D0793635

EARTH SCIENCE

EARTH SCIENCE
VOLUME I

THE PHYSICS AND CHEMISTRY OF EARTH

EDITOR
JAMES A. WOODHEAD
Occidental College

EDITORIAL BOARD

RENÉ DE HON, VOLUME I
University of Louisiana at Monroe

DENNIS G. BAKER, VOLUME IV
University of Michigan

ANITA BAKER-BLOCKER, VOLUME IV
Applied Meteorological Services

DAVID K. ELLIOTT, VOLUME II
Northern Arizona University

CHARLES W. ROGERS, VOLUMES III & V
Southwestern Oklahoma State University

SALEM PRESS, INC.
Pasadena, California   Hackensack, New Jersey

Managing Editor: Christina J. Moose
Project Development: Robert McClenaghan
Manuscript Editors: Doug Long, Amy Allison
Acquisitions Editor: Mark Rehn
Research Supervisor: Jeffry Jensen
Photograph Editor: Philip Bader
Assistant Editor: Andrea E. Miller
Indexers: Melanie Watkins, Lois Smith
Research Assistant: Jeffrey Stephens
Production Editor: Cynthia Beres
Page Design and Layout: James Hutson
Additional Layout: William Zimmerman
Graphics: Electronic Illustrators Group

Copyright © 2001, by Salem Press, Inc.

All rights in this book are reserved. No part of this work may be used or reproduced in any manner whatsoever or transmitted in any form or by any means, electronic or mechanical, including photocopy, recording, or any information storage and retrieval system, without written permission from the copyright owner except in the case of brief quotations embodied in critical articles and reviews. For information address the publisher, Salem Press, Inc., P.O. Box 50062, Pasadena, California 91115.

∞ The paper used in these volumes conforms to the American National Standard for Permanence of Paper for Printed Library Materials, Z39.48-1992 (R1997).

Library of Congress Cataloging-Publication Data
Earth science / editor, James A. Woodhead.
    p. cm.
Expands and updates Magill's survey of science: earth science series.
Includes bibliographical references and indexes.
Contents: v. 1. The physics and chemistry of earth — v. 2. The earth's surface and history — v. 3. Earth materials and earth resources — v. 4. Weather, water, and the atmosphere — v. 5. Planetology and earth from space.
ISBN 0-89356-000-6 (set : alk. paper) — ISBN 0-89356-001-4 (v. 1 : alk paper) —
ISBN 0-89356-002-2 (v. 2 : alk. paper) — ISBN 0-89356-003-0 (v. 3 : alk. paper) —
ISBN 0-89356-004-9 (v. 4 : alk. paper) — ISBN 0-89356-005-7 (v. 5 : alk. paper)
1. Earth sciences. I. Woodhead, James A. II. Magill's survey of science. Earth science series.

QE28 .E12 2001
550—dc 21

00-059567

First Printing

PRINTED IN THE UNITED STATES OF AMERICA

CONTENTS

PUBLISHER'S NOTE

Earth Science—comprising 495 lengthy essays on basic topics in geology, geophysics, atmospheric sciences, oceanography, environmental science, paleontology, and planetology—is designed to provide an introduction to our latest understanding of the Earth's physical features and dynamic processes, as well as those of the solar system in which it resides. Librarians, students, and general readers alike will turn to this reference work for both basic information and current developments, from plate tectonics through the latest discoveries on Mars, presented in easy-to-understand language with copious ready-reference aids.

These five volumes heavily expand upon and update the original 377 articles of *Magill's Survey of Science: Earth Science Series* (1990), as well as the 55 essays of that publication's one-volume supplement, published in 1998. Among these original 432 essays, 4 were dropped, with coverage provided in newer and updated articles; 11 were combined into 5 more concise essays; and the remainder were thoroughly updated by area experts. Additionally, 19 essays—18 from Salem's *USA in Space* and 1 from Salem's *Environmental Issues*—were updated and added to round out the publication's coverage of Earth in the context of space and the solar system. All essays' bibliographies have been updated with recently published sources for further study. Finally, the editors determined the need for 64 entirely new entries, bringing the set's total to 495 in-depth essays, along with more than 500 illustrations—tables, charts, drawings, and photographs—which display basic principles, phenomena, and geological features of our world and the solar system.

Another major feature of this publication is its categorized organization: Each volume focuses on a broad area of Earth science. Students seeking information on topics that tend to be covered together in the course curricula will find these topics grouped together. This aid to accessing related information is supplemented by cross-references at the end of every essay, which direct readers to related essays, as well as two lists of the essays at the end of each volume: the Alphabetical List of Contents, which arranges the essays with their page references alphabetically, and the Categorized List of Contents, which arranges the essays in detailed groupings of all essays related to a particular subtopic.

The volumes are broken into different areas of Earth science as follows: Volume I, *The Physics and Chemistry of Earth*, discusses the planet's structure from crust to core, including its dynamics and physical processes: from plate tectonics to mantle convection, geochemical processes such as water-rock interactions to geochronology and dating methods, rotation to polar wander, deep-focus earthquakes to soil liquefaction. This volume also delves into the explorations geologists are undertaking, such as the Ocean Drilling Program, that are revolutionizing our understanding of the Earth's structure, climate, and mineral resources.

Volume II, *The Earth's Surface and History*, surveys structures and surface features, from continents and moutains to fold belts, glaciers, and volcanoes. Many of these are specific: The reader will find not only "Geysers and Hot Springs" but also "Yellowstone National Park"; not only "Stratovolcanoes" but also "Mount Pinatubo," "Mount St. Helens," and other specific volcanoes; not only "Mountain Belts" but also the Andes, the Sierra Nevada, the Himalaya, and other ranges. In addition, Volume II addresses the geological history that gave rise to these structures and the life-forms that evolved on Earth, in essays that cover the different periods of Earth's history, the evolution of life from "Archaebacteria" to "Dinosaurs" to "Human Evolution," and subdisciplines such as paleogeography, paleoclimatology, and paleoseismology.

Volume III, *Earth Materials and Earth Resources*, the "rocks" volume, will be familiar to many beginning students of geology. Minerals, igneous rocks, metamorphic rocks, sedimentary rocks, land and soil, and their economic resources, from gems to petroleum, are covered here. Special sections address energy resources, petroleum resources, and the environmental implications of resource exploitation.

Volume IV, *Weather, Water, and the Atmosphere*, examines Earth as a living planet and therefore also considers many processes that shape, change, and threaten to deteriorate the environment. Here, Earth's atmosphere extends beyond the troposphere to consider such influences as radiation from beyond the planet. The processes of the planet's water bodies, intimately involved with these dynamics, are also discussed, including the hydrologic cycle, El Niño and La Niña, and many of the Earth's major seas, rivers, and oceans. Weather and climatic phenomena—from the dramatic, such as hurricanes and tornadoes, to the subtle processes of desertification and erosion—are addressed as well.

Volume V, *Planetology and Earth from Space*, considers our world from the perspective, and in the context, of space. Scientists are hourly learning more about Earth as they learn from other planets and moons what may have transpired in its past and what may lie in the future. Both earlier and later missions, from the Voyagers to Magellan, Cassini, Galileo, and the Mars expeditions, have cast light on how our planet works. Likewise, satellite-borne technology has focused its remote-sensing devices on Earth from beyond the atmosphere: From the Air Density Explorers to the latest Landsat and the Global Positioning System, we are learning more, with greater precision, about the physical layout, dynamics, and fragility of our relatively small globe.

Each essay retains the familiar Magill format: All the entries begin with ready-reference top matter, including a summary statement in which the contributing author explains why the topic is important to the study of the Earth and its processes. A listing of principal terms and their definitions helps to orient the reader to the essay. The text itself is broken by informative subheadings that guide readers to areas of particular interest. An annotated and updated Bibliography closes each essay, referring the reader to external sources for further study that are of use to both students and nonspecialists. Finally, a list of Cross-References directs the reader to other essays in the five volumes that offer information on related topics.

At the end of every volume, several appendices are designed to assist in the retrieval of information. As mentioned above, the Alphabetical List of Contents lists all essays alphabetically by title, followed by page numbers. The Categorized List of Contents, more detailed than the general Table of Contents, breaks the essays into useful categories to offer readers access to related essays; essays are often listed in more than one category to display a subcategory's full coverage in these volumes. Volume V, the final volume, contains several additional tools: an Alphabetical List of Elements, the Periodic Table of Elements, a Time Line of Earth History (which incorporates the most authoritative and updated version of the Geologic Time Scale), a Glossary of Earth science terminology, a Bibliography containing standard and useful sources by category, and a listing of useful Web sites on Earth sciences. A comprehensive subject index rounds out the set.

Salem Press in indebted to the many specialists who contributed their time and expertise to this publication. James A. Woodhead of Occidental College edited the original 1990 edition and returned to provide his valuable counsel in the development of both the table of contents and its arrangement. René De Hon of the University of Louisiana at Monroe read and updated the current edition's essays in Volume I, *The Physics and Chemistry of Earth*. David K. Elliott of Northern Arizona University updated Volume II, *The Earth's Surface and History*. Charles W. Rogers of Southwestern Oklahoma State University lent his expertise to the updating of two volumes: Volume III, *Earth Materials and Earth Resources*, and Volume V, *Planetology and Earth from Space*. Dennis G. Baker of the University of Michigan, with Anita Baker-Blocker of Applied Meteorological Services, updated Volume IV, *Weather, Water, and the Atmosphere*. Special mention must be made of Roger Smith, who played a principal role in shaping the 1998 supplement's contents. Project editors Robert McClenaghan and Doug Long were principally responsible for the volumes' organization and manuscript editing, respectively. Hans J. Neuhart of Electronic Illustrators Group prepared the illustrations. Finally, thanks are due to the many academicians and professionals who worked to communicate their expert understanding of the Earth sciences to the general reader; a list of these individuals and their affiliations appears in the following pages, and their contributions are gratefully acknowledged.

CONTRIBUTORS

Stephen R. Addison
University of Central Arkansas

Mary D. Albanese
*American Institute of Professional
  Geologists*

Arthur L. Alt
College of Great Falls

Rajkumar Ambrose
Monmouth College

Michael S. Ameigh
St. Bonaventure University

Valentine J. Ansfield
University of South Dakota

Richard W. Arnseth
Science Applications International

George K. Attwood
Maharishi International University

N. B. Aughenbaugh
University of Mississippi

Dennis G. Baker
University of Michigan

Victor R. Baker
University of Arizona

Anita Baker-Blocker
Applied Meteorological Services

John M. Bartley
University of Utah

Thomas W. Becker
Webster University

Alvin K. Benson
Brigham Young University

John L. Berkley
State University of New York, Fredonia

Elizabeth K. Berner
Yale University

David M. Best
Northern Arizona University

Richard J.Boon
Independent Scholar

Danita Brandt
Eastern Michigan University

Walter Bressette
NASA Langley Research Center

Alan Brown
Livingston University

Michael Broyles
Collin County Community College

David S. Brumbaugh
Northern Arizona University

James A. Burbank, Jr.
Western Oklahoma State College

Scott F. Burns
Louisiana Tech University

Byron D. Cannon
University of Utah

Roger V. Carlson
*Jet Propulsion Laboratory,
  California Institute of Technology*

Robert S. Carmichael
University of Iowa

Robert E. Carver
University of Georgia

Dennis Chamberland
Independent Scholar

Dennis W. Cheek
Rhode Island Department of Education

D. K. Chowdhury
*Indiana University and Purdue
  University at Fort Wayne*

Habte Giorgis Churnet
University of Tennessee at Chattanooga

Mark Cloos
University of Texas at Austin

Raymond A. Coish
Middlebury College

John H. Corbet
Memphis State University

Robert G. Corbett
Illinois State University

William C. Cornell
University of Texas at El Paso

James R. Craig
*Virginia Polytechnic Institute and
  State University*

William W. Craig
University of New Orleans

Ralph D. Cross
University of Southern Mississippi

Robert L. Cullers
Kansas State University

E. Julius Dasch
*National Aeronautics and Space
  Administration*

Loralee Davenport
Mississippi University for Women

Larry E. Davis
Washington State University

Ronald W. Davis
Western Michigan University

Dennis R. Dean
University of Wisconsin—Parkside

René A. De Hon
University of Louisiana at Monroe

Albert B. Dickas
University of Wisconsin

James A. Dockal
*University of North Carolina at
  Wilmington*

Bruce D. Dod
Mercer University

Dave Dooling
Independent Scholar

Walter C. Dudley, Jr.
University of Hawaii at Hilo

Dean A. Dunn
University of Southern Mississippi

Steven I. Dutch
University of Wisconsin—Green Bay

David K. Elliott
Northern Arizona University

John J. Ernissee
Clarion University of Pennsylvania

David G. Fisher
Lycoming College

Dennis R. Flentge
Cedarville College

Richard H. Fluegeman, Jr.
Ball State University

George J. Flynn
*State University of New York,
    Plattsburgh*

Robert G. Font
Strategic Petroleum, Inc.

Annabelle M. Foos
University of Akron

John W. Foster
Illinois State University

Dell R. Foutz
Mesa State College

Robert C. Frey
Independent Geologist

A. Kem Fronabarger
College of Charleston

Charles I. Frye
Northwest Missouri State University

Stephanie Gallegos
Jet Propulsion Laboratory

Roberto Garza
San Antonio College

Joyce Gawell
Del Valle High School

Soraya Ghayourmanesh
Nassau Community College

Karl Giberson
Eastern Nazarene College

Gail G. Gibson
*University of North Carolina at
    Charlotte*

Billy P. Glass
University of Delaware

Douglas Gomery
University of Maryland

Gregory A. Good
West Virginia University

Pamela J. W. Gore
De Kalb Community College

Daniel G. Graetzer
University of Washington, Seattle

Hans G. Graetzer
South Dakota State University

Martha M. Griffin
Columbia College

William R. Hackett
Idaho State University

William J. Hagan, Jr.
St. Anselm College

Edward C. Hansen
Hope College

Clay D. Harris
Middle Tennessee State University

Jasper L. Harris
North Carolina Central University

Charles D. Haynes
University of Alabama

Paul A. Heckert
Western Carolina University

Sara A. Heller
College of Charleston

Thomas E. Hemmerly
Middle Tennessee State University

Charles E. Herdendorf
Ohio State University

David F. Hess
Western Illinois University

Carl W. Hoagstrom
Ohio Northern University

William Hoffman
Independent Scholar

Earl G. Hoover
*American Institute of Professional
    Geologists*

Robert M. Hordon
Rutgers University

Robert A. Horton, Jr.
*California State University at
    Bakersfield*

Louise D. Hose
Westminster College

Ruth H. Howes
Ball State University

Stephen Huber
Beaver College

Samuel F. Huffman
University of Wisconsin—River Falls

Pamela Jansma
Jet Propulsion Laboratory

Albert C. Jensen
Central Florida Community College

Jeffrey A. Joens
Florida International University

Brian Jones
Independent Scholar

James O. Jones
University of Texas at San Antonio

Richard C. Jones
Texas Woman's University

Pamela R. Justice
Collin County Community College

Karen N. Kähler
Independent Scholar

Kyle L. Kayler
Richard H. Gorr and Associates

Christopher Keating
Angelo State University

John P. Kenny
Bradley University

Diann S. Kiesel
*University of Wisconsin Center—
    Baraboo/Sauk County*

Michael M. Kimberley
North Carolina State University

Richard S. Knapp
Belhaven College

David R. Lageson
Montana State University

Ralph L. Langenheim, Jr.
University of Illinois at Urbana-Champaign

Gary G. Lash
State University of New York College, Fredonia

Joel S. Levine
NASA Langley Research Center

Leon Lewis
Appalachian State University

W. David Liddell
Utah State University

J. Lipman-Boon
Independent Scholar

M. A. K. Lodhi
Texas Tech University

James Charles LoPresto
Edinboro University of Pennsylvania

Donald W. Lovejoy
Palm Beach Atlantic College

Gary R. Lowell
Southeast Missouri State University

Spencer G. Lucas
New Mexico Museum of Natural History

David N. Lumsden
Memphis State University

Michael L. McKinney
University of Tennessee, Knoxville

Paul Madden
Hardin-Simmons University

David W. Maguire
C. S. Mott Community College

Mehrdad Mahdyiar
Leighton and Associates

Nancy Farm Mannikko
Independent Scholar

Carl Henry Marcoux
University of California, Riverside

Glen S. Mattioli
University of California, Berkeley

Michael W. Mayfield
Appalachian State University

Paul S. Maywood
Bridger Coal Company

Lance P. Meade
American Institute of Mining Engineers

Nathan H. Meleen
American Geophysical Union

Robert G. Melton
Pennsylvania State University

Randall L. Milstein
Oregon State Unniversity

Joseph M. Moran
University of Wisconsin—Green Bay

Otto H. Muller
Alfred University

Phillip A. Murry
Tarleton State University

John E. Mylroie
Mississippi State University

John Panos Najarian
William Paterson College

Brian J. Nichelson
U.S. Air Force Academy

Bruce W. Nocita
University of South Florida

Edward B. Nuhfer
University of Wisconsin—Platteville

Divonna Ogier
Oregon Museum of Science and Industry

Steven C. Okulewicz
City University of New York, Hunter College

Michael R. Owen
St. Lawrence University

Susan D. Owen
Bonneville Power Administration

David L. Ozsvath
University of Wisconsin—Stevens Point

Donald F. Palmer
Kent State University

Robert J. Paradowski
Rochester Institute of Technology

Gordon A. Parker
University of Michigan—Dearborn

Donald R. Prothero
Occidental College

C. Nicholas Raphael
Eastern Michigan University

Donald F. Reaser
University of Texas at Arlington

Jeffrey C. Reid
North Carolina Geological Survey

Gregory J. Retallack
University of Oregon

Mariana Rhoades
St. John Fisher College

J. A. Rial
University of North Carolina at Chapel Hill

J. Donald Rimstidt
Virginia Polytechnic Institute and State University

Raymond U. Roberts
Pacific Enterprises Oil and Gas Company

Charles W. Rogers
Southwestern Oklahoma State University

James L. Sadd
Occidental College

Neil E. Salisbury
University of Oklahoma

Virginia L. Salmon
Northeast State Technical Community College

Cory Samia
Oregon Museum of Science and Industry

Panagiotis D. Scarlatos
Florida Atlantic University

Elizabeth D. Schafer
Independent Scholar

David M. Schlom
California State University, Chico

Kenneth J. Schoon
Indiana University Northwest

Rose Secrest
Independent Scholar

John F. Shroder, Jr.
University of Nebraska at Omaha

Stephen J. Shulik
Clarion University of Pennsylvania

R. Baird Shuman
*University of Illinois at Urbana-
Champaign*

William C.Sidle
Ohio State University

Dorothy Fay Simms
Northeastern State University

Paul P. Sipiera
Harper College

Clyde Curry Smith
University of Wisconsin—River Falls

Philip A. Smith
*Southern Illinois University of
Edwardsville*

Roger Smith
Independent Scholar

John Brelsford Southard
Massachusetts Institute of Technology

Joseph L. Spradley
Wheaton College, Illinois

Kenneth F. Steele
*Arkansas Water Resources Research
Center*

Robert J. Stern
University of Texas at Dallas

David Stewart
Southeast Missouri State University

Dion Stewart
Adams State College

Toby Stewart
Independent Scholar

Ronald D. Stieglitz
University of Wisconsin—Green Bay

Bruce L. Stinchcomb
*St. Louis Community College at
Florissant Valley*

Anthony N. Stranges
Texas A&M University

Locke Stuart
Goodard Space Flight Center

Frederick M. Surowiec
Society of Professional Journalists

Eric R. Swanson
University of Texas at San Antonio

Stephen M. Testa
California State University, Fullerton

Donald J. Thompson
California University of Pennsylvania

Keith J. Tinkler
Brock University

Leslie V. Tischauser
Prairie State College

D. D. Trent
Citrus College

Ronald D. Tyler
Independent Scholar

Robert Veiga
*National Aeronautics and Space
Administration*

Nan White
Maharishi International University

James L. Whitford-Stark
Sul Ross State University

Thomas A. Wikle
Oklahoma State University

Ian Williams
University of Wisconsin—River Falls

Shawn V. Wilson
Independent Scholar

Dermot M. Winters
*American Institute of Mining
Engineers*

James A. Woodhead
Occidental College

Grant R. Woodwell
Mary Washington College

Lisa A. Wroble
Redford Township District Library

Jay R. Yett
Orange Coast College

David N. Zurick
Eastern Kentucky University

EARTH SCIENCE

1
EARTH'S STRUCTURE AND INTERIOR PROCESSES

EARTH'S CORE

The core is the Earth's densest, hottest region and its fundamental source of internal heat. The thermal energy released by the core's continuous cooling stirs the overlying mantle into slow, convective motions that eventually reach the surface to move continents, build mountains, and produce earthquakes.

PRINCIPAL TERMS

CONVECTION: the process in liquids and gases by which hot, less dense materials rise upward to be replaced by cold, sinking fluids

MAGNETIC FIELD: a force field, generated in the core, that pervades the Earth and resembles that of a bar magnet

P WAVES: seismic waves transmitted by alternating pulses of compression and expansion; they pass through solids, liquids, and gases

S WAVES: seismic waves transmitted by an alternating series of sideways movements in a solid; they cannot be transmitted through liquids or gases

SEISMIC WAVES: elastic oscillatory disturbances spreading outward from an earthquake or human-made explosion; they provide the most important data about the Earth's interior

STRUCTURE AND COMPOSITION

The Earth's core extends from a depth of 2,900 kilometers to the center of the Earth, about 6,300 kilometers below the surface. The core is largely liquid, although toward the center, it becomes solid. The liquid part is known as the outer core; the solid part, the inner core. Ambient pressures inside the core range from 1 million to nearly 4 million atmospheres, and temperatures probably reach more than 5,000 degrees Celsius at the Earth's center.

Being almost twice as dense as the rest of the planet, the core contains one-third of the Earth's mass but occupies a mere one-seventh of its volume. Surrounding the core is the mantle. The boundary between the solid mantle and the underlying liquid core is the core-mantle boundary (CMB), a surface that demarcates the most fundamental compositional discontinuity in the Earth's interior. Below it, the core is mostly made of iron-nickel. Above it, and all the way to the surface, the mantle is made of silicates (rock-forming minerals). The solid inner core contains 1.7 percent of the Earth's mass, and its composition may simply be a frozen version of the liquid core. The boundary between the liquid and the solid cores is known as the inner core boundary (ICB); it appears sharp to seismic waves, which easily reflect off it.

The core has lower wave-transmission velocities and higher densities than the mantle, a consequence of its being of a different chemical composition. The core is probably composed of 80 to 90 percent (by weight) iron or iron-nickel alloy and 20 to 10 percent sulfur, silicon, and oxygen; it therefore must be a good electrical and thermal conductor. The mantle, in contrast, is composed mainly of crystalline silicates of magnesium and iron and is therefore a poor conductor of electricity and a good thermal insulator.

FORMATION

This sharp contrast in physical properties is a major end product of the way in which the Earth evolved thermally, gravitationally, and chemically. It is difficult, however, to tell whether the Earth's core formed first and the Earth was accreted from the infall of meteorites and other gravitationally bound materials or, alternatively, the core differentiated out of an already formed protoearth, in which silicates and iron were separated after a cataclysmic "iron catastrophe." This event may have occurred when iron, slowly heated by radioactivity, suddenly melted and sank by gravity toward the Earth's center, forming the core. Unfortunately, the two scenarios are equally likely, and both give the same end result; moreover, there probably are other scenarios. Calculations show, however, that

3

iron sinking to the core must have released great amounts of energy that would have eventually heated and melted the entire Earth. Cooling of the outer parts proceeded rapidly, by convection, but as the silicate mantle solidified, it created a thermal barrier for the iron-rich core, which, not being able to cool down as readily, remained molten. The inner core began then to form at the Earth's center, where the pressure was greatest and solidification was (barely) possible.

GEOMAGNETIC FIELD

The most tangible consequence of the existence of a fluid, electrically conducting core is the presence of a magnetic field in the Earth that has existed for at least 3.5 billion years with a strength not very different from what it has today. The process that generates and maintains the geomagnetic field is attributable to a self-exciting dynamo mechanism—that is, an electromagnetic induction process that transforms the motions of the conducting fluid into electric currents, which in turn induce a magnetic field that strengthens the existing field. (For the system to get started, at least a small magnetic field must be present to initiate the generation of electric currents.) The increased magnetic field in turn induces stronger currents, which further strengthen the field, and so on. As the magnetic field increases beyond a certain high value, it begins to affect the fluid flow; there is a mechanical force, known as the Lorentz force, that is induced in a conductor as it moves across a magnetic field. The stronger the magnetic field, the stronger the Lorentz force becomes and the more it will tend to modify the motion of the fluid so as to oppose the growth of the magnetic field. The result is a self-regulating mechanism which, over time, will attain a steady state.

A dynamo mechanism is needed to explain the geomagnetic field, because there can be no permanently magnetized substances inside the Earth. Magnetic substances lose their magnetism as their temperature increases above the so-called Curie temperature (around 500 degrees Celsius for most magnetic substances), and most of the mantle below the depth of 30 kilometers and all the core is at temperatures well above the Curie point. The basic problem, then, is to find a source of energy that can maintain the steady regime of flow in the core against decay by somehow maintaining the

fluid currents that induce the field. A favored view is that the necessary energy to maintain the flow is provided by the growth of the inner core as it is fed by the liquid core. According to some researchers, this process would provide enough gravitational energy to stir the core throughout. Thermally and compositionally driven flows can also be invoked as possible models of core fluid dynamics, but there is still no evidence that decides the question. Measurements based on variation of the magnetic field confirm circulation of the fluid outer core much as the dynamo theory predicts.

A most extraordinary feature of the core-generated magnetic field is that, at least over the past few hundred million years, it has reversed its polarity with irregular frequency. For example, it is known that at times the field has reversed as frequently as three times every million years, but in other cases, more than 20 million years went by without a noticeable reversal. A reversed geomagnetic field means simply that the magnetic needle of a compass would point in the opposite direction as it does today. (For convenience, the present orientation of the needle is considered normal.) The important point is that the rocks that form (for example, lavas that cool below the Curie point as they become solid rock) during either a reverse or a normal period acquire and preserve that magnetism. Unlike the swinging compass needle, a rock keeps the magnetic field direction that existed at the time of its formation forever frozen in its iron-bearing minerals. Therefore, rocks formed throughout geologic time have recorded the alternating rhythms of normal and reverse Earth magnetism. This sequence of magnetic reversals contains the clue to the core's nature.

CONVECTION CURRENTS

Geophysicists are eager to learn whether the core is vigorously convecting as a consequence of the inner core's growth. If that were the case, the core would be delivering a great amount of heat to the mantle, whose low thermal conductivity would create a barrier to the upcoming heat. As a result, the local temperature gradient at the base of the mantle would probably be very high, so that a layer 100 kilometers thick, say, at the base of the mantle, would be gravitationally unstable. From this layer, thermal inhomogeneities would rise through the mantle in the form of plumes of buoyant, hot, lower-mantle material. Several such plumes might reach the upper mantle or set the entire mantle into convection. These convection currents would be responsible for the motion of the tectonic plates on the Earth's surface and, consequently, for the uplifting of mountain ranges, the formation of oceanic basins, and the occurrence of volcanic eruptions and earthquakes. Continental drift and plate tectonics, the most visible effects of the internal cooling of the Earth, would thus be linked to the growth of the inner core and to Earth's earliest history. This view of the Earth is very speculative, but it is favored by many geoscientists, who recognize its beauty and simplicity.

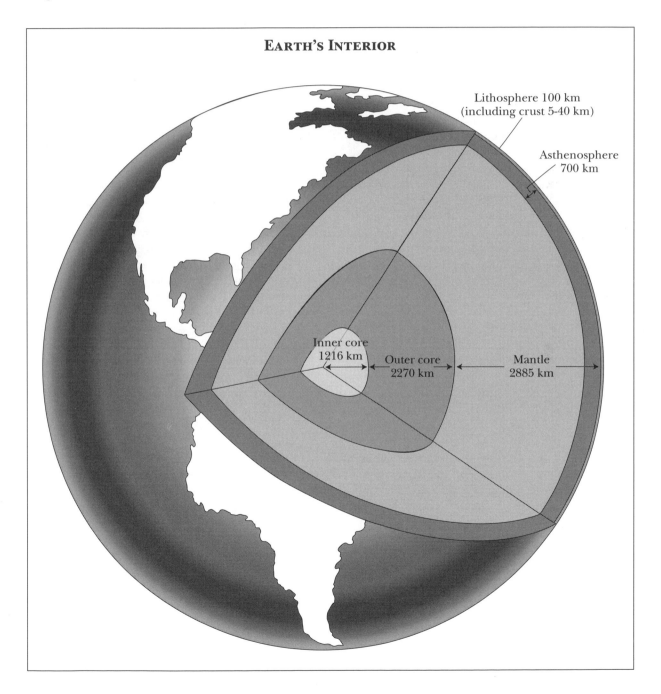

EARTH'S INTERIOR

Lithosphere 100 km
(including crust 5-40 km)

Asthenosphere
700 km

Inner core
1216 km

Outer core
2270 km

Mantle
2885 km

SEISMIC WAVE ANALYSIS

Knowledge of the structure, physical properties, and composition of the core is entirely based on indirect evidence gathered largely from analyses of seismic waves and the study of the Earth's gravitational and magnetic fields. The first evidence for the existence of the core was presented in a paper suggestively entitled "The Constitution of the Interior of the Earth, as Revealed by Earthquakes," published in 1906 by Richard D. Oldham, of the geological survey of India. Thirty years later, Inge Lehmann, from the Copenhagen seismological observatory, presented seismic evidence for the existence of the inner core. In the past few decades, with the advent of high-speed computers and technological advances in seismometry, seismologists have developed increasingly sensitive instrumentation to record seismic waves worldwide and sophisticated mathematical theories that allow them to construct models of the core that explain the observed data.

Seismic waves provide the most important data about the core. Earthquakes or large explosions generate elastic waves that propagate throughout the Earth. These seismic waves may penetrate deep in the Earth and, after being reflected or transmitted through major discontinuities such as the CMB and ICB, travel back to the surface to be recorded at the seismic stations of the global network. The most direct information that seismic or elastic waves carry is their travel time. Knowing the time it takes for elastic waves to traverse some region of the Earth's interior allows the calculation of their velocity of propagation in that region. The velocity of seismic waves strongly depends on the density and rigidity, or stiffness, of the material through which they propagate, so estimates of the mechanical properties of the Earth can in principle be derived from seismic travel time analyses.

P AND S WAVES

Seismic waves that propagate through the deep interior of the Earth are of two types: compressional waves (also called P waves) and shear waves (also called S waves). Compressional waves produce volume changes in the elastic medium; shear waves produce shape distortion without volume change. If the medium has some rigidity, both P and S waves can be transmitted. If the medium has no rigidity, it offers no resistance to a change in shape; no elastic connection exists that can communicate shearing motions from a point in the medium to its neighbors, so S waves cannot propagate, although P waves can.

After many years of careful observations, it has been determined that S waves are not transmitted through the outer core. Therefore, the outer core material has no rigidity, but behaves as a fluid would. Similar observations suggest that the inner core is solid; the actual rigidity of the inner core is very difficult to estimate, however, since shear waves inside the inner core are isolated from the mantle by the outer core and can only travel through it as P waves converted from S waves at the ICB. Nevertheless, when the whole Earth is set into vibration by a very large earthquake, the average rigidity of the inner core can be estimated by comparing the observed frequencies of oscillation with those theoretically computed for models of the Earth that include a solid inner core. Model studies have indicated that the inner core is indeed solid, because a totally liquid core model does not satisfy the observations.

The average velocity of P waves in the Earth is about 10 kilometers per second, whereas the average P-wave velocity in the rocks accessible to measurement at the Earth's surface is 4 to 5 kilometers per second. The S-wave velocity is nearly half that of P waves in solids and zero in perfect fluids.

The velocity of P waves drops abruptly from 13.7 kilometers per second at the base of the mantle to 8.06 kilometers per second across the CMB, at the top of the core. From this point down, the velocity steadily increases to 10.35 kilometers per second at the ICB, where it jumps discontinuously to 11.03 kilometers per second at the top of the inner core. From there to the center of the Earth, the velocity of P waves increases slowly to reach 11.3 kilometers per second. The S-wave velocity increases from zero at the ICB to around 3.6 kilometers per second at the Earth's center. The core's density abruptly increases from 5,500 kilograms per cubic meter at the base of the mantle to nearly 10,000 kilograms per cubic meter just underneath the CMB. From there, the density increases slowly to nearly 13,100 kilograms per cubic meter at the Earth's center. In comparison, the density of mercury at room temperature and ambient pressure is 13,600 kilograms per cubic meter.

GRAVITATIONAL AND MAGNETIC FIELD STUDY

That the core is mostly iron is consistent with iron's being cosmically more abundant than other heavy elements and with the high electrical conductivity the core needs to have in order to generate the Earth's magnetic field. The fluidity of the outer core has been demonstrated by measurements not only of seismic wave transmission but also of the oscillation period of gravitational waves in the core excited by the lunisolar tides. The existence of a sustained, steady magnetic field is also consistent with a fluid outer core.

New views of the Earth's interior are produced, sometimes unexpectedly, by the analyses of data collected by satellite missions. Data from orbiting satellites that measure tiny variations of the Earth's gravitational field, combined with computer-aided seismic tomography of the Earth's interior, have revealed large-density anomalies at the base of the mantle and a large relief of more than 2 kilometers on the CMB. Seismic tomography uses earthquake-generated waves that penetrate the mantle in a multitude of directions to map the three-dimensional structure of its deep interior, just as computerized medical tomography uses multiple X-ray images to create a three-dimensional view of internal organs of the body. Essential to the success of these studies, however, is the installation of dense networks of seismic sensors all over the surface of the Earth; this installation, however, is a very expensive procedure.

CONTINUING RESEARCH

Any study of the Earth's physical environment is likely to provide insight into the nature and future of the planet and, consequently, the future of humankind. If geophysicists come to understand how the Earth's core works, they will be able to predict the geomagnetic field's activity for years to come. Thus, they will be able to predict an upcoming reversal. According to the best estimates, a reversal does not occur suddenly but takes about ten thousand years. That means that during a reversal,

there is a time of very small or even zero field intensity. Under such conditions, the magnetic shielding that prevents the highly energetic solar wind particles from reaching the Earth's surface will disappear, leaving the Earth directly exposed to lethal radiations.

Seismic data can probe the inner core only partially from the Earth's surface, unless the source of the seismic waves and the receivers are located antipodally to each other. Such an arrangement would allow scientists to measure seismic waves that had penetrated the center of the Earth. It would be possible to construct a global experiment to investigate the inner core by deploying an array of highly sensitive seismic sensors antipodal to either a seismically active region or an underground nuclear explosion testing ground. Despite the wealth of unique data that would be obtained from such an experiment, it would be a very expensive endeavor, and not devoid of risk.

The inner core has not yet been sufficiently explored. It is the remotest region of the Earth and therefore the most difficult to reach; furthermore, it is hidden beneath the "seismic noise" created by the crust, mantle, and outer core. The inner core, however, holds the key to the understanding of the Earth's early history and its subsequent development as a planet.

J. A. Rial

CROSS-REFERENCES

Earth Tides, 101; Earth's Core-Mantle Boundary, 9; Earth's Crust, 14; Earth's Differentiation, 20; Earth's Lithosphere, 26; Earth's Magnetic Field, 137; Earth's Magnetic Field: Origin, 144; Earth's Magnetic Field: Secular Variation, 150; Earth's Mantle, 32; Earth's Origin, 2389; Earth's Rotation, 106; Earth's Shape, 111; Earth's Structure, 37; Geothermal Phenomena and Heat Transport, 43; Heat Sources and Heat Flow, 49; Lithospheric Plates, 55; Magnetic Reversals, 161; Mantle Dynamics and Convection, 60; Nickel-Irons, 2718; Plumes and Megaplumes, 66.

BIBLIOGRAPHY

Bolt, Bruce A. *Inside the Earth: Evidence from Earthquakes.* San Francisco: W. H. Freeman, 1982. An elementary treatment of what is known about the Earth's interior, mostly through the study of seismic waves, the author's major field of research. The book contains abundant diagrams that illustrate accurately important results of the investigation of the core and mantle. For readers with some knowledge of mathematics, the book includes brief derivations of important formulas, separated by "boxes" from the main text. It is well written and includes anecdotal descriptions of great scientific discoveries along with personal views of the history and development of seismology. Illustrated.

Brush, Stephen G. *Nebulous Earth: The Origin of the Solar System and the Core of the Earth from Laplace to Jeffreys.* Cambridge: Cambridge University Press, 1996. Brush's book, volume 1 in the History of Modern Planetary Physics series, contains useful information on the nebular hypothesis, the origin of the solar system, and the Earth's core. Includes a bibliography and index.

Clark, Sydney P. *Structure of the Earth.* Englewood Cliffs, N.J.: Prentice-Hall, 1971. Although slightly out of date, this short review of the Earth's structure and composition is an excellent first reading to gain a global perspective on geology and geophysics. Illustrations are abundant and very clear. The text is simply written, yet the author manages to convey complex concepts about tectonics, wave propagation, and ray theory with ease. The chapter dedicated to seismology is the best and most carefully written section of the book.

Gurnis, Michael, et al., eds. *The Core-Mantle Boundary Region.* Washington, D.C.: American Geophysical Union, 1998. This collection of articles is one volume of the American Geophysical Union's Geodynamics series. Although intended for the specialist, the essays contain plenty of information suitable for the careful college-level reader. Bibliography.

Hamblin, William K. *Earth's Dynamic Systems.* 8th ed. Upper Saddle River, N.J.: Prentice Hall, 1998. This geology textbook offers an integrated view of the Earth's interior not common in books of this type. The illustrations, diagrams, and charts are superb. Includes a glossary and laboratory guide. Suitable for high school readers.

Jacobs, J. A. *The Earth's Core.* 2d ed. New York: Academic Press, 1987. This is a highly technical text, but it is perhaps the best reference for a detailed description of the most accepted core models. The tables—which give the numerical values of the density, temperature, rigidity, and wave velocity distributions within the Earth—are of interest to anyone wanting a quantitative description of the core. A long list of research articles is included.

Jeanloz, Raymond. "The Earth's Core." *Scientific American* 249 (September, 1983): 56-65. The best elementary treatment of the structure and composition of the core. Jeanloz is a leading expert in the field. In this article, the origin, evolution, and present state and composition of the core are discussed in detail. The language is precise but not too specialized. The entire issue is dedicated to the Earth and Earth dynamics, so it should be of great interest to some readers.

Olson, Peter. "Probing Earth's Dynamo." *Nature* 389 (September, 1997): 337-338. Provides confirming evidence of circulation of the fluid outer core much as the dynamo theory predicts. A westward circular motion is detected beneath the North Pole.

Press, Frank, and Raymond Siever. *Understanding Earth.* 2d ed. New York: W. H. Freeman, 1998. This comprehensive physical geology text covers the formation and development of the Earth. Readable by high school students, as well as by general readers. Includes an index and a glossary of terms.

Vogel, Shawna. *Naked Earth: The New Geophysics.* New York: Plume, 1996. Vogel's book, written to be easily understood by students with little background in the Earth sciences, is a good introduction to the physics of the Earth's core and internal structure. Index.

EARTH'S CORE-MANTLE BOUNDARY

The core-mantle boundary is a pronounced discontinuity separating the outer core from the mantle of the Earth. It is a chemical and mineralogical as well as a thermal boundary. The topography of the core-mantle boundary is believed to be controlled by the dynamic processes in the mantle and the outer core.

PRINCIPAL TERMS

CORE-DIFFRACTED PHASES: those elastic waves incident at the outer core at a grazing angle that are diffracted and arrive within the shadow zones for direct waves

CORE-REFLECTED PHASES: elastic waves that are reflected from the core-mantle boundary

CORE-TRANSMITTED PHASES: elastic waves that travel through the Earth's core

CRUST: the thin, surface layer of the Earth, with an average thickness of 34 kilometers; it consists of low-density, silicate rocks

EPICENTER: the region at the Earth's surface directly above the focus, or hypocenter, of an earthquake

FOCUS: the region within the Earth from which earthquake waves emanate; also called hypocenter

LONGITUDINAL (P) WAVES: elastic waves in a medium where particle displacements are parallel to the direction of wave propagation

MANTLE: a zone, located at depths of approximately 34-2,885 kilometers, consisting of silicate minerals

OUTER CORE: a zone, located at depths of approximately 2,885-5,144 kilometers, that is in a liquid state and consists of iron sulfides and iron oxides

SHEAR (S) WAVES: elastic waves in a solid medium where displacements are perpendicular to the direction of wave propagation

A COMPOSITIONAL BOUNDARY

The core-mantle boundary (CMB) is a prominent discontinuity within the Earth. The mantle above the boundary is largely solid, of relatively low temperature, and primarily composed of magnesium and iron silicates. The outer core below the boundary is liquid, of higher temperature, and composed of dense materials such as iron oxides and iron sulfide alloys. This boundary separates two dynamic systems, one operating in the mantle as hot spots and convection cells, the other—in the outer core—consisting of convection currents and eddies of the core fluid. The motions of the core fluid appear to be responsible for the Earth's magnetic field. The approximate depth of the CMB is 2,885 kilometers.

It appears that the lower mantle is primarily composed of magnesium-iron silicates present in the perovskite structure. Although some amount of aluminum-calcium silicates and magnesium-iron oxides may also be present, their relative abundance is not known. Measurement of the melting point of silicate perovskite led scientists to estimate the temperature of the D″ zone (the lowermost part of the mantle) in the range of 2,600-3,100 Kelvins. Similar studies of outer core materials, which are primarily iron sulfides and iron oxides, indicate that the temperature of the outermost core is at least 3,800 Kelvins. Thus, the temperature increases by about 700 Kelvins in the D″ zone, resulting in partial melting of some minerals and thereby making the zone soft with anomalous characteristics.

EARTH ZONES

Detailed studies of the seismic velocities in the Earth led K. E. Bullen to divide the interior of the Earth into seven concentric, spherical zones termed A through G. For an earthquake occurring in the crust or the upper mantle, the downgoing waves pass through the different zones of the Earth before they emerge at the surface. To study the CMB, the effects of these zones on seismic propagation need to be considered. It has been found that the Earth is laterally heterogeneous, particularly the crust (zone A), the low-velocity zone of the upper mantle (B), and the lowermost part of the mantle (D″). The D″ zone, known as the core-mantle transition zone, is approximately 200-300 kilometers thick and is located just above the CMB. A suggestion was made that the D″ zone was thinly layered, which resulted in variations in the reflected amplitude of waves from the CMB. However, accurate analysis of the short-period seismic waves recorded by seismological arrays indicated that such is not the case. Studies utilizing inversions of seismic waveforms and travel times indicate large-scale variations (more than 1,000 kilometers) in the velocities in the D″ zone. The P-wave velocity varies as much as 1.5 percent, which is three to four times more than in the middle part of the mantle. The longitudinal (P) waves appear to travel faster in the portions of this zone that are located below North America, China, the eastern part of the Indian Ocean, and off the Pacific coast of Chile. Lower P-wave velocities are observed below the southern part of Africa, the New Hebrides Islands, the South Pacific Ocean, and the Argentine Basin.

Similar large-scale variations have also been observed for shear (S) waves. The D″ zone under the American continents, Asia, the northern Indian and Pacific oceans, and Antarctica are characterized by higher S-wave velocities. Lower S-wave velocities are observed underneath the Central and South Pacific Ocean, the Atlantic Ocean, the major parts of Africa, and the southern part of the Indian Ocean. These long wavelength velocity variations in the D″ zone appear to continue upward in the mantle. Thus, the thermally induced convection currents and hot spots in the mantle appear to be related to large-scale, lateral velocity variations in the D″ zone. The lateral heterogeneity of the D″ zone is also evident in the short-scale (less than 100-kilometer) and the intermediate-scale (100-1,000-kilometer) lengths, determined primarily through studies of scattered core phases and waveform modeling techniques.

BOUNDARY UNDULATIONS

Seismologists studying the core-mantle boundary (CMB) by means of reflected waves from the core have long been frustrated by the strong scatter of the reflected amplitudes. A major part of this scatter is believed to be the result of undulations of the core-mantle boundary. The lateral ex-

tent of these undulations is of the order of thousands of kilometers. The elevation of the boundary may change as much as 5-8 kilometers above or below its normal depth. Topographic highs of the CMB have been observed beneath the Indian Ocean, the Pacific Ocean, and the Atlantic Ocean (particularly in the North Atlantic). The CMB is depressed below the Tonga-Karmadec area, the China-Japan region, Central Africa, and the area off the west coast of South America. Because most of these areas are associated with the subduction of oceanic plates, the CMB structure is believed to be caused by the dynamic processes in the mantle, which may again be related to the convection processes in the outer core. Subduction of a crustal plate is caused by the downwelling convective flow in the mantle. When the convective flow reaches the core boundary, it depresses the CMB into the hot, liquid core. Core fluids may partially invade the "topographic low" of the CMB, altering the chemical composition of the D″ zone. Similarly, beneath an upwelling zone of mantle flow, liquid core material may be "sucked" up into the mantle, creating a topographic high of the CMB. The "lows" of the CMB, being subjected to the higher temperatures of the outer core, may melt and recrystallize at the topographic "highs" of the CMB. Thus, the overall effect is to smooth the CMB, which is continually disturbed by the convective circulations in the mantle. With heat dissipation in the mantle, the outer core slowly cools and the core materials crystallize and underplate the mantle. Thus, the outer core slowly shrinks with time, and the CMB gets deeper.

A THERMAL BOUNDARY

Although near the Earth's surface the temperature increases quickly with depth, the rate of increase slows below 200 kilometers depth. The core-mantle boundary not only is a compositional boundary but also appears to be a thermal boundary, where the temperature increases by at least 700 Kelvins. Thermal coupling between the mantle and the outer core, however, may change laterally, resulting in a variable heat flow across the boundary. Although no consensus has been reached among scientists, it is possible that the mantle dynamics are at least partially responsible for controlling the heat flow.

The Earth behaves like a large magnet. The

magnetic field of the Earth—that is, the geomagnetic field—undergoes a slow change known as the secular variation. The origin of the geomagnetic field appears to be related to motions of the outer core fluid. Studies suggest that the deep mantle and the outer core play a significant role in shaping the secular variations. Upwellings in the outer core material are associated with the hot and seismically slow regions of the D″ zone; downwellings are associated with the cold and higher-velocity portions of D″. Cold regions in the mantle transmit greater amounts of heat from the outer core, thereby setting up mantle circulations. The topography of the CMB is controlled by the circulations in the mantle as well as in the outer core. The topographic relief of the CMB may also set up a lateral temperature gradient which may be responsible for the secular variation of the magnetic field.

SEISMOLOGIC STUDY

Various subdisciplines of geophysics are being utilized to determine the nature and the structure of the core-mantle boundary. They include, among others, seismology, geodesy and geodynamics, high-temperature and high-pressure mineral physics, geothermometry, and geomagnetism. Seismology has been the most important among all these subdisciplines and has contributed most of the information about the Earth's interior.

Seismology deals with earthquakes and the propagation of earthquake waves through the Earth. Whenever an earthquake occurs, different types of waves are generated. Surface waves travel along the Earth's surface, and longitudinal (P) waves and shear (S) waves travel through the interior of the Earth. It is often helpful to visualize P and S waves in the form of rays originating from an earthquake focus, or hypocenter, and radiating in all directions through the Earth. Because of the increased rigidity and incompressibility of rocks downward, the velocities of these waves increase with depth. As a result, the downgoing seismic rays (except for vertical or near-vertical rays) are curved back to the surface. Thus, the seismographic stations that are farther away from the epicenter record direct seismic rays, which penetrate through the deeper layers in the mantle.

The outer core has no rigidity, as it is liquid. Consequently, the velocities of seismic waves decrease abruptly as they cross the CMB. P waves decrease from 13.5 to 8.5 kilometers per second and are steeply refracted in the outer core. S waves do not propagate through the outer core. As a result, shadow zones are produced for both direct P and S waves recorded at the ground surface beyond 12,000 kilometers from the epicenter. The presence of a liquid outer core was discovered through the existence of the shadow zones and the absence of core-transmitted shear waves.

Seismic rays emerging at steep angles from the hypocenter encounter the CMB. Part of the incident energy is reflected back from the boundary, and the rest is refracted through the outer core. A P wave can be reflected back as a P and as an S wave, called PcP and PcS waves (or phases) respectively. Similarly, an S wave reflected back from the CMB as a P or an S wave is called ScP or ScS. These core-reflected phases have been important in the study of the nature, shape, and depth of the CMB. Because S waves cannot transmit through liquid, the refracted energy in the outer core propagates in the form of P waves. These waves are designated as a K phase. Thus, PKP is a phase that travels from the hypocenter in the mantle as a P wave, propagates as a P (that is, K) in the outer core, and reemerges in the mantle as a P wave. Similarly, SKS and other combinations, such as PKS and SKP, are often observed in the seismic records. A joint study of the core-reflected phases (for example, PcP) and the core phases (for example, PKP) is often important in resolving the depths and topography of the CMB. Seismic rays incident at a large angle on the CMB are diffracted. Study of these diffracted waves provides important information on the D″ zone above the CMB. Using the waveform modeling techniques, scientists are determining the thickness and fine structures of the D″ zone.

SEISMIC TOMOGRAPHY AND FREE OSCILLATIONS

Another important tool is seismic tomography. It utilizes the same principle used in CAT-scan X rays of humans. In a CAT scan, the X-ray source and the camera are rotated around the body and a large number of images are produced. A computer processes these images and forms a three-dimensional image of the internal organs of the subject. The seismological data collected worldwide can similarly be processed to form a three-dimensional image of the Earth's interior. Seismic tomography is providing valuable information on the CMB as well as the Earth's mantle.

A large earthquake sets the Earth vibrating like a bell. If the Earth were perfectly spherical with uniform layering, it would produce a pure tone, vibrating at a preferred frequency. Departures from the spherical shape of the Earth, as well as depth-related discontinuities, produce additional tones involving distortions of the Earth. Thus, recordings of these various modes of the Earth's vibrations, known as free oscillations, can furnish information about the shape of the CMB.

GRAVITY MEASUREMENTS AND GEODETIC OBSERVATIONS

Satellite measurements of the Earth's gravity field and the geodetic observations of the geoid can also provide us with information on the CMB. Theoretical models of the Earth's interior, particularly the mantle, the D″ zone, and the CMB, can be constructed to match observed geoidal undulations and the gravity anomalies. It appears that a 2-3-kilometer change in elevation of the CMB can explain 90 percent of the observed large-scale gravity anomalies. Astronomic observations of the Earth's wobble furnish additional constraints on the shape of the CMB. The Earth has an equatorial bulge caused by its rotation. The Moon pulls at the bulge and attempts to align it along the orbital plane of the Moon, generating a wobble, or a nutational motion, of the Earth's axis. (This motion is similar to the wobble of a spinning top or a gyroscope.) Deformation of the CMB, which is not formed by rotation, produces certain irregularities in the nutational motion. Studies of these irregularities indicate that the undulations of the CMB are less than 1 kilometer in height.

ROCK AND MINERAL STUDIES

Major developments in instrumentation have made it possible to simulate in the laboratory the temperature and pressure conditions of the deep mantle. Scientists can now study how the crystal structures of minerals change with increased temperature and pressure. Measurements on the electrical properties of rocks under high pressure, and possible alloying of iron by sulfur and oxygen that may occur in the outer core, are also being studied. These investigations are important for com-

plete understanding of the mineral compositions, structure, temperature, and pressure environment of the Earth's deep interior.

GEOLOGIC SIGNIFICANCE

The study of the core-mantle boundary is important from several perspectives. The CMB is believed to be associated with deep mantle plumes, the mantle convection currents that drive the lithospheric plates, and may be responsible for secular variations of the geomagnetic field. As the most pronounced discontinuity within the Earth, the undulations at the CMB may also cause regional gravity anomalies and can affect the transmission of seismic waves. The transmission effects of seismic waves crossing the CMB should be determined in order to study the geometry and the physical and chemical parameters of materials at the boundary as well as the outer and inner core of the Earth. Furthermore, because core-reflected phases travel along vertical or near vertical paths in the mantle, they are often utilized to study heterogeneity and the seismic behavior in the mantle. Knowledge of the nature of the CMB is necessary to determine these mantle characteristics.

Scientists from various geophysical subdisciplines have made a concerted effort to investigate the structure and nature of the CMB and the deep interior of the Earth. A committee on Studies of the Earth's Deep Interior (SEDI), under the auspices of the International Union of Geodesy and Geophysics (IUGG) and the American Geophysical Union (AGU), was formed to stimulate exchange of scientific information about the Earth's interior. The first meeting of SEDI was held in Spain in 1988; starting with the fall of 1987, special sessions on the Earth's deep interior have been held at most of the AGU meetings.

D. K. Chowdhury

CROSS-REFERENCES

Earthquakes, 316; Earth's Core, 3; Earth's Crust, 14; Earth's Differentiation, 20; Earth's Lithosphere, 26; Earth's Mantle, 32; Earth's Shape, 111; Earth's Structure, 37; Elastic Waves, 202; Geothermal Phenomena and Heat Transport, 43; Geothermometry and Geobarometry, 419; Gravity Anomalies, 120; Heat Sources and Heat Flow, 49; Lithospheric Plates, 55; Mantle Dynamics and Convection, 60; Plumes and Megaplumes, 66.

BIBLIOGRAPHY

Bolt, Bruce A. *Earthquakes.* New York: W. H. Freeman, 1988. This volume presents information on the Earth's interior obtained from seismological studies. Suitable for high school and college levels.

_____. *Inside the Earth.* San Francisco: W. H. Freeman, 1982. A good introduction to seismology for the nonscientist, this well-illustrated, concise book summarizes the seismological methods and the results.

Eiby, G. A. *About Earthquakes.* New York: Harper & Row, 1957. Lucidly written, this book provides the historical perspective of seismological discoveries. Suitable for high school and undergraduate students.

Gurnis, Michael, et al., eds. *The Core-Mantle Boundary Region.* Washington, D.C.: American Geophysical Union, 1998. This collection of articles is one volume of the American Geophysical Union's Geodynamics series. Although intended for the specialist, the essays contain plenty of information suitable for the careful college-level reader. Bibliography.

Hemley, Russel J., ed. *Ultrahigh-Pressure Mineralogy: Physics and Chemistry of the Earth's Deep Interior.* Washington, D.C.: Mineralogical Society of America, 1998. Although intended for the reader with a background in Earth sciences, this book provides excellent descriptions of geophysical and geochemical processes in the Earth's interior.

Hodgson, J. H. *Earthquakes and Earth Structures.* Englewood Cliffs, N.J.: Prentice-Hall, 1964. This volume summarizes the important seismological observations prior to 1960. Suitable for high school and undergraduate students.

Jackson, Ian, ed. *The Earth's Mantle: Composition, Structure, and Evolution.* Cambridge: Cambridge University Press, 1998. Intended for the college student, *The Earth's Mantle* provides a clear and complete description of the elements that make up the Earth's mantle and the process of change that it has undergone since its formation. Includes bibliography and index.

Jacobs, John. *Deep Interior of the Earth.* London: Chapman and Hall, 1992. This introductory geophysics textbook is formidable for the average student because there is considerable mathematics in some chapters, but it does cover many useful topics. It contains a minimum of equations but many figures and graphs.

Kerr, Richard A. "Continents of the Core-Mantle Boundary." *Science* 233 (August 1, 1986): 523-524. This short, well-written article provides some of the results of current research about the core-mantle boundary. Appropriate for readers at any level.

Lay, T. "Structure of the Core-Mantle Transition Zone: A Chemical and Thermal Boundary." *EOS: Transactions of the American Geophysical Union* 70, no. 4 (1989): 44-59. This is an important review article on the transition zone at the core-mantle boundary. Suitable for a university-level audience with some background in the geosciences.

_____. "Structure of the Earth: Mantle and Core." *Reviews of Geophysics* 25 (June, 1987): 1161-1167. This important review article summarizes major works on the mantle and the core. Appropriate for university-level readers.

Young, C. J., and T. Lay. "The Core-Mantle Boundary." *Annual Review of Earth Planetary Sciences* 15 (1987): 25. A summary of recent research on the core-mantle boundary. The treatment is at a university level.

EARTH'S CRUST

Humankind's existence and modern society depend upon the crust of the Earth. The dynamic changes involved in the creation and destruction of crustal rock also liberate gases and water that form oceans and the atmosphere, cause earthquakes, and create mineral deposits essential to society.

PRINCIPAL TERMS

ANDESITE: a volcanic igneous rock type intermediate in composition and density between granite and basalt

BASALT: a dark-colored igneous rock rich in iron and magnesium and composed primarily of the mineral compounds calcium feldspar (anorthite) and pyroxene

DENSITY: the mass per unit of volume (grams per cubic centimeter) of a solid, liquid, or gas

GRANITE: a silica-rich igneous rock light in color, composed primarily of the mineral compounds quartz and potassium- and sodium-rich feldspars

ISOSTASY: the concept that suggests that the crust of the Earth is in or is trying to achieve flotational equilibrium by buoyantly floating on denser mantle rocks beneath

P WAVE: the fastest elastic wave generated by an earthquake or artificial energy source; basically an acoustic or shock wave that com-

presses and stretches solid material in its path

PLATE TECTONICS: the theory that the crust and upper mantle of the Earth are divided into a number of moving plates about 100 kilometers thick that meet at trench sites and separate at oceanic ridges

REFLECTION: the bounce of wave energy off a boundary that marks a change in density of material

REFRACTION: the change in direction of a wave path upon crossing a boundary resulting from a change in density and thus seismic velocity of the materials

SNELL'S LAW: a statement of the fact that refraction of seismic waves across a boundary will occur such that the ratio of the two velocities of the material on either side of the boundary is equal to the size of the two angles on either side of the boundary formed by the ray path and a line perpendicular to the boundary

ROCKS OF EARTH'S CRUST

The crust of the Earth is the outermost layer of rock material of the Earth. It is distinct from the region of rocks lying beneath it, called the mantle, in that the rock materials composing the crust are of a different composition and a lower density. Density may be described as the weight per unit of volume of solid materials. Therefore, if a cubic centimeter of granite, which makes up much of the crust of the Earth of continents such as North America, could be weighed, it would total 2.7 grams. Deeper crustal rocks under continents have higher densities, some approaching the 3.3 grams per cubic centimeter characteristic of upper mantle rocks. A sample of crustal rock underlying the ocean basins would reveal that it is a rock type known as basalt, with a density of about 2.9

grams per cubic centimeter.

Compared with the rocks of the mantle, the rocks of the Earth's crust are quite varied. The rocks of the crust can be classified as belonging to one of three broad groups: igneous, sedimentary, and metamorphic. Both granite and basalt are igneous rocks. Such rocks are formed by cooling and crystallization from a high-temperature state called magma or lava. Other igneous rock types of the Earth's crust that are intermediate in rock composition and density between granite and basalt include andesite and granodiorite. Igneous rocks may form by melting of other igneous and metamorphic rocks in the crust or upper mantle, or by melting of sedimentary rocks.

Metamorphic rocks are formed from other rocks that have been subjected to pressures and temperatures high enough to cause the rock to respond by change in the crystalline structure of the rock materials. These temperatures are not high enough to melt the rock. Such changes often occur in the deep parts of the crust, where heat is trapped and great pressure occurs from the weight of the overlying rock. As a consequence of this high pressure, densities of metamorphic rocks of the lower crust average about 2.9 grams per cubic centimeter.

Sedimentary rocks of the Earth's crust are formed by chemical change and physical breakdown into fragments of other rocks exposed to the atmosphere and water of the Earth's surface. The density of sedimentary rocks is generally less than that of igneous rocks, ranging from about 2.2 to as high as 2.7 grams per cubic centimeter.

THICKNESS OF EARTH'S CRUST

The boundary between the rocks of the crust and the mantle is known as the Mohorovičić discontinuity, or simply Moho. The nature of this boundary varies from place to place. Under parts of the crust that have recently been stretched or compressed by mountain-building forces, such as under the great desert basins of the western United States, the Moho is a very sharp, distinct boundary. Elsewhere, in the interior of continents that have not been deformed for long time periods, the Moho appears to be an area of gradual density change with increasing depth rather than a distinct boundary. The position of the Moho, and thus the thickness of the crust, varies widely. The crust is thickest under the continents, reaching a maximum of 70 kilometers beneath young mountain chains such as the Himalaya. Under the ocean basins, the crustal thickness varies from 5 to 15 kilometers.

OCEANIC CRUST

Thickness of the crust is directly related to its formation and evolution through geologic time. Only within the last twenty-five years have geoscientists understood this relationship. The crustal rocks of the Earth are constantly being created, deformed, and destroyed by a process known as plate tectonics. Plate tectonics theory suggests that the crust and upper mantle of the Earth are divided into a number of separate rock layers that resemble giant plates. These plates are in motion, driven by heat from the Earth's interior. Where the heat reaches the surface along boundaries between plates on the ocean floor, new rocks are formed by rising lava, creating new ocean basin crust. Because new crust is being created, crust must be consumed or destroyed elsewhere so that the Earth's volume will remain constant. The sites where crust is consumed also lie on the ocean floor. Topographically, such sites are deep trenches where the crust bends down into the mantle to be heated and remelted. Such a process of recycling ocean-basin crust means, first of all, that ocean-basin crust is never geologically very old. The oldest sea-floor crust in the western Pacific is 175 million years old as compared to about 4.5 billion years for the age of the Earth. Second, it suggests that since ocean-basin crust goes through a geologically short life and uncomplicated history, it has a rather uniform thickness of about 5 to 15 kilometers, unchanged between the time it is born and the time it is destroyed.

CONTINENTAL CRUST

Continental crust has a much longer life and a more complicated history, reflected in a highly variable crustal thickness. It is created at the sites

CHEMICAL COMPOSITION OF EARTH'S CRUST

Element	Weight (%)	Volume (%)
Oxygen (O)	46.59	94.24
Silicon (Si)	27.72	0.51
Aluminum (Al)	8.13	0.44
Iron (Fe)	5.01	0.37
Calcium (Ca)	3.63	1.04
Sodium (Na)	2.85	1.21
Potassium (K)	2.60	1.88
Magnesium (Mg)	2.09	0.28
Titanium (Ti)	0.62	0.03
Hydrogen (H)	0.14	—

SOURCE: Data are from William C. Putnam, *Geology*, 2d ed., revised by Ann Bradley Bassett, 1971, and Sybil P. Parker, ed., *McGraw-Hill Concise Encyclopedia of Science and Technology*, 2d ed. 1989.

where oceanic crust is consumed, also known as subduction zones. As the crust and upper mantle, or lithospheric plate, are bent back down into the Earth, this material heats up. Eventually, melting of part of this rock material occurs, creating volcanoes near trench sites. Such volcanoes have lavas rich in elements such as calcium, potassium, and sodium. When these lavas cool to form rock, the rock type that results is an andesite, named for volcanic rocks abundant in the Andes of South America. These continental volcanic rocks are less dense than basalts and, once created, remain on the top of a lithospheric plate, where they are carried along by the motion of the sea floor and underlying lithospheric plate as it moves away from the ocean ridge boundaries. Eventually, the sea-floor motion may cause pieces of this continental crust to collide and weld together, forming larger pieces of continental crust. Thus, continents grow with time by two processes: volcanism above subduction zones and collision. The process of collision causes rocks to pile up like a throw rug pushed against a wall, creating high mountains that also extend downward with roots that increase crustal thickness. Continents thus grow along their edges where young mountain belts, called orogenic belts, are found, such as the Andes and the mountain systems of the western United States. The crust is relatively thick under young mountain belts, piling upward and sinking downward simultaneously to form a thick wedge of rock. In this sense, it is much like a buoyant iceberg, with the majority of its mass below the water or, in this case, below sea level. The buoyancy of the lighter continental rocks above the denser mantle rocks is known as the principle of isostasy, or flotational equilibrium. Just as the iceberg must reach a flotational level by displacing a volume of water equal to its mass, so must the continental crustal rocks displace a volume of denser mantle rocks to reach their buoyancy level. Thus, under higher mountainous terrain thicker crust is found, whereas at lower elevations, such as under the ocean basin, the thinnest crust is found.

Toward the center of continental landmasses are core areas of older rocks known as cratons. The age of rocks found in the cratons ranges from about 500 million to an extreme of 3.8 billion years. The cratons of the world compose about one-half of the area of the continental crust and have been free of deformation and mountain-building forces for long periods of time. Consequently, their surfaces tend to be relatively flat as a result of surface processes such as weathering and stream-cutting acting on the exposed rocks over a geologically long period of time. The thickness of the continental

PRIMARY ROCKS AND MINERALS IN EARTH'S CRUST

Rocks	% Volume of Crust	Minerals	% Volume of Crust
Sedimentary		Quartz	12
Sands	1.7	Alkali feldspar	12
Clays and shales	4.2	Plagioclase	39
Carbonates (including		Micas	5
salt-bearing deposits)	2.0	Amphiboles	5
		Pyroxenes	11
Igneous		Olivines	3
Granites	10.4	Clay minerals (and	
Granodiorites, diorites	11.2	chlorites)	4.6
Syenites	0.4	Calcite (and aragonite)	1.5
Basalts, gabbros,		Dolomite	0.5
amphibolites, eclogites	42.5	Magnetite (and	
Dunites, peridotites	0.2	titanomagnetite)	1.5
		Others (garnets, kyanite,	
Metamorphic		andalusite, sillimanite,	
Gneisses	21.4	apatite, etc.)	4.9
Schists	5.1		
Marbles	0.9	**Totals**	
		Quartz and feldspars	63
Totals		Pyroxene and olivine	14
Sedimentary	7.9	Hydrated silicates	14.6
Igneous	64.7	Carbonates	2.0
Metamorphic	27.4	Others	6.4

SOURCE: Michael H. Carr et al., *The Geology of the Terrestrial Planets*, NASA SP-469, 1984. Data are from A. B. Ronov and A. A. Yaroshevsky, "Chemical Composition of the Earth's Crust," American Geophysical Union Monograph 13.

crust in cratonic areas is variable, which is a reflection of their long and complex histories. These areas were at one time thickened because of mountain-building activity, but long and varying periods of stability have caused them to lose some crustal thickness as well. Figures for central Canada and the United States show a range of from 30 to 50 kilometers for thickness of the craton.

SEISMIC WAVE VELOCITIES

Elastic waves are created by both earthquakes and artificial sources and may be used to study the crust of the Earth. This branch of Earth science is called seismology. When energy is released in rock by a source, the rock is set in motion with an up-and-down or back-and-forth wavelike motion. These waves force the rock to respond like a rubber band, stretching and compressing it without permanently deforming it. Such a response is called elastic. This response can be used as a key to studying the physical properties of the rocks because of a wave generated by a seismic relationship between velocity and rock properties. The fastest elastic wave is the primary, or P, wave. This wave is basically an acoustic wave or sound wave traveling in rock, compressing and stretching rock materials in its path. Therefore, the density, rigidity, and compressibility of a material determine wave velocity. A simple example would be to compare the velocity of an acoustic wave in air, called the speed of sound, to that in rock. In air near sea level, an acoustic wave travels at about 0.3 kilometer per second, whereas in rocks near the Earth's surface, the same kind of wave travels at about 5 kilometers per second. Air is much less dense than rock and has no rigidity (no permanent shape).

Rocks of the lower crust beneath the continents have P-wave velocities of between 6.8 and 7.0 kilometers per second. It can be shown in the laboratory that metamorphic rocks known as granulites, when placed under the pressures and temperatures of the lower crustal depths, have velocities in this range. Other rocks under the same pressures and temperatures (600 to 900 degrees Celsius, 5,000 to 10,000 atmospheres) may have similar velocities. Samples of lower crustal rocks known as xenoliths, however, exist at the surface, having been brought up by volcanic activity. These samples also suggest that granulite is a good choice.

Rocks of the upper crust in continental areas have P-wave velocities of around 6.2 kilometers per second. Here, rocks at the surface of a granite to granodiorite composition suggest the appropriate choice of rock. When such rocks are velocity-tested in the laboratory under the appropriate range of pressures and temperatures, there is a good match between rock type and velocity.

The composition of the oceanic crust is well known. Here, basalts yield a velocity of around 6.7 kilometers per second, reflecting the rather uniform composition of the geologically simpler oceanic crust. Finally, part of the continental crust is mantled by sedimentary rocks. This material has among the lowest velocities, ranging from less than 2 kilometers per second up to an extreme of about 6 kilometers per second and reflecting a wide range of compositions as well as the presence of open space and fluids contained therein.

SEISMIC WAVE REFRACTION AND REFLECTION

The thickness of continental crust has been determined by the study of seismic waves that bounce off (reflect) or bend (refract) when they cross the Moho. The density contrast and resulting change in velocity of seismic waves when they cross the Moho from crust to mantle cause the wave path to change angle or bend. The same phenomenon occurs when light crosses from air to liquid in a glass. This can be shown by placing a straight straw in the liquid and gazing along its length. The straw will appear bent even though it is actually the light wave that has bent.

Waves that leave the source at one critical angle will cross the boundary and travel along beneath it, radiating energy back to the surface at the same angle. This is the critically refracted ray path, and the sine of the critical angle can be predicted from Snell's law of refraction to be the ratio of the crust and mantle velocities. The geometry of this wave path is determined by two factors, the ratio of the crust and mantle velocities, and the thickness of the crust. The thicker the crust, the longer the travel time of the wave for a particular pair of velocities. Using critically refracted P waves, thicknesses have been estimated for much of the crust.

It has been possible in many areas to check the crustal thickness determined by critically refracted waves by using information from reflected waves. This has been applied with particular success to the study of Earth's crust in continental areas. The technique is similar to that of depth sounding in ships, in which an acoustic wave is sent down from a ship, bounces off the bottom, and returns. The depth is proportional to the time of travel of the wave, also called the two-way time. The depth can be found by multiplying water velocity by travel time. The same basic procedure has been used under the continents with artificial acoustic wave sources such as explosives and vibrator trucks.

DYNAMIC EVOLUTION OF EARTH'S CRUST

An understanding of the geometry, evolution, and composition of the Earth's crust increases humankind's knowledge of the nature of the world. It is easy to show that humankind's very existence, as well as the material wealth of modern societies, is totally dependent on the crust of the Earth. The crustal state is one of dynamic evolution, with rock materials being created, deformed, and destroyed at plate tectonic boundaries. The process that makes creation and destruction of rocks possible is that of crystallization and melting of the mineral compounds that compose rock, a process known as volcanism. Volcanic activity over the billions of years of the Earth's existence has, by the expulsion of gases trapped in lavas that reach the surface, provided the water vapor and other gases necessary to form the oceans and atmosphere, which are necessary to support life.

An understanding of volcanoes, of the how, why, and where they occur, requires an understanding of the Earth's crust and of crustal dynamics. Certainly, this can be important as viewed from the perspective of Mount St. Helens and other volcanoes of the northwestern United States. Mount St. Helens is a volcano formed by remelting of part of the oceanic crust that is slowly being taken back into the interior of the Earth. As this process of subduction and remelting of the oceanic crust will continue into the future for millions of years, so will eruptions continue to occur at Mount St. Helens, as well as at other volcanoes of the Cascade Mountains. Thus, an understanding of the crust of the Earth shows that the disastrous May 18, 1980, eruption of Mount St. Helens

was not a onetime event.

The dynamic evolution of the Earth's crust is also accompanied by the movement of large plates of the crust and upper mantle, up to 100 kilometers thick, against one another. The San Andreas fault of California is one place where two of these plates of crustal material rub against each other. The forces created by this motion are released as energy in large earthquakes, posing a threat to life and property. Eventually, knowledge of how crustal rocks change and respond to these forces before an impending earthquake may allow their prediction.

Exploration for important economic minerals is guided by knowledge about the evolution and composition of the crust. The creation of valuable metal deposits, such as gold and copper, during volcanic activity at ocean ridge sites where new oceanic crustal rocks are also being created is occurring in the Red Sea between Africa and Asia. Consequently, exploration efforts for such metallic ores can be directed toward identifying ancient ridge site deposits. The formation of continental sedimentary rocks in the Gulf of Mexico is trapping organic materials that will be turned into oil and natural gas. Looking for similar types of sedimentary rocks in the appropriate crustal environment would be worthwhile for explorers in the petroleum and natural gas industries.

David S. Brumbaugh

CROSS-REFERENCES

Andesitic Rocks, 1263; Basaltic Rocks, 1274; Continental Crust, 561; Continental Growth, 573; Continental Structures, 590; Earthquake Distribution, 277; Earth's Core, 3; Earth's Core-Mantle Boundary, 9; Earth's Differentiation, 20; Earth's Lithosphere, 26; Earth's Mantle, 32; Earth's Structure, 37; Elastic Waves, 202; Evolution of Earth's Composition, 386; Geothermal Phenomena and Heat Transport, 43; Granitic Rocks, 1292; Heat Sources and Heat Flow, 49; Isostasy, 125; Lithospheric Plates, 55; Mantle Dynamics and Convection, 60; Mountain Belts, 841; Plate Margins, 73; Plate Motions, 80; Plate Tectonics, 86; Plumes and Megaplumes, 66.

BIBLIOGRAPHY

Bally, A. W. *Seismic Expression of Structural Styles.* Tulsa, Okla.: American Association of Petroleum Geologists, 1983. An excellent visual treatment of the structure and layering of primarily the upper crust throughout the world. Sections into the crust of offshore Scotland and northwest Germany show the Moho. Suitable for a broad audience from general readers to scientific specialists.

Bott, M. H. P. *The Interior of the Earth.* New York: Elsevier, 1982. This book was intended for undergraduate and graduate students of geology and geophysics as well as for other scientists interested in the topic. The plate tectonic framework of the outer part of the Earth is strongly emphasized.

Brown, G. C., and A. E. Mussett. *The Inaccessible Earth.* London: Allen & Unwin, 1981. A good general introduction geared toward the undergraduate college student. The primary topics are the internal state and composition of the Earth. Included are background material on seismology and three chapters discussing the Earth's crust.

Crossley, David J., ed. *Earth's Deep Interior.* Amsterdam: Gordon and Breach Science, 1997. A highly technical but extremely detailed guide to geophysics and the Earth's crust, *Earth's Deep Interior* is intended for the careful college student with an Earth science background. The volume provides illustrations that prove useful to all levels.

Fountain, Daviv M., R. Arculus, and R. W. Kay, eds. *Continental Lower Crust.* Amsterdam: Elsevier, 1992. One volume of the Developments in Geotectonics series, *Continental Lower Crust* provides an in-depth description of the composition and evolution of the Earth's lower crust. Includes bibliographical references and index.

Jackson, Ian, ed. *The Earth's Mantle: Composition, Structure, and Evolution.* Cambridge: Cambridge University Press, 1998. Intended for the college student, *The Earth's Mantle* provides a clear and complete description of the elements that make up the Earth's mantle and the process of change that it has undergone since its formation. Includes bibliography and index.

Phillips, O. M. *The Heart of the Earth.* San Francisco: Freeman, Copper, 1968. An excellent and well-written book intended for a general college and noncollege audience with no background in geophysics. The book has an excellent chapter on seismology and the way in which earthquake waves are used to determine physical properties from velocity and to infer crustal structure by refracted waves.

Smith, David G., ed. *The Cambridge Encyclopedia of Earth Sciences.* Cambridge, England: Cambridge University Press, 1981. This general reference provides an excellent overview of the Earth sciences. Chapter 10 is an extensive discussion of the Earth's crust, including useful illustrations and diagrams. Contains a glossary, an index, and recommendations for further reading.

Taylor, Stuart R., and Scott M. McLennan. *The Continental Crust: Its Composition and Evolution.* Boston: Blackwell Scientific, 1985. A text aimed at undergraduate and graduate geology and geophysics students as well as general Earth scientists. It is clearly written and up to date and has excellent, well-rounded scientific references.

EARTH'S DIFFERENTIATION

Earth's differentiation describes the formation of "layers" within the early Earth when it originated more than 4 billion years ago. A core surrounded by a mantle overlain with a crust, on which humans now live, was created by chemical and physical processes as the Earth cooled from a hot molten sphere.

PRINCIPAL TERMS

CRUSTAL DIFFERENTIATION: origin of continental and oceanic crust through remelting of original, heavier crust

DIFFERENTIATION: origin of layers (core, mantle, and crust) through differential settling of material in the molten Earth, based on densities

HOMOGENEOUS ACCRETION THEORY: one of two major theories on Earth's differentiation: that differentiation occurred after the Earth had formed through accretion of debris

INHOMOGENEOUS ACCRETION THEORY: one of two major theories on Earth's differentiation: that differentiation occurred while the Earth was accreting debris because denser debris formed first

PARTIAL MELTING: melting of rock that results in a magma concentrated in some minerals and depleted in others as compared to the original unmelted rock

SEISMIC WAVES: vibrational waves caused by earthquakes that are a major tool in analyzing the layers of rock within the Earth

EARTH'S "LAYERS"

The Earth today is a spherelike body composed of layers arranged according to density. The highest-density ("heaviest") material, mainly nickel and iron, is at the core, which is about 3,400 kilometers (2,100 miles) thick with an average density of about 10-13 grams per cubic centimeter. The lightest elements are dominant in the crust, the outermost layer at the surface of the Earth. This layer is very thin—only 5-60 kilometers (3-37 miles) thick. The average density of the crust is about 2.8 grams per cubic centimeter. The common elements of the crust include silicon, aluminum, calcium, potassium, and sodium. They combine to form various silicate minerals, especially the feldspars and quartz. These minerals are major components of granites and other abundant rocks. Between the crust and the core is the mantle, which is intermediate in density (4.5 grams per cubic centimeter). It is very thick, about 2,900 kilometers (1,800 miles).

HOMOGENEOUS ACCRETION THEORY

There are two major theories on how the Earth became differentiated in this way: the homogeneous accretion theory and the inhomogeneous accretion theory. The homogeneous accretion theory states that the Earth formed by randomly sweeping up debris (meteors, dust) in its orbit around the Sun. This process, which occurred about 4.6 billion years ago, caused the debris to be added on (accreted) to the early Earth in a sort of "snowball" effect that made the Earth progressively larger. The bigger it became, the more its gravity increased and the more debris it accumulated. According to this theory, because the debris was accreted at random, the Earth at this time was undifferentiated; that is, it was homogeneous, meaning it was roughly the same throughout. At some time (scientists are uncertain exactly when) during or after the accretion, differentiation occurred when this body became molten. The liquid state allowed heavier (denser) elements to sink to the core so that it became enriched in iron and nickel. The lighter elements rose to the surface to form the crust, while elements of intermediate density stayed below them. This process would also explain why radioactive elements such as uranium and thorium are common in the crust, because they would have tended to combine with the low-density crustal minerals at this time.

For the homogeneous accretion theory to be correct, there must have been some process which heated up the originally solid body, causing it to become molten liquid. Scientists agree that there were probably three processes that could have caused this heating. First, radioactive isotopes were much more common in the early Earth; radioactive decay of these isotopes would have produced much heat. This heat source has greatly decreased through time as the isotopes have decayed to more stable states. Second, during accretion of debris, the energy released as the debris impacted the Earth was converted to heat energy. Third, much heat is created by gravitational compression. As more and more material was added to the Earth, progressively greater temperatures were generated at the core. Some scientists have estimated that the combination of these three processes would have raised the temperature of the Earth by as much as 1,200 degrees Celsius. While this theory has many supporters, it is far from conclusively proven. Current research based on computer models of physical laws indicates that even with all three of these processes operating, there might not have been enough heat generated to warm a cold planetary body to a molten state.

INHOMOGENEOUS ACCRETION THEORY

To provide an alternative to the problems of the homogeneous accretion theory, the inhomogeneous accretion theory was proposed. The main difference is that this latter theory states that the differentiation of layers occurred as accretion was occurring. Instead of accreting as an originally homogeneous body, which then became layered through density gradients, the Earth is thought to have accreted the layers in sequence. First, a naked core developed from dense matter in the debris of the orbit. Later, a less dense mantle and an even lighter crust accreted around the core as lighter debris in the orbit was encountered. Yet why should progressively lighter matter be encountered in such a cloud of debris? Calculations show that in a cooling cloud of hot gas and dust, such as that in the early Earth's orbit, iron and nickel would condense first to form the core. As the cloud cooled further, silicates of progressively lighter elements would condense so that they would be accreted in that order.

POINTS OF AGREEMENT BETWEEN THEORIES

In spite of their differences, the two theories agree on a number of major points about the Earth's differentiation, the most basic being the

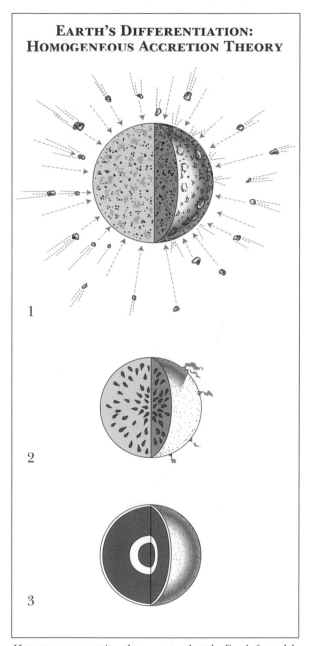

EARTH'S DIFFERENTIATION: HOMOGENEOUS ACCRETION THEORY

Homogeneous accretion theory states that the Earth formed by randomly sweeping up debris (meteors, dust) in its orbit around the Sun. (1) This process, which occurred about 4.6 billion years ago, caused the debris to be added on (accreted) to the early Earth in a sort of "snowball" effect that made the Earth progressively larger and still undifferentiated, or homogeneous. (2) Gravitational compression, increasing temperatures in the interior, initiated differentiation, as iron, nickel, and other denser elements sank to the core. (3) The result was that lighter elements rose to the surface to form the crust, and elements of intermediate density stayed between, forming today's core, mantle, and crust.

assumption that the Earth originated from condensation of a large cloud of dust and debris around the Sun. The theories also agree that the differentiation must have been well under way by 3.8 billion years ago. The oldest rocks on Earth date to this age, and they contain remanent magnetism, which indicates that the Earth had a magnetic field by then; therefore, scientists know that the core had formed. Rotation of the core produces the field. Another major point of agreement is that the different melting points of the various elements played a key role in the origin of layering. In both theories, undifferentiated matter formed separate layers because the denser materials tended to solidify first. Thus, nickel and lead condense early in the homogeneous accretion theory and then sink to the core, or condense early in the dust cloud in the inhomogeneous theory and are accreted first as the core.

CRUSTAL DIFFERENTIATION

Scientists are currently debating the evidence for the two models and, presumably, a consensus will someday be reached. Whichever of the two models is correct, it must be noted that the differentiation of the Earth did not end with the formation of merely these three layers. The early crust almost certainly consisted mainly of ultramafic minerals, which are relatively heavy minerals high in magnesium and iron. Nevertheless, today's crust, composing the continents, contains much lighter sialic minerals, which are high in silicon and aluminum. Experimental evidence shows that in order for this continental crust to differentiate, the ultramafic crust would need to undergo remelting, which would cause the lighter minerals to separate from the heavier ones.

There are two theories as to how this remelting came about. One relies on convection currents within the very hot early Earth itself. For millions of years after it formed, the Earth stayed extremely hot compared to today's interior temperatures. Therefore, as the crust began to harden and surface cooling began, there was much volcanism as hot magma from the interior sought release through the crust. Volcanoes created some local density differences in the crust and also led to erosion as some of the higher areas were exposed to weathering. It is thought by many scientists that these density differences together with accumulating sediment along the volcano margins led to remelting as convection in the underlying magma carried up great amounts of heat. In addition, convection currents return to deeper depths when they cool off, so they may have carried ultramafic crust and sediment down with them to be remelted. This recycling of rock by convection cells in hot magma was probably the beginning of plate tectonics. This same recycling process continues, although at a much slower pace because the convection currents are driven by heat and the Earth is much cooler today.

The second theory about the differentiation of the crust states that meteorite impacts were involved. Scientists know that the early Earth was (with all the planets) heavily bombarded with debris. This theory says that large meteors penetrated the original ultramafic crust, remelting some of the crust and causing a large rim to form around the crater. The remelting, along with the rim formation, might have led to a continental "nucleus" that formed the center onto which later continental material was accreted. This theory is not exclusive of the first theory because the nuclei created by the impacts would then participate in convection recycling, helping to create still more continental crust. Whether impacts were involved or not, it is clear that the amount of continental crust has continued to increase since the original differentiation to form the land masses as they are known today. At the same time, the amount of ultramafic crust has diminished because it is readily destroyed by the erosion and remelting process. Instead there remains only oceanic crust, underlying the ocean basins. The oceanic crust is more mafic than the continental crust but not as high in magnesium and iron as the original ultramafic crust. It consists of crust that has not been as completely recycled as is continental crust and is therefore somewhat denser. Today, the crust is composed entirely of the lighter continental crust, with a density of about 2.7 grams per cubic centimeter, and the denser oceanic crust, with a density of about 3.0 grams per cubic centimeter.

STUDY OF SEISMIC WAVES

It is not possible to study directly the differentiation of the Earth because the process occurred at least 4 billion years ago. If technology ever becomes adequate, it may be possible to study the differentiation of other planets in other systems. Astronomers have recently discovered debris around other stars that appears to represent a solar system in the process of formation. Until then, there are two less direct methods of studying Earth's differentiation: observations of the current internal structure of the Earth to see what changes continue to occur and laboratory and field study of the Earth and meteorites to make inferences about those processes involved in differentiation.

The core, mantle, and even the deeper crust are far too deep to be reached by conventional drilling; therefore, no one has ever seen rocks from those parts of the Earth. Instead, all scientists know of them comes from the study of seismic waves produced by earthquakes, which provide a kind of X ray of the interior. For example, if an earthquake occurs on one side of the planet, the seismic vibrations traveling to stations on more distant parts of the planet not only travel rapidly but follow refracted (curved) paths. This high speed indicates that the Earth's interior becomes denser with depth, because vibrations travel faster in denser material—which refraction confirms, because the changing density will cause waves to travel at different speeds depending on where they are. Even more telling is the behavior of different kinds of seismic waves. P (primary) waves move back and forth in the direction of travel and will go through solid or liquid material. S (secondary) waves, which move at right angles to the travel direction, will go through solids but not through liquids. Observations show that S waves are not received at locations directly on the other side of the Earth from an earthquake. This "shadow zone" indicates that the Earth has a core that is composed of liquid. On the other hand, P waves are received on the opposite side of the Earth, but they are refracted in some parts but not in others. Calculations show that this phenomenon occurs because the core is divided into two layers: a solid inner core and a liquid outer core. Similar refractions are used to locate the top of the mantle, where "seismic discontinuities" cause waves to change speed beween the mantle and crust.

LABORATORY STUDIES

Laboratory studies used to model differentiation show that iron and nickel will differentially separate from silicate minerals at an early stage in cooling. This finding is confirmed by the study of metallic meteorites, which are composed mostly of nickel and iron. These meteorites appear to have formed under extremely high temperatures and pressures such as exist in planetary cores. In fact, such meteors are thought to have originated from early planets that were shattered in collisions shortly after the solar system formed. The next major group of minerals to separate are olivine and pyroxene. These are silicates that are high in iron and magnesium and that form the rock called peridotite. This observation, plus density estimates of the mantle itself using seismic velocity data, indicates that the mantle is composed largely of that rock. (The rate of seismic wave travel can be related to what the medium is composed of chemically.) Laboratory studies of peridotite also indicate how oceanic and continental crust would differentiate from it. As peridotite rises up from the mantle at divergent plate boundaries, the pressure and temperature of the magma's environment begin to decrease dramatically. This temperature decrease leads to partial melting of the magma, which produces minerals of basaltic composition. These minerals become accreted onto the oceanic basin floor to become oceanic crust. At convergent plate boundaries where subduction is occurring, the oceanic crust, along with many sediments on it, is being pushed underneath the other plate and is being remelted. Laboratory studies of magmas of this composition show that its partial melting will create a new magma that is rich in silica and other minerals found in granites and in other common rocks making up continental crust.

SHORT- AND LONG-TERM APPLICATIONS

It is difficult to appreciate the importance of an event that occurred at least 4 billion years ago, but there are two good reasons to do so. On a personal level, each human being needs to have a firm grasp of how he or she and the planet arrived at this moment. Many of the problems facing humankind today arise because decision makers fail to consider long-term consequences.

On a more immediate, practical level, the events involved in Earth's differentiation very much shaped the Earth as humans now live on it; the better those events are understood, the better the planet will be understood and the more wisely its resources used. For example, humankind lives on only a tiny fraction of the crustal thickness, yet the crust itself is only a small fraction of the total planet. How far down will mineral resources be found? A project that had been planned to drill to the mantle to sample it directly was abandoned because it was too expensive. At present, industrialized society is running low on many materials, such as chromium, which is mined from ores naturally enriched by differentiation. If it were not for such differentiation, there would be no civilization, because it costs too much (uses too much energy) to separate the usable minerals in rocks that have not undergone natural enrichment.

For example, any average rock (like a granite) contains many valuable elements, such as gold. They are so dilute, however, that the cost of extracting them is too great to be economical. Is petroleum formed by magmatic processes of differentiation as some (only a few) geologists say? It is not known for certain. Of particular relevance is the origin of earthquakes. The processes that formed the mantle and crust played a major role in establishing the processes of plate tectonics. Scientists now know that most earthquakes are directly associated with plate tectonism. Some earthquakes originate at great depths, while others begin nearer to the surface.

On a longer time frame, our knowledge of Earth's differentiation will make it much easier to understand the formation and to utilize the resources of other planets. The other terrestrial planets—Mercury, Venus, and Mars—also underwent planetary differentiation at about the same time as did the Earth. Mars is the most habitable by humans, but all of them, especially Mercury, are rich in minerals and in other materials that humans can use. Comparative planetology can also tell scientists about the geological future of the Earth. It is important to note that Earth's differentiation is not truly complete; plate tectonism means that the crust continues to evolve. Furthermore, physiochemical changes continue to occur in the core and mantle as the interior slowly cools. By looking at other worlds that have already gone through these changes, scientists can draw conclusions about Earth's fate. Mercury, for example, because of its small size (about that of Earth's moon), lost its internal heat billions of years ago, and any movement of crust or volcanic activity has long since ceased.

Michael L. McKinney

CROSS-REFERENCES

Continental Growth, 573; Earth's Core, 3; Earth's Core-Mantle Boundary, 9; Earth's Crust, 14; Earth's Lithosphere, 26; Earth's Mantle, 32; Earth's Origin, 2389; Earth's Structure, 37; Evolution of Earth's Composition, 386; Geothermal Phenomena and Heat Transport, 43; Heat Sources and Heat Flow, 49; Lithospheric Plates, 55; Mantle Dynamics and Convection, 60; Nickel-Irons, 2718; Plate Tectonics, 86; Plumes and Megaplumes, 66; Solar System's Origin, 2607.

BIBLIOGRAPHY

Brush, Stephen G. *Nebulous Earth: The Origin of the Solar System and the Core of the Earth from Laplace to Jeffreys.* Cambridge: Cambridge University Press, 1996. Brush's book, volume 1 in the History of Modern Planetary Physics series, contains useful information on the nebular hypothesis, the origin of the solar system, and the Earth's core. Includes a bibliography and an index.

Gurnis, Michael, et al., eds. *The Core-Mantle Boundary Region.* Washington, D.C.: American Geophysical Union, 1998. This collection of articles is one volume of the American Geophysical Union's Geodynamics series. Although intended for the specialist, the essays contain plenty of information suitable for the careful college-level reader. Bibliography.

Head, James W., Charles A. Wood, and Thomas A. Mutch. "Geologic Evolution of the Terrestrial Planets." *American Scientist* 65 (January/February, 1977): 21. Comparative evolution of Mars, Venus, Mercury, and Earth in a widely read review article. Technical but very informative to the motivated layperson or student.

Jackson, Ian, ed. *The Earth's Mantle: Composition, Structure, and Evolution.* Cambridge: Cambridge University Press, 1998. Intended for the college student, *The Earth's Mantle* provides a clear and complete description of the elements that make up the Earth's mantle and the process of change that it has undergone since its formation. Includes bibliography and index.

Jacobs, John. *Deep Interior of the Earth.* London: Chapman and Hall, 1992. This introductory geophysics textbook is formidable for the average student because there is considerable mathematics in some chapters, but it does cover many useful topics. It contains a minimum of equations but many figures and graphs.

Kaufmann, William J., III. *Planets and Moons.* New York: W. H. Freeman, 1979. One of the standard and most-respected texts on comparative planetology, this book includes excellent discussions of planetary origins and differentiation. Some parts of this college text are highly advanced, but much is suitable for the interested layperson and the advanced high school student because chapters begin with the basics.

Levin, H. *The Earth Through Time.* New York: Saunders, 1988. A summary of the Earth's growth and differentiation is found in this highly respected and widely used basic freshman text. Very well illustrated and clearly written; an excellent introduction to the subject. Technical references for further research.

Moorbath, Stephen. "The Oldest Rocks and the Growth of Continents." *Scientific American* 236 (March, 1977): 92. A well illustrated discussion of differentiation processes of the continental and oceanic crusts, including the role of plate tectonism. Especially interesting discussion of earliest known rocks. Suitable for the interested layperson or the advanced high school student.

Ozima, Minoru. *The Earth: Its Birth and Growth.* Translated by J. F. Wakabayashi. New York: Cambridge University Press, 1981. An excellent overview of Earth's differentiation from the beginning of planetary condensation to the present. Suitable for interested laypersons and advanced high school students. Technical in parts, but many basic concepts, too.

Short, Nicholas M. *Planetary Geology.* Englewood Cliffs, N.J.: Prentice-Hall, 1975. A complete introduction to planetary evolution, with detailed description on an elementary level. Comprehensible to interested laypersons and high school students.

Wetherill, George W. "The Formation of the Earth from Planetesimals." *Scientific American* 244 (June, 1981): 162. A well-illustrated account of Earth's origin and subsequent differentiation. Very readable by the motivated layperson or high school student.

Wicander, R., and J. Monroe. *Historical Geology.* St. Paul, Minn.: West, 1989. An up-to-date survey of Earth history, with a good summary discussion of Earth's differentiation. A basic college-level text, but readable for the layperson and the advanced high school student.

EARTH'S LITHOSPHERE

Within the lithosphere, earthquakes occur, volcanoes erupt, mountains are built, and new oceans are formed. An understanding of the lithosphere's structure is needed in the search for oil and gas, for the prediction of earthquakes, and for the verification of nuclear test ban treaties.

PRINCIPAL TERMS

ASTHENOSPHERE: the partially molten weak zone in the mantle directly below the lithosphere

BASALT: a dark-colored igneous rock containing minerals, such as feldspar and pyroxene, high in iron and magnesium

CRUST: the rocky, outer "skin" of the Earth, made up of the continents and ocean floor

GRANITE: a light-colored igneous rock containing feldspar, quartz, and small amounts of darker minerals

MANTLE: the thick, middle layer of the Earth between the crust and the core

MOHOROVIČIĆ DISCONTINUITY (MOHO): the boundary between the crust and the mantle, named after the Yugoslavian seismologist Andrija Mohorovičić, who discovered it in 1909

PERIDOTITE: an igneous rock made up of iron- and magnesium-rich olivine, with some pyroxene but lacking feldspar

REFLECTED WAVE: a wave that is bounced off the interface between two materials of differing wave speeds

REFRACTED WAVE: a wave that is transmitted through the interface between two materials of differing wave speeds, causing a change in the direction of travel

DEFINING TERMS

The lithosphere is the rigid outer shell of the Earth. It extends to a depth of 100 kilometers and is broken into about ten major lithospheric plates. These plates "float" upon an underlying zone of weakness called the asthenosphere. The phenomenon is somewhat like blocks of ice floating in a lake: As lake currents push the ice blocks around the lake, so do currents in the asthenosphere push the lithospheric plates. The plates carry continents and oceans with them as they form a continually changing jigsaw puzzle on the face of the Earth.

The word "lithosphere" is derived from the Greek *lithos*, meaning stone. Historically, the lithosphere was considered to be the solid crust of the Earth, as distinguished from the atmosphere and the hydrosphere. The words "crust" and "lithosphere" were used interchangeably to mean the unmoving, rocky portions of the Earth's surface. Advances in the understanding of the structure of the Earth's interior, resulting mostly from seismology, have forced the redefinition of old terms. "Crust" presently refers to the rocky, outer "skin" of the Earth, containing the continents and ocean floor. "Lithosphere" is a more comprehensive term that includes the crust within a thicker, rigid unit of the Earth's outer shell. To appreciate the reason for this redefinition, it is necessary to learn about the nature of the Earth's interior.

EARTH'S INTERIOR

Except for the upper 3 or 4 kilometers, the Earth's interior is inaccessible to humans. Therefore, indirect methods, such as studying earthquakes and explosions, are used to learn about the inside of the Earth. Earthquakes and explosions, both conventional and nuclear, generate two types of energy waves: compressional (P) waves and shear (S) waves. P waves travel faster than do S waves and are generally the first waves to arrive at an observation station. The speed of a wave, however, depends on the rock through which it travels. When seismic waves encounter a boundary between two different rocks, some energy is reflected back, and some is transmitted across the boundary. If the rock properties are very different, the transmitted waves travel at a different speed and their travel path is bent, or refracted. This phenomenon can be illustrated by placing a pencil in a glass of water. Light in water travels at a speed different from that of light in air, so light is refracted, or bent, as it travels from water to air. Thus, the pencil appears to be bent. P and S waves are reflected and refracted as they travel through the Earth. Waves following different paths travel at different speeds.

Since 1900, seismologists have studied P and S waves arriving at different locations from the same earthquake. They discovered three distinct layers in the earth: the crust, the mantle, and the core. The boundaries separating these layers show abrupt changes in both P- and S-wave speeds. These changes in wave speeds provide information about the Earth's interior. Scientists studying the theory of traveling elastic waves, such as earthquake waves, related the speed of waves to the physical properties of the material through which they travel. It was found that S waves do not travel through liquids. From this finding, scientists concluded that the Earth's core had a liquid outer region and a solid inner region. Other scientists measured the P- and S-wave speeds of many different rocks and provided clues to the kind of rocks found inside the Earth.

The continental crust averages 30-40 kilometers thick and is divided into two main seismic layers. One layer, the upper two-thirds of the crust, has P- and S-wave speeds corresponding to those of granitic rocks. The speeds increase slightly in the bottom third of the continent, corresponding to rocks of basaltic composition. The average oceanic crust is 11 kilometers thick and is of basaltic composition. Beneath both continental and oceanic crust, the P- and S-wave speeds increase sharply. This boundary between the crust and mantle is called the Mohorovičić discontinuity, or Moho. The Moho marks a compositional change to a dense, ultramafic rock called peridotite.

THE ASTHENOSPHERE

At an average depth of 100 kilometers, the S-wave speed decreases abruptly. It remains low for about 100-150 kilometers. This region is called the low velocity zone (LVZ). Laboratory experiments have shown that seismic-wave speeds, particularly those of S waves, decrease in rocks containing some liquid. The LVZ in the mantle indicates a zone of partial melting, perhaps 1-10 percent melt. The presence of the melt reduces the overall strength of the rock, giving the region its name, "asthenosphere," from the Greek *asthenes*, meaning "without strength."

The partially molten asthenosphere is very mobile, allowing the more rigid lithosphere above it to move about the Earth's surface. The boundary between the lithosphere and the asthenosphere does not mark a change in composition; it marks a change in the physical properties of the rocks. The lithosphere defines this region of crust and mantle from the mantle region below by its seismic-wave speeds and its physical properties.

LITHOSPHERIC PLATES

Seismic-wave speeds and earthquake distribution provide information about the lithospheric plates and the boundaries between them. Like the Earth's crust, lithosphericic plates are not the same everywhere. For example, the Pacific plate contains primarily oceanic crust, the Eurasian plate is mostly continental, and the North American plate contains both continental and oceanic crust. The lithosphere is thinnest at spreading centers, or regions where two plates are moving away from each other, such as the Mid-Atlantic Ridge and the East Pacific Rise. Here, the asthenosphere is close to the surface and the melt portion pushes upward, separating the plates and creating new lithosphere. Shallow earthquakes occur as the new crust is cracked apart. In areas such as western South America or southern Alaska, two plates are coming together, with the oceanic lithosphere being thrust under the continental plate. Earthquakes occur as deep as 700 kilometers as one plate slides under the other. Along the California coast, two plates slide past each other along faults that cut through the lithosphere. Earthquakes are common, and the faults can move several meters at a time. Where two continental plates, India and Eurasia, have collided, the crust is highly faulted and 65 kilometers thick. Earthquakes in and near the Himalaya are numerous, often occurring along deep fault zones.

Although earthquakes are most common along plate boundaries, they can also occur within lithospheric plates. Some earthquakes are related to newly forming boundaries. The Red Sea is believed to be a recently formed spreading center pushing the Arabian Peninsula and Africa apart.

Some earthquakes result from the movement along ancient geologic faults buried within the crust. The causes of some earthquakes, however, such as the one in 1886 in Charleston, South Carolina, remain unknown.

UPPER AND LOWER LITHOSPHERE

Structural details within plate regions cannot be determined by earthquake studies alone. P and S waves generated by explosions are reflected and refracted by layers within the lithospheric plates. Regional studies show the upper lithosphere to be highly variable. In mountainous regions, such as the Appalachians or the Rocky Mountains, the continental crust is thicker than average and shows much layering. In the midcontinent and the Gulf of Mexico regions, the crust consists of thick layers of sediments and sedimentary rocks. Oil companies, combining the data from many controlled explosions, discovered petroleum and natural gas within these layers from the changes in P- and S-wave speeds. Other regional seismic studies have found ancient geological features deep within the crust. Similarities in the seismic structure between these and other known features can uncover potential sites of much-needed natural resources. The discovery of the oil fields of northern Alaska was prompted by the area's structural similarity to the Gulf of Mexico, a known source of oil and gas.

The seismic structure of the lower lithosphere is less well known. Early studies show that it is also highly variable and that crustal structures are often related to features deep in the lithosphere. Much work, however, remains in unraveling the details of the lithosphere.

STUDY OF SEISMIC WAVES

Scientists use a number of seismic techniques to study the lithosphere. They use P and S waves generated by earthquakes that travel through the Earth (body waves) and along the Earth's surface (surface waves). Reflection and refraction seismology use seismic waves generated by explosions to study the continental and oceanic lithosphere. Data from experimental studies of rocks are used to relate seismic speeds to specific kinds of rocks. Computers help analyze the vast amounts of seismic data and are used to develop models to aid in the understanding of the Earth.

The use of P and S waves from earthquakes is the oldest method of studying Earth's structure. The times at which P and S body waves, reflected and refracted by the layers in the Earth, arrive at different distances from the same earthquake are related to the average speed at which the waves travel. The arrival times of surface waves also depend on the layer speeds. Using seismic waves from many earthquakes, seismologists can determine the seismic structure of the lithosphere.

In regions with numerous earthquakes, seismologists record P and S waves using many portable seismographs, instruments that record seismic-wave arrivals. The scientists can then determine a more detailed regional structure. Earthquakes, however, do not occur regularly everywhere on the Earth. Until an average regional structure is known, it will be difficult to determine the precise location and time of an earthquake.

Explosions as a source of seismic waves to study crustal structure have been developed and used extensively by the oil industry. With an explosive source, the location and time of detonation can be precisely controlled. Two basic techniques using artificial sources are reflection seismology and refraction seismology. Refraction seismology studies the arrivals of waves that are refracted, or bent, by the layers in the crust. The scientist determines an average velocity structure for an area by recording the time the first waves arrive at receivers located varying distances from the explosion. To determine deep structure, the distance between the explosion and the receivers must be very large. Reflection seismology allows a deeper look into the crust by studying reflections from many different layers. The seismic-wave receivers do not need to be placed as far from the source as they must in refraction studies. The reflection technique combines the results from many explosions, producing a picture of the Earth's layers. This method is used extensively in the search for oil and gas. The techniques of reflection and refraction seismology have been applied to the lithosphere. Long reflection and refraction profiles have been acquired over geologically interesting but little-understood regions.

STUDY OF ROCK PROPERTIES

Seismic waves are vibrations traveling around and through the Earth. Because of friction, these vibrations eventually stop, and seismic waves no longer travel. Earthquakes and explosions generate waves that vibrate at many frequencies. The Earth slows each frequency differently. As a seismic wave travels through different rocks, the shape of its vibrations recorded on a seismograph is related to the properties of the rocks through which it travels. The analysis of seismic waveforms has shown differences between waves generated by earthquakes and by explosions.

To understand the lithosphere, it is necessary to know about rocks. Using a hydraulic press, scientists squeeze rocks in the laboratory to pressures and heat them to temperatures present deep within the Earth. They then measure the rocks' physical properties at these conditions. Experimentally measured P and S speeds are compared to wave speeds determined from earthquakes and explosions to infer the kind of rocks and the conditions that exist within the Earth. The complexity of the lithosphere, however, does not allow simple answers.

COMPUTER MODELING

To aid the scientists in their studies, computers are used to develop models—simplified representations—of the Earth. By making changes in the model, the scientist can study changes in computed seismic properties and compare them to the observed Earth properties. Changes in the model are made to resemble the Earth more closely. In modeling the lithosphere, scientists incorporate data from a wide range of sources, such as earthquake studies, experimental rock studies, and geologic maps. The computer allows the Earth scientist to test more complex models in an effort to provide a better understanding of the lithosphere.

SIGNIFICANCE

For the Earth scientist, increased knowledge of the seismic structure of the lithosphere helps in unraveling the processes by which geologic features are formed. The movement of the lithospheric plates about the Earth creates mountain ranges, causes earthquakes, and devours or creates ocean basins. Because much of the Earth is inaccessible, seismic waves generated by earthquakes and explosions are used to look deep within the Earth to provide a picture of the Earth's structure.

Increased knowledge of the lithosphere is important to the average person for three reasons: First, earthquakes are caused by movements between and within the lithospheric plates. Every year, lives are lost and millions of dollars in damage occur because of earthquakes and earthquake-related phenomena. Detailed knowledge of the lithosphere helps scientists understand where and how earthquakes occur. This information can lead to regional assessment of the potential for earthquakes and earthquake-related damage. Knowledge of the earthquake potential of a region can result in the improvement of local building codes and the evaluation of existing emergency preparedness plans. Earthquake-hazard assessment can also aid in prediction by determining the probability of future earthquake occurrence. Some success in long-term predictions has been seen in Japan and China. Eventually, the increased understanding of the lithosphere may lead to the short-term prediction of earthquakes.

Second, detailed knowledge of lithospheric structure will lead to the discovery of potential sites of needed natural resources, such as oil, gas, and coal; metals, such as iron, aluminum, copper, and zinc; and nonmetal resources, such as stone, gravel, clay, and salt. Scientists are beginning to unravel the relationship of tectonic features to the formation of many mineral deposits. Detailed knowledge of the structure of the lithosphere from seismic studies can uncover deeply buried features that may provide new sources for critically needed resources.

Finally, scientists require detailed information on the seismic structure of the lithosphere to locate and identify earthquakes and nuclear explosions. More structural information will also lead to better identification of the differences between these two types of seismic wave sources. An accurate and reliable means of distinguishing between earthquakes and nuclear explosions is critical for the verification of any nuclear test ban treaty.

Pamela R. Justice

CROSS-REFERENCES

BIBLIOGRAPHY

Bakun, William A., et al. "Seismology." *Reviews of Geophysics* 25 (July, 1987): 1131-1214. A series of articles summarizing research in seismology in the United States from 1983 to 1986. Reviews recent findings and unresolved problems in all areas of seismology. Articles are somewhat technical but suitable for the informed reader. Extensive bibliographies.

Bolt, Bruce A. *Earthquakes*. New York: W. H. Freeman, 1988. A popular, illustrated book on the many features of earthquakes. Chapter topics include the use of earthquake waves to study the Earth's interior and earthquake prediction. A bibliography and an index are included. Suitable for the layperson.

Bullen, K. E., and B. A. Bolt. *An Introduction to the Theory of Seismology*. 4th ed. New York: Cambridge University Press, 1985. Introductory sections of most chapters provide historical and nonmathematical insight into the subject, suitable for the general reader. Contains a selected bibliography, references, and an index. (Designed as a text for the advanced student with a mathematics background.)

Dahlen, F. A. *Theoretical Global Seismology*. Princeton: Princeton University Press, 1998. Intended for the college-level reader, this book describes seismology processes and theories in great detail. The book contains many illustrations and maps. Bibliography and index.

Doyle, Hugh A. *Seismology*. New York: John Wiley, 1995. A good introduction to the study of earthquakes and the Earth's lithosphere. Written for the layperson, the book contains many useful illustrations.

Langel, R. A. *The Magnetic Field of the Earth's Lithosphere: The Satellite Perspective*. Cambridge, England: Cambridge University Press, 1998. Focusing on remote sensing, Langel's book describes what has been learned about geomagnetism through the use of artificial satellite technology. Includes illustrations.

Mutter, John C. "Seismic Images of Plate Boundaries." *Scientific American* 254 (February, 1986): 66-75. An article on the application of explosion seismology to the study of plate boundaries. Summarizes the method of seismic reflection profiling. Shows results of studies across different plate boundaries. Well illustrated. Suitable for the general reader.

Pitman, Walter C. "Plate Tectonics." In *McGraw-Hill Encyclopedia of the Geological Sciences*. New York: McGraw-Hill, 1978. A brief summary of plate tectonics, discussing evidence for the theory and an explanation of causes of present-day features. Cross-referenced, illustrated, with bibliography. Suitable for the general reader.

Press, Frank, and Raymond Siever. *Earth*. 4th ed. San Francisco: W. H. Freeman, 1986. A book for the beginning reader in geology. Of interest are chapter 17, "Seismology and the Earth's Interior," and chapter 19, "Global Plate Tectonics: The Unifying Model," for an overall understanding of the importance of the lithosphere. Illustrated and supplemented with numerous marginal notes. Chapter bibliographies and glossary.

_____. *Understanding Earth*. 2d ed. New York: W. H. Freeman, 1998. This comprehensive physical geology text covers the formation and development of the Earth. Readable by high school students, as well as by general readers. Includes an index and a glossary of terms.

Reynolds, John M. *An Introduction to Applied and Environmental Geophysics*. New York: John Wiley, 1997. An excellent introduction to seismology, geophysics, tectonics, and the lithosphere. Appropriate for those with minimal scientific background. Includes maps, illustrations, and bibliography.

Smith, Peter J., ed. *The Earth*. New York: Macmillan, 1986. A well-illustrated, comprehensive guide to the Earth sciences for the general reader. Chapter 3, "Internal Structure," describes historical development of the current view of the Earth's lithosphere. Chapters 1, 2, and 5 provide related material. Includes glossary of terms.

Thomson, Ker C. "Seismology." In *McGraw-Hill Encyclopedia of the Geological Sciences*. New York: McGraw-Hill, 1978. A brief summary of the principles of seismology. Discusses methods of determining Earth structure and of detecting nuclear explosions and describes related research. Cross-referenced and illustrated, with bibliography. Suitable for the interested general reader.

EARTH'S MANTLE

The mantle is that portion of the inner Earth that lies between the crust and the outer core. It is composed of rocks that are of greater density than those of the crust. The mantle contains a zone in which the rock is under such great temperature and pressure that it exists in a plastic state. It is upon this zone that the major plates of the Earth's crust slide.

PRINCIPAL TERMS

DIFFERENTIATION: layering within rock that results from differences in density; the lighter material rises to the surface while the heaviest material sinks to the bottom of a mixture of substances

DISCONTINUITY: a rapid change in the properties of rock with increasing depth

FOCUS: the point within the Earth that is the center of an earthquake and the point of origin of seismic waves

IGNEOUS ROCKS: a family of rocks which have solidified from the molten state

POLYMORPHISM: the characteristic of a mineral to crystallize into more than one form

PRIMARY WAVE: a compressional type of earthquake wave, which will travel in any medium and is the fastest wave

SECONDARY WAVE: a transverse type of earthquake wave, slower than a primary wave, which will not travel in a liquid

DIFFERENTIATION

The mantle is that portion of the interior of the Earth that extends from the base of the crust to the boundary of the outer core. This distance is approximately 2,900 kilometers, roughly 45 percent of the radius of the Earth. Since the thickness of the Earth's crust is not uniform, the distance from the ground surface to the upper boundary of the mantle varies significantly. It has been determined that the thickness of the crust in continental areas is approximately 40 kilometers, while in the ocean basins the thickness is only some 5 kilometers.

Evidence for the existence of differentiation, layering within the Earth caused by density differences, was first observed from the study of earthquake waves. In 1906 a seismologist, Andrija Mohorovičić, studied the records of an earthquake that had taken place in Yugoslavia. At a certain distance from the actual focus of the earthquake, two types of earthquake waves were received. Those types were the primary waves, P waves, and the secondary waves, S waves. Although there was only a single shock, a short time later another set of P and S waves were received by the same seismograph. Mohorovičić concluded that the second set of waves were actually reflections of the original waves. When the rock was stressed and broken at the focus, P and S waves were sent out in all directions. The P and S waves initially received by the seismograph traveled by the most direct route. Waves directed downward into the Earth were reflected back from a surface and were recorded. This reflecting surface is called a discontinuity. A discontinuity is a rapid change in the properties of rock with increased depth. Knowing the velocities of seismic waves in the rocks nearer to the surface, Mohorovičić calculated the distance from the surface to the discontinuity. This boundary between the crust and the mantle is known as the Mohorovičić discontinuity, or Moho, named in his honor.

MANTLE MATERIALS

Unlike rocks of the crust, rocks of the mantle have never been directly observed, and therefore only indirect evidence of their nature or composition exists. At one time in the mid-1960's, a project to drill down through the Earth's crust to the mantle was begun. The undertaking was appropriately named Project Mohole. Unfortunately, because of lack of funding, the idea was abandoned.

The greatest source of information on the nature of mantle materials comes from the study of earthquake waves. Since the velocity of seismic waves through the mantle is known, the types of rock that conduct waves at this known velocity are primary candidates for being mantle materials. These rocks are peridotite and eclogite, which both occur to some extent in the crust. Peridotite is a heavy, dark green rock from the igneous rock family. Igneous rocks are those that have cooled and solidified from a molten state. Peridotite consists of the elements magnesium, oxygen, and silicon. The second possibility, eclogite, is composed of the minerals garnet and jadeite. Eclogite is very similar chemically to basalt, which is a lava commonly associated with worldwide volcanic activity. Since the source of volcanic activity is believed to be in the mantle, eclogite might well be the material that is transformed into basalt as pressures are reduced. Since the crust is far less dense than the mantle, as the molten material moves upward toward the surface, the lithostatic pressure exerted by overlying rock layers would be significantly less.

A third type of material that is believed to originate in the mantle is the rare substance known as kimberlite. Kimberlite occurs in pipe-shaped deposits and is mined extensively for diamonds. Diamonds are a form of carbon that has been placed under great pressure. These pressures have been calculated, and it has been concluded that pressures of this magnitude could occur only 100 kilometers or more within the Earth. The diamond pipes must have originated within the mantle.

By the use of seismic wave information, it has been shown that the mantle is not uniform throughout. It is assumed that greater depths in mantle rock would produce greater pressures and, therefore, greater rock density. If this is true, seismic wave velocity would also increase. Primary wave velocities in the upper mantle are approximately 8 kilometers per second and gradually increase with depth to a velocity of roughly 8.3 kilometers per second. At this depth, the velocity of the waves begins to drop to a value of somewhat less than 8 kilometers per second. It has been concluded that the rock composition at that depth does not change but that its physical state does. Because of the geothermal gradient, temperatures at this depth and pressure have risen to near the partial melting point of the rock. The material then assumes plastic or flow properties. This low-velocity layer was first identified by Beno Gutenberg in 1926.

CONTINENTAL DRIFT

According to the theory of continental drift, the Earth's surface consists of pieces called tectonic plates. These eighteen or so lithospheric plates slide over a plastic zone in the mantle. Apparently the low-velocity layer of the mantle is the asthenosphere, the plastic zone upon which the plates move. The asthenosphere has been found to vary significantly in depth from the surface of the Earth. It has been found to be as close as 20 kilometers in depth near an ocean ridge; the asthenosphere averages some 100 kilometers in depth under continents.

Beneath the asthenosphere the wave velocity begins to increase again. Sharp increases at depths of 400 kilometers and 650 kilometers have been noted. Scientists believe that these increased ve-

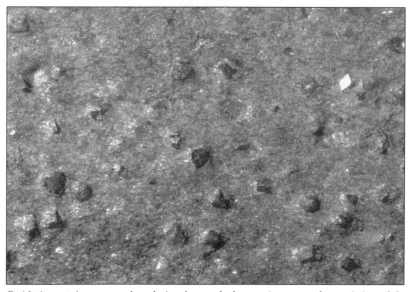

Peridotite, a primary mantle rock, is a heavy, dark-green igneous rock, consisting of the elements magnesium, oxygen, and silicon. This sample is composed of pyroxene omphacite with small garnets. (© William E. Ferguson)

locities are caused by polymorphism. Polymorphism is a term that means "many different forms." When a rock or a mineral is subjected to increasing temperature and pressure, it may rearrange its internal structure to compensate for this added stress. As a result, the density of the substance is increased, and therefore the velocity of the seismic waves passing through it is also increased. From the 650-kilometer anomaly, the wave velocities gradually increase until reaching the boundary of the outer core, where the S wave is no longer conducted.

Critical to the modern explanation of how the continents move is the topic of heat flow in the mantle. Although the mechanism of heat flow in the mantle is not completely understood, it is known that the Earth's interior heat was left over both from its original formation and from the decay of radioactive elements. It is believed that this heat causes rock to rise in the form of a current from the depths. It cools nearer to the surface and then plunges back deep within the mantle. This process is known as convection and is easily explained with the heating of a beaker of water or a container of a gas. A convection current is the density flow of a liquid or a gas. The hot material is less dense and rises, and cooler material moves in to replace the rising material. As surface material cools, it plunges below to be replaced by more rising material. This concept of a convection current is the modern explanation of heat flow in the mantle and of the mechanism that drives continental movements.

It has been known since the nineteenth century that the age of the Hawaiian islands increases from southeast to northwest. In the early 1960's, when the theory of continental drift was becoming more acceptable to Earth scientists, it was suggested that these volcanic islands recorded the movement of the seafloor. It was postulated that there existed a magma, or molten rock, source deep within the mantle. This hot spot, or plume, was a long-lived source of magma. During an eruption of the hot spot, volcanic material would be extruded out onto the seafloor. Eventually the material would break the surface of the ocean and an island would be formed. Since the Earth plates were in constant motion, the newly created island would then move away from the hot spot. Further eruption would create new islands and therefore a chain of islands like Hawaii. More than one hundred hot spots have been found worldwide.

STUDY OF SEISMIC WAVE VELOCITIES

Although the Earth's mantle cannot be directly observed, it can be studied indirectly by various techniques. The primary method of study is by use of seismic waves. These waves may be generated by an earthquake or a large explosion such as that produced by a nuclear test. At the point of rock fracture, energy is released in the form of several types of waves traveling outward in all directions and at velocities that depend upon the density of the conducting medium. The waves that travel deep into the Earth increase in velocity as they encounter denser material. Since the mantle is much denser than the crust, wave velocities in the mantle are higher than those velocities in the crust. The study of velocities of seismic waves helps scientists determine the type of rock through which the waves are traveling. When calculated velocities of waves are compared with known velocities in various types of rock, the subsurface material can be identified. The study of wave velocities can also be used to identify discontinuities in the subsurface.

When an earthquake or a large explosion occurs, energy waves travel outward from the point of energy release in all directions. These energy waves are of three main types: P waves, S waves, and L waves (surface waves). The P waves are similar to sound waves in that they are compressional in nature. The particles in a compressional wave vibrate back and forth parallel to the direction in which the wave is traveling. Primary waves will pass through any type of material. The secondary waves are transverse types of waves similar to the wave form of electromagnetic radiation. The particles of matter that make up an S wave travel perpendicular to the direction of wave propagation. The S waves are considerably slower than P waves, so at a recording station the P waves always arrive first. The L waves are also transverse waves that travel along the surface; these types are the slowest of the three waves.

As these waves travel through rock, their velocity depends on the density of the material. Since pressure increases with depth, and increased pressure results in rocks of greater density, the velocity of seismic waves is in general greater with depth.

As waves encounter boundaries between rock layers of different density or composition, some of the waves are reflected back toward the surface. It is this reflection of waves that allows scientists to determine the depth of various parts of the Earth's interior, including the mantle.

Not all waves are reflected at a discontinuity. Some waves are refracted into the newly encountered material. If the material is denser, the velocities are higher. Waves reaching the surface after being refracted through an area of greater density may arrive in a shorter time interval than those that traveled a shorter distance but through a less dense medium.

SEISMIC TOMOGRAPHY

Another modern method of studying the subsurface was first used in the field of medicine. The CAT (computerized axial tomography) scan is a composite image of X rays taken from a number of different angles. The computer assembles these images into a three-dimensional representation of the particular organ under study. The Earth science equivalent of the CAT scan is known as seismic tomography. Seismic data from all over the world are analyzed. These data provide pictures of the Earth from many different angles. The goal of seismic tomography is to assemble these pictures into a three-dimensional image of the interior of the Earth. Seismic tomography makes it possible to construct a three-dimensional image of the interior of the Earth. Using this method, seismologists have constructed images of convective circulation within the mantle. Hotter, rising materials are distinguished from cooler, descending materials by differing densities and changing velocities of seismic wave propagation.

EARTHQUAKES, VOLCANOES, AND DIAMONDS

By studying the change in velocities of waves, scientists are able to determine the nature and the composition of the Earth's mantle. The behavior of the material of the mantle has a direct bearing on much of the activity of the Earth's surface; consideration of the role of the mantle is essential in studying such concerns as earthquakes, volcanism, and even diamond mining.

A low velocity zone in the mantle is referred to as the asthenosphere. It is upon the plastic rock of this zone that the tectonic plates of the Earth's crust move. It is these moving plates that make the active fault zones and areas of extensive volcanism that exist on the Earth's surface. For example, Southern California is prone to earthquakes because it lies upon two different plates. The city of Los Angeles lies on the Pacific plate while the city of San Francisco lies upon the North American continental plate. As the Pacific plate moves to the north, it rubs against the boundary of the continental plate. The result is an earthquake.

In areas where one plate is moving below another plate, volcanism is common. The plate that is being subducted into the mantle undergoes remelting. This molten material then finds its way up to the surface through cracks and fissures. The result is a volcano. The eruption of Mount St. Helens is an example of this type of action.

Of economic importance is the mining of diamonds. Diamonds are a polymorphic form of carbon and are found in deposits of kimberlite, an igneous rock that originates deep within the mantle. As molten material it works its way to the surface through cracks and fissures; thus, knowledge of the activity of the mantle and its action on the crust aids in finding likely areas for diamond mining.

David W. Maguire

CROSS-REFERENCES

BIBLIOGRAPHY

Cailleux, André. *Anatomy of the Earth.* Translated by J. Moody Stuart. New York: McGraw-Hill, 1968. A complete, well-illustrated volume describing the Earth's interior and how it is studied. The book also treats such topics as the origin of the Earth and continental drift. The book is for general readers.

Compton, R. R. *Interpreting the Earth.* New York: Harcourt Brace Jovanovich, 1977. This well-illustrated volume discusses the geology of the Earth's surface. It also offers chapters on tectonics and continental drift. The book is suitable for general readers.

Gurnis, Michael, et al., eds. *The Core-Mantle Boundary Region.* Washington, D.C.: American Geophysical Union, 1998. This collection of articles is one volume of the American Geophysical Union's Geodynamics series. Although intended for the specialist, the essays contain plenty of information suitable for the careful college-level reader. Bibliography.

Hemley, Russel J., ed. *Ultrahigh-Pressure Mineralogy: Physics and Chemistry of the Earth's Deep Interior.* Washington, D.C.: Mineralogical Society of America, 1998. Although intended for the reader with a background in Earth sciences, this book provides excellent descriptions of geophysical and geochemical processes in the Earth's interior.

Jackson, Ian, ed. *The Earth's Mantle: Composition, Structure, and Evolution.* Cambridge: Cambridge University Press, 1998. Intended for the college student, *The Earth's Mantle* provides a clear and complete description of the elements that make up the Earth's mantle and the process of change that it has undergone since its formation. Includes bibliography and index.

Jacobs, John. *Deep Interior of the Earth.* London: Chapman and Hall, 1992. This introductory geophysics textbook is formidable for the average student because there is considerable mathematics in some chapters, but it does cover many useful topics. It contains a minimum of equations but many figures and graphs.

Jacobs, John A., Richard D. Russell, and J. T. Wilson. *Physics and Geology.* 2d ed. New York: McGraw-Hill, 1974. A technical volume covering such topics as composition of the Earth, geochronology, isotope geology, thermal history of the Earth, magnetism, and seismic studies. The text is intended for college-level students of geology or physics. Some differential equations are used in the book.

Phillips, Owen M. *The Heart of the Earth.* San Francisco: Freeman Cooper, 1968. A technical volume covering various topics in geophysics, such as gravitation, mass, earthquakes and seismic waves, volcanism, continental drift, and Earth magnetism. The volume is well illustrated with drawings and numerical tables. The reader should have a working knowledge of college algebra.

Skinner, B. J., and S. C. Porter. *The Dynamic Earth.* New York: John Wiley & Sons, 1989. A well-written, well-illustrated, and very colorful volume on the geology of the Earth. It would be suitable for the college student beginning geology.

Tennissen, A. C. *The Nature of Earth Materials.* 2d ed. Englewood Cliffs, N.J.: Prentice-Hall, 1983. A complete, well-illustrated volume covering the nature and structure of rocks and minerals. The volume would be suitable for the college student of geology, mineralogy, or petrology.

Weiner, Jonathan. *Planet Earth.* New York: Bantam Books, 1986. A colorful, well-illustrated, well-written book describing the Earth and how it is studied. This volume is the companion to the PBS television series of the same name. It is suitable for general readers.

EARTH'S STRUCTURE

Processes that are occurring in the interior of the Earth have profound effects upon the surface of the Earth and its human population. The results of processes operating in the interior include earthquakes, volcanic activity, and the shielding of life-forms from solar radiation.

PRINCIPAL TERMS

ASTHENOSPHERE: a region of the upper mantle that has less rigid and probably plastic rock material that is near to but below its melting temperature

CORE: the central spherical region of the Earth consisting of an outer liquid layer and a solid inner core

CRUST: the thinnest, outermost layer of the Earth

DENSITY: the mass per unit of volume (grams per cubic centimeter) of a solid, liquid, or gas

GRANITE: a silica-rich igneous rock light in color composed primarily of the mineral compounds quartz and potassium- and sodium-rich feldspars

LITHOSPHERE: the outer layer of the Earth, including the outer mantle and the crust

MANTLE: a layer of dense silicate rock that lies between the crust and core and comprises the majority of the Earth's volume

P WAVE: the fastest elastic wave generated by an earthquake or artificial energy source; basically an acoustic or shock wave that compresses and stretches solid material in its path

PERIDOTITE: a silicate igneous rock consisting largely of the mineral compound olivine

PLATE TECTONICS: the theory that the crust and upper mantle of the Earth are divided into a number of moving plates about 100 kilometers thick that meet at trench sites and separate at oceanic ridges

REFLECTION: the bounce of wave energy off of a boundary that marks a change in density of material

REFRACTION: the change in direction of a wave path upon crossing a boundary resulting from a change in density and thus seismic velocity of the materials on either side of the surface

EARTH'S CRUST

Evidence that comes primarily from the study of earthquake waves reveals that the interior of the Earth is not homogeneous. It is instead divided into a number of layers of varying thickness, some of which show a change in composition. The thinnest layer is the outermost one known as the crust. The crust of the Earth varies in thickness from about 5 kilometers under parts of the ocean basins up to about 70 kilometers under the highest mountain ranges of the continents. The rock materials of the crust are composed of a number of different rock types, but if an average continental rock could be chosen, it would probably be best represented by a granite. Granite is an igneous rock, formed by crystallization from a hot liquid known as a magma. It characteristically is rich in the element silicon, which constitutes about 68 percent of its composition. The ocean basin areas

of the crust, on the other hand, are characterized by an igneous rock type known as basalt. Basalt is not as rich in silicon (48 percent of its composition) but does have a greater abundance of the elements magnesium and iron.

The base of the crust is marked by a boundary known as the Mohorovičić discontinuity, or Moho. In places, the Moho is quite sharp, such as under the Basin and Range Province of the western United States. It is marked by a change in density of rock types on either side of the boundary. Density is the weight per unit of volume of materials. Thus, if a cubic centimeter of rock of a granitic composition under the continents and just above the Moho could be sampled, it would weigh 2.9 grams. Rocks below the Moho, however, would weigh 3.3 grams per cubic centimeter. This suggests a change in composition to a denser type of material. It is believed this material below the

Moho is probably a rock type known as peridotite. Peridotite is similar to basalt in composition, but the former is richer in magnesium while having slightly less silicon than basalt.

EARTH'S MANTLE

Peridotite is believed to represent the basic composition of the layer of the Earth underlying the crust known as the mantle. The mantle comprises the bulk of the Earth, representing about 80 percent by volume. The mantle is also heterogeneous. In the upper mantle at depths beneath the surface ranging from 100 to 350 kilometers is a zone of less rigid and more plastic, perhaps even partially melted, material. This zone has been termed the asthenosphere. The mantle and crust above it, acting as a more rigid unit or plate, are known collectively as the lithosphere. The change in physical properties in the asthenosphere occurs because at about 100 kilometers, temperatures in the upper mantle are close to the melting point of peridotite. Although temperature continues to increase below 350 kilometers, the tremendous pressures at those depths are high enough to keep melting of peridotite from occurring.

The asthenosphere has been suggested to play an important role in changes taking place in the lithosphere above. The theory of plate tectonics suggests that the lithosphere is divided into a number of plates about 100 kilometers thick that are in constant motion, driven by hot, convective currents of material moving slowly in the plastic asthenosphere. The heat rises along plate boundaries marked at the surface by volcanic mountain ranges in the ocean basins known as mid-ocean ridges. The slowly moving currents in the asthenosphere then move laterally away from the ocean ridges beneath the lithospheric plates, perhaps helping to carry the plates above away from the ridges. As they move laterally, these asthenospheric convection currents cool, eventually becoming denser and sinking back downward. The sites where the convection currents sink are also sites where lithospheric plates dive into the mantle, perhaps pulled by the sinking currents. At these sites, marked at the surface by gashes or trenches in the ocean-basin floor, crustal rocks may be carried into the upper mantle as deep as 670 kilometers.

Two other changes in properties occur within the mantle. At 400 and 670 kilometers below the surface, increases in density occur. Although one might suspect a change in composition to account for the jump in density, laboratory studies of rocks under pressure suggest a simpler explanation. The primary constituent of peridotite is olivine. At pressures that exist at 400 kilometers and again at 670 kilometers, there is a change that occurs that causes collapse of the crystalline structure and, as a result, produces a denser mineral compound with the same composition of iron and magnesium silicate. At pressures existing at 400 kilometers olivine converts to the denser mineral compound spinel. At the even higher pressures at 670 kilometers, spinel will convert to yet a denser mineral compound with the same composition, known as perovskite.

Thus, the changes occurring in the mantle to produce the asthenosphere and the 400- and 670-kilometer boundaries or discontinuities are not related to changes in composition but to changes related to temperature and pressure. Recall that crustal materials may be carried downward into the mantle no deeper than 670 kilometers. Although some difference of opinion exists on this point, if it is true, as most believe, it may suggest that the rock below this level is simply too dense for the lithospheric plates to penetrate.

EARTH'S CORE

The next layer beneath the mantle is called the outer core. This layer begins at a depth of about 2,900 kilometers beneath the surface and continues to a depth of 5,100 kilometers. There is a large density increase across the core-mantle boundary. At the base of the mantle, density has increased to a value of 5.5 grams per cubic centimeter, compared to about 3.3 grams per cubic centimeter at the Moho. At the top of the outer core, the density is estimated to be 10 grams per cubic centimeter. Iron is the only abundant element that would have the required density at the tremendous pressure of millions of atmospheres at these depths. Thus, the core-mantle boundary represents a composition change from the silicate perovskites of the lower mantle. Pure iron would give too high a density, so an iron alloy with silicon or possibly sulfur has been suggested.

At the pressures and temperatures that must exist at the depths in the outer core, iron compounds would be in a liquid state. Complex currents of metallic iron alloy, generated in the fluid outer core by the Earth's rotation, give rise in some complex and, as yet, poorly understood way to the Earth's main magnetic field. Some of the changes in the Earth's magnetic field, such as a slow westward drift, are a direct consequence of this rotation-generated magnetic field.

The outer core mantle boundary is a sharp one, but whether it is a smooth, spherical shape or irregular with hills or peaks on its surface is not well known. There is also some evidence from seismology that the lower mantle within 100 kilometers of the core boundary repr... zone with a chan...

ity to maintain a fixed shape, or the rigidity of the material, is one of these physical properties, while the resistance to squeezing or a change in volume is another. The S wave, however, which moves material in the path from side to side, is sensitive only to rigidity. Therefore, an S wave cannot travel across a substance that is not rigid, such as a gas or liquid, while a P wave can, but with reduced velocity. It has been found for the interior of the Earth that both P waves and S waves cross the asthenosphere of the upper mantle, but with reduced velocity, suggesting lower rigidity but not a liquid ..e, since the S wave is propagated through the ... Therefore, it seems that the asthenosphere ..represent a plastic but still solid region.

..sewhere in the Earth's interior, only the ..'s outer core shows a sharp drop in velocity ... P wave as it crosses the mantle outer core ..ary. At this point, the S wave disappears, ..ing no rigidity of the material of the outer ..nce it is known that gases cannot exist at ..s at the depth of the outer core, the mate-.. region must consist of a liquid.

..er locations, the interior of the Earth is .. increases in velocity for both the P .. S wave. There is a sharp increase in ve-.. base of the crust. Above the Moho, ..velocities are around 7 kilometers per ..le below it they jump to 8 kilometers ..r more. Below the asthenosphere, the ..ty increases gradually to a depth of ..lometers. At this depth, there is a ..ncrease in P-wave and S-wave veloci-.. wave velocities in rocks under pres-..ratory indicate that peridotite is an ..ice for upper-mantle rocks. At 400 ..sures are such that peridotite col-..inel. This change in mineral com-..ount for the increased velocity in

..ers depth in the mantle, a sec-..r increase in wave velocities oc-..es may cause a second collapse .. denser mineral compound, ..rain, P and S waves passing ..mposed of perovskite would ..elocity. Moreover, the veloci-.. he mantle match those ob-..tories from waves passing ..vskite samples placed under the

PLEASE CHECK (✓) REASON FOR RETURN. THANK YOU.

10 Wrong content level
11 Ordered in error
12 Poor quality
13 Specialized interest
14 Peripheral by YBP
15 Duplicated by YBP
16 Reprint
17 Too popular/low level
18 Narrow geographic area

1 Reject–no reason
2 Duplicate from another source
3 Damaged/defective
4 No to series (specify if subseries)
5 Out of scope/subject excluded
6 Sufficient coverage
7 Not on press list
8 Too expensive
9 Billing/shipping error

Your reasons for returning this book will help us to assess your profile and recommend potential revisions to your approval plan.

Comments:

Init.

kinds of pressures found at 670 kilometers. A final increase in velocity may be observed at the outer to inner core boundary. This could be explained by a phase transition from liquid to solid iron. Such a proposal is supported by the reappearance of the S wave in the inner core.

SEISMIC WAVE REFRACTION AND REFLECTION

Another way in which the existence of structural boundaries within the Earth can be shown from a study of seismic waves is to examine the behavior of the waves when they encounter the boundaries. Depending on the angle at which the waves approach the boundary, as well as on the properties of materials on both sides of the boundary, a seismic wave may bounce off or reflect from such a boundary, or refract or bend as it crosses the boundary. The same bending occurs when light crosses from air to liquid in a glass. This can be shown by placing a straight straw in the liquid and looking down along it. The straw will appear bent when it is actually the light wave that has bent. P waves are reflected off the Moho, mantle-core, and outer-inner core boundaries, providing clear evidence that there are sharp boundaries between these layers. Waves have also been detected bouncing off of the 670-kilometer discontinuity.

The bending or refraction of waves yields further evidence. As P waves cross the mantle-core boundary, they are refracted away from the center of the Earth because of the change in density. This lenslike focusing action for waves passing through the outer core leaves a gap on the other side of the Earth where the earthquake waves emerge. This gap is known as the P-wave shadow zone because no P waves will reach the surface in this area. This shadow zone is also evidence of the existence of a liquid outer core.

SEISMIC TOMOGRAPHY

Advances in computer science have allowed the identification of even subtler details about the Earth's interior. Computerized tomography is a technique used in medicine, in which X rays from all directions are analyzed in a computer to give a three-dimensional picture of the human body. Seismic tomography is an analogous approach that uses seismic waves that travel from earthquakes to seismographs around the world to map the Earth's interior. This includes both P and S waves in the interior as well as the results of the study of surface waves, which can also move rock material at great depths. By looking at the time of travel between two points, scientists are able to compare velocities along different paths. Such an approach has already resulted in maps of slow and fast regions of the mantle that probably represent warmer (less rigid) and colder (more rigid) regions.

SIGNIFICANCE OF EARTH'S INTERIOR PROCESSES

The interior of the Earth has profound effects on humans and their environment. The interior acts as a complex great engine. The heat energy released affects the crust of the Earth. This release of energy is the driving force behind plate tectonics that results in the formation and evolution of oceanic and continental crustal rocks. In the process, earthquakes and volcanic activity occur that create hazards for the human population on the Earth's surface. Complete acceptance of the plate tectonic theory could not occur without the discovery of the asthenosphere, which makes the movement of the lithospheric plates more plausible.

We are fortunate that the Earth has a magnetic field. Without it, the age of discovery and exploration would not have been possible, for navigation by magnetic compasses allowed voyages across uncharted oceans. Modern-day navigation is equally dependent on the magnetic field. It is now known that the Earth's magnetic field is generated from deep inside, in the region of the outer core. Another important implication of the core-generated magnetic field is the changes it undergoes through time. In particular, at rather irregular intervals the magnetic poles switch places between north and south. The details of such a switch are uncertain; however, it is known that the magnetic field decreases in strength. Since the magnetic field shields life-forms on the Earth's surface from extremes of solar radiation, there is some concern for the effect on the human population. Some scientists suspect that genetic changes occur during polar reversal periods that aid the process of biologic evolution. Thus, the surface of the Earth as well as the life-forms on it depend upon and are strongly affected by changes occurring within the Earth, in its mantle and core.

David S. Brumbaugh

BIBLIOGRAPHY

Bolt, Bruce A. "Fine Structure of the Earth's Interior." In *Planet Earth*. San Francisco: W. H. Freeman, 1974. An extremely well-illustrated review of how seismic waves have been used to discover and define the various layers of the Earth's interior. This collection of articles from *Scientific American* is written at a general-interest college level. Little background or expertise in mathematics is required.

_____. *Inside the Earth: Evidence from Earthquakes*. San Francisco: W. H. Freeman, 1982. This book is written for undergraduate college students in physics and the Earth sciences and for nonspecialists interested in a more detailed summary of knowledge of the Earth's interior. The text is relatively free of mathematics and is clearly and well illustrated. It is a rather concise, up-to-date, and readable treatment of the use of seismic waves to discover and interpret the Earth's interior. A large list of useful references is included.

Brown, G. C., and A. E. Mussett. *The Inaccessible Earth: An Integrated View to Its Structure and Composition*. London: Chapman and Hall, 1993. This book deals well with geophysics topics as they relate to the structure, chemical composition, and evolution of the Earth. Its excellent line drawings helpfully illustrate complex ideas.

Brush, Stephen G. *Nebulous Earth: The Origin of the Solar System and the Core of the Earth from Laplace to Jeffreys*. Cambridge: Cambridge University Press, 1996. Brush's book, volume 1 in the History of Modern Planetary Physics series, contains useful information on the nebular hypothesis, the origin of the solar system, and the Earth's core. Includes a bibliography and index.

Cromie, W. J. "Windows to the Earth." *Mosaic* 15, no. 6 (1984): 28-37. The articles in this journal are written for the nonspecialist, providing a very readable review of the latest developments in research of the interior of the Earth. A good summary of seismic tomography.

Heppenheimer, T. A. "Journey to the Center of the Earth." *Discover* 8 (November, 1987): 86-92. A very well-illustrated treatment of planet Earth with color illustrations and clear diagrams. Includes a short treatment of the latest advances in understanding of the Earth's interior, as well as an excellent explanation of seismic tomography. This article describes the development of the relationships between the Earth's interior and its processes and the dynamic changes occurring on the surface of the Earth.

Jackson, Ian, ed. *The Earth's Mantle: Composition, Structure, and Evolution*. Cambridge: Cambridge University Press, 1998. Intended for the college student, *The Earth's Mantle* provides a clear and complete description of the elements that make up the Earth's mantle and the process of change that it has undergone since its formation. Includes bibliography and index.

Jeanloz, Raymond. "The Earth's Core." *Scientific American* 249 (September, 1983): 46. This article is geared toward general science and undergraduate college audiences. It is well illustrated with excellent color photographs and diagrams. The references are restricted to a few key ones. Emphasis is on the Earth's magnetic field and on the physical state and chemical composition of the inner and outer core.

McKenzie, D. P. "The Earth's Mantle." *Scientific American* 249 (September, 1983): 67. An excellent companion article to "The Earth's Core." Again, the color illustrations are excellent and helpful. There is an emphasis on the physical state and composition of the mantle and a strong development of the relationship between processes in the mantle and the dynamics of the crust.

GEOTHERMAL PHENOMENA AND HEAT TRANSPORT

The temperature of rock at any particular location and depth in the Earth's subsurface is naturally fixed by heat generated in the Earth's core, the supplemental heat generated from the radioactive decay of unstable elements in Earth materials above the core, and the thermal conductivity of Earth material. Rock temperatures may be increased locally by a fourth factor, convective heat flow resulting from the upward movement of hot fluids.

PRINCIPAL TERMS

CURIE ISOTHERM: the depth in the Earth's subsurface at which magnetic minerals suddenly lose their magnetic properties as a result of high temperature

FUSION ENERGY: heat derived from the natural or human-induced union of atomic nuclei; in effect, the opposite of fission energy

GEOTHERMAL ANOMALIES: locales where the temperature of rock at depth is uncommonly high

RADIOGENIC HEAT: heat that results from the radioactive decay of isotopes; such decay always results in loss of atomic mass weight

THERMAL CONDUCTIVITY: the ability to permit heat to flow by induction

UNSTABLE ELEMENTS: those elements that have isotopes that decay by natural fission and thus yield heat

GEOTHERMAL GRADIENT

The high temperatures of the Earth's interior, at least in places, have been apparent for centuries. Geothermal energy gives rise to the famous hot springs and hot-water spas of the world, found on nearly all continents and on islands such as Iceland. The Earth's interior heat was also obvious from active volcanoes, spewing forth molten rock at temperatures of 600 degrees Celsius or higher.

Because of the relative youthfulness of geology as a science, it was not until the nineteenth century that persons trained in scientific observation and theory were able to make the logical deduction that, indeed, the entirety of the Earth's deep interior must be hot. With the advent of drilling technology in the late nineteenth century, it became known that all water wells more than a hundred meters deep produce groundwater at a temperature higher than the average annual surface temperature. With the advent of deep drilling technology for the petroleum industry in the 1930's, a geothermal gradient became apparent: The deeper the rock, the warmer its temperature, with no exceptions. Finally, as a result of wire-line probes developed for the oil industry for the purpose of acquiring physical and chemical data from measurements within the drill holes, it was learned that the temperature gradient with depth is not a constant—that the gradient is dependent upon the nature of the geology of the location.

Geophysical research led to the conclusion that, quantitatively, the geothermal gradient is a function of the heat flow, which is different from place to place, and the thermal conductivity of the rock itself. Some rock (for example, rock salt) is very conductive of heat, while other rock (for example, sandstone) is quite nonconductive. As a rock's thermal conductivity increases, the geothermal gradient decreases. Therefore, a chart of borehole temperatures (temperature log), plotted as a curve showing depth temperature, almost always appears as an irregular curve, even though the heat flow is a constant. The irregularities are attributable solely to the differing values of thermal conductivity of rocks in the sequence of strata that have been drilled in the creation of the borehole. The temperature log of a 7,000-meter-deep borehole in west central Texas, for example, would reveal one hundred or more "breaks" in the plotted temperature curve. Each "break" in the otherwise gradually increasing curve toward higher temperatures represents a different contact between one flat-lying geologic formation and the next. "Bottom-hole" temperature in that locale would be approximately 190 degrees Celsius at the 7,000-meter depth.

The irregular curve of temperature with depth is analogous to the warmth a person feels when lying under a wool blanket. The body generates internal heat, and the blanket, being of low thermal conductivity, deters the heat from escaping. As a result, there is a damming of calories within the blanket's full thickness—exactly the effect of geologic formations that happen to have low values of thermal conductivity. Similarly, the temperature gradient through the glass wool insulation in an attic is far higher than through the clay brick of a wall. If one drilled through massive, homogeneous granite, the temperature log would appear as a consistent curve toward ever high temperature at depth, because the thermal conductivity of the rock would be virtually constant. Any inconsistency in the curve would reflect local radiogenic heat.

FUELING EARTH'S HEAT ENGINE

Everywhere on the Earth there is geothermal flux (Earth-derived heat flow). Of the total flux, the largest component in most places is the heat generated within the core of the planet. Surprisingly, geologists and geophysicists remain uncertain as to the nature of the enormous heat engine that keeps the interior of the planet "alive" with geothermal energy regardless of the supplemental heat provided from within the Earth's crust. The Los Alamos National Laboratory has calculated that the temperature of the Earth's solid inner core, below a depth of about 5,100 kilometers, is 7,000 degrees Celsius. That level of heat is twice the melting temperature of diamond, which has the highest melting temperature of all minerals. Since heat flows from higher heat to lower heat, the core's enormous heat engine makes itself felt even at the Earth's surface. For example, the thickness of permafrost in the arctic is limited at depth by the heat flux from the inner Earth, especially so in regions where very thick glacial ice provided a thermal blanket in the ice ages.

In the past, it was widely thought that the great heat engine of the Earth's core was fueled by the decay of isotopes, particularly uranium and thorium. There are strong suggestions that the solid inner core is composed of iron and lesser amounts of nickel. No one can certify the core's actual composition, except that it has a very high density. One could argue that the Earth is too old to have

sustained the heat engine of the core to the present. The geologic Earth is approximately 4.6 billion years of age, about the half-life of uranium 238. This means that in the core the rate of radioactive decay from uranium 238, if any exists, would presently be less than half the original rate. There is no good evidence that the heat flux has reduced by half over geologic time.

Another possible fuel for the core's heat engine is fusion, yielding heat from a natural union of atomic nuclei—the same process which has been studied in the United States as a potential and safer substitute for the fueling of nuclear power plants. If the Earth's solid inner core is indeed producing heat by fusion, it is doing so with heavy elements, unlike scientific fusion experiments, in which elements of low atomic mass weight are logically used. Little is known about the actual physical environment of the inner core. Aside from its extremely high temperature, which scientists cannot reproduce in a sustained fashion, the elements making up the core remain of uncertain identity, and the compressive stresses of the central core can only be guessed at.

Within the relatively thin lithospheric crust of the planet, the compressive stress on rock is easily approximated for any given depth. In the very deep Earth, however, the forces of gravity weaken markedly as a result of the negative gravity imposed by the mass of overlying material. That is to say, as the central core is approached, the conventional gravitational attraction in a downward direction lessens, and negative gravity in an upward direction increases. These complexities give rise to the uncertainties regarding the core's actual physical environment as well as its chemistry. It would appear that fusion must be seriously considered as the mechanism by which the planet stays hot through long geologic time.

RADIOGENIC HEAT

Whatever the fuel may be that drives the Earth's great heat engine, the core's thermal production flows ever upward in keeping with the upward direction of the thermal gradient. Once that heat arrives at the solid-rock crust, the "primal" heat becomes supplemented by radiogenic heat, especially under the world's continents, where the crustal thickness is greater than under the ocean basins and where radioactive minerals are more abundant. Radiogenic heat derives from the spontaneous decay of radioactive elements, which are part of the crystal structure of many minerals. Each time an isotope decays to become another element, it loses atomic mass weight, and the lost weight manifests itself by the production of calories. The greater the weight loss, the more calories result from the fission process. This natural heat generation is, in effect, duplicated by controlled fission in all nuclear power plants. In the Earth's crust, however, the heat does not readily escape. It simply enters the massive, existing flow of heat from the core and thereby supplements it.

In most places under the world's continents, supplemental radiogenic heat is quantitatively less than the "primal" heat, perhaps contributing 10-25 percent of the total heat flow in the shallow crust. Under certain geologic conditions, however, radiogenic heat amounts to more than 50 percent of the total flow. For example, granite, as an intrusive igneous rock, is commonly composed of numerous minerals made up of radioactive, elemental isotopes. These include uranium 238, thorium 232, potassium 40, and others. The actual quantity of heat generated from these isotopes can be calculated or at least approximated. While granite is by no means the only rock capable of yielding radiogenic heat, it serves as a classic example and is a major contributor in locales where massive granite occurs. One of the most famous granite bodies, in these respects, is the Conway granite of northeast central New Hampshire, which contains extraordinary amounts of uranium and thorium as well as potassium. The total heat flow in shallow rock under that area is uncommonly high, and more than half of the heat derives from the radiogenic contribution.

GEOTHERMAL ANOMALIES

Along the United States' eastern seaboard, extending from the panhandle of Florida to New Jersey, the coastal plain sediments overlie and bury dozens of intrusive granite bodies never observed by the human eye. Their existence has been demonstrated by geophysical means, including the measurement of high levels of geothermal heat flow in the sedimentary strata directly above the granitic bodies. In the American West there are numerous areas mapped as geothermal anomalies, places where shallow rock temperatures are extraordinarily high. Some of these anomaly areas may in fact be a result of high levels of inductive heat flow. Others, however, appear to have resulted from the transport of heat from deep zones to shallow zones by convection, where groundwater at depth is able to circulate upward, carrying with it the calories the waters collected. Such a geohydrologic situation results in the water heating the rock. That phenomenon is one of the complicating factors in geothermal assessments, which usually target those anomalies caused by an unusually shallow Curie isotherm.

Beneath all points on the Earth's surface there is a specific depth at which magnetic minerals suddenly lose their magnetic properties. That specific depth is known as the Curie isotherm. Rock temperature principally controls the depth of the Curie isotherm—about 650 degrees Celsius. The significance of the Curie isotherm as a geothermal "marker" cannot be overstated. If one knows the depth to that isotherm beneath a particular spot, the true average geothermal gradient is easily calculated, simply by taking the rock temperature at a depth of 10 meters and dividing the depth of the Curie isotherm by the difference in temperature, resulting in the gradient in terms of degrees per kilometer.

The entire world could be quickly mapped for geothermal, high heat-flow anomalies, were it possible to measure the depth of the Curie isotherm by some geophysical technique. Unfortunately, that technique remains to be discovered. The Curie isotherm may lie at an average depth of about 26 kilometers, far too deep and too hot for even the most advanced drilling technology. In places, the Curie isotherm may lie only 6 kilometers deep—for example, under the Imperial Valley of southern California—shallow enough to reach by drilling, except for the fact that temperatures associated with the Curie isotherm prohibit drilling with present-day materials and equipment.

Geothermal anomalies exist under the deep ocean basins as well as under the continents. Is it possible that variations in geothermal heat flow under the oceans could render significant influence on world weather patterns? Ocean temperatures and the behavior of ocean currents are known to affect weather. The warm region of the Pacific Ocean off the coast of South America, known as El Niño, geographically coincides with a known and mapped geothermal anomaly in the subocean. Opportunities abound for serious research in the little-known by-products of the geothermal phenomenon.

Geothermal Measurements

Most geophysical research on the geothermal phenomenon and on heat flow focuses on specific determinations of heat flow units (HFU), thermal conductivity of rock or sediment (K), the geothermal gradient (G), and heat generation units (HGU). For purposes of quantitative analysis, a system of numerical values is applied to each of the above physical conditions and properties.

Heat flow units are a measure of the quantity of heat that is flowing upward through rock or sediment beneath a certain geographic place. This quantity would be a constant if the Earth were homogeneous and if all the heat were of "primal" origin—that is, generated in the Earth's core. The Earth is not, however, of homogeneous material; additional heat is added to the total heat flow from radiogenic sources, the decay of isotopes, at least in the Earth's hard crust. One HFU is one millionth (0.000001) of a calorie per square centimeter per second. This means that where there is an HFU value of 1,000,000, for example, the quantity of heat flow would be 1 calorie passing upward through each horizontally oriented square centimeter in 1 second.

Thermal conductivity is a physical property of the particular rock or sediment through which the heat is trying to flow. This property, or K, is expressed as a velocity, in keeping with the scientific method of assigning a value to other categories of conductivity. (For example, hydraulic conductivity is expressed as centimeters per second or meters per day.) Thermal conductivity is expressed as 0.001 calorie across a distance of 1 centimeter in 1 second with a thermal gradient of 1 degree Celsius. The following are typical values of thermal conductivity in various rocks and sediments: granite, 7.8; gneiss, 6.4; rhyolite, 8.2; limestone, 6.2; shale, 3.5; dolomite, 11.0; quartzite, 11.8; marl, 3.5; schist, 5.7; sandstone, 4.7; rock salt, 14.0; and basalt, 4.1. It will be noted that the range of typical K values is substantial, so that there is a multitude of changes in the geothermal gradient in a drill hole that penetrates many different rock types. The actual value of K in a rock is of complex ori-

gin, involving the K value of individual minerals of which the rock is constituted as well as the rock's density, porosity, and temperature. As the temperature increases, the K value of solids usually decreases. At room temperature, pure water has a K value of about 1.4. Therefore, rock that has porosity (void spaces) and is saturated with groundwater has a relatively low thermal conductivity so long as the heat flow does not become convective. (See the values of shale and sandstone above.)

The geothermal gradient (G) is numerically identified in terms of degrees Celsius change in rock temperature across a vertical distance of 1 kilometer.

Heat generation units (HGUs) constitute that portion of the total heat flow that is contributed largely within the Earth's crust as a result of radiogenic process—that is, the decay of isotopes. HGUs are referenced in the same terms as HFUs. Units of HGU are difficult to derive with any degree of precision. The heat production decay of uranium 238, thorium 232, and potassium 40, which are the principal contributors of radiogenic heat, is determined by the content of these elements in a rock in parts per million, except for potassium, for which percentage by volume is used. The actual concentrations of these three elements by laboratory testing requires either continuous drill cores of the rock or (less accurately) drill cuttings.

Algebraic Relationship of Measurements

The algebraic relationship of HFU, K, and G is HFU = K × G × 0.01. It thus becomes possible to calculate the HFU value where K and G are known. K can be measured in a laboratory from a rock core obtained in a drilling operation. G can be measured in the field, usually by means of a temperature log by conventional down-hole geophysical methods used in most deep borings. There is no known method, however, by which units of HFU can be measured directly in the field. HFU values obtained by the equation above do not always have the precision one might wish, since accurate average values of K are difficult to obtain.

It may well be possible to measure the depth of the Curie isotherm from ground surface. That achievement would represent a giant step forward in human comprehension of the geothermal phenomenon and in the economic potential of geothermal energy.

SIGNIFICANCE OF EARTH'S INTERNAL HEAT

Humankind pays insufficient respect to the remarkable natural phenomenon of the Earth's internal heat, which is generated primarily by a source within the planet's core. The process by which this gigantic heat engine produces thermal energy has never been identified, yet it produces hundreds of times more heat during any given moment than does the civilized world with all its burning of fossil fuels and all its fission-based nuclear power plants.

Cold planets are those that are devoid of internal heat. If the Earth were such a planet, without its interior oven, life would not exist in the form it has taken on Earth. The oceans would be solid ice, except for shallow surface water in the low latitudes and shallow water in summer in the moderate climates. Groundwater resources would not exist, because rock temperatures below a few meters would instantly freeze any fluid. There would be no crude oil available for production, because natural heat is required for the chemical process by which oil is formed; furthermore, if oil did exist it would be too viscous to flow. All the continents would have permafrost, including the tropics. The atmosphere would have far less average water vapor, the total oceanic evaporation being inhibited by the predominance of ice rather than of water. Thus, the world's precipitation patterns would be drastically diminished to the point of enabling agriculture only in the tropics, if at all. Most of the world's great rivers would be either dry or frozen, except possibly in the tropics. Many of the strategic minerals would not exist, because for their formation, most require either metamorphism (change of form by heat and pressure) or very hot subsurface water. The world would suffer from the absence of bituminous coal, as the processes of its creation require moderate warmth in the subsurface environment in which the transformation of organic debris to coal takes place. In all likelihood, the Earth would have no magnetic field. One blessing, however, might be cited: Earthquakes would be unknown.

A RENEWABLE RESOURCE

As it stands, to the good fortune of human beings, the quantity of heat that is both constantly generated in the solid inner core of the planet and contained in all the rock and fluids of the Earth is quantitatively beyond imagination. Furthermore, geothermal energy is renewable as a resource. When heat is extracted from a certain zone at depth, a thermal gradient is quickly established in that local sphere, and more heat begins to flow to replace the heat that has been removed. In fact, geothermal energy and groundwater are the only geologic resources that are classified as renewable.

More than 99 percent of all geothermal energy in the Earth's solid crust is, in effect, locked up in the form of hot, dry rock. Electric power by geothermal energy is produced from natural hot water or steam, but these occur only where fortuitously favorable geologic conditions are found.

One day humankind may have the ingenuity to find an economic method by which the great thermal resource of the planet, contained in hot, dry rock down to about 10 kilometers, can be extracted for conversion to electric power. That heat resource lies everywhere beneath the surface as if waiting to be tapped for its clean energy—nothing burned, no residue, no atmospheric contamination, and renewable.

John W. Foster

CROSS-REFERENCES

BIBLIOGRAPHY

Albu, Marius, David Banks, and Harriet Nash, et al., eds. *Mineral and Thermal Groundwater Resources.* London: Chapman and Hall, 1997. Well illustrated, this book looks at geothermal resources and groundwater and their relationship to geothermal processes. Suitable for college-level students with some background in the Earth sciences. Contains bibliographical references and index.

Birch, R. F., et al. *Heat Flow and Thermal History in New England and New York.* New York: Interscience, 1968. One of the early publications that reveal the physical-thermal relationship between heat flow that derives from the Earth's heat engine at the core and the heat derived from radiogenic sources within the crust.

Clark, S. P., Jr., ed. *Handbook of Physical Constants.* New York: Geological Society of America, 1966. A compendium of data vital to research in almost any field of geophysics and geology. Section 21, entitled "Thermal Conductivity," is the source of typical values of thermal conductivity of various rocks and sediments listed under "Geothermal Measurements" (above).

Combs, J., and G. Simmons. "Terrestrial Heat Flow Determinations in the North Central United States." *Journal of Geophysical Research* 78 (January, 1979). A classic report of research on geothermal conditions in the United States midlands, concluding that even benign-appearing regions have substantial ranges of geothermal flux.

Costain, John K., et al. *Evaluation and Targeting of Geothermal Energy Resources in Southeastern United States.* Blacksburg: Virginia Polytechnic Institute, 1979. This report on geothermal research is considered to be an excellent source for the basics of geothermal investigatory procedures.

Foster, John W. *Method for Producing a Geothermal Reservoir in a Hot Dry Rock Formation for the Recovery of Geothermal Energy.* Patent 4,223,729.
Washington, D.C.: Patent and Trademarks Office, September 23, 1980. The first patent that describes a method of creating a man-made fracture complex in massive hot, dry rock at depth by simultaneous hydraulic pressurization of two separate boreholes. The process applies a new concept in rock mechanics: mutually converging stress fields.

Hodgson, Susan F. *A Geysers Album: Five Eras of Geothermal History.* Sacramento: California Department of Conservation, Division of Oil, Gas and Geothermal Resources, 1997. Hodgson's book provides an easy understanding of geothermal history in the United States. It includes useful graphics, illustrations, and maps.

Kron, Andrea, and Grant Heiken. *Geothermal Gradient Map of the Conterminous United States.* LA: 8476-MAP. Springfield, Va.: National Technical Information Service, 1980. A well-researched map that shows the patterns of geothermal gradients in degrees Celsius per kilometer of depth. The map does not, however, distinguish between thermal gradients attained by true conductive heat flow and those where convection by hot fluids has greatly influenced the gradients.

Plate, Erich J., et al., eds. *Buoyant Convection in Geophysical Flows.* Boston: Kluwer Academic Publishers, 1998. Although highly technical at times, this collection of papers provides good descriptions and discussions about heat convection theories, geophysics, and heat flows. The collection also offers illustrations and a bibliography.

HEAT SOURCES AND HEAT FLOW

Several significant sources contribute to the internal heat of the Earth. Radioactive decay of isotopes of thorium, uranium, and potassium in the crust produces most of the heat observed at the surface of the continents. Terrestrial heat flow is readily observed in deep wells, mines, and tunnels which penetrate below the narrow zone on the surface which is heated by daily and seasonal radiation changes. Another major source is heat convected outward from the Earth's core and through the mantle. This heat produces the majority of the heat flow measured in the oceans, especially at the mid-oceanic ridges.

PRINCIPAL TERMS

ASTHENOSPHERE: the semimolten portion of the outer mantle (ranging to a depth of 250 kilometers) which lies at the base of the lithosphere

BASALT: a fine-grained, dark extrusive igneous rock

CONDUCTION: the transfer of heat or the transportation of matter caused by temperature differences

CONVECTION: the transfer of heat by the movement or circulation of the heated parts of a liquid or gas

CORE: the center portion of the Earth which is divided into a liquid outer portion and a solid, denser inner section

CRUST: the outermost layer of the Earth, ranging in thickness from 5 to 60 kilometers; it consists of rocky material which is less dense than the mantle

LITHOSPHERE: the outer, rigid portion of the Earth which extends to a depth of 100 kilometers; it includes the crust and uppermost portion of the mantle

MANTLE: the portion of the Earth's interior extending from about 60 kilometers in depth to 2,900 kilometers; it consists of relatively high-density minerals which consist primarily of silicates

TERRESTRIAL HEAT FLOW

The surface of the Earth receives heat from several different sources. Solar radiation provides the largest amount of heat, an amount that is approximately five thousand times greater than that moving outward from the subsurface. However, solar radiation is almost all reradiated into space, so it has very little effect on the Earth's temperature deeper than a few meters. The amount of terrestrial heat flow is so small that it would take several months for a dish of water placed on the surface to heat up by an additional 1 degree Celsius. During that period of time, the surface temperature on the Earth could easily fluctuate 20 to 30 degrees Celsius, depending on the time of year. Because of the large variation in surface temperatures, it is necessary to measure terrestrial heat flow at depths that lie well below those that are affected by these relatively short-term changes.

The fact that subsurface temperatures increase with depth has been known from deep mines and wells which penetrate several kilometers into the Earth's crust. The temperature at a given depth cannot be determined with the degree of accuracy that pressures at depth can be calculated. Variations in temperature are attributable to several important variables, particularly radioactive heat production and the coefficient of heat transfer. Our inability to measure these parameters at great depths within the crust makes it difficult to establish a well-defined temperature versus depth graph.

In general, the amount of outward heat flow on the surface is small. The mean heat flow for the Earth ranges between 60 and 70 milliwatts per square meter. Average values for oceanic and continental areas are essentially equal. Significant differences for values measured in the two general areas depend on the specific geologic setting within the oceanic or continental area. Observed values also differ significantly for measurements taken at specific locations in either the oceans or the continents. Heat-flow measurements in continental regions vary from 41 milliwatts per square meter in Precambrian shields (large masses of igneous and/or metamorphic rock) to 74 milliwatts per square meter for more active, younger Mesozoic and Cenozoic areas. Within the ocean basins, heat-flow values range from 49 to 80 milliwatts per square meter or more. Many researchers provide a multitude of different values for the various geologic regions.

OCEANIC HEAT FLOW

Molten basaltic rock formed by extrusion at the mid-oceanic ridges cools as it moves out from this primary heat source. This heat is being transferred by both convection from the mantle and conduction of heat produced by radioactive isotopes within the rocks in the oceanic layer. As the molten basalt cools, it contracts and undergoes a small density increase, which allows the newly formed layer to sink deeper into the asthenosphere. The depth to which the cooler basaltic layer sinks has been shown to be roughly proportional to the square root of its age. Oceanic crust that is 2 million years old is covered by about 1.5 kilometers of water, whereas crust that is 20 million years old would be located at a considerably greater depth below the ocean surface.

Subduction trenches have the lowest heat-flow values of measurement sites in the oceans. These trenches are deep depressions on the ocean floor usually adjacent to the boundary of continental-oceanic or oceanic-oceanic plate collisions. Several reasons exist for lower values in those regions. As discussed earlier, most of the heat observed in the oceans results from volcanic activity associated with the mid-oceanic ridges. This heat is lost once the hotter volcanic material moves away from the ridge. Trenches are very distant from the ridges, so the oceanic floor in the trench areas has long since been removed from the heat source as the floor has spread outward.

Trenches associated with continental-oceanic plate collisions are also areas that receive large volumes of sediment that are shed from the continents through normal erosion processes. In addition, the deposition of organic and fine-grained clastic material from seawater fills the trenches.

All these sediments produce an insulating blanket that can be as thick as 1 kilometer overlying the basaltic oceanic floor. Any heat produced by the basalts is held in by this sedimentary layer. Heat-flow measurements taken in the oceans are made in this soft sediment layer, not in the underlying basalt.

RADIOACTIVE DECAY OF ISOTOPES

Heat is produced by the natural radioactive decay of isotopes of thorium, uranium, and potassium. The long-lived radioactive isotopes which contribute most to the overall heat production are thorium 232, uranium 238, potassium 40, and uranium 235 (listed in order of decreasing importance). All these isotopes have half-lives roughly the same as the age of the Earth. These isotopes are not readily incorporated into the internal structure of the most common minerals. They do tend to become concentrated in minerals that have lower melting points, such as those occurring in granites. A granitic crust 20-25 kilometers thick can produce the amount of terrestrial heat flow observed over these areas. If a more intermediate composition for the continental crust is used, it is still possible to account for at least two-thirds of the heat flux of the continents as being generated from the crust. The remaining heat flow (about 20 milliwatts per square meter) is assumed to come from the mantle.

Oceanic crust which has an average thickness of 5 kilometers produces less than 3 percent of the heat flow observed in oceanic measurements. As previously mentioned, the oceans and continents exhibit near-equality in their total heat-flow measurements. The implication is that the mantle underlying oceanic rocks has a higher temperature and greater amounts of radioactive elements than continental mantle and thus serves as the primary heat source in the oceans. This higher heat flow is conducted through the basaltic layer and also convected upward at the mid-oceanic ridges. It must be remembered, however, that, as a result of the mobility of the lithosphere, this higher-temperature mantle material underneath the oceans eventually moves under the continents and helps compound the interpretation of actual heat flow being produced by the mantle. Almost 75 percent of the Earth's total heat loss is through the ocean floors, a percentage similar to the amount of surface covered by the oceans (70 percent). The vast percentage of this oceanic heat loss is related to the formation of new oceanic crust at the mid-oceanic ridges.

HEAT FLOW VARIABLES

The amount of heat flow observed in the oceans and on the continents is dependent on a number of variables. The age of the rock is of primary importance in that older rocks produce less heat, because the radioactive isotopes in the rocks have had a longer time to break down and hence less of the original heat-producing parent isotope is present. Older basalts on the ocean floor are also farther removed from their primary heat source associated with volcanism along the mid-oceanic ridge, where they were first extruded.

Continental heat-flow values are dependent on this same concept of age of the parent rock. On the continents, the lowest heat-flow values are associated with Precambrian shield areas, those portions of the continents which represent the stablest and oldest central core of the landmasses. These areas have also been subjected to the most erosion, which has removed a significant portion of the rocks and minerals containing the radioactive constituents. Some portions of the continents have undergone extensional stresses, which have pulled the landmasses apart and thus thinned the continents. In these places, such as the Basin and Range area of Nevada, the crust has been thinned just as a rubber band becomes thinner when it is extended. The result is that the mantle is much closer to the surface and its influence is enhanced with respect to the amount of heat being moved upward toward the surface. Areas on the continents which have experienced Mesozoic and Cenozoic mountain-building activity also show high levels of heat flow, but they also display the greatest amount of variation in observed values. However, these areas of tensional tectonics and mountain-building activity make up a small percentage of the continents and thus do not contribute much to the average values observed on land.

Surface temperatures can also be affected by groundwater flow, soil moisture, slope orientation, vegetative cover, topography, and Sun angle. These contributing factors must be removed in order to obtain heat-flow readings that are representative of rocks at depth.

HEAT FLOW CONDUCTION AND TRANSFER

Scientists have found that the Earth is losing heat. The heat that is being lost is produced partly by higher rates of radioactive heat production in the past and also by heat generated by the formation of the Earth. The inner core of the Earth has a temperature estimated to be between 4,000 and 5,000 degrees Celsius. If the Earth were a perfect conductor of heat, the rate of heat loss would offset the rate of heat generation. The mantle is a poor conductor of heat, however, and thus it stores heat, which is slowly released as the core and lower mantle cool. Calculations have shown that heat in the lower mantle is not completely transferred to the upper mantle and later to the surface. If large-scale convection cells existed within the mantle, there would be a more efficient transfer of this deep-seated heat. This is not observed. The best models of the internal transfer of heat seem to point to two levels of convection cells, one lying in the lower mantle and a second, separate level in the upper mantle. This latter level serves as an additional insulator from the deeper heat, which is trying to rise to the surface. The lateral motion of these convection cells in the upper mantle also serves as the primary mechanism to move the continental masses around. The continents are of a lower density than the underlying material; hence they "float" on the lithosphere and asthenosphere. These convection cells raft the continents around on the Earth's surface, although at a slow rate (several centimeters per year). It must be remembered that heat-transfer processes in the mantle and core are not directly observable. Therefore, geophysicists do not totally agree on the mechanisms which explain the production and transfer of heat at great depth.

Heat flow itself is controlled by the second law of thermodynamics. This law states that in order for thermal equilibrium to be attained, heat must move from warmer to colder material. This means that in the Earth heat flows from the warmer interior to the colder lithosphere and crustal-atmosphere boundary. Heat produced within the Earth is transferred in two ways. Deep-seated sources in the mantle and core transfer heat to the upper mantle and crust by convection. Heat rises slowly up through the mantle until it encounters the base of the lithosphere. Some heat is conducted into the relatively cooler lithosphere, while the remainder moves laterally along the boundary of the lithosphere and mantle. As it does so, the temperature of the upper mantle lessens and eventually the cooler rock material sinks back into the mantle, where it is reheated and returned into the convective cycle. Within the oceanic and continental masses, heat is also conducted by the rock. Although rocks are generally poor conductors of heat, the vast quantities being generated by radioactive decay and moved by convection are conducted toward the surface.

HOT SPOTS

Heat sources mentioned so far are large, often extending hundreds or thousands of kilometers across and through the Earth and displaying a broad horizontal and vertical expanse within the crust and mantle. Mantle plumes or hot spots represent much more localized heat sources. Only several dozen of these features have been recognized to date. Their spatial distribution is widespread and generally dispersed, with a slight concentration located along portions of the mid-oceanic ridges. Hot spots are usually only several tens of kilometers in diameter. They are thought to be conduits of heat rising from the mantle and intersecting the Earth's surface. Several well-known examples include Iceland, the Hawaiian Islands, and Yellowstone National Park in Wyoming.

Iceland is the result of a hot spot which lies directly on the Mid-Atlantic Ridge. Its volcanic nature is direct evidence of the heat and type of rock produced by the upward movement of heat from beneath the ocean floor. Geothermal energy produced by the volcanism is used as the primary heat source for the island. An obvious hazard of the geologic setting of the island is that volcanic eruptions can adversely affect everyday life in the area.

The Hawaiian Islands are volcanic mountains which rise from the deep ocean floor to elevations of more than 4,200 meters above sea level. When considered in total, they are the highest mountains on Earth. The entire string of islands in the Hawaiian Islands chain formed as the result of the Pacific plate having moved in two separate stages in a northwesterly direction over a hot spot. The present position of the Pacific plate has the hot spot centered on the southeastern corner of the island of Hawaii. This area has experienced very

active volcanism in the past few centuries and has been the site of numerous eruptions since 1983. Ocean-floor reconnaissance has detected the formation of a new island to the southeast of the island of Hawaii. This submarine volcanic feature has been named Loihi, and it will continue to grow until it breaks through the ocean surface to produce another major island in the chain.

The Yellowstone caldera, located in and around Yellowstone National Park in the western United States, is an excellent example of a hot spot that has risen through the continental lithosphere. A topographic high is centered on the caldera. Elevations decrease as the cooler lithosphere moves out from the center of the hot spot. Evidence for the heat exists in geologically recent volcanic activity and the present-day hot springs and geysers found in the park.

HEAT-FLOW MEASUREMENT

Heat-flow determinations depend on two separate measurements: the rate of increase of temperature with depth (which is termed the vertical temperature gradient, r) and the thermal conductivity (K) of the rocks in which the temperatures are being measured. The flux, or rate, of heat flow (q) is calculated using the formula: $q = {}^{-}Kr$. The units of q are watts per square meter, those of K are watts per meter per degree, and those of r are per degree per meter (deg $C^{-1}m^{-1}$). Absolute temperatures (in Kelvins) can also be used to express these parameters. The minus sign denotes the fact that heat flows down the temperature gradient, from the warmer spots to the colder ones. It must be noted that it is standard practice to consider heat-flow values as positive numbers even though the values obtained from the equation are indeed negative. The temperature gradient is fairly linear within certain depth ranges in the Earth. David G. Smith, in *The Cambridge Encyclopedia of Earth Sciences* (1981), provides a figure which roughly defines these different gradients for varying depths in the crust and upper mantle. The rate of increase of the temperature gradient, also referred to sometimes as the geothermal gradient, is greatest in the outer 1,500 kilometers. The rate of increase decreases significantly at a depth of 2,900 kilometers, where the core-mantle boundary exists. This is attributable to a change in state of the minerals present at that depth.

From the above equation, it is clear two variables must be measured in order to determine the rate of heat flow. The thermal conductivity (K) is usually measured at discrete points along a core sample taken from the borehole. Errors can be introduced because of temperature contamination as the cores are brought to the surface. Average values are obtained by taking a series of measurements along the retrieved core. Measurements of the temperature gradient (r) are obtained by placing a probe into the borehole (or driving it into the ocean sediment). The probe has a series of thermal sensors attached to it which record the temperatures at various distances along the probe. The gradient is calculated by dividing the temperature differences by the known distance of separation between the respective probes.

Observed heat-flow values of ocean bottoms are much less variable than those measured on the continents. In water depths exceeding several hundred meters, it is necessary only to measure temperatures in the upper few meters of the sediments and to establish the thermal conductivity over this same interval. These more shallow probe depths in the oceans are permitted because of the more stable heat regime at the boundary of the cold seawater and the ocean sediments. On the continents, however, seasonal variations resulting from solar radiation can affect heat-flow measurements to varying depths. For an average continental rock, the daily change in temperature affects rocks only to a depth of about 15 centimeters; annual variations extend down about 3 meters, while longer-term variations can reach to more than 8 meters. In addition, the flow of groundwater in the upper 50-100 meters alters the heat regime in the subsurface. In some areas, such as highly fractured rocks, groundwater effects can penetrate to depths of as much as 1 kilometer. Therefore, heat-flow measurements must be taken at depths great enough to remove these effects.

David M. Best

CROSS-REFERENCES

Basaltic Rocks, 1274; Earth's Core, 3; Earth's Core-Mantle Boundary, 9; Earth's Crust, 14; Earth's Differentiation, 20; Earth's Lithosphere, 26; Earth's Mantle, 32; Earth's Structure, 37; Elemental Distribution, 379; Geothermal Phenomena and Heat Transport, 43; Granitic Rocks, 1292; Hawaiian Islands, 701; Hot Spots, Island Chains, and Intraplate Volcanism, 706; Lithospheric Plates, 55; Magmas, 1326; Mantle Dynamics and Convection, 60; Ocean Ridge System, 670; Oceanic Crust, 675; Plate Margins, 73; Plate Tectonics, 86; Plumes and Megaplumes, 66; Radioactive Decay, 532; Spreading Centers, 727; Subduction and Orogeny, 92; Yellowstone National Park, 731.

BIBLIOGRAPHY

Cook, A. H. *Physics of the Earth and Planets*. New York: Halsted Press, 1973. This book provides succinct presentations describing the physical properties of the Earth observed both on the surface and in the interior. A relatively advanced text, suitable for college-level readers.

Francheteau, Jean. "The Oceanic Crust." *Scientific American* 249 (September, 1983): 114. A clearly written article in an issue which addresses all facets of the geology of the Earth. This article has good diagrams showing the role the oceanic crust plays in the overall geologic setting of the Earth. Suitable for high-school-level readers.

Jacobs, J. A. *The Earth's Core*. London: Academic Press, 1975. Deals in detail with all aspects of the Earth's inner and outer core. The origin of the core, its constitution, and its thermal and magnetic properties are discussed in detail. Well suited to the serious science student.

Jacobs, John. *Deep Interior of the Earth*. London: Chapman and Hall, 1992. This introductory geophysics textbook is formidable for the average student because there is considerable mathematics in some chapters, but it does cover many useful topics. It contains a minimum of equations but many figures and graphs.

Lowell, Lindsay, William G. Hample, et al., eds. *Geology and Geothermal Resources of the Imperial and Mexicali Valleys*. San Diego: San Diego Association of Geologists, 1998. This detailed account of the geothermal resources and processes of the Imperial and Mexicali Valleys in Calfornia contains many helpful illustrations and maps to help clarify the subject. Includes a bibliography.

Plate, Erich J., et al., eds. *Buoyant Convection in Geophysical Flows.* Boston: Kluwer Academic Publishers, 1998. Although highly technical at times, this collection of papers provides good descriptions and discussions about heat convection theories, geophysics, and heat flows. The collection also offers illustrations and a bibliography.

Skinner, Brian J., and Stephen C. Porter. *Physical Geology.* New York: John Wiley & Sons, 1988. A well-written, clearly illustrated text which provides all the basic concepts of geology at a level which advanced high school and college science students can understand.

Smith, David G., ed. *The Cambridge Encyclopedia of Earth Sciences.* New York: Crown Publishers, 1981. Chapter 9, "The Energy Budget of the Earth," offers a good summary of the external and internal energy sources affecting the Earth, including a table showing the contributions that geologic regions and isotopes make to the total heat-flow regime on Earth. An excellent glossary and good graphics augment the text, which is aimed at college-level readers.

Stacey, Frank D. *Physics of the Earth.* 2d ed. New York: John Wiley & Sons, 1977. Although slightly outdated, this advanced text still provides the mathematical basis for understanding many global geophysical processes. Also provides thorough explanations of the processes which involve the physics and geology both on and in the Earth. Contains an excellent bibliography listing several hundred references on specific topics.

LITHOSPHERIC PLATES

Lithospheric plates are large, distinct, platelike segments of brittle rock. They are composed of upper mantle material and oceanic or continental crust. The seven major and numerous minor plates fit together to form the outer crust of the earth.

PRINCIPAL TERMS

ASTHENOSPHERE: a layer of the mantle in which temperature and pressure have increased to the point that rocks have very little strength and flow readily

DENSITY: the mass of a given volume of material as compared to an equal volume of water

FELSIC: rocks composed of the lighter-colored feldspars, such as granite

ISOSTASY: the balance of all large portions of the Earth's surface when floating on a denser material

LITHOSPHERE: the outermost portion of the globe, including the mantle above the asthenosphere

MAFIC: rocks composed of dark, heavy, iron-bearing minerals such as olivine and pyroxene

MANTLE: the layer of the Earth between the crust and the outer core

PLATE STRUCTURE AND COMPOSITION

The lithosphere is the sphere of stone or outer crust of the Earth. It is composed of seven major platelike segments and numerous smaller ones. These lithospheric plates fit together in jigsaw-puzzle fashion. The recognition of the existence of these plates and their distinct boundaries has led to the theories of plate tectonics and seafloor spreading.

Lithospheric plates are layered. The bottom layer is the rigid upper portion of the mantle. The upper mantle is composed of dense, grayish green, iron-rich rock. Some plates have another solid layer of oceanic crust. This crustal rock is composed primarily of basalt. Some plates consist of only upper mantle and a thin covering of oceanic crust, while other plates have mantle material, oceanic crust, and continental crust. The continental crust is primarily granitic and is less dense than the basalt of the oceanic crust. Until recently, it was assumed that all plates had a continuous layer of oceanic crust and that continental crust was an additional layer, riding on the top. That no longer appears to be the case. The continental crust may be underlain by areas of oceanic crust in a discontinuous fashion, but the two crustal types are actually complexly intermingled.

The upper crustal rocks range from 12 kilometers thick over the ocean plains to more than 30 kilometers thick on the continental masses. The Mohorovičić discontinuity defines the boundary between the crust and upper mantle. This boundary is recognized because seismic waves suddenly accelerate at it. The lithospheric plates, including the rigid upper mantle, are 75 to 150 kilometers thick. They float on the asthenosphere, which is a deeper portion of the mantle. The rock of the asthenosphere is under such pressure and increased temperature that it has little strength and can readily flow in much the same fashion as warm candle wax. The contact between the plates and the asthenosphere is marked by a sudden decrease in the speed of seismic waves.

The ability of the lithospheric plates to float on the asthenosphere is a key to understanding them. In much the same way that ice floats on water, the plates float on the material below them. Ice is able to float because it is less dense than water. As ice forms, it crystallizes and expands to fill more space. A given volume of ice has less density than the same volume of water. The density of the different layers of the Earth increases toward the solid iron and nickel core. The lower mantle floats on the outer core, the asthenosphere floats on the lower mantle, and the lithospheric plates float on the asthenosphere.

PLATE MARGINS

The plates fit together along margins. There are generally considered to be only three types of plate margin: ridges, trenches, and transform faults. Ridges, such as the Mid-Atlantic Ridge, are characterized by rifts or spreading centers. Trenches are margins where one plate is being forced below another and are the deepest areas of the ocean floor. Transform faults, such as the San Andreas fault in California, are areas where two plates are sliding alongside each other. The complex interactions of the lithospheric plates have led to the formation of the continents as they now exist. The plate margins do not necessarily follow the continental outlines. Continents may be composed of more than one plate. All the rocks and minerals that are on or near the surface are located on these plates. Geologic processes such as mountain building, earthquakes, and volcanism can be observed at or near the plate margins.

FORMATION OF CRUSTAL MATERIAL

Both types of crustal material, oceanic and continental, form through crystallization. This process is dependent on time, temperature, and pressure. As a molten material cools, a complex series of reactions occurs. The denser minerals crystallize early in the cooling of a molten material. If there is sufficient time in the cooling process, these early-formed dense minerals will gradually react with the remaining molten materials to form less dense minerals.

At the divergent plate margin, melting of the upper mantle gives rise to a silicate magma rich in iron and magnesium. This magma intrudes along fractures to be emplaced in the ocean floor as dikes and erupts to the surface as lava flows. The mafic material thus formed is called "basalt" and makes up the ocean floor. As new material is added by injection into the basalt of the ocean floor, it must push the existing material out of the way. Thus, a new seafloor is added at the spreading centers, and the seafloor is made up of progressively older material away from the spreading center. On the other hand, continents are composed of mostly granitic material which is formed from silicate magmas that are low in iron and magnesium and high in alkali-elements such as sodium and potassium. These granites are less dense than basalt.

If a basaltic oceanic plate collides with lighter continental crustal material, the continental crust will ride up over the oceanic plate, and the oceanic plate will be pushed down into the hotter mantle, where it will be assimilated back into the mantle. This convergent margin is marked by a deep oceanic trench on the ocean side of the collision zone. The descending oceanic plate is known as a subduction zone.

When two plates of continental material collide, neither can be subducted. If they do not begin to slide alongside each other, the compressive forces will form mountains. These mountains cannot rise higher than their isostatic balance. They must either be eroded by wind and water or sink back into the asthenosphere. The eroded pieces of rock, called sediment, are transported to lower areas called basins. As the sediment becomes more deeply buried, the pressure of overlying sediments causes them to lithify or become sedimentary rock. These sedimentary rocks have considerable pore space between the individual grains of sediment or silt and therefore are not very dense. They become additional continental crust material.

If the sedimentary rocks are buried deep enough, the increased temperature and pressure will begin a process known as metamorphism. During metamorphism, the original minerals in the sedimentary rock react with each other to form new minerals that are stable in the new environment. As temperature and pressure increase with depth, the rock becomes more and more like granite. If the temperature reaches high enough, the rock may melt and become magma.

MULTIDISCIPLINARY STUDY OF LITHOSPHERIC PLATES

Lithospheric plates fit together to form the crust or rock surfaces of the Earth. The study of the surface of the Earth and its composition is an extremely broad subject. It includes many of the subdisciplines of geology and oceanography. The study of the Earth's surface and extraction of economic minerals have existed since humankind's earliest times. Flint and obsidian used in toolmaking were early trade items. Mining geology and mineralogy are almost as old. The early Greeks and Romans wrote books on geology.

Humans have used minerals and the metallic minerals since prehistory. Much knowledge of the Earth is essentially a by-product of what was learned during the search for minerals, mineral ores, and gems. Something as simple as the formation of a nail requires iron ore and carbon. Mining geologists assay ores looking for economic deposits, and mineralogists study minerals.

Geophysicists bounce sound waves through the Earth to determine subsurface structures. With the use of seismographs, they listen to earthquakes to pinpoint their locations. They also measure the gravity and magnetic field of specific areas of the Earth. Petroleum geologists search for oil and gas by drilling into the Earth's surface. Their interpretation of drill cuttings and core samples provides information about ancient environments. Volcanologists study volcanoes. They employ lasers to measure any minute movements on the surface of a volcano. They also use seismographs to detect the earthquakes that may signal an onset of volcanic activity. Because of the potential devastation of volcanoes, prediction has become increasingly important. Petrologists examine rocks to understand the Earth processes that formed them. Their primary tools are the scanning electron microscope and X-ray diffraction machines.

Geochemists analyze the chemical composition of rocks and minerals and the reactions that may have caused their formation and dissolution. Paleontologists study fossilized life-forms, while paleoecologists study ancient environments. Planetary scientists investigate meteorites and Moon rocks to increase understanding of the Earth and its lithospheric plates. Much of what is known about the mantle material is a result of the study of meteorites.

STUDY OF OCEANIC CRUST

Much early geologic work was done on the more readily accessible continental crust. Recently, scientists have made considerable progress in the study of the oceanic crust. Early exploration of the ocean floor was through simple depth measurements from ships. Sailors lowered a weighted line over the side of a ship and physically measured the depth to the seafloor. The echo sounders developed in the early 1900's allowed for more rapid measurements of the ocean depths. In time, continuous profiles of the seafloor were made. Instead of the featureless plain that was expected, oceanic ridges, deep trenches, and numerous submerged volcanoes appeared.

Dredging is an old but ongoing method of sampling the surface of the ocean floor. The deep-diving bathysphere paved the way for bathyscaphes and other high-technology submersibles. Much recent work has been done with television cameras. A major find was made by a geologist in the late 1970's. A seafloor volcanic vent actually had life-forms subsisting on the chemically rich waters near it. Until this time, it had been assumed that all life on the Earth was dependent on photosynthesis. This initial television discovery of chemosynthetic life-forms shocked the scientific community.

Drilling on the ocean floor has been accomplished by drill ships such as the *Glomar Challenger.* The cores of the deep ocean floor indicated a much younger oceanic crust than had been expected. Much that was learned about the oceanic crust simply did not fit with the scientific theories of the day. Serious rethinking had to be done, and many theories had to be radically changed.

SIGNIFICANCE OF PLATE INTERACTIONS

Lithospheric plates are the brittle rocks that float on the hot, plastic asthenosphere. They fit together to form the crust of the Earth. Each step people take is either on the surface of a lithospheric plate or on something that is directly or indirectly made from one. Weathered surface rock provides the soil in which plants grow. The plants provide a breathable atmosphere and sustain animal life. The interaction of lithospheric plates leads to earthquakes, volcanic activity, and tidal waves. These impressive geologic displays have caught the human imagination since earliest times.

Since the lithospheric plates form the solid surface of the Earth, in a real sense everything humans touch is related to them. Even something as unlikely as plastic is made from petroleum products extracted from the Earth's crustal rocks. Coal, oil, and gas are burned to provide heat and electricity. Minerals extracted from lithospheric plates become the gold that makes jewelry, crowns teeth, and is part of circuit boards and computer chips. The minerals and compounds extracted from the lithospheric plates provide the iron for skyscrapers, cars, and car fuel. Coal, oil, and gas are formed by complex interactions of ancient plant life during the rock-forming processes that have occurred during the formation of upper portions of the lithospheric plates. The surface of the Earth and its ongoing geologic processes also affect the weather.

The study of the composition and motion of lithospheric plates has created nearly all the body of knowledge in the field of geology. Numerous subdisciplines have arisen to study specific areas of geology. Study of the oceanic crust is relatively new, and recent discoveries are changing commonly accepted views of the Earth. As views change, more discoveries seem to become possible. As more is learned about the Earth's surface, views must be altered and upgraded to explain the phenomena observed.

Raymond U. Roberts

BIBLIOGRAPHY

Condie, Kent C. *Plate Tectonics and Crustal Evolution.* 4th ed. Oxford: Butterworth Heinemann, 1997. An excellent overview of modern plate tectonics theory that synthesizes data from geology, geochemistry, geophysics, and oceanography. A very helpful tectonic map of the world is enclosed. The book is nontechnical and suitable for a college-level reader. Useful "suggestions for further reading" follow each chapter.

Davies, Thomas A. *Glaciated Continental Margins: An Atlas of Acoustic Images.* London: Chapman and Hall, 1997. Written for the college student, this book explores glacial landforms and their relationship to sedimentation, plate tectonics, continental drift, and the lithosphere. Filled with helpful maps and illustrations.

Glen, William. *Continental Drift and Plate Tectonics.* Columbus, Ohio: Charles E. Merrill, 1975. A college-level introductory text, this volume covers the concepts of lithospheric plates and their formation and motion. Although technical, the material is introduced in a fashion that does not require a background in geology. Includes a very good index and an extensive supplementary-reading reference section.

Gross, M. Grant. *Oceanography.* 2d ed. Columbus, Ohio: Charles E. Merrill, 1971. Designed as an introductory-course text in oceanography for the college student. The first three chapters discuss oceanic plates and the seafloor, or oceanic crust. The historical section on seafloor study and current methodology is valuable, as is the index and an extensive supplementary reading list.

Marvin, Ursula B. *Continental Drift.* Washington, D.C.: Smithsonian Institution Press, 1973. Taking a historical approach, Marvin provides considerable discussion of plates and plate theory, covering old theories and explaining their progression toward new ones; includes discussion of the views that disagree with current theory. Index and extensive bibliography. For college-level readers.

Miller, Russell. *Planet Earth: Continents in Collision.* Alexandria, Va.: Time-Life Books, 1983. A clear and excellently illustrated introduction to lithospheric plates and plate tectonics. Extensive historical background is provided, and the concepts are introduced in a logical fashion. Good index and extensive bibliography. For advanced high school and college readers.

Olsen, Kenneth H., ed. *Continental Rifts: Evolution, Structure, Tectonics.* Amsterdam: Elsevier, 1995. The various essays provide good explanations of plate tectonics and continental rifts. Slightly technical but suitable for the careful reader. Illustrated.

Reynolds, John M. *An Introduction to Applied and Environmental Geophysics.* New York: John Wiley, 1997. An excellent introduction to seismology, geophysics, tectonics, and the lithosphere. Appropriate for those with minimal scientific background. Includes maps, illustrations, and bibliography.

_____. *Planet Earth: Earthquake.* Alexandria, Va.: Time-Life Books, 1982. Offers some exceptional illustrations of lithospheric plates. Chapter 5, "Dreams of Knowing When and Where," contains a good discussion of equipment and methodology used to determine plate movement. Well indexed, with a good bibliography for additional reading. For high school and introductory college students.

_____. *Planet Earth: Volcano.* Alexandria, Va.: Time-Life Books, 1982. Although much of this book is about the surface effects of volcanoes, there are several technical sections. Focus is primarily on the formation of the basaltic portions of lithospheric plates. The methodology discussed in chapter 5, "Monitoring the Earth's Heartbeat," is particularly valuable. Good index and bibliography.

Walker, Bryce. *Geology Today.* 10th ed. Del Mar, Calif.: Ziff-Davis, 1974. An excellent introductory text for the study of lithospheric plates, crustal rocks, plate movement, and geologic processes. The progression of concepts is clear and logical, and the volume is exceptionally well illustrated and indexed. The bibliography includes listings of other technical reference books.

MANTLE DYNAMICS AND CONVECTION

Mantle dynamics is the study of the motion of the Earth's mantle, which is primarily generated by convection. Convection within the mantle causes the transfer of heat from one region of the Earth to another. This convection facilitates the movement of the lithospheric plates of the Earth, resulting in mountain building, earthquakes, volcanism, and the evolution of continents and ocean basins. Understanding mantle dynamics and convection is a major component of the framework for explaining how the Earth developed, how it works, and why it is constantly changing.

PRINCIPAL TERMS

ASTHENOSPHERE: the weak zone directly below the lithosphere, from 10 to 200 kilometers below the Earth's surface, believed to consist of soft material that yields to viscous flow

CONVECTION CELL: a pattern of movement of mantle material in which the central area is uprising and the outer area is downflowing because of density changes produced by heat variations

CORE-MANTLE BOUNDARY: the seismic discontinuity 2,890 kilometers below the Earth's surface that separates the mantle from the outer core

LITHOSPHERE: the relatively rigid outer zone of the Earth, which includes the continental crust, the oceanic crust, and the part of the

upper mantle lying above the weaker asthenosphere

LOWER MANTLE: the seismic region of the Earth between 670 and 2,890 kilometers below the surface, consisting of the D' and D'' layers

MANTLE PLUME: a vertical cylindrical distribution of material in the mantle within which abnormal amounts of heat are conducted upward to form a hot spot at the Earth's surface

SEISMIC TOMOGRAPHY: a processing technique for constructing a cross-sectional image of a slice of the subsurface from seismic data

UPPER MANTLE: the part of the mantle that lies above a depth of about 670 kilometers, consisting of the B layer and the C layer

CHEMICAL AND MECHANICAL PROPERTIES OF THE MANTLE

The Earth's interior consists of a series of shells of different compositions and mechanical properties. Based on chemical composition, the outermost layer is the crust, consisting of both continental and oceanic crust. The next major compositional layer of the Earth is the mantle, which is approximately 2,890 kilometers thick and constitutes about 82 percent of the Earth's volume and 68 percent of its mass. By studying fragments of the mantle that have been brought to the surface by volcanic eruptions, it is deduced that the mantle is chemically composed of silicate rocks containing primarily silicon, oxygen, iron, and magnesium.

Based upon physical, or mechanical, properties, the solid, strong, rigid outer layer of the Earth is termed the lithosphere ("rock sphere"). The lithosphere includes the crust and the uppermost part of the mantle. The Earth's lithosphere varies

greatly in thickness, from as little as 10 kilometers in some oceanic regions to as much as 300 kilometers in some continental areas. The Earth's lithosphere is broken up into a series of large fragments, or rigid plates. Seven major plates and a number of smaller ones have been distinguished, and they grind and scrape against one another as they move independently, similar to chunks of ice in water. Much of the Earth's dynamic activity occurs along plate boundaries, and the global distribution of associated tectonic phenomena, particularly earthquakes and volcanism, delineates the boundaries of the plates.

Within the upper mantle, there is a major zone where the temperature and pressure are such that part of the material melts, or nearly melts. Thus the rocks in this region of the Earth lose much of their strength, becoming soft and plasticlike, so that they can slowly flow as a viscous liquid. This zone of easily deformed mantle is termed the asthenosphere ("weak sphere"). Seismic velocities are about 6 percent lower in the asthenosphere than in the lithosphere. Although there is no fundamental change in chemical composition between the two regions, the lithosphere and the asthenosphere are mechanically distinct.

The lithosphere rides over the plastic, partly molten asthenosphere. As the lithosphere moves, the continents split, and the large plates drift thousands of kilometers across the Earth's surface. All the major structural features of the Earth are the result of a system of moving lithospheric plates. Movement in the plate tectonic system is driven by the loss of internal heat energy, primarily from within the mantle. This heat-driven internal movement is responsible for the creation of ocean basins and continents, as well as deformations of the Earth's solid outer layers that generate earthquakes, mountain belts, and volcanic activity at the plate boundaries. The primary source of heat within the mantle appears to be the radioactive decay of uranium, thorium, and potassium.

Below the asthenosphere, the rock becomes stronger and more rigid. The higher pressure below the asthenosphere offsets the effect of higher temperatures, making the rock stronger than in the overlying asthenosphere. The layer from the base of the Earth's crust to a depth of approximately 670 kilometers is designated the upper mantle, which includes the asthenosphere and the lower part of the lithosphere. Unraveling all the layers in the upper mantle has proved rather difficult. One model of the structure of the upper mantle designates a B layer, nearly 400 kilometers thick and of fairly uniform composition, and a C layer, between 200 and 300 kilometers thick, in which the chemical composition appears to be quite variable.

The transition from the upper to the lower mantle is quite gradual, with the depth of the separating, or transition, boundary usually thought to be between 600 and 670 kilometers. One of the fundamental questions about the seismic discontinuity that separates the upper mantle from the lower mantle is whether it is a barrier to lower mantle convection. One theory is that this boundary temporarily prevents the penetration of mantle material but ultimately will allow passage. Since seismic wave velocities appear to be very steady throughout the D′ region, a layer between 670

and 2,700 kilometers deep, the lower mantle is assumed to have a less complex structure than the upper mantle. Variations of properties inside the D′ region appear to be predominantly caused by the effects of simple compression. However, inside the D″ region, from 2,700 kilometers deep down to the outer core, seismic velocity falls continuously, indicating some continuous changes in physical properties and chemical composition that could produce deep mantle convection.

EVIDENCE FOR MANTLE CONVECTION

Better understanding of plate tectonics from analysis and interpretation of vast amounts of seismic data, as well as the application of improved observational and experimental techniques to the study of the properties of mantle materials, has confirmed the existence of mantle convection. Apparent episodic material exchanges between the upper and lower mantle, the genesis of plume-like upwellings, and the ultimate fate of subducted slabs within the Earth are all aspects of the mantle's dynamic convection system. One of the most significant questions is whether mantle convection is isolated in the upper mantle or involves the whole mantle.

The advent of seismic tomography in the mid-1980's, coupled with new laboratory and computational capabilities in the 1990's, has profoundly impacted the understanding of mantle convection. Global seismic tomography has helped to resolve the parameters that characterize mantle dynamics, particularly a viscosity increase from the upper to the lower mantle, an endothermic phase transition at the upper mantle-lower mantle boundary, heat flow across the core-mantle boundary (CMB), and the effect of the motion of rigid surface plates on convection patterns. Using seismic tomography, images of three-dimensional variations in seismic velocities revealed the deep structure of the underlying surficial plates. Patterns of high and low seismic velocity have been revealed throughout the mantle, with the strongest variations found in the upper 300 kilometers.

Unexpected features, such as deep roots of high-velocity material extending 300 to 400 kilometers below continental cratons, have greatly enhanced the understanding of plate tectonics and continental formation. Deep-seated upwellings under the ocean ridges and beneath major volcanic hot spots are indicated by low-velocity regions and by deflections of transition-zone discontinuities. While some tomographic images show descending lithospheric slabs that are apparently blocked at the upper mantle-lower mantle boundary, others show lithospheric slabs sinking nearly to the CMB. Numerical models show that reasonable simulations of subducted lithospheric slabs and plumes will penetrate the upper mantle-lower mantle boundary, although they may be temporarily blocked. Laboratory models of descending sheets interacting with contrasting density and viscous interfaces support this conclusion. Some slabs of subducting lithosphere have been seismically imaged as high-velocity tabular downwellings extending throughout the mantle.

UPPER MANTLE DYNAMICS AND CONVECTION

Because of its ability to flow, the asthenosphere figures prominently in the dynamic theories on the causes of vertical motion observed at the Earth's surface, such as postglacial rebound. Periodic compensatory adjustments that take place in the interior of the Earth in response to changing mass distributions at the surface that arise from erosion, sedimentation, glaciation and deglaciation, and volcanism are thought to occur through flow in the asthenosphere.

Likewise, the asthenosphere plays a prominent role in models of the large horizontal movements of the lithosphere as observed in continental drift and plate tectonics. As it slowly churns in large convection cells, the asthenosphere is the lubricating layer over which the plates glide. Thermal convection in the asthenosphere is thought to be a fundamental force in driving the tectonic plates. According to this scenario, hot mantle material rises at the mid-oceanic spreading ridges (divergent boundaries), where it escapes as magma, cools, and generates new oceanic crust. The seafloor moves in conveyor-belt fashion, ultimately to be destroyed at convergent plate boundaries, where it is subducted, or carried down, into the asthenosphere and eventually remelted. The rest of the hot mantle material spreads out sideways beneath the lithosphere, slowly cooling in the process. As it flows outward, it drags the overlying lithosphere outward with it, thus continuing to open the ridges. When the hot mantle material

cools, the flowing material becomes dense enough to complete the convection cycle by sinking back deeper into the mantle, and tomographic images indicate that this is happening under subduction zones at converging boundaries.

Some debate continues as to whether convection is confined to the upper mantle in a thin asthenosphere or whether it occurs throughout the mantle. The convection cells need not be confined to the asthenosphere. Seismic velocity data indicate that oceanic lithosphere can be subducted to depths of approximately 700 kilometers. Thus convection cells may operate at least down to those depths. In addition, in some places in the asthenosphere, the temperature may reach the rock-melting temperature and produce magma, thus giving the asthenosphere another dynamical role as the source region of many types of igneous rocks.

LOWER MANTLE DYNAMICS AND CONVECTION

A serious ongoing debate among Earth scientists about the size of convection cells in the mantle began in the early 1980's. Because of the different chemical compositions of the upper and lower mantle, geochemists have argued that the upper and lower mantles must have isolated convection cells with virtually no mixing between them. Thus slabs of the lithosphere that sink below the surface at the edge of tectonic plates should stay within the upper mantle, with their material being recycled there. In addition, geochemical evidence also exists for distinct reservoirs in the mantle, emphasizing the importance of plume flows from internal boundary layers as distinct from large-scale flows associated with the oceanic plates.

On the other hand, many geophysicists have maintained that convection involves the entire mantle. In the late 1990's, numerous three-dimensional seismic tomographic studies mapped seismic speeds in the Earth's mantle, and the interpretations provide strong evidence that the lithosphere is sinking well into the lower mantle. The tomographic images of seismic wave speeds at different depths are a rough indication of the temperature distribution in the mantle. Waves travel more quickly in regions that are colder, and more slowly in hotter regions. In numerous locations beneath the Earth, tomographic images show cold regions to be a continuous function of depth far into the lower mantle, suggesting the descent of some slabs of oceanic lithosphere into the lower mantle at the edge of tectonic plates. One interpretation of the tomographic images shows that the mantle's heterogeneity is dominated by large-scale structures, which support a mantle convection system dominated by large flow patterns. Evidence is not conclusive as to whether the mantle is only a layered convective system, but the evidence does suggest that significant material transport occurs across the upper mantle-lower mantle boundary.

In addition, increasing geophysical evidence supports the conjecture that the Earth's core interacts with the surrounding mantle. Images from seismic tomography reveal that the lowermost 200 to 400 kilometers of the mantle is one of the most heterogeneous regions of the Earth. Above the CMB, seismic images indicate the presence of two laterally variable seismic discontinuities, one at 130 to 400 kilometers and another at 5 to 50 kilometers above the CMB. According to the tomographic images, both regions have anisotropic properties, and the complexities of physical and dynamic processes are as sophisticated as those present in the lithosphere and shallow asthenosphere, which supports the idea of a dynamic lower mantle involved in convection.

In the late 1990's, the discovery that the D″ layer is associated with dramatically reduced seismic velocities at 5 to 50 kilometers above the CMB changed the persisting idea that the deep region of the lower mantle was solid. From tomographic images, some of the most unusual anomalies seen in the lower mantle are thin patches, less than 40 kilometers thick, in which seismic velocities are locally reduced by 10 percent or more. Such ultralow-velocity zones are not seen anywhere else in the mantle. Explaining their presence requires massive local melting within the lowermost mantle, meaning that this region is most likely convecting. The significant heterogeneity, which likely involves locally hot and partially molten zones near the CMB, is indicative of the dynamical behavior of the D″ layer. Some Earth scientists reason that many of the volcanic plumes associated with hot spots at the Earth's surface, such as the Hawaiian island chain, represent upwelling jets of hot rock in the mantle that are preferentially lined up above the ultralow-velocity molten patches in the D″ layer.

INTEGRATED MANTLE CONVECTION

In the late 1990's, many Earth scientists believed that it was necessary to make a compromise between whole-mantle convection and isolated-mantle convection. With regard to the existing geochemical evidence, the main requirement is that there exist chemically distinct reservoirs that do not have to be totally confined to the lower mantle. Three suggested models have emerged. Some Earth scientists support a model of the mantle that contains isolated, discontinuous volumes of material dispersed throughout. Other Earth scientists suggest that a reasonable model is one in which the mass transport in the mantle does not occur in a steady, continuous fashion but rather in an intermittent, nonsteady state. Still others believe that lithospheric slabs descending into the mantle lose geochemically monitored elements in the upper mantle as they sink to lower depths. The best model may contain aspects of all three of these alternatives.

Laboratory experiments in the mid-1990's have shown that the oxides of the Earth's deep mantle react vigorously when placed in contact with liquid iron alloys, thought to exist in the outer core, at the high pressures and temperatures at the CMB. These experiments suggest that the rocky mantle is slowly dissolving, over geological time spans, into the liquid metal of the outer core. The slow dissolution appears to be related to a fundamental change in the bonding character of oxygen at high pressures. Whereas oxygen forms insulating compounds at low pressures, it can become a metal-alloying component at high pressures. Thus, when coupled with seismic tomography, experimental and theoretical investigations of high pressures point to the CMB as perhaps being the most chemically active region in the Earth's interior. Numerical models of the mantle suggest that the Earth may have undergone a transition from layered to whole-mantle convection caused by a combination of secular cooling and a decrease in heat production in the mantle from radioactive decay.

The products of the chemical reactions at the CMB, where insulating oxides meet metallic alloys, may well explain the seismologically observed heterogeneity of the D″ layer in the mantle. In addition, piles of oceanic crust that have settled toward the bottom of the mantle may further contribute to the heterogeneity of the region.

The possible occurrence of varying amounts of metal alloys at the base of the mantle is particularly important because metal conducts heat much more readily than insulating oxides do. Consequently, heat may be emerging from the CMB in a spatially variable manner that determines the pattern of convection throughout the Earth's mantle.

SIGNIFICANCE

Many of the geophysical and geological phenomena of the Earth's crust are consequences of the dynamics and thermal convection within the underlying mantle. The major features of mantle convection are deduced from seismic data, laboratory investigations, and computational modeling. The emerging picture is that the tectonic plates of the lithosphere are the most active component of mantle convection, whereas mantle plumes are an important secondary component. Direct consequences of mantle dynamics and convection include the relative motions of the lithospheric plates, the spreading of the seafloor and formation of new crust, volcanism in its various tectonic settings, much of the Earth's seismic activity, and the majority of the observed heat flow through the Earth's surface.

Unraveling the complexities of the mantle continues to be a challenge for seismology, but the results play a crucial role in answering questions regarding the composition, dynamics, and evolution of the Earth. Better understanding of the present-day mantle is providing a much more complete picture of the evolution and interaction of the Earth's thermal and tectonic regimes. Plausible arguments indicate that the mantle was episodically layered in Precambrian times and that plate tectonics would not have worked when the mantle was more than about 50 degrees Celsius hotter than at present. Whether plumes were more or less important in the past is being studied. Based on the present model of the mantle's dynamic and convection patterns, plumes do not offer an alternative to plate tectonics because they are derived from a different thermal boundary layer (the D″ layer).

Insight into the dynamic interactions in the lower part of the mantle near the CMB is very important for better understanding past geological phenomena of significant magnitude. In particular, there is evidence for periods of massive volcanic eruptions (superplumes) that were hundreds of times greater than anything the Earth has experienced in recent geological time. Models of the deep mantle based on three-dimensional tomographic images and laboratory observations indicate that superplume events could be the surface manifestation of fluid-dynamical instabilities triggered at the CMB. Such models may be generating the first glimpses of how such massive instabilities are initiated deep inside the Earth.

Alvin K. Benson

CROSS-REFERENCES

BIBLIOGRAPHY

Brown, G. C., and A. E. Mussett. *The Inaccessible Earth*. 2d ed. New York: Chapman and Hall, 1993. Provides an understanding of mantle convection and how the lithospheric plates are an essential part of it. Contains insights concerning whether the mantle's D layer is a barrier to whole-mantle convection. Basic groundwork for general readers is included in a series of notes at the end of the book.

Hamblin, W. K., and E. H. Christiansen. *Earth's Dynamic Systems*. 8th ed. Upper Saddle River, N.J.: Prentice Hall, 1998. Very readable description of the Earth's two major dynamic systems: the hydrologic system and the tectonic system. Basic explanations of mantle composition, dynamics, and convection. Many excellent color photos and illustrations.

Jackson, I., ed. *The Earth's Mantle*. Cambridge, England: Cambridge University Press, 1998. Comprehensive overview of the composition, structure, and evolution of the mantle layer. Reviews the evolution of the Earth. Draws on perspectives from isotope geochemistry, cosmochemistry, fluid dynamics, petrology, seismology, geodynamics, and mineral and rock physics. Written for more advanced readers.

Jeanloz, R., and B. Romanowicz. "Geophysical Dynamics at the Center of the Earth." *Physics Today* 50 (August, 1997): 22. Description of what occurs in the Earth's core and at the core-mantle boundary. Based on geophysical observations and laboratory and computational models. Explanation of interactions at the D″ layer in the lower mantle.

Lay, T., and Q. Williams. "Dynamics of Earth's Interior." *Geotimes* 43 (November, 1998): 26. Excellent overview of the present understanding of the thermal, chemical, and dynamical state of the Earth's deep interior. Colored cross sections of the Earth's interior at various depths generated from seismic data.

Lillie, R. J. *Whole Earth Geophysics*. Upper Saddle River, N.J.: Prentice Hall, 1999. Introductory book that illustrates how different types of geophysical observations, especially seismic, have helped determine the Earth's gross structure and composition. Good explanation of the theory of plate tectonics and the mantle's role in it.

Strahler, A. N. *Plate Tectonics*. Cambridge, Mass.: GeoBooks, 1998. A textbook of basic principles and important plate tectonic data. Generally descriptive presentation, with supportive quantitative data. Basic explanations of mantle convection and its role in plate tectonics.

PLUMES AND MEGAPLUMES

A plume is a pipe that extends into the mass of hot rocks that exist in the mantle of the Earth and brings them to the surface, forming a "hot spot." A megaplume is a supermass of extremely hot rocks that moves very slowly under the surface of the Earth and influences the breakup of tectonic plates.

PRINCIPAL TERMS

CRUST: the rock and other material that make up the Earth's outer surface

GUYOT: a formation made by plume activity in the ocean that has a flat top wholly under water

HOT SPOT: a heat source fed by a plume that reaches deep into the Earth and produces molten rock

MAGMA: molten rock generated deep within the Earth that is brought to the surface by volcanoes and plumes

MANTLE: the part of the Earth below the crust and above the core composed of dense, iron-rich rocks

PLATE TECTONICS: the theory that accounts for the major features of the Earth's surface in terms of the interactions of the continental plates that make up the surface

SEAMOUNT: an isolated dome formed under the sea by plumes reaching a height of at least 2,300 feet

TECTONICS: the history of the larger features of the Earth, rocks and mountains, islands and continents, and the forces and movements that produce them

HOT SPOTS

There are more than one hundred regions of the world known as "hot spots," which are fed by plumes of hot rock rising from deep in the Earth's mantle. These hot spots are responsible for a particular type of volcanic activity, which, unlike other active volcanoes, has its origins deep in the interior of the Earth. Plumes are found far away from the most active centers of volcanic activity and are usually most active in flat landscapes or at the bottom of oceans rather than in mountainous regions, as is true of more-typical volcanoes.

The hot spots come from material found deep within the Earth's mantle, the solid layer of rocks that extends to more than 3,000 miles below the Earth's surface, just above the core. Plumes apparently arise in regions of the mantle that are stirred about as the large continental plates that cover part of the Earth move slowly across the surface. This movement of plates has been going on since early in the Earth's history, beginning at least 4.6 billion years ago, when the Earth's crust was just forming. As the huge plates of rock that make up the continents formed, they were originally one giant mass, but they began breaking up and moving apart at a fraction of an inch per year. "Plate tectonics," as the study of these movements is called, describes how the continents reached their present locations; they are still moving apart, and still only by fractions of an inch every year. The plate movement helps explain the building of mountain ranges, for as the plates crash into one another, they push their margins up into mountains such as the Himalayas and the Andes. Plate movement also helps scientists to understand the activities of volcanoes. Most volcanic activity occurs in those areas where the major plates that make up the Earth's surface (including the Eurasian, American, African, Pacific, Indian, and Antarctic plates) come together. At these margins of plate contact, the pressure of the plates pressing against one another creates fissures and breaks in the Earth's surface, through which magma—the hot, molten material coming from deep in the mantle—can flow. The plates that form the continents are called "continental plates." Other plates are found at the bottom of the oceans; these are called "oceanic plates."

HOT SPOT VOLCANOES

As the plumes of mantle material move upward toward the surface, they feed and create what are called "hot spot" volcanoes. These range in size according to how deep the plume has reached into the depths of the Earth. The deepest plumes create the largest volcanoes. The material coming up through the plumes (magma) consists of gigantic blobs of melted rock. Plumes and the hot spots connected to them move much more slowly than the continents above them. When one of the continental plates crosses over a plume, the magma flowing upward creates a large structure that looks like a dome. Such domes are usually about 125 miles wide and can be hundreds of miles long. Approximately 10 percent of the Earth's surface is covered with these domelike structures. As the magma continues to burst upward, the dome increases in size; as it does, cracks and small openings appear, and the hot magma flows through these openings onto the surface. The most well-known domes created by plumes are the Hawaiian Islands. Geologists believe that all the islands in the Hawaiian chain were created from a single plume. As the Pacific plate passed over the hot spot, the islands popped up and out from the ocean floor, and the plume pumped huge quantities of magma into and out of the resulting dome.

In the Atlantic, volcanic islands formed from plumes are found along the mid-ocean ridge. The Azores and Ascension Islands appear on this ridge. The location of these island chains and hot spots can help scientists understand the movement of tectonic plates. The plumes appear to be fixed in relationship to one another and move at velocities of only a few millionths of an inch each year. Sophisticated measuring instruments can account for this movement, however. They also move in what appear to be well-established tracks, and as they move, their heat weakens the rocks above. Over time, these weakened surfaces begin to crack, causing rifts, or giant cracks in the Earth. Some of these rifts become huge valleys, such as that found in the East African nation of Ethiopia.

HOT SPOT DISTRIBUTION

Of the hundred or so hot spots, more than half are found on the continental plates, with about twenty-five found in Africa. The African plate has remained over these hot spots for millions of years. The shape of the continent, which is covered by hundreds of basins, domes, and ridges, was greatly influenced by the slow movement of the continental plate over these plume-fed hot spots. Hot spots are also found in great numbers under the Antarctic and Eurasian plates. It seems likely that hot spots are more likely to be found under slow-moving plates, since those continents moving more rapidly, such as North and South America, have only a very few areas of volcanic activity caused by hot spots.

One area in North America located over a hot spot is Yellowstone National Park. The hot spot below the park creates the many geysers, including Old Faithful, that are found in the region. Geysers are created when surface water seeps into the ground. When it comes in contact with boiling magma, the water is heated rapidly; it then boils upward until it explodes through cracks in the Earth's crust.

Hot spots do have a limited life span. Typically, a plume feeding a hot spot cools off and disappears after about 100 million years. Their positions also change. The plume feeding the Yellowstone geysers originated farther to the north, around the Snake River in Idaho, 400 miles away, around 15 million years ago. Over time, the North American plate has slipped across it, putting the hot spot in its present location. Yet Yellowstone, too, is only a temporary home for this hot spot, and its slow movement to the southwest continues. The slow drift accounts for the volcanic activity in the area. Scientists believe that at least three major volcanic eruptions have taken place in this region during the last two million years. They predict that another massive explosion, hundreds of times greater than the huge Mount St. Helens eruption in 1981, will hit the area sometime in the next few thousand years.

ICELAND

The hot-spot theory is an important contribution to the science of plate tectonics. Few geologists doubt the existence of these hot spots and the plumes that create them. One of the most intensely studied hot spots is the dome that makes up the North Atlantic island of Iceland. This megaplume, which was raised above the ocean floor more than 16 million years ago, lies across a formation known as the Mid-Atlantic Ridge. The dome is actually about 900 miles long, but only about 350 miles of it—Iceland—lies above sea level. To the south of the island, the dome tapers off gradually and dips below the sea. In 1918, a volcanic eruption under a glacier to the north of Iceland melted enough ice to create an oceanic flood of water that was twenty times greater than the yearly flow of the Amazon River, the world's largest river. Luckily, the floodwater did not hit land, or many islands and the European coastline would have been devastated.

Within the next few million years, a very short period in geologic time, the Mid-Atlantic Ridge will have moved away from the hot spot, carrying Iceland with it. This will dry up the source of magma supplying the volcanoes on Iceland, and they will no longer erupt.

SEAMOUNTS AND GUYOTS

Most volcanic hot spots never rise above sea level and remain as underwater volcanoes. Magma erupting from these plumes forms structures called "seamounts." These are isolated, though they form into long chains along the surface of the oceanic plate. A few seamounts are found with extensive fissures and cracks; in these cracks, magma has cooled over hundreds of thousands of years, piling up thousands of magma flows, one on top of the other. A few of these are high enough to break through the ocean's surface. These seamounts become volcanic islands and dot various ridges in the crust, forming island chains such as the Galapagos Islands off the west coast of Ecuador. The tallest seamounts rise more than two and one-half miles above the sea floor and are found to the east of the Philippine Islands, where the crust is over 100 million years old. Generally, the older the crust, the larger the number of undersea volcanoes. The majority of seamounts are found in the Pacific Ocean, where there are between five and ten volcanic hot spots in every 5,000 miles of ocean floor. The Hawaiian Islands were created as the oceanic plate passed over a plume and formed the Emperor Seamounts, the name used by students of plate tectonics to refer to the Hawaiian Island chain.

Other plumes, in ancient geological times, formed undersea volcanoes called guyots (pronounced GHEE-ohs). Dozens of these once rose high above the Pacific Ocean. Over millions of years, however, constant wave action eroded the tops of these guyots below the sea's surface. Guyots, over time, moved away from their sources of magma, the hot spots, and this also helps to account for their disappearance beneath the ocean.

MID-CRETACEOUS MEGAPLUME ACTIVITY

The plumes described above are moderate in size and are considered to be a normal part of the Earth's mantle. Some geologists are convinced, however, that once, millions of years ago, the Earth went through an extremely intense period of volcanic eruptions. During this period, giant megaplumes exploded from deep within the Earth, expelling huge quantities of molten material. These structures spread across the Earth's surface, becoming ten times larger than average plumes. These "superplume" explosions were responsible for the volcanic activity that affected the ocean floor during the mid-Cretaceous period about 90 million to 100 million years ago. One result of this unusually violent activity was the creation of hundreds of seamounts in the western Pacific. Another area affected by these megaplumes was the Parana River basin in Brazil, where hundreds of rift valleys were created. It was also during this time that the Andes Mountains in South America and the Sierra Nevada in the western United States were formed.

Megaplume activity during the mid-Cretaceous led to a 100-foot rise in sea level and a 10-degree increase in the temperature of the Earth's air. This increase was caused by the release of huge amounts of carbon dioxide into the air during volcanic explosions. A key result of this activity was an enormous increase in plankton, the microscopic organisms that drift in the oceans and are the first link in the ocean's food chain. As the plankton died, they devolved into huge deposits of oil. Perhaps 50 percent of the world's known oil supply dates to this period of megaplume activity. The volcanic activity of the giant plumes also brought large quantities of diamonds from the Earth's interior closer to the surface, from which they are mined.

ROLE IN EARTH'S HISTORY

Plumes and megaplumes have had a dramatic influence on the history of the Earth. Plume activity has created islands, volcanoes, valleys, and mountains. The superheated rocks brought from far within the interior of the Earth have helped to form geysers, oil deposits, and diamonds. Tracking the slow movement of hot spots can help scientists to predict future events, such as the possibility of volcanic eruptions or the creation of new rift valleys. The study of past volcanic explosions can help scientists to keep the public informed about the potential harm that can be expected from future eruptions of plumes. Volcanic eruptions have occurred periodically throughout the Earth's history. From a geological point of view, periods of intense volcanic activity last for relatively brief spans of time, perhaps from 2 million to 3 million years. There are particularly intense periods of major activity every 32 million years. These latter periods—and the mid-Cretaceous period might have been one of them—coincide with mass extinctions of life, as volcanic gases flow into the atmosphere, releasing thousands of tons of sulphur and other dangerous chemicals. Some of the released gases are converted into acids that also have a devastating impact on living things. Hot spots and megaplumes expel huge amounts of ash, dust, and molten rock from their cracks and fissures. These materials absorb the Sun's radiation and can cause intense heating or cooling of the atmosphere. The dust can also shade out the Sun's light for long periods of time. The re-

duced sunlight can, again, cause mass extinctions of plants and animals because of the extreme cold produced. Intense volcanic activity can also produce acid rain, which could kill the leaves of plants and make the oceans and lakes unlivable. Scientists believe that the Earth has been victimized by such violent activity at least three times in the past, the last time being about 65 million years ago. Hot spots and megaplumes were responsible for much of this violent volcanic activity.

Leslie V. Tischauser

CROSS-REFERENCES

Acid Rain and Acid Deposition, 1803; African Rift Valley System, 611; Continental Drift, 567; Earth's Core, 3; Earth's Core-Mantle Boundary, 9; Earth's Crust, 14; Earth's Differentiation, 20; Earth's Lithosphere, 26; Earth's Mantle, 32; Earth's Structure, 37; Eruptions, 739; Geothermal Phenomena and Heat Transport, 43; Geysers and Hot Springs, 694; Hawaiian Islands, 701; Heat Sources and Heat Flow, 49; Hot Spots, Island Chains, and Intraplate Volcanism, 706; Island Arcs, 712; Lithospheric Plates, 55; Magmas, 1326; Mantle Dynamics and Convection, 60; Mount St. Helens, 767; Ocean Ridge System, 670; Plate Motions, 80; Plate Tectonics, 86; Recent Eruptions, 780; Seamounts, 2161; Volcanic Hazards, 798; Yellowstone National Park, 731.

BIBLIOGRAPHY

Ballard, Robert D. *Exploring Our Living Planet.* Washington, D.C.: National Geographic Society, 1983. A well-illustrated guide to modern theories of continental drift, plate tectonics, and the activities of volcanoes. A good place to begin an investigation of the history of the Earth's formation and the various forces that have created the Earth's features. Includes pictures, maps, and an index.

Condie, Kent C. *Plate Tectonics and Crustal Evolution.* 4th ed. Oxford: Butterworth Heinemann, 1997. An excellent overview of modern plate tectonics theory that synthesizes data from geology, geochemistry, geophysics, and oceanography. A very helpful tectonic map of the world is enclosed. The book is nontechnical and suitable for a col-

lege-level reader. Useful "suggestions for further reading" follow each chapter.

Eicher, Don L., A. Lee McAlester, and Marcia L. Rottman. *The History of the Earth's Crust.* Englewood Cliffs, N.J.: Prentice-Hall, 1984. A brief introduction to plate tectonics and geological history. A good beginning for those unfamiliar with the topic. Useful illustrations, charts, and an index.

Erickson, Jon. *Plate Tectonics: Unraveling the Mysteries of the Earth.* New York: Facts on File, 1992. An excellent, well-written, easily understandable description of the forces shaping the Earth's geology, including a detailed and illustrated discussion of plumes and hot spots. Megaplumes, however, are not described. A very good introduction to the subject. Illustrations, bibliography, index.

Kearey, Philip, and Frederick J. Vine. *Global Tectonics.* 2d ed. Cambridge, Mass.: Blackwell Science, 1996. A textbook written in somewhat technical language; nevertheless contains some good illustrations and a detailed discussion of megaplumes. Designed for college courses in geology. Index and bibliography.

Olsen, Kenneth H., ed. *Continental Rifts: Evolution, Structure, Tectonics.* Amsterdam: Elsevier, 1995. The various essays provide good explanations of plate tectonics and continental rifts. Slightly technical but suitable for the careful reader. Illustrated.

Reynolds, John M. *An Introduction to Applied and Environmental Geophysics.* New York: John Wiley, 1997. An excellent introduction to seismology, geophysics, tectonics, and the lithosphere. Appropriate for those with minimal scientific background. Includes maps, illustrations, and bibliography.

Seyfert, Charles K., and L. A. Sirkin. *Earth History and Plate Tectonics: An Introduction to Historical Geology.* New York: Harper & Row, 1973. A textbook for geology students, but easy to understand and well illustrated. Read this book after consulting some of the briefer descriptions of the formation of the Earth's mantle and crust.

Sullivan, Walter. *Continents in Motion: The New Earth Debate.* New York: McGraw-Hill, 1974. A somewhat dated but still useful summary of the differing points of view of various theorists of the Earth's formation and how such views have changed over time. Popularly written; easily understood without a technical background in geology.

2
PLATE TECTONICS

PLATE MARGINS

The outer 70-200 kilometers of the Earth comprise a number of rigid plates that move about independently of one another. Each plate interacts with the adjacent plate along one of three types of plate margin: convergent, divergent, or conservative. The interactions of the moving plates along these margins cause most earthquakes as well as much of the volcanic activity on the surface of the earth.

PRINCIPAL TERMS

ACCRETIONARY PRISM: a complex structure composed of fault-bounded sequences of deep-sea sediments mechanically transferred from subducting oceanic lithosphere to the overriding plate; it forms the wall on the landward side of a trench

ASTHENOSPHERE: the soft, partially molten, layer below the lithosphere

CONVECTION CELL: a single circular path of rising warm material and sinking cold material

LITHOSPHERE: the outer, rigid shell of the Earth that contains the oceanic and continental crust and the upper part of the mantle

RIFTING: the process whereby lithospheric plates break apart by tensional forces

SEAFLOOR SPREADING: the concept that new ocean floor is created at the ocean ridges and moves toward the volcanic island arcs, where it descends into the mantle

SUBDUCTION ZONE: a region where a plate, generally oceanic lithosphere, sinks beneath another plate into the mantle

TRANSFORM FAULT: a fault connecting offset segments of an ocean ridge along which two plates slide past each other

VOLCANIC ISLAND ARC: a curving or linear group of volcanic islands associated with a subduction zone

WADATI-BENIOFF ZONE: the inclined band of earthquake focus points interpreted to delineate the subducting oceanic lithosphere; better known as the Benioff zone

PLATE TECTONICS

The surface of the Earth is a mosaic of several large and more numerous smaller plates that move about laterally. The boundaries, or margins, of these plates are the sites of most of the volcanic and earthquake activity on the Earth. The geological concept of the plate and plate margins ultimately has its origins in the theory of continental drift. The first comprehensive theories of continental drift were independently proposed around 1910 by the American geologists F. B. Taylor and H. H. Baker and the German meteorologist Alfred Wegener. Wegener's work was particularly thorough, and he is generally considered to be the person who first made the theory of continental drift an important scientific issue. Wegener devoted much time and effort to matching geological features and fossil types on both sides of the Atlantic Ocean. He argued, based on this work, that approximately 230 million years ago, all the continents were joined as one supercontinent that he named Pangaea. Furthermore, Wegener suggested that Pangaea broke apart approximately 170 million years ago. Since then, the various parts of the supercontinent—the modern continents—have moved to their present positions. Wegener also postulated that as the continents move through the oceans, their leading edges become crumpled and often collide with other continents, thereby forming the mountain belts.

In the period after World War II, several important discoveries made by oceanographers studying the ocean basins lent support to Wegener's scoffed-at theory. This postwar research ultimately led to the formulation of the theory of seafloor spreading proposed by Harry Hammond Hess of Princeton University. According to Hess's theory, new ocean floor is continuously forming at the ocean ridges, or the large subsea mountain ranges that traverse the Earth, in a conveyor belt fashion. As this process continues, the newly formed seafloor moves laterally away from the ocean ridge on both sides of the ridge. The opening in the Earth's surface created by the spreading of the seafloor at the ridge is filled with magma, or molten rock, from the mantle, which cools to form new seafloor. Hess and other scientists suggested that if the seafloor is continuously moving, the continents must also be moving with it. Thus, the concepts of continental drift and seafloor spreading were combined into the more comprehensive theory of plate tectonics. This revolutionary theory in the Earth sciences describes the movement of rock in the Earth's 70- to 200-kilometer-thick outer brittle shell, the lithosphere, as it moves over the deeper, more ductile, partially molten asthenosphere. The lithosphere, which includes continental and oceanic crust and the upper part of the mantle, comprises a number of large and small, rigid lithospheric plates that move independently of one another. As these plates move, they interact along one of three types of plate margin: divergent or accreting margins, convergent or destructive margins, and conservative or neutral margins.

DIVERGENT PLATE MARGINS

Divergent or accreting plate margins are tensional plate boundaries that correspond to the ocean ridges. According to plate tectonic theory, ocean ridges, also referred to as spreading centers, are sites on the Earth's surface where new oceanic lithosphere is formed by the process of rifting, or the tensional separation of plates. Rifting can occur in the oceans as well as on the continents, as in the case of the East African rift zone. If rifting begins on a continent and continues for an extended period of time, a new ocean basin will ultimately form—as is now occurring in the Middle East, where the Red Sea rift is creating a new ocean between Saudi Arabia and northeastern Africa.

The crests of ocean ridges are characterized by deep valleys believed to have been caused by tensional faulting in response to oppositely directed lateral movement of the plates bound by the ocean ridge. This valley is referred to as the rift valley and is marked by a high degree of earthquake activity. The rift valley is also the site of voluminous outpourings of basaltic magma, or molten rock enriched in iron and magnesium.

Some of the most dramatic evidence of the processes occurring at divergent plate margins has

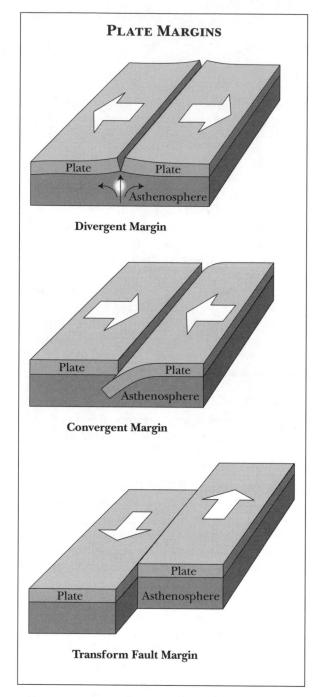

PLATE MARGINS

Divergent Margin

Convergent Margin

Transform Fault Margin

come from a series of observations made from submersibles on the Galápagos spreading ridge near the equator, just west of South America and that part of the East Pacific Rise south of the Gulf of California. Researchers observed undersea hot springs and mounds of iron-rich clay minerals and manganese dioxide precipitated from the hot ore-carrying springs. Similar submersible dives to the Mid-Atlantic Ridge southwest of the Azores permitted observation of submarine volcanism and yielded abundant evidence of tensional faulting within the narrow rift valley.

CONVERGENT PLATE MARGINS

Convergent or destructive plate margins are those areas of the Earth's surface where the lithospheric plates grind together head-on and are then recycled back into the asthenosphere. Thus, these margins are characterized by compressive tectonic forces. Submarine features typical of convergent plate margins are long, narrow troughs on the seafloor referred to as trenches. Hess postulated that the trenches mark the positions where ocean lithosphere created at ocean ridges is drawn down or sinks into the mantle. The trenches, which are the deepest points of the oceans, are closely associated with volcanic island arcs, or linear or arcuate groups of volcanic islands such as Japan or the Aleutian Islands.

Convergent margins are characterized by a high incidence of earthquakes, many of which originate at depths greater than 600 kilometers within the Earth. Scientists have demonstrated that the focus points of these earthquakes (those points within the Earth from where the seismic energy is first generated) are generally found within a band inclined from the trench toward the volcanic island arc or continent called the Wadati-Benioff zone (better known as the Benioff zone). Realization of this distribution of earthquakes at convergent margins led scientists to speculate that the Wadati-Benioff zone delineates a slab of dense lithosphere formed at an ocean ridge that, according to plate tectonic theory, is sinking at the trench into the mantle. This process of lithospheric sinking at convergent margins is referred to as subduction. In general, the most powerful earthquakes at convergent margins are the shallow focus earthquakes generated close to the trenches. These earthquakes occur when a sinking lithospheric plate moves beneath an island arc or a continent and drags the overriding plate down a bit. Eventually, this process reaches a critical point, and sudden slip occurs along the boundary of the sinking plate and the plate beneath which it is moving, thereby creating the earthquake.

SUBDUCTION

The volcanic island arcs, like the distribution of earthquake focus points, can be considered in terms of the process of subduction. The basic question concerns how the magma that spewed from the volcanic islands forms beneath the island arcs. It is generally agreed that island arc volcanism is caused by melting of the subducting plate as it descends into the hot mantle. Generation of magma may also be assisted by frictional melting along the upper surface of the subducting plate as it moves beneath the island arc.

A final point regarding convergent margin processes concerns the fate of marine sediments on subducting oceanic lithosphere. Initially, it was postulated that all the marine sediments on descending oceanic lithosphere should be piled up and folded on the bottom of the trench. Studies of modern trenches, however, indicate that trenches generally contain only minor amounts of sediment. A more complete understanding of this problem was gained when more sophisticated geophysical techniques were applied to the study of convergent margins. Results of these investigations suggest that marine sediments carried on a subducting plate are stripped off the plate at the trench. These sediments are attached to the leading edge of the overriding plate to form a complexly deformed sequence of sedimentary rocks called an accretionary prism that builds the landward wall of the trench.

In summary, convergent plate margins display a series of features that have been interpreted in terms of subduction, the dominant plate tectonic process occurring at these margins. Trenches mark the locations on the Earth where oceanic lithosphere, formed at a divergent plate margin, sinks into the mantle. At least a portion of the deep marine sediment carried atop this plate is mechanically transferred to the leading edge of the overriding plate to form an accretionary prism. As the plate moves deeper into the mantle, it fractures and generates earthquakes along its length. Additionally, it begins to melt, thereby producing magma that rises to the surface of the Earth to form a volcanic island arc. If the oceanic plate is attached to a continent, that continent eventually reaches the trench. Because, however, continental lithosphere is less dense and therefore more buoyant than oceanic lithosphere, the continent cannot be subducted, and instead it collides with the volcanic island arc or the overriding continent. This collision results in the formation of a mountain belt, as in the case of the collision of the Indian subcontinent and the Tibetan plateau to the north, which is still forming the Himalaya belt.

CONSERVATIVE PLATE MARGINS

The third type of plate margin is the conservative, or neutral, margin. These margins, along which lithosphere is neither created nor destroyed, are characterized by oppositely directed horizontal movement of adjacent plates. The actual boundaries of the moving plates are marked by transform faults, or faults along which plates slide horizontally past one another. The San Andreas fault of California, perhaps the best-known example of a transform fault, marks the boundary between the northwest-moving Pacific plate and the North American plate. This plate boundary is characterized by contrasting geology on both sides of the fault, by little if any volcanic activity, and by powerful shallow focus earthquakes such as the kind that devastated San Francisco in 1906.

Transform faults separate offset segments of ocean ridges. Although ocean ridges extend continuously for thousands of kilometers, they are actually broken into much smaller segments separated by transform faults that are oriented at nearly right angles to the ridge segments. The relative movement of the two plates along a transform fault is caused by creation of new seafloor at the two offset ridge segments. As ocean lithosphere forms at and moves away from one ridge segment, it slides in the opposite direction past lithosphere forming at the other ridge segment. The transform fault, therefore, marks the contact of the oppositely moving plates and is situated between the ridge segments.

TRIPLE JUNCTIONS

Plate tectonic theory requires that there be single points, called triple junctions, at which three lithospheric plates meet. In the Middle East, for example, three divergent plate margins—the Gulf of Aden, the East African rift, and the Red Sea rift—meet at what is referred to as a ridge-ridge-ridge triple junction. Almost any combination of the three plate margins—ridge, trench, and trans-

form fault—can form triple junctions. Some types of triple junctions move with the plates, and they may even be subducted.

ECHO SOUNDING AND SEISMIC REFLECTION PROFILING

Much of what is known about plate margins, particularly convergent and divergent margins, has come about through detailed study of the surface and interior structure of the ocean floor. One of the most important techniques developed in this regard is echo sounding. In echo sounding, a sound pulse generator-receiver system mounted on the hull of a ship emits sound pulses at regular intervals. Each pulse travels to the ocean floor at a known velocity and echoes back to the ship, where its return is detected by the pressure-sensitive receiver. The recording apparatus, a precision depth recorder, indicates the travel time of the sound pulse to and from the ocean bottom on an advancing paper chart. As the ship moves across the ocean, travel time marks for a succession of pulses detected by the receiver are displayed on the chart profile. The depth to the sea bottom is then calculated by multiplying the velocity of the sound pulse by one-half its travel time. This method was most instrumental in defining the ocean ridges and trenches on the ocean floor.

A somewhat more sophisticated approach using seismic waves allows scientists to study the internal structure of the upper part of oceanic lithosphere. In this approach, referred to as seismic reflection profiling, a ship emits sound waves powerful enough to penetrate the bottom of the ocean and then to reflect back to the ship. More specifically, these sound waves, which may be generated either by undersea explosions or by compressed air, reflect from the surface of the ocean floor and from internal sediment layers and faults back to the ship, where they are picked up by a receiver, or hydrophone, towed behind the ship. The travel times of the waves reflected off and from within the seafloor are recorded on charts by a sparker-profiler. Seismic reflection profiling has helped scientists to understand better the interior structure of the ocean bottom. For example, the faults bounding the slivers of marine sediments in accretionary prisms at convergent margins were recognized through the use of this technique.

MULTINATIONAL RESEARCH PROJECTS

The Deep Sea Drilling Project (DSDP), a multinational research program initiated in 1968, attempted to understand better the evolution and geologic history of the modern oceans by drilling through the deep-sea sediments into the underlying igneous floor of the ocean. A specially designed ship, the *Glomar Challenger,* was used in the drilling. Among other things, results of the DSDP indicated that the age of the igneous ocean floor increases away from ocean ridges, thereby substantiating the major tenet of seafloor spreading: that oceanic lithosphere is produced at, and moves away from, the ridges. The Deep Sea Drilling Project was superseded by a new program of research with many of the same goals, the Ocean Drilling Program (ODP).

Many details of plate margins, particularly divergent ocean ridges, have been revealed through direct observation of the seafloor. The French-American Mid-Ocean Undersea Study (project FAMOUS) of 1973 and 1974, for example, concentrated on a small area of the Mid-Atlantic Ridge southwest of the Azores. Several deep-sea submersible submarines were used to dive to the ridge to map the shape of the rift valley and to collect samples of the ocean floor. This project permitted observation of the tensional faults that formed the rift valley and extrusion of basaltic magma. In several submersible dives to transform faults, scientists recovered igneous rock samples that showed evidence of the shearing associated with horizontal plate movement along the transform faults. A number of submersible dives to the East Pacific rise in the eastern Pacific Ocean have allowed marine geologists to observe submarine hot springs associated with lava extrusion at a divergent margin.

HEAT FLOW AND ROCK SEQUENCE STUDIES

Measurement of terrestrial heat flow (the amount of heat that escapes from the Earth's interior through the seafloor) yielded particularly valuable information regarding the nature of plate margins. Results of heat flow studies indicated that the ocean ridges, once thought to be dormant submarine mountain ranges, are actually sites where large amounts of heat from the interior of the Earth reach the Earth's surface. This finding fit in well with Hess's convection-cell interpretation of ocean ridges. In addition, heat-flow studies demonstrated extremely low heat-flow values in the trenches, an observation consistent with Hess's proposal that convection cells sink into the mantle at convergent margins.

Finally, studies of modern plate margins have been supplemented by investigations of rock sequences exposed on land and interpreted to have formed at ancient plate margins. This approach is particularly useful to the study of convergent plate margins. For example, highly deformed or chaotic rock units, referred to as mélanges and exposed in the Appalachian belt, along coastal California, and elsewhere, have been interpreted as marine sediments that were incorporated into ancient accretionary prisms. By studying these exposed sedimentary rocks, geologists can understand better the processes occurring at modern convergent plate margins.

DANGERS AND BENEFITS

Plate margins are generally not a major concern of most people unless they happen to live or work near one. Nevertheless, one has only to pick up a newspaper to see the effects of plate margins and their attendant processes on human life. Convergent margins, for example, are characterized by frequent earthquakes and are generally prone to volcanic activity, some of which may be violent. The powerful earthquakes and deadly volcanoes of the Aleutian Islands, Central and South America, Japan, and Indonesia attest the potentially dangerous conditions of convergent plate margins. Conservative plate margins, like the San Andreas fault, are susceptible to powerful earthquakes, although volcanic activity is not likely. The instability of plate margins must be kept in mind by community planners so that proper building codes can be created and followed to reduce the potential for catastrophe in these areas.

Despite the obvious dangers of living close to or along plate margins, there can be some benefits. In Iceland, for example, the heat emanating from the Mid-Atlantic Ridge is used as geothermal energy. Indeed, Reykjavík, the capital of Iceland, is heated entirely by geothermal energy.

Understanding the relation of plate tectonics and metal deposits is of paramount importance given the growing global need for various metals. At ocean ridges, for example, marine geologists and oceanographers have observed the formation of metallic sulfide ores. These deposits, which form in association with the basalt magma extruded at the ridge, precipitate out of the hot water that circulates through the newly erupted basalt. Convergent plate margins are characterized by various types of metal deposits formed in association with magma generated during subduction. In the Andes belt of South America, iron, copper, and gold ores accumulated in response to subduction of the Pacific ocean floor beneath the western coast of South America.

Gary G. Lash

BIBLIOGRAPHY

Bonatti, E., and K. Crane. "Ocean Fracture Zones." *Scientific American* 250 (May, 1984): 40. Excellent discussion of transform faults and associated oceanic fractures. Suitable for the college-level reader.

Condie, Kent C. *Plate Tectonics and Crustal Evolution.* 4th ed. Oxford: Butterworth Heinemann, 1997. An excellent overview of modern plate tectonics theory that synthesizes data from geology, geochemistry, geophysics, and oceanography. A very helpful tectonic map of the world is enclosed. The book is nontechnical and suitable for a college-level reader. Useful "suggestions for further reading" follow each chapter.

Dewey, J. F. "Plate Tectonics." *Scientific American* 226 (May, 1972): 56. A good overview of the theory of plate tectonics and plate margins. Can be read by the high school or college student.

Heirtzler, J. R. "Seafloor Spreading." *Scientific American* 219 (December, 1968): 60. This article provides an excellent discussion of the theory of seafloor spreading. Suitable for high school students.

Heirtzler, J. R., and W. B. Bryan. "The Floor and the Mid-Atlantic Rift." *Scientific American* 233 (August, 1975): 78. Excellent discussion of divergent plate margins, with the Mid-Atlantic Ridge as the example. Can be read by high school and college students.

Kearey, Philip, and Frederick J. Vine. *Global Tectonics.* 2d ed. Cambridge, Mass.: Blackwell Science, 1996. This college text gives the reader a solid understanding of the history of global tectonics, along with current processes and activities. The book is filled with colorful illustrations and maps.

Kious, Jacquelyne W. *This Dynamic Earth: The Story of Plate Tectonics.* Washington, D.C.: U.S. Department of the Interior, United States Geological Survey, 1996. Kious is able to explain plate tectonics in a way suitable for the layperson. The book deals with both historic and current theory. Illustrations and maps are plentiful.

Marsh, B. D. "Island-Arc Volcanism." *American Scientist* 67 (March/April, 1979): 161. A detailed discussion of volcanic activity at convergent margins. Suitable for college students.

Sutherland, Lin. *The Volcanic Earth: Volcanoes and Plate Tectonics, Past, Present, and Future.* Sydney, Australia: University of New South Wales Press, 1995. Although Sutherland focuses on volcanic activity in Australia, the book provides an easily understood overview of volcanic and tectonic processes, including the role of igneous rocks. Includes color maps and illustrations, as well as a bibliography.

Tokosoz, M. N. "The Subduction of the Lithosphere." *Scientific American* 233 (November, 1975): 88. This article describes the process of subduction at convergent margins and can be read by high school and college students.

Uyeda, Seiya. *The New View of the Earth: Moving Continents and Moving Oceans.* San Francisco: W. H. Freeman, 1971. An excellent presentation of the evolution of the theory of plate tectonics from continental drift. Convergent and divergent plate margins are particularly well discussed, with numerous examples from the Pacific Ocean. Probably most suitable for college-level readers.

PLATE MOTIONS

In order to trace the geological history of the Earth, it is necessary to know how the tectonic plates have moved around upon its surface. Using geological evidence, scientists can determine their relative locations at various times in the past. Such information can help in understanding the distribution of geological provinces and also in locating economically important formations.

PRINCIPAL TERMS

DECLINATION: the angle in the horizontal plane between true north and the direction that the magnetization of a rock points

EULER POLE: the point on the surface of the Earth where an axis, about which a rotation occurs, penetrates that surface

FRAME OF REFERENCE: a part of the planet, with respect to which all velocities are quoted

HOT SPOT: a point on the Earth's surface, unrelated to plate boundaries, where volcanic activity occurs

INCLINATION: the angle in the vertical plane be-

tween horizontal and the direction of magnetization of a rock

PANGAEA: a supercontinent consisting of all the present continental fragments; it existed approximately 200 million years ago

RELATIVE VELOCITY: the velocity of one object measured relative to another

TRIPLE JUNCTION: a point where three plate boundaries meet

VECTOR: a quantity that is defined by both magnitude and direction

VELOCITY: speed and direction of motion

RELATIVE VELOCITY

A central tenet of the plate tectonic theory is that the plates are moving across the surface of the Earth. Though all motion is relative in the context of plate tectonics, motion must be defined with respect to a given frame of reference. There is also the difficulty of treating an enormous period of time over which the plate motion has taken place. Some of the geological methods available to Earth scientists can be utilized to find the position of a plate millions of years ago, while other methods can yield its present velocity. It may be difficult, however, to resolve these two pieces of information into a consistent pattern describing the history of the plate's motion.

In order to measure a plate's motion, the first step is to find its velocity with respect to an adjoining plate; that is termed a relative velocity. Such a relative velocity is actually a linear velocity and, like any velocity, is a vector, which means that it is described not only by the speed of the plate (the magnitude of the vector) but also by the direction in which the plate moves. Some of the methods that geologists employ to find plate velocities give

both magnitude and direction; others provide only one of these quantities.

EULER POLES

Yet, plates do not move in straight lines, as they are constrained to be on the surface of a globe. In fact, the plates are moving along curved paths, so their velocities should be described as angular velocities, strictly speaking. If one is considering only a very small area on the Earth's surface, then linear velocities are an acceptable approximation. While linear velocities are quoted in units of millimeters per year, angular velocities are quoted as degrees per year, or radians per year. Furthermore, angular velocities are described as a change in angle per unit time around a pivot point (or axis). An everyday analogy might be a door: When a door opens, it pivots at the hinge, and the entire door moves at a particular angular velocity around this pivot. Note that the linear velocity of various parts of the door varies. Near the hinge, the distance moved in the time taken to open the door is small, so the linear velocity here is small too. The door handle moves a much greater distance in the same time, so its linear velocity is greater. Note also that as the door opens the linear velocity of any point on the door changes continually, as any point on the door is constantly changing direction. Considering the door handle, at every instant during the opening its direction of motion is changing (even though the speed may be constant); hence, the velocity is also changing.

Now consider a plate on the surface of the Earth: The linear velocity of the plate has a small magnitude near the pivot point around which it moves. This pivot point is called an Euler pole (for a Swiss mathematician, Leonhard Euler, who developed these concepts). Farther away from the Euler pole, the magnitude of the linear velocity increases. Suppose that two plates are spreading apart and that the pivot point is the north geographic pole. The mid-ocean ridge between the plates would lie on a line of longitude. The linear velocity of one plate wth respect to the other (at any instant) would be zero at the Euler pole and increase to a maximum at the equator. On the other side of the equator, the linear velocity would decrease until it reached zero again at the south pole, where another Euler pole would be located. In fact, the two Euler poles are just the points where the axis around which the rotation is taking place penetrates the Earth's surface.

RECONSTRUCTING ANCIENT LANDMASSES

Knowledge of relative velocities and Euler poles enabled the reconstruction of the position of the continental landmasses in the past. Approximately 200 million years ago, the continents were grouped in a single supercontinent called Pangaea. Pangaea then split into a northern fragment (Laurasia) and a southern part (Gondwanaland) separated by the Tethys Sea. Since then, the fragmentation has continued, and the plates have shifted such that the continents have drifted to the positions they occupy today. Some of the continental fragments have drifted quite rapidly, such as the Indian subcontinent, which broke from Africa and Antarctica and drifted north until colliding with Asia to form the Himalayan mountains.

Economic deposits, such as oil or coal, may have originally formed before the breakup of Pangaea. Therefore, if the location of one such deposit is known, then, by reconstructing the ancient landmasses, it may be possible to determine the full paleogeographic extent of the environment that gave rise to the deposit in the past. By using this method, geologists can predict where further economic sites may lie, even if these sites are currently thousands of miles from the known deposit on a separate continent.

PREDICTING FUTURE PLATE MOVEMENT

Using the known present relative velocities, how the plates will move in the future can be predicted. For example, the Atlantic Ocean will continue to open, mostly at the expense of the shrinking Pacific Ocean. Australia and Africa will continue to move north, as will Baja California and parts of southern California as the San Andreas fault lengthens. In roughly 10 million years, Los Angeles and San Francisco will become neighbors.

Although plate motions are very slow, the consequences of those motions can often be very abrupt and dramatic. Study of plate tectonics has led to an understanding of why certain regions of the Earth are prone to such hazards as earthquakes and volcanic eruptions. It would be to everyone's benefit to be able to predict when these events will take place, and some of the methods that are used to determine plate motions can give direct information concerning these phenomena. In a region such as Southern California, measurements of relative velocities along the San Andreas fault can be applied to the forecasting of earthquakes, and are, therefore, of direct interest to the local population.

ABSOLUTE PLATE MOTIONS

Although relative velocities are clearly very useful, the absolute motions of plates can be defined from a frame of reference that is geologically determinable. What is needed is a frame of reference fixed with respect to the interior of the planet, beneath the lithosphere; this region of the interior is termed the mesosphere. It appears that there are locations on the Earth's surface that are in some way tied to the mesosphere: hot spots. Hot spots are places where there is volcanic activity that is apparently unrelated to plate boundary activity. These regions are often typified by lavas that are geochemically dissimilar to those formed at either mid-ocean ridges or island arcs, and the suggestion is that the dissimilarity results from the fact that their magma source is much deeper.

Perhaps the best example of a hot spot trace is the Hawaii-Emperor chain of seamounts, which is basically a chain of extinct volcanoes except for the island of Hawaii itself. As one moves away from Hawaii along this seamount and island chain, the lava flows become progressively older. In the hot spot hypothesis, this is explained by the postulation that each island (or seamount) formed over the hot spot, but that the motion of the Pacific plate over the hot spot continually moved the islands away from the magma source— rather like a conveyer belt moving over a static Bunsen burner, leaving a progressively lengthening scorch mark. Interestingly, the Hawaii-Emperor chain has a bend in it, at about the location of Midway Island, which is interpreted as meaning that the Pacific plate motion changed direction at the time that that island formed (some 37 million years ago). Because it is possible to date the lava flows on these islands, the velocity of the Pacific plate relative to this particular hot spot can be calculated (its direction being obtained from the bearing of the seamount chain). The same can be done for other hot spots, too, and thus geologists can ultimately find the velocities of the hot spots relative to one another. The result of this procedure is the discovery that the hot spots move with respect to one another but at rates much slower than do the plates. Assuming that these relative motions are insignificant and that the hot spots are in reality fixed with respect to their proposed source, the mesosphere, then the mean hot spot frame of reference can be defined and all plate motions calculated with respect to that. In fact, this method essentially determines the velocities of the plates with respect to a mantle velocity that best simulates all the known hot spot traces. Absolute plate velocities determined by this method are commonly given in contemporary global plate motion analyses.

THEORETICAL IMPLICATIONS

The analysis of absolute plate motions has a bearing on theories concerning why the plates move. One group of plates is apparently moving quite slowly, with velocities of between 5 and 25 millimeters per year. This group includes the Eurasian plate, the North and South American plates, and the African and Antarctic plates. In contrast, the Indian, Philippine, Nazca, and Pacific plates move much more rapidly, and the Cocos plate has a velocity of roughly 85 millimeters per year. This observation has led to the realization that it is the plates with actively subducting margins that move the fastest. None of the slower group has a significant percentage of its margin being subducted, whereas all in the faster group do; the implication may be that the subduction process itself plays an important role in driving plate motions. This idea is in contradiction to the earlier hypothesis that the lithospheric plates rode on the back of giant convection cells within the mantle. If this latter view were correct, one might expect the larger plates to move faster (although that is debatable, depending on the geometry of the convecting cells). At the least, a passive plate theory such as that would not produce the correlation noted above. It seems that the plates are not passive players in the plate tectonic cycle but are an active part of convection.

DETERMINING EULER POLE LOCATIONS

Geologists have a variety of ways to determine how plates have moved in the past and how they may move in the future. Crucial to this endeavor is determining the location of Euler poles, but finding the location of an Euler pole for the relative motion between two plates can be difficult. As indicated in the previous section, however, if one follows a line of longitude along which a ridge lies, one must eventually arrive at the Euler pole. Unfortunately, ridges do not always lie on the geographic longitude lines of the Earth. The ridge system separating two plates describes its own set of longitude lines, which may not correspond with geographic longitude lines. To distinguish them, these longitude lines can be referred to as great circles. (In fact, any circle that is drawn around the Earth is a great circle. All great circles would be identical in length on a perfectly spherical Earth. Latitude lines, on the other hand, are not great circles, with the exception of the equator, and vary considerably in length. They are referred to as small circles.) Fortunately, mid-ocean ridge segments are offset by transform faults; therefore, a set of great circles can be drawn through the various segments of the ridge system and hence reveal the Euler pole. The transform faults can also be used; small circles drawn through these also define the position of the Euler pole. This latter case has the added advantage that the fracture zones on either side of the transform fault effectively extend their length and make the geometric construction easier, as it is advantageous to have as long a feature as is possible to which to fit the circle in order to cut down on the errors inevitably involved with any line-fitting method.

VECTOR ADDITION

When a plate does not have a mid-oceanic ridge system separating it from a neighboring plate, other methods must be employed. Such is the case with the Philippine plate, which is surrounded entirely with subduction zones. In this instance, finding its velocity with respect to its neighboring plates and the Euler pole around which the rotation occurs is much more difficult. The motion of the Philippine plate is usually found by adding the velocities of all the other plates on the Earth's surface and finding the resultant. The velocity that exactly cancels this resultant is taken to be the velocity of the Philippine plate. Locations where three plates meet at a single point (triple junctions) can be analyzed by vector addition also, and if the velocities of two of the plates are known, then the velocity of the third can be determined. Significantly, the relative velocity of the triple junction itself can be determined; from that number it can be determined if any of the plate boundaries is lengthening or shortening. In this fashion geologists were able to determine that the San Andreas fault is lengthening. At its southern end, the triple junction (a convergence of all three types of plate boundaries) migrates south, and at its northern end, the triple junction (two transform faults meet a subduction zone) moves north. This deduction led to the realization that part of the ancient Pacific seafloor, the Farallon plate and part of the East Pacific Rise, had been subducted down a trench that used to lie offshore of western North America. The two remnants of this older plate are the Cocos plate to the south and the Gorda (or Juan de Fuca) plate to the north.

INSTANTANEOUS VELOCITIES

The relative motions of two plates can also be measured by more direct approaches. One technique is to try to measure directly the changes in positions occurring over a few years, which, from a

geological point of view, is instantaneous. These are referred to as instantaneous velocities. One example of how that may be done is using geodetic measurements, essentially surveying the region across a plate boundary at regular intervals and, therefore, observing the motion. This technique does not lend itself to the examination of mid-ocean ridges but has been extensively used in studying the San Andreas fault in California. The results of these measurements give the magnitude of the relative linear velocity between the Pacific plate and the North American plate to be between 50 and 75 millimeters per year, the direction of this relative velocity being known from the bearing of the fault line. These numbers agree quite well with other estimates based upon geological evidence, such as the separation of once-continuous geological features that has taken place over much longer time periods. Another example is the use of satellite laser ranging (SLR); this technique employs a laser beam bounced off a satellite, which affords a method to calculate the distance between two points on the surface of the Earth with great accuracy. The distance between two points on separate plates is regularly found, and hence the velocity between the points is calculated. By this method, the relative velocity between North America and Europe has been found to have a magnitude of approximately 15 ± 5 millimeters per year. Once again, that is in agreement with geological data for much longer time periods. In some cases, however, the agreement between the results of SLR and geological evidence is not as good. In the Zagros mountains of Iran, the two methods do not agree, implying that the instantaneous velocity indicates a change in the relative motion of the two plates on either side of this plate boundary.

FINITE VELOCITIES

The velocities calculated by geological means over much longer time spans are referred to as finite velocities. One major technique used to determine finite velocities depends upon the Vine and Matthews theory of seafloor spreading. The geomagnetic polarity record is now well established and the dates of the geomagnetic reversals known (although it is still undergoing refinement and short polarity episodes are sometimes added to the known record). This time scale can be used to identify marine magnetic anomalies caused by the magnetization of the sea floor and affords a method by which to date a point on the sea floor. If one measures the distance between two locations of the same age, located on either side of a mid-ocean ridge, then it is quite simple to calculate the relative velocity between the two plates (the direction of the motion being, in most cases, perpendicular to the ridge or parallel to the transform faults). If the separation of two points on the ocean floor, on the same plate but of different ages, is measured, then geologists can still find the "half spreading rate" of the ridge (the amount of new crustal material added per year at the ridge), but this is not a relative velocity.

REMANENT MAGNETIZATION

Because the oceanic crust is quite young, the oldest approximately 160 million years old (as compared to 4.6 billion years of Earth history), the techniques described above are not applicable to the majority of the history of the Earth. In order to work out plate motions for older periods, other methods must be employed. The most prevalent of these methods is the use of the remanent magnetization of rocks. Remanent magnetization is acquired when rocks form, and it is oriented parallel to the geomagnetic field at the time and place at which they are forming. This magnetization is retained in much the same way that a bar magnet retains its magnetization. If the rock is subsequently moved, by being carried along with a moving plate, the rock may end up at a location where the direction of the geomagnetic field is substantially different from that of its magnetization. It is this difference between the field and magnetization directions that was critical in proving that continents do indeed drift across the Earth's surface and that was a contributing factor in the acceptance of this theory by geologists. The angle that the geomagnetic field makes with the horizontal varies considerably, from vertically up at the south magnetic pole to horizontal at the equator to vertically down at the north magnetic pole. This angle is referred to as the inclination. By calculating the inclination of the magnetization of the rock, the latitude at which the rock formed, called the paleolatitude, can be ascertained. If a series shows paleolatitudes for successively older rocks, the latitudinal motion of the plate over time can be traced. Unfortunately, the same cannot be done for longitude, for the simple reason that while latitude is an inherent property of a spinning planet, longitude is not.

Plates not only shift in latitude but also rotate as they move with respect to one another. The angle between true north and a rock's magnetization is referred to as the declination, and it is this angle that allows such rotations to be determined. It is interesting that in recent years geologists have been able to delineate rotations of small blocks near the edges of the major plates, which means that a considerably more complex story unravels concerning the interactions at plate boundaries. In both Southern California and Southeast Asia, there are numerous microplates that may have rotated between larger plates.

Ian Williams

CROSS-REFERENCES

Continental Rift Zones, 579; Earth's Core, 3; Earth's Core-Mantle Boundary, 9; Earth's Crust, 14; Earth's Differentiation, 20; Earth's Lithosphere, 26; Earth's Mantle, 32; Earth's Structure, 37; Geothermal Phenomena and Heat Transport, 43; Gondwanaland and Laurasia, 599; Heat Sources and Heat Flow, 49; Hot Spots, Island Chains, and Intraplate Volcanism, 706; Lithospheric Plates, 55; Mantle Dynamics and Convection, 60; Ocean Basins, 661; Plate Margins, 73; Plate Tectonics, 86; Plumes and Megaplumes, 66; Subduction and Orogeny, 92.

BIBLIOGRAPHY

Condie, Kent C. *Plate Tectonics and Crustal Evolution.* 4th ed. Oxford: Butterworth Heinemann, 1997. An excellent overview of modern plate tectonics theory that synthesizes data from geology, geochemistry, geophysics, and oceanography. A very helpful tectonic map of the world is enclosed. The book is nontechnical and suitable for a college-level reader. Useful "suggestions for further reading" follow each chapter.

Cox, Allan, and R. B. Hart. *Plate Tectonics: How It Works.* Palo Alto, Calif.: Blackwell Scientific, 1986. A well-illustrated and detailed account of the methodology of plate tectonics. Includes information on many different aspects of plate tectonic theory and supplies explanations of the mathematical techniques utilized in solving plate tectonic problems. Suitable for those with a good mathematical background.

Dewey, J. F. "Plate Tectonics." In *Continents Adrift and Continents Aground.* San Francisco: W. H. Freeman, 1976. This article appears in a book of articles reprinted from *Scientific American.* The first part of the article gives a succinct explanation of plate rotations and includes several excellent diagrams. While not giving a full mathematical treatment, the article does approach some complex ideas in an understandable fashion. Suitable for high school readers who have some prior knowledge of the subject.

Kearey, Philip, and Frederick J. Vine. *Global Tectonics.* 2d ed. Cambridge, Mass.: Blackwell Science, 1996. This college text gives the reader a solid understanding of the history of global tectonics, along with current processes and activities. The book is filled with colorful illustrations and maps.

Press, Frank, and Raymond Siever. *Earth.* 4th ed. New York: W. H. Freeman, 1986. A general geology text. The chapter on global plate tectonics is quite thorough and contains "boxes" that explain the motions of the plates. Hot spots are explained elsewhere in the text and not related to plate motions. The diagrams, although only two-tone, are quite detailed. The text is appropriate for advanced high school readers.

_____. *Understanding Earth.* 2d ed. New York: W. H. Freeman, 1998. This comprehensive physical geology text covers the formation and development of the Earth. Readable by high school students, as well as by general readers. Includes an index and a glossary of terms.

Prichard, H. M. *Magmatic Processes and Plate Tectonics.* London: Geological Society, 1993. Although fairly technical, this special publication has relevent information about plate motions and plate tectonics. The maps and graphics help to illustrate the ideas presented.

Uyeda, Seiya. *The New View of the Earth.* San Francisco: W. H. Freeman, 1978. A very readable account of the development of plate tectonics up to the early 1970's. Does not go into mathematical detail concerning plate motions but does give many examples. Suitable for high school readers.

Wyllie, Peter J. *The Way the Earth Works.* New York: John Wiley & Sons, 1976. A good introductory geology text written from the point of view of plate tectonics. The author does not go into detail concerning the mathematics involved with determining plate motions. Well illustrated and easy to read. Information concerning plate motions is disseminated throughout the text. Suitable for high school readers.

PLATE TECTONICS

Plate tectonics is the theory that the Earth's surface is composed of major and minor plates that are being created at one edge by the formation of new igneous rocks and consumed at another edge as one plate is thrust, or subducted, below another. This elegant theory accounts for the formation of earthquakes, volcanoes, and mountain belts; the growth and fracturing of continents; and many types of ore deposits.

PRINCIPAL TERMS

ANDESITE: a volcanic rock that occurs in abundance only along subduction zones

BASALT: a dark-colored, fine-grained igneous rock

CONTINENTAL RIFT: a divergent plate boundary at which continental masses are being pulled apart

CONVERGENT PLATE BOUNDARY: a compressional plate boundary at which an oceanic plate is subducted or two continental plates collide

DIVERGENT PLATE BOUNDARY: a tensional plate boundary where volcanic rocks are being formed

EARTHQUAKE FOCUS: the area below the surface of the Earth where active movement occurs to produce an earthquake

OCEANIC RISE: a type of divergent plate boundary that forms long, sinuous mountain chains in the oceans

SUBDUCTION ZONE: a convergent plate boundary where an oceanic plate is being thrust below another plate

TRANSFORM FAULT: a large fracture transverse to a plate boundary that results in displacement of oceanic rises or subduction zones

PLATE BOUNDARIES

Plate tectonics is the theory that the Earth's crust is composed of seven major rigid plates and numerous minor plates with three types of boundaries. The divergent plate boundary is a tensional boundary in which basaltic magma (molten rock material that will crystallize to become calcium-rich plagioclase, pyroxene, and olivine-rich rock) is formed so that the plate grows larger along this boundary. The rigid plate, or lithosphere, moves in conveyer-belt fashion in both directions away from a divergent boundary across the ocean floor at rates of 0-18 centimeters per year. The lithosphere consists of the crust and part of the upper mantle and averages about 100 kilometers thick; it is thicker over continental than over oceanic crust. The lithosphere seems to slide over an underlying plastic layer of rock and magma called the asthenosphere. Eventually, the lithosphere meets a second type of plate boundary, called a convergent plate margin. If lithosphere-containing oceanic crust collides with another lithospheric plate containing either oceanic or continental crust, then the oceanic lithospheric plate is thrust or subducted below the second plate. If both intersecting lithospheric plates contain continental crust, they crumple and form large mountain ranges, such as the Himalaya or the Alps. Much magma is also produced along convergent boundaries. A third type of boundary, called a transform fault, may develop along divergent or compressional plate margins. Transform faults develop as fractures transverse to the sinuous margins of plates, in which they move horizontally so that the plate margins may be displaced many tens or even hundreds of kilometers.

OCEANIC RISES

Divergent plate margins in ocean basins occur as long, sinuous mountain chains called oceanic rises that are many thousands of kilometers long. The rises are often discontinuous, as they are displaced long distances by transform faults. The two longest oceanic rises are the East Pacific Rise, running from the Gulf of California south and west into the Antarctic, and the Mid-Atlantic Ridge, running more or less north-south across the middle of the Atlantic Ocean. The oceanic rises are deep-sea mountain ranges, and there is a rift valley that runs down the middle of the highest part of the mountain chain. The rift valley apparently forms along the ocean rises as the plates move outward from the rises in both directions and pull apart the lithosphere. The oceanic floor descends from a maximum elevation at the oceanic rises to a minimum in the deepest trenches along subduction zones. Thus, the lithosphere moves downhill from the oceanic rises to the convergent plate margins. It is thought that the lithosphere gradually cools and contracts as it moves from the oceanic rises to the convergent margins.

The oceanic rises are composed of piles of basalts forming gentle extrusions. There is high heat flow out of oceanic rises because of the large volume of magma carried up toward the surface. The magnetic minerals in the lavas are frozen into alignment with the Earth's magnetic field. Half the magnetized lavas move out from the oceanic rises in one direction, and the other half move out in the opposite direction. The magnetic field of the Earth appears to reverse itself periodically over geologic time. The last magnetic reversal occurred about 730,000 years ago. This last reversal can now be observed at the same distance in both directions away from the oceanic rises. A series of such magnetic reversals can be traced back across the Pacific ocean floor for a period of about 165 million years. Many shallow-focus earthquakes occur at depths of up to 100 kilometers below the surface, along the rises and transform faults. Presumably they result from periodic movement that releases tension in the lithosphere.

CONTINENTAL RIFTS

A second type of divergent plate margin, called a continental rift zone, occurs in continents. Ex-

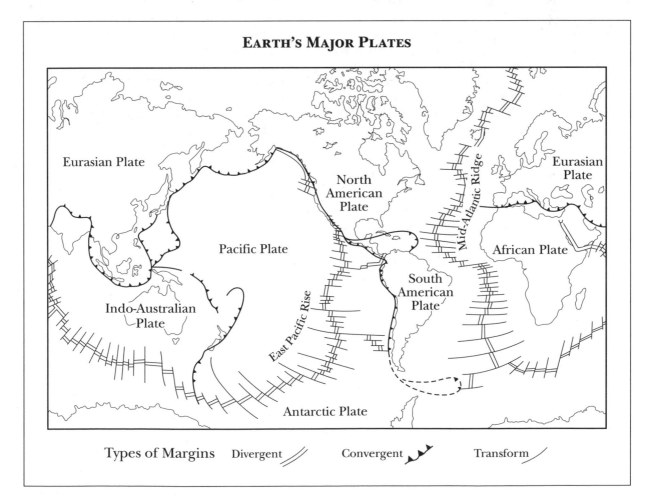

EARTH'S MAJOR PLATES

Eurasian Plate

North American Plate

Pacific Plate

Indo-Australian Plate

East Pacific Rise

South American Plate

Mid-Atlantic Ridge

Eurasian Plate

African Plate

Antarctic Plate

Types of Margins   Divergent   Convergent   Transform

amples are the Rio Grande Rift, occurring as a sinuous north-south belt in central New Mexico and southern Colorado, and the East African Rift, occurring as a sinuous north-south belt across eastern Africa. These rift zones occur as down-dropped blocks forming narrow, elongate valleys that fill with sediment. The rift valleys often contain rivers or elongate lakes. They are characterized by abundant basalts with high potassium contents and, often, smaller amounts of more silica-rich rocks called rhyolites. Rhyolites are light-colored volcanic rocks containing the minerals alkali feldspar (potassium, sodium, and aluminum silicate), quartz (silica), sodium-rich plagioclase, and often minor dark-colored minerals. Shallow-focus earthquakes result in these areas from the tension produced as the continental crust is stretched apart, much as taffy is pulled.

Many rift valleys never become very large. Others grow and may actually rip apart the continents to expose the underlying oceanic crust and rise, as is occurring in the Red Sea. There the oceanic crust is near enough to the continents that it is covered with sediment. Eventually, the continents on both sides of the Red Sea may be pulled apart so far that the underlying oceanic floor will be exposed, with no sediment cover. About 240 million years ago, the continents of North and South America, Europe, and Africa were joined in an ancient landmass called Pangaea. They slowly broke apart along the north-south Mid-Atlantic Ridge from about 240 to 70 million years ago. At first, only a rift valley similar to the East African Rift was formed. Later it opened, much like the area of the Red Sea today. Finally, the continents drifted far enough apart during the last 70 million years to form a full-fledged ocean basin, the Atlantic Ocean.

HOT SPOTS

As the lithosphere moves slowly across the ocean floor, minor volcanic activity is generated over hot spots on the ocean floor. The Hawaiian Islands are situated over one of these hot spots. The basalts produced there are much richer in potassium than are those formed over oceanic rises. The Hawaiian Islands are part of a linear, northwest-trending chain of islands, about 2,000 kilometers long, that extends to the island of Midway. The volcanic rocks become progressively older from the Hawaiian Islands to Midway Island. Presumably, Midway Island formed first as the plate slid over the hot spot. As the plate moved to the northwest, the source of magma was removed from Midway, and newer volcanoes began progressively to form over the same hot spot.

SUBDUCTION ZONES

Eventually, the lithospheric plate with oceanic crust reaches a compressional plate boundary and may be subducted below other oceanic crust; one result is the island arcs in the western Pacific Ocean, such as Japan. Or they may be pulled below continental crust, often at angels of 20-60 degrees to the horizontal (the Andes in western South America are the result of such movement). The intersection of the two colliding plates is marked by a sinuous, deep trench forming the deepest portions of the ocean floors. Sediment collects along the slopes of the trench, carried down from the topographic highs of the upper plate. Mountain belts are built up on the nonsubducted plate, as a result of the tremendous amounts of igneous rock that form and of the compressional forces of the plate collision, which throw much sediment and metamorphic rock in the nonsubducted plate to higher elevations.

The subducted plate can be traced to depths as great as 700 kilometers. Some of the sediments collecting along the trench are carried rapidly to great depths, where they undergo a very high-pressure and low-temperature metamorphism. (Metamorphism is the transformation of minerals in response to high temperatures and pressures deep within the Earth.) Some rocks are carried more slowly to great depths and have a more normal, higher-temperature metamorphism. During metamorphism, many minerals containing water along the subducted plate gradually break down and give off water vapor, which moves up into the overlying plate. The water vapor is believed to lower the melting point of these rocks within the subducted and overlying plates so that widespread melting takes place, producing the abundant basalts and andesites that build up island arcs or continental masses above the subducted plate. In addition, much rhyolitic magma is formed in the continental crust, presumably through the melting of some of the higher-silica rocks in the continents.

EARTHQUAKE ZONES

Sometimes a continent is carried by an oceanic plate into another continent at a subduction zone, which happened when India collided with the Asian continent. Such a collision crumples the continents into very high mountains; the Himalaya were formed in this way. This process produces an earthquake zone that is more diffuse (with foci to depths up to 300 kilometers) than are those along subducted plates. No volcanic rocks are produced in these continental-continental plate collisions. Instead, abundant granites crystallize below the surface. Granites contain the same minerals as do rhyolites. Rhyolites form small crystals by quick cooling when they crystallize rapidly in volcanic rocks; granites form larger crystals from magma of the same composition by slow cooling below the Earth's surface.

DEVELOPMENT OF PLATE TECTONIC MODEL

Plate tectonics is a major, unifying theory that clarifies many large-scale processes on the Earth: the formation of volcanoes, earthquakes, mountain belts, and many types of ore deposits, as well as the growth, drift, and fracturing of continents. The major concepts to support the theory were put together only in the late 1950's and the 1960's, yet many of the keys to developing the theory had been known for many years. Beginning in the seventeenth century, a number of people noticed the remarkable "fit" in the shape of the continents on opposing sides of the Atlantic Ocean and suggested that the continents could have been joined at one time. It was not until the early twentieth century that Alfred Wegener put many pieces of this puzzle together. Wegener noticed the remarkable similarity of geological structures, rocks, and especially fossils that were currently located on opposite sides of the Atlantic Ocean. Most notably, land plants and animals that predated the hypothesized time of the breakup of the continents, at about 200 million years before the present, were remarkably similar on all continents. Subsequently, their evolution in North and South America was quite different from their development in Europe and Africa. Climates could also be matched across the continents. For example, when the maps of the continents were reassembled into their predrift positions, the glacial deposits in southern Africa, southern South America, Antarctica, and Australia could be explained as having originated as one large continental glacier in the southern polar region.

One of the biggest problems with the concept of continental drift at that time was the lack of understanding of a driving force to explain how the continents could have drifted away from one another. Then, in 1928, Arthur Holmes proposed a mechanism that foreshadowed the explanation geologists later adopted. He suggested that the mantle material upwelled under the continents and pulled them apart as it spread out laterally and produced tension. The basaltic oceanic crust would then carry the continents out away from one another much like rafts. When the mantle material cooled, Holmes believed, it descended back into the mantle and produced belts along these areas. From the 1920's to the early 1960's, however, continental drift theories had no currency, for there was no real evidence for driving forces that might move the continents. It was not until the ocean floors began to be mapped that evidence was found to support a plate tectonic model. The topography of the ocean floor was surveyed, and large mountain ranges, such as the Mid-Atlantic Ridge with its rift valleys, and the deep ocean trenches were discovered. Harry Hess suggested in the early 1960's that the oceanic ridges were areas where mantle material upwelled, melted, and spread laterally. Evidence for this seafloor spreading hypothesis came from the mirror-image pattern of the periodically reversed magnetic bands found in basalts on either side of the ridges. The symmetrical magnetic bands could be explained only by the theory that they were originally produced at the ridges, as the Earth's magnetic field periodically reversed, and then were spread laterally in both directions at the same rate.

Supporting evidence for plate tectonics began to accumulate during the 1960's. Further magnetic pattern surveys on ocean floors confirmed that the symmetrical pattern of matching magnetic bands could be found everywhere around ridges. Also, earthquake, volcanic rock, and heat-flow patterns were discovered to be consistent with the concept of magma upwelling along rises and seafloor material being subducted along oceanic trenches. Oceanic and lithospheric plates could then be defined, and the details of the interaction of the plate boundaries could be understood. With this overwhelming evidence, most geologists became convinced that the plate tectonic model was valid.

ECONOMIC APPLICATIONS

Plate tectonics is important economically because of the theory's usefulness in predicting and explaining the occurrence of ore deposits. Plate boundaries such as the mid-oceanic rises are areas of high temperature in which hot waters are driven up toward the surface. These hot waters are enriched in copper, iron, zinc, and sulfur, so sulfide minerals such as pyrite (iron sulfide), chalcopyrite (copper and iron sulfide), and sphalerite (zinc sulfide) form along oceanic rises. One such deposit in Cyprus has been mined for many centuries. Tensional zones sometimes formed in basins behind subduction zones may form deposits similar to those at oceanic rises. In addition, ferromanganese nodules form in abundance in some places by chemical precipitation from seawater. These nodules are enriched in cobalt and nickel, as well as in iron and manganese as complex oxides and hydroxides. They could potentially be mined from ocean floors.

Deposits enriched in chromium occur in folded and faulted rocks on the nonsubducted plate next to the oceanic trench in subduction zones. This deposit is found in some peridotites (olivine, pyroxene, and garnet rocks) or dunites (olivine rock) that have been ripped out of the upper mantle and thrust up into these areas. The ore mineral chromite (magnesium and chromium oxide) is found in pods and lenses that range in size from quite small to massive. Many intrusions of silica-rich magma above subduction zones contain water-rich fluids that have moved through the granite after it solidified. The water-rich fluids deposit elements such as copper, gold, silver, tin, mercury, molybdenum, tungsten, and bismuth throughout a large volume of the granite in low concentrations. Hundreds of these deposits have been found around subduction zones in the Pacific Ocean.

Robert L. Cullers

CROSS-REFERENCES

Continental Growth, 573; Continental Rift Zones, 579; Earthquake Distribution, 277; Earth's Core, 3; Earth's Core-Mantle Boundary, 9; Earth's Crust, 14; Earth's Differentiation, 20; Earth's Lithosphere, 26; Earth's Magnetic Field, 137; Earth's Magnetic Field: Origin, 144; Earth's Magnetic Field: Secular Variation, 150; Earth's Mantle, 32; Earth's Structure, 37; Faults: Transform, 232; Geothermal Phenomena and Heat Transport, 43; Gondwanaland and Laurasia, 599; Heat Sources and Heat Flow, 49; Hot Spots, Island Chains, and Intraplate Volcanism, 706; Island Arcs, 712; Lithospheric Plates, 55; Magmas, 1326; Magnetic Reversals, 161; Mantle Dynamics and Convection, 60; Mountain Belts, 841; Ocean Basins, 661; Ocean Ridge System, 670; Oceanic Crust, 675; Plate Margins, 73; Plate Motions, 80; Plumes and Megaplumes, 66; Subduction and Orogeny, 92.

BIBLIOGRAPHY

Condie, Kent C. *Plate Tectonics and Crustal Evolution.* 4th ed. Oxford: Butterworth Heinemann, 1997. An excellent overview of modern plate tectonics theory that synthesizes data from geology, geochemistry, geophysics, and oceanography. A very helpful tectonic map of the world is enclosed. The book is nontechnical and suitable for a college-level reader. Useful "suggestions for further reading" follow each chapter.

Kearey, Philip, and Frederick J. Vine. *Global Tectonics.* 2d ed. Cambridge, Mass.: Blackwell Science, 1996. This college text gives the reader a solid understanding of the history of global tectonics, along with current processes and activities. The book is filled with colorful illustrations and maps.

Kious, Jacquelyne W. *This Dynamic Earth: The Story of Plate Tectonics.* Washington, D.C.: U.S. Department of the Interior, United States Geological Survey, 1996. Kious is able to explain plate tectonics in a way suitable for the layperson. The book deals with both historic and current theory. Illustrations and maps are plentiful.

Motz, Lloyd M., ed. *The Rediscovery of the Earth.*

New York: Van Nostrand Reinhold, 1979. An unusual book, as it is written by many of the experts who developed the plate tectonic model. Begins at an elementary level so that someone without much background in plate tectonics should be able to understand the discussion; the discussion progresses, however, to an advanced level. Beautifully illustrated with photographs and diagrams.

Press, Frank, and Raymond Siever. *Understanding Earth*. 2d ed. New York: W. H. Freeman, 1998. This comprehensive physical geology text covers the formation and development of the Earth. Readable by high school students, as well as by general readers. Includes an index and a glossary of terms.

Seyfert, Carl K., and L. A. Sirkin. *Earth History and Plate Tectonics*. New York: Harper & Row, 1973. This book integrates the plate tectonic concept with the evolution of plants and animals through geologic time. Some under- standing of plate tectonics, rocks, and minerals would be helpful before using this source. Written as an introductory text in historical geology, so important concepts are reviewed. Good illustrations.

Skinner, Brian J., et al. *Resources of the Earth*. Englewood Cliffs, N.J.: Prentice-Hall, 1988. A good book for the layperson who is interested in the history, use, production, environmental impact, and geological occurrence of ore deposits. Technical terms are kept to a minimum. Well illustrated; contains a glossary. Suitable for someone who is taking a course in geology.

Windley, Brian F. *The Evolving Continents*. 2d ed. New York: John Wiley & Sons, 1984. A more advanced source than the others listed. Summarizes how the continents have evolved through geologic time. The reader should understand plate tectonic processes well before attempting to read this book.

SUBDUCTION AND OROGENY

Subduction and orogeny are fundamental consequences of plate tectonics and are the two processes that build mountains on the edges of continents. Through the recognition of subduction, scientists have been better able to determine regions where risks of earthquakes and volcanic explosions are significant.

PRINCIPAL TERMS

CONTINENTAL MARGIN: the edge of a continent that is both exposed on land and submerged below the water that marks the transition to the ocean basin

CRUST: the outermost layer of the Earth, which consists of materials that are relatively light; the continental crust is lighter than oceanic crust, which allows it to float while oceanic crust sinks

FAULTING: the process of fracturing the Earth such that rocks on opposite sides of the fracture move relative to one another; faults are the structures produced during the process

FOLDING: the process of bending initially horizontal layers of rock so that they dip; folds are the features produced by folding and can be as small as millimeters and as big as kilometers long

GEOSYNCLINES: major depressions in the surface of the Earth where sediments accumulate; geosynclines lie parallel to the edges of continents and are long and narrow

INTRUSION: the process of forcing a body of molten rock generally derived from depths of tens of kilometers in the Earth into solidified rock at the surface

MAGMA: molten rock that is the source for volcanic eruptions

THEORY OF GEOSYNCLINES

Subduction and orogeny are two processes that are fundamental to the evolution of continents. All continents contain long, narrow mountain chains near their edges that are composed of folded and faulted rocks that are younger than the rocks in the continental interiors. The event that formed the mountains is termed an orogeny, and the mountain chain itself is called an orogenic belt. Because of the proximity of mountain chains to the edges of continents, scientists have believed for centuries that orogenies reflected movements localized along continental margins. It is only recently, however, that orogeny was coupled with subduction, the process in which seafloor descends below a continent or another piece of seafloor. Earlier views of orogeny were part of the theory of geosynclines. Geosynclines are linear basins that form on subsiding regions of the Earth's surface adjacent to continental margins, fill with sediment, and evolve into mountains composed of folded and faulted sedimentary strata. The origin of the compressive forces responsible for the creation of the mountains was not known.

Erosion of the newly created mountains provides sediment for new geosynclines that develop seaward of the mountain belt, thereby completing one geosynclinal cycle, which typically lasts on the order of a few hundred million years. The advent of the theory of plate tectonics in the 1960's led the majority of the scientific community to abandon the geosynclinal cycle in favor of the subduction process as an explanation for orogenies. Subduction was attractive because it readily provided a mechanism by which the large compressive forces needed to form mountains could be produced.

Orogenic belts are characterized by the folding and faulting of layers of rock, by the intrusion of magma, and by volcanism. Folds and faults form parallel to the continental margin and extend hundreds of kilometers toward the continental interior. Folding bends layers of rocks, whereas faulting takes rocks that were side by side and stacks them on top of each other in sheets up to 20 kilometers thick. Both processes significantly shorten the horizontal and thicken the vertical dimensions of the continents. At the same time as they are folded and faulted, the rocks are intruded by magmas derived from tens of kilometers below the surface. Some of the magmas eventually erupt, building volcanoes on the deformed rocks. An additional feature of orogenic belts is the juxtaposition of sequences of rock that have nothing in common with each other. The rocks in the two sequences may be different in age, composition, or style of folding. The origin of this juxtaposition was unreconciled by the theory of geosynclines, which holds that all rocks in a mountain belt were originally deposited near one another and were derived from the same source. The theory of plate tectonics, however, easily explains the juxtaposition.

THEORY OF PLATE TECTONICS

In order to understand subduction and orogeny, one must have a clear grasp of the theory of plate tectonics. The theory states that the surface of the Earth is composed of about twelve rigid plates which are less than 100 kilometers thick. Plates are either oceanic or continental. Below the plates is a partially molten layer that allows the rigid plates to float and move relative to each other at speeds between 2 and 10 centimeters per year. The motions are defined primarily by the oceanic plates; the continental plates drift passively. The relative motions of the plates define three types of boundaries: convergent, where plates move toward one another; divergent, where plates spread apart; and transcurrent, where plates slide smoothly past each other. Convergent boundaries are frequently along the margins of continents, and divergent boundaries are commonly in the ocean basins. For example, the west coast of South America is a convergent boundary, and the Mid-Atlantic Ridge, the mountain range that runs down the middle of the Atlantic Ocean, is a divergent boundary. Divergent boundaries are zones along which two plates separate. This type of plate boundary is typically demarcated by a linear ridge system in an ocean basin where magma rises from deep in the Earth to fill the gap created by the diverging plates. When the hot magma contacts the cold seawater, it solidifies into new oceanic crust. As the plates continue to separate, additional magma wells up from the Earth's interior, allowing the continuous creation of oceanic crust at the ridge. This process is known as seafloor spreading and is responsible for the drifting of the continents on the surface of the Earth. Convergent boundaries are where two plates move toward each other and one plate subducts, or descends below, the other. The subducting plate is always oceanic, but the overriding plate may be either oceanic or continental. This reflects the greater density of oceanic crust relative to continental crust, which allows the oceanic plates to sink readily into the Earth's interior whereas the continental plates remain afloat. When two continental plates collide, neither plate subducts—they are too light—but the plates push against each other with tremendous force such that their edges buckle and huge mountain ranges grow. This process built the world's tallest mountains, the Himalaya, which are the result of the collision between the subcontinent of India and the continent of Asia.

SUBDUCTION ZONES

Subduction zones are characterized by a progression from the subducting to the overriding plate of deep trenches, high mountains, and many volcanoes that occupy an area hundreds of kilometers wide and thousands of kilometers long. The deep trench, frequently filled with sediments eroded from the adjacent mountains, marks the point in the ocean floor where the subducting plate bends to descend below the overriding plate. As the oceanic plate descends, these sediments are scraped onto the overriding plate. Slivers of oceanic crust may also scrape off and mix with the sediments. The offscraped rocks form an intricately folded and faulted region tens of kilometers wide and several kilometers high at the edge of the overriding plate. These complexly deformed mixtures of sediments and slivers of oceanic crust are called mélanges and are characteristic of most ancient subduction zones now exposed on land.

Another important feature of subduction zones is the linear belt of volcanoes on the overriding plate that parallels the plate boundary. The volcanoes grow from the eruption of magma that is generated at the interface between the subducting and overriding plates at depths between 100 and 200 kilometers. At these depths, the temperature of the Earth is high enough to melt small areas of either the subducting or the overriding plate. The magma rises, intruding the rocks at the surface and eventually erupting to build the volcanic belt. Some of the magma, however, may solidify between the top of the oceanic plate and the surface.

The similarity of features in orogenic belts and subduction zones is striking and forces the obvious conclusion that subduction leads directly to orogeny. An orogeny can occur either during subduction of an oceanic plate below a continental plate, such as on the west coast of South America, or during the collision of two continental plates, such as in the Himalaya. Because continents do not subduct, the compressive forces are much greater in a continent-continent collision than in seafloor subduction. The mountains produced during collision (Himalaya), therefore, are much taller than those generated during subduction (Andes).

CONSEQUENCES OF PLATE TECTONICS

The theory of plate tectonics elucidates important differences between the oceans and the continents and provides a mechanism by which different rock sequences can be juxtaposed in orogenic belts. The ocean basins are transient features that are constantly modified by the growth and destruction of new seafloor at divergent and convergent boundaries, respectively. In contrast, the continents are too light to be subducted and are permanent features of the Earth's surface. This consequence of plate tectonics is supported by the 200 million year age of the oldest seafloor and the 4 billion year age of the most ancient rocks on the continents. Continents, therefore, drift, fragment, and collide as relative plate motions change through geologic time. The collision of continents that were once widely separated allows the bringing together of rocks that have had very different histories. As the collision leads to orogeny, these different sequences of rock may be juxtaposed in the same mountain belt.

The difference between the age of orogenic belts and the interiors of continents implies that the continents have evolved through time by the addition of material at their edges during orogenies. Orogenic belts are also of different ages, ranging from a billion years to zero (actively forming). Two or three belts whose ages decrease away from the continental interior may define one edge of a continent. This suggests that orogenies have occurred repeatedly through geologic time and that continents have added material continuously to their margins since the formation of their interiors. Because the ocean floor is so young, orogenic belts are the only record of subduction and collision events prior to 200 million years ago.

If subduction is the only mechanism responsible for orogeny, plate tectonics must have been active since early in the history of the Earth.

ANALYSIS OF EARTHQUAKES

Subduction and orogeny are studied by hundreds of scientists, each of whom looks at only a small part of the picture. One may determine the composition of volcanic rocks that are characteristic of subduction zones; another may examine the styles of folds and faults in orogenic belts. Three techniques, however, are dominant in the study of subduction and orogeny: the analysis of the locations and sizes of earthquakes, the discrimina-

The Big Sur coast of California. The California coastline lies at a plate boundary where the Pacific plate is subducting under the North American plate. (© William E. Ferguson)

tion of relationships between different types of rocks in the field, and the investigation of features in deep-sea trenches and in the submerged region of folded and faulted rocks. The first defines where subduction and orogeny occur today, whereas the second determines what the physiographic expressions of these processes are, how they are preserved in the rocks, and where they were active in the past. The third technique provides a direct link between subduction and orogeny and illustrates the early stages of development of a mountain belt.

One of the most important discoveries of plate tectonics was that earthquake zones define plate boundaries. Earthquakes occur when a fracture, or fault, forms in the Earth's crust, and the two pieces on either side of the fault move, or slip, past each other. For large earthquakes, the slip is on the order of 10-20 meters. The forces responsible for faulting are simply the result of the relative motions of the plates at the plate boundaries. The motion can accumulate in the rocks for hundreds of years prior to causing a rupture. When the crust finally breaks, the energy stored by the rocks is released suddenly as waves that travel through the Earth and generate the intense vibrations associated with an earthquake. The rupture continues for as much as 1,000 kilometers and moves at speeds in excess of 10,000 kilometers per hour.

The energy carried by the waves is recorded on seismographs, which are instruments that monitor ground motion. Seismographs are composed of a mass attached to a pendulum. The mass remains still during an earthquake, measuring the amount the Earth moves around it. The motion is recorded on a chart as a series of sharp peaks and valleys that deviate from the background value measured during times of no earthquake activity. The arrival of the waves at different times at different places allows the geophysicist to calculate the location, or epicenter, of the earthquake. The amount of the deviation of the peaks and valleys from the background noise is an estimate of the magnitude of the earthquake.

Earthquakes near mountain belts define zones that extend at an angle from the surface of the Earth at the deep-sea trench to depths of hundreds of kilometers below the continents. This zone corresponds to the subducting plate at a convergent boundary. As a result, the locations of sub-

duction zones that are currently active are very well known. The descent of a subducting plate below an overriding continent has triggered some of the deepest and largest earthquakes ever recorded. Continued motion of the plate and rupturing of the Earth's crust in response translate into mountain ranges on the Earth's surface.

STUDY OF ROCKS AND DEEP-SEA TRENCHES

Analysis of earthquakes is essential to evaluate the modern plate tectonic setting of the Earth but reveals nothing about the geologic past. Information about plate tectonics of the past must be obtained from looking at ancient mountain belts. Recognition of relationships among rocks in the field involves determining the ages, compositions, and histories of the rocks. This process led to the discovery that mountain belts on different continents contained rock sequences that were very similar. For example, rocks in the Appalachian Mountains of the east coast of North America were found to match closely those in the Atlas Mountains of the west coast of Africa. Conversely, recognition of relationships determined that dissimilar rock sequences frequently are adjacent to each other in the same orogenic belt. Both phenomena are most readily explained by continental drift, seafloor spreading, and subduction.

The critical link between ancient orogenic belts and modern subduction zones identified by earthquake activity was provided by deep-sea trenches. Using highly sophisticated techniques to "see" the ocean floor, scientists discovered the region of offscraped rocks that lies on the overriding plate in a subduction zone. These regions sometimes continue to the continental margin, where they are exposed on land as mountains. Thus, subduction was observed to cause folding and faulting in rocks and to build mountains, both important processes in orogenic belts.

EARTHQUAKE AND VOLCANO HAZARD ASSESSMENT

The theory of plate tectonics provides scientists with a process that can be observed—subduction—to explain the origin of mountains. Because young mountain chains are the locus of most of the large earthquakes that occur today, understanding subduction yields insight into the potential for destructive earthquakes in any given area. This is extremely important because most of the global population lives along convergent plate boundaries. The identification of subduction zones at the margins of the Pacific Ocean has explained the so-called "Ring of Fire," a region of abundant earthquakes and volcanoes that had long puzzled the scientific community. Restriction of most earthquakes to plate boundaries allows the assessment of earthquake hazards anywhere in the world if the locations of plate boundaries are known. For example, the city of Santiago in Chile, which is above a subduction zone, has a high risk, whereas the city of Chicago in the United States, which is in the continental interior, has a low risk.

Additional information can also be gathered about the type of earthquakes that may occur. In subduction zones, the piece of the crust that is above the rupture typically moves upward relative to the piece below, which generates waves that shake the ground in certain directions. At divergent plate boundaries, however, the piece of crust that is above the rupture moves downward relative to the piece below. This motion produces waves that vibrate the ground in directions different from those generated by earthquakes in subduction zones. Additional differences between convergent and divergent boundaries that may affect ground motion include the depth and size of the earthquakes. subduction zones generate the deepest and largest earthquakes; earthquakes at divergent plate boundaries are more frequent, smaller, and shallower. Knowledge of the way the ground may move helps civil engineers to design and construct buildings able to withstand large earthquakes.

Eruptions of volcanoes that lie above subduction zones can be devastating. These volcanoes typically erupt violently and explosively in contrast to volcanoes near mid-ocean ridges, which erupt quietly and smoothly. This reflects the greater viscosity (resistance to flow) of magmas at convergent boundaries relative to those at divergent boundaries. Because of their greater viscosity, the magmas above subduction zones tend to plug the volcanoes at the surface, preventing any eruptions. Finally, when the pressure below the plug is great enough, the volcano erupts with such force that cities nearby are damaged considerably. For example, in 79 C.E., the entire city of Pompeii, Italy, was destroyed, and hundreds of people were killed by the volcano, Vesuvius. Clearly, the investigation of subduction and orogeny is beneficial to understanding the forces of nature that are harmful to humankind. Perhaps someday in the future, large earthquakes and violent volcanic eruptions may be predicted far enough in advance that precautions can be taken to prevent the loss of human life.

Pamela Jansma

CROSS-REFERENCES

Andes, 812; Continental Growth, 573; Earthquake Engineering, 284; Earthquake Prediction, 309; Earthquakes, 316; Earth's Core, 3; Earth's Core-Mantle Boundary, 9; Earth's Crust, 14; Earth's Differentiation, 20; Earth's Lithosphere, 26; Earth's Mantle, 32; Earth's Structure, 37; Geosynclines, 630; Geothermal Phenomena and Heat Transport, 43; Heat Sources and Heat Flow, 49; Himalaya, 836; Lithospheric Plates, 55; Mantle Dynamics and Convection, 60; Mountain Belts, 841; Ocean Basins, 661; Plate Margins, 73; Plate Motions, 80; Plate Tectonics, 86; Plumes and Megaplumes, 66; Seismometers, 258; Volcanic Hazards, 798.

BIBLIOGRAPHY

Bebout, Gray E. *Subduction Top to Bottom*. Washington, D.C.: American Geophysical Union, 1996. Bebout's book gives clear definitions and explanations of subduction, folding, faults, and orogeny. Illustrations and maps help to clarify some difficult concepts.

Kearey, Philip, and Frederick J. Vine. *Global Tectonics*. 2d ed. Cambridge, Mass.: Blackwell Science, 1996. This college text gives the reader a solid understanding of the history of global tectonics, along with current processes and activities. The book is filled with colorful illustrations and maps.

Osterihanskay, Lubor. *The Causes of Lithospheric Plate Movements*. Prague: Charles University, 1997. This college-level text examines geography and geoecology in relation to plate tectonics and the Earth's lithosphere. Many theories are illustrated with graphics and maps.

Press, F., and R. Siever. *The Earth*. 4th ed. New York: W. H. Freeman, 1986. An excellently illustrated introductory text on geology, the book has five chapters that deal with folding, faulting, plate tectonics, earthquakes, and orogeny. A map of the major plates is on the inside back cover. The glossary is huge and indispensable. Recommended for senior high school and college students.

Shelton, J. S. *Geology Illustrated*. San Francisco: W. H. Freeman, 1966. This book has a superb collection of photographs and sketches drawn from the photographs that illustrate specific geologic features such as folds, faults, and volcanic landforms. Although it does not talk about subduction and orogeny directly, this source does help the reader in visualizing the features representative of convergent plate boundaries.

Short, Nicholas M., and Robert W. Blair. *Geomorphology from Space: A Global Overview of Regional Landforms*. Washington, D.C.: National Aeronautics and Space Administration, 1986. This book contains beautiful pictures taken by various satellites that orbit the Earth. Text accompanies each picture and explains the tectonic setting. Many mountain belts and volcanic chains are shown. Although the text is fairly technical, the photographs are worth examining by anyone. Recommended for college-level students.

Uyeda, Seiya. *The New View of the Earth: Moving Continents and Moving Oceans*. Translated by Masako Ohnuki. San Francisco: W. H. Freeman, 1978. Although slightly dated, the book discusses the historical context of the theory of plate tectonics in addition to explaining the theory very well. Nontechnical and designed for the nonscientist. Interesting stories about the people responsible for the theory abound. Suitable for anyone interested in plate tectonics.

Wilson, J. T., ed. *Continents Adrift*. San Francisco: W. H. Freeman, 1972.

_____. *Continents Adrift and Continents Aground*. San Francisco: W. H. Freeman, 1977. These two volumes are collections of articles originally printed in *Scientific American* magazine. The amount of overlap is very small, which amply illustrates the rapid advances in plate tectonics in the late 1960's and early 1970's. The articles are very well illustrated, and the introductions in each volume are very helpful. Suitable for anyone interested in any aspect of plate tectonics.

3
GEODESY AND GRAVITY

EARTH TIDES

Earth tides are deformations of the crust of the Earth as a result of gravitational interaction with the Moon and the Sun. Knowledge of the effects of these tidal forces is important to Earth scientists who search for natural resources.

PRINCIPAL TERMS

DEFORMATION: the alteration of an object from its normal shape by a force

GRAVIMETER: a device that measures the attraction of gravity

HOMOGENEOUS: having uniform properties throughout

OBLATE SPHEROID: a spherically shaped body that is flattened at the polar regions

OSCILLATE: to fluctuate or to swing back and forth

PENDULUM: a mass suspended in such a way that it can swing freely

PERTURB: to change the path of an orbiting body by a gravitational force

SYNCHRONIZED ROTATION-REVOLUTION: a situation in which the rotation rate of a body is equal to its rate of revolution

GRAVITATIONAL ATTRACTION

Earth tides are the deformation of the solid portion of the Earth by the combined gravitational forces of the Moon and the Sun. Although other bodies within and beyond the solar system gravitationally attract the Earth, the distances are great enough to make their tidal effect upon the Earth negligible. Consider the Earth-Moon system. According to Sir Isaac Newton's law of gravity, every particle of mass in the universe is attracted to every other particle of mass by a force that is directly proportional to the product of the masses and inversely proportional to the square of the distance between them—which means that gravity is always an attractive force, but its magnitude depends to a great extent upon the distance between the two bodies in question. Since gravity is an inverse square law, the following relationship holds true: If the distance between two bodies is doubled, the attraction of gravity becomes one-fourth as great; if the distance between the bodies is tripled, the attraction becomes one-ninth as great, and so on. According to this law, each particle of the Moon attracts each particle of the Earth. Because these particles are not all equidistant from one another, the force of gravity varies in intensity. Gravitational attraction is greatest between the particles that are closest; therefore the surface of the Earth nearest the position of the Moon is subjected to more attraction than is the surface of the Earth opposite the Moon. It is this difference in relative position that causes the tidal force and thus the deformation of the Earth.

Albert Michelson measured the Earth tides in 1913 by observing water tides in long horizontal pipes. He had assumed that the Earth was rigid, but he did not observe the tidal values that the theory indicated he should. The difference could be accounted for when the Earth was assigned a rigidity so that it was able to respond to lunar gravitational forces by raising crustal tides to a height of several centimeters.

OCEAN TIDES

The ocean tides may be considered as being analogous to the Earth tides. Like Earth tides, ocean tides are caused by the gravitational forces of both the Sun and the Moon. Because of its relative closeness, the Moon is the greater factor. Its gravitation causes the water in the oceans to bulge outward a distance of one meter or so. There are two water bulges on the surface of the Earth: one in the direction of the Moon and one in the direction opposite the direction of the Moon. This latter bulge forms because of the reduced amount of gravity at that position on the Earth's surface. Another way of looking at it might be as follows: The Earth is being attracted toward the Moon or, in a sense, is falling toward the Moon. Therefore, the water on the lunar side is falling toward the Moon and is actually ahead of the Earth's surface. The water on the opposite side of the Earth is also falling toward the Moon but cannot quite keep up with the Earth's surface and so forms a bulge. Theoretically, as the Earth rotates with respect to the Moon, the water level rises and falls as these bulges of water are swept around the Earth. In reality, the height and timing of tides may vary considerably. In some bays, the tidal water may accumulate to heights of 10 meters and greater. Because there are two tidal bulges, there are two high tides per day.

The Sun also exerts a tidal force on the Earth, but because of its greater distance, its influence is only about one-half as great as the Moon's. Extremely large high tides are generated when the Sun, the Moon, and the Earth lie along a straight line. The tidal forces of the Sun and the Moon then act in the same direction. These tides are known as spring tides, though they have nothing to do with the season. The nature of the ocean tides provides an immediate observation and a fairly simple observation of the nature of tidal forces.

EARTH'S SHAPE

Tidal forces also have an affect on gravity, as does the shape of the planet. The ancient Greeks taught that the Earth is a sphere. The philosopher Plato reasoned that all heavenly bodies are perfect and therefore must be spherical; because the Earth was a heavenly body, its shape was thus spherical. In about the year 230 B.C.E., Eratosthenes calculated the circumference of the Earth to be 12,560 kilometers, which is only 112 kilometers less than the current estimate. During the seventeenth century, several measurements were made on the Earth's surface. The size of one degree of arc in the Northern Hemisphere proved to be somewhat smaller than a degree of arc farther south. It was concluded from these studies that the Earth is flattened toward the poles and thus is not spherical. The shape of the Earth is rather an oblate spheroid, as explained by Sir Isaac Newton in his famous work of 1687, *Principia*.

If the Earth were a perfect sphere and homogeneous in composition, the gravity measurements at all points on the surface would be identical and the orbits of Earth satellites would be perfectly circular or elliptical. Because the Earth's gravitational field is uneven, resulting from the fact that the Earth is neither perfectly spherical nor homo-

geneous, the orbits of satellites are somewhat perturbed. The paths of satellites can be observed and plotted with a high degree of precision. The data indicate that the Earth is an oblate spheroid, its radius 21 kilometers longer at the equator than at the poles. It behaves as though it were a fluid balanced between gravitational forces, which tend to make it spherical, and centrifugal forces resulting from its rotation, which tend to flatten it.

ACCELERATION OF GRAVITY

The acceleration of gravity near the Earth's surface is measured in gals, in honor of Galileo. A gal is the amount of force that will accelerate a mass 1 centimeter per second per second, or 1 centimeter per second squared. The total value for the acceleration of gravity is 980 gals, which is equivalent to the more familiar value of 9.8 meters per second squared. It is known that when the Moon is directly overhead, at a position known as the zenith, the value for the acceleration of gravity at that point on the Earth's surface is slightly less than if the Moon were in any other position. This phenomenon is a result of the gravitational influence or tidal force that the Moon exerts on the Earth. The attraction of the Moon's gravity causes a point on the Earth's surface to be distended slightly. Values for the amount of distension have been found to be about 0.073 meter. The fact that this point on the Earth's surface has been gravitationally pulled away from the center of the Earth will result in a slightly reduced value in the acceleration of gravity toward the center of the Earth. These values have been found to be in the vicinity of 0.2 milligal. (A milligal is one thousandth of a gal.)

SYNCHRONIZED ROTATION-REVOLUTION

Subtle effects of tidal forces on the Earth exist. When the Earth and the oceans are subjected to tide-raising forces, energy caused by friction is dissipated. The result of this friction is the reduction in the period of the Earth's rotation. In the case of a binary system such as the Earth and the Moon, the result of tidal forces produces a synchronized state of rotation-revolution. In other words, the rate that the Moon rotates on its axis is the same as the rate at which it revolves around the Earth in its orbit—which is the reason that the same face of the Moon always points toward the Earth. This particular phenomenon occurs elsewhere in the solar system; for example, the Sun and Mercury, as well as Pluto and its moon Charon, form other such binary systems.

There is a law in physics that states that angular momentum is conserved. If the rotation rates of the Earth and the Moon are slowing but their masses stay the same, the distance between them must be increasing. Evidence from paleontological studies indicates that at one time, the Earth had a faster rotation rate and the Moon was much closer than it is today. It is now known that the Moon is moving away from the Earth 3.2 centimeters per year.

VERTICAL AND LINEAR DEFORMATION STUDIES

At the beginning of the nineteenth century, the concept that the Earth was not perfectly rigid but in fact was somewhat deformable began to be accepted. The first studies of the deformation of the Earth's crust were conducted in France in the early 1830's. These early studies were accomplished by using containers of mercury and comparing the motion of the liquid metal with the rise and fall of the ocean tides. The horizontal pendulum was the first instrument to record the effect of Earth tides with scientific precision. It consisted of a rigid bracket whose base contained three leveling screws. At both the top and the bottom of the bracket (which resembled a C-clamp), two metal wires were attached. These wires were all attached to a metal arm in such a way as to suspend it in position. At the end of the arm was attached a small mass. The slightest vibration caused by changes of the ground would cause the pendulum arm to begin oscillating back and forth. This instrument was but the first of many types and variations of the pendulum.

In the 1900's, the gravimeter came into use in the field of exploration geophysics and was later used to detect the minute changes in gravity brought about by Earth tides. Gravimeters are designed to measure the differences in the acceleration of gravity. There are several different types of these instruments, most of which consist of a mass suspended by springs. The greater the force, such as gravity, pulling on the mass, the more the spring stretches. The upward force is a function of the strength of the spring, or the spring constant. When the mass is in balance (not oscillating) the spring constant is equal to the force of gravity. Any

change in gravity will then produce a corresponding change in the stretch of the spring. During a period of a maximum Earth tide, gravity will be slightly reduced, resulting in a slight upward drift of the mass.

The pendulums and the gravimeter are used to study the vertical deformation of the Earth's surface. The linear deformation may be measured by means of a device called an extensometer. The first results from the use of this device were reported in the early 1950's. The extensometer consists of a wire 1.6 millimeters in diameter that is held nearly horizontal between two fixed supports about 20 meters apart. A mass of 350 grams is suspended from the center of the wire by a smaller wire with a diameter of 0.2 millimeter. Variations in the 20-meter distance between the two fixed supports as a result of linear deformations of the Earth's surface can cause variations in the tension of the main wire. These variations cause the suspended mass to oscillate vertically. By methods of calibration, the oscillation can be translated into values of linear deformation.

ECONOMIC AND GEOLOGIC APPLICATIONS

The knowledge of how Earth tides function is necessary for an understanding of the deformable nature of the Earth and of the Earth's gravitational interaction with the Moon and the Sun. This knowledge is important to those who explore for the oil, gas, groundwater, and minerals that are necessary for life in the modern world. To the geophysicists who use the technique of gravity surveying, it is necessary to know whether the change in the value of gravity indicated by their instruments is caused by a subsurface geological structure or by the gravity of the Moon or the Sun.

For this reason, gravity surveyors must make what is known as a tide correction, which accounts for the time-varying gravitational attraction of the Sun and the Moon. The attraction is cyclic because the positions of the Sun and Moon are constantly changing with regard to a fixed position on the surface of the Earth. To those Earth scientists who use the technique of searching for magnetic anomalies, or areas where the Earth's magnetism is greater or less than expected, the Sun's effect on the Earth's magnetic field is very important. The Sun's tidal force produces wind currents in the Earth's ionosphere in the same way that it produces ocean tides. Since these winds in the ionosphere consist of waves of charged particles, there is an associated electric current. With this current comes a fluctuating magnetic field. The geophysicist, therefore, needs to know if the Sun's tidal force is causing deviations in the equipment being used.

David W. Maguire

CROSS-REFERENCES

Earth's Rotation, 106; Earth's Shape, 111; Earth-Sun Relations, 1851; Experimental Rock Deformation, 208; The Geoid, 115; Gravity Anomalies, 120; Isostasy, 125; Ocean Tides, 2133; Stress and Strain, 264.

BIBLIOGRAPHY

Baugher, Joseph F. *The Space-Age Solar System.* New York: John Wiley & Sons, 1988. A well-illustrated, very readable volume on the planets, moons, and other bodies that make up our solar system. Suitable for the layperson.

Davidson, Jon P., Walter E. Reed, and Paul M. Davis. *Exploring Earth: An Introduction to Physical Geology.* Upper Saddle River, N.J.: Prentice Hall, 1997. An excellent introduction to physical geology, this book explains the composition of the Earth, its history, and its state of constant change. Intended for high-school-level readers, it is filled with colorful illustrations and maps.

Hamblin, William K. *Earth's Dynamic Systems.* 8th ed. Upper Saddle River, N.J.: Prentice Hall, 1998. This geology textbook offers an integrated view of the Earth's interior not common in books of this type. The illustrations, diagrams, and charts are superb. Includes a glossary and laboratory guide. Suitable for high school readers.

Howell, Benjamin F. *Introduction to Geophysics.* New York: McGraw-Hill, 1959. A technical volume dealing extensively with various areas in the study of geophysics. Topics such as seismology and seismic waves, gravity, isostasy, tectonics, continental drift, and geomagnetism are covered. The reader should have a working knowledge of differential and integral calculus. Suitable for college students of physics or geophysics.

Lutgens, Frederick K., and Edward J. Tarbuck. *Earth: An Introduction to Physical Geology.* 6th ed. Upper Saddle River, N.J.: Prentice Hall, 1999. This college text provides a clear picture of the Earth's systems and processes that is suitable for the high school or college reader. In addition to its illustrations and graphics, it has an accompanying computer disc that is compatible with either Macintosh or Windows. Bibliography and index.

Melchior, Paul. *The Earth Tides.* Elmsford, N.Y.: Pergamon Press, 1966. A highly detailed, highly technical volume on the discovery and the observation of Earth tides. Goes into great detail on the evolution of the instrumentation used for Earth tide detection. Suitable for college students of geophysics or engineering.

Robinson, Edwin S., and Cahit Coruh. *Basic Exploration Geophysics.* New York: John Wiley & Sons, 1988. A well-illustrated volume dealing with the science of geophysics both in theory and in applications. Contains well-developed chapters on seismic, gravity, and magnetic exploration techniques. The reader should have a working knowledge of algebra and trigonometry. Suitable for college students of geology, geophysics, or physics.

Spencer, Edgar W. *Dynamics of the Earth.* New York: Thomas Y. Crowell, 1972. An introduction to the principles of physical geology. Covers all aspects of geology, from introductory mineralogy through a study of the agents that shape the planet's surface. Concludes with units on global tectonics and geophysics. These later chapters tend to be somewhat technical, requiring the use of algebra. Suitable for college-level geology students.

Stacy, Frank D. *Physics of the Earth.* New York: John Wiley & Sons, 1969. A technical volume dealing with such topics as Earth rotation, gravity, seismology, and internal structure, magnetism, and radioactivity. Advanced mathematics is used throughout. Suitable for college students of physics or geophysics.

Taff, Laurence G. *Celestial Mechanics*. New York: John Wiley & Sons, 1985. A technical volume dealing with Newtonian gravitation, how it manifests itself, and how gravitational bodies interact with one another. Calculus and differential equations must be understood prior to reading. Suitable for college-level students of physics, astrophysics, and mathematics.

Wilhelm, Helmut, Walter Zuern, Hans-Georg Wenzel, et al., eds. *Tidal Phenomena*. Berlin: Springer, 1997. A collection of lectures from leaders in the fields of Earth sciences and oceanography, *Tidal Phenomena* examines Earth's tides and atmospheric circulation.

Complete with illustrations and bibliographical references, this book can be understood by someone without a strong knowledge of the Earth sciences.

Zeilik, Michael, and Elske Smith. *Introductory Astronomy and Astrophysics*. New York: Saunders College Publishing, 1987. A technical volume having to do with such topics as celestial mechanics, interactions of gravitational bodies, the planets, the origin of the solar system and the universe, stars, and cosmology. Some advanced mathematics is used. Suitable for college students of astronomy or astrophysics.

EARTH'S ROTATION

The rotation of the Earth produces the days and nights that provide the daily rhythm of life. Rotation causes the Earth to be flattened at the poles and to bulge at the equator. The Coriolis force, which influences the circulation of the atmosphere and oceans, is a result of this motion.

PRINCIPAL TERMS

CENTRIFUGAL FORCE: a pulling-apart force

CORIOLIS FORCE: an effect on free-moving bodies on the Earth because of the planet's rotation

DAY: the interval of time between successive passages of the Sun or star over a meridian of the planet Earth

OBLATE SPHERE: the shape of the Earth that is the result of rotation

PLANE OF THE ECLIPTIC: the plane of the orbit of the Earth around the Sun

PRECESSION: the wobbling of the Earth's rotational axis

REVOLUTION: the yearly orbit of the Earth around the Sun

AXIS OF ROTATION

The spinning of the Earth on its polar axis is called rotation. The ancient Greeks who studied the motions of the universe considered the Earth to be a motionless body in the center of a geocentric (Earth-centered) universe. An exception was Heraclides (fourth century B.C.E.), who thought that the Earth did rotate. In general, the Greeks reasoned that a stationary object (in their experience, the Earth) tends to remain at rest, and they had no physical theory of gravitation to account for planetary motions. Later, the work of Nicolaus Copernicus, Galileo Galilei, and Johannes Kepler, culminating with Sir Isaac Newton's law of universal gravitation, explained how moving bodies tend to remain in motion. In other words, the Earth's rotational motion is imparted to the objects that move along with the Earth as it rotates and revolves around the Sun. Today, it is commonly accepted that the Earth rotates around a heliocentric (Sun-centered) system.

The polar axis of rotation, which is inclined 23½ degrees from the perpendicular (90 degrees) to the plane of the orbit of the Earth, passes through the center of the Sun and the Earth. The axis of the Earth, therefore, makes an angle of 66½ degrees with the plane of the ecliptic, or path of the Earth. The inclination has an effect on the length of days and nights as the Earth orbits the Sun. The Earth rotates from west to east, or in a counterclockwise direction, thus making the Sun, Moon, planets, and stars appear to move from east to west across the sky. Rotation produces the nights and days of the year and the sunrises and sunsets of each day. People refer to the Sun as rising and setting because it appears that the Sun is moving around the Earth, when in fact, the Earth is moving around the Sun.

VELOCITY AND RATE OF ROTATION

The velocity of rotation is 1,038 miles per hour (mph) at the equator; it decreases to 899 mph at 30 degrees latitude, to 519 mph at 60 degrees latitude, and to 0 mph at 90 degrees latitude (the polar axis of rotation). Because the Earth's velocity is 1,038 mph at the equator and because 1 degree of longitude at this latitude equals approximately 69 miles, division of 1,038 mph by 69 miles gives 15 degrees. The Earth then rotates through 15 degrees of longitude every hour; consequently, the Earth's rotation is also important from the standpoint of time.

The Earth is not a good timepiece. Its rate of rotation varies through the course of the year. A day may be defined as the interval of time between successive passages of a meridian, or line of longitude from the North to South Pole, under a body of reference (the Sun or a star). A day with reference to the Sun is called a solar day, and a day with reference to a star is called a sideral day.

(The solar day is approximately four minutes longer than the sidereal day.) Therefore, we have solar and sidereal time. The difference between the two is a result of the Earth's revolution around the Sun in a counterclockwise manner. This motion makes the Sun appear to shift eastward about 1 degree per day. The Earth in fact makes one complete revolution (360 degrees) once every 365¼ days; the rotation velocity varies because of the elliptical orbit of the Earth, which brings it closer to the Sun (perihelion) around January 3 and takes it farthest from the Sun (aphelion) on July 4. The apparent solar day, which is not constant, measures the time interval between successive passages of the Sun over the same meridian. This is the time indicated by a sundial. Mean solar time is the average solar day, or the twenty-four-hour clock day that consists of 86,400 seconds. The solar day would coincide with the Earth's if the Earth were not inclined and moved at a constant velocity as it revolved.

CORIOLIS FORCE

If the Earth were a true sphere, it would have a constant diameter of 7,927 miles and a circumference of 24,901 miles. Because of rotation, however, the Earth is flattened in the polar regions and it bulges at the equator, thus making it slightly elliptical. This form is known as the ellipsoid of rotation, or the oblate spheroid (flattened sphere), and results in objects located at the equator being at a greater distance from the center of the Earth than those at the poles. The centrifugal force developed as a result of rotation is consequently greater at the equator than at the poles. The gravitational pull is 289 times greater than the centrifugal force and is increased slightly at the polar axis due to rotation. Because the Earth is slightly flattened, the length of a degree of latitude changes from 68.70 miles at the equator to 69.40 miles at the poles. The Coriolis force or effect, named after a nineteenth century French mathematician who studied this phenomenon, is caused by the Earth's rotation. It is an apparent force that affects free-moving bodies (wind, water, missiles). For example, if a missile were fired in a northerly direction, it would be seen veering to the right in the Northern Hemisphere because the observer, along a certain meridian, has rotated westward with reference to the missile's path. In the Southern Hemisphere,

objects would veer to the left. At the equator, where an object would be farthest from the axis of rotation, the Coriolis deflective force is zero, and therefore free-moving bodies would not be deflected. At 30 degrees latitude, the deflective force is 50 percent greater, and at 90 degrees, at the axis of rotation, the deflective force is 100 percent.

VARIATIONS IN ROTATION

Overall, the rotation velocity of the Earth is decreasing. Astronomers agree that the length of the day is increasing by milliseconds per century, as the rotational velocity is decreasing by that amount. Thus the amount of days in a year has decreased through geologic time. The decrease in velocity is a result primarily of the tidal friction caused by the gravitational pull of the Moon. Evidence for a slowing rotational velocity comes from the study of fossils. Paleontologists have found that living corals produce approximately 365 growth lines on their shells versus about 400 growth lines for Devonian fossil corals about 370 million years old. There are numerous detailed studies that show that the Earth's rotation varies constantly in minute proportions for various reasons. The reasons for these variations include the speeding up of the rotation of the mantle (this is more likely in reference to the asthenosphere, which is part of the mantle that behaves in a plastic manner), the transfer of momentum (motion) between the mantle and outer core, movements of air masses, changes in wind patterns, periodic exchange of angular momentum between the atmosphere and the Earth's mantle, transfer of angular momentum between the atmosphere and the oceans, monsoons, earthquakes, volcanoes, and the plate tectonics (crustal plate movements). The periods of variation on the Earth's rotation vary. There are short-term and seasonal variations attributed to wind velocities. Length-of-day variations are mainly attributed to atmospheric changes. Decade variations may be attributed to climatic variations, which may be related to volcanism. Length-of-day changes caused by earthquakes amount to only 1 percent of the total causes of variation. Monthly changes in the length of day result from changes in the distance of the Moon with reference to the Earth because of changes in the Earth's tidal bulge. It is evident from many studies that the Earth's rotational velocity has varied throughout geologic time and continues to change on a daily, weekly, monthly, seasonal, yearly, or even longer-term basis.

PRECESSION

It may be assumed that the Earth's axis of rotation points towards Polaris—our present North Star; however, because of the slowing of the Earth's rotational speed, the Earth's axis is becoming more and more inclined from the perpendicular to the plane of the ecliptic. Like a spinning top, as it begins to slow down, its axis begins to tilt and wobble. Because the axis of the Earth's equinoxes precess (shift) by 50 seconds of arc a year along the ecliptic, it takes the axis of the Earth 25,800 years to complete one precession, or wobble. In other words, the rotation axis of the Earth will again point to the same position in the sky in 25,800 years. Changes in the orientation of the Earth's axis as a result of precession cause the polar axis to inscribe a large imaginary upside-down cone in the sky. As the axis points to different parts of the sky, other stars serve as north stars. Because of precession, star charts have to be changed every 50 years. The seasons of the year will be reversed, because as the Earth revolves, the polar axis of rotation will be pointing to a different position in the sky. Superimposed upon the precessional motion are two other motions. One of these small oscillating motions is called nutation, which has a semiamplitude of 9.2 seconds of arc and a period of 18.6 years. This motion is associated with the periodic variation in the angle that the Moon's motion makes with the Earth's equator. The other motion, called Chandler's wobble, has two oscillations. One of the oscillations, the Chandler component, has a period of twelve months. The twelve-month component is a result of meteorological effects associated with seasonal changes in air masses. The second oscillation of the Chandler wobble, the 14.2-month component, is caused by shifts in the Earth's interior mass. Thus, the precession circle is not a smooth, round circle but a squiggly one. These two motions and others caused by various forces affect the Earth's polar motion.

OBSERVATION AND TIME MEASUREMENT

Scientists have used methods ranging from visual observation to satellites and lasers to fossils to learn about the rotation of the Earth. Visual ob-

servations by the early Greeks led them to conclude that the Earth did not rotate. Through the much later contributions of Copernicus, Galileo, Kepler, and Newton, a new physics evolved that explained the planetary motions of the universe. Kepler derived three basic laws of planetary motion. The first two are important with reference to the Earth's rotation. The first law states that the planets move around the Sun in elliptical orbits, with the Sun located at one focus. The second law states that as a planet revolves, a line connecting it to the Sun sweeps over equal areas in equals period of time. The telescope first used by Galileo in astronomical observations allowed scientists to see that heavenly bodies rotate. Telescopic observations showed that other planets were slightly flattened. This led observers to question whether the Earth was similarly shaped. Evidence indicated that the Earth's gravity varied with latitude; when the physical laws of gravity were considered, it became evident that the Earth was oblate in form.

In 1671, Jean Richer, a French astronomer, made time measurements with a pendulum clock both in Paris (49 degrees north) and in Cayenne, French Guiana (5 degrees north) and compared the two. In French Guiana, the clock "lost" two and one-half minutes per day. He attributed this loss to a decrease in gravitational pull toward the equator. It was later determined that, as a result of the Earth's rotation, the polar area of the Earth was flattened and that it bulged at the equator because of centrifugal force. Measurements to confirm the Earth's oblateness were made in the eighteenth century in Lapland (67 degrees north) and near Quito, Ecuador (4 degrees north); it was found that the length of a degree of arc near the equator was less than that in France and even less than that in Lapland. In 1851, the French physicist Jean-Bernard-Léon Foucault hung an iron ball with a 200-foot-long wire—adding a pin at the bottom to make marks in the sand underneath—from the dome of the Panthéon, in Paris. It was observed that the path of the ball moved toward the right. Since the pendulum kept swinging in the same direction, this meant the building underneath the pendulum was rotating. The apparent movement to the right of the pendulum also demonstrated the apparent force called the Coriolis effect. Sundials also help to demonstrate that the Earth rotates as the shadow of the gnomon (stick) moves across the face of the dial. In the 1950's, atomic clocks began to be used to measure time accurately over long periods. When time kept by the clocks was compared to time determined by the rotation of the Earth, new variations in the Earth's rotation were found. An interesting technique used to determine the changing rotational velocity of the Earth through time has been the study of growth lines located between coarse (presumably annual) bands in fossil and modern-day corals.

SATELLITES AND LASERS

Until the 1970's, telescopes were used to observe stars to determine length of day and polar motion. This technique was limited by bending of starlight by the atmosphere. New techniques used to determine length of day and polar motion involve the use of satellites and lasers. One method called Lunar Laser Ranging (LLR) involves the emission of light pulses from a laser on Earth to reflectors left on the Moon by Apollo and Soviet spacecraft. The returning pulses of light are received by a telescope. The total travel time is calculated to determine the Earth-to-Moon distance. By observing the time the Moon takes to cross a meridian during successive passages, this method has provided very good length-of-day measurements. Another technique involves the use of the Laser Geodynamics Satellite (Lageos). This satellite is covered by prisms that reflect light from pulsed lasers on Earth. Again, the returned beam is received by a telescope and the round-trip travel time is used to infer the one-way distance from the Earth to the satellite. Scientists hope that this method, which includes a network of stations on Earth, will be able to tell something about the yearly movement of crustal plates, which is believed to cause variations in the Earth's rotation. A very accurate technique known as Very-Long Baseline Interferometry (VLBI) is also being used to plot continental drifts as well as variations in Earth's rotation and the position of the poles. In this method, radio signals (typically quasars) from space are received by two radio antennas and are tape-recorded. The tapes are compared, and the difference in arrival times of the signals at the two radio antennas is used to calculate the distance between the two. If the distance between the two antennas has changed, the crustal plates have moved. All of these techniques are important to scientists who need accurate data about the Earth's rotation. The layperson, however, needs only to observe the daily east-west motion of the Sun, Moon, planets, and stars as the world turns from west to east on its polar axis of rotation.

GEOLOGIC SIGNIFICANCE

The spinning of the Earth on its polar axis once every twenty-four hours is very much a part of the daily rhythm of life. It gives to human reckoning days and nights and daily time references; because there are 360 degrees in a circle and the Earth rotates through 15 degrees of longitude every hour, there are 24 time zones in the world. The rotational velocity of the Earth varies from day to day because of the elliptical orbit of the Earth around the Sun. For day-to-day time-keeping purposes, however, the mean solar day of twenty-four clock hours is used. In addition, the shape of the Earth is affected by the Earth's rotation. The spinning of the Earth on its polar axis sets up a centrifugal force that causes the Earth to bulge at the equator and the polar area to be flattened. Because of this phenomenon, the distance to the center of gravity in the Earth varies, and the gravitational pull on objects on the Earth also varies: Objects weigh slightly less in the equatorial area than in the polar regions of the world. The Coriolis force, an apparent force caused by the rotation of the Earth, causes free-moving bodies to be deflected to the right in the Northern Hemisphere and to the left in the Southern Hemisphere. This force controls the circulation patterns of the atmosphere and the oceans. It governs the direction of winds as they flow in or out of pressure systems. It has set up the wind belts of the world. The force also influences the flow of water, as evidenced by the ocean currents of the world. Consequently, these patterns help to produce certain climatic regions along the borders of continents. Because the Earth's crust is broken into several moving crustal plates, it has been suggested that the shape and location of the continents have been influenced by the Coriolis force. The Coriolis force is believed to be important in producing the convection currents in the liquid outer core, which in turn give rise to the internal magnetic field of the Earth.

Roberto Garza

BIBLIOGRAPHY

Davidson, Jon P., Walter E. Reed, and Paul M. Davis. *Exploring Earth: An Introduction to Physical Geology.* Upper Saddle River, N.J.: Prentice Hall, 1997. An excellent introduction to physical geology, this book explains the composition of the Earth, its history, and its state of constant change. Intended for high-school-level readers, it is filled with colorful illustrations and maps.

Gould, S. G. "Time's Vastness." *Natural History* 88 (April, 1979): 18. This article summarizes the reasons for the slowing down of the Earth's rotation. It discusses the use of corals as a proof that the length of the day is increasing and that the number of days in a year are decreasing. Suitable for high-school-level readers.

Hamblin, William K. *Earth's Dynamic Systems.* 8th ed. Upper Saddle River, N.J.: Prentice Hall, 1998. This geology textbook offers an integrated view of the Earth's interior not common in books of this type. The illustrations, diagrams, and charts are superb. Includes a glossary and laboratory guide. Suitable for high school readers.

Hoyle, Fred. *Astronomy.* Garden City, N.Y.: Doubleday, 1962. This book traces astronomical discoveries through time. Of particular value to the study of rotation are the chapters on planetary motion and ancient astronomy, Copernicus and Kepler, and the theory of gravitation. An excellent book for anyone interested in astronomy. It includes an index and numerous illustrations.

King-Hele, D. "The Shape of the Earth." *Scientific American* 217 (October, 1967): 17. This article begins with the historical views of the shape of the Earth. It then discusses, with a good set of illustrations, how satellites have helped scientists to learn more about the shape of the Earth. Suitable for high school and other readers interested in a nontechnical approach.

McDonald, G. E. "The Coriolis Effect." *Scientific American* 186 (1952): 72. The article takes a nontechnical approach to the study of how objects move on the Earth as a result of the Coriolis effect. Suitable for high school readers.

Markowitz, W. "Polar Motion: History and Recent Results." *Sky and Telescope* 52 (August, 1976): 99. This article reviews studies of polar motion. It takes an indepth look at how the Earth's rotation and precession motions are affected by various forces.

Monroe, James S. *Physical Geology: Exploring the Earth.* 2d ed. Minneapolis/St. Paul: West Publishing Company, 1995. This college text is intended for the college student with some Earth science background, but can also be used as an introduction to physical geology. It provides a nice explanation of gravity and the Earth's rotation complete with illustrations.

Mulholland, J. D. "The Chandler Wobble." *Natural History* 89 (April, 1980): 134. The article discusses how small movements affecting the Earth's axis may be associated with other terrestrial phenomena. Suitable for high school readers.

Plummer, Charles C., and David McGeary. *Physical Geology.* Boston: McGraw-Hill, 1999. A college-level introductory geology textbook that is clearly written and wonderfully illustrated. An excellent sourcebook of basic information on geologic terminology and fundamentals of geologic processes. Contains CD-ROM. An excellent glossary.

Rosenburg, G. D., and S. K. Runcorn, eds. *Growth Rhythms and the History of the Earth's Rotation.* New York: John Wiley & Sons, 1975. This book is a good compilation of studies that can serve as an introduction to the methods of determining the history of the Earth's rotation. The text is suitable for college-level readers not intimidated by technical language. Each study includes a bibliography, and the book is carefully indexed by author, taxonomy, and subject.

Rothwell, Stuart C. *A Geography of Earth Form.* 2d ed. Dubuque, Iowa: Wm. C. Brown, 1973. This book is an introduction to the basic topics of the Earth's shape, map projections, planetary motions, and time and its measurements. It is suitable as a supplement to a college-level physical geography course.

Smylie, D. E., and L. Mansinha. "The Rotation of the Earth." *Scientific American* 225 (December, 1971): 80. This article analyzes measurements indicating that the Earth's wobble may be due to earthquakes. It is a well-illustrated article that can be read by high school readers.

Strahler, A. N. *Physical Geography.* 2d ed. New York: John Wiley & Sons, 1960. This book is written for a freshman college-level course in physical geography. It is an excellent textbook as well as a reference book. Each chapter provides a bibliography. The book has an index and excellent illustrations.

Zharkov, Vladimir, et al., eds. *The Earth and Its Rotation: Low-Frequency Geodynamics.* Heidelberg, Germany: Wichmann, 1996. Although Zharkov does focus heavily on complex mathematical models of the Earth and the Earth's rotation, this book offers a good understanding of geophysics and geodynamics for the person with some background in the Earth sciences. Bibliography and index.

EARTH'S SHAPE

It has been known for centuries that Earth is not a perfect spheroid. The circumference of the planet is significantly greater at the equator than in the dimension of the meridians, the so-called polar circumference. This oblateness is the result of the substantial centrifugal force generated by Earth's daily rotation around its axis.

PRINCIPAL TERMS

CENTRIFUGAL FORCE: an upward or outward force that results from mass in motion along a curved path

DEFORMATION: the warping of Earth materials as a result of strain in response to stress

EARTH TIDE: the slight deformation of Earth resulting from the same forces that cause ocean tides, those that are exerted by the Moon and the Sun

GEODESY: a branch of applied mathematics that determines the exact positions of points on the Earth's surface, the size and shape of the Earth, and the variations of terrestrial gravity and magnetism

GEOID: the figure of the Earth considered as a mean sea level surface extended continuously through the continents

LITHOSPHERIC CRUST: the relatively thin outer portion of Earth's "onion" structure, composed of solid rock

PERFECT SPHEROID: a three-dimensional body that has the same circumference regardless of the direction by which it is measured; that is, perfectly "round"

OBLATENESS

The surface of the Earth is irregular. Any attempt to map the surface requires a zero reference datum upon which to locate positions and from which to measure elevations. This reference surface is called the "geoid." The exact configuration or shape of the geoid is a function of centrifugal force and distribution of density within the Earth.

A view of Earth from satellite distance in space would, to the naked eye, suggest that the planet is a perfect spheroid. Yet, finer measurements from space and on Earth itself reveal that the planet is an oblate spheroid, meaning that it is deformed. In other words, Earth is distended at its "waistline," the equator. Earth's flattening at the poles, or its oblateness, is about three one-thousandths in terms of its diameter, which, as measured at the equator, is 12,756 kilometers and, pole to pole, is 12,714 kilometers.

The discovery that Earth is not a perfect spheroid dates to the seventeenth century, when measurements of the distance of 1 degree of latitude (one-ninetieth of the distance from the equator to a pole) demonstrated inconsistencies from one place to another. It eventually became evident that the distance from one latitude to another becomes less the farther from the equator the measurements were taken. Extreme precision in the measurements of Earth's oblateness was not possible until the advent of twentieth century instrumentation.

Earth turns on its axis, making one complete rotation every twenty-three hours, fifty-six minutes, and four seconds, which means that on the equator at sea level, there is rotational velocity of about 1,670 kilometers per hour. At 45 degrees latitude (north or south), however, the surface velocity is approximately one-half that speed, and at the poles there is no rotational velocity. This differential in rotational velocities means that the centrifugal force at work in the low latitudes is great, while in the polar regions the centrifugal force is small or nonexistent. From this differential derives the equatorial deformation and thus Earth's oblate spheroidal shape.

PRINCIPLE OF GRAVITATIONAL EQUILIBRIUM

It is necessary at this point to consider why all planets are essentially spheroidal while the hundreds of objects that comprise the asteroid belt lying between Mars and Jupiter are not, which is known as the principle of gravitational equilibrium. Any object that has density also develops its own gravitational force. When the mass of an object is sufficiently large, it cannot sustain any shape other than spheroidal; the fragments of the asteroid belt are not sufficiently large to have met that requirement. On the other hand, Earth, all the other planets of the solar system, and most of the satellites of planets (the Moon, for example) are indeed spheroids. An object with large mass (density times volume) develops its own gravitational force in a direction downward from the surface of the object. This force results in unit weight or specific gravity; the greater the force of gravity, the greater an object's weight. Only a spheroidal shape will permit physical equilibrium in a large mass in which the forces of gravity exceed the strength of the material that makes up the mass.

One is easily misled as to the rigidity of planet Earth. Its solid lithospheric crust has a thickness that approximates an eggshell in relative dimensions. Therefore, it is understandable that the Earth's "shell" has been broken, warped, and distorted through geologic time and into the present. It also becomes comprehensible that Earth is extremely sensitive to distortion and deformation from forces imposed upon it, upsetting gravitational equilibrium, as in the case of centrifugal force created by rotational velocity, especially in the equatorial latitudes where the velocities are highest.

A comparison of the oblateness of Earth with that of other planets in the solar system demonstrates that centrifugal force from rotational velocity is the principal cause of the Earth's equatorial "bulge." Oblateness is not directly proportional to the surface velocity of rotation at the planet's equator; the strength or rigidity of a planet's material composition affects the degree of warping. Planets Mercury and Mars, however, each with very slow rotational velocities, have no discernible oblateness and are nearly perfect spheroids. Mars has three times more oblateness than Earth, yet with only

about one-half of the rotational velocity. The average density of Mars is about 30 percent less than that of Earth, which suggests that its materials have less rigidity, perhaps accounting for Mars' excessive equatorial bulge.

EARTH TIDES

Lesser forces imposed upon Earth also affect the shape of the planet, a prime example being Earth tides. The oceans of the world have two huge bulges, or regions where the ocean surface is relatively high. Sea level rises as the bulge is approached and falls as it is passed. Therefore, two high tides and two low tides are recorded each day. The ocean bulges are caused primarily by the gravitational attraction of the Moon, which slightly counters Earth's controlling force, its own gravitational attraction. The Sun also imposes an attraction, although less than that of the Moon. When both the Sun and the Moon are positioned in line with the Earth, the oceans display the highest tides, as the negative attractions of both bodies are imposed collectively. The solid Earth distorts very slightly from the same external force imposed by the Moon and the Sun, but the distortion is so slight as to render great difficulties in actual measurement. The cycle of these Earth tides is much longer than that of ocean tides; they are sufficiently long to be called static tides.

CHANGES IN CONTROLLING PHENOMENA

The passing of geologic time has brought startling changes in the phenomena which control

PLANET SHAPES AND ROTATIONAL PERIODS AND VELOCITIES			
Planet	*Oblateness*	*Rotational Period*	*Surface Velocity of Rotation*
Mercury	0.0	59 days	10.8
Venus	0.0	–243 days	6.5
Earth	0.003	24 hours	1670.0
Mars	0.009	24.5 hours	866.0
Jupiter	0.06	10 hours	45,087.0
Saturn	0.1	10.5 hours	36,887.0
Uranus	0.06	–17 hours	14,794.0
Neptune	0.02	16 hours	9,794.0
Pluto	?	–6.5 days	128.0

NOTES: Oblateness, or ellipticity, is a measure of the planet's flatness, or deviation from a perfect sphere. Rotational periods are rounded to the nearest half day or half hour; a minus sign signifies retrograde rotation.

Earth's shape. The distance from Earth's surface to the Moon was far shorter than at present, and Earth's rotational velocity was much faster back in geologic time. Therefore, Earth's shape does not remain constant. About 350 million years ago, Earth appears to have had about 405 days in the year. Fossil coral of that age shows microscopic diurnal growth lines, and the year is measured by seasonal changes in patterns. On the basis of 405 days per year, the planet's rotational velocity must have been at least 10 percent faster. Such a rotational velocity would have generated significantly greater centrifugal force, and Earth's oblateness would have been more severe than at present. One can speculate as to what effect that condition might have had on ancient dynamic processes; couple that thought with the Moon's nearer proximity to Earth, which would have resulted in far more severe ocean tides in the Paleozoic era (about 250-575 million years ago). The two phenomena are scientifically linked. It is the Moon's relatively strong negative gravitational attraction on Earth that is believed to be the principal force causing the slowing of Earth's rate of rotation.

GEODETIC SURVEYS AND SATELLITE MEASUREMENTS

Geodesy is the investigation of the Earth's shape. Typically the shape has been determined by geodetic surveys which employ a combination of terrestrial surveying techniques and astronomical measurements to determine the location of points on the terrestrial spheroid. Laborious field-surveying techniques are being replaced by satellite measurements. A growing network of global positioning satellites (GPS) and radar ranging from orbit make it possible to measure location and elevation whether on land or sea with increasing precision.

Until recently, it was generally assumed that, other than the two tidal bulges, the surface of the oceans represents the smooth curvature of the planet. Geodesists, however, have hypothesized for decades that the ocean surface should theoretically have highs and lows conforming to the peaks and deeps of the ocean floor. Actual measurements in the early 1980's proved those theories to be correct.

It has become possible to map the topography of an ocean surface to a precision of a few centimeters. The instrument used is the satellite-mounted radar altimeter, which makes continuous measurements of the distance from the satellite to the water surface. Assuming the satellite itself maintains a consistent orbital path, a perfectly curved ocean surface should reflect an equally consistent distance. The fact is, with wave crests and troughs averaged out, ocean surfaces show substantial deviations from a smooth curve. Major seamounts and suboceanic ridges are clearly represented by corresponding high places of ocean surface. Likewise, the major deep-sea troughs, as are common in the western Pacific, the Indian Ocean, and the Caribbean, reveal themselves with troughs in the ocean surface above. This phenomenon derives from geographic variations in the acceleration of gravity. In the case of a deep-sea trough, for example, the space within the trough is filled with seawater instead of with rock. The density or specific gravity of seawater is slightly greater than 1.0, whereas the density of suboceanic rock is typically about 2.85—which means that the acceleration of gravity at sea level directly over a deep-sea trough is less than normal. A compensating rise in the ocean surface maintains nature's equilibrium.

ROLE IN EARTH'S HABITABILITY

Earth's inhabitants suffer little or no effect from the planet's distortion. It cannot be observed with the naked eye, and it does not appear to play a role in weather patterns and climate. The principal cause of deformation, however, affects the entire habitability of the planet. The daily rotation of Earth around its polar axis and its 23.5-degree tilt, with respect to the plane of orbit around the Sun, results in the seasons of the year. If the Earth rotated today at the rate it rotated 350 million years ago, days and nights would be 10 percent shorter. An interesting meteorological challenge is to calculate exactly how shorter days and concurrently shorter nights would affect weather, climates, agriculture, and human health. Conversely, were rotation to slow appreciably, as surely it must in geologic time, the effect would be disastrous. The daytimes would be far hotter, and the nighttimes would be far colder because of longer exposure to solar radiation in daylight and longer nocturnal radiation at night. It is doubtful that the agricultural systems of today's society could be sustained.

Earth's daily turn on its axis results in other familiar phenomena. The ocean currents of the Northern Hemisphere always flow clockwise, while those of the Southern Hemisphere flow counterclockwise. Witness the Gulf Stream of the north Atlantic and the Japanese (Alaskan) current of the north Pacific, always turning to the right, while the south Atlantic flow is to the left in rotation. This phenomenon is caused by Earth's west-to-east rotation. If Earth rotated in the opposite direction, the Sun would rise in the west and set in the east, and the phenomenon would also be reversed. The planet's oblateness, however, would be unchanged, as the equatorial bulge results from centrifugal force. That force has no regard to compass direction, deriving as it does solely from the difference in surface velocity of the land surface: very high speed in the equatorial belt, diminishing to zero speed at the poles.

John W. Foster

CROSS-REFERENCES

Earth-Sun Relations, 1851; Earth Tides, 101; Earth's Rotation, 106; Earth's Structure, 37; The Geoid, 115; Gravity Anomalies, 120; Isostasy, 125; Polar Wander, 172; Solar System's Origin, 2607.

BIBLIOGRAPHY

Davidson, Jon P., Walter E. Reed, and Paul M. Davis. *Exploring Earth: An Introduction to Physical Geology.* Upper Saddle River, N.J.: Prentice Hall, 1997. An excellent introduction to physical geology, this book explains the composition of the Earth, its history, and its state of constant change. Intended for high-school-level readers, it is filled with colorful illustrations and maps.

Dott, R. H., and R. L. Batten. *Evolution of Earth.* 3d ed. New York: McGraw-Hill, 1981. Describes the physical and paleontological evolution of Earth from pre-Paleozoic to the present. Designed as an introduction to the natural history of the planet and to the fossil evidence of the past.

Hamblin, William K. *Earth's Dynamic Systems.* 8th ed. Upper Saddle River, N.J.: Prentice Hall, 1998. This geology textbook offers an integrated view of the Earth's interior not common in books of this type. The illustrations, diagrams, and charts are superb. Includes a glossary and laboratory guide. Suitable for high school readers.

Heacock, John G., ed. *The Structure and Physical Properties of the Earth's Crust.* Geophysical Monograph 14. Washington, D.C.: American Geophysical Union, 1971. Contributors were drawn from pertinent disciplines of geophysics, physics, geochemistry, and geology. Serves primarily as a reference for advanced students of Earth science.

Melchior, Paul. *The Earth Tides.* Oxford, England: Pergamon Press, 1966. A sophisticated treatment of the physical phenomenon of small distortions of Earth resulting from gravitational forces imposed by the Moon and the Sun.

Munk, W. H., and G. J. F. MacDonald. *The Rotation of Earth: A Geophysical Discussion.* New York: Cambridge University Press, 1960. A detailed analytical treatment of the physics of Earth's rotation. Designed for professional geophysicists. Includes discussion of the small fluctuations in rotation as a result of redistribution of angular momentum, thought to be caused by dynamics in the fluid outer core.

Plummer, Charles C., and David McGeary. *Physical Geology.* Boston: McGraw-Hill, 1999. A college-level introductory geology textbook that is clearly written and wonderfully illustrated. An excellent sourcebook of basic information on geologic terminology and fundamentals of geologic processes. Contains CD-ROM. An excellent glossary.

Siever, Raymond. *The Solar System.* San Francisco: W. H. Freeman, 1975. Although dated, this comprehensive and readable compendium of twelve parts by twelve authors, including part 6, "The Earth," still contains many concepts pertinent to the Earth's shape and gravity. This work is unique in that it could serve as a reference, as a textbook, or simply as reading for the inquiring mind.

Smith, James. *Introduction to Geodesy: The History and Concepts of Mode Geodesy.* New York: Wiley, 1997. Geared toward the college student, this book provides a nice introduction to the study of the Earth's shape and rotation. Includes many illustrations and maps that add clarity to key concepts. Bibliography and index.

Stacey, Frank D. *Physics of Earth.* 2d ed. New York: John Wiley & Sons, 1977. A reference volume on solid-Earth geophysics, including radioactivity, rotation, gravity, seismicity, geothermics, magnetics, and tectonics. Carries detailed numerical tabulations on dimensions, properties, and unit conversions.

THE GEOID

The shape of the sea-level surface, over the oceans and under the continents, is given by the geoid. This shape differs from the best-fitting ellipsoid by amounts ranging up to approximately one hundred meters, and these variations provide valuable information concerning models of the convection and tectonics of the planet.

PRINCIPAL TERMS

ELLIPSOID OF REVOLUTION: a three-dimensional shape produced by rotating an ellipse around one of its axes

EQUIPOTENTIAL SURFACE: a surface on which every point is at the same potential, used here to include gravitational and rotational effects; no work is done when moving along an equipotential surface

GLOBAL POSITIONING SYSTEM (GPS): a group of satellites that go around Earth every twenty-

four hours and that send out signals that can be used to locate places on Earth and in near-Earth orbits

LEGENDRE POLYNOMIALS: mathematical functions used to describe equipotential surfaces on spheres

MANTLE CONVECTION: thermally driven flow in the Earth's mantle thought to be the driving force of plate tectonics

APPROXIMATING EARTH'S SHAPE

The geoid is an imaginary surface that is at sea level everywhere on the Earth. Over the oceans, it is generally at mean sea level; under the continents, it is the elevation the sea would have if all of the continents were cut by narrow sea-level canals. It is usually represented as the difference in elevation between sea level and some ellipsoid representing the average shape of the Earth, and its relief is on the order of 100 meters. It is important in surveying and geodesy because elevations are measured above or below this surface, and it is important in geology and geophysics because its departures from a perfect ellipsoid reveal information about the Earth's interior.

The shape of the Earth can be approximated, with varying degrees of complexity and with different levels of success, by different mathematically defined shapes. If represented as a sphere with a radius of 6,371 kilometers, the shape is very simple, but it will have a radius that will be 7 kilometers too small at the equator and 15 kilometers too large at the poles. Nonetheless, this is adequate for many purposes.

For some purposes, however, a spherical shape is entirely inadequate. Gravity varies with the radius of the Earth. Surveys seeking to detect density variations beneath the surface using sensitive

measurements of gravity need a way to account for the gradual increase in radius from the poles to the equator. This change in radius, usually called the Earth's "flattening," is obtained by dividing the difference between the equatorial radius and the polar radius by the equatorial radius. A modification to the spherical shape is obtained by letting the radius vary slightly with latitude, using a straightforward function that includes the value for flattening. This shape is the spheroid, sometimes known as the "niveau spheroid," and for many years was used in gravity surveys. It permitted data reduction at a time when computers filled rooms, if not buildings. This simple formula is actually an approximation of a slightly more complex shape, the ellipsoid.

EQUIPOTENTIAL SURFACES

An ellipsoid of revolution is the shape of a solid produced by rotating an ellipse about one of its axes. If an ellipse with a minor axis equal to the polar diameter and a major axis equal to the equatorial diameter is rotated about the poles, the resulting shape is the Earth's ellipsoid. This ellipsoid is used in studies when the sphere or spheroid is inadequate, and it forms the basis for the geoid.

The geoid is an example of an equipotential surface. If there was some way of sliding a mass around on its surface without any friction, no work would be done in moving that mass from place to place, because the mass would stay at the same potential. It is not difficult to calculate the shape of this surface for various idealized situations. It is also possible to measure the shape of the geoid. Much can be learned by comparing the observed shape with the shapes generated by the models.

If the Earth were a stationary sphere of uniform density, the geoid would also be a sphere. If the Earth were a rotating sphere of uniform density, the geoid would become an ellipsoid. This is because the rotation produces centrifugal force. (Rotation involves an acceleration, which, when multiplied by a mass, must be balanced by a centripetal force, in this case one supplied by gravity. For these purposes, using the non-Newtonian centrifugal force will prove simpler.) This centrifugal force acts in a direction perpendicular to the rotation axis. At the equator, it would be directly opposed to the gravitational pull of the Earth. An equipotential surface would need to be higher there to make up for this. The equator would be farther from the center of the Earth, and the poles would be closer to the center of the Earth, but because the geoid is an equipotential surface, traveling from the equator to one of the poles would not involve going downhill. If the Earth formed a rigid sphere, oceans would be much deeper at the equator than at the poles. The scale of the Earth and its billions of years of existence, however, have allowed it to deform much as if it were a fluid.

EVOLVING KNOWLEDGE OF GEOID

Suppose this model is allowed to assume the equilibrium shape of a fluid with the Earth's mean density, rotating in space once a day. In 1686, Sir Isaac Newton determined that such a model would form an ellipsoid with a flattening of 1 part in 230. His solution piqued considerable interest. This much flattening would result in differences in the length of a degree of latitude between the equator and the poles, which should have been measurable using the techniques available in the early part of the eighteenth century. Expeditions were made to Lapland and Peru to do just this. The results showed a flattening, but of only about 1 part in 300. The current value is 1 part in

298.257, and many geoids are presented in terms of elevation above or below this reference ellipsoid.

As additional geodetic surveys, gravity surveys, and satellite orbit determinations were done, knowledge of the geoid evolved. It is now known that the Indian Ocean just off the southern tip of India is about 100 meters beneath the ellipsoid. Other ocean lows exist in the western North Atlantic Ocean (⁻50 meters), the eastern North Pacific Ocean (⁻50 meters), and the Ross Sea near Antarctica (⁻60 meters). On the continents, lows are present in central Asia (⁻60 meters) and northern Canada (⁻50 meters). High areas of the geoid occur over New Guinea (⁺75 meters), southeast of Africa halfway to Antarctica (⁺50 meters), in the North Atlantic Ocean (⁺60 meters), in western South America (⁺40 meters), and in southern Alaska (⁺20 meters). These highs and lows dominate the geoid. Their existence and locations have been known since the 1970's.

UNDULATIONS AND GEOID ANOMALIES

The huge areas over which individual highs and lows extend require that they be produced by large, deep-seated density variations. Their existence suggests an Earth that is not in hydrostatic equilibrium, which in turn suggests that they result from density variations that do not persist for more than a few hundred million years. Therefore, mantle convection seems to be the most likely cause of these undulations, and many geophysical studies of the long-wavelength undulations of the geoid have concentrated on determining what they tell us about this convection. In general, lows on the geoid are above areas of rapid spreading, and highs are above subducted slabs. Although still evolving and hence subject to change, most of these investigations seem to suggest that mantle convection is driven by descending, not ascending, plumes; that convection involves the whole mantle, not just the seismically active (less than 670 kilometers deep) mantle; and that viscosity increases by a factor of about ten at some depth, probably 670 kilometers.

Other research involves the smaller-wavelength geoid anomalies, particularly over ocean areas. The data are filtered to remove the larger effects, and what remains is usually an excellent indicator of sea-bottom topography. The additional mass produced by a mountain on the seafloor attracts extra water, which piles up and causes a high on the geoid. The geoid is depressed above trenches because trench areas have less mass than the normal seafloor, so they attract less water. This small-wavelength low, with an amplitude of ten meters or so, is usually superimposed on a much larger wavelength high produced by the huge mass excess of the cold slab descending into the mantle nearby.

Undulations of the geoid give some of the best evidence for lateral density variations in the mantle. Seismic data also reveal lateral variations within the mantle, but these are variations in seismic wave velocities, which may or may not correspond directly to density variations. Eventually, the two lines of investigation promise to reveal the inner workings of the planet.

LAPLACE'S EQUATION

Because gravity depends on the shape of the geoid, if there were enough accurate determinations of gravity at sea level, mathematical manipulations could be performed to find the shape of the geoid. However, gravity measurements taken above sea level can be adjusted in ways that convert them to equivalent values for sea-level readings. By the middle of the twentieth century, great progress had been made in mapping the gravity field over much of North America and Europe. This gave some indication of how the geoid undulated locally, but accurate determinations of geoid heights actually require considerable knowledge of gravity from around the entire Earth. When the first satellite was launched in 1957, a new technique suddenly became available that permitted global gravity—and geoid height—to be calculated.

A satellite's orbit is influenced by the distribution of mass beneath it. The motion of the satellite in its orbit and the precession (gyration) and nutation (wobble) of the orbit are all influenced by the gravitational field it experiences. If the orbit is carefully tracked, it reveals much about the Earth's shape and gravity. This requires a considerable mathematical effort, seeking solutions to a partial differential equation called Laplace's equation. These solutions take the form of coefficients of Legendre polynomials and associated polynomials. To see how they can describe the geoid,

consider a surface suspended in space above a chessboard.

To describe the topography of this surface, each square could be designated by its row and column number and its elevation listed. However, there is another way to do this. The description can begin with the average elevation for the whole surface. Once that is determined, the surface is divided into quarters labeled "1" through "4"; the difference between the average elevation for each quarter and the average for the whole surface is then listed. Each quarter is then divided into four parts labeled "A" through "D," and the difference between the average elevation for each part and the average for the whole quarter is again listed. If this is done once more, labeling the divisions "a" through "d," elevation data will exist for all sixty-four squares of the chessboard. If the quarters are numbered clockwise from the upper left, then the square in row 3, column 7 would be designated 2Ca, and its elevation would equal the total surface average plus the difference for quarter 2, plus the difference for sixteenth 2C, plus the difference for sixty-fourth 2Ca.

One advantage this system has is that the earlier results do not change as the description gets more and more detailed. The geoid uses a similar system; however, it divides a spherical grid rather than a square one, so instead of repeatedly quartering, it uses Legendre polynomials. Instead of dividing the board horizontally, the geoid goes to the next degree; instead of dividing the board vertically, the geoid goes to the next order. A geoid determined to degree and order 12 will have much more detail than one determined to degree and order 6, but the coefficients for the first six degrees and orders should be the same.

SATELLITE DATA

As satellites have been tracked through more orbits by better technology, the geoid has become known to increasingly better precision. It was known to degree and order 8 by 1961, 16 by 1971, and 20 by 1985. Current models use supercomputers to work through 10,000 parameters, yielding some results to degree and order 360. There are differences between them; however, typically they vary on the order of half a meter at degree and order 50 or so. Much of the difficulty lies in correctly merging data from more than thirty satellites and tens of thousands of individual gravity measurements made on land.

Satellite data are influenced by factors such as atmospheric drag and refraction effects, and considerable thought and engineering have gone into finding ways around such problems. Geodetic satellites must be in near-Earth orbits in order to be influenced significantly by the undulations in the geoid, and in such orbits atmospheric drag is significant. One approach has been to design satellites out of uranium, which is denser than lead, giving them a great mass for a small cross-sectional area. Another solution has been to suspend a massive core inside a spherical shell. Atmospheric drag will act on the shell, pushing it closer to the core on the side experiencing the drag. Appropriate use of thrusters built into the shell can then ensure that these effects are perfectly canceled. Refraction effects also result from the atmosphere and can be minimized by determining the satellite's position relative to other satellites in higher orbit. Lasers are used for this, reflecting off corner cubes embedded in the outer shell of the satellite.

Ground-based gravity data are influenced by instrument design, operator expertise, and the care with which topographic effects have been removed. Accurate elevations are essential, and much of the Earth's land surface has not been mapped with sufficient topographic precision to provide this control. This, combined with political and economic realities, has resulted in a data set that includes very little information from Asia and not enough from South America. The Global Positioning System (GPS) constellation of satellites promises great improvement in the data set by providing better satellite tracking as well as better elevation control in poorly mapped areas.

PREDICTIVE VALUE

Better knowledge of the geoid will lead to better understanding of plate motions and convection in the mantle, permit more accurate placement of satellites in orbit, and provide an enhanced base for surveys on Earth. One of the outstanding questions in geophysics concerns the scale of convection in the mantle. No one knows if the convection cells extend throughout the entire mantle or if there are two convection regimes, one above a depth of 670 kilometers, the other below it. Unequivocal answers have yet to be found, but

the study of the geoid offers great promise. Models for most of the highs of the geoid use the mass of the descending slabs to generate them. It appears that a slab reaching only to depths of 670 kilometers may not have sufficient mass. This somewhat tentative result favors deep mantle convection. As knowledge of the geoid improves, particularly in continental areas for which only sparse gravity data exist, the validity of this result may be established. An improved grasp of convection in the mantle will increase understanding of the plate-tectonic theory. As that develops, it may well produce a capability to predict earthquakes and volcanoes and to mitigate the death and destruction they cause.

As communications technology develops, the need for many more satellites to relay the ever-increasing traffic will grow. Just as minor perturbations in satellite orbits help us to define the geoid, a better understanding of the geoid will permit much better predictions of just how a satellite will behave in orbit. This is true even for the very high communications satellites. As the population of satellites grows, the need for such refinements in orbit calculations will become necessary to avoid collisions.

Ocean currents such as the Gulf Stream are powered by differences in elevation of the sea surface. These differences, on the order of a meter or so, occur because huge, warm lenses of water float on denser water below. Water at the surface, which in this case is not an equipotential surface, tries to flow down the slope, is affected by the Earth's rotation, and ends up going around the lens of warmer water instead. Measuring such tiny variations in the level of the sea is difficult. As our knowledge of the geoid improves, however, we should be able to observe these discrepancies. This should provide important data from which changes in ocean currents can be predicted. Finally, because all surveys on land measure elevations with respect to sea level, improvements in charting sea level will make the elevation measures on land much better.

Otto H. Muller

CROSS-REFERENCES
Earth Tides, 101; Earth's Rotation, 106; Earth's Shape, 111; Earth's Structure, 37; Gravity Anomalies, 120; Isostasy, 125; Plate Margins, 73; Plate Motions, 80; Plate Tectonics, 86; Sea Level, 2156.

BIBLIOGRAPHY

Davidson, Jon P., Walter E. Reed, and Paul M. Davis. *Exploring Earth: An Introduction to Physical Geology.* Upper Saddle River, N.J.: Prentice Hall, 1997. An excellent introduction to physical geology, this book explains the composition of the Earth, its history, and its state of constant change. Intended for high-school-level readers, it is filled with colorful illustrations and maps.

Fowler, C. M. R. *The Solid Earth.* Cambridge, England: Cambridge University Press, 1990. This college textbook contains a brief but easily understood treatment of the geoid and examples of geoid height-anomaly calculations.

Garland, G. D. *The Earth's Shape and Gravity.* Oxford, England: Pergamon Press, 1977. An excellent, easy-to-read general treatment that describes the geoid and the techniques and data manipulations needed to obtain it. Although somewhat dated (the geoid map included is from 1961), the treatment is thorough. Potential theory is discussed in an appendix, and the book tries to avoid higher math as much as possible, making this the best source for a quantitative introduction to the subject.

Hamblin, William K. *Earth's Dynamic Systems.* 8th ed. Upper Saddle River, N.J.: Prentice Hall, 1998. This geology textbook offers an integrated view of the Earth's interior not common in books of this type. The illustrations, diagrams, and charts are superb. Includes a glossary and laboratory guide. Suitable for high school readers.

Heiskanen, Weikko A., and Helmut Moritz. *Physical Geodesy.* San Francisco: W. H. Freeman, 1967. The standard text on this subject for many years, this book presents a thorough treatment of the techniques and theoretical bases for classical geodesy. Somewhat heavy with equations.

King-Hele, Desmond. *Scientific American*, October, 1967. This easy-to-read paper captures the excitement and sense of discovery shared by scientists working with the early satellites. Although the data set has grown tremendously since this paper was written, most of the major features of the geoid can be seen on the map.

Lambeck, Kurt. *Geophysical Geodesy.* Oxford, England: Clarendon Press, 1988. The best book available on the subject, it includes a detailed treatment of the geoid and how it is obtained. Considerably better than other references in explaining satellite techniques and dynamic ocean topography. Although rich in equations, many containing Legendre polynomials, most of the relevant concepts and results are also described in words. Contains many maps, including one of global gravity and one of geoid heights from 1985.

Murthy, I. V. *Gravity and Magnetic Interpretation in Exploration Geophysics.* Bangaloree: Geological Society of India, 1998. This book is an excellent source of information about gravity and geomagnetism. In addition to useful illustrations, the book comes with a CD-ROM that complements the information in the chapters. Bibliography and index. Intended for the reader with some Earth science knowledge.

Smith, James. *Introduction to Geodesy: The History and Concepts of Mode Geodesy.* New York: Wiley, 1997. Geared toward the college student, this book provides a nice introduction to the study of the Earth's shape and rotation. Includes many illustrations and maps that add clarity to key concepts. Bibliography and index.

Stacey, Frank D. *Physics of the Earth.* 2d ed. New York: John Wiley & Sons, 1977. A standard geophysics textbook for many years, this book examines the geoid and its implications for geophysical models of Earth. Contains a map of a geoid of 1971.

Tsuboi, Chuji. *Gravity.* London: George Allen & Unwin, 1983. A remarkable book by an accomplished Japanese scientist. The geoid is given unusual treatment, using Cartesian, cylindrical, and spherical approaches. Although quantitative and rigorous, many of the treatments are practical, drawn from work done by the author in the first half of the twentieth century. Almost entirely devoted to presatellite work.

GRAVITY ANOMALIES

Gravity anomalies represent variations in measured gravity from the values that are expected for specific locations and elevations on the Earth. Gravity anomalies reveal changes in the density of rocks in the subsurface and were the first evidence for the existence of mountain roots and the differences between oceanic and continental crust.

PRINCIPAL TERMS

AMPLITUDE: the positive or negative value (intensity) of an anomaly as measured against the background values in the region

BOUGUER GRAVITY: a residual value for the gravity at a point, corrected for latitude and elevation effects and for the average density of the rocks above sea level

DENSITY: the mass of a specific volume of a given material

FREE-AIR GRAVITY: a residual value for the gravity at a point, corrected for latitude and elevation effects; this value allows the scientist to determine differences in the densities of subsurface rocks

HALF-WIDTH: the distance over which the amplitude of an anomaly falls from its maximum value to half the maximum amplitude

ISOSTASY: the concept of balance by which continental and oceanic crust are "floating" on the denser substrate of the mantle

MILLIGAL: the basic unit of the acceleration of gravity, used by geophysicists in measurement of gravity anomalies equal to 0.001 centimeter per second squared

WAVELENGTH: the distance over which an anomaly rises to its maximum amplitude and falls again to background values

GRAVITY VARIATION FACTORS

A gravity anomaly represents a departure from the expected value of the acceleration of gravity at any point on the Earth's surface. In general, such departures are small compared to the total gravity of the Earth, which averages 980 centimeters per second squared. The actual value varies as a function of latitude and elevation. This variation occurs because the Earth is not a perfect sphere but a spheroid of revolution. The equatorial radius is 6,378 kilometers, 21 kilometers longer than the polar radius of 6,357 kilometers. Since gravity decreases over distance, the gravity at the equator is less than that at the pole. Added to this is the effect of the Earth's rotation. Together, these effects lead to gravity values of 978.0490 centimeters per second squared at the equator and 983.2213 centimeters per second squared at the pole, with a value for any latitude between predicted by a simple formula. Latitude explains the largest variation in gravity values of the Earth. A second major effect results from elevation, which brings about a decrease in gravity of approximately 0.094 centimeter per second squared for every thousand feet of elevation above sea level.

The gravity at two points on the Earth's surface will depend on latitude and elevation effects and on the densities of the rocks beneath the two points. The densities are of particular interest to the geophysicist. In order to evaluate these densities, the gravity values measured at two points on the Earth's surface must be corrected for the latitude and elevation effects.

FREE-AIR AND BOUGUER ANOMALIES

The correction for the shape and rotation of the Earth is made with the simple formula mentioned above and has the effect of reducing the effect of latitude difference to zero. After this correction has been made, differences in gravity measured at two points may be attributed to differences in elevation and variations in the density of the underlying rocks. The correction for elevation is made relative to sea level and may be divided into two parts. The first involves the effect of being farther from the center of the Earth as a result of the elevation. This correction is called the free-air correction, because it corrects the effect of distance as if there were only air between the point on the surface of the Earth and sea level. The second part of the elevation correction involves the subtraction of the effect of the slab or rock between the surface and sea level. This latter correction is termed the Bouguer correction.

Gravity values corrected for latitude and incorporating the free-air correction are termed the free-air gravity or free-air anomaly. Gravity values corrected by latitude and by free-air and Bouguer corrections are called the Bouguer gravity values or Bouguer anomaly. Bouguer gravity values may also involve corrections for irregular topography and the curvature of the Earth.

ROCK DENSITY

After the measured gravity values have been corrected, all differences caused by latitude, elevation, and topography have been removed mathematically, and the residual gravity anomalies reflect the lateral changes in rock density at depth. These gravity anomalies are typically small, usually less than ±0.2 centimeter per second squared or

±200 milligals—representing the differences between measured and ideal gravity values.

The value of the anomaly is related directly to a surplus or deficiency in mass in the subsurface. The effect of this mass difference is related to the acceleration of gravity by Isaac Newton's equation $\delta g = G^{\delta M} r2$, where δg is the difference in gravity values over what is expected, G is the universal gravitational constant, r is the distance from the anomalous mass to the point on the surface of the Earth where the gravity is measured, and δM is the increased or decreased mass.

This mass is usually expressed as the product of the change in density ($\delta\sigma$) times volume, or $\delta M = \delta\sigma \times V$. The density difference is the physical property of the rocks in the subsurface that causes the gravity anomaly. The amplitude and wavelength of the anomaly are caused by the size of the density contrast, the size and shape of the body, and the depth of the body.

This relationship can be illustrated by use of the simplest shape that may cause a gravity anomaly: a sphere. The figure shows spheres of differ-

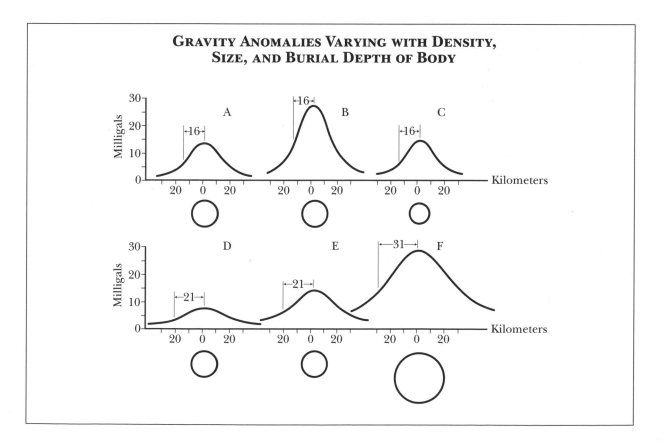

GRAVITY ANOMALIES VARYING WITH DENSITY, SIZE, AND BURIAL DEPTH OF BODY

ent densities and sizes that are buried at different depths and the gravity anomalies associated with them. The anomalies have a maximum amplitude directly above the centers of the buried spheres, and the amplitudes decrease away from the bodies. In the figure, the horizontal axis marks distance in kilometers right and left of the center of the spheres, and the vertical axis marks the anomaly amplitude in milligals. In the first panel, three spheres are buried with their centers at the same depth of 20 kilometers. Most of the rocks in the area have a density of 2.7 grams per cubic centimeter; this value represents the background value of density in the example. Sphere A has a radius of 10 kilometers and a density of 2.9 grams per cubic centimeter, giving a positive density contrast, δσ, of 0.2 gram per cubic centimeter. The excess mass causes a positive gravity anomaly with a maximum amplitude of 13.97 milligals. The second sphere, B, has a density of 3.12 grams per cubic centimeter, a density contrast of 0.42 gram per cubic centimeter, and a 10-kilometer radius, causing a maximum anomaly value of 27.24 milligals. The third sphere, C, has a density of 3.12 grams per cubic centimeter but is smaller, with a radius of only 8 kilometers. This sphere causes an anomaly of 13.95 milligals. Even though sphere C has the same density as sphere B, its size, and therefore its mass, is less, and the resulting anomaly is less.

Comparison of anomalies A and C shows them to be nearly identical. This occurs even for bodies of different densities when the product of volume times density yields the same mass. Note that the half-width of anomalies A, B, and C are all the same, with a value of 16 kilometers. This occurs because the center of mass of all three spheres is buried at 20 kilometers.

In the figure's second panel, spheres D and E are the same sizes and densities as A and B, respectively, but are buried at 28 kilometers depth. The amplitudes of each anomaly are diminished as a result of the greater distances to the bodies. Furthermore, the wavelengths and the half-widths of the anomalies are increased. Here, the half-widths are approximately 21 kilometers. This example shows that a pair of anomalies such as A and E with approximately the same amplitude will have different wavelengths or half-widths if the bodies causing the anomalies are at different depths.

The last sphere is buried at a depth of 40 kilo-

meters with a radius of 20 kilometers and a density contrast of 0.2 gram per cubic centimeter. The maximum amplitude of the anomaly is 27.94 milligals, close to the amplitude of sphere B, but the half-width of the anomaly is 31 kilometers, nearly twice that of anomaly B.

While most rock bodies do not approximate spherical shapes, the above relationships show several general principles of gravity interpretation. The anomaly occurs because of an excess (or deficit) of mass in the subsurface. The amplitude is controlled by the density, size, and depth of occurrence of the rock body. The wavelength and half-width are related to the depth of the body. The longer the wavelength and half-width, the greater is the depth of the source of the anomaly.

More complexly shaped geologic bodies cause more complex anomalies, and more complex equations are used to describe them. Shapes often used to simulate geologic bodies include cylinders, slabs, and three-dimensional prisms. The power of the gravity method involves the use of equations to calculate an anomaly that matches as closely as possible the gravity values measured in the field. The match allows the scientist to infer much about the character of rocks at depth.

DETERMINING ROCK DENSITY

Gravity anomalies show the density variations in the subsurface related to the occurrence of specific rock types. Practically, these variations must involve a lateral change in density from one place to another. Geologists and geophysicists try to understand the variation in terms of specific rock types. The densities of rocks vary as a function of composition, mineralogy, the occurrence of open spaces, and physical conditions such as temperature and pressure. Among sedimentary rocks, sandstone has a density range of 2.35-2.55 grams per centimeter cubed; shale, 2.25-2.45; limestone, 2.45-2.65; and loose sand, 1.90-2.00. Igneous rocks include granite, with a density range of 2.60-2.80 grams per centimeter cubed; gabbro, 2.85-3.10; and peridotite, 3.15-3.25. Metamorphic rocks include the following density ranges: granite gneiss, 2.60-2.70; schist, 2.70-2.90; amphibolite, 2.80-3.10; eclogite, 3.30-3.45; and marble, 2.70-2.75.

Using these data, one can get an idea of the rocks that are likely to cause a positive or negative gravity anomaly. If one were to rely only on these figures in interpreting rock types from gravity data, however, one would find a great many possibilities because of the overlapping of density ranges for different rock types. Thus, a limestone may have a density similar to that of a granite or a gneiss. For this reason, the geophysicist must be guided by what is known about the geology in the area being studied and by careful measurements of the actual densities for rocks from outcrop or drill holes in the region.

GRAVITY MODELING

The boundaries of the shapes of the rock bodies, which are mathematically determined by gravity modeling, are interpreted as having a geologic significance. Thus, these boundaries, representing the contact between rocks of different densities, may be interpreted as faults, intrusive contacts, unconformities, or normal depositional contacts, depending on what else is known about the geology.

By tracking changes in acceleration as a satellite orbits a planetary body, space scientists are able to map variations in the gravity field. These observations led to the discovery of lunar mascons (apparent mass concentrations in the lunar near-surface rocks). Mascons were eventually determined by gravity modeling to be caused by relatively thin layers of dense basalt pooled in large lunar basins mirrored by upward migration of dense mantle material. Gravity maps have been constructed for Mars and Venus by satellite measurements as well.

GRAVITY ANOMALY PATTERNS

The largest differences in gravity occur as gravity anomalies between the ocean basins and the large continental masses. Ocean basins have positive gravity values, related to the dense rocks that underlie the oceans. Continents, on the other hand, have negative Bouguer gravity values, which reveal that thick sections of low-density rock occur beneath the continents.

The relationship between the density of underlying rock and elevation is a very general one that applies to most areas of the Earth. As early as 1850, measurements had shown that mountains were underlain by rocks less dense than those underlying the surrounding lowlands. The relationship was explained by two models. The first was that continents and mountains were high because

they were underlain by thicker sections of low-density rocks. Seismic data on the depths to the base of the continental crust have confirmed this model in most places. The second model suggested that mountains were high relative to lowlands because mountains are less dense than are the rocks under the lowlands. This relationship has also been verified in a number of areas and does explain the lower elevations found in some rift valleys.

Scientists have found that the patterns of gravity anomalies in a large area may give a distinctive grain or "fabric" to regions on the continents. The patterns in a particular region may involve a series of positive and negative anomalies of a certain amplitude aligned in a particular direction. Adjacent regions may have groups of anomalies of different amplitudes, wavelengths, or orientations that contrast with one another in the same way that the different sections of a quilt stand out against one another. The distinctive regional character of gravity in many places allows geologists to divide the crust into provinces. Other geologic observations and age determinations have shown these provinces to be pieces of crust that were assembled over time to make the continents.

GEOPHYSICAL APPLICATIONS

Gravity anomalies are a major source of geophysical understanding of the Earth, but there are uncertainties in the interpretation and modeling of gravity data. These uncertainties are lessened by the use of geologic information and other geophysical survey techniques. Magnetic, electrical, or seismic data can provide additional information of the depth, size, and shape of the body in the subsurface. Knowledge of these variables improves the scientist's ability to define rock density and the rock type.

Gravity anomalies have helped to define the compositional and structural differences between oceanic and continental crust and the crustal thickening that occurs under mountain ranges. The patterns of gravity anomalies also reveal the internal structure of continents and the stages of continental development. The occurrence of gravity anomalies associated with the oceanic trench systems and island areas also attested the dynamic character of these features and constituted evidence for the theory of plate tectonics.

ECONOMIC APPLICATIONS

Of major importance is the application of gravity anomalies to natural resource discovery and evaluation. Many economically important features are related to changes in rock density and are detectable using gravity measurements. Lateral changes in density may occur in areas where anticlines or faulting have formed traps for oil and gas. These features may cause either positive or negative anomalies. One of the most successful of these applications is in the energy industry, where a large number of producing fields are related to thick, intrusive masses of salt called salt domes. The low-density salt moves upward in the sedimentary section, creating folds and faults that are excellent traps for oil and gas. The negative gravity anomalies associated with the salt domes have been used to locate pools of oil and gas in this geologic environment.

A second application of economic importance is in the location and evaluation of groundwater aquifers in certain parts of the country. Aquifers are often found where Quaternary (about 2 million years to the present) sand and gravel deposits are especially thick. Since the density of sand and gravel is much less than that of rock, thick sections cause negative gravity anomalies proportional to aquifer thickness. Information from gravity surveys may thus be useful in land-use planning and development.

Donald F. Palmer

CROSS-REFERENCES

Aquifers, 2005; Continental Crust, 561; Continental Structures, 590; Earth Tides, 101; Earth's Rotation, 106; Earth's Shape, 111; The Geoid, 115; Isostasy, 125.

BIBLIOGRAPHY

Bott, M. H. P. *The Interior of the Earth: Its Structure, Constitution, and Evolution.* London: Edward Arnold, 1982. This text emphasizes the use of geophysics in the interpretation of the major compositional and structural components of the Earth. The treatment is generally nonmathematical. The text integrates different geophysical methods pertaining to the different parts of the crust, mantle, and core. Discussion of gravity anomalies is found in sections dealing with general crustal structure, mountain ranges, rift systems, the mid-ocean ridges, continental margins, island arcs, and global gravity variations resulting from mantle inhomogeneities.

Brown, G. C., and A. E. Mussett. *The Inaccessible Earth: An Intergrated View to Its Structure and Composition.* London: Chapman and Hall, 1993. This book deals well with geophysics topics as they relate to the structure, chemical composition, and evolution of the Earth. Its excellent line drawings helpfully illustrate complex ideas.

Dobrin, M. B., and C. H. Savit. *Introduction to Geophysical Prospecting.* 4th ed. New York: McGraw-Hill, 1988. The newest edition of one of the most popular books in the application of geophysics to natural resource investigations. The majority of the book deals with seismic exploration. Three chapters on gravity prospecting treat the subject quite adequately. The mathematical treatment is good. Equations for gravity modeling of simple shapes are given.

Hamblin, William K. *Earth's Dynamic Systems.* 8th ed. Upper Saddle River, N.J.: Prentice Hall, 1998. This geology textbook offers an integrated view of the Earth's interior not common in books of this type. The illustrations, diagrams, and charts are superb. Includes a glossary and laboratory guide. Suitable for high school readers.

Jacobs, John A. *Deep Interior of the Earth.* London: Chapman and Hall, 1992. An informative, mathematical treatment of the global variations of gravity as a function of latitude, elevation, and topography. Also provides an overview of the Earth's structure and composition.

Lutgens, Frederick K., and Edward J. Tarbuck. *Earth: An Introduction to Physical Geology.* 6th ed. Upper Saddle River, N.J.: Prentice Hall, 1999. This college text provides a clear picture of the Earth's systems and processes that is suitable for the high school or college reader. In addition to its illustrations and graphics, it has an accompanying computer disc that is compatible with either Macintosh or Windows. Bibliography and index.

Robinson, Edwin S., and Cahit Coruh. *Basic Exploration Geophysics*. New York: John Wiley & Sons, 1988. The text includes three chapters on gravity, with clear descriptions and excellent diagrams that deal with the geologic interpretations that can be made based on gravity modeling.

Smith, P. J. *Topics in Geophysics*. Cambridge, Mass.: MIT Press, 1973. Contains an excellent, mathematically simple explanation of the free-air and Bouguer corrections, the principles of gravity modeling, and continental structures. This text is especially good for students at the high school level.

Stacey, F. D. *Physics of the Earth*. 2d ed. New York: John Wiley & Sons, 1977. A mathematically rigorous treatment of the gravity of Earth and isostasy. This text is best used by those who have a background in calculus and basic physics.

ISOSTASY

Isostasy is a principle that describes the vertical positioning of segments of the Earth's lithosphere relative to one another in terms of the elevation of the land and depth to the top of the asthenosphere. It is, in effect, a restatement of Archimedes' principle or an application of that principle to the outer layers of the Earth.

PRINCIPAL TERMS

ARCHIMEDES' PRINCIPLE: the notion that a solid, floating body displaces a mass of fluid equal to its own mass

ASTHENOSPHERE: the layer immediately underneath the lithosphere, which acts geologically like a fluid

COLUMN: a cylindrical segment of the Earth oriented on a line from the center of the Earth to any point on its surface, beginning somewhere in the asthenosphere and ending somewhere within the atmosphere

DENSITY: a term referring to the amount of mass per unit volume of a substance

LITHOSPHERE: the outermost solid layer of the Earth

SEA LEVEL: the position of the surface of the ocean relative to the surface of land

SUBSIDENCE: the sinking of the Earth's surface or a decrease in the distance between the Earth's surface and its center

UPLIFT: the rising of the Earth's surface or the increase in distance between the Earth's surface and its center

VISCOSITY: the ability of a fluid to flow

ARCHIMEDES' PRINCIPLE

Isostasy, sometimes called the doctrine or principle of isostasy, is a fundamental principle of the Earth sciences that describes the spacial positioning of lithospheric mass within the Earth. Isostasy requires that the total mass of air, water, and rock within any vertical column extending from within the asthenosphere, through the lithosphere, to within the atmosphere is equal to the total mass of any other column in the same area of the Earth, extending from the same depth in the asthenosphere to the same elevation in the atmosphere. The concept of isostasy is analogous to the concept of buoyancy in physics. Buoyancy was first explained by Archimedes, who, as legend tells it, lowered himself into bathwater, observed the level of the water rise against the wall of the bath pool, and thus realized that ships float because they displace a mass of water equal to the mass of the ship. This discovery came to be known as Archimedes' principle.

Many centuries later, scholars realized that Archimedes' principle could be used to explain why the Earth has both high mountain ranges and deep ocean basins. The main obstacle to the acceptance of the principle was the belief of early scholars that the Earth was a solid, rigid body. The idea that the ground on which they stood could be compared to a boat floating on the sea was totally beyond their comprehension. The knowledge needed to draw that analogy did not become available until the mid-nineteenth century, when British surveyors under the direction of Sir George Everest were engaged in the trigonometrical survey of India near the Himalaya. The surveyors noted that the distance between the towns of Kalianpur and Kaliana, when measured by triangulation methods, differed by 5.236 seconds of arc, or about 160 meters from the distance when measured by astronomical methods. Two British scholars, George Biddell Airy and John Henry Pratt, realized the cause of this apparent error, though each provided different interpretations of the geologic conditions that gave rise to the difference in distances. Their interpretations later came to be known as the Airy hypothesis and the Pratt hypothesis of isostasy.

AIRY AND PRATT HYPOTHESES

While both Airy and Pratt applied the Archimedes' principle to explain the elevation of the Himalaya and the discrepancy in distance between

128

the two survey methods, their hypotheses differed in the way they explained how the mass is distributed below the mountains. Airy viewed that apparent mass deficiency below the mountains as a result of the mountains having a root of low-density rock that extends well into a lower-lying, denser, fluid layer upon which the mountains and all other surficial rock layers float. This lower-density mountain and mountain root combination were envisioned as being like a boat floating upon a denser fluid, which Airy thought was lava. The "boat" was thus made buoyant by the root's displacing a mass of the fluid equal in mass to the combined mountain and mountain root. To Airy, the higher the mountain, the deeper the root must extend to compensate for the elevated mass. By analogy, of two vessels of the same areal extent, one tall and the other of low profile, the tall vessel projects deeper into and rises higher out of the water. Pratt saw the situation somewhat differently. He maintained that the position of the base of the solid crust must be the same everywhere. The differences in surficial elevations, Pratt thought, arise from some areas having experienced less "contraction" than other areas during the cooling of the Earth. These areas of less contraction are also of less density and float higher in accordance with Archimedes' principle. Regional variations in surface elevation, according to the Airy hypothesis, result from variations in the thickness of the solid outer layer of the Earth. Airy thought the density of the outer layer was the same everywhere, but the position of the base varies according to the magnitude of surface elevation. According to the Pratt hypothesis, the regional variations in surface elevation result from variations in density of the solid outer layer, with the base of that layer being of equal position everywhere. The Pratt model has a flat-bottomed crust.

Elevated terrains in both the Airy and the Pratt models have less mass near the surface than low-lying terrains. Therefore, according to these hypotheses, the plumb bob was pulled by gravity away from the area of the mountains and toward the lower-lying plains of India. Clarence Edward Dutton in 1889 recognized the significance of this variation in the amount of mass near the surface and concluded, "Where the lighter matter was accumulated there would be a tendency to bulge, and where the denser matter existed there would be a tendency to flatten or depress the surface." Dutton coined the term "isostasy" for this definition between land surface elevation and rock mass as mandated by Archimedes' principle.

ISOSTATIC COMPENSATIONS

Earth scientists acknowledge the validity of both the Airy and the Pratt hypotheses. They consider large-scale or regional land surface elevation variation to result from variations in density and thickness of the lithosphere and also from variations in density of the asthenosphere. Furthermore, Earth scientists recognize that the density and/or thickness of the lithosphere at any particular place can change through time and thus result in vertical movements of lithospheric plates to compensate for these changes. If one were to heat a solid object, such as a steel pipe, it would expand and thus decrease its density and increase its length. If a segment of the lithosphere of the Earth were to be heated, the rock within that segment would also become less dense, the thickness of the lithospheric segment would increase, and the elevation of the land surface of that segment would rise in accordance with both the Airy and the Pratt hypotheses. This rising of the land surface is known as uplift. Similarly, if a segment of the lithosphere were to cool, the rock would increase in density, the thickness would be reduced, and the elevation of the surface would be reduced. This reduction of land surface elevation through time is referred to as subsidence. If cargo were added to the deck of a boat, it would be seen to ride lower in the water. The top of the cargo, however, would be at a greater distance above the water. Removing deck cargo has the opposite effect. If a segment of the lithosphere were to have sediment deposited on its surface, its base would project a greater distance into the asthenosphere, and its top would be described as being at a greater elevation. If material is removed from the lithosphere by erosion, the base of the lithosphere rises and the land surface elevation decreases.

The vertical adjustments in the position of the lithosphere to maintain equilibrium are referred to as isostatic compensations. To be in equilibrium, the total amount of mass within a column of the Earth that extends from within the atmosphere, through the hydrosphere and lithosphere, and into the asthenosphere must be equal to the total mass of any other column of the same areal range that extends from the same elevation in the atmosphere to the same depth in the asthenosphere. A change in the lithosphere within a column in terms of mass or density will be compensated by changes in the mass of the atmosphere, hydrosphere, and asthenosphere.

When sediment or rock is deposited or eroded from the top of the lithosphere, mass is added or subtracted from the lithosphere, and isostatic adjustments are made to compensate for this change. If sediment is deposited upon the surface of the lithosphere, the added load displaces some of the asthenosphere; thus, there is less asthenospheric mass in the column. If the top of the lithosphere were below sea level, then hydrospheric mass would also be displaced; if sediment accumulated until it were stacked above sea level, then mass within the atmosphere would be displaced also. If sediment or rock is eroded from the top of the lithosphere, the base of the lithosphere rises, and mass is added to the asthenospheric portion of the column. If the top of the column were initially above sea level, then the mass of the atmospheric portion of the column would increase; if erosion cut below sea level, then hydrospheric mass would be added to the column. Depositional isostatic subsidence is seen along continental margins such as the Gulf coast of Texas, where there are great accumulations of sediment. Isostatic rebound is associated with erosion (melting) of the Pleistocene ice sheets. Such glacio-isostatic rebounds have been measured in eastern North America and Northern Europe.

THERMO-ISOSTATIC UPLIFT AND SUBSIDENCE

When the lithosphere is warmed or cooled, the situation becomes more complex. The warming or cooling of the lithosphere is geologically accomplished by changes in the temperature of the asthenosphere. Therefore, the density of both the lithosphere and asthenosphere would be expected to vary with temperature changes. If temperature change is the only process operating, then the mass of the lithosphere is constant regardless of its temperature; its thickness and density, however, would have changed. If the lithosphere is warmed, the mass in the hydrosphere and/or atmospheric portions of the column would have decreased in an amount equal to the increase in mass of the asthenospheric portion. The net effect is an increase in the elevation of the land surface, or thermo-isostatic uplift. If the lithospheric portion of the column were to cool, there would be an increase in the mass of the hydrospheric and/or atmospheric portions of the column and a decrease in the asthenospheric portion. The net effect would be a decrease in land elevation, or thermo-isostatic subsidence. Isostatic uplift is seen in the area of the Mid-Atlantic Ridge, the greatest mountain range on the surface of the Earth. It also may explain why the continent of Africa has such a greater average elevation relative to sea level than do the other continents. Isostatic subsidence has been suggested to be the underlying mechanism for the formation of the thick sediment accumulations within continental areas. The Michigan Basin and the Williston Basin in North America are examples of these accumulations.

The processes that give rise to isostatic adjustments take millions of years. The resulting isostatic adjustments are also very slow to occur. When a person steps onto a boat, it instantly rides lower in the water, because the compensation of the boat for the additional load is immediate. The medium upon which the boat floats, water, has a very low viscosity. If the boat were afloat in a more viscous fluid, such as cold molasses, the adjustment to the added mass would be noticeably slower, perhaps taking a minute or more. The asthenosphere is very viscous. Consequently, isostatic adjustments to lithospheric changes may take tens of thousands of years.

EVALUATION OF GLACIO-ISOSTATIC REBOUND

Isostasy is a principle or law of the Earth sciences, and, as such, it cannot be collected, observed, or quantified. What can be observed or quantified are the results of lithospheric segments satisfying or attempting to establish isostatic equilibrium. If a geologic process changes the mass or density of a segment of the lithosphere, vertical adjustments in the position of the lithosphere are necessary to reestablish the equilibrium. These vertical adjustments are slow; 1 centimeter per year would be considered fast. To measure the changes in lithosphere position caused by isostatic compensation, one needs a hypothetical measuring stick and a clock. The "stick" in nearly all cases measures the distance between the top of the lithospheric segment and sea level. Because the time over which the adjustment process occurs is quite long, the clock that is used is the decay of radioisotopes, such as carbon 14, potassium 40, and uranium 238.

The application of these tools to the study of isostatic compensation can be illustrated with the evaluation of the phenomenon of glacio-isostatic rebound. During the last glaciation, the Wisconsin, vast sheets of ice covered portions of Antarctica, North America, Europe, and the southern tip of South America. That ice constituted a load on the decks of several lithospheric boats. From 18,000 to 6,500 years ago, most of the ice sheets in North America, Europe, and South America melted. The meltwater increased the volume of water in the oceans. Consequently, the level of the oceans rose 100 meters relative to a fixed point on a landmass that was not glaciated, such as the island of Cuba. From 6,500 years ago to the present, little additional ice has melted. Thus, the amount of water in the oceans has been constant. Sea level, therefore, should have been constant worldwide.

During this period of time, however, sea level has not been constant in those areas where glacial ice had once loaded the lithosphere. In those areas, fixed points on the land surface are rising relative to sea level. Some areas are currently rising at the rate of 2 centimeters per year; other areas have already risen nearly 140 meters. Scientists can determine how far and how fast the lithosphere has rebounded or is rebounding by examining the locations of exposed shoreline sediments or marine terraces. The sediments would have been deposited and the terraces formed by waves on a beach when sea level was at that land point. Part of that sediment would have been the remains of plants and animals that were alive at the time of deposition of the sediment. By surveying the current difference in elevation between the ancient shoreline sediments and the present sea level, scientists can determine the amount of vertical uplift since the sediment was deposited. By determining the radiometric age of the remains of organic life using carbon 14 dating methods, scientists can calculate the length of the time over which that amount of rebound occurred. Several different shoreline deposits or terraces in the same region can reveal different land positions relative to sea level and how the rate of rebound has changed with time.

Geologists can therefore determine the viscosity of the asthenosphere, project how much rebound will occur in the future, and estimate how much rebound will have occurred when isostatic equilibrium is established. This estimate can be translated into how thick the ice was when the glaciers were present. Ice thickness equals the product of the total rebound times the ratio of the density of the asthenosphere to the density of the ice.

INTEREST TO GEOLOGISTS AND HISTORIANS

The relationship between isostasy and the surface of the land is analogous to the relationship between buoyancy and the deck of a ship. Humans can overload the deck of a ship and sink it into the sea, but they cannot overload the lithosphere and sink it into the asthenosphere. This area of nature is one of the few that is not heavily influenced by human activity. If all the engineers of the world used all the earthmoving equipment in the world to pile soil, sediment, and rock in one huge mound, they could not in their lifetimes cause a segment of the lithosphere to ride 1 millimeter lower in the asthenosphere. Nature, however, in a few hundred millennia can pile enough snow and ice on Antarctica to sink land surface so substantially that most of the subice rock surface (the preglaciation top of the lithosphere) now lies below sea level, several hundred meters below where it originally was. Besides geologists and geophysicists, isostasy touches the lives of very few people directly. The notable exceptions are those historians who ponder why certain Viking harbors in Scandinavia are now situated above sea level: The answer pertains to glacio-isostatic rebound.

James A. Dockal

CROSS-REFERENCES

BIBLIOGRAPHY

Condie, Kent C. *Plate Tectonics and Crustal Evolution.* 4th ed. Oxford: Butterworth Heinemann, 1997. An excellent overview of modern plate tectonics theory that synthesizes data from geology, geochemistry, geophysics, and oceanography. A very helpful tectonic map of the world is enclosed. The book is nontechnical and suitable for a college-level reader. Useful "suggestions for further reading" follow each chapter.

Davidson, Jon P., Walter E. Reed, and Paul M. Davis. *Exploring Earth: An Introduction to Physical Geology.* Upper Saddle River, N.J.: Prentice Hall, 1997. An excellent introduction to physical geology, this book explains the composition of the Earth, its history, and its state of constant change. Intended for high-school-level readers, it is filled with colorful illustrations and maps.

Dockal, J. A., R. A. Laws, and T. R. Worsley. "A General Mathematical Model for Balanced Global Isostasy." *Mathematical Geology* 21 (March, 1989): 147. A comprehensive mathematical treatment of isostasy. Discusses the connection between isostatic adjustments and global sea-level changes. Suitable for college-level students with a working knowledge of algebra.

Hamblin, William K. *Earth's Dynamic Systems.* 8th ed. Upper Saddle River, N.J.: Prentice Hall, 1998. This geology textbook offers an integrated view of the Earth's interior not common in books of this type. The illustrations, diagrams, and charts are superb. Includes a glossary and laboratory guide. Suitable for high school readers.

Hart, P. J., ed. *The Earth's Crust and Upper Mantle.* Washington, D.C.: American Geophysical Union, 1969. A somewhat technical book that gathers together many aspects of the crust and mantle or lithosphere and asthenosphere. A chapter by E. V. Artyushkov and Y. U. A. Mescherikov deals quite well with recent isostatic movements and provides a good bibliography of the foreign literature on isostasy. Suitable for college-level students.

Jordan, Thomas H. "The Deep Structure of the Continents." *Scientific American* 240 (January, 1979): 92-107. Discusses new ideas on the makeup and nature of the lithosphere and asthenosphere. Provides considerable insight into how knowlege of the deep Earth is obtained. Suitable for college-level students.

Lutgens, Frederick K., and Edward J. Tarbuck. *Earth: An Introduction to Physical Geology.* 6th ed. Upper Saddle River, N.J.: Prentice Hall, 1999. This college text provides a clear picture of the Earth's systems and processes that is suitable for the high school or college reader. In addition to its illustrations and graphics, it has an accompanying computer disc that is compatible with either Macintosh or Windows. Bibliography and index.

Mather, K. F., ed. *A Source Book in Geology, 1900-1950.* Cambridge, Mass.: Harvard University Press, 1967. A collection of major landmark geologic works dating from 1900 to 1950. Included in this collection is Joseph Barrell's "The Status of the Theory of Isostasy," which summarizes much of the early thinking on isostasy. Other relevant works include the two studies of the properties of the asthenosphere, one by Felix Vening Meinesz and the other by Beno Gutenberg. Suitable for high school and college-level students.

Mather, K. F., and S. L. Mason, eds. *A Source Book in Geology, 1400-1900.* Cambridge, Mass.: Harvard University Press, 1970. A collection of major landmark geologic works dating from 1400 to 1900. Each paper is condensed from its original length. The collection includes the works of G. B. Airy, J. H. Pratt, and C. E. Dutton. Brief biographic sketches are given for the authors. Suitable for high school and college-level students.

Plummer, Charles C., and David McGeary. *Physical Geology.* Boston: McGraw-Hill, 1999. A college-level introductory geology textbook that is clearly written and wonderfully illustrated. An excellent sourcebook of basic information on geologic terminology and fundamentals of geologic processes. Contains CD-ROM. An excellent glossary.

Walcott, R. I. "Late Quaternary Vertical Movements in Eastern North America: Quantitative Evidence of Glacio-Isostatic Rebound." *Review of Geophysics and Space Physics* 10 (November, 1972): 849-884. A review paper that collects and evaluates from published sources the evidence of vertical movements of the lithosphere that are attributed to glacio-isostatic rebound. Charts present accumulated data for sea-level changes in North America. Maps portray the magnitude of rebound. Contains an excellent bibliography on glacio-isostatic rebound. Suitable in part for advanced high school and college-level readers.

4
EARTH AS A MAGNET

EARTH'S MAGNETIC FIELD

The study of the Earth's magnetic field—its origin, its history, and other characteristics—has vast implications for humanity, ranging from the location of ore deposits to the disruption of communication systems. The Earth is the only planet of the inner solar system with a strong magnetic field, which has implications for the formation of the Earth and the solar system as well as for the existence of life.

PRINCIPAL TERMS

DECLINATION: for a particular location on the Earth's surface, the horizontal angle between true north and the compass needle direction

DIPOLE FIELD: the field shape produced by two electrically charged particles, such as a proton and electron, or by two magnetic poles, such as north and south

INCLINATION: the vertical angle between the horizontal plane and the magnetic field

MAGNETIC ANOMALIES: distortions in the magnetic field, produced by an object such as an iron ore body

MAGNETIC STORM: rapid changes in the Earth's magnetic field as a result of the bombardment of the Earth by electrically charged particles from the Sun

MAGNETOPAUSE: the outer boundary of the Earth's magnetic field

MAGNETOSPHERE: the volume of space in which the magnetic field of the Earth is located

SECULAR VARIATION: a change in the Earth's magnetic pole position on the Earth's surface over hundreds of years

A DIPOLE FIELD

The ultimate source of a magnetic field is a moving electrical charge, which is an electrical current as in the electron flow in home wiring. Approximately 90 percent of the Earth's magnetic field found at any location is produced by electrical currents in the Earth's outer core. A dynamo effect is thought to be the origin of this main field.

The main field is predominantly a dipole field, or a field resulting from two charges (the prefix "di" is derived from the Greek word meaning "two")—in this case, two magnetic charges. A bar magnet has a dipole field because it has a north end, representing one charge, and a south end, representing the other charge. A magnetic field line leaves the Earth's surface in the Southern Hemisphere, arcs over the Earth, and enters the Earth in the Northern Hemisphere. On the Earth's surface, many of these lines, emerging and penetrating at angles from 0 to 90 degrees to the horizontal, make up the dipole field of the Earth. The magnetic south pole of the Earth, which lies in the Northern Hemisphere, is located where the lines penetrate the Earth at 90 degrees, and the magnetic north pole (in the Southern Hemisphere) is

where they leave at 90 degrees. The field's strength is 0.6 gauss (the unit of magnetic induction) at the poles and 0.3 gauss at the magnetic equator. (A small magnet has about a 1-gauss field strength.) The difference results from the field lines bunching together at the poles and spreading apart around the magnetic equator. The number of field lines present in a particular area determines the magnetic field strength.

The magnetic poles are not located at the geographical poles of the Earth, which are the points where the imaginary spin axis of the Earth penetrates the surface. They are not stationary; rather, they apparently wander around the polar regions. The south magnetic pole is located approximately 11 degrees from the geographic pole in the islands north of Canada.

Magnetic anomalies contribute another portion of the interior field. They distort the dipole shape of the main field. Some of these anomalies are associated with the outer core. Other anomalies result from rock bodies, with the strongest known located at Kursk in the Soviet Union, 400 kilometers south of Moscow. An anomaly can arise from an igneous rock body such as basalt, which has a

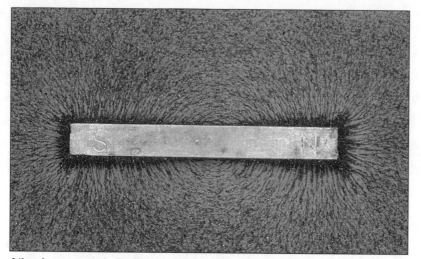

Like a bar magnet, the Earth has a dipole field because it has a north end, representing one charge, and a south end, representing the other charge. A magnetic field line leaves the Earth's surface in the Southern Hemisphere, arcs over the Earth, and enters the Earth in the Northern Hemisphere. (© William E. Ferguson)

SOLAR WIND

Other portions of the field are external to the Earth and tend to vary rapidly in direction and magnitude with a time period of hours to days. The gravitational fields of the Sun and the Moon cause shifts in the atmosphere of the Earth, which move charged particles within the atmosphere, thus producing a magnetic field that changes with the motion of the Sun and Moon.

The Sun pushes electrons, protons, and, to a lesser extent, other particles, outward from its surface. These particles are collectively known as the solar wind, and they hit the Earth's magnetic field at speeds of hundreds of kilometers per second. Because they are charged as well as in motion, they have their own magnetic fields that interact with the Earth's magnetic field and contribute to it. Normally, a magnetic field extends to infinity, but because of these interactions, the Earth's field has a boundary known as the magnetopause that surrounds the magnetosphere, or the magnetic field of the Earth. The solar wind also changes the shape of

large amount of the magnetic iron oxide, magnetite, or, for smaller anomalies, from human-made objects such as a plowshare. An important set of anomalies associated with the oceanic ridges are known as the magnetic seafloor stripes. They support the concept that the Earth's magnetic field has reversed polarity many times in the past and also the concept that plates, or segments of the Earth's outer rock layer, the lithosphere, have moved.

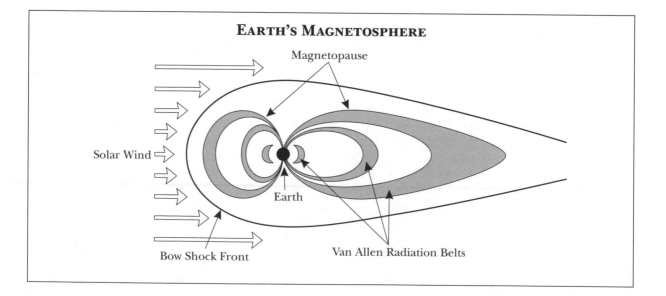

the field: The area pointing toward the Sun is pushed in toward the Earth so that the magnetopause is ten Earth radii from the Earth, and the field pointing away from the Sun is elongated into a teardrop shape that is more than one thousand Earth radii long.

Some of the solar wind particles, particularly electrons and protons, are trapped in the Earth's magnetic field. These form the Van Allen belts, which were discovered in the late 1950's by Dr. James Van Allen while he was analyzing magnetic data from satellites. The inner belt is a broad ring, about 3,000 kilometers above the magnetic equator; the outer ring is about 14,000 kilometers above the equator and is much larger in volume. The belt's particles are traveling at high speeds, and if they hit an object, they disturb its atomic structure.

SOLAR FLARES

The Sun follows a sunspot cycle, in which the number of sunspots increases and decreases over a period of eleven years. The sunspots are associated with the magnetic field of the Sun. During times of maximum sunspot activity, the Sun is very active, and solar flares erupt from its surface. These flares are large plumes of hot gases traveling along the magnetic field lines of the Sun. Some of these particles travel out from the Sun and hit the Earth's magnetic field, producing magnetic storms that cause wild variations in the Earth's field. They also cause disruptions in various communication networks on Earth as a result of the changes in the Earth's upper atmosphere.

In the late winter and early spring of 1988-1989, many solar flares were produced, and they were so violent that they were off the scale used to rank solar flare intensity. It is at these times that auroras are common, as charged particles from the Sun enter the Earth's atmosphere near the magnetic poles and hit air molecules, causing them to glow. The particles enter near the poles because they are moving parallel to the field and it does not affect them. If they enter farther away from the magnetic poles, they are moving across the field lines, and therefore the field changes their direction of travel so that they are scattered. Auroras were observed in the northern sky as far south as Florida in solar flare events in the late 1980's.

LIGHTNING STRIKES

Another external field is a very rapidly changing one: the magnetic field associated with a lightning strike. Lightning is caused by the flow of electrical charges from the ground to clouds or vice versa. Locally, this current produces a very large increase and then decrease to normal field strength. The field can induce electrical currents in metal objects and can cause their destruction as a result of the heating produced by the current. After a strike, wire segments a foot long may be found lying on the ground; the heat caused changes in their structure that caused them to become very brittle so that they easily broke.

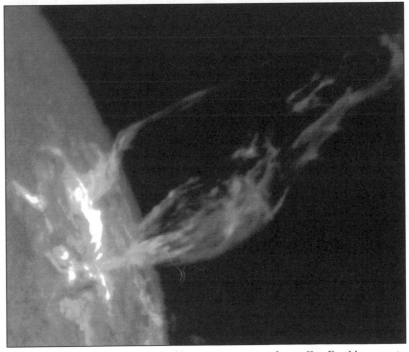

A solar flare. Eruptions during 1999 were strong enough to affect Earth's magnetosphere, and it was predicted that a huge energy release in the next year would disrupt electronic activity on Earth. (AP/Wide World Photos)

SECULAR VARIATION

These variations are of an external nature, but the internal field also varies, though over a period of thousands of years. This phenomenon is referred to as secular variation and is caused by changes in the outer core of the Earth, where the Earth's main field is produced. The variation is also linked to the shifting of the magnetic pole position around the geographic pole position. The magnetic pole wanders around the geographic pole along a looping and twisting path, but, recorded over a long period of time, the pole's position averages out to be the location of the geographic pole.

An interesting change in the magnetic field occurred in 1969: a rapid, worldwide change in the field known as a jerk. Geomagnetists still do not understand its cause, and some believe that, rather than a "real" change, it is a glitch in the processing of the data. A small error in a computer program or a malfunctioning component of an instrument can generate faulty data. Scientists repeat experiments and process data by different methods in order to eliminate these difficulties.

In addition to these various field changes, measurements of the field's strength since the mid-nineteenth century indicate that it is decreasing at a rate of 5 percent per century. Intensity readings obtained from rock samples reveal that the field peaked at an intensity of one and a half times the present value around eleven hundred years ago, and before that it was one-half the present strength at 3500 B.C.E. With these changes in mind, scientists cannot predict what the field will do in the future. It may continue to decrease for a few years or hundreds of years and thereafter increase. It may change polarity, as it has been known to have done many times in the past.

STUDY OF EARTH'S MAGNETIC FIELD

The study of the Earth's magnetic field is a subdiscipline of geophysics, which is the study of various physical characteristics of the Earth. Geomagnetism is the study of the field as viewed over the past several centuries, when measurements were taken of the field. Prior to that time, the study known as paleomagnetism examines various rock layers to determine the field's history. Subdivisions of geomagnetism include rock magnetism, magnetic reversals, and magnetic stratigraphy.

In order to study the Earth's magnetic field scientifically, measurements must be made, as they enable scientists to observe changes in a phenomenon and, therefore, to develop theories about it. In order to determine the field's orientation, scientists obtain two angles and a magnitude. A compass gives one angle, as it measures the horizontal angle between geographic north, once true north is determined for a location, and the needle direction. This angle is called declination. The vertical angle between the horizontal plane and the field line is the inclination. For example, at longitude 80 degrees west and 40 degrees north latitude (the location of Pittsburgh, Pennsylvania), the declination is 9 degrees west of geographic north, and the inclination is approximately 70 degrees from the horizontal into the Earth. In the Southern Hemisphere the inclination is upward. The magnetic equator is located where the inclination is 0 degrees. At the magnetic poles, the inclination is 90 degrees up for the pole in the Southern Hemisphere and 90 degrees down for that in the Northern Hemisphere.

STUDY ON A PERMANENT BASIS

There are two categories for studying the Earth's magnetic field: either on a permanent or on a temporary basis. Around the world, 130 permanent magnetic observatories were established to observe the everyday changes in the magnetic field and to report any significant changes, such as the apparent jerk of 1969, in the scientific literature. Per area, Europe has the most observatories and Asia the fewest. Japan has nine, and Antarctica has twenty-four. A number of others are scattered around the globe on islands. It was at observatories in London and Paris, established centuries ago, that the secular variation of the field was first observed. These early observatories could measure only the vertical and horizontal directions of the field. The horizontal was determined by a compasslike device and the vertical by a small, magnetized rod placed so that it could pivot in the vertical direction.

Starting in the mid-nineteenth century, magnetometers were developed for the measurement of the field's intensity. There are a number of different magnetometers used to measure the field. One is the rubidium magnetometer, which applies the fact that a magnetic field affects the frequency of light given off by an electron moving from a higher to lower energy level around the nucleus of the rubidium atom. The change in the light is proportional to the strength of the magnetic field. The change is measured by the electronics of the magnetometer, and the strength of the field is displayed. In conjunction with the stationary posts, satellites carry magnetometers for the measurement of the field at various heights above the surface, and they provide readings for virtually the entire globe.

STUDY ON A TEMPORARY BASIS

The study of the field on a temporary basis is performed on a small area of Earth's surface. In this context, scientists are interested in what the field has to reveal about objects under the surface. A magnetic iron ore deposit affects the magnetic field near it, and thus scientists can discover the deposit's location. In historic archaeology, researchers are interested in determining where to dig for artifacts. A magnetic survey of the area may provide interesting magnetic anomalies worthy of further investigation.

A common magnetometer used in survey work is a proton precession magnetometer. This formidable name belies a rather simple procedure for measuring a magnetic field. A closed cylinder, filled with water and surrounded with coils of wire, is attached by wires to a box containing the electronics and the power supply of the magnetometer. A current is passed through the coil, and this produces a very strong magnetic field in the water-filled cylinder. The magnetic field forces the protons in the hydrogen atoms to align with the field. The field is then removed, but the Earth's weaker field is still present to affect the protons. The protons naturally spin so that it is difficult to change their orientation with a weaker magnetic field; thus, they start to precess around the Earth's magnetic field. (The wobble seen in a spinning top is precession that is caused by the pull of the Earth's gravity on the top.) The proton precesses at a frequency that depends on the strength of the magnetic field. This frequency is measured by the magnetometer electronics, and the strength of the field is displayed. This value is recorded along with the position of the magnetometer on the grid laid out in the survey area. The data are plotted on a piece of paper and patterns of high- and low-intensity field strength are noted. High-intensity areas indicate a buried magnetic object such as iron shot for a cannon. Using this method, the archaeologist can concentrate on areas that have a higher potential for containing artifacts, instead of digging by trial and error, with subsequent savings in time and funds.

MAGNETIC STORMS AND COSMIC RAYS

When magnetic storms strike the Earth, communication networks, such as radio, are disrupted. When these storms occur, auroras are very prominent in the polar regions, and nature's version of painting with light can be observed. When lightning strikes, radio waves are produced that spiral along the magnetic field lines and bounce into the Southern Hemisphere, where they can be heard on a radio receiver as a "whistler."

The magnetic field also affects cosmic rays, or charged particles moving through space that enter the Earth's atmosphere and even the solid Earth. Some of the particles are prevented from entering the atmosphere because of their interaction with the magnetic field. Scientists are not sure if a decrease in the field would lead to more cosmic rays reaching the surface and possibly producing greater numbers of genetic mutations or even death. Changes in the field strength have been posited as the cause of the mass extinctions that have occurred in the distant past.

SUPPORT FOR PLATE TECTONICS THEORY

Perhaps most important to Earth scientists is the role the Earth's magnetic field has played in lending support for the theory of plate tectonics, which describes the lithosphere as composed of shifting, or moving plates; thus, the continents have shifted position on the globe. Geophysicists have determined that rocks record the position of the Earth's magnetic field for various periods of the Earth's history. The results of their studies provided direct evidence, previously lacking, that the continents have shifted position over hundreds of millions of years and are continuing to do so. The theory of plate tectonics has led to a better understanding of how the Earth operates, how various ores are produced, how mountains originate, why certain areas are good oil producers and others are not, and why earthquakes occur in patterns.

Stephen J. Shulik

CROSS-REFERENCES

BIBLIOGRAPHY

Chapman, Sydney, and Julius Bartels. *Geomagnetism.* 2 vols. Oxford, England: Clarendon Press, 1940. Although outdated, this book contains virtually all there was to know about the Earth's magnetic field at the time it was published. It is of value to browse through to gain historic perspective on the discipline of geomagnetism.

Courtillot, V., and J. L. Le Mouel. "Time Variations of the Earth's Magnetic Field: From Daily to Secular." In *Annual Review of Earth and Planetary Science,* vol. 16, edited by G. G. Wetherill. Palo Alto, Calif.: Annual Reviews, 1988. This source covers the geomagnetic field in general and then launches into an in-depth study of its variations, from short-term to very longterm. Very little mathematics, and many figures. At the end of the chapter is a listing of over a hundred references.

Garland, G. D. *Introduction to Geophysics.* 2d ed. London: W. B. Saunders, 1979. Used as a text for introductory geophysics, this book contains in chapter 17, "The Main Field," very readable material on the main field and its generation. Time variations are discussed in chapter 18; chapter 19 covers the external field; chapter 16, methods of measurement. Some equations. Many figures and graphs of interest to the less well informed reader. References are placed at the end of each chapter.

Hamblin, William K. *Earth's Dynamic Systems.* 8th ed. Upper Saddle River, N.J.: Prentice Hall, 1998. This geology textbook offers an integrated view of the Earth's interior not common in books of this type. The illustrations, diagrams, and charts are superb. Includes a glossary and laboratory guide. Suitable for high school readers.

Jacobs, John. *Deep Interior of the Earth.* London: Chapman and Hall, 1992. This introductory geophysics textbook is formidable for the average student because there is considerable mathematics in some chapters, but it does cover many useful topics such as geomagnetism, the Earth's present magnetic field, auroras, and the magnetosphere. It contains a minimum of equations but many figures and graphs.

Knecht, D. J., and B. M. Shuman. "The Geomagnetic Field." In *Handbook of Geophysics and Space Environment,* edited by A. S. Jursa. Springfield, Va.: National Technical Information Service, 1985. This source covers the geomagnetic field and its various aspects: terminology, coordinate systems, sources of the field, measurements, the main field, and sources of geomagnetic data. Some sections have no mathematics, but others have a small amount. Many figures help the reader to understand the authors' narratives. A number of references are listed at the end of the chapter. Excellent.

Lutgens, Frederick K., and Edward J. Tarbuck. *Earth: An Introduction to Physical Geology.* 6th ed. Upper Saddle River, N.J.: Prentice Hall, 1999. This college text provides a clear picture of the Earth's systems and processes that is suitable for the high school or college reader. In addition to its illustrations and graphics, it has an accompanying computer disc that is compatible with either Macintosh or Windows. Bibliography and index.

Motz, Lloyd, ed. *Rediscovery of the Earth.* New York: Van Nostrand Reinhold, 1979. As a collection of articles for the nonscientist by scientists renowned in their respective fields, the text makes very interesting reading, augmented with many colorful illustrations. The chapter "The Earth's Magnetic Field and Its Variations" is written by Dr. Takesi Nagata, who has authored hundreds of articles on diverse aspects of geophysics besides the Earth's magnetic field.

Plummer, Charles C., and David McGeary. *Physical Geology.* Boston: McGraw-Hill, 1999. A college-level introductory geology textbook that is clearly written and wonderfully illustrated. An excellent sourcebook of basic information on geologic terminology and fundamentals of geologic processes. Contains CD-ROM. An excellent glossary.

Smith, David G., ed. *The Cambridge Encyclopedia of Earth Sciences.* New York: Cambridge University Press, 1982. Chapter 7, "The Earth as a Magnet," contains information about the Earth's present-day magnetic field, geomagnetic field changes, and magnetic anomalies. The term "encyclopedia" does not characterize this publication well, as the book is organized by chapters that are grouped into six parts pertaining to general categories of the Earth sciences. The text is well written at a nontechnical level, with many colorful diagrams and figures. Excellent.

Stacey, F. D. *Physics of the Earth.* New York: John Wilcy & Sons, 1977. In section 8.1, "The Main Field," the author provides a short, technical description of the main field that is of interest to the more advanced student. As a textbook for geophysics, it covers many other areas on a technical level.

EARTH'S MAGNETIC FIELD: ORIGIN

The Earth has a magnetic field that scientists believe did not appear spontaneously. The process of discovery leads into other areas of geophysics and into physics and chemistry. A dynamo effect in the outer core of the Earth is perhaps the most likely source of most of the magnetic field.

PRINCIPAL TERMS

CONDUCTOR: a material that permits the easy flow of electrical charge as opposed to an insulator, such as glass

CONVECTION: the movement of fluid as a result of density differences caused by heating

CURIE TEMPERATURE: the temperature above which a permanently magnetized material loses its magnetization

DIPOLE FIELD: the field shape produced by two electrically charged particles, such as a proton and electron, or by two magnetic poles—north and south

DYNAMO EFFECT: the movement of a conductor through a magnetic field, producing an electrical current that in turn generates a magnetic field

MAGNETIC POLE: the location on the Earth's surface where the Earth's magnetic field is perpendicular to the surface

S WAVES: waves, similar to those in a rope when it is shaken, that are produced in the Earth by an earthquake and that cannot travel through fluids

SECULAR VARIATION: any long-term change in the position of the Earth's magnetic pole location

MAGNETIC BAR THEORY

Central to study about the Earth's magnetic field are questions concerning its origin. Is it the result of a great bar magnet similar in concept, if on a larger scale, to an ordinary, store-bought magnet, or is a more complex process involved? The hypothesis that the source of the Earth's field is a bar magnet can be tested by comparing characteristics of the Earth's field with those of a huge bar magnet. Because the Earth is immense, the magnet must be correspondingly large. The Earth has a strong field that a small bar magnet could not account for anyway, as a magnet's field strength decreases with distance from the magnet. A small magnet at the center of the Earth would produce a field with too low a strength at the surface of the Earth, given the distance involved: 6,371 kilometers.

A bar magnet and the Earth both possess north and south magnetic poles that produce a dipole field. The "di" prefix is from a Greek term for "two": A dipole field is a field produced by two poles. If a bar magnet is located somewhere inside the Earth, the magnetic field at the Earth's surface should not shift position. It is difficult to imagine how a structure such as a large bar magnet could physically move in order to account for the movement of the field. Yet the field at the surface does change. Some of the alteration is the result of outside influences, such as magnetic storms caused by energetic particles from the Sun, and some motion is from some changing conditions within the Earth. One change is difficult to explain: the secular variation of the Earth's field. This variation is the slow shift in the magnetic pole position over periods of years because of the changing of some interior condition that alters the field's position. A bar magnet cannot account for the observed secular variation.

Scientists do not accept an explanation based on one line of evidence, and so they search for other ways to support or refute a hypothesis. In this case, the next step is to look at the temperature change in the Earth's interior. Moving below the Earth's surface, depths are rapidly reached where the temperature is so high that rocks melt. The temperature is above the Curie temperature of all known permanently magnetized materials. The Curie temperature (or Curie point) is the temperature above which a material is no longer permanently magnetic. Heating a magnet above its Curie temperature, for example, causes it to lose its ability to attract iron objects such as pins. When the magnet cools, it once again becomes permanently magnetized but not to the same strength as before.

An alternative theory is that the Earth's outer layer of cool material contains enough magnetic material to account for the magnetic field. There are two problems with this idea: It does not explain the movement of the field, and there is not enough permanently magnetized material to account for more than a minimum percentage of the Earth's magnetic field. As the magnetic bar theory cannot be substantiated, perhaps a look at the manner in which magnetic fields are produced will establish another hypothesis.

DYNAMO EFFECT

The ultimate source of any magnetic field is the movement of electric charges, such as electrons, protons, and ionized atoms. Wires, for example, have magnetic fields around them because electric charges (electrons) move through the wires in the form of the electric current. In a magnet, the electrons surrounding the nucleus of an atom are moving, which denotes electrical current, however minute. All these small currents taken together produce the magnetic field of the material. The conclusion is, therefore, that some electric current within the Earth, other than atomic current, is producing the magnetic field of the Earth.

The current cannot be a passive current, one that was started sometime in the past and allowed to flow, generating a passive magnetic field. A process that could have started such a passive current is unknown. An electrical current encounters resistance that causes a passive current to diminish with time. The resistance converts electrical energy into heat. To sustain the current requires a conductor and a source of energy. The layer of the Earth under the thin surface crust, the mantle, is composed of rock that is an insulator, and, thus, very little current can flow. The innermost layer, the core, is roughly 3,500 kilometers in radius and metallic in composition, most probably iron with a small percentage of nickel and an even smaller percentage of other elements. Therefore, the core is a possible conductor through which the charge can flow.

Some process must actively generate the current.

The most probable is a dynamo effect, comparable to the method employed to generate electrical energy for home and industrial use. ("Generate" is not a very accurate term—the process is actually one of converting one form of energy, such as the chemical energy in coal, into electrical energy and not creating the electrical energy from nothing, as "generation" implies.) A length of wire moved across a magnetic field will induce a small current in the wire. One end of the wire is now electrically positive, and the other end is negative. Take a longer length of wire, form it into loops, and turn these loops in a magnetic field, and the result is an electrical generator, or dynamo, with electrical energy available at the ends of the wire. The actual design of a dynamo is not so simple, but the energy conversion process is very straightforward and well understood.

CONDUCTION AND CONVECTION

The requirements for producing the Earth's field include a source of energy that causes a conductor to move through a magnetic field, which causes an electric current that in turn produces a magnetic field that adds to the original field. Movement is also essential; energy and conductors cannot produce even the smallest magnetic field without free, unimpeded movement between the conductor and the initial magnetic field. These necessary conditions are met by resources within the Earth.

The heat energy of the Earth's interior is more than adequate for the generation of the field. The temperature at the Earth's center is between 4,000 and 6,000 degrees Celsius. Some of this heat energy escapes from the interior through the Earth's surface and is detected as heat flow. It is much smaller than the amount of energy that the surface receives from sunlight, but it is enough. The heat energy is transferred from the interior to the surface by conduction and convection and produces such phenomena as earthquakes, volcanoes, and mountain building.

Convection is important where material is fluid. It is seen in a container of boiling water—heated water expands and rises, and cooler water sinks to replace it. This reaction produces a convection cell that transfers heat from a hot area to a cold one. In molten regions of the Earth, convection transfers heat outward from the hot interior to outer, cooler areas. Earthquake studies indicate that the inner core, with a radius of 1,200 kilometers, is solid, but the outer core, with a thickness of 2,300 kilometers, is molten. An earthquake produces waves known as S waves that cannot travel through liquids, and they do not travel through the outer core of the Earth. This molten material is free to move and transfer heat, or convect.

The Earth has an average density 5.5 times that of water, and it is much denser than rock. The Earth is not entirely rocky material and must have a denser portion to give a high average density. Therefore, the Earth has a mantle of lower-density rock and a core of much higher-density conductive metal that is most probably iron in composition. (The metallic nature of the core can be deduced from the unique high-density composition of metal.)

MAGNETIC FIELDS

With a source of energy—heat—that causes molten iron to move in convection cells, all that is needed is a magnetic field to start the process. Magnetic fields, however weak, are always present in the universe. The outer core's liquid iron moves through a weak field and produces a weak electrical current. When electric charge flows, it produces a magnetic field. The weak electric current can produce a magnetic field that reinforces the initial field. Molten iron moves through this stronger field and generates more current, which then yields a stronger magnetic field.

Although it may seem like a perpetual-motion, or closed, system, that is not the case. The energy comes from the Earth's heat supply, which is a finite source of energy. Also, the process does not continue generating an increasingly stronger field. The process reaches a steady-state condition at which the current and field achieve stable levels. It becomes harder to produce more magnetic field because the electric and magnetic field interact to slow the process to a steady state. A bicycle with an electric light powered by a generator spun by one of the bicycle's wheels demonstrates this principle: It is harder to pedal with the light operating than without. A similar effect is operating in the outer core.

This process can explain features seen in the Earth's field. The shifting of the field's position is caused by alterations in the convection cells within the core. The core is free to move inside the Earth, and changes in the Earth's motion through space can cause the mantle to shift position relative to the core, which will affect the convection currents of the outer core and the generation of the field. Curie temperature does not pertain to electromagnets, and in fact a high temperature in the core is required as an energy source for the convection. Studies indicate that the Earth's magnetic field has reversed polarity many times in the past, the last occurring about 700,000 years ago. The dynamo process is unstable over long periods of time and can change polarity. Models of dynamos can be constructed that are simple versions of the Earth's dynamo, which, when set in operation, reveal changes in the field's intensity and polarity. Evaluations of changes in the magnetic field from 1870 to 1990 tend to confirm that a whirlpool slowly spins clockwise in the liquid core beneath the North Pole.

This dynamo process can be used to explain the presence or absence of a magnetic field for the planets, their moons, and the Sun. Mercury, Venus, Mars, and the moons do not have substantial magnetic fields; they do not possess a molten conductive core. The Earth, the Sun, and the larger planets have substantial fields. The Earth has a molten iron core; the Sun is a fluid, conductive plasma; and Jupiter and the other outer planets contain thick layers of liquid, metallic hydrogen. The dynamo effect explains the origin of about 95 percent of the Earth's magnetic field; the rest comes from the magnetic fields associated with magnetic minerals in the Earth's crust and from external sources such as electrical charges flowing in the atmosphere. They, however, are minor considerations in comparison to the main field generated by the dynamo in the Earth's outer core.

STUDYING EARTH'S INTERIOR

The study of the magnetic field's origin involves many areas of science. Earthquakes may be destructive, but their waves act as probes of the Earth's interior. Various types of earthquake waves travel from the starting point of the earthquake, the focus, into the Earth's interior. Their speed and direction of travel are determined by the density and elastic properties of the material through which they are traveling. The waves reflect off the core-mantle boundary and return to seismic stations, where the waves are recorded on a seismograph. Calculations, using data from these seismograms, reveal the thickness of the mantle and the size of the outer core. They also reveal that the seismic wave known as an S wave cannot travel through the outer core of the Earth. S waves cannot travel through a liquid, and, therefore, it can be inferred that the outer core of the Earth is liquid.

In determining the volume of the Earth and its mass, the Earth's density can be ascertained by dividing the mass by its volume; the average density of the Earth is five and a half times that of water. This figure may not seem very significant, but rocks are only three times the density of water. A portion of the Earth's interior must be much denser in order to yield such a high average. Only metals have the required density, but some metals, such as aluminum, are too low in density, and others, such as uranium, are too high. Still others are closer to the required density but are too rare, as is the case with gold or silver. Iron is a good candidate, as it has the right density and is fairly common. Therefore, the core of the Earth is concluded to be mostly iron, with some nickel and other elements: a conclusion reached mainly with simple mathematics and previously established scientific knowledge.

STUDYING MAGNETIC FIELDS

Studies of magnetic characteristics of materials determine the Curie temperatures and types of magnetism the materials possess. Heating a permanently magnetized material in a magnetic field and recording the temperature when the material is no longer strongly pulled by the field establishes the Curie temperature. Other studies investigate the manner in which magnetic fields are generated, such as moving charges producing magnetic fields. These are areas of study for physicists or engineers, whose work can have great implications in a discipline such as geophysics.

Those studying the Earth's heat flow, a subdiscipline of geophysics, also advance study on the connection between heat and magnetism. In the ocean, the heat flow is determined by placing two temperature probes several meters apart on a tube that is lowered into the sediment on the ocean bottom. The system is allowed to come to equilibrium with its surroundings, and the temperature

of each probe is read. There is a mathematical formula that is used to ascertain the heat flow from the temperature difference, the distance between the probes, and the heat-conduction characteristics of the surrounding sediment. The calculated heat flow permits the determination of the temperature change with depth for the Earth and the fact that there is enough heat energy to produce the magnetic field.

Magnetometers such as the spinner magnetometer are used to determine the characteristics of the Earth's magnetic field in the past, and this information places constraints on the methods by which the field is produced. Some igneous and sedimentary rocks are good recorders of the field. Igneous rock samples are collected and taken to the laboratory for analysis. Basalt, an igneous rock, has a strong magnetization that is measured using a spinner. The sample is placed in the instrument and spun at a high speed. Surrounding the sample are coils of wire in which an electric current is induced by the spinning magnetic field of the sample. The current is proportional to the magnetic field strength, and the electronics of the magnetometer convert it into a reading of that strength. This work has established that the magnetic field has changed polarity many times in the past.

SIGNIFICANCE OF EARTH'S MAGNETIC FIELD

For civilization in general, the simple knowledge of how to generate electric current from a moving magnetic field or a magnetic field from a moving electrical charge is fundamental. Without the laws governing electric and magnetic fields, most of the world's standard of living would collapse. There would be no electric motors for drills, mixers, fans, and automobile starters. No electric generators would produce electricity, with no electric heaters for water and house, no electric lights, and no television. Moreover, beyond creature comforts, the electromagnetic force is one of the four basic forces in the universe, along with gravity and the two nuclear forces (strong and weak). A universe without magnetic fields is incomprehensible.

The fact that the Earth has a magnetic field indicates that the Earth is a dynamic planet with many geological processes taking place. Without this activity, there would be no earthquakes, no volcanoes, no moving plates, and no mountains. It is doubtful that the varied life-forms that presently exist could survive on such a planet, and they most certainly could not have originated on such an Earth. The active generation of the Earth's magnetic field thus symbolizes the vitality of the Earth.

Stephen J. Shulik

CROSS-REFERENCES

BIBLIOGRAPHY

Busse, F. H. "Recent Developments in the Dynamo Theory of Planetary Magnetism." In *Annual Review of Earth and Planetary Sciences*, edited by W. W. Wetherill, vol. 11. Palo Alto, Calif.: Annual Reviews, 1983. This article gives the reader the opportunity to see the advances a particular discipline has made in seven years when compared with the article by Levy, cited below. An outline of the dynamo theory is given, with models of the dynamo for various planets. The observation evidence is discussed, along with the paleomagnetic data, geomagnetic reversals, and secular variation. References are listed, along with a number of figures.

Butler, Robert F. *Paleomagnetism: Magnetic Domains to Geologic Terranes*. Boston: Blackwell Scientific Publications, 1992. Butler's exploration of the Earth's magnetic fields begins with basic descriptions of what paleomagnetism is and how it occurs. The book is filled with illustrations to back up difficult concepts covered in the text.

Garland, G. D. *Introduction to Geophysics*. 2d ed. Philadelphia: W. B. Saunders, 1979. Used as a text for introductory geophysics, this book covers, in sections 17.4 and 17.5, the cause of the main field and the dynamo theory. A few equations, but many figures and graphs that are of interest to the less technically informed reader. At the end of the chapter is a listing of thirty-two references.

Gubbins, D., and T. G. Masters. "Driving Mechanisms for the Earth's Dynamo." In *Advances in Geophysics*, edited by B. Saltzman, vol. 21. New York: Academic Press, 1979. This article looks at such topics as the physical and chemical properties of the core and the power sources for the magnetic field. References are located at the end of the article. Mathematics and numerous figures and tables are included.

Hamblin, William K. *Earth's Dynamic Systems*. 8th ed. Upper Saddle River, N.J.: Prentice Hall, 1998. This geology textbook offers an integrated view of the Earth's interior not common in books of this type. The illustrations, diagrams, and charts are superb. Includes a glossary and laboratory guide. Suitable for high school readers.

Kennett, J. P. *Marine Geology*. Englewood Cliffs, N.J.: Prentice-Hall, 1982. Kennett devotes pages 21-23 to a brief discussion of the subdivision of geophysics called paleomagnetism, the study of the Earth's magnetic field, from generation to magnetization methods such as TRM. Not exhaustive but very readable, with no mathematics and some figures. The text is a veritable encyclopedia on marine geology: plate tectonics, oceanic structure, sediments, margins, and history.

Lapedes, D. N., ed. *McGraw-Hill Encyclopedia of Geological Sciences*. New York: McGraw-Hill, 1978. Pages 704-708, under the heading of "Rock Magnetism," provide a concise description of many aspects associated with rock magnetism: how rock magnetization occurs, the present field, magnetic reversals, field generation, secular variation, and apparent polar wandering, among other subjects. Very readable, with no mathematics and a fair number of graphs, tables, and figures.

Levy, E. H. "Generation of the Planetary Magnetic Fields." In *Annual Review of Earth and*

Planetary Sciences, edited by F. Donath, vol. 4. Palo Alto, Calif.: Annual Reviews, 1976. This article presents an overview of how the generation of magnetic fields within the various planets affects the dynamo effect. Some mathematical equations and a number of figures, along with many references at the end of the book.

Lutgens, Frederick K., and Edward J. Tarbuck. *Earth: An Introduction to Physical Geology*. 6th ed. Upper Saddle River, N.J.: Prentice Hall, 1999. This college text provides a clear picture of the Earth's systems and processes that is suitable for the high school or college reader. In addition to its illustrations and graphics, it has an accompanying computer disc that is compatible with either Macintosh or Windows. Bibliography and index.

Merrill, R. T., and M. W. McElhinney. *The Magnetic Field of the Earth: Paleomagnetism, the Core, and the Deep Mantle*. San Diego: Academic Press, 1998. The authors cover the basic material associated with the Earth's magnetic field. Chapters deal with the origin of the magnetic field, as well as the origin of secular variation and field reversals. Bibliography and index. Numerous tables, figures, and mathematical equations.

Motz, L., ed. *Rediscovery of the Earth*. New York: Van Nostrand Reinhold, 1979. As a collection of articles for the nonscientist by scientists renowned in their respective fields, the text makes very interesting reading, augmented with many colorful illustrations. The chapter "The Earth's Magnetic Field and Its Variations" is written by Dr. Takesi Nagata, who has authored hundreds of articles on diverse aspects of geophysics besides the Earth's magnetic field, and covers a wide range of magnetic field topics. Two pages are devoted to the origin of the field. A small amount of mathematics.

Murthy, I. V. *Gravity and Magnetic Interpretation in Exploration Geophysics*. Bangaloree: Geological Society of India, 1998. This book is an excellent source of information about gravity and geomagnetism. In addition to useful illustrations, the book comes with a CD-ROM that complements the information in the chapters. Bibliography and index. Intended for the reader with some Earth science knowledge.

Olson, Peter. "Probing Earth's Dynamo." *Nature* 389, no. 6649 (September 25, 1997): 337-338.

Olson, Peter, and Jonathan Aurnou. "A Polar Vortex in the Earth's Core." *Nature* 402, no. 6758 (November 11, 1999): 170-173. Brief technical reports on the finding of evidence for circulation of the Earth's liquid core and its significance for the dynamo hypothesis.

Opdyke, Neil D. *Magnetic Stratigraphy*. San Diego: Academic Press, 1996. Intended for someone with little scientific background, *Magnetic Stratigraphy* examines the magnetic fields of the Earth, focusing largely on paleomagnetism. Contains fifty pages of bibliographical resources, as well as an index.

Plummer, Charles C., and David McGeary. *Physical Geology*. Boston: McGraw-Hill, 1999. A college-level introductory geology textbook that is clearly written and wonderfully illustrated. An excellent sourcebook of basic information on geologic terminology and fundamentals of geologic processes. Contains CD-ROM. An excellent glossary.

Smith, D. G., ed. *The Cambridge Encyclopedia of Earth Sciences*. New York: Crown Publishers, 1981. Chapter 7, "The Earth as a Magnet," contains a discussion of the field's origin. The term "encyclopedia" is not the best term for this publication because it is broken into chapters that are grouped into six parts pertaining to general categories of the Earth sciences. The text is well written at a nontechnical level, with many colorful diagrams and figures.

Stacey, F. D. *Physics of the Earth*. New York: John Wiley & Sons, 1977. Under section 8.4, "Generation of the Main Field," the author provides a short, technical description of the origin of the field, which will be of interest to the more advanced student. Equations are rather formidable, but several figures illustrating the dynamo effect are included. A large number of references at the end of the text. Many other areas of geophysics are covered at a technical level.

EARTH'S MAGNETIC FIELD: SECULAR VARIATION

The Earth's magnetism manifests itself at every point on the Earth and above it. It has a direction, as indicated by a freely suspended compass needle, and an intensity. This direction and intensity change slowly over decades and centuries, a process known as secular variation.

PRINCIPAL TERMS

CORE: the region of the inner Earth beneath the crust and mantle, discovered by seismologists; the outer core, between 1,000 and 3,500 kilometers from the Earth's center, is the source of geomagnetism and secular variation

GEODYNAMO THEORIES: theories that explain the cause of the Earth's magnetic field and its secular variation in terms of movements of fluid and electricity in the Earth's outer core

GEOMAGNETIC ELEMENTS: measurements that describe the direction and intensity of the Earth's magnetic field

ISOMAGNETIC CHARTS: maps on which are traced curves, all the points of which have the same value in some magnetic element

MAGNETIC SURVEY: measurements of the magnetic elements at many points, on or above the Earth's surface, carried out by field

teams, airborne magnetometers, ships at sea, or satellites

MAGNETOMETER: a scientific instrument used to measure the elements of the Earth's magnetic field

PALEOMAGNETISM: the study of the record of remanent or fossil magnetism in rocks, indicative of past states of the Earth's magnetic field and very useful in explaining secular variation

REPEAT STATION: a location where the magnetic elements have been measured more than once, for the purpose of determining secular variation

SELF-EXCITING DYNAMO: also called a self-sustaining dynamo, a model of the Earth's outer core in which the magnetic field produced by convection is in the same direction as the field through which the motion occurs

GEOMAGNETIC ELEMENTS

The secular variation of the Earth's magnetic field is a long-term change in the magnetic forces produced in the Earth's core. The phenomenon is observable anywhere on the Earth's surface. Its most familiar example is a gradual alteration of the direction in which an ordinary compass needle points. It also is seen in the change of the inclination; that is, the angle at which a magnetic needle suspended by its center of gravity dips below the horizon. Finally, the intensity of the force which returns a magnetized needle to its rest position is changing. Secular variation differs from other variations in the Earth's magnetism. For example, the diurnal (or daily) variation repeats cyclically. The secular variation, on the other hand, is a constant drift of direction and intensity which never repeats. These slow changes are termed "secular" changes, in analogy with similarly named grad-

ual drifts in astronomical variables. Although its effects are seen best over periods of decades or even centuries, with modern methods secular variation can be detected across a much shorter interval.

The needle of a magnetic compass points generally north and south, but it does not point exactly so. The angle between geographic and magnetic north was originally called magnetic variation, meaning the variation from true north. Declination, as it is now called, was discovered around the twelfth century, although it was originally believed to be caused by abnormalities in the needle's magnetization or in its suspension. By the sixteenth century, Europeans had accepted declination as a phenomenon of the Earth's magnetism. Inclination was also discovered during that century. Hence William Gilbert could write in 1600 of both declination and inclination as natural phenomena. The needle indicated the direction of the Earth's magnetic field. By the early sixteenth century, Europeans had noticed that the declination varies from place to place, though this discovery can be attributed to no single observer. The discovery arose in the practices of navigation, chart making, and "dialing," or the crafting of sundials and magnetic compasses—all activities connected with exploration. Perhaps Christopher Columbus, and certainly Sebastian Cabot, noted that while the compass pointed east of north near Europe, it pointed west of north in the New World.

ISOGONIC MAPS AND ISOPORIC CHARTS

Indeed, all three magnetic elements—declination, inclination, and intensity—vary over the planet. If places with the same declination are connected by curved lines, maps of equal declination, called isogonic maps, can be produced. The first such printed map was produced by Edmond Halley in about 1701. Similar maps displaying equal inclination or equal intensity are also drawn. One line which early attracted much attention was the agonic line, or "line of no variation," that is, a line along which the compass pointed to geographic north. It engendered hope for determining longitude by compass. Meanwhile, Henry Gellibrand announced in 1635 that declination changed over time as well as space. In his case, the declination had shifted from 11.3 degrees east for London in 1580 to 4.1 degrees east in 1634. Later investigators discovered that the inclination and the intensity of the magnetic field also gradually change. Between 1700 and 1900, the inclination at London decreased from almost 75 to 67 degrees. The intensity of the dipole field is decreasing at the rate of 8 percent per century.

Secular variation has been characterized by the westward drift of the agonic line. This drift can be visualized another way. As one can map the magnetic elements, one can map how these elements change. Curved lines connect points that are changing at a certain rate. For example, all points where the declination is shifting, say, eastward at 1 degree per century, might be connected together. These charts, known as isoporic charts, came into wide use in the mid-twentieth century. Areas of most rapid change are called isoporic foci. These isoporic foci are drifting westward, just as the agonic line is. While this drift has been a prominent feature of secular variation since Gellibrand, some evidence exists for eastward drifts during prehistoric times.

THEORIES OF SECULAR VARIATION

Secular variation is distant from direct experience. First, its very description presumes a system of mathematical analysis. The geomagnetic field and its secular change are often described in terms of dipoles, that is, idealized bar magnets superimposed on each other, or one dipolar field superimposed on a nondipolar field. Second, the effects produced by causes inside the Earth are always discussed in isolation from those caused by external processes, even though measurements cannot separate the two types of effects. Last, any discussion of the cause of secular variation is necessarily indirect, as is any discussion of the cause of the main field. Ultimately, separation of the description of geomagnetic secular variation from its mathematical analysis and theory is difficult.

Theories of geomagnetic secular variation have been extremely diverse. That is predictable, given the inaccessibility of the cause of the Earth's magnetism. Gilbert had suggested that the Earth behaves as if it had an ordinary bar magnet of extraordinary intensity at its center. From worldwide declination data, Halley discerned in 1683 a pattern dependent on four magnetic poles, not merely two. Robert Hooke proposed in 1674 that the magnetic axis of the Earth is tilted about 10 degrees from the axis of rotation and that this axis revolves around the rotational axis every 370 years. In 1692, Halley suggested that his four poles could explain secular variation. Two of these poles he assigned to the Earth's outer crust and the other two to a nucleus, which rotated slightly more slowly than the crust, on the same axis. The crustal magnetic poles were, he said, fixed in place, though as the nucleus drifted slowly westward under the crust, so did its magnetic poles. This motion explained, he thought, drift of the agonic line. The idea that the core is permanently magnetized was later ruled out by its temperature; it is too hot to be magnetic. Current discussion of dipoles in the core do not assume permanent magnets.

GEODYNAMO THEORIES

Two alternative theories are that the Earth's very rotation causes its magnetism and that the

Earth acts as an electromagnet. The former hypothesis reached its highest state of development around 1950 in work by Patrick M. S. Blackett, but the latter approach has proven more fruitful. In its rudimentary early nineteenth century form, this model assumed that the Earth's magnetism was produced by electrical currents flowing inside the Earth, from east to west, as in an electromagnet made up of a coil of wire. The model of electric currents flowing inside the core became adequately sophisticated to address the data only with the theoretical investigations of Walter Elsasser, from 1939 on, and of Sir Edward Crisp Bullard, starting in 1948. Elsasser proposed that the combination of the movement of molten materials and the simultaneous flow of electricity in these materials produced both the Earth's main magnetic field and the secular variations in it. This dynamo was driven, he suggested, by the heat generated by radioactive materials. Convection of hotter materials upward and of colder materials downward, he said, produced the dynamo. Various models have been advanced to show that such convection cells can produce the observed effects, and many debates are still waged over the character and cause of the dynamo.

There is significant agreement that the Earth's field and its secular variation are the result of motions in the core. It is also agreed that, for the most part, the magnetic field lines (or lines of force) travel with the moving fluid; that is, the lines are frozen. Similarly, the dynamo is a self-exciting (or better, self-sustaining) dynamo. Most scientists accept the two most probable energy sources for the dynamo to be heat from radioactive materials and convection caused by the settling of denser materials to the inner core. As any other dynamo must be driven—perhaps by a waterfall or by steam from coal or by a nuclear reactor—so, too, the geodynamo requires a source for its power. The Earth does not create its magnetic field; rather, it converts some other form of energy into it. The Earth's magnetic field is not a perpetual motion machine.

In the end, one must emphasize that theories of the geomagnetic dynamo and of its power sources are tentative. Their connection to secular variation is also exploratory. This area of geophysical theory is a most active and challenging one, and it is in rapid flux.

MAGNETIC INSTRUMENTS AND SURVEYS

Because detection of geomagnetic secular variation requires measurement of an invisible force, special apparatus is required. The simplest way to detect this slow change is to observe the declination of a magnetic compass over some decades; until the twentieth century, that was the only way. All the instruments employed by famous students of geomagnetism, from Gilbert in 1600 to C. F. Gauss in the 1830's, used adaptations of the compass to measure the magnetic elements. Among other goals, these scientists aimed to measure these elements more accurately, so as to reveal secular change in a shorter interval. During the twentieth century, however, there was a sustained trend to replace magnetic needle instruments with ones based on new principles. Around 1900, research-quality Earth inductors were developed to replace the dip circle in measuring inclination. The idea behind this first electrically based geomagnetic instrument is simple: If a coil of wire is rotated in a magnetic field, and if its axis differs from the direction of that field, an electric current circles through the coil. Let the axis coincide with the field, however, and the current will cease. Magnetic scientists used this "null" method to determine inclination more accurately and more conveniently. The Earth inductor was followed in the 1930's by the flux-gate magnetometer. This instrument is based on a high-permeability alloy, that is, one that magnetizes readily. Around two cores of such material are wound two coils of wire, in opposite directions, that carry the same alternating current. When placed in the Earth's field, the changes in the magnetic fields of these two cores do not cancel each other. The net effect indicates the Earth's magnetic element in one direction. When, however, this magnetometer is oriented parallel to the Earth's field, no current is produced. The flux-gate magnetometer has seen wide use in aerial geomagnetic surveys.

Other generations of magnetic instruments have appeared since the flux-gate. Some of the most useful are the proton procession, the Rubidium vapor, and the superconducting magnetometers. These devices take advantage of principles of quantum physics. Some of them, like the proton precession instrument, measure only the total intensity of the Earth's field. Others, like the superconducting magnetometer, are directional. Both types are many times more sensitive than older instruments and also perform much faster.

Magnetic surveys have been an essential part of the method of studying secular variation. All over the world, teams of observers have established "repeat stations," or places for careful observation of the magnetic elements at times separated by, say, several decades. These global data have then been analyzed according to one of several mathematical approaches. Magnetic surveys have been greatly facilitated not only by the new instruments mentioned above but also by the way those instruments are used. Surveys are now often conducted very quickly with instruments carried by airplanes and satellites (such as MAGSAT). Data that once took decades to gather are now collected in months. Moreover, calculations have been accelerated unimaginably by computers. That is critical, as the utility of the data depends on extensive calculations. Worldwide magnetic charts are produced much more frequently now than in 1900, and the study of secular variation is thus much more fine-grained.

PALEOMAGNETISM

Equally impressive changes have been wrought by the use of geomagnetic methods to study the magnetic properties of rocks. Many rocks provide a record of the Earth's magnetic field at the time they were deposited. The phenomenon is called remanent (or fossil) magnetism, and the science is called paleomagnetism. Until this development, mostly since 1950, secular variation studies were limited to data obtained by direct measurement of the Earth's field. Little was known of the magnetic field before 1600. New phenomena that have been revealed by these methods include reversals of the magnetic field.

The deep interior of the Earth is more inaccessible than the surfaces of Mars and Venus. People can fly to these planets but can only scratch at the Earth's surface. Thus, scientists must watch closely the faint but distinct signals received at the Earth's surface for clues to the processes that drive this planet. Magnetic secular variation provides one of the few methods by which such information can be obtained. Perhaps the most exciting implications of secular variation are related to reversals of the Earth's main magnetic field. Reversals have happened scores of times over geological history.

They tie geomagnetic secular variation in with the chronology of the planet and with seafloor spreading and plate tectonics.

The magnetism impressed upon minerals as they formed and on sedimentary beds as they settled has been preserved over hundreds of millions of years. When these records are gathered from around the world, they provide a unique look at the past state of the Earth's magnetic field. As scientists expand this record, it becomes a key to solving more geochronology problems and may even aid in archaeological investigations.

TOPOGRAPHIC MAP INTERPRETATION AND NAVIGATION

One cannot sense the magnetic field of the Earth directly, unaided by scientific instruments. Even so, most people are familiar with the magnetic compass, and many know roughly how to use it. Two areas which require greater familiarity with geomagnetic phenomena are the reading of topographic maps and navigating at sea. These activities demand close attention to declination and its secular variation.

In the lower left-hand corner of almost any topographic map produced by the U.S. Geological Survey there are three arrows, which point to true north, to magnetic north, and to "grid" north, respectively. With this declination information, one can adjust field readings to correspond to the map. In areas where secular variation of geomagnetism proceeds at a high rate, it may also be necessary to know when declination readings were last taken and the rate of their change. For example, near Tay River in the Canadian Yukon Territory, the declination was listed as longitude 33°25′ east in 1979 and decreasing at 3.3 feet per year. Thus, if secular variation continued at its current rate, in eighteen years the declination would decrease to longitude 32°25′ east. In a century this could mean a change of more than 5 degrees. Secular variation cannot be predicted reliably over so long a period, however, and maps are therefore updated regularly in magnetic surveys.

Information regarding declination at sea and especially near the coast is of even greater importance. Every vessel is sometimes beset by fog, and thus an essential bit of navigational data is the declination. With it and a known starting point, one can at least know the direction toward which the ship is pointing. Up-to-date charts are, again, the best means to circumvent secular variation. As the date of the magnetic declination data recedes into the past, however, reliable information concerning its secular change becomes more important.

Gregory A. Good

CROSS-REFERENCES

Earth's Core, 3; Earth's Core-Mantle Boundary, 9; Earth's Magnetic Field, 137; Earth's Magnetic Field: Origin, 144; Earth's Rotation, 106; Geobiomagnetism, 156; Magnetic Reversals, 161; Magnetic Stratigraphy, 167; Plate Tectonics, 86; Polar Wander, 172; Rock Magnetism, 177.

BIBLIOGRAPHY

Brush, Stephen G. *Nebulous Earth: The Origin of the Solar Sytem and the Core of the Earth from Laplace to Jeffreys.* Cambridge: Cambridge University Press, 1996. Brush's book, volume 1 in the History of Modern Planetary Physics series, contains useful information on the nebular hypothesis, the origin of the solar system, and the Earth's core. Includes a bibliography and index.

Carnegie Institution. *The Earth's Core: How Does It Work?* Washington, D.C.: Author, 1984. This 32-page pamphlet, available from the Carnegie Institution of Washington (1530 P Street NW, Washington, D.C. 20005), discusses active research undertaken by the institution's scientists in the study of the Earth's core. A nontechnical account of the core in many perspectives: seismological, geochemical, and geomagnetic. Suitable for high-school-level readers.

De Bremaecker, Jean-Claude. *Geophysics: The Earth's Interior.* New York: John Wiley & Sons, 1985. This well-written text is intended for college-level students with some calculus and some physics background. Nevertheless, the author is careful to explain difficult concepts or mathematical statements. Chapter 10, "Magnetostatics," and chapter 11, "The Earth's Magnetic Field," can be read separately from the rest of the book to provide an in-depth survey of geomagnetism, its measurement, and its secular variation. Especially useful are the technical appendices on mechanical quantities, magnetic quantities, data about the Earth, notation, and some relevant mathematics. One of the best treatments available.

Gurnis, Michael, et al., eds. *The Core-Mantle Boundary Region.* Washington, D.C.: American Geophysical Union, 1998. This collection of articles is one volume of the American Geophysical Union's Geodynamics series. Although intended for the specialist, the essays contain plenty of information suitable for the careful college-level reader. Bibliography.

Hamblin, William K. *Earth's Dynamic Systems.* 8th ed. Upper Saddle River, N.J.: Prentice Hall, 1998. This geology textbook offers an integrated view of the Earth's interior not common in books of this type. The illustrations, diagrams, and charts are superb. Includes a glossary and laboratory guide. Suitable for high school readers.

Hoffman, Kenneth A. "Ancient Magnetic Reversals: Clues to the Geodynamo." *Scientific American* 258 (May, 1988): 76-83. This article provides a fine update on how paleomagnetic data are providing a new understanding of the geodynamo. A nonmathematical account, with clear graphical representations of the main field, the timetable of magnetic reversals, and the wandering of the virtual (apparent) geomagnetic poles during several important reversals of the Earth's magnetic field. The discussion of how the geodynamo can or could reverse the field it produces is an especially useful summary of ideas of the 1980's.

McConnell, Anita. *Geomagnetic Instruments Before 1900.* London: Harriet Wynter, 1980. This short book provides one of the clearest expositions of the basics of geomagnetism for the lay reader. Includes illustrations of many of the basic early forms of instrumentation, especially European.

Multhauf, Robert P., and Gregory Good. "A Brief History of Geomagnetism and a Catalog of the Collections of the National Museum of American History." *Smithsonian Studies in History and Technology* 48. Washington, D.C.: Smithsonian Institution Press, 1987. This monograph, available in many libraries, has two parts. The first section sketches discoveries in geomagnetism from about 1600 to World War II, with an emphasis on work done in America. The second section illustrates and discusses more than fifty magnetometers, dip circles, Earth inductors, and other magnetic instruments in the collection of the Smithsonian Institution.

Thompson, Roy, and Frank Oldfield. *Environmental Magnetism*. London: Allen & Unwin, 1986. This book captures the broad range of possible applications of knowledge of magnetism in the study of the Earth that have appeared since the 1950's, from the study of magnetic minerals to biomagnetism. This is an introductory, nonmathematical, college-level text. Although its chapters on basic magnetic principles are valuable, the most unusual feature of the book is the many application chapters. Especially relevant to secular variation are chapter 5, "The Earth's Magnetic Field"; chapter 6, "Techniques of Magnetic Measurements"; chapter 13, "Reversal Magnetostratigraphy"; and chapter 14, "Secular Variation Magnetostratigraphy." The schematic illustrations of apparatus and experiments are another useful feature of this book.

GEOBIOMAGNETISM

Geobiomagnetism refers to the interaction of living organisms with the Earth's magnetic field. Many animals, plants, and even bacteria have displayed in laboratory experiments the ability to sense and to use the Earth's magnetic field in various ways, notably in navigation.

PRINCIPAL TERMS

BIOMAGNETISM: the magnetic fields generated by living organisms

GEOMAGNETISM: the magnetic field generated by the Earth

MAGNETITE: an isometric mineral, an oxide that is sensitive to magnetic fields

MAGNETOMETER: a device used to detect and measure magnetic fields

SQUID: an extremely sensitive magnometer capable of detecting and measuring very weak magnetic fields

NAVIGATION ABILITY OF LIVING ORGANISMS

The ability of living organisms to navigate accurately over great distances has long fascinated and baffled naturalists and life scientists. How are many species of birds able to migrate thousands of miles annually, often across open seas, and unerringly reach their destinations? How can homing pigeons find their way back to their coops after having been taken many miles from them in enclosed containers? How can honeybees, after having located a desirable food source miles away from their hives, not only return to the food source but also communicate its location to other honeybees without actually taking them there? These are only three examples of the remarkable direction-finding abilities displayed by living organisms. Systematic research into methods by which living organisms navigate got under way only during World War II. In experiments in the 1940's and 1950's, researchers showed that living organisms use a variety of means to find directions. These means include celestial navigation (use of the Sun and the stars to find directions), used by several species of migratory birds and some crustaceans; navigation by sound reflection, used by bats and many forms of sea-dwelling mammals; navigation by electricity, used by many species of fish; and navigation by using the Earth's magnetic field. Many life-forms as diverse as bacteria, butterflies, fish, and birds have built-in compasses, in the form of minute, magnetic, mineral grains, that enable them to orient to the Earth's magnetic field.

BIRDS' USE OF EMF

Hans Fromme was conducting observations of several robins in a cage at the Frankfurt Zoological Institute in 1957 at a time that robins in Germany were preparing for their annual migration to Spain. Fromme was not satisfied with the then-accepted theory that the robins found their way to Spain by celestial navigation, because radar trackings of the birds had shown that they flew straight toward their destination on nights when heavy cloud cover precluded their being able to see the stars. Fromme caged his birds in a room from which they could not see the heavens. Nevertheless, when the robins outside began their southwestward migration, Fromme's birds became restless and fluttered up to the southwestern corner of their cage. Obviously, Fromme reasoned, they must be responding to some stimulus other than the stars or the Sun. He guessed that this stimulus might be the Earth's magnetic field. Scientists had long known that the Earth acts in many ways as a giant electromagnet of considerable power. Until Fromme's experiments, however, few scientists suspected that geomagnetism affected living organisms or could be used by them for various purposes. In order to test his theory, Fromme put his birds into a special steel chamber which reduced the power of the Earth's magnetic field (EMF) to 0.14 gauss (a unit of measure for magnetic force). The average strength of the EMF at Frankfurt is 0.41 gauss. In this enclosure, the robins still became restless at their normal migration time, but their flutterings were random, no longer toward the southwest corner of the cage. Further research showed that over a period of days the robins adjusted to the reduced magnetic field and once again flew toward the southwest corner of their cage. Fromme and his colleagues were able to "fool" the robins by creating an artificial magnetic field that created a false southwest. The robins rapidly adjusted to the artificial field and fluttered toward the southwest of the artificial field. After Fromme's experiments proved that robins used the EMF to navigate, life scientists began investigating the effects of geomagnetism on a great variety of life-forms, ranging from bacteria through higher vertebrates to human beings. These experiments resulted in a series of dramatic and unexpected discoveries.

In addition to Fromme's experiments with robins, other experiments have demonstrated conclusively that many species of birds rely on the EMF to navigate. The homing pigeon provides perhaps the best example. Carefully conducted experiments showed that a simple bar magnet attached to the back of a homing pigeon's head completely disrupts its navigational ability. Other experiments showed that homing pigeons are remarkably sensitive to the most minute local fluctuations (anomalies) in the EMF, which may perhaps explain their remarkable homing ability.

OTHER ORGANISMS' USE OF EMF

One group of scientists observed anaerobic bacteria of the *Spirillum* type, which are usually found in aquatic mud. When taken into open water, the bacteria swim along magnetic field lines, natural or artificial, which take them toward the magnetic north pole in the Northern Hemisphere and the magnetic south pole in the Southern Hemisphere. This reaction takes them directly to their natural habitat, the mud of the seafloor.

Other scientists have shown that many different species of insect use the EMF in a number of ways and for a variety of purposes: The common honeybee, for example, can communicate by use of the EMF to its hive mates the location of a food source which none but it has visited. The honeybee accomplishes this direction-giving by performing a complicated series of movements (called a "waggle dance") on the honeycomb in which its movements are oriented by the EMF. The so-called

compass termite of Australia uses the EMF in an entirely different way from the honeybee. The compass termites build large nests, sometimes 13 feet high and 10 feet long but only about 3 feet wide, the temperature of which is regulated by use of the EMF. The long axis of the nest always runs due north and south. This magnetic orientation has the advantage of exposing the long sides of the nest to the direct warming rays of the Sun during the early morning and late afternoon. In the middle of the day, when the Sun's rays might be too hot, however, only the relatively thin top edge is exposed to its direct rays.

Many species of fish also use the EMF. Sharks and rays, for example, are apparently able to detect changes in the EMF to locate potential prey. Scientists have shown that the fish interact with the EMF by introducing electrical fields into their environment, which they use to orient themselves in the EMF and to register fluctuations therein caused by magnetic anomalies or by other living creatures. The organ involved in the fishes' ability to interact with the EMF appears to be the electroreceptive ampullae of Lorenzini, which respond to very low electrical voltage gradients. Freshwater eels, both the European and the American varieties, apparently use the EMF to guide them from the rivers where they spend their adolescence to the Sargasso Sea, to which they migrate for purposes of reproduction once they have reached maturity.

MAGNETIC PROPERTIES OF HIGHER ORGANISMS

Scientists investigating the magnetic properties of higher organisms have also made spectacular and totally unexpected discoveries. Researchers in this area, called biomagnetism, have found that most organs in higher vertebrates (including man) produce weak magnetic fields that can be detected and measured using the very sensitive instruments made available by modern technology. The organs producing such magnetic fields include the liver, the brain, and the heart. Magnetic measurements of these organs provide information about them that no other sort of test, including X rays and electroencephalograms, can yield. A number of researchers in the field of biomagnetism suspect that the magnetic fields produced by some living organisms allow them somehow to use the EMF for direction-finding. This relationship, however, has not yet been scientifically demonstrated.

INSTRUMENTS FOR STUDYING GEOBIOMAGNETISM

The sensitive instruments necessary to study geobiomagnetism emerged from weapons research conducted by both sides during World War II. In their efforts to discover ever more efficient ways to detect enemy submarines and aircraft, both Allied and Axis scientists investigated various applications of electromagnetism. Governments invested huge sums of money into scientific research projects that offered even the most tenuous hope of producing revolutionary weapons. Some of the better-known results of these military-oriented scientific projects include radar, sonar, and nuclear fission. After the war, a part of the research conducted by military research projects led to the development of instruments capable of detecting the very weak magnetic fields produced by living organisms: biomagnetism.

Biomagnetic fields are very faint, usually less than one-tenth that of the Earth, and cannot be measured with the magnometers used to measure the EMF. Magnetic fields stronger than 1 gauss are measurable by a simple but sensitive magnetometer called a fluxgate. Measurement of weaker fields requires the use of an extremely sensitive cryogenic magnetometer called a SQUID (acronym for superconducting quantum interference device). No instrument yet devised, however, has been able to show how organisms interact with the EMF or which device or organ is involved in that interaction, although some clues have been discovered. Nevertheless, abundant evidence exists that such interaction does take place.

In the experiments with bacteria mentioned earlier, researchers introduced an artificial magnetic field pointing at right angles to the sea bottom. The bacteria invariably aligned themselves with the new field. When scientists cultured these bacteria in a largely iron-free medium, the bacteria lost their ability to orient themselves along the EMF. Upon examination, the researchers found the bacteria that were cultured in a natural environment contained 1.5 percent (dry weight) iron, which almost certainly is the agent which allows their interaction with the EMF. Exactly how this interaction occurs, however, is unknown.

THEORIES OF GEOBIOMAGNETISM

Geophysicists have proposed several theories as to the mechanisms at work in geobiomagnetism. The paramagnetic molecule theory states that molecules with unpaired magnetic spins are present in all living cells, which may line up with external magnetic fields, although this has yet to be demonstrated. Even if such alignment does occur, there is no evidence or even theory as to how an organism's nervous system could use the information to deduce the direction of the field. The electrodynamic theory states that if a force of electrically charged particles is introduced into a magnetic field, the field exerts a force which influences their direction of motion. Whether detectable effects can be produced in living organisms, allowing them to detect the weak EMF, is debatable and has yet to be demonstrated. According to the magnet hypothesis, the ingestion of magnetic material or the formation of magnetic material within specialized cells by living organisms allows them to sense the Earth's magnetic field. Magnetotactic bacteria produce intercellular iron sulfide, greigite, which is magnetic. Magnetite, an iron oxide, in the trigeminal nerve cells of trout and other fish enables them to detect changes in the magnetic field.

APPLICATIONS BENEFITING HUMANKIND

If scientists could learn the methods by which living organisms sense and use the EMF for navigation, the benefits would be enormous. Faint local variations in the EMF could be used to steer planes and ships to their destinations without recourse to the expensive and complicated navigational equipment presently employed for that purpose.

The new field of biomagnetism has already yielded unexpected results in medical technology. In the very near future, biomagnetism will almost certainly play a major role in the detection and diagnosis of human maladies, and perhaps in their treatment as well. It is not inconceivable that, as we learn more about the magnetic properties of living organisms and their interaction with the EMF, ways of treating malfunctions of bodily processes through manipulating these magnetic fields will evolve.

Paul Madden

Cross-References

Bibliography

Barnothy, Madeleine F., ed. *Biological Effects of Magnetic Fields.* 2 vols. New York: Plenum Press, 1964, 1969. The articles in this older but still valuable work cover the entire spectrum of research into the effects of the EMF on living organisms. Many of the articles use very technical language, but the average reader will nevertheless be able to gain insights into the scope of research being done and the possibilities presented by further investigation into the field of biophysics.

Blakemore, R. P., and R. B. Frankel. "Magnetic Navigation in Bacteria." *Scientific American* 245 (June, 1981): 42-49. This article presents incontrovertible evidence that some forms of bacteria are able to use the EMF for purposes of navigation. The authors also make a compelling case that magnetite ingested by the bacteria is the agent that allows them to use this unique form of navigation. The article is written in such a way as to be intelligible to readers without advanced degrees in either physics or biology.

Dubrov, A. P. *The Geomagnetic Field and Life.* New York: Plenum Press, 1978. Dubrov surveys the entire spectrum of the effects of the EMF on living organisms, both proven and possible. Written for a general rather than a professional reading audience, the book is accessible to anyone with a moderate background in science. It is probably the most comprehensive book on the subject to date.

Fenwick, Peter. "The Inverse Problem: A Medical Perspective." *Physics in Medicine and Biology* 32 (April, 1987): 5-10. Valuable to anyone wishing an understanding of the new science of biomagnetism. The author explains how new techniques for measuring the magnetic fields produced by living organisms aid in solving perplexing problems in medical diagnoses.

Gulrajani, Ramesh M. *Bioelectricity and Biomagnetism.* New York: Wiley, 1998. A thorough look at the effects of biomagnetism on the Earth and its life-forms. Suitable for people with little scientific background, the book is clearly written and filled with helpful illustrations.

Hamblin, William K. *Earth's Dynamic Systems.* 8th ed. Upper Saddle River, N.J.: Prentice Hall, 1998. This geology textbook offers an integrated view of the Earth's interior not common in books of this type. The illustrations, diagrams, and charts are superb. Includes a glossary and laboratory guide. Suitable for high school readers.

Ioannides, A. A. "Trends in Computational Tools for Biomagnetism: From Procedural Codes to Intelligent Scientific Models." *Physics in Medicine and Biology* 32 (January, 1987): 77-84. Ioannides' article is an imaginative guide to the seemingly unlimited future applications of biomagnetic technology in medicine. The nonspecialist should read the article with a dictionary close at hand.

Jungreis, Susan A. "Biomagnetism: An Orientation Mechanism in Migrating Insects?" *The Florida Entomologist* 70 (1987): 277-283. Jungreis makes a convincing case that a number of insect species use the EMF as a navigational tool. The author also tested a number of migratory insects for significant levels of magnetic particles in their bodies that might help explain the mechanisms involved. Only one of five species tested displayed evidence of such particles. Readers with a moderate background in science will be able to follow this article with little difficulty.

Kholodov, E. A. *Magnetic Fields of Biological Objects.* Translated by A. N. Taruts. Moscow: Nauka, 1990. Although slightly technical, this book does provide great insight into the relationships among biological organisms, the Earth, and magnetic fields.

Malmivuo, Jaakko. *Bioelectromagnetism: Principles*

and Applications of Bioelectric and Biomagnetic Fields. New York: Oxford University Press, 1995. Malmivuo does a fine job describing the basic principles of bioelectromagnetism in terms that a person with little to no scientific background can grasp. Numerous charts and graphs help illustrate important points.

Markl, Hubert. "Geobiophysics: The Effect of Ambient Pressure, Gravity and of the Geomagnetic Field on Organisms." Translated by B. P. Winnewisser in *Biophysics*, edited by Walte Hoppe et al. New York: Springer-Verlag, 1983. Markl's article is written for the reader with substantial background in the sciences; nevertheless, it is worth the effort necessary to understand the author's arguments, because he addresses the problem involved in geobiomagnetism from a number of perspectives and offers insights not available elsewhere.

Plummer, Charles C., and David McGeary. *Physical Geology*. Boston: McGraw-Hill, 1999. A college-level introductory geology textbook that is clearly written and wonderfully illustrated. An excellent sourcebook of basic information on geologic terminology and fundamentals of geologic processes. Contains CD-ROM. An excellent glossary.

Reite, M., and J. Zimmerman. "Magnetic Phenomena of the Central Nervous System." *Annual Review of Biophysics and Bioengineering* 7 (1978): 167-188. This article suggests that the understanding of the functions of the central nervous systems of humans will be greatly enhanced as the study of biomagnetism proceeds. It gives one example of the vistas opened up by geobiophysical research.

Street, Philip. *Animal Migration and Navigation*. New York: Charles Scribner's Sons, 1976. Street's book contains only one short section on geobiomagnetism, but there are strong suggestions in the sections examining animal navigation and migration that the explanation for the navigational abilities of many species of life may be explained by their interaction in some manner with the EMF. Written for a general readership.

Walker, Michael M., Carol E. Diebel, et al. "Structure and Function of the Vertebrate Magnetic Sense." *Nature* 390 (November, 1997): 371-376. This article describes the function of magnetic crystals within the various sensing organs of vertebrates.

MAGNETIC REVERSALS

The investigation of Earth's magnetic field history, as recorded by diverse rock types, has disclosed that the magnetic field changes position relative to the surface of the Earth. The information accumulated from the study of these reversals is used to explain many of the events that have occurred in the history of the Earth, such as continental collisions.

PRINCIPAL TERMS

BASALT: dark-colored, fine-grained igneous rock frequently found beneath the sediment covering the ocean floor

DETRITAL REMANENT MAGNETIZATION (DRM): sedimentary rock magnetization acquired by magnetic sediment grains aligning with the magnetic field

NORMAL POLARITY: orientation of the Earth's magnetic field so that a compass needle points toward the Northern Hemisphere

POLARITY: orientation of the Earth's magnetic field relative to the Earth

REVERSE POLARITY: orientation of the Earth's magnetic field so that a compass needle points toward the Southern Hemisphere

THERMAL REMANENT MAGNETIZATION (TRM): magnetization acquired as a magma's magnetic material becomes permanently magnetized

NORMAL AND REVERSE POLARITIES

Research into the history of the Earth's magnetic field has revealed that the field has flipped polarity many times in the past. Presently, the field is oriented so that a compass needle points towardthe Northern Hemisphere of the Earth. This orientation is known as normal polarity. If a compass needle were to point toward the south, that would indicate a reverse polarity. The flipping, or reversal, of the field involves the exchange of pole positions from Northern Hemisphere to Southern, either normal to reverse or reverse to normal.

To determine whether the polarity change is a real field change or simply a modification in a rock's magnetic-recording mechanism, numerous rocks were analyzed to ascertain their magnetic characteristics and to determine whether these alter over time. Only a very small percentage of the rocks studied, including an igneous rock from Japan, displayed a self-reversing tendency. This find persuaded geophysicists that self-reversing tendencies in rocks do not need to be considered in the study of the field's history. Therefore, geophysicists do not have to test every rock to determine whether it self-reverses.

THERMAL REMANENT MAGNETIZATION

Geologists must still verify that the polarity changes are real phenomena that are consistent from one region of the Earth to another. They make use of the fact that when magnetic grains form in magma, they magnetically align themselves with the magnetic field present at that time. This type of rock recording of magnetic direction is known as thermal remanent magnetization (TRM), and the best recorder is rock of basaltic composition.

The Hawaiian Islands are the site of basaltic rock formed from magma that has sporadically erupted from the Hawaiian volcanoes over a period of millions of years. The island of Hawaii is a large volcano that sits several kilometers below sea level and rises several kilometers above sea level. Measured from base to summit, Hawaii is the highest mountain in the world.

A detailed polarity history of the island is difficult to develop because volcanic eruptions are intermittent, with several thousands of years between eruptions; however, an overall appreciation of the field changes can be acquired by sampling the distinct layers located in the eroded sides of the volcanoes. Back in the laboratory, an "abso-

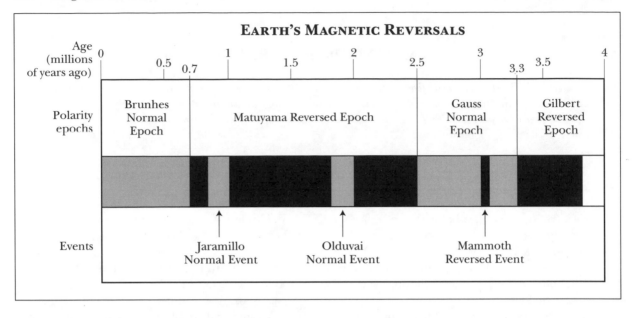

lute" date for the rocks can be obtained using radioactive-dating procedures. Relative dates—the sequence of occurrence for the samples—can also be established, as sample A is from a layer that lies below sample B, and so on. Relative dating helps assure that the absolute dates are correct. If sample A has an absolute date of 120,000 years and sample B a date of 140,000 years, but sample B is located physically above sample A in the volcano, then something is wrong. Accurate dating is an important aspect in the establishment of the polarity time scale.

By using a magnetometer, the polarity and the field direction of the sample can be determined. If the field points down, the polarity is normal. A field that points up indicates a reversed polarity. Once enough data have been collected (several hundred samples), the sample polarities can be plotted against the sample date. In this manner, the polarity history can be determined for the past 4 million years.

The polarity scale shows that from 4 million to 3.3 million years ago, the field was reversed. This period is called the Gilbert reverse epoch (the major periods are named for scientists who have advanced the discipline of magnetism). The field was normal until 2.5 million years ago during the Gauss normal epoch, except for a brief period of reversed polarity around 3 million years ago, known as the Mammoth reverse event. The Matu-

yama reverse epoch continued until 700,000 years ago. This epoch contained two normal events: the Olduvai, around 2 million years ago; and the Jaramillo, about 1 million years ago. The Brunhes is the present-day normal epoch. Other normal or reverse events may have been present in these epochs.

DETRITAL REMANENT MAGNETIZATION

Geophysicists are compelled to find other methods that verify the validity of the polarity scale and to extend and add more detail to the existing scale. One technique utilizes the sediment layer covering most of the ocean basin. This sediment can record magnetic field direction by the mechanism of detrital remanent magnetization (DRM). Long sediment cores are obtained from various areas in the ocean, and the magnetic polarity of areas along the length of the individual cores is measured. Again, a pattern of polarity changes is evident.

Radioactivity cannot efficiently date the layers of the core. Fortunately, the sediment is laid down very slowly, and this rate is measurable. The rate is on the order of millimeters per one thousand years; thus, a layer 10 millimeters from the surface was deposited approximately 15,000 years ago. Polarity is plotted against calculated age, and analysis shows that the sediment-based data correspond well with the land-based scale.

MAGNETIC SEAFLOOR STRIPES

In the 1950's, magnetometers were towed behind ships that sailed over the oceanic ridge to the south of Iceland. The data were plotted on a map of the research area, and something strange became evident: The recorded magnetic field varied over the area. The map revealed a striped pattern of weaker and stronger field intensities that was aligned parallel to the ridge, now known as magnetic seafloor stripes. Fred Vine and Drummond Matthews, working together, and Lawrence Morley, working alone, realized that polarity changes caused the stripes.

In the mid-1960's, a revolution in the Earth sciences was occurring with the development of the theory of plate tectonics. Scientists theorized that the Earth's surface rock was split into plates of thin but considerable area. These plates had boundaries that interacted in several possible ways: They could move together, or converge; they could move apart, or diverge; or they could slide past each other in an area known as a transform fault. At the diverging boundary, the motion should produce a breach between the plates, but none was found. Investigation disclosed that the volcanically active oceanic ridge was the diverging boundary and that basaltic magma quickly filled any gap. New plate material is formed at this diverging boundary, and the cooling magma records the magnetic field present at the time of cooling by thermal remanent magnetization (TRM). The cooled magma moves away parallel to the ridge as the plates diverge. The magnetic field of basaltic rock that recorded the Earth's magnetic field during a period of the reversed polarity cancels some of the Earth's present-day field. This cancellation produces an area of lower-intensity field parallel to the ridge. The rock recording normal polarity adds to the Earth's field, resulting in a strong intensity stripe.

The last polarity change 700,000 years ago was represented by rock that was located many kilometers from the center of the ridge. By dividing that distance by 700,000 years, the rate at which the plate is forming can be calculated. That value is approximately 2-5 centimeters per year depending on which portion of the ridge is being measured. This rate is comparable to how fast human fingernails grow. As a result of this movement, North America, which is west of the ridge, is now about 25 meters farther west of Europe, which is

east of the ridge, than when Christopher Columbus sailed in 1492.

IDENTIFYING TRENDS

Other research has extended the polarity scale back hundreds of millions of years. This extension permits the identification of long-term trends depicting the manner in which the field has changed polarity. The field has remained very stable, with few polarity changes, several times in the past. In the Pennsylvanian period of the late Paleozoic era, for example, the field was predominantly in the reversed position; the field was normal for a long time in the Cretaceous period of the Late Mesozoic era. At other times, including the present Cenozoic era, the field has flipped many times.

ANALYZING ROCK SAMPLES

The procedure used to study magnetic reversals depends on the area of investigation. Land-based investigations are straightforward: The researcher chooses a likely site and conducts preliminary research to ascertain whether others have studied the area and whether the site will yield samples appropriate for study. If an area displays promise, the required rock samples are obtained. The rock should not be severely weathered, as that may alter the magnetic or radioactive components of the rock, which could lead to incorrect results. The sample is collected using a gasoline-powered drill with a tube-shaped, diamond-tipped drill bit. The resulting sample is a cylinder of rock still attached at its base to the original rock. A brass tube, the size of the drill bit, is slipped over the cylinder. Brass is used because it is nonmagnetic and will not alter the sample's magnetic characteristics. Attached to the tube is a small platform on which is placed a Brunton compass, used to measure the orientation of the sample. The compass is very important, as sample orientation, the sampling site latitude and longitude, and the magnetic field direction of the sample are needed to calculate the sample's pole position. An orientation mark is made on the sample with a brass rod, and the sample is broken from the original rock. An identification number is assigned, which is carefully recorded along with all other pertinent information.

In the laboratory, the samples are prepared for the measurement of the rock's magnetic field direction by cutting them into lengths of 2.5 centimeters; thus, several small samples are obtained from each core, which can be used for dating purposes and for verifying the sample's polarity. Scientists do not rely on one measurement but make multiple assessments of a characteristic to ensure sample integrity. Because they are basalt samples, the rocks' magnetic directions are measured by a spinner magnetometer. A spinner works similarly to an electrical generator in that the sample is spun at high speed near coils of wire. The magnetic sample induces an electric current in the coils that is proportional to the strength of the sample's magnetic field. The rock's magnetic field direction is determined from these signals.

Usually, a computer performs the calculation of the final pole position using sample orientation, site latitude and longitude, and rock field direction. This pole position is plotted on a graph known as a stereonet, which is a two-dimensional representation of the Earth's surface. Normal polarity poles plot in the Northern Hemisphere of the stereonet and those of reversed polarity in the Southern Hemisphere.

The basalt is also analyzed to determine its age. The amount of a suitable radioactive element (the parent isotope) and the amount of the element into which the radioactive element decays (the daughter isotope) are measured. The sample's age is calculated from these measurements.

Obtaining Sediment Polarity Scale

The sediment polarity scale is more difficult to obtain, because the cores come from the ocean bottom—3 kilometers below sea level, in some places—using a coring device that is dropped from a ship. Back in the laboratory, samples are taken along the length of the core. Their position on the core is measured, as their distance from the top of the core determines the age of the sample. The original core is not oriented as it is taken from the seafloor, because the scientist is not interested in the sample's pole position. The scientist is interested in the sample's polarity, and an unoriented core will yield that information.

The sediment has a weak magnetic field, so a superconducting magnetometer is used for the measurements. When some materials are cooled close to absolute zero, near the boiling point of liquid helium, they have no resistance to electrical current. Superconducting magnetometers that can detect small magnetic fields, such as those of sediment, employ these materials. Again, the polarity of the sediment sample is determined from the magnetometer readings.

Detecting Seafloor Stripes

The detection of the seafloor stripes is a simple but tedious endeavor that requires the towing of a magnetometer "fish" several hundred meters behind a ship along parallel tracks across the area of interest. The magnetometer is towed to prevent the detection of ship-related magnetic fields. The readings—signal strength and ship's position—are plotted on a map and the stripe patterns are observed.

Geologic Applications

The fact that field reversals are a worldwide occurrence and instantaneous from a geological perspective permits the establishment of the field of magnetostratigraphy for correlating rock layers from various continents. In piecing together the Earth's history as revealed in the rocks, it is necessary to know what is occurring around the world at approximately the same time. In order to accomplish this feat, ways have been developed to correlate rocks in one area with rocks in another area. One way is by using index fossils, which are fossils that are widespread and common but which lived for only a short time. The problem is that index fossils are limited in distribution and cannot be used for correlating between continents. Magnetostratigraphy bypasses this difficulty.

Magnetic reversals require that the strength of the magnetic field decrease to near zero and then build back with reversed polarity. This change in polarity can take anywhere from 2,000 to 20,000 years. The magnetic field strength of the past has been both greater and weaker than present. The magnetic field of the Earth shields living organisms from damaging cosmic radiation. Periods of reduced magnetic field strength allow increased exposure of life-forms to radiation. Periods of mass extinctions or accelerated genetic changes may accompany these periods of weakened magnetic field strength.

The fact that the Earth's magnetic field has reversed many times in the past is important because it verifies the theory of plate tectonics. Before the plates rupture, the rock layers along the potential rift area dome upward; to relieve the pressure, the dome splits in a three-armed rift. Two of the arms expand in length and join arms from other domed areas. These joined arms form the boundary between the two plates; the third arm fails to enlarge but forms a wedge-shaped basin that fills with sediment. Over time, organic material in the sediment is converted to petroleum. These sources of petroleum were unknown until the development of the theory of plate tectonics. Understanding plate tectonics also increases the understanding of earthquakes and their origin, which could lead to their prediction and even to their control.

Stephen J. Shulik

CROSS-REFERENCES

Basaltic Rocks, 1274; Earth's Magnetic Field, 137; Earth's Magnetic Field: Origin, 144; Earth's Magnetic Field: Secular Variation, 150; Geobiomagnetism, 156; Geologic Time Scale, 1105; Geothermal Phenomena and Heat Transport, 43; Magnetic Stratigraphy, 167; Ocean Ridge System, 670; Oceanic Crust, 675; Plate Tectonics, 86; Polar Wander, 172; Radioactive Decay, 532; Radioactive Minerals, 1255; Rock Magnetism, 177; Spreading Centers, 727; Stratigraphic Correlation, 1153.

BIBLIOGRAPHY

Butler, Robert F. *Paleomagnetism: Magnetic Domains to Geologic Terranes.* Boston: Blackwell Scientific Publications, 1992. Butler's exploration of the Earth's magnetic fields begins with basic descriptions of what paleomagnetism is and how it occurs. The book is filled with illustrations to back up difficult concepts covered in the text.

Cox, Allan, ed. *Plate Tectonics and Geomagnetic Reversals.* San Francisco: W. H. Freeman, 1973. Cox was a leader in establishing the magnetic polarity scale for the past 4 million years. In this book, he provides fascinating introductions to chapters that are composed of seminal papers concerning magnetic reversals and their contribution to the development of the theory of plate tectonics. The papers are advanced for the average reader, but have many graphs, diagrams, and figures that merit attention. The introductions are good, as they reveal the human side of scientists in their quest for knowledge.

Garland, G. D. *Introduction to Geophysics.* 2d ed. London: W. B. Saunders, 1979. Used as text for introductory geophysics, this book contains in sections 21.4 through 21.6 very readable material on magnetic reversals, magnetic anomalies, and the magnetic character of the oceans and continents. The many figures and graphs are of interest to the general reader.

Glen, William. *The Road to Jaramillo: Critical Years of the Revolution in Earth Science.* Stanford, Calif.: Stanford University Press, 1982. A history of the plate tectonics revolution of the mid-1950's to the mid-1960's. Chapter 6 is devoted to the evolution of the magnetic polarity scale. Chapter 7 deals with magnetic seafloor stripes and their interpretation. Other chapters are also important, such as chapter 5, with its discussion of the participants in the Earth science revolution.

Jacobs, John. *Deep Interior of the Earth.* London: Chapman and Hall, 1992. This introductory geophysics textbook is formidable for the average student because there is considerable mathematics in some chapters, but does cover many useful topics such as geomagnetism, the Earth's present magnetic field, auroras, and the magnetosphere. It contains a minimum of equations but many figures and graphs.

Kennett, J. P. *Marine Geology.* Englewood Cliffs, N.J.: Prentice-Hall, 1982. Pages 71-78 are devoted to magnetostratigraphy and magnetic reversals. The text is a veritable all-you-need-to-know book about marine geology: plate tectonics, oceanic structure, sediments, margins, and history. Various dating methods are also described. Very readable, with no mathematics and some figures.

_____, ed. *Magnetic Stratigraphy of Sediments.* Benchmark Papers in Geology, vol. 54. Stroudsburg, Pa.: Dowden, Hutchinson, and Ross, 1980. A memorial edition to Norman

Watkins, who was intimately involved with the research aimed at the development of plate tectonic theory via the study of magnetization of sediments. A collection of seminal papers concerned with the magnetic stratigraphy of sediments, which relies on the fact of magnetic reversals. As a collection, the reading is uneven, as some papers are very involved, while others are less technical. The editor provides short essays at the beginning of each chapter. Graphs, tables, figures, and diagrams abound.

Lapedes, Daniel N., ed. *McGraw-Hill Encyclopedia of Geological Sciences*. New York: McGraw-Hill, 1978. Pages 704-708, under the heading "Rock Magnetism," provide a concise description of many aspects associated with rock magnetism: how rock magnetization occurs, the present field, magnetic reversals, secular variation, and apparent polar wandering, among others. Very readable, with no mathematics and a fair number of graphs, tables, and figures.

Merrill, R. T., and M. W. McElhinney. *The Magentic Field of the Earth: Paleomagnetism, the Core, and the Deep Mantle*. San Diego: Academic Press, 1998. The authors cover the basic material associated with the Earth's magnetic field. Chapters deal with the origin of the magnetic field, as well as the origin of secular variation and field reversals. Bibliography and index. Numerous tables, figures, and mathematical equations.

Monastersky, Richard. "The Flap over Magnetic Flips." *Science News* 143 (June, 1993): 378-380. In an article written for the general public, the author reviews recent data and hypothesis as to what happens during the time that the magnetic field is reversing polarity. The simple dipole field becomes more complex, but the field configuration returns to similar configuration each time it reverses.

Motz, Lloyd, ed. *Rediscovery of the Earth*. New York: Van Nostrand Reinhold, 1979. As a collection of articles for the nonscientist by renowned scientists in their respective fields, the text makes very interesting reading and is augmented with many colorful illustrations. The chapter "The Earth's Magnetic Field and Its Variations" was written by Takesi Nagata, who has written hundreds of articles on diverse aspects of geophysics besides the Earth's magnetic field.

Plummer, Charles C., and David McGeary. *Physical Geology*. Boston: McGraw-Hill, 1999. A college-level introductory geology textbook that is clearly written and wonderfully illustrated. An excellent sourcebook of basic information on geologic terminology and fundamentals of geologic processes. Contains CD-ROM. An excellent glossary.

MAGNETIC STRATIGRAPHY

The Earth's magnetic field has fluctuated between a polarity like that of today's field ("normal") and one completely opposite ("reversed") thousands of times in the last 600 million years. The magnetic minerals in erupting lavas and in settling sediments align with the prevailing field at the time the rock forms and thus record the Earth's polarity history. The pattern of polarity changes in a thick sequence of rock and can be matched from area to area, providing scientists with a very powerful tool of correlation.

PRINCIPAL TERMS

CORRELATION: matching the sequence of events (distinctive layers, fossils, magnetic polarity intervals) between two stratigraphic sections

CURIE POINT: the temperature at which a magnetic mineral locks in its magnetization

MAGNETIC DOMAIN: a region within a mineral with a single direction of magnetization; mineral grains smaller than about 100 microns contain only one domain, while larger grains can contain several domains

MAGNETIC POLARITY TIME SCALE: the geologic history of the changes in the Earth's magnetic polarity

MAGNETIC REMANENCE: the ability of the magnetic minerals in a rock to "lock in" the magnetic field of the Earth prevailing at the time of their formation

PALEOMAGNETISM: the study of the ancient magnetic field of the Earth, as recorded by magnetic minerals in rocks

RADIOMETRIC DATING: the estimation of the numerical age of a rock by measuring the decay of radioactive minerals, such as uranium, rubidium, or potassium

STRATIGRAPHY: the study and interpretation of geologic history from layered rock sequences (usually sedimentary)

MAGNETIC REMANENCE

A compass shows that the Earth's magnetic field lines point toward the North Pole, but 800,000 years ago, a compass needle would have pointed to the South Pole. The Earth's magnetic field has apparently changed polarity thousands of times in the geologic past, and an excellent record of its history extends back over the last 150 million years. This history is recorded in the magnetic minerals of rocks that were deposited or erupted in the geologic past.

Several minerals common in the Earth's crust are known to be magnetic, but the most important are the iron oxides magnetite and hematite. Magnetite contains three atoms of iron and four of oxygen; hematite contains two atoms of iron and three of oxygen. When a magma cools, the magnetic domains (areas within a crystal that have the same magnetic direction) within crystals align with the field at that time and lock in that direction as the rest of the rock crystallizes. This process is known as thermal remanent magnetization (TRM). Since TRM is formed by cooling, it is found only in igneous and metamorphic rocks. The only igneous rocks that are commonly layered and capable of stratigraphic study are lava flows. Stacked sequences of lava flows were the source of the first discovery that the Earth's magnetic field had reversed. The temperature at which this magnetization is locked in is known as the Curie point. For magnetite, the Curie point is about 578 degrees Celsius, but for hematite, it can be as high as 650 degrees Celsius. The actual Curie point varies with the variation in iron and titanium content in the mineral.

When rocks with magnetic minerals are eroded, the magnetic grains become sedimentary particles that are transported by wind and water. As these particles settle, they too align with the prevailing field. As the rest of the sediment is hardened into rock, the sedimentary rock records the direction of the field at the time it was formed. This is known as detrital (or depositional) remanent magnetization (DRM). Since most stratigraphic sequences are sedimentary, most magnetic stratigraphy is concerned with the DRM of sediments.

After a rock is formed, it is possible for water seeping through it to oxidize the iron and precipitate new minerals (particularly hematite and iron hydroxides, such as goethite). Since these new minerals are formed by chemical activity, the magnetic field they lock in is known as chemical remanent magnetization (CRM). These minerals lock in a magnetic field that records the time of chemical alteration rather than the time of the formation of the rock. This magnetization is usually a secondary, "overprinted" one that obscures the original magnetization, which is the most interesting to the paleomagnetist.

MAGNETIC POLARITY TIME SCALE

Thick sequences of lava flows or layers of sediments that span long periods of time record the changes in the Earth's magnetic field through that time interval. By sampling many levels through such a sequence, the paleomagnetist can determine the magnetic sequence, or magnetic stratigraphy, of that local section. Under the right conditions, the magnetic pattern of a section is distinctive. It can be matched to the pattern in a number of other sections of approximately the same age, and these sections can be correlated by the polarity changes. If the pattern is long and distinctive enough and its numerical age can be estimated (usually by radiometric dating), then it is possible to match the pattern to the worldwide magnetic polarity time scale and to estimate an even more precise age.

The worldwide magnetic polarity time scale was first developed in the 1960's, when a group of scientists found that all lava flows with potassium-argon dates less than about 700,000 years were normally magnetized (like the present Earth's field), and those older than 700,000 years were usually reversely magnetized (opposite the present Earth's field). They began to seek out more and more lava flows around the world, sampling them for both their magnetism and their potassium-argon age. In about five years of sampling, they found a consistent pattern: All rocks of the same age had the same magnetic polarity, no matter where they were located. This immediately suggested that their magnetic properties were caused by worldwide magnetic field reversals rather than by local peculiarities of the rocks themselves.

SEAFLOOR SPREADING

Continuous sequences of lava flows that could be dated, however, were not available for time periods older than about 13 million years. What was needed was a terrestrial process that continuously recorded the Earth's magnetic field behavior and could be dated. Such a process was discovered in the early 1960's at the same time that magnetic polarity reversals were documented. The crust of the ocean floor is constantly pulling apart, and the gap is filled by magma from the mantle below. When the magma cools it locks in the magnetic polarity prevailing at the time. Continual seafloor spreading pulls apart this newly cooled crustal material and carries it away from the mid-ocean ridge, causing new magma to fill in the rift, to cool, and to lock in a new polarity. This process of cooling, magnetization, and spreading acts as a "tape recorder" that produces a magnetic record of the present field at the center of the ridge and progressively older fields away from the ridge crest. In a few places in the ocean basin, this ocean-floor "tape recording" goes back about 150 million years.

The steady spreading of oceanic plates provides the only continuous record of the changes in the Earth's magnetic field between 13 million and about 150 million years ago. In 1968, the first attempt was made to construct a magnetic polarity time scale. Using the known rates of spreading of several mid-ocean ridges, scientists extrapolated several oceanic spreading records back to about 100 million years and placed tentative dates on all the polarity events that were recorded. Since 1968, many attempts have been made to date this polarity time scale more precisely. Ironically, most of the new dates have shown that the original 1968 extrapolation was remarkably good, and new versions of the time scale differ very little from the first version. This proves the assumption that seafloor spreading is a relatively steady, constant process.

EVALUATION AS CORRELATION TOOL

Magnetic stratigraphy has proven to be one of the most powerful tools of correlation and dating available. It has many features that other methods of correlation do not. Unlike correlation by distinctive rock units or by the changes in fossils through time, magnetic polarity changes happen on a worldwide basis and can be recorded in any type of rock (lavas or marine or nonmarine sediments) formed at that time. No rock type is formed worldwide, and fossils are restricted by the environments in which they lived. Thus, rocks formed in both the oceans and land can be directly correlated by magnetic stratigraphy, even though the rock sequences are different and they do not share the same fossils.

Another unique feature of magnetic stratigraphy is that polarity changes take place within about 4,000-5,000 years, which is considered instantaneous in a geological sense for any event that occurred more than about a million years ago. Thus, a polarity zone boundary represents a worldwide, geologically instantaneous "time plane" that can be used as a very precise marker wherever it is found. By contrast, the changes in fossil assemblages in a stratigraphic section can seldom resolve events down to a few thousand years, and radiometric dates typically have analytical errors that are anywhere from hundreds of thousands to millions of years.

The major limitation of magnetic stratigraphy is that most magnetic patterns are not unique. When paleomagnetists sample a rock, they get only normal or reversed polarity, not a numerical age. To date a sequence, some other form of dating must be used to place the magnetic pattern on the magnetic polarity time scale. For example, a sequence of "normal-reversed-normal" is not unique by itself; it has occurred many times in the geologic past. If, however, a distinctive set of fossils or a radiometric date can constrain that pattern to a certain period in Earth history, then there may be only one part of the magnetic polarity time scale that matches that pattern at that particular point in time. This match gives a more precise age estimate than does the fossil or radiometric date alone.

ROCK SAMPLING AND MEASUREMENT METHODS

Paleomagnetists study ancient magnetic fields by sampling rocks of the proper age and rock type. If it is a lava or other very hard rock, they use a portable drill that collects a short core about 2 centimeters in diameter. Lavas tend to be strongly magnetized compared to other rock types. If the rock to be sampled is a softer sedimentary rock

that might break up while drilling, then simple chisels and scrapers are used to extract a hand sample. Sediments tend to have magnetizations that are weaker than those of lavas by a factor of one hundred to one thousand. In addition, only fine-grained sediments (siltstones, claystones, fine sandstones, and limestones) record a remanence; coarse sandstones have more than one magnetic domain within each magnetic grain, which cancel one another out. In both cases, the direction of the present Earth's magnetic field is marked on the sample, so it can be compared with the direction recorded in the rock.

The samples are then measured in a device called a magnetometer, which determines the direction and intensity of the field recorded by the sample (its natural remanent magnetization, or NRM). Some magnetometers are portable, but they are only suitable for measuring strongly magnetized lavas. Most labs now use a superconducting cryogenic magnetometer. Its sensing area is kept at 4 degrees Celsius above absolute zero (−269 degrees Celsius) so that it is superconducting, or has almost no resistance to electrical current. When a sample is lowered into the sensing area, even weak magnetic fields in the sample cause changes in electrical current, which are then converted into a magnetic signal.

Typically, the field direction found in the sample (NRM) is a composite of several different magnetic fields. For example, if the rock were deposited during a period of reversed polarity, it may still have a young magnetic overprint acquired during the normal polarity that is seen today. The interaction of these two directions may give an NRM that is neither normal nor reversed, but some intermediate direction. To get rid of unwanted overprinting, the samples must be treated with high temperatures (thermal demagnetization) or high external magnetic fields (alternating field demagnetization), which destroys the less stable (and presumably young overprinted) component of the magnetization. After each treatment at progressively higher temperatures or progressively higher applied fields, the sample is measured again. Interpreting the change of direction and strength of the magnetic component during this stepwise demagnetization enables the paleomagnetist to decide which magnetic mineral is the carrier of the magnetic remanence and also which temperature or field is best for magnetically "cleaning" samples.

After magnetic cleaning, each sample produces a direction that presumably represents the field direction at the time the rock was formed. This remanence is known as the primary, or characteristic, remanence. Because several samples are taken of each lava flow or of each sedimentary bed, the directions of all of the samples from a given site are averaged to omit random "noise." The more tightly all the directions from a site cluster, the more reliable they are likely to be. There are statistical methods that measure this clustering and allow the paleomagnetist to determine the quality of the data. Data that cluster poorly or give nonsensical results can be rejected.

ROLE IN UNDERSTANDING GEOLOGIC HISTORY

Magnetic stratigraphy has become one of the most powerful tools of dating geologic events. It is critical to understanding geologic history and provides a much greater understanding of certain aspects of the geological past than was previously possible. For example, there has been great controversy over how fast evolution takes place or when mass extinctions occurred. By more precisely dating the sequences in which these events are recorded, scientists can determine rates of evolution much more precisely or determine a much more accurate date for the timing of a mass extinction, which may, in turn, allow the determination of the causes of these events and resolve many long-standing controversies. Magnetic stratigraphy has been used to date the long history of evolution of fossil mammals and dinosaurs in the terrestrial environment and the details of the evolution of the world ocean in marine sections. In many marine sections, the use of magnetic stratigraphy has allowed precise dating of climatic changes, particularly the glacial-interglacial fluctuation of the last ice age. This precise dating, in turn, has allowed scientists to determine that the glacial-interglacial cycles were controlled by changes in the Earth's orbital motions, and they thus deciphered the cause of the ice ages. A better understanding of how some of these events (climate change, ice ages, mass extinctions) occurred in the past will help scientists to decide if such events are likely to happen again in the near future.

Donald R. Prothero

CROSS-REFERENCES

Earth's Magnetic Field, 137; Earth's Magnetic Field: Origin, 144; Earth's Magnetic Field: Secular Variation, 150; Geobiomagnetism, 156; Geologic Time Scale, 1105; Magnetic Reversals, 161; Polar Wander, 172; Rock Magnetism, 177; Stratigraphic Correlation, 1153.

BIBLIOGRAPHY

Bebout, Gray E. *Subduction Top to Bottom*. Washington, D.C.: American Geophysical Union, 1996. Bebout's book gives clear definitions and explanations of subduction, folding, faults, and orogeny. Illustrations and maps help to clarify some difficult concepts.

Boggs, Sam, Jr. *Principles of Sedimentology and Stratigraphy*. Columbus, Ohio: Charles E. Merrill, 1986. A college-level textbook on stratigraphy and sedimentology, which devotes a chapter (chapter 15) to magnetic stratigraphy. This chapter, although brief and not concerned with the practical aspects of magnetic stratigraphy, does give one of the few up-to-date accounts available in any stratigraphy textbook.

Brush, Stephen G. *Nebulous Earth: The Origin of the Solar System and the Core of the Earth from Laplace to Jeffreys*. Cambridge: Cambridge University Press, 1996. Brush's book, volume 1 in the History of Modern Planetary Physics series, contains useful information on the nebular hypothesis, the origin of the solar system, and the Earth's core. Includes a bibliography and index.

Cox, Allan, ed. *Plate Tectonics and Geomagnetic Reversals*. San Francisco: W. H. Freeman, 1973. A collection of the classic papers that led to the plate tectonics revolution, edited by a man who was responsible for the paleomagnetic data that propelled it. It includes many of the pioneering papers that first described the reversals of the Earth's magnetic field as well as the discovery of the magnetic polarity time scale and seafloor spreading. One of its best features is the editorial introductions, which place the papers in historical context.

Cox, Allan, and R. B. Hart. *Plate Tectonics: How It Works.* Palo Alto, Calif.: Blackwell Scientific, 1986. A college-level textbook that explains many facets of plate tectonics, with examples and problem sets. Several chapters give an excellent discussion of paleomagnetism.

Glen, William. *The Road to Jaramollo: Critical Years of the Revolution in Earth Sciences.* Stanford, Calif.: Stanford University Press, 1982. A history of the plate tectonics revolution, recounting the important individuals and their discoveries that led to the discovery of continental drift and seafloor spreading. The development of the magnetic polarity time scale was a key part of this, and the rivalry between various labs in discovering and dating the magnetic reversals is described in detail.

Gurnis, Michael, et al., eds. *The Core-Mantle Boundary Region.* Washington, D.C.: American Geophysical Union, 1998. This collection of articles is one volume of the American Geophysical Union's Geodynamics series. Although intended for the specialist, the essays contain plenty of information suitable for the careful college-level reader. Bibliography.

Hamblin, William K. *Earth's Dynamic Systems.* 8th ed. Upper Saddle River, N.J.: Prentice Hall, 1998. This geology textbook offers an integrated view of the Earth's interior not common in books of this type. The illustrations, diagrams, and charts are superb. Includes a glossary and laboratory guide. Suitable for high school readers.

Kearey, Philip, and Frederick J. Vine. *Global Tectonics.* 2d ed. Cambridge, Mass.: Blackwell Science, 1996. This college text gives the reader a solid understanding of the history of global tectonics, along with current processes and activities. The book is filled with colorful illustrations and maps.

Kenneth, J. P., ed. *Magnetic Stratigraphy of Sediments.* Stroudsburg, Pa.: Dowden, Hutchinson, and Ross, 1980. An anthology of classic papers on magnetic stratigraphy. Most of the papers deal with magnetic stratigraphy of marine sediments and their application to paleo-oceanographic problems, but some also cover terrestrial magnetic stratigraphy. The editor wrote introductions that place all the papers in historical context.

Lindsay, E. H., et al. "Mammalian Chronology and the Magnetic Polarity Time Scale." In *Cenozoic Mammals of North America,* edited by M. O. Woodburn. Berkeley: University of California Press, 1987. This chapter contains one of the best reviews of the practical aspects of magnetic stratigraphy as applied to terrestrial sections. Although it is written on the professional level, it assumes little or no background in rock magnetism.

McElhinny, M. W. *Paleomagnetism and Plate Tectonics.* New York: Cambridge University Press, 1973. One of the most popular college textbooks on paleomagnetism. Although some of the text is outdated, it contains an excellent discussion of the field as it was at that time.

Prothero, D. R. *Interpreting the Stratigraphic Record.* New York: W. H. Freeman, 1989. A college-level textbook on stratigraphy, containing a chapter on magnetic stratigraphy. The discussion on magnetic stratigraphy is considerably clearer, more thorough, and more up-to-date than that by Boggs.

_____. "Mammals and Magnetostratigraphy." *Journal of Geological Education* 36 (1988): 227. A nontechnical article detailing the practical aspects of terrestrial magnetostratigraphy. It also reviews the progress on the terrestrial, fossil-mammal-bearing record up to the time of the article.

Tarling, D. H. *Palaeomagnetism: Principles and Applications in Geology, Geophysics, and Archaeology.* London: Chapman and Hall, 1983. One of the best books available on paleomagnetism. It discusses magnetic stratigraphy on a much more general level than does McElhinny. A good first resource in reading about the subject.

POLAR WANDER

Evidence from several of the Earth sciences clearly demonstrates that the Earth's magnetic and geographic poles have been located at widely separated places relative to its surface during the planet's geological history.

PRINCIPAL TERMS

ASTHENOSPHERE: a hypothetical zone of the Earth that lies beneath the lithosphere and within which material is believed to yield readily to persistent stresses

ICE AGES: periods in the Earth's past when large areas of the present continents were glaciated

LITHOSPHERE: the outer layer of the Earth

NORTH GEOGRAPHIC POLE: the northernmost region of the Earth, located at the northern point of the planet's axis of rotation

NORTH MAGNETIC POLE: a small, nonstationary area in the Arctic Circle toward which a compass needle points from any location on the Earth

PALEOMAGNETISM: the intensity and direction of residual magnetization in ancient rocks

PLATE TECTONICS: the study of the motions of the Earth's crust

CONTINENTAL DRIFT

Shortly before World War II, geophysicists discovered a method of determining the location of rocks on the Earth's surface at the time they were formed, relative to the north magnetic pole. Thus began the study of paleomagnetism. Paleomagnetic studies quickly yielded very puzzling and often contradictory results. The new science produced evidence that the north magnetic pole has changed its location by thousands and even tens of thousands of miles hundreds of times during the Earth's geologic history. Since Earth scientists are generally agreed that the north magnetic pole has always corresponded closely with the north geographic pole, this evidence seemed to indicate that the Earth's axis of rotation must have changed, a highly unlikely occurrence.

As the paleomagnetic evidence for different locations of the poles in the past accumulated through measurements of rock formations from around the world, more and more Earth scientists began to accept the theory of continental drift. This theory offered an explanation of the paleomagnetic evidence without the necessity of postulating that the Earth's axis of rotation had changed in the past. Alfred Wegener, early in the twentieth century, had drawn attention to the theory that the continents moved in relation to one another. Most geologists initially greeted his theories with derision, but many others agreed with him, causing an often bitter controversy in the Earth sciences that lasted almost half a century. The ever-growing body of paleomagnetic evidence could be explained by postulation that the surface areas of the Earth move in relationship to the planet's axis of rotation. This explanation proved to be more acceptable to geologists than the idea that the axis of rotation changed.

With the growing acceptance of the theory of continental drift in the 1940's, geologists began trying to explain the mechanism that caused it. They postulated that the Earth has a stable and very dense core overlain by an area called the asthenosphere, which is made up of rock rendered plastic by heat and pressure. Floating on the asthenosphere is the Earth's outer crust, or lithosphere. Dislocation within the Earth caused by the action of heat and pressure result in the movement of the lithosphere relative to the core and to the axis of rotation. The initial attempts to explain continental drift have been considerably revised and refined into the modern theories of plate tectonics and ocean-bed spreading, but the basic premise remains the same: The surface areas of the Earth move in relationship to its core and to its axis of rotation. The result of the movement of the Earth's lithosphere is that the surface area located at the axis of rotation does not remain the same over long periods of time. This shifting accounts for the apparent "wandering" of the poles as well as for several other puzzling aspects of Earth's geologic history.

EVIDENCE FROM GLACIERS AND PALEOCLIMATOLOGY

Striking evidence that the surface areas of the Earth have moved enormous distances during geologic history relative to its axis of rotation comes from the study of glaciers. Observations from around the globe show that almost all the land areas of the Earth have been glaciated at some time in the past, including parts of Africa, India, and South America presently located on or near the equator. Without postulating either a substantial shifting of the Earth's surface relative to its axis of rotation or a change in the axis, equatorial glaciation is inexplicable. If global temperatures dropped to levels sufficient to glaciate even the equator at some time in the past, all life on Earth would have been destroyed. If, however, the areas of Africa, India, and South America which are presently located in tropical locales once shifted to the polar regions and shifted from there to their present locations, their ancient glaciation is not at all mysterious.

The study of paleoclimatology has also produced evidence supporting the proposition of the shifting of the Earth's crust relative to its axis of rotation. Paleoclimatologists study the climates of past ages on the various parts of Earth's surface. They have found that Antarctica once supported rich varieties of plant and animal life, many of which could only have lived in temperate and even subtropical climates. Explorations in the far northern regions of Canada, Alaska, and Siberia have revealed that those areas also supported multitudes of animals and luxurious forests in the past, as did many of the islands presently located within the Arctic Circle. Obviously, those regions must have had much warmer climates at the times when the plant and animal life flourished there, which can be explained in only one of two ways: Either the climate of the entire world was much warmer in the past, or those areas now located near the poles were once located in much more temperate latitudes. If the entire world had warmed to the point that the polar areas had temperate climates, the tropical and subtropical areas of the Earth

would have been much too hot to support life, which is demonstrably untrue according to the fossil record. Thus, the areas now near the poles must have been located in temperate climatic latitudes in the past.

APPROXIMATE CHRONOLOGY

Earth scientists, using the evidence discussed above and paleomagnetism, have established an approximate chronology showing which areas of the Earth's surface were located at its north rotational axis during past ages. At the beginning of the Cambrian period (roughly 600 million years ago), the area of the Pacific Ocean now occupied by the Hawaiian Islands was at or near the Earth's north rotational axis. By the Ordovician period 100 million years later, the surface of the Earth had shifted in such a manner that the area approximately 1,000 miles north and east of modern Japan was on or near the North Pole. Fifty-five million years later, during the Silurian period, modern Sakhalin Island north of Japan was within the Arctic Circle. During the next 20 million years, the area of modern Kamchatka in eastern Siberia shifted to a position very near the pole. Earth scientists have identified ninety-nine separate locations that occupied the polar regions at one time or another during the ensuing 395 million years from the Silurian to the Pleistocene. During the past million years, forty-three different areas of the Earth's surface have been on or near the north geographic poles, averaging over 1,500 miles distance from each other.

REMAINING CONTROVERSIES

Although contemporary Earth scientists have reached a consensus that the surface of the Earth has shifted relative to the planet's axis of rotation many times in the past, several problems remain. One area on which there is no unanimity of opinion is the mechanism responsible for crustal shift. The answer most likely lies in high-pressure physics and the nature of the asthenosphere. Another, more controversial, problem concerns the speed of crustal shifts. During most of the twentieth century, almost all the geologists who were daring enough to accept the theory of continental drift assumed that the movement of surface features of the Earth relative to the axis of rotation and relative to one another was very slow, on the order of a few inches per year at most. Then an increasing number of Earth scientists began arguing for short periods of relatively rapid movement of the Earth's crust and long periods of stability.

These problems notwithstanding, there can no longer be any doubt that the surface of the Earth has shifted many times relative to its rotational axis. The phenomenon has led to the mistaken assumption that the rotational axis has moved relative to the Earth's surface—thus the term "polar wander." The rotational axis of the Earth has remained constant throughout its history; apparent polar wander is caused by the shifting of the Earth's crust.

STUDY OF PALEOMAGNETISM

The study of paleomagnetism has yielded irrefutable evidence that many different areas of the Earth's surface have occupied polar positions during the history of the planet. Scientists studying paleomagnetism measure the weak magnetization of rocks. Virtually all rocks contain iron compositions that can become magnetized. In the study of paleomagnetism, the most important of these compositions are magnetite and hematite, which are commonly found in the basaltic rocks and sandstones. Paleomagnetism may also be measured in less common rocks that contain iron sulfide. In igneous rocks, magnetization takes place when the iron compositions within the rocks align themselves with the Earth's magnetic field as the rocks cool. In sedimentary rocks, small magnetic particles align with the magnetic field as they settle through the water and maintain that alignment as the sediments into which they sink solidify.

Magnetized rocks not only indicate the direction of the north magnetic pole at the time they were formed but also show how far from the pole they were at formation by the angle of their dip. Scientists call their horizontal angle the variation and their dip the inclination. Variation reveals the approximate longitude of the rock sample at the time of its formation, relative to the north magnetic pole, and inclination gives its approximate latitude. By ascertaining the date at which the rock sample being examined was formed, using well-known dating methods, scientists are able to establish the area of the Earth's surface relative to the north magnetic pole that was occupied by the rock at the time of its formation.

PROBLEMS IN PALEOMAGNETIC STUDIES

There are, however, many pitfalls for the unwary scientist investigating paleomagnetism. A rock whose magnetism is being studied may have moved considerable distances from its place of formation by glacial action or by crustal movement along a major fracture in the Earth's surface, such as the San Andreas fault on North America's west coast. High temperatures, pressure, and chemical action can distort or destroy the magnetization of a rock. Folding and the movement of the continents relative to one another may also alter the original orientation of the rocks whose magnetism is being studied. All these pitfalls may be avoided through the expedient of basing estimates of the relative position of the north magnetic pole on a great number of rock samples of the same age, gathered from many different locations on all the continents.

Another problem in paleomagnetic studies involves the constant movement of the north magnetic pole relative to the north geographic pole. Recent studies show that the north magnetic pole moved from 70 degrees to 76 degrees north latitude (approximately 345 miles, or 576 kilometers) during the period 1831-1975. This phenomenon might accurately be called true polar wander, though it does not involve any alteration either of the Earth's axis of rotation or of the surface of the planet relative to its axis of rotation. Most geophysicists studying this movement have concluded that over a period of several thousand years, the average position of the north magnetic pole coincides closely with that of the north geographic pole. Thus, when scientists learn that the north magnetic pole was located near Hawaii 600 million years ago, it is a virtual certainty that modern Hawaii was at that time located near the north geographic pole.

RECONSTRUCTION OF CONTINENTAL POSITIONS

The position of the magnetic field as determined for rocks of different ages in North America yields a wandering path from the eastern Pacific Ocean, looping across the ocean basin, through western Asia, to its present position. The apparent path of polar wandering as determined from Europe matches that of the North American path in general shape but does not coincide except at the present pole position. The curves are separated by about 40 degrees of longitude which is about the width of the Atlantic Ocean. These paths do coincide if the Atlantic Ocean is removed and the continents are shifted to allow a fit along opposing coastlines. Analysis of apparent pole positions from all of the continents indicate that each continent has its own series of magnetic poles.

Recognition of polar wander supports the basic concept of continental drift which is now imbedded in the phenomenon of seafloor spreading. Repositioning landmasses so that rocks of the same age agree in their apparent North Pole position allows reconstruction of continental positions through time. Thus, paleomagnetic data indicate that the continents were once assembled into a single landmass, known as Pangaea which was split

apart as the Atlantic Ocean opened. Once the position of landmasses throughout time is fully understood, paleomagnetic data from deformed or displaced rocks within that landmass may be used to reconstruct the terrain prior to deformation.

Paul Madden

Cross-References

Continental Glaciers, 875; Earth's Crust, 14; Earth's Magnetic Field, 137; Earth's Magnetic Field: Origin, 144; Earth's Magnetic Field: Secular Variation, 150; Earth's Rotation, 106; Fossil Record, 1009; Geobiomagnetism, 156; Ice Ages, 1111; Magnetic Reversals, 161; Magnetic Stratigraphy, 167; Mass Extinctions, 1043; Paleoclimatology, 1131; Plate Tectonics, 86; Rock Magnetism, 177.

Bibliography

Brooks, C. E. P. *Climate Through the Ages.* New York: McGraw-Hill, 1949. This work synthesizes data from many Earth sciences to demonstrate clearly that every surface area of the Earth has at different times in the planet's history been subjected to every extreme of climate, from arctic to equatorial and everything in between.

Daly, R. A. *Our Mobile Earth.* New York: Scribner's, 1926. Daly was one of the first Earth scientists to propose that the Earth's surface has shifted over long distances relative to its axis of rotation and over relatively brief periods of time; this book explains Daly's views and his theory on the mechanism that causes the shifts. Suitable for anyone with a high school education.

Doell, Richard R., and Allan Cox. *Paleomagnetism.* Vol. 8, Advances in Geophysics. New York: Academic Press, 1961. Despite its extensive use of technical terms, this book can prove informative to the layperson interested in the scientific underpinnings of paleomagnetism and associated problems.

Hapgood, Charles H. *Earth's Shifting Crust.* Philadelphia: Chilton, 1958. Revised and reissued by Chilton in 1970 as *The Path of the Poles.* Hapgood's pioneering work presents a sound and clear argument that the Earth's surface has shifted rapidly many times relative to its axis of rotation. Includes a foreword by Albert Einstein. Readily understandable to the layperson.

Hibben, Frank C. *The Lost Americans.* New York: Thomas Y. Crowell, 1946. Hibben provides a wealth of information concerning the great animal extermination at the end of the Pleistocene. His evidence shows clearly that the extermination was the result of one or more natural catastrophes of enormous proportions. Material is presented in nontechnical language accessible to any reader with a high school education.

Hooker, Dolph Earl. *Those Astonishing Ice Ages.* New York: Exposition Press, 1958. Hooker's book is designed to make information concerning past ice ages accessible to a general audience. Includes evidence that areas now on or near the equator were once glaciated.

Irving, E. "Pole Positions and Continental Drift Since the Devonian." In *The Earth: Its Origins, Structure, and Evolution*, edited by M. W. McElhinny. New York: Academic Press, 1980. Irving uses the results of half a century of magnetic measurements to establish which areas at the Earth's surface were located at its magnetic pole during the various geological periods. A layperson can follow the gist of his arguments.

King, Lester C. *Wandering Continents and Spreading*

Sea Floors on an Expanding Earth. New York: John Wiley & Sons, 1983. Most of this book is written in language much too technical for the general public. It does contain, however, a chapter on paleomagnetism with a good explanation of the techniques and pitfalls of that science and a chapter of plate tectonics with considerable evidence that shiftings of the Earth's crust have occurred rapidly and at irregular intervals over geological history.

Lutgens, Frederick K., and Edward J. Tarbuck. *Earth: An Introduction to Physical Geology.* 6th ed. Upper Saddle River, N.J.: Prentice Hall, 1999. This college text provides a clear picture of the Earth's systems and processes that is suitable for the high school or college reader. In addition to its illustrations and graphics, it has an accompanying computer disc that is compatible with either Macintosh or Windows. Bibliography and index.

Merrill, R. T., and M. W. McElhinney. *The Magnetic Field of the Earth: Paleomagnetism, the Core, and the Deep Mantle.* San Diego: Academic Press, 1998. The authors cover the basic material associated with the Earth's magnetic field. Chapters deal with the origin of the magnetic field, as well as the origin of secular variation and field reversals. Bibliography and index. Numerous tables, figures, and mathematical equations.

Munyan, Arthur C., ed. *Polar Wandering and Continental Drift.* Tulsa, Okla.: Society of Economic Paleontologists and Mineralogists, 1963. The articles contained in this publication range in topic from ancient climates through the study of paleomagnetism. Although a background in geology is necessary to understand all the nuances in the articles, most of them can be followed by the general reader.

Opdyke, Neil D. *Magnetic Stratigraphy.* San Diego: Academic Press, 1996. Intended for someone with little scientific background, *Magnetic Stratigraphy* examines the magnetic fields of the Earth, focusing largely on paleomagnetism. Contains fifty pages of bibliographical resources, as well as an index.

Tarling, Donald H., Peter Turner, et al., eds. *Paleomagnetism and Diagenesis in Sediments.* London: Geological Society, 1999. This collection of essays, written by leading scientists in their respective fields, deals with key paleomagnetic concepts and principles. Although technical at times, there are many charts and illustrations that help to clarify complicated subjects. Bibliography and index.

Tauxe, Lisa. *Paleomagnetic Principles and Practice.* Boston: Kluwer Academic Publishers, 1998. Tauxe offers a clear definition of paleomagnetism, then explores its causes and its effects on the Earth and the Earth's systems. The book comes with a CD-ROM that complements the material covered in the chapters.

Whitley, D. Gath. "The Ivory Islands of the Arctic Ocean." *Journal of the Philosophical Society of Great Britain* 12 (1910). Whitley describes in detail the myriad bones of large land mammals which lived during the Pleistocene, stacked to heights of more than 100 feet on many of the islands within the present Arctic Circle. That these animals could not have lived in those areas given present climatic conditions is axiomatic.

ROCK MAGNETISM

Rock magnetism is the subdiscipline of geophysics that has to do with how rocks record the magnetic field, how reliable the recording process is, and what conditions can alter the recording and therefore raise the possibility of a false interpretation being rendered by geophysicists.

PRINCIPAL TERMS

BASALT: a very common, dark-colored, fine-grained igneous rock

BLOCKING TEMPERATURE: the temperature at which a magnetic mineral becomes a permanent recorder of a magnetic field

CURIE TEMPERATURE: the temperature above which a permanently magnetized material loses its magnetization

DAUGHTER PRODUCT: an isotope that results from the decay of a radioactive parent isotope

DETRITAL REMANENT MAGNETIZATION: the magnetization that results when magnetic sediment grains in a sedimentary rock align with the magnetic field

FERROMAGNETIC MATERIAL: the type of magnetic material, such as iron or magnetite, that retains a magnetic field; also called permanent magnet

GRANITE: a low-density, light-colored, coarse-grained igneous rock

MAGNETITE: a magnetic iron oxide composed of three iron atoms and four oxygen atoms

RADIOACTIVITY: the spontaneous disintegration of a nucleus into a more stable isotope

THERMAL REMANENT MAGNETIZATION: the magnetization in igneous rock that results as magnetic minerals in a magma cool below their Curie temperature

MAGNETIC FIELD PRODUCTION

The direct study of the Earth's magnetic field is only four centuries old. This study involves the measurement of the field with scientific instruments and subsequent analysis of the resulting data. Four centuries is a very small fraction of the 4.6 billion years that the Earth has existed; thus, direct study affords scientists very little understanding of the nature of the field over long periods of time. It is useful to know what happened to the Earth's magnetic field in those billions of years before the present, because the field can be a source of information about conditions on the Earth's surface and its interior. Magnetic minerals in rocks serve as recording devices, giving scientists clues regarding the nature of the ancient magnetic field.

A moving electric charge, such as an electron, produces a magnetic field that is the ultimate source of any larger magnetic field. An atom is composed of a nucleus, with its protons and neutrons, and the electrons that surround the nucleus. The protons do not move within the nucleus, and therefore they do not produce a magnetic field. The electrons, however, orbit the nucleus, and this movement produces a weak magnetic field. In addition, the electrons spin on their axes, and this activity also gives rise to a small magnetic field.

TYPES OF MAGNETISM

Because all atoms have electrons orbiting and spinning, one might think that all materials should have a permanent magnetic field, but the situation is more complicated. Strictly speaking, every material is magnetic, but there are different types of magnetism. Some materials are paramagnetic: When they are placed in an external magnetic field, the atoms align with the field. The atoms act as small compasses, orienting with the field, and the material is magnetized; the magnetic fields produced by the atom's electrons add to the intensity of the external field. When the external field is removed, however, the atom's orientation becomes randomized because of vibrations caused by heat, and the material is consequently demagnetized. Many materials, such as quartz, are paramagnetic and are not able to record the Earth's magnetic field.

177

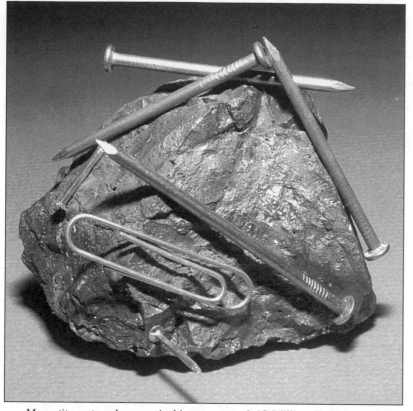

Magnetite, a strongly magnetized iron compound. (© William E. Ferguson)

A much smaller number of minerals are ferromagnetic. There are various types of ferromagnetism, but the underlying principle is the same. In ferromagnetic materials, an external magnetic field again aligns the atoms parallel to the field, and the material is magnetized. When the field is removed, however, the atoms remain aligned, and the substance retains its magnetization; it is "permanently" magnetized. Actually, the substance can be demagnetized by heating or stress. Dropping a bar magnet on the floor or striking it with a hammer will demagnetize it slightly. The shock randomizes some of the atoms so that they cease to contribute to the overall magnetic field. The heating of a magnet above its Curie temperature also destroys its magnetization by randomizing the atoms and making the material paramagnetic. As the temperature drops below the Curie point, the material becomes slightly remagnetized, because the weak field of the Earth aligns some of the atoms.

In ferromagnetic materials, atoms are not all aligned in one direction; rather, they are found in aligned groups, called domains. Under a microscope, the domains are barely visible. Within a particular domain, the atoms are aligned, but all the domains are not aligned in the same direction. A "permanent" magnetic material that is unmagnetized has all the domains randomly aligned, and the overall field cancels to zero. When placed in a magnetic field, some of the domains realign parallel to the direction of the field and stay aligned after the field is removed. It is these domains that give the material its overall magnetization. If a high enough magnetic field is applied, all the domains align with the field, and the magnetization has reached its saturation point; the strength of the material's magnetic field is at a maximum. One of the areas of research for physicists is the quest for materials that have high magnetic field strengths but with less material. Such materials are useful in making small, but powerful, electric motors.

MAGNETIC MINERALS

Rocks are classified into three main groups: igneous, formed from crystallized molten rock; sedimentary, formed from weathered rock material; and metamorphic, produced when other rock is modified with heat, pressure, and fluids. Most magnetic minerals occur in igneous and sedimentary rocks.

Materials such as iron, cobalt, and nickel are ferromagnetic; for this reason they are used in making various permanent magnets. These metals are not found naturally on the Earth's surface in the uncombined state, so they do not contribute to rocks' recording ability. Most of the minerals that make up rocks, such as quartz and clay, are not ferromagnetic. These minerals are useless as recorders, but many rocks contain magnetite or hematite, which are good recorders. These common magnetic minerals are oxides of iron.

Hematite is Fe_2O_3, which means that there are two iron atoms for every three oxygen atoms. Hematite is red in color, similar to rust on a piece of iron. Most reddish-brown hues in sedimentary rock are caused by hematite. This magnetic mineral is not a very strongly magnetized compound, but it is a very stable recorder in sedimentary rocks. Unfortunately, in many cases, its formation postdates that of the rock in which it occurs, so it does not necessarily record the magnetic field at the time of the rock's formation. Magnetite (Fe_3O_4) has been known as lodestone for several millennia. It is a strongly magnetized iron compound that makes some igneous rocks very magnetic and supplies some of the recording ability of sedimentary rocks. The magnetite in rocks can record the field direction by one of several methods.

THERMAL REMANENT MAGNETIZATION

In igneous rocks, magnetite crystals form as the magma cools. As the crystals grow, they align themselves with any magnetic field present. This process is called thermal remanent magnetization, or TRM. If the crystals are quite small or quite large, they cannot permanently record the field direction; after a short time the recording fades and becomes unreadable. The magnetism of such small grains is called superparamagnetism: They do align with a magnetic field, but they easily lose their orientation. The larger grains contain many magnetic domains that become misaligned over time so that the recording fades.

Grains the size of fine dust are good recorders. Unfortunately, not all igneous rocks have grains of the proper size. The size of the mineral crystal depends on the rate of cooling: When magma is cooled very slowly, large crystals are produced, while a rapid cooling results in smaller crystals. Granite is coarse-grained and thus is not the best recorder. The best igneous recorder is basalt, a black, fine-grained rock. Basalt is fairly common on the surface of the Earth, particularly in the ocean basins, where nothing but basalt underlies the sediment on the basin floor.

A useful magnetic recorder must provide information about how old it is. Basalt again fills this requirement, for its crystallization can be dated by measuring the amount of radioactive elements and their daughter products it contains. Clearly, basalt is an ideal source of information on the magnetic field. Unfortunately, it does not occur everywhere on the Earth; moreover, as a recorder, it covers only times of eruptions of magma. Some other recorder must be used to fill in the blanks.

DETRITAL REMANENT MAGNETIZATION

Sedimentary rock is formed from the products of the rock weathering that accumulate mostly in watery environments, such as rivers, lakes, and oceans. Clastic sedimentary rocks are formed from fragments of rock and mineral grains, such as grains of quartz in sandstone. Chemical sedimentary rock is derived from chemical weathering products, such as calcium carbonate or calcite, which is the major constituent of limestone. Most of the material in sedimentary rocks is not ferromagnetic, but there are a few grains of magnetite and other ferromagnetic compounds. As the grains fell through the water, they aligned with the magnetic field present at that time. When they hit the bottom, they retained the orientation, for the most part, and were subsequently covered by more sediment. This process is termed detrital remanent magnetization, or DRM.

An interesting aspect of DRM is the role that organisms play in its formation. The grains of magnetic minerals that fall through the water are oval-shaped, and when they strike the surface of the sediment they become misaligned with the field. Organisms such as worms disturb the sediment in a process known as bioturbation, which moves the sediment around and realigns the magnetic grains with the field. In the mid-1980's, it was discovered that certain varieties of bacteria have small grains of magnetite in their bodies. The bacteria use the grains like compasses to find their way down into the sediment on which they feed. The bacteria eventually die, and the magnetite grains become part of the sediment, aligned with the magnetic field; this phenomenon is known as biomagnetism.

The grain-size problem also occurs in DRM, given that a sediment particle can be the size of a particle of clay, a boulder, or anything in between. Conglomerate, a rock composed of rounded pebbles and other large particles, is not a good recorder, nor is coarse sandstone. Finer sandstones, shales, siltstones, and mudstones are much better. Most chemical rocks, such as halite (common table salt), are poor recorders; limestone may or may not be good, depending on the conditions of formation.

The magnetization in sedimentary rocks is generally between one thousand and ten thousand times weaker than is the magnetization in a basalt. Very sensitive magnetometers are needed to measure the magnetic field in these specimens. To be useful in geomagnetic studies, sedimentary rocks must be dated, but this is a difficult task, as they cannot be dated using radioactive methods. By a complex method of determination, fossils can act as indicators of the age of the rock in which they are found. If igneous rock layers are located above and below the rock layer of interest, and if these igneous rock layers can be dated, an intermediate age can be assigned to the sedimentary layer.

STUDY OF ROCK MAGNETISM

A magnetometer useful in the study of rock magnetism is the superconducting rock magnetometer, or SCM. Superconductivity is the phenomenon of a material's losing its resistance to electric current at low temperatures. Liquid helium is used to cool a portion of the magnetometer, composed of a cylinder of lead closed at one end. As the lead cools, it becomes superconducting, and if done in a region of low magnetic-field intensity, this low field is "trapped" inside the cylinder. Magnetic field sensors known as SQUIDS, or superconducting quantum interference devices, are very sensitive to low-intensity magnetic fields. The sample is lowered into the device, and its electronic display shows the intensity of the sample's magnetization. Such devices are useful in studying the rock magnetism of low-intensity sedimentary rocks.

The Curie temperature is important for establishing the thermal remanent magnetization for igneous rocks. A sample of a particular ferromagnetic material in a magnetic field is heated and the temperature is measured; the sample's Curie temperature is determined when the pull of the magnetic field on the sample weakens. The Curie point for various ferromagnetic materials is established by this method. Once that is done, the procedure is reversed. A sample of an unknown ferromagnetic material can be heated in a magnetic field to determine its Curie point, which can then be compared with the established table of values

to identify the magnetic mineral. This method does not establish the exact composition of the material, but it does narrow down the possibilities, which is of value because other methods for determining composition are more expensive. In addition, it has been discovered that Curie temperature is not the only factor critical to the recording process. At the Curie point, the material is ferromagnetic but the recording ability is weak. The material has to cool through the blocking temperature for recording stability. Thereafter, magnetic minerals are magnetically stable for periods of billions of years.

Another area of study is the determination of the best grain size and shape for magnetic recording. Researchers experiment with different sizes and shapes of magnetic grains in magnetic fields of various strengths and directions and measure their responses to changes. It was found that crystals of magnetic materials such as magnetite develop features known as domains. These are areas where the atoms are aligned in one direction and produce the unified magnetic field for the domain. A small crystal has only one domain that can easily shift to another direction; therefore small crystals are poor recorders. If the crystal is quite large, it has many domains in which it is again easy to shift direction. Crystals with one large domain or several small domains are magnetically "hard" in that it is more difficult to shift the magnetic alignment. For magnetite, these are dust-sized particles, around 0.03 micron in diameter.

METHODS OF "MAGNETIC CLEANING"

Other research reveals that the rock's recording of the field is not as "neat and clean" a process as portrayed in the previous paragraphs. Many events can lead to the alteration of the magnetic alignment. If the rock is heated above the Curie point and then cooled, the magnetic alignment is that of the field present at that time, and the old alignment is erased. The rock may be changed chemically, and old magnetic minerals may be destroyed and new ones produced. This process is referred to as chemical remanent magnetization, or CRM.

These secondary magnetizations can be removed in some cases, and they can even provide more information on the rock's history. One method of magnetic cleaning or demagnetization involves subjecting the rock sample to an alternating magnetic field while other magnetic fields are reduced to zero. This "cleaning" will remove that portion of the mineral's magnetization that is magnetically "softer" than is the maximum alternating field. The magnetization above the level is unaffected and should represent the original magnetization. Heating a sample to a certain temperature is another method of demagnetization.

ROLE IN STUDY OF EARTH'S HISTORY

The study of the Earth's magnetic field history, and all the inferences about the Earth drawn from that study, depends on the ability of rocks to record information about the magnetic field at the time of the rocks' formation. That ability, in turn, is dependent upon the magnetic characteristics of a few permanently magnetized minerals, such as magnetite.

The study of rock magnetism is rather esoteric; only a few individuals worldwide are involved in this subdiscipline of geomagnetism. Yet, such study has shown that rocks can faithfully record the history of the Earth's magnetic field. This record is used to infer conditions on the Earth hundreds of million years ago. Such studies have lent support to the idea that the continents have actually moved over the surface of the globe, and thus the theory of plate tectonics was born, with all its implications for the formation and location of petroleum and ore deposits, the origin of earthquakes and volcanoes, and the formation of mountain ranges such as the Himalaya. Such is an example of the odd twists and turns that science can take. Seemingly inconsequential findings can lead to a theory with great potential for making the Earth and its workings much more understandable.

Stephen J. Shulik

CROSS-REFERENCES

Basaltic Rocks, 1274; Earth's Age, 511; Earth's Magnetic Field, 137; Earth's Magnetic Field: Origin, 144; Earth's Magnetic Field: Secular Variation, 150; Geobiomagnetism, 156; Magmas, 1326; Magnetic Reversals, 161; Magnetic Stratigraphy, 167; Oxides, 1249; Plate Tectonics, 86; Polar Wander, 172; Radioactive Decay, 532.

BIBLIOGRAPHY

Butler, Robert F. *Paleomagnetism: Magnetic Domains to Geologic Terranes.* Boston: Blackwell Scientific Publications, 1992. Butler's exploration of the Earth's magnetic fields begins with basic descriptions of what paleomagnetism is and how it occurs. The book is filled with illustrations to back up difficult concepts covered in the text.

Cox, Allan, ed. *Plate Tectonics and Geomagnetic Reversals.* San Francisco: W. H. Freeman, 1973. Cox provides fascinating introductions to chapters that are composed of seminal papers concerning magnetic reversals and their contribution to the development of the theory of plate tectonics. Information on rock magnetism is scattered throughout the book in discussions on baked sediments, magnetization of basalt, magnetic intensity, and self-reversals in rocks. The papers are advanced for the average reader, but there are many graphs, diagrams, and figures that merit attention.

Glen, William. *The Road to Jaramillo: Critical Years of the Revolution in Earth Science.* Stanford, Calif.: Stanford University Press, 1982. This book gives a history of the plate tectonics revolution of the mid-1950's to the mid-1960's. Rock magnetism is specifically covered on pages 103-109. Other aspects of rock magnetism are discussed in various portions of the book, for example, those dealing with the magnetic minerals associated with rock magnetism and deep-sea core work. Of particular interest are the sections devoted to instruments used to measure rock magnetism.

Hamblin, William K. *Earth's Dynamic Systems.* 8th ed. Upper Saddle River, N.J.: Prentice Hall, 1998. This geology textbook offers an integrated view of the Earth's interior not common in books of this type. The illustrations, diagrams, and charts are superb. Includes a glossary and laboratory guide. Suitable for high school readers.

Hargraves, R. B., and S. K. Banerjee. "Theory and Nature of Magnetism in Rocks." In *Annual Review of Earth and Planetary Sciences*, vol. 1, edited by F. Donath. Palo Alto, Calif.: Annual Reviews, 1973. The article covers the theories of the natural remanent magnetization of rocks (NRM). NRM is the combined magnetization of all magnetic minerals in a rock, such as those resulting from TRM. The various carriers of remanence are also covered, with a table that lists each mineral and its composition, crystal structure, magnetic structure, and other pertinent information. The paleomagnetic potential of rocks is also discussed. A few mathematical equations, but nothing too formidable. Numerous figures and a long list of references.

Lapedes, D. N., ed. *McGraw-Hill Encyclopedia of Geological Sciences*. New York: McGraw-Hill, 1978. Pages 704-708, under the heading "Rock Magnetism," provide concise descriptions of how rock magnetization occurs, the present field, magnetic reversals, secular variation, and apparent polar wandering. The text is very readable, with no mathematics and a fair number of graphs, tables, and figures.

Merrill, R. T., and M. W. McElhinney. *The Magentic Field of the Earth: Paleomagnatism, the Core, and the Deep Mantle*. San Diego: Academic Press, 1998. The authors cover the basic material associated with the Earth's magnetic field. Chapters deal with the origin of the magnetic field, as well as the origin of secular variation and field reversals. Bibliography and index. Numerous tables, figures, and mathematical equations.

O'Reilly, W. *Rock and Mineral Magnetism*. New York: Chapman and Hall, 1984. O'Reilly covers the atomic basis for magnetism, the magnetization process, the various remanent magnetizations such as TRM, the magnetic properties of minerals, and, finally, the applications of rock and mineral magnetism. Many tables, figures, and photographs of minerals. Some mathematics and chemistry are also included but should not be too difficult. References are included at the end of each chapter.

Plummer, Charles C., and David McGeary. *Physical Geology*. Boston: McGraw-Hill, 1999. A college-level introductory geology textbook that is clearly written and wonderfully illustrated. An excellent sourcebook of basic information on geologic terminology and fundamentals of geologic processes. Contains CD-ROM. An excellent glossary.

Stacey, F. D. *Physics of the Earth*. New York: John Wiley & Sons, 1977. Under section 9.1, "Magnetism of Rocks," the author provides a short, technical description of rock magnetism. Several figures show the various types of magnetic alignments. The domain structure of ferromagnetic materials is also discussed. Many other aspects of the Earth's magnetic field are also covered at a technical level.

Tarling, D. H. *Paleomagnetism: Principles and Applications in Geology, Geophysics, and Archeology*. New York: Chapman and Hall, 1983. A very good resource on the subject of paleomagnetism, or the ancient magnetic field. Chapter 2 is devoted to the "physical basis" for the magnetization of material, with a discussion of the atomic level and the resulting magnetic domains. The various remanent magnetizations are covered in detail. Chapter 3 deals with the various magnetic minerals and their identification. The magnetization of the various rock types are covered in chapter 4. Chapter 5 discusses instruments used in paleomagnetic work. The remainder of the text deals with mathematical analysis used in paleomagnetic work and, finally, paleomagnetic applications.

5
SEISMOLOGY

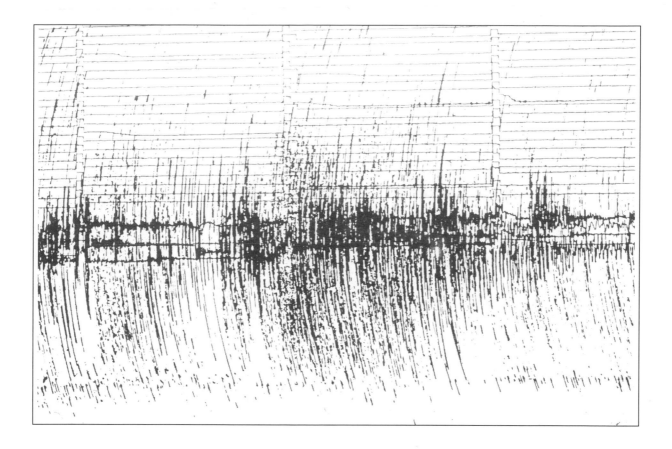

CREEP

Creep involves small deformations under small stresses acting over long periods of time. The effect of time on rock properties is important in understanding geologic processes as well as deformation and failure. In general, creep results in a decrease in strength and an increase in ductile or plastic flow.

PRINCIPAL TERMS

CREEP TESTS: experiments that are conducted to assess the effects of time on rock properties, in which environmental conditions (surrounding pressure, temperature) and the deforming stress are held constant

DISLOCATION: a linear defect or imperfection in the atomic structure (arrangement) of rock-forming minerals; virtually all minerals and crystals contain dislocations

DUCTILITY: the rock property that expresses total percent deformation prior to rupture; the maximum strain a rock can endure before it finally fails by fracturing or faulting

ELASTIC DEFORMATION: a nonpermanent deformation that disappears when the deforming stress is removed

PLASTIC DEFORMATION: a nonrecoverable deformation that does not disappear when the

deforming stress is removed

STRAIN: the deformation resulting from the stress, calculated from displacements; it may involve change in volume, shape, or both

STRAIN RATE: the rate at which deformation occurs, expressed as percent strain per unit time

STRESS: the force per unit area acting at any point within a solid body such as rock, calculated from a knowledge of force and area

STRESS-STRAIN TEST: a common laboratory test utilized in the study of rock and soil deformation; stress is plotted versus strain throughout the test along the vertical and horizontal axes

ULTIMATE STRENGTH: the peak or maximum stress recorded in a stress-strain test

ENVIRONMENTAL FACTORS

Creep is an important geologic process related to rock deformation. It involves small displacements that occur under the influence of small but steady stresses that act over long periods of time. Scientists and engineers involved in experimental rock and soil deformation and the assessment of creep commonly perform stress-strain tests. These experiments are designed to deform Earth materials in the laboratory under controlled conditions.

The effect of environmental factors such as surrounding (confining) pressure, temperature, pore-fluid pressure, and strain rate (or time) have been documented through the years based on countless tests. In essence, these factors dictate whether rocks will fracture as brittle substances or considerable ductile flow and creep strain will occur prior to rupture. The effect of increasing confining pressure on dry rocks (containing no appreciable amounts of liquid pore fluid) at room

temperature is to increase both the ultimate strength and the ductility. Rocks tested under constant confining pressures tend to weaken and become more ductile as temperature increases. An increase in confining pressure on rocks saturated with pore fluids generally results in a decrease in both ultimate strength and ductility. This result is caused by the fact that part of the load (or stress) is carried by the pore fluid and less by grain-to-grain contacts. Decreasing the strain rate (or increasing the time during which the stress is applied) lowers ultimate strength and increases ductility—which basically defines the influence of creep strain on rock properties.

DEFORMATION STAGES

The mechanism of creep may be expressed as follows: Rocks subjected to the steady action of small stresses first undergo elastic deformation. After a given period of time, the elastic limit is exceeded. (The elastic limit is the point of no return beyond which deformation is permanent or nonrecoverable.) Following elastic deformation, rocks undergo strain hardening, a phenomenon characterized by a continuous rise in stress with increasing strain because of dislocations moving within individual mineral grains, interfering with one another and causing a literal "traffic jam" at the interatomic level. This initial stage of deformation comprising elastic behavior and strain hardening is termed transient creep; following the transient creep stage, steady-state creep is achieved. During this stage, rocks deform by plastic or ductile flow under a constant strain rate. Deformation mechanisms are characterized by gliding flow (intracrystalline movements) and by recrystallization. Gliding flow may take the form of translation or twin gliding. In translation gliding, layers of atoms slide one interatomic distance or a multiple thereof relative to adjacent layers. The overall mineral grain changes shape, but the interatomic lattice (arrangement) remains unchanged. In twin gliding layers, atoms slide a fraction of an interatomic distance relative to adjacent layers, distorting the interatomic lattice. Recrystallization involves rearrangements of the deforming minerals at the molecular scale through solution and redeposition by local melting or by solid diffusion. A common type of recrystallization occurs by local melting at those grain contacts experiencing the greatest stress and by precipitation (or redeposition) along grain contacts subjected to low stress. Recrystallization can also occur through mixing and rearrangement of the atoms and molecules in mineral grains by "spreading" into each other, analogous to the mixing of gases and liquids through the process of diffusion. Beyond steady-state creep, the final stage, known as accelerated creep, is reached. During accelerated creep, strain rate increases rapidly, ending in rock failure by fracturing or faulting. Deformation mechanisms during this final stage are characterized by cataclasis and formation of voids or pores. Cataclasis involves mechanical crushing, granulation, fracturing, and rotation of mineral grains. It results in intergranular movements.

Creep strain is equal to the sum of all of the stages of deformation, starting with elastic strain, followed by the transient stage, and culminating with steady-state and accelerated flows prior to failure by rupture. The rate at which creep strain occurs is very sensitive to temperature, with creep rates increasing rapidly as temperature rises. In fact, increasing temperature has been used as an alternative to experiments involving low strain rates or deformations over long periods of time. Increasing temperature or lowering strain rates affects rocks in a similar fashion by decreasing the ultimate strength and increasing the overall ductility.

LABORATORY AND FIELD STUDY

The study of creep is conducted in the laboratory in special experiments under controlled conditions. Environmental factors such as confining pressure, temperature, pore pressure, and strain rate are closely monitored and regulated. Among the methods that have been used to study creep are tension, bending, uniaxial compression, and triaxial compression. Pure tension has been utilized mainly to study creep in metals but has not been common in testing rocks. Bending is a simple method that has been used in creep studies of coal. By far, however, uniaxial and triaxial compression experiments have been utilized most often in testing creep behavior in rocks. In uniaxial compression, rock samples are loaded with the stress directed vertically. The sample itself is generally unconfined laterally. The vertical or axial load is maintained constant, and percent strain is plotted as a function of time. In triaxial compression, a rock sample is loaded in a pressure chamber, and an all-around confining pressure is applied. The magnitude of the confining pressure can be significant, simulating pressure conditions expected several kilometers below the Earth's surface. An axial or vertical load is then applied and maintained constant. The total stress along the vertical axis of the specimen is the sum of the axial load plus the confining pressure. The deforming stress therefore equates to the axial load. The latter is often referred to as differential stress or deviatoric stress because it is the stress that deviates from the all-around confining (or hydrostatic) pressure. The deviatoric stress is maintained constant until failure occurs. Some pressure vessels are equipped with heating elements in order to increase the surrounding or ambient temperature; others have the additional capability of recording the increase in fluid pressure for samples saturated with pore fluids. Some of the recent designs have the capacity to subject samples to confining pressures of 20 kilobars (20,000 atmospheres), temperatures of 1,000 degrees Celsius, and strain rates as low as 10^{-10} per second.

Creep tests are not easy to run from a purely mechanical point of view. For example, the choice of magnitude of the deviatoric stress is a matter of difficulty and importance because each experiment may occupy an apparatus for a considerable time. (It is not unusual for creep tests to last for a period of one year.) In addition, if the stress is too low, little effect is produced; if too high, failure may occur too quickly. Temperature effects must be closely controlled because they can accelerate creep rates. Also, with many rocks, absorption of water produces effects similar to creep, so that humidity must be monitored and regulated.

Earth and soil creep that may eventually result in landsliding or damage to foundations and retaining structures can be studied in the field and laboratory. Evidence of creep strain along slopes may be detected by direct observation; bent or distorted tree trunks are common indicators. The rate of creep strain is recorded through installation and monitoring, or strain (displacement), gauges. The magnitude of pressures exerted on human-made structures resulting from creep flow can be predicted through laboratory experiments designed to record shear strength and shrink-swell (potential volume change) of soils and argillaceous (clay-rich) rocks. Specialized laboratory experiments simulating pressure-temperature conditions expected in the Earth's mantle have been designed to study the effects of creep as a mechanism for releasing stored strain energy resulting in earthquakes.

PETROGRAPHIC STUDY

Mechanisms common to creep (such as translation and twin gliding, recrystallization, and cataclastic flow) are routinely documented by studying thin sections of deformed rock specimens using the petrographic (polarizing) microscope and the universal stage. The petrographic microscope differs from a conventional model in that it is equipped with two polarizing elements and other accessories. When both polarizers (or nicols) are engaged, a ray of light transversing a mineral grain is generally refracted into two rays that vibrate in

planes at right angles to each other. In contrast, when the lower polarizer is the only one engaged, the light impinging on the mineral grain is plane-polarized. The universal stage allows the rock-forming minerals in the thin sections to be studied at different inclinations from vertical and horizontal axes.

APPLICATIONS FOR STRUCTURAL AND ENGINEERING GEOLOGY

Understanding creep, or the effect of time on rock properties, is important to the structural geologist studying rock mechanics as well as to the engineering geologist concerned with landslide prediction and control or with the stability of Earth retaining structures. Deformation of Earth materials may occur as brittle failures or after considerable ductile or plastic strain has resulted. Intuitively, it is easy to understand rock failure through fracturing or cracking, since one tends to think of rocks as brittle substances. Under the influence of high confining pressures, elevated temperatures, and stresses acting over long periods of time, however, rocks can and do undergo considerable ductile or plastic deformation. Entire mountain chains of visibly folded rock are common throughout the planet.

Studying the process of folding, creep, and rock flowage has a number of practical applications. For example, it is common to find commercial quantities of oil and gas in folded structures known as anticlines. Therefore, understanding how and where rocks fold and which rock types are likely to develop the best porosity and permeability during the process is of critical significance in the search for new petroleum reserves. Similarly, quantifying creep strain and rates is very important in predicting, preventing, and correcting Earth hazards such as landslides and in the proper design of foundations and retaining walls. On the subject of slope instability, creep can play a key role. Earth creep, or the slow, imperceptible downslope movement of soil and argillaceous rocks is the main cause of a specific type of landslide recognized worldwide. In this form of creep, considerable volumes of Earth move as the sum of a very large number of minute displacements of individual particles and grains that do not necessarily strain at the same rate. This motion may be caused by expansion and contraction of clay-rich rocks in response to fluctuations in moisture content, which is especially critical in Earth materials containing minerals from the smectite or montmorillonite family that expand considerably when wet and contract when dry. The end result of this creep strain is mass flow or landsliding. Similarly, soil creep can exert enormous stresses on retaining walls and foundations. Pressures exceeding 207,000 kilopascals or 2.1 kilobars (where one bar is basically equivalent to one atmosphere of pressure) have been recorded in north-central Texas.

Finally, creep is important in understanding earthquake mechanisms. Earthquakes are classified as shallow, intermediate, and deep based on their focal depth. Shallow earthquakes have focal depths not exceeding 70 kilometers. Intermediate earthquakes occur within a range of 70-300 kilometers. Deep earthquakes occur between 300 and 700 kilometers. The elastic rebound theory and the brittle failure of rock are accepted as the main mechanism giving rise to earthquakes—but only of the shallower types, because at depths where intermediate and deep earthquakes occur, the environmental conditions are conducive to ductile behavior. Convection currents in the Earth's mantle and the thermal instability of creep have been proposed as the major mechanism responsible for the deeper earthquakes.

Robert G. Font

CROSS-REFERENCES

Cross-Borehole Seismology, 190; Discontinuities, 196; Earth's Mantle, 32; Earthquake Distribution, 277; Earthquake Prediction, 309; Elastic Waves, 202; Expansive Soils, 1479; Experimental Rock Deformation, 208; Faults: Normal, 213; Faults: Strike-Slip, 220; Faults: Thrust, 226; Faults: Transform, 232; Fold Belts, 620; Folds, 624; Landslides and Slope Stability, 1501; Petrographic Microscopes, 493; Plate Motions, 80; Rocks: Physical Properties, 1348; San Andreas Fault, 238; Seismic Observatories, 246; Seismic Tomography, 252; Seismometers, 258; Soil Chemistry, 1509; Stress and Strain, 264.

BIBLIOGRAPHY

Dennen, William H., and Bruce R. Moore. *Geology and Engineering.* Dubuque, Iowa: Wm. C. Brown, 1986. A complete reference dealing with the subject of engineering geology, this text is well illustrated and readable. Chapter 13 includes a discussion of slope stability and creep flow.

Font, Robert G. *Engineering Geology of the Slope Instability of Two Overconsolidated North-Central Texas Shales.* Vol. 3, Reviews in Engineering Geology. Washington, D.C.: Geological Society of America, 1977. A review of three distinct types of landslides, their causes, occurrence, and prevention along northcentral Texas; one type is a classical example of creep strain and ductile flow. Although technical, the article is well illustrated and should be relatively easy for the nonscientist to follow.

Griggs, David T. "Creep of Rocks." *The Journal of Geology* 47 (April/May, 1939): 225-251. A classic reference on the subject of creep, this article is thorough and well illustrated. Covers the most important aspects of the subject as related to rock mechanics and structural geology. Although technical, it should not be too difficult for the nonscientist.

Heard, Hugh C. "Effect of Large Changes in Strain Rate in the Experimental Deformation of Yule Marble." *The Journal of Geology* 71 (March, 1963): 162-195. A very thorough and well-written article on the effects of time on rock properties. Technical, but definitely readable material for the nonscientist.

Hobbs, Bruce E., Winthrop D. Means, and Paul F. Williams. *An Outline of Structural Geology.* New York: John Wiley & Sons, 1976. Chapter 1 is a good review of mechanical properties of rocks, the concepts of stress and strain, and the response of rocks to stress. A fine discussion of ductile flow and creep strain is presented in the chapter, which is well illustrated.

Nabarro, Frank. *The Physics of Creep: Creep and Creep-Resistant Alloys.* London: Taylor and Francis, 1995. Nabarro's book introduces the physics of creep in language comprehensible to readers with little scientific background. The text also deals with ways to combat slope instability.

Spencer, Edgar W. *Introduction to the Structure of the Earth*. New York: McGraw-Hill, 1977. Chapter 4 is a good review of experimental study of rock deformation and is well illustrated. The effects of time on rock properties and the subject of creep are reviewed.

CROSS-BOREHOLE SEISMOLOGY

Cross-borehole seismology is a geophysical exploration technique involving the acquisition of data that can be used to image subsurface geology and determine the spatial distribution of the physical properties of geological materials. The data are acquired by placing seismic detectors in one borehole and a seismic source in an adjacent borehole. Seismic energy that propagates from the source travels through or reflects from subsurface geological materials. The total time of travel and amplitude of this energy is recorded by the detectors and is used to construct an image of the subsurface geology between the boreholes.

PRINCIPAL TERMS

AMPLITUDE: the maximum departure (height) of a wave relative to its average value

IMAGING: a computer method for constructing a picture of the subsurface geology from acquired seismic data

INVERSION (INVERSE PROBLEM): using measured data to construct a geological model that describes the subsurface and is consistent with the measured data

LITHOLOGY: the description of rocks, such as rock type, mineral makeup, and fluid in the rock pores

REFLECTIVITY: the ratio of the amplitude of the reflected wave to that of the incident wave

RESOLUTION: the ability to separate two features that are very close together

SEISMIC REFLECTION PROFILING METHOD: measurements made of the travel times and amplitudes of events attributed to seismic waves that have been reflected from interfaces where seismic properties change

SEISMIC TOMOGRAPHY: a processing technique for constructing a cross-sectional image of a slice of the subsurface from seismic data

SEISMOLOGY: the study of seismic waves

TRAVELTIME: the amount of time it takes seismic energy to travel from the source into subsurface geology and arrive back at a seismic detector

CROSS-BOREHOLE VERSUS SURFACE EXPLORATION

Various methods exist that allow geoscientists to produce reasonable images of the Earth's subsurface geology. Seismic data can be acquired using surface seismic exploration and cross-borehole seismology. When these data are properly processed and interpreted, an image of the subsurface is constructed.

In seismic reflection profiling, geophysicists typically arrange seismic detectors along a straight line at or near the surface of the Earth and then generate sound waves by vibrating the ground. Seismic waves can be generated by detonating charges of dynamite, by dropping a weight on the ground, or by pounding the ground with a sledgehammer. To eliminate environmental risks associated with using explosives, a system called vibroseis is often used, in which a huge vibrator mounted on a special truck repeatedly strikes the Earth to produce sound waves.

A seismograph records how long it takes sound waves to travel through or reflect from rock layers and arrive at seismic detectors. In seismic reflection profiling, the recorded data display the amplitudes of the reflected sound waves as a function of traveltime. Such a graphic record is called a seismogram. The seismic source, detectors, and seismograph are then moved a short distance along the line, and the experiment is repeated. For seismic reflection profiling, the frequency range of investigation is limited to between 10 and 300 cycles per second, depending on the seismic velocity in the subsurface materials and the depth to the target of interest. This frequency range does not provide the necessary resolution of subsurface features that is typically required when making production decisions.

To help solve this problem, cross-borehole seismology was developed. In this method, a source of seismic energy is placed in one borehole, and appropriate seismic detectors are placed in an adjacent borehole. Boreholes can be vertical or horizontal. Between seven and twenty detectors are placed 2 to 5 meters apart, and adjacent wells are at least 100 to 300 meters apart but always less than 1 kilometer apart. The source is fired, and the resulting seismic energy propagates through or reflects from the rock and is detected in the adjacent borehole. The travel times and amplitudes of seismic waves that have been transmitted or reflected through the rock mass between the drill holes are recorded. The source and detectors are then moved to another position in their respective boreholes, and the process is repeated. This procedure is continued until the region of interest is adequately covered by the propagating energy.

In contrast to seismic investigations conducted at the surface, cross-borehole surveys are performed at target depths; therefore, they do not suffer from the low-frequency resolution problems that are pervasive in surface seismic recordings. For cross-borehole seismology, the operating frequency range is from 400 to 30,000 cycles per second. Such data can provide high resolution of the subsurface geology between boreholes.

CROSS-BOREHOLE TECHNOLOGY

The idea of cross-borehole seismology has existed for many years. Field experiments were discussed as early as 1953 by Norman H. Ricker. Seismic surveys between boreholes were carried out in France by the Institute Français du Petrole in the early 1970's. Initially, the images produced from the data were fuzzy, filled with artifacts, and generally not worth the cost to acquire. Some geophysicists, however, saw in these images the proof of cross-borehole concepts and knew that with additional processing effort, coupled with advancements in seismic source and detector technology, the clarity and resolution would make the acquisition and processing of cross-borehole seismic data cost-effective.

Few disciplines affecting exploration geophysics developed more rapidly in the 1980's and 1990's than cross-borehole methods. Interaction among various scientific disciplines, along with advancements in seismic field data acquisition, imaging and inverse-problem theory, and computing speed, accelerated the development of cross-borehole

seismology. Borehole source and detector technology, as well as computer-based data analysis algorithms, advanced to the point that routine application of the method became feasible by the late 1990's. Much of the progress in cross-borehole seismology has been driven by the development of powerful, nondestructive borehole seismic sources and by the increasing need for enhanced definition of oil-producing reservoirs.

Many borehole sources are made of piezoelectric ceramic materials that convert varying voltages in the material into mechanical vibrations, typically generating seismic signals in the range of 400 to 2,000 cycles per second. The same materials are also used in the seismic detectors because of their controllability, their high-frequency response, and their good impedance match to hard rock. The sources and detectors are packaged in metallic housings to form borehole probes that can operate in fluid-filled boreholes more than 1,000 meters deep. In the simplest borehole arrangement, the source and the detectors hang freely in the boreholes, but for better coupling or for operation in dry holes, more sophisticated sources and detectors with electrically powered clamping mechanisms lock the seismically active parts firmly against the rock in the borehole. For clamping detectors, the piezoelectric material is oriented to respond to seismic waves in three dimensions, allowing both compressional and shear waves to be recorded.

Another important borehole source was developed by Texaco in the 1990's. A nondestructive, broadband air-gun array provided energy transmission over distances exceeding 600 meters. The broadband nature of the source provided the necessary spatial resolution of the subsurface.

Although the source waveform may be a pulse, transmitting a continuous, coded signal is a better way to achieve maximum transmission range. In the mid-1990's, a high-energy, broadband, clamped borehole vibrator was designed and successfully implemented to accomplish this task. Inside the borehole vibrator, a 114-kilogram mass is suspended below a hydraulic piston that is set into axial motion by a hydraulic valve. The vibrator clamp is coupled to the motion of the piston, which transmits stress to the borehole wall and sends seismic energy into the formation. This is analogous to surface vibroseis. The vibrator source transmits a continuous, controlled signal over a long period of time, keeping the stress low while transmitting considerable energy. Thus, a large amount of seismic energy is transmitted into the formation without harming the cement casing in a cased borehole.

SEISMIC TOMOGRAPHY

By applying the techniques of tomographic imaging, cross-borehole seismic data can be used to image the subsurface and to estimate subsurface physical properties. Tomographic imaging is a highly effective way of condensing and organizing the large amount of information contained in high-density cross-borehole seismograms. The medical community has used computed tomography (CT) scanning since the 1960's to generate imaged cross sections of different portions of the human body. In medical imaging techniques, such as X-ray tomography, the source and receiver rotate all the way around the object to be imaged. Cross-borehole seismology is similar to the medical case, except the angular coverage is not nearly as great.

In the early 1980's, geophysicists began applying techniques similar to those of medicine to geophysical problems. These problems range from estimating the internal velocity structure of the subsurface to formulations that provide a complete three-dimensional image of the subsurface geology. In the early 1990's, tomographic reconstruction became a standard technique for analyzing cross-borehole and surface seismic data.

Obtaining information about subsurface geology from cross-borehole data constitutes a type of inverse problem. That is, measurements are first made of energy that has propagated through and reflected from within the subsurface. The received travel times and amplitudes of this energy are then used to estimate values of the physical parameters of the medium through which it has propagated. The parameters that are typically extracted are velocities and depths, from which a gross model of subsurface structure can be derived. Initially, gross subsurface structure was considered the ultimate goal of seismic tomography, but it became obvious that an accurate set of velocities versus depth can effectively be used to constrain other types of seismic inversion, including the velocity control necessary for constructing an accurate depth image of the subsurface. During the 1990's, the goal for cross-borehole seismic data evolved into obtaining a reasonable estimate of the properties of the subsurface geology, particularly density, compressibility, shear rigidity, porosity (pore space in rocks), and permeability (ability for fluids to flow in rocks). To accomplish this goal, both compressional and shear wave data must be recorded.

CROSS-BOREHOLE DATA PROCESSING

Since seismic waves propagating in the Earth's subsurface readily spread, refract, reflect, and diffract, algorithms for processing cross-borehole seismic data had to be developed to produce realistic subsurface images. Effective software and interactive graphics are required to pick and process the acquired cross-borehole data into a reasonable image. The received borehole signals are first filtered and digitized. By adding up many waveforms per detector, signal-to-noise ratios are enhanced. When using a vibrator source, impulse seismograms are obtained by cross-correlating the received data with the vibrator waveform, which further reduces random noise in the data. By implementing these signal enhancement techniques, the transmitted power in the borehole can be kept low enough to avoid damage to expensive boreholes and still record high-quality seismograms across distances of more than a few hundred meters in most rock types.

Using a variety of computer programs, cross-borehole seismograms are processed to yield seismic sections that represent the Earth's reflectivity in time. However, since wells are drilled in depth, not in travel time, the seismic data need to be converted to depth in the imaging process. Furthermore, since reflectivity is a property associated with subsurface interfaces, a rock sample in the laboratory does not have any intrinsic reflectivity. Therefore, reflectivity is not an actual rock property, and this parameter must be converted to another parameter that really describes the rock. Typically, the chosen parameter is the internal velocity of seismic wave propagation through subsurface materials.

Internal velocity structure is estimated from cross-borehole and surface seismic data using seismic tomographic methods. Because subsurface velocities vary according to the physical properties of the rock through which the wave travels, geophysicists can use these velocities to determine the depth and structure of rock formations. In addition, since seismic waves change in amplitude when they are reflected from rocks that contain gas and other fluids, the fine details of amplitude changes in the seismograms can be used to infer the type of rocks in the subsurface. Thus, conventional seismic sections can yield far more information about the subsurface geology by using cross-borehole data and tomographic techniques to produce subsurface images as a function of depth and to estimate rock properties from the images.

The basic procedure for processing both cross-borehole and surface seismic data using tomography is iterative seismic tomography. The procedure involves the following steps: picking transmitted or reflected events on raw seismograms; associating these events with the structure of a proposed subsurface geological model; dividing the section of rock to be imaged into a grid of pixels with local rock and fluid parameter values; using the laws of physics to trace raypaths of seismic waves through the proposed geological model from the seismic source down through or from a subsurface boundary and back to the seismic detectors; comparing the ray-traced travel times and amplitudes from the model with the travel times and amplitudes recorded on the seismogram; and updating the geological model (medium parameters or geometry) to make the ray tracing consistent with the observed data. Corrections are made to the medium parameters or geometry systematically to reduce the differences between the observed and modeled travel times and amplitudes. After several iterations, the differences between the recorded and modeled data become acceptably small, and a "best" image of the subsurface geology that is self-consistent with the acquired data emerges.

Since the source-detector coverage of the object or area of interest is far from complete in cross-borehole seismology, nonuniqueness of solutions and lowered resolution of images result. Thus, tomographic images based on limited view angles must be interpreted with care. However, the tomographic parameter determination is still very useful, especially in areas of significant lateral velocity variations in the subsurface. By including cross-borehole, seismic reflection profiling, well-

log, and any other available geophysical data, such as gravity, electrical, or radar, in the tomographic process, the resolution and certainty in describing the subsurface geology greatly improves.

APPLICATIONS

Cross-borehole seismology has been used in a number of different applications. The greater the degree of angular coverage around the rock mass of interest, the greater will be the reliability of the constructed subsurface image. By making numerous measurements for various source-detector positions in adjacent boreholes and analyzing the travel times and amplitudes for these source-detector locations, the velocity, elastic parameters, and attenuation of the intervening rock can be estimated from the transmitted and reflected energy. Cross-borehole seismology has been used for hydrocarbon exploration, mineral exploration, fault detection, stress monitoring in coal mines, delineation of the sides of a salt dome, investigation of dams, mapping dinosaur bone deposits, and nuclear waste site characterization.

In 1985, a research team from the Southwest Paleontology Foundation began excavating a giant sauropod dinosaur (45 to 60 meters long), later named *Seismosaurus*. The *Seismosaurus* skeleton was discovered after eight tailbones were exposed by weathering and erosion. By 1987, almost the entire tail had been excavated. To remove the rest of the skeleton, the location of the rest of the skeletal remains needed to be determined. In 1989, a number of boreholes were drilled in the area, and cross-borehole seismology provided data to construct vertical cross sections of where the *Seismosaurus* was located in the subsurface.

The Engineering Geoscience group at the University of California, Berkeley, used cross-borehole seismology in the 1990's to search for buried treasures, particularly the Victorio Peak treasure (Spanish weapons, coins, refined gold bars) in southeastern New Mexico and Yamashita's treasure (gold and jewels) in the Philippines. Their goal was to detect and delineate underground voids or cavities that could possibly contain the buried treasures. Although neither study resulted in a discovery, the ability of cross-borehole seismology to detect buried channels and voids was clearly demonstrated at both sites.

One of the primary applications of cross-borehole seismology is the enhanced characterization of petroleum reservoirs in existing fields. Fewer frontier fields remain to be discovered, and existing fields may still hold up to two-thirds of their petroleum. The geologic detail needed to properly exploit most hydrocarbon reservoirs substantially exceeds the detail required to find them. For effective planning, drilling, and production, a complete understanding of the lateral extent, thickness, and depth of the reservoir is absolutely essential. This can only be accomplished from detailed seismic interpretation of three-dimensional cross-borehole data integrated with three-dimensional seismic reflection profiling.

A common practice in three-dimensional seismic reflection profiling is to place the seismic detectors at equal intervals and collect data from a grid of lines covering the area of interest. In addition, since many adjacent boreholes are typically available in existing petroleum fields, cross-borehole seismic data can be collected between numerous adjacent boreholes across the site. Integration of cross-borehole and surface seismic data using tomography-based imaging algorithms yields seismic depth sections and parameter characterization of the subsurface geology. The resulting depth sections assist in interpreting the structure (geometry), stratigraphy (depositional environment), and lithology (rock and fluid types) of established hydrocarbon reservoirs. A repeated sequence of cross-borehole surveys as a function of time can aid in monitoring the effectiveness of enhanced oil-recovery methods. Based upon integrated tomographic models of the subsurface geology generated from the cross-borehole and surface seismic data, more wells can be drilled in the field at strategic locations, allowing a three-dimensional view of the subsurface to eventually emerge. These data provide petroleum companies with a continuously utilized and updated management tool that impacts reservoir planning and evaluation for years after the surface and cross-borehole seismic data were originally acquired and processed.

SIGNIFICANCE

The concept of estimating material properties from data collected around an object has a broad variety of applications. Cross-borehole seismology provides spatially continuous, high-resolution data that are necessary to image reservoir-scale features large distances from a well, including faults, stratigraphic boundaries, unconformities, porosity, and fracturing. Coupled with seismic tomography, cross-borehole seismology provides an important methodology for investigating subsurface geology.

In addition to measuring one-way transmitted energy in the subsurface, seismic reflection waves can also be recorded in cross-borehole surveys. These reflections are significantly better in resolving rock layers than surface seismic reflections because of the fact that the cross-borehole frequencies are one to two orders of magnitude greater than those of surface surveys. Integration of cross-borehole and surface seismic surveys is especially valuable, particularly for analyzing subsurface velocity structure. Furthermore, the seismic problem of imaging subsurface geology in depth and estimating physical rock parameters can be solved by correlating and integrating cross-borehole and surface seismic data using tomographic algorithms. Numerous geophysical, engineering, and environmental applications have emerged.

Evaluation and exploitation of existing petroleum reservoirs is a major application of cross-borehole seismology. Reservoir complexity produced by spatial heterogeneities in porosity, permeability, clay content, fracture density, overburden pressure, pore pressure, fluid phase behavior, and other related factors leads to large uncertainties in estimated total recovery. The most feasible approach for mapping these spatial variabilities comes from surface and cross-borehole geophysical measurements that are integrated through seismic tomographic processing. There is little doubt that cross-borehole seismology will play a major role in helping to solve not only exploration problems but also production and recovery problems in the petroleum industry.

In many areas of natural resource exploration and exploitation, numerous drill holes often exist, making cross-borehole seismology an ideal investigation tool. At these sites, a full suite of well-log data, rock-core analyses, and three-dimensional cross-borehole and surface seismic data can be acquired. Employing seismic tomography, these data can be processed, integrated, and interpreted to yield a geological model of subsurface lithology that closely approximates reality. The data acquired using cross-borehole seismology have been successfully processed and interpreted to yield three-dimensional images and lithological estimates of the subsurface geology.

Alvin K. Benson

CROSS-REFERENCES
Creep, 185; Discontinuities, 196; Earth Resources, 1741; Earth's Crust, 14; Elastic Waves, 202; Experimental Rock Deformation, 208; Faults: Normal, 213; Faults: Strike-Slip, 220; Faults: Thrust, 226; Faults: Transform, 232; Nuclear Waste Disposal, 1791; Oil and Gas Exploration, 1699; Petroleum Reservoirs, 1728; Rocks: Physical Properties, 1348; San Andreas Fault, 238; Seismic Observatories, 246; Seismic Reflection Profiling, 371; Seismic Tomography, 252; Seismometers, 258; Stress and Strain, 264; Well Logging, 1733.

BIBLIOGRAPHY

Lines, L. R. "Cross-Borehole Seismology." *Geotimes* 40 (January, 1995): 11. Applications of cross-borehole seismology and seismic tomography to the shallow subsurface.

_____, ed. *The Leading Edge* 17 (July, 1998): 925-959. Contains five excellent papers that conceptually discuss advanced technology and a variety of applications of cross-borehole seismology.

Russell, B. H. *Introduction to Seismic Inversion Methods.* Tulsa, Okla.: Society of Exploration Geophysicists, 1988. Russell discusses techniques used for the inversion of cross-borehole and surface seismic data, including principles of seismic tomography. Contains some good data illustrations.

Sheriff, R. E., ed. *Reservoir Geophysics.* Tulsa, Okla.: Society of Exploration Geophysicists, 1992. Describes applications and shows examples of using cross-borehole seismology and seismic tomography to evaluate and exploit existing petroleum reservoirs. Reviews borehole source and detector technology.

Stewart, R. R. *Exploration Seismic Tomography.* Tulsa, Okla.: Society of Exploration Geophysicists, 1991. Includes some of the historical development of cross-borehole methods and tomography. Reviews the fundamentals of seismic tomographic techniques, and discusses applications of cross-borehole seismology, surface seismic profiling, and seismic tomography to exploration geophysics.

Tarantola, A. *Inverse Problem Theory: Methods for Data Fitting and Parameter Estimation.* Amsterdam: Elsevier, 1987. An in-depth treatise on determining subsurface information from surface and cross-borehole seismic data. Includes qualitative discussions, as well as technical details.

Telford, W. M., L. P. Geldart, and R. E. Sheriff. *Applied Geophysics.* 2d ed. Cambridge, England: Cambridge University Press, 1990. Contains a basic overview of cross-borehole seismology and how it relates to other seismic methods of exploration.

DISCONTINUITIES

Discontinuities are boundaries within the Earth that divide the crust from the mantle, the mantle from the core, and the outer core from the inner core. The term is also used to describe the less dramatic boundaries within layers.

PRINCIPAL TERMS

CRUST: the top layer of the Earth, composed largely of the igneous rock granite; it ranges from 3 to 42 miles in thickness

EARTHQUAKE: a tremor caused by the release of energy when one section of the Earth rapidly slips past another; earthquakes occur along faults or cracks in the Earth's crust

EARTHQUAKE WAVES: vibrations that emanate from an earthquake; earthquake waves can be measured with a seismograph

INNER CORE: the innermost layer of the Earth; the inner core is a solid ball with a radius of about 900 miles

MANTLE: the largest layer of the Earth, about 1,800 miles in thickness; the mantle is within 3 miles of the Earth's surface at some locations

OUTER CORE: the outer portion of the core, about 1,300 miles in thickness; it is believed to be composed of molten iron

SEISMOGRAPH: a device that measures earthquake waves

EARTH'S INTERIOR

Discontinuities are underground boundaries between layers of the Earth. The closest discontinuity to the Earth's surface is the Mohorovičić, which divides the Earth's crust from the mantle underneath. Other discontinuities divide the mantle from the outer core and the outer core from the inner core. Minor discontinuities are found within these layers.

The interior of the Earth has been the object of much speculation and interest for thousands of years. Because direct observation of the Earth's interior is usually impossible, however, inferences about its structure and characteristics must be made from phenomena seen or felt at or near the Earth's surface. Several phenomena do give indications of the subsurface earth: caves that are often cool and damp, cool water emanating from springs and artesian wells, hot water spewing upward from geysers, and volcanoes from which extremely hot lava erupts. These phenomena give a mixed and incomplete picture of the Earth beneath the surface.

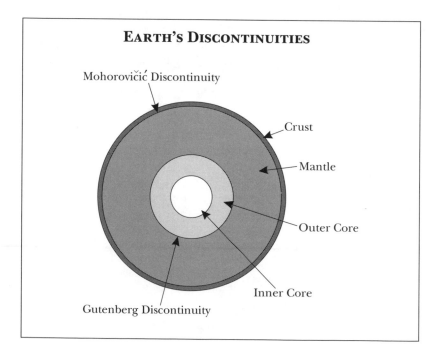

EARTH'S DISCONTINUITIES

Mohorovičić Discontinuity

Crust

Mantle

Outer Core

Inner Core

Gutenberg Discontinuity

SEISMIC WAVES

The structure and composition of the interior of the Earth can, however, be inferred from the study of earthquake waves. Seismographs can detect three types of vibrations: surface waves (the ones that can cause damage when there is an earthquake), P (primary) waves, and S (secondary) waves, which are also generated by every quake. P waves are compressional (pushing) waves, in which Earth or rock particles move forward in the direction of wave movement; S waves are shear waves, in which the particle motion is sideways or perpendicular to the direction of wave movement. The more efficient P waves travel twice as fast as S waves and thus are always detected first by a seismograph. Seismographs record these waves on charts, called "seismograms," attached to moving drums. By noting the arrival times of the various waves, seismologists can determine the distance to an earthquake and can see the effects on these waves caused by the type of rock through which the waves have moved.

Seismic waves travel through rock layers at specific speeds, which are different for each type of mineral or rock. For example, waves travel through basalt at 5 miles a second and through peridotite at 8 miles a second. Seismogram study has shown that the Earth's interior is not homogeneous, but rather is composed of several major layers and many sublayers.

GUTENBERG DISCONTINUITY

In 1906, Richard Oldham discovered that S waves are never detected on the opposite side of the Earth from any earthquake. As he already knew that S waves cannot travel through liquid substances, Oldham postulated that the center of the Earth must be composed of a molten core and that the materials above this core are not molten. The depth of the boundary between this core and the material above it was discovered eight years later by Beno Gutenberg. Now called the Gutenberg discontinuity, it is located about 1,800 miles beneath the Earth's surface.

MOHOROVIČIĆ DISCONTINUITY

When Oldham made his discovery of a central core, Andrija Mohorovičić was the director of the Royal Regional Center for Meteorology and subduction at Zagreb, one of the leading seismological observatories in Europe. In 1909, his meticulous study of a Croatian earthquake showed that some of the P waves from that quake had traveled faster than others. He already knew that other waves speed up or slow down when they move from one medium into another (as when light moves from air into water) and that this change in speed can result both in reflection, a bouncing back of waves, and in refraction, or a change in wave direction through the new medium. He deduced that the faster-moving P waves had traversed down through the Earth, through a discontinuity to a material of a different density, and then had come back up to the surface. Deep in the Earth was a material that allowed for faster

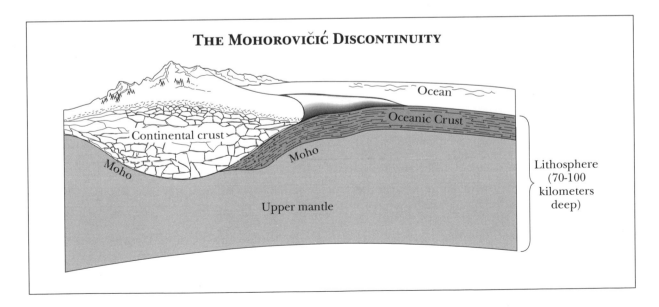

THE MOHOROVIČIĆ DISCONTINUITY

Ocean

Oceanic Crust

Continental crust

Moho

Moho

Upper mantle

Lithosphere (70-100 kilometers deep)

transmission of P waves. Above this discontinuity, seismic waves travel at about 4.2 miles a second; below the boundary, they travel at about 4.9 miles a second.

When Mohorovičić's results were replicated by other seismologists, it was concluded that the discontinuity was a global phenomenon. Data from these studies showed that there were two very distinct layers of the earth: an upper, less-dense layer now called the "crust," and a denser layer below called the "mantle." Thus Andrija Mohorovičić had discovered what is now called the Mohorovičić discontinuity, the boundary between the Earth's crust and mantle (it is often called the "Moho").

The crust of the Earth is made up of continents and ocean basins that are very different from one another. Continental crust is made primarily of granite. Covering this granite over much of the Earth's continents may be found layers of younger sedimentary rock such as sandstone, limestone, and shale. Ocean basins, on the other hand, are composed of the dark, heavy rock basalt.

Mohorovičić believed the discontinuity between the crust and mantle to be about 30 to 35 miles below the surface of the Earth. Subsequent studies have shown that it is usually at a depth of about 21 miles. However, the Moho has an irregular shape that is roughly a mirror-image of the surface of the Earth. Under the continents, the Moho is much deeper; under the oceans, the crust is very thin, and the Moho is as close as from 3 to 5 miles from the surface. The greatest depth of the Moho is probably beneath the Tibetan Plateau, where it reaches a depth of 42 miles.

EARTH'S MANTLE

The continents are higher because they are composed of granite, which is a lower-density rock than basalt or the materials of the mantle. Even though the mantle is composed of solid rock, under long-term stresses, the rock moves slowly like a liquid. Thus, just as ice floats in water, the continents actually are floating upon the heavier mantle rock. The Moho is the boundary between continental granitic rocks and the denser peridotite rock of the mantle.

The mantle extends from the bottom of the crust to a depth of 1,800 miles. It appears to be made of the rocks somewhat similar chemically to those in the Earth's crust but more "basic,"—that is, having more of the heavy iron and magnesium minerals such as olivine, and less lightweight aluminum. The mantle also appears to be composed of layers with discontinuities about 220 and 400 miles beneath the Earth's surface. Although mantle rock is solid, it can, under certain conditions, behave somewhat like a liquid. Under long-term pressures, the molecules of this solid rock can move like liquids, but under sharp, short-term stresses, mantle rock fractures like a brittle solid.

In 1957, a project was conceived to drill a hole through the thin oceanic crust down past the Mohorovičić discontinuity to bring up rock from the mantle. Although the "Mohole" project was approved and funded by the National Science Foundation, funds for it were cut off by the U.S. Congress in 1966.

MANTLE CONVECTION CURRENTS

Heat within the Earth is created through the decay of radioactive isotopes. Although this generated heat is very small when compared to the heat received from the Sun, it is well insulated and is enough to create volcanoes and the convection currents of the mantle. Mantle rock is extremely hot; because of the pressure on it from the crust above, however, it cannot melt, except where there is a decrease in this pressure.

Studies at the surface of the Earth have revealed areas where great heat flow comes from the mantle. Near the center of the Atlantic Ocean, the basaltic ocean bottom has split; the two sides are being pulled away from each other as Europe and Africa move away from the American continents. At this split, a decrease in pressure allows the hot mantle rock to melt and well upward, filling the gap between the dividing ocean bottoms. Thus the new ocean basin is made of material directly from the mantle. Within the mantle are large, slow-moving convection currents where hot mantle rock moves upward, cools off, and slowly sinks. These currents are believed to be the driving forces of continental drift.

EARTH'S CORE

At the bottom of the mantle, beneath the Gutenberg discontinuity, is the Earth's core. Seismic studies have shown that the outer core, which extends from roughly 1,800 to 3,100 miles beneath the Earth's surface, is not a perfect sphere. The core rises in areas where hot mantle rock is moving upward and is depressed where cooler mantle rock is moving downward. The density of the core is much greater than that of the mantle. It is believed that this core is made of molten nickel and iron and that its motion generates the Earth's magnetic field and aurora borealis.

In 1936, Danish seismologist Inge Lehmann discovered evidence for a solid core within the molten center of the Earth by detecting seismic waves that had been deflected back to the surface from within the core. When she realized that these waves, though very weak, travel faster through this most-central part of the Earth than through the rest of the molten core, she was able to infer that this inner core was composed of solid material completely surrounded by the molten outer core. This most central layer of the Earth extends from 3,100 miles beneath the Earth's surface to the center of the Earth, 4,000 miles down.

Sesimographs

The seismoscope is an ancient instrument that shows Earth movements. A Chinese seismoscope of the second century C.E. had eight dragon figures each with a ball in its mouth. When the Earth trembled, a ball would fall from the dragon's mouth into the mouth of a frog figure underneath it. European seismoscopes often used bowls of water that would spill when agitated. In 1853, Luigi Palmieri designed a seismometer that used mercury-filled tubes that would close an electric circuit and prompt a recording device to start moving when the Earth vibrated.

In 1880, British seismologist John Milne invented the first modern seismograph, which employed a heavy mass suspended from a horizontal bar. When the Earth would quake, the bar would move, and that movement would be recorded on light-sensitive paper beneath. Most seismographs employ a pendulum, which, because of inertia, remains still as the Earth moves underneath it.

When seismographs measure shock waves from nearby earthquakes, they first receive the P waves, which vibrate in the direction in which the waves are moving. S waves, which vibrate perpendicular to the direction in which the waves are moving, then arrive, followed by surface waves. When seismographs record more distant earthquakes, the

results are complicated by the reflection and refraction of seismic waves resulting from the various discontinuities underground. As the complications were deciphered, seismologists realized that the recordings described the rock layers below and between the quake and the seismograph.

Once geologists realized that they could learn about the Earth by examining seismograms, some researchers became impatient when they wanted to study a particular area but had to wait for an earthquake to occur. This became particularly difficult in areas where earthquakes did not occur frequently. Milne solved the problem by dropping a one-ton weight from a height of about 25 feet. The impact of this weight on the ground generated seismic waves that were weaker than, but similar to, those generated by earthquakes. To create stronger waves, seismologists explode charges of dynamite. These artificially induced shock waves have enough energy to reach deep into the planet. Since the 1970's pistons on large trucks have been used to strike the Earth and create artificial seismic waves.

When charges are exploded and the vibrations recorded by several nearby seismographs, a detailed description of rock layers can be detected. Since 1923, when a seismograph was first used to locate a large underground pool of petroleum, seismology has played a large part in the oil and gas industries. Earthquake waves artificially produced by explosions are also able to determine the location of underground geologic structures that may contain mineral deposits.

With the advent of the space age, seismographs connected to radio transmitters have been placed on the surfaces of the Moon and Mars. There are more than a thousand seismographs in constant operation gathering seismic data around the world. Data from the National Earthquake Information Center is updated daily and is available on the Internet.

Earthquake Study and Prediction

The same technology that has indicated the location of discontinuities deep within the Earth has also provided a greater knowledge about the crust. Whereas ancient civilizations feared earthquakes as manifestations of angry gods, quakes are now seen as results of energy released when plates of the Earth's crust move past one another.

Although earthquakes do occur in many places on the Earth's crust, they are most common in certain areas such as the "Ring of Fire" around the Pacific Ocean. Most earthquakes are linked directly to the movement of the Earth "plates," or sections of the crust. The Pacific plate and the North American plate meet along the San Andreas fault, which runs from western Mexico through California to the Pacific Ocean. The two plates are moving past each other along this fault. Each time there is movement along the fault, tremendous amounts of energy are released, and the Earth quakes. Quakes along this fault and others have caused untold damage.

One of the primary goals of seismologists is to determine a way to predict exactly when earthquakes will occur. If this information were known in advance, people could prepare for quakes, and far fewer deaths would occur. Many phenomena have been observed before quakes, such as increased strains upon bedrock, changes in the Earth's magnetic field, changes in seismic wave velocity, strange movements of animals, changes in groundwater levels, increased concentrations of rare gases in well water, geoelectric phenomena, and changes in ground elevation. However, none of these dependably occurs before every quake, and thus these signs have not become reliable indicators.

Seismologists cannot prevent earthquakes from occurring, nor can they yet predict the exact time of a major quake, but they can predict where earthquakes are likely to occur. It is believed that certain active faults where there has been no earthquake activity for thirty years or so are about ready for an earthquake. With this information, urban and regional planners can provide for quake-resistant roads, bridges, and buildings.

SEARCH FOR OIL AND GAS RESERVOIRS

Seismology and the search for minor discontinuities play a great part in the search for oil, gas, and mineral resources. Since much petroleum is retrieved from off-shore locations where the crust of the Earth is thinner, knowing the location of the Mohorovičić discontinuity sets the lower boundary for exploration.

Seismic studies are used regularly to assist in the search for oil and gas reservoirs. Natural gas and petroleum both can become trapped under some geologic formations. Seismologists routinely create a survey of an area before drilling to find minor discontinuities or boundaries between two different rock types, such as shale and sandstone. These surveys are made by measuring the reflection of seismic waves from the underlying rock layers. Geologic structures that can contain petroleum, natural gas, or mineral deposits can be identified from these surveys. Seismic surveys can show the distance and direction to these structures.

Kenneth J. Schoon

CROSS-REFERENCES

BIBLIOGRAPHY

Calder, Nigel. *The Restless Earth: A Report on the New Geology.* New York: Viking Press, 1972. A companion book to the television program "The Restless Earth," this book emphasizes how geologists came to the conclusion that the continents are moving. Illustrated with black-and-white and color photographs and diagrams. Indexed.

Cromie, William J. *Why the Mohole?* Boston: Little, Brown, 1964. A 1960's view of the never-finished American and Soviet plans to drill holes through the entire crust of the Earth in order to reach the mantle. The author was public-information officer for the American project. Several diagrams and photographs, a bibliography, and an index.

Dahlen, F. A. *Theoretical Global Seismology*. Princeton: Princeton University Press, 1998. Intended for the college-level reader, this book describes seismology processes and theories in great detail. The book contains many illustrations and maps. Bibliography and index.

Doyle, Hugh A. *Seismology*. New York: John Wiley, 1995. A good introduction to the study of earthquakes and the Earth's lithosphere. Written for the layperson, the book contains many useful illustrations.

Emiliani, Cesare. *Planet Earth: Cosmology, Geology, and the Evolution of Life and Environment*. Cambridge, England: Cambridge University Press, 1992. A large, comprehensive book containing basic information about matter and energy, many aspects of the physical and historical Earth, and a large section about the Earth's relationship to the universe. The last section is a brief history of the Earth sciences.

Erickson, Jon. *Rock Formations and Unusual Geologic Structures*. New York: Facts on File, 1993. An easy-to-read description of the Earth's crust, including the creation, deformation, and erosion of rock. Clear black-and-white photographs, diagrams, and maps along with a large glossary, bibliography, and index.

Jackson, Ian, ed. *The Earth's Mantle: Composition, Structure, and Evolution*. Cambridge: Cambridge University Press, 1998. Intended for the college student, *The Earth's Mantle* provides a clear and complete description of the elements that make up the Earth's mantle and the process of change that it has undergone since its formation. Includes bibliography and index.

Lambert, D., and the Diagram Group. *Field Guide to Geology*. New York: Facts on File, 1988. A profusely illustrated book about the Earth, its seasons, rocks, erosional forces, and geological history. Contains a list of "great" geologists and a list of geologic museums, mines, and spectacular geologic features. Indexed.

Miller, Russell. *Continents in Collision*. Alexandria, Virginia: Time-Life, 1983. A thorough text describing how Earth motions have created geologic features. Profusely illustrated with color and black-and-white illustrations. Bibliography and index.

Plummer, Charles C., and David McGeary. *Physical Geology*. Boston: McGraw-Hill, 1999. A college-level introductory geology textbook that is clearly written and wonderfully illustrated. An excellent sourcebook of basic information on geologic terminology and fundamentals of geologic processes. Contains CD-ROM. An excellent glossary.

Reynolds, John M. *An Introduction to Applied and Environmental Geophysics*. New York: John Wiley, 1997. An excellent introduction to seismology, geophysics, tectonics, and the lithosphere. Appropriate for those with minimal scientific background. Includes maps, illustrations, and bibliography.

Tarling, D., and M. Tarling. *Continental Drift: A Study of the Earth's Moving Surface*. Garden City, N.J.: Anchor Press, 1971. A small paperback book with black-and-white photographs and diagrams that help the reader to understand the principles of Earth structure and plate tectonics.

Vogt, Gregory. *Predicting Earthquakes*. New York: Franklin Watts, 1989. A good text on the Earth's interior and on how earthquakes are generated, detected, and measured. The last chapter discusses the prediction of earthquakes and efforts to control their effects. Black-and-white photographs and diagrams, glossary, and index.

Weiner, Jonathan. *Planet Earth*. New York: Bantam Books, 1986. A companion volume to the television series "Planet Earth," this book is well illustrated with both black-and-white and color pictures and diagrams. No glossary, but a comprehensive bibliography and index.

ELASTIC WAVES

The vibrations of the Earth, felt as earthquakes, are elastic waves in soil and solid rock. These waves are similar to sound waves, which travel through the air, and sonic waves, which travel through water.

PRINCIPAL TERMS

BODY WAVE: a seismic wave that propagates interior to a body; there are two kinds, P waves and S waves, that travel through the Earth, reflecting and refracting off of the several layered boundaries within the Earth

ELASTIC MATERIAL: a substance that, when compressed, bent, stretched, or deformed in any way, undergoes a degree of deformation that is proportional to the applied force and returns back to its original shape as soon as the force is removed

HOMOGENEOUS: having the same properties at every point; if elastic waves propagate in exactly the same way at every point, they are homogeneous

IDEAL SOLID: a theoretical solid that is isotropic, is homogeneous, and responds elastically under applied forces, stresses, compressions, tensions, or shears

ISOTROPIC: having properties the same in all directions; if elastic waves propagate at the same velocity in all directions, they are isotropic

REFLECTION: when an elastic wave strikes a boundary between two substances or between two rock layers of different seismic velocities, part of the incident ray bounces back (reflects)

REFRACTION: when an elastic wave passes through a boundary between two rock layers of different seismic velocities, the rays passing through are bent (refracted) in another direction

SURFACE WAVE: a seismic wave that propagates parallel to a free surface and whose amplitudes disappear at depth; there are two kinds, Rayleigh waves and Love waves, that travel at the surface around the Earth

ELASTIC BEHAVIOR

Elastic waves are experienced frequently every day: Everything heard is an elastic wave in the air; every vibration felt, in the ground as a truck passes or in the floor from vibrations in a building, is from elastic waves in solid matter. Although the experience of elastic wave energy is familiar, the exact nature of this phenomenon is not something visible to the eye or easily described in a visual way. Ripples resulting from dropping an object in a still body of water move in ever-increasing circles away from the splash; such waves are not elastic; rather, they are gravity waves. Yet, elastic waves are analogous to this example in that they originate from a disturbance and propagate outward and away in concentric circles or spheres.

An elastic wave moves in an elastic medium, which can be a solid, liquid, or gas. A substance is said to respond elastically if when it is compressed, stretched, bent, or submitted to shear forces it deforms in proportion to the applied force and then immediately returns to its original unstressed state when the force is removed. A good illustration of this property is a spring scale. A 1-kilogram weight placed on the scale will cause the spring inside the scale to be stretched (deformed) into a longer length, such as dropping 1 centimeter. A 2-kilogram weight on the scale would cause it to move down 2 centimeters. Hence, the displacement of the spring is proportional to the force applied. When the weights are removed from the scale, it immediately returns to zero, its original unstressed length and shape. This is elastic behavior. If the spring of the scale stretched 1 centimeter for 1 kilogram and then stretched more or less than a centimeter for the second kilogram, it would not be elastic because it would not be a proportional response. Also, if the spring did not return to zero after the weight was removed but retained some permanent deformation, it would not be elastic.

In the case of wave propagation in the Earth, consider the effect of striking the ground with a sledgehammer. When struck, the ground would be suddenly compressed, which would be transmitted to the neighboring soil and rock around and beneath the strike. Except in the immediate vicinity of the blow, where permanent deformation (nonelastic) may occur, the response of the neighboring soil and rock would be elastic. It would be temporarily compressed by the force of the blow and then immediately relax back into the former condition. A compression wave would irradiate spherically away from the blow, traveling across the surface (like the ripples on a pond) and down into the Earth. In the passing of an elastic wave, the medium passing the wave is restored to its original unstressed state as if no wave had ever come through at all.

ELASTIC WAVE VELOCITIES

Elastic waves travel at certain velocities depending on the density and elastic stiffness or compressibility of the medium. If a substance is soft, elastic waves move more slowly; if it is very stiff, elastic waves move fast. In air, sound waves move at approximately 300 meters per second; in water, sonic waves move at roughly 1,200 meters per second; in rock, compressional waves move at a rate of from 3,000 to more than 10,000 meters per second, depending on the rock's hardness and how deeply it is buried.

If within a medium through which elastic waves can move the velocity is the same everywhere, that medium is called "homogeneous." If, in addition, at any given point in that medium the velocities are the same in all directions, the substance is termed "isotropic." In most rocks of the Earth, which occur in layers, the deeper below the surface, the higher the velocity becomes. The increasing weight of the overburden acting on rocks found deeper in the Earth causes their density and stiffness to change, and, generally, in the vertical direction the velocity of a seismic wave is different from its velocity in horizontal directions. Thus, many rocks of the Earth are not isotropic; neither are they homogeneous. By analyzing seismograms from earthquakes, quarry blasts, and underground nuclear explosions, the inhomogeneities and anisotropies of the Earth have been described to give a picture of what the Earth's interior is like.

ELASTIC WAVE TYPES

There are two basic kinds of elastic wave: P waves, or compressional waves, and S waves, or shear waves. P waves are sometimes called "push-pull" waves because they consist of a series of pushes (compressions) and pulls (rarefactions), where the motion of a particle of matter as the wave passes by is parallel to the direction the wave passed. S waves are sometimes called "shake" or "shear" waves, because they consist of shearing or shaking motions where the movement of a particle of matter as the wave passes by is transverse, or perpendicular, to the direction the wave passed. A "Slinky" toy spring, held in two hands, can provide an illustration of sending waves back and forth. The alternate stretched and compressed parts of the spring move from one end to the other. If a long rope is tied to a post and the end is shaken up and down, a wave will move from the shaken end to the post, but the motion of the particles of the rope are up and down, transverse to the wave motion. P waves can move in all substances, solid, liquid, or gas. S waves can move only in solids. Compressional and shear waves are the only types that can propagate anywhere interior to a solid, like the rocks of the Earth. These are called "body waves." Earthquakes generate both compressional and shear waves at the source where the fault moves.

There are two other important kinds of elastic waves, but these travel only parallel to free surfaces, like the surface of the Earth, and have amplitudes that decay with depth. They are called "surface waves." The two kinds are "Rayleigh waves" and "Love waves," each named after the scientist who discovered and described it.

When a Rayleigh wave passes by on the surface of the Earth, a particle of soil or rock is first moved forward, then up, then backward, and then down to its starting point in an elliptical path. For Rayleigh waves, when the displaced particle is at the top of its elliptical motion, it is moving in the opposite direction of the Rayleigh wave front. This is called an "elliptic retrograde" motion. When a Love wave passes by on the surface of the Earth, a particle of soil or rock is moved from side to side perpendicular to the direction of the wave front. Love waves are horizontally polarized shear waves traveling parallel with the surface.

With regard to velocity, compressional waves

are the fastest; next are shear waves, which move at roughly six-tenths the speed of the compressional wave; slowest are the surface waves, which move at approximately nine-tenths the speed of shear waves.

Elastic Wave Forms

One final aspect needs to be described in talking about elastic waves, and this is the form of the wave. Regardless of the type of wave (P, S, Rayleigh, or Love), they all consist of trains of disturbances that move through the Earth. A wave that has only one vibration is a pulse. In elastic waves, even ones that sound like sharp pops or explosions, there is a train of several cycles of vibra-

tion—sometimes a few seconds in duration and sometimes for many minutes or even hours. (Rayleigh and Love waves can be recorded for an hour or more on seismographs when generated by a very strong earthquake.)

The form of a wave is described by its frequency and its amplitude, as well as by its particle motion. Frequency is merely the number of times per second that the vibrations occur as the wave passes. Earthquake waves have frequencies of from several cycles per second down to several seconds per cycle. In addition to the time between peaks of an elastic wave's passage, there is also a distance that can be measured between peaks. This is called the "wavelength." For waves in the Earth,

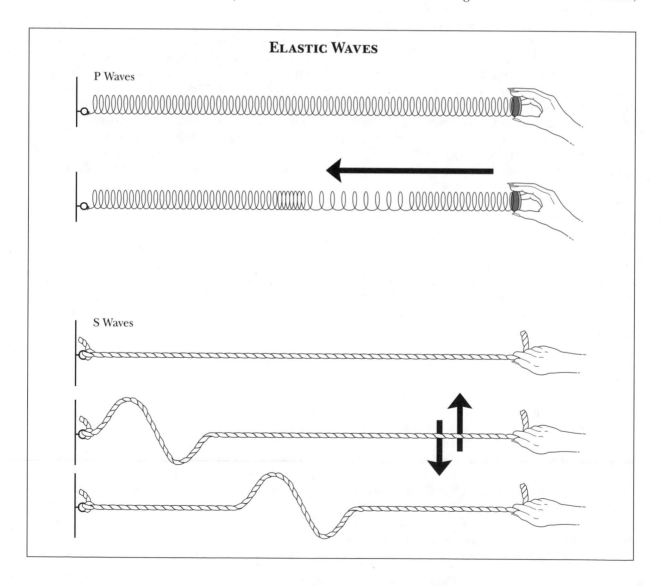

ELASTIC WAVES

P Waves

S Waves

the wavelengths can vary from a few meters (for high-frequency P and S waves) to a kilometer or more (for low-frequency surface waves). Hence, in an earthquake one part of a railroad track can be under compression, being sheared to the left, while a few hundred meters away another part is under tension, being sheared to the right, all at the same instant.

ELASTIC WAVE PROPAGATION COMPLEXITY

Even though a source of elastic waves may be simple, generating only one kind of wave, as soon as boundaries between differing layers are encountered, other kinds of waves result, reflecting and refracting in many directions. Those that eventually find themselves back at the surface can be recorded. When P and S waves arrive at the surface, it is their complex interaction at the surface that produces the Rayleigh and Love waves. With regard to earthquake-generated waves, not only do waves reflect from the source back to the surface off the boundaries of crust, mantle, and core, but some waves can pass completely through the Earth and be recorded on the other side. During an exceptionally strong earthquake, waves can refract through to the other side and then return again through the core and mantle to be recorded again on the original side. Surface waves generated by large earthquakes have also been known to circumnavigate the globe, sometimes circling several times before their amplitudes become too small to detect. In very large earthquakes (those of 8.6 or more on the Richter scale), these waves have been measured to complete as many as ten or more passages around the world, taking approximately 3 hours for each trip.

A wave train of a single type can change from one type to another repeatedly along its ray path. This is of great interest to seismologists. For example, a P wave may start from where it is generated at an earthquake fault and travel down until it hits the Mohorovičić, or Moho, discontinuity, the boundary between the Earth's upper crust and mantle below. There it can refract through, turning into an S wave, bending its direction of travel slightly, and taking on a new velocity. As it propagates farther and farther downward, it speeds up until it hits the boundary of the outer core, where it must either change again or reflect back toward the surface. If it passes through the boundary, it

must transform back into a P wave because the outer core of the Earth acts like a plastic liquid and will not permit the passage of S waves. Continuing on past the center of the Earth, it would strike the boundary between core and mantle on the other side and, again, it could change back into an S wave. As it traveled up toward the other side of the Earth, it would gradually slow in speed until it hit the Moho on the other side. There it could turn, again, into a P wave and move through the crust until it emerged at the ground surface. There it would be reflected back toward the Earth's interior or, perhaps, follow a curved ray path that would skip back to the surface at another location.

During this long and varied path, the ray would travel at P-wave velocities when in a compressional phase and at S-wave velocities (roughly 40 percent slower) when in a shear phase.

SEISMOGRAPHS AND SEISMIC STATIONS

Elastic waves in the Earth are measured and recorded by various kinds of seismographs. Some measure vertical motions only, some horizontal; some measure compressional waves only, such as those that move through water. To describe the particle motion of a train of passing waves requires a set of three seismographs: one for vertical motion and two for horizontal—one for east-west motion and one for north-south.

Seismic stations permanently installed to monitor earthquakes are built in a variety of ways, depending on what is to be measured. A given seismic sensor can only detect a given band of frequencies; outside that band it is insensitive. Since earthquakes generate a wide range of frequencies from high to ultralow, some stations measure high frequencies (also called "short periods") while others measure low frequencies (called "long periods"). Moreover, seismic sensors respond only to a given range of amplitudes. Hence, some earthquake observatories have extremely sensitive instruments that detect and magnify even the tiniest vibrations 100,000 times or more. Other stations also have so-called strong motion instruments that do not record at all, unless a real jolt passes through. This diversity in equipment is necessary, because when strong high-amplitude seismic waves hit a high-magnification instrument, the readings go off the scale and cannot be deciphered. On the other end, strong-motion equipment is insensitive to smaller tremors.

REFLECTION SEISMIC PROFILING

Because the interaction of seismic waves with the details (inhomogeneities) within the Earth's interior enables what is there to be described, buried out of sight, artificially generated seismic waves are sometimes used to find oil and other things of interest belowground. "Reflection seismic profiling," a method used by oil companies the world over, usually entails the use of an explosive to send elastic waves into the ground. Then, by an array of seismic sensors called geophones, deployed to catch the reflections at the land surface, geophysicists can deduce the structures of the subsurface. Some seismic profiling methods employ only P waves, while some have been successful with artificial S-wave sources.

When P and S waves hit a boundary between two rock layers, they each split into four parts. A P wave, for example, will reflect back both a P and an S wave and will also transmit a portion of its energy through the boundary into the next layer down that also splits into a P wave and an S wave. The wave bouncing back is the "reflected" portion of the incident wave, while the part that passes through is the "refracted" portion. An incident S wave similarly splits into four parts, a reflected P and S and a refracted P and S.

PRACTICAL APPLICATIONS

Understanding elastic wave propagation within the Earth not only has been the means by which seismologists have been able to define the inner structure of the Earth but also has enabled the discovery of almost all of the oil deposits found since the mid-twentieth century. Other practical applications include the monitoring of underground nuclear testing to verify that countries are living up to their treaties. Also, because submarine earthquakes are the cause of seismic sea waves, and because the seismic waves passing through the Earth travel several times faster than the seismic sea waves (or tsunamis) do through water, these destructive waves from the sea can be predicted, sometimes hours before they strike a coastline, thus saving many lives.

David Stewart

CROSS-REFERENCES

BIBLIOGRAPHY

Aki, Keiiti, and Paul G. Richards. *Quantitative Seismology: Theory and Methods.* 2 vols. San Francisco: W. H. Freeman, 1980. This is an advanced text on elastic waves and earthquake seismology. A modern version of Ewing, Jardetzky, and Press (cited below) and the treatise by Love (also cited below), it is written at the graduate university level.

Bedford, A. *Introduction to Elastic Wave Propagation.* New York: Wiley, 1994. This text is an excellent introduction to seismology and the behavior of elastic waves within the Earth.

Bolt, Bruce A. *Inside the Earth.* San Francisco: W. H. Freeman, 1982. This book is an elementary but thorough treatment of seismic waves in the Earth, useful to lower-level college or advanced high school students.

Bullen, K. E., and Bruce A. Bolt. *Introduction to Theory of Seismology.* 4th ed. Cambridge, England: Cambridge University Press, 1985. A thorough treatment of elastic wave theory, it is readable at the undergraduate science-major level.

Dahlen, F. A. *Theoretical Global Seismology.* Princeton: Princeton University Press, 1998. Intended for the college-level reader, this book describes seismology processes and theories in great detail. The book contains many illustrations and maps. Bibliography and index.

Ewing, W. Maurice, Wenceslas S. Jardetzky, and Frank Press. *Elastic Waves in Layered Media.* New York: McGraw-Hill, 1957. This is a definitive text on elastic waves in layered media as they occur within the Earth. Mathematical, it is written for graduate-school-level or advanced physics students.

Graff, Karl F. *Wave Motion in Elastic Solids.* New York: Dover, 1991. This is an advanced text on elastic waves and earthquake seismology. Some sections may be suitable for the general reader, but it is written at the college level. Bibliographical references and index included.

Love, A. E. H. *A Treatise on the Mathematical Theory of Elasticity.* Mineola, N.Y.: Dover, 1944. First published in 1884 in Cambridge, England, this comprehensive work remains important. The Love waves of earthquake seismology were first described and discovered by the author. Best suited for graduate students.

Richter, Charles F. *Elementary Seismology.* San Fransisco: W. H. Freeman, 1958. This is a classic work by the seismologist who devised the Richter scale. While dated in some subject areas, it is very readable and thorough. The chapter on elastic waves and their propagation through the Earth is excellent.

EXPERIMENTAL ROCK DEFORMATION

In order to understand how and why rocks in the Earth have deformed, experiments are done using laboratory apparatus that simulate some of the conditions found in the crust and mantle of the Earth. These experiments have shown that the mechanical behavior of rocks is complex but can be deciphered. The results help to develop an intuition which leads to meaningful interpretations of field situations.

PRINCIPAL TERMS

BRITTLE BEHAVIOR: the sudden failure of a sample by catastrophic loss of cohesion

CONFINING PRESSURE: pressure acting in a direction perpendicular to the major applied stress in a rock deformation experiment

DISLOCATION: a defect in a crystal caused by misalignment of the crystal lattice; the presence of dislocations greatly reduces the stress necessary to produce permanent deformation

DUCTILE BEHAVIOR: permanent, gradual, non-recoverable deformation of a solid; sometimes called plastic deformation

ELASTIC BEHAVIOR: recoverable deformation where the strain is proportional to the stress

PORE PRESSURE: the pressure in the fluid within the pores of a rock

STRAIN: a measure of deformation including translation, rotation, dilatation, and distortion; it is usually measured as a percent or ratio and results from stress

STRESS: the intensity of forces (force per unit area) acting within a body; may refer to a particular stress acting in a particular direction on a particular plane or to the collection of all stresses acting on all planes at that point

BRITTLE VS. DUCTILE BEHAVIOR

Experiments in rock deformation aim to develop an understanding of how rocks behave mechanically. Under conditions found at the surface of the Earth, most rocks behave as brittle solids, but when subjected to the conditions found at depth within the Earth, those same rocks behave in a very ductile manner. If a chalk-sized piece of rock were squeezed between the jaws of a vise, it would fail in a sudden fashion, almost like an explosion. The pieces of rock recovered from such an experiment would be sharp shards similar to the remnants of a rock that was smashed with a hammer. The breaking strength, pore pressure, and internal angle of friction determine the failure criterion for a given type of rock. The failure criterion encompasses the combinations of factors that can cause the rock to fail. Engineers can apply such criteria to determine the stability of a slope, for example, or the spacing required for pillars in an underground mine.

Often a rock will deform without breaking. Rock layers are sometimes buckled into folds that can vary from angular kink bands to smoothly undulating surfaces. Other rocks seen in roadcuts and outcrops show, by the patterns of their banding and textures, that they have deformed in a very ductile way, flowing much like a fluid. To simulate this behavior in a laboratory experiment, the conditions under which deformation occurred must be considered.

DEFORMATION CONDITIONS

The most striking examples of ductile deformation come from deep crustal or upper mantle depths. What are the pressure, temperature, and strain rate like at depths of 20 kilometers? The pressure produced by 200 kilometers of rock with a density of approximately 2.7 grams per cubic centimeter is about 5.2 kilobars, which is approximately 75,000 pounds per square inch, or 5,000 atmospheres of pressure. By studying minerals found in volcanic rocks that come from deep sources, scientists have learned that the temperatures at a depth of 20 kilometers are 250-500 degrees Celsius. If an object deforms so that its length changes by 1 percent in a second, it is undergoing a deformation with a strain rate of 0.01

per second. Geologists have found examples of deformation that incorporate datable features, and from these have learned that strain rates of 10^{-13} to 10^{-14} per second are typical for geological processes. At such rates, the length of an object would change by about one part per million per year. This is too slow to study in the laboratory, so experiments are run at strain rates on the order of 10^{-5} per second. The results are extrapolated to estimate how rocks would behave at very low strain rates. An extrapolation over nine orders of magnitude is risky but is supported by theoretical considerations.

Experiments have been designed to study the effects of pressure, temperature, and strain rate, as well as pore pressure, anisotropy, and water content. The emerging picture reveals that rocks exhibit a complex mechanical behavior. At low confining pressures, low temperatures, and high strain rates, they behave as elastic solids when subjected to stresses up to their breaking strength, then fail in a brittle fashion. At high confining pressures, high temperatures, and low strain rates, they deform in a ductile fashion, with yield strengths and viscosities that are functions of temperature and strain rate. One result of this behavior is that when subjected to the very low strain rates associated with convection, mantle rocks flow easily; when subjected to the high strain rates resulting from the passage of those seismic waves called shear waves, mantle rocks respond like elastic solids.

DISLOCATIONS

Results from flow experiments indicate that at high temperatures and pressures and low strain rates, rocks deform by the movement of offsets in crystal lattices called dislocations. This is similar to the way a caterpillar moves forward—only a few of its legs are in motion at any one time, but the movements propagate along as waves, and the whole animal moves forward. As dislocations move through a crystal, only a few bonds are broken at a time, but the entire crystal deforms as a result.

The study of dislocations and how they move has resulted in a better understanding of ductile deformation of rocks and other materials. Several different mechanisms, such as power law creep and diffusion creep, have been found to be active

in different substances under different conditions. Flow laws have been formulated, and maps have been constructed which show, for a given mineral, which flow law will dominate the deformation for a given stress difference and temperature. Because rocks are aggregates of different minerals, their behavior is more complex than that of any single mineral. Progress is being made, however, and eventually the behavior of the material of the crust and upper mantle will be better understood.

STUDYING LOWER MANTLE AND CORE BEHAVIOR

To study lower mantle and core behavior, experiments have been designed using diamonds as platens (flat plates that exert or receive pressure). Diamonds can withstand very high pressures and are transparent to visible light. This transparency permits visual observations of phase changes and allows samples to be heated to very high temperatures using lasers. Conditions similar to those within the Earth's core can be simulated in such experiments, but the measurements that are possible are limited by the need to use small samples.

STUDYING FRACTURE AND FLOW

The methods used to study rock deformation in the laboratory depend on whether fracture or flow is the subject of investigation. Many studies of fracture are motivated by the desire to understand how earthquakes occur, and, if possible, to develop means of predicting them. Because damaging earthquakes frequently occur in rock near the surface, these experiments are done at low temperatures and confining pressures. Studying the flow of rock at high temperatures and high confining pressures develops insights into mantle convection and plate tectonics. The general procedure is to prepare a sample of the rock to be studied; attach strain gauges to the sample to monitor changes in strain during the course of the experiment; insert the sample between the platens of the press; adjust confining pressure, temperature, and pore pressure to the conditions of interest; and then squeeze the sample while recording data from the sensors.

An experiment with no confining pressure is called a "uniaxial" test, because all the stress is applied along one axis. A hydrostatic confining pressure can be applied to the sample by immersing it within a medium and then compressing that medium. Although the terminology is not quite correct, this kind of experiment is usually called a "triaxial" experiment. The confining medium can be a solid (such as talc), a fluid (commonly kerosene), or a gas.

During experiments exploring the fracturing process, the sample may be instrumented by gluing small microphones to it. Just before the rock fails catastrophically, several small acoustic events often can be located within the sample by triangulation from a number of microphones. These noises are thought to be produced by the extension of small, naturally occurring fractures as they grow in response to the increasing stress. Determining the relationship between these events and the fracture that finally forms offers a promising means of earthquake prediction: Foreshocks are commonly recorded in the vicinity where a large earthquake is about to occur and may be similar to the acoustic events observed in the laboratory.

High confining pressures, high temperatures, and low strain rates are needed to study the flow of rock. Such experiments are technically difficult, particularly if they run for long periods of time. Commonly, the temperature is increased above that typically expected to occur within the Earth at the pressures being studied, in order to increase the rate of deformation. At the completion of the experiment, the sample may be recovered, sliced into thin sections, and studied under a microscope. If the textures produced in the laboratory resemble those observed in samples from the crust and mantle, it is likely that similar processes are active. The flow laws operative during the experiment can be determined and then adjusted for differences in temperature and strain rate, which permits the extrapolation to the conditions present within the Earth to be conducted with more confidence.

UNDERSTANDING FORMATION OF GIANT STRUCTURES

A topographic map from the Valley and Ridge Province of the Appalachian Mountains shows sinuous ridges tracing out elaborate folds in a coherent pattern extending for hundreds of miles. The landscape south of San Francisco is dominated by long, linear valleys parallel to the San Andreas fault. Roadcuts near the Thousand Islands reveal swirling, flowing patterns which appear to have formed as if the marble there behaved like a fluid.

Each of these phenomena is a striking demonstration of how rocks deform when subjected to the mammoth stresses involved in mountain building and plate tectonics, yet they are all very different from one other. To understand how such giant structures are formed, scientists have performed experiments in the laboratory on small samples of the rocks from which the structures are made. They have learned that the behavior of rock is a function of its environment at the time it is being deformed, the size of the stresses applied to it, and the rate at which those stresses are applied.

FRACTURE ORIENTATIONS AND PATTERNS

Some of this behavior can be compared to that of three familiar materials: modeling clay, beeswax, and Silly Putty. Modeling clay shows a behavior that varies with its environment, particularly temperature. A piece of cold modeling clay is difficult to work with. Most people spend a few minutes kneading it in their warm hands; its behavior changes noticeably as it warms. A piece of very cold modeling clay may shatter if it is dropped on the floor, unlike a piece that has been warmed. If the pieces were reassembled, there would be a pattern of fractures related to the orientation of the rock within the vise. Much larger, but similar, fractures occur within the crust of the Earth, which are called joints or faults. Theoretical considerations and data from laboratory experiments are used to interpret the orientations and patterns produced by these brittle fractures.

If a tennis ball were put in the vise, it would shorten in the direction it was squeezed and would get fatter in the plane of the jaws of the vise. A rock deforms elastically, just like a tennis ball, but at a much smaller scale. Sensors, called strain gauges, attached to the rock sample will record these tiny changes in shape. Careful monitoring of the stress applied by the vise and the strain experienced by the rock would help to determine the elastic constants that describe the behavior of the rock before failure begins. These constants, called Young's modulus and Poisson's ratio, can be used to calculate seismic wave velocities. As failure occurs, fractures grow across the sample. In the rock-in-a-vise example, these fractures will usually form perpendicular to the jaws of the vise, corresponding to what are called extension joints. If the sample were enclosed in a jacket that provided pressure on its sides, the experiment would be conducted with a confining pressure present. Under these conditions, many fractures might form at an angle to the jaws of the vise, producing a set of what are called conjugate shear joints. Alternatively, one fracture might develop, and the sample might slip in opposite directions on both sides of this fracture. Such a fracture corresponds to a fault in the field. Measuring the angles at which these fractures form would show that they are somewhat constant for fractures produced in the same material. By increasing the confining pressure, the stress needed to break the sample also increases. Graphing the results permits the determination of another material constant, called the internal angle of friction. A comparison of this angle with the angle at which conjugate fractures and faults form shows that they are simply related.

PORE PRESSURE

These results characterize some of the mechanical behavior of the rock from which the sample was taken. Young's modulus, Poisson's ratio, breaking strength, and the internal angle of friction are material constants that vary little among different samples from the same rock. Different types of rocks have different elastic constants and strengths, just as they have different densities.

Fluids within the pores of a rock play a significant role in its brittle behavior. Experiments that control the pressure of such pore fluids show that the strength of the rock decreases as the pore pressure increases. Some of a rock's resistance to failure is provided by the pressure of one grain against the next. Pore pressures reduce this pressure and so weaken the rock, which helps to explain why most catastrophic landslides have occurred after heavy rainfalls. Slopes that are stable when dry can weaken as the pore pressure within

them increases to become unstable and to fail. High pore pressure may also facilitate movements on thrust faults deep within the Earth.

SIZE AND RATE OF STRESSES

The behavior of beeswax varies with the size of the stresses applied to it. A chunk of beeswax feels hard and makes a sharp, rapping sound when struck against a table. The fact that hives and statues in wax museums maintain their shape for years attests to the ability of beeswax to resist the forces of gravity over long periods of time. Yet it is easy to stick a thumbnail into a chunk of beeswax. The stress produced by the edge of a nail is greater than the strength of the beeswax, and it deforms, whereas the stresses produced by gravitational force are less than the strength of the beeswax and are unable to cause it to deform.

Silly Putty shows a behavior which varies with the rate at which stresses are applied to it. Throw a sphere of Silly Putty onto the floor, and it bounces. Smash it with a sledgehammer, and it shatters. But pull on it slowly, and it will stretch. In response to a rapid application of stress, Silly Putty behaves like a brittle, elastic solid. But when subjected to a slowly applied force, its behavior is much more like that of a fluid.

Otto H. Muller

CROSS-REFERENCES

Creep, 185; Cross-Borehole Seismology, 190; Discontinuities, 196; Earthquake Engineering, 284; Earthquake Prediction, 309; Earthquakes, 316; Elastic Waves, 202; Faults: Normal, 213; Faults: Strike-Slip, 220; Faults: Thrust, 226; Faults: Transform, 232; Geothermometry and Geobarometry, 419; Land-Use Planning, 1490; Landslides and Slope Stability, 1501; Mining Processes, 1780; Phase Changes, 436; Rocks: Physical Properties, 1348; San Andreas Fault, 238; Seismic Observatories, 246; Seismic Tomography, 252; Seismometers, 258; Stress and Strain, 264.

BIBLIOGRAPHY

Billings, Marland P. *Structural Geology.* 3d ed. Englewood Cliffs, N.J.: Prentice-Hall, 1972. Chapter 2, "Mechanical Principles," includes an elementary review of experimental rock deformation. Descriptive, with little prior knowledge assumed, this book is suitable for the general reader.

Dahlen, F. A. *Theoretical Global Seismology.* Princeton: Princeton University Press, 1998. Intended for the college-level reader, this book describes seismology processes and theories in great detail. The book contains many illustrations and maps. Bibliography and index.

Davis, George H. *Structural Geology of Rocks and Regions.* New York: John Wiley & Sons, 1984. Chapter 5, "Dynamic Analysis," discusses experimental rock deformation in a manner that is suitable for the general reader. It takes the reader step by step through an experiment involving the compression of a limestone, with the procedures, results, and interpretations of those results carefully described. The treatment is the most descriptive and least technical of the references listed here.

Doyle, Hugh A. *Seismology.* New York: John Wiley, 1995. A good introduction to the study of earthquakes and the Earth's lithosphere. Written for the layperson, the book contains many useful illustrations.

Hobbs, Bruce E., Winthrop D. Means, and Paul F. Williams. *An Outline of Structural Geology.* New York: John Wiley & Sons, 1976. Section 1.4, "The Response of Rocks to Stress," provides a nice summary of the results obtained from experiments in rock deformation. Chapter 2, "Microfabric," presents excellent discussions of the development of microfabric and crystallographic preferred orientations in deformed rocks. This treatment includes many striking microphotographs, line drawings, and stereonet plots. Suitable for technically oriented college students.

Jackson, Ian, ed. *The Earth's Mantle: Composition, Structure, and Evolution.* Cambridge: Cambridge University Press, 1998. Intended for the college student, *The Earth's Mantle* provides a clear and complete description of the elements that make up the Earth's mantle and the process of change that it has undergone since its formation. Includes bibliography and index.

Jaeger, John Conrad, and Neville George Wood Cook. *Fundamentals of Rock Mechanics.* New York: John Wiley & Sons, 1976. This 585-page book is a standard text in the field of experimental rock deformation. Although parts of it become so technical that they are suitable only for experts in the field, much of it is of interest to technically minded college students. Chapter 6, "Laboratory Testing," is thorough, lucid, and contains fewer complex equations than most of the book.

Marshak, Stephen, and Gautam Mitra. *Basic Methods of Structural Geology.* Englewood Cliffs, N.J.: Prentice-Hall, 1988. Chapter 10, "Analysis of Data from Rock-Deformation Experiments," by Terry Engelder and Stephen Marshak, begins by providing a good overview of the subject, followed by results from several experiments which the reader is invited to interpret. This technique probably results in a better understanding and a firmer intuitive grasp of what is entailed in doing experiments on the mechanical behavior of rocks than any other reference listed here. Suitable for the general reader.

Suppe, John. *Principles of Structural Geology.* Englewood Cliffs, N.J.: Prentice-Hall, 1985. Chapter 4, "Deformation Mechanisms," and chapter 5, "Fracture and Brittle Behavior," discuss the results of experiments in rock deformation. The emphasis is on flow laws and failure criteria and how these are related to the microstructures and chemical bonds within the rock. Suitable for college students, but a background in physical chemistry or material science would be helpful.

FAULTS: NORMAL

Normal faults are common features that occur when the Earth's crust is subjected to tensional forces. The sense of movement is primarily vertical and results in an extension of the crust. These faults are generally associated with broad-flexed or uplifted areas and are an integral part of the modern concept of plate tectonics.

PRINCIPAL TERMS

DIP: the angle of inclination of a fault, measured from a horizontal surface; dip direction is perpendicular to strike direction

FAULT: a break in the Earth's crust that is characterized by movement parallel to the surface of the fracture

FAULT DRAG: the bending of rocks adjacent to a fault

FOOTWALL: the crustal block underlying the fault

GRABEN: a long, narrow depressed crustal block bounded by normal faults that may form a rift valley

HANGING WALL: the crustal block that overlies the fault

HORST: a long, narrow elevated crustal block bounded by normal faults that may result in a fault-block mountain

SLICKENSIDES: fine lines or grooves along a faulted body that usually indicate the direction of latest movement

STRESS: the forces acting on a solid rock body within a specified surface area

THROW: the vertical displacement of a rock sequence or key horizon measured across a fault

NORMAL FAULT FORMATION

A normal fault is a fracture that separates two crustal blocks, one of which has been displaced downward along the fractured surface. Some workers use the term "gravity fault" to indicate an apparent normal fault if genesis, rather than geometry, is implied. Crustal blocks overlying or underlying a normal or reverse fault are commonly designated as the hanging wall and footwall blocks, respectively. These are old descriptive terms that were used in the early English coal-mining districts. Faults that were inclined toward the down-dropped side were common in the area and the term "normal fault" was applied. At places where the movement was in the opposite direction, the breaks in the rock were designated as reverse faults. The displacement of normal faults, which can be intermittent, ranges from less than a meter up to thousands of meters. The inclination or dip of the fault can be from nearly horizontal to vertical but generally ranges from 45 to 60 degrees. In some areas, the angle of dip decreases with depth and results in a curved surface that is concave upward. This curved surface is termed a listric normal fault and is a common type of fracture in the Gulf coast region of the United States.

Normal faults are the product of a dynamic process that results in conditions of changing stress (force per unit area) along a plane of weakness in the Earth's crust. The fault develops from a point along this plane. According to Lamoraal de Sitter, the stress is minimum at the starting point along the surface and maximum at the edges. Because of the edge conditions, the plane steepens and splits into several divergent smaller faults, or splays. These small segments may join to form a larger normal fault with a scalloped trace. Subparallel normal faults with smaller displacements generally accompany the large faults. At places, the adjacent beds are systematically fractured without significant displacement. These fractures are termed joints and generally have a high density (close spacing) near the fault.

The deforming forces can be related to three mutually perpendicular but unequal axes designated as maximum principal stress (σ_1), intermediate principal stress (σ_2), and minimum principal stress (σ_3). In the case of normal faults, the primary deforming force (σ_1) is vertical or nearly vertical. The least stress (σ_3) is horizontal. The normal faults are actually steeply inclined shear fractures that formed in response to forces promoting the sliding of adjacent blocks past each other. These fractures generally form at an angle of 30 degrees from the maximum principal stress. The orientation of the maximum principal stress is horizontal for thrust faults and for wrench (transform) faults.

Normal Fault Classification

Normal faults are classified according to the type of displacement of fault blocks relative to a known point. Based on the slip or actual movement of formerly adjacent points on opposing fault blocks, three types are commonly designated: strike slip, or movement along the trend of the fault; dip slip, or movement directly down the fault surface; and oblique slip, or diagonal movement down the fault surface. The movement along all these examples is translational; consequently, no rotation of the blocks in respect to each other has occurred outside a disturbed zone adjacent to the fault. If the actual displacement is not known, the term "separation" is used by most geologists to indicate the apparent movement on a map or cross section. Heave and throw are the

horizontal and vertical components, respectively, of the dip separation as measured along a vertical profile that is at right angles to the trend of the fault.

There are several varieties of normal faults. Detachment (denudation) faults have a low angle of dip (usually less than 30 degrees) and are common features in the western United States. Growth or contemporaneous faults are listric normal faults that are active during sediment accumulation. Layered rocks on the downthrown side of the fault are thicker than equivalent beds on the upthrown side. Smaller subsidiary or antithetic faults commonly form on the downthrown side of the main fault but dip in a direction opposite to the master fault. These are common features along the Gulf coast of the United States.

Some special faults may result from the same stress orientation as normal faults; that is, the maximum principal stress is vertical. These closely related faults, however, are characterized by rotational movement between blocks. For example, hinge faults increase in displacement along the length of the fault; linear features that were parallel before faulting are not all parallel after faulting. A pivotal (scissor) fault is another example of a rotational fault. In this type, the fault blocks pivot about an axis that is at right angles to the fault surface; the movement on the downthrown side is in opposite directions (up and down) along the length of the fault.

Associated Features

Major steeply dipping normal faults occur in the Colorado Plateau (Arizona and New Mexico) where these features are closely associated with regional flexures called monoclines. The western part of the plateau along the Colorado River is divided into large structural blocks by three north-trending faults. One of these faults, the Hurricane fault of Arizona and Utah, dips to the west and has a maximum displacement of 3,048 meters.

At some places, normal faults bound narrow blocks that have been displaced up or down. An uplifted or elevated block is called a horst; the depressed block is termed a graben. Topographically, these structural features may be represented by a series of mountain ranges and intervening valleys, respectively. The Basin and Range Province of the western United States is a good example of this

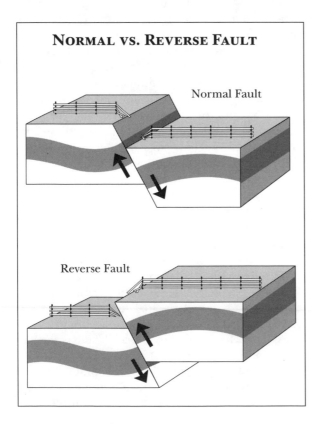

NORMAL VS. REVERSE FAULT

Normal Fault

Reverse Fault

A normal fault in sandy shale in the Chilhowee Mountains, Tennessee. (U.S. Geological Survey)

the part of the rock column that is missing. Caution, however, must be exercised to make sure that the strata have not been eroded, have not thinned, or have not changed character laterally.

There are many distinctive geometric, mineralogic, or physiographic features that are associated with large normal faults. Some of these features, however, are also characteristic of other types of faults and should thus merely be considered as "clues" in recognizing normal faults. Many normal faults are expressed at the surface by low cliffs or scarps that reflect the minimum displacement along the fault; however, these straight slopes are usually modified by erosion to form faultline scarps. The scarps may be notched by streams crossing the upthrown block at a high angle to the fault trend; continued erosion by these side streams may result in triangular-shaped bedrock facets on the footwall with fan-shaped stream deposits on the downthrown hanging wall block. Movement along the fault usually disturbs rocks adjacent to the break and results in beds along the fault being bent up or down; in other words, the rocks bend before rupture takes place. This phenomenon is called fault drag. Normal drag occurs when rocks on the upthrown side are bent down into the fault and rocks on the downthrown side are bent upward. Reverse drag occurs when beds on both sides of the fracture are bent down. Because movement along the fault produces an irregular surface between the fault blocks at some places, subsurface fluids are provided an avenue to the surface. Both hot- and cold-water springs are common occurrences along large normal faults. Solutions moving along the fault may also deposit minerals such as calcite or quartz between the blocks. These fillings are usually stained yellow or reddish brown by iron oxide. Movement along the fault is usually recorded by fine lines or by narrow grooves on the fault surface called slickensides. These features, however, may only indicate the latest movement along the fault. Impressions of the slicken lines

horst-and-graben type complex. In this region, both low- and high-angle normal faults have shaped an area that extends from southern Oregon southward to northern Mexico; the area has been broadly uplifted and the crust stretched in an east-west direction by the normal faulting. Some estimates of the total extension across the region are more than 100 percent. The displacement along these large faults ranges up to 5,486 meters. In Europe, the Rhine graben is a classic example of a well-developed rift system. This narrow structural trough trends northward for nearly 300 kilometers through West Germany and controls the path of the upper Rhine River.

CLUES TO IDENTIFICATION

Normal faults are recognized in the field or on vertical aerial photographs and satellite images by identifying features characteristic of this type of fault. On the Earth's surface, these faults occur as geological lines that are revealed by a sharp, curvilinear line in the bedrock that is usually accentuated by vegetation, a sharp contact in adjacent rock types in section or map view, a marked change in structural style, an abrupt change in topography, or an anomalous drainage pattern. Normal faults are usually recognized in vertical drill holes by an omission of rock layers; comparison of rock samples (drill cuttings) or mechanical logs from adjacent borings will generally reveal

A thrust fault showing fault drag in Atacam Province, Chile. (U.S. Geological Survey)

are sometimes preserved on the outer surfaces of the mineral fillings. A series of larger scale (several centimeters or more of relief), parallel grooves and ridges may produce an undulating fault surface. The movement of large blocks along the fault usually produces low, steplike irregularities on the surface that are steeply inclined in the direction of movement. These features can be used to identify the direction of movement when a fault surface is poorly exposed. As the fault blocks slide past each other, angular rock fragments are dislodged and may accumulate to form a tectonic breccia; a microscopic breccia, or mylonite, may also result from movement. In some cases, the dislodged rock may be ground to a pliable, claylike substance called gouge. At places, a large fragment of bedrock, called a horse, is caught along a normal fault.

CHARACTERISTIC MAP PATTERNS

Normal faults generally occur in definite region patterns that are easily represented on geo-logic maps. Zones of overlapping, or *en echelon*, normal faults are common in the Gulf coast region of the United States. In Texas, individual faults within the Balcones and Mexia-Luling fault systems are not continuous along strike but overlap with adjacent faults that have a similar trend. These fault zones generally follow the path of the buried Ouachita fold belt and mark the boundary between the geologically stable area of Texas and the less stable Gulf coast region. Parallel or subparallel faults in an area also form a distinctive map pattern; if most of the faults are downthrown in the same direction, these structural features are designated as step faults. Radial fault patterns are common over or around central uplifts or domed areas of the crust. These faults are generally associated with local stretching of the crust that results from the emplacement of salt masses (plugs) or igneous intrusions.

Some normal faults are also closely related to the development of plunging (inclined) folds and form characteristic map patterns. According to de Sitter, steeply dipping normal cross faults may form nearly at right angles to the trend of concentric (formed of parallel layers) folds. These faults are parallel to the principal deforming force and occur during the folding process. The maximum displacement occurs along the highest part (crest) of the fold; these faults usually die out along the flanks. Longitudinal crest faults may occur parallel to the trend of the folds. These normal faults are perpendicular to the principal deforming force and probably form as the compressional forces diminish.

ROLE IN SHAPING EARTH'S SURFACE

Faults have played a significant role in shaping the Earth's surface throughout geologic time. The occurrence of normal faults is closely tied to modern plate tectonics, a unifying concept for the geological sciences. The faults are generally associated with modern and ancient divergent lithospheric plate boundaries, both on continents and in the ocean basins. The regions adjacent to modern plate margins, which are characterized by high heat flow and shallow-focus earthquake activity, are places where new oceanic-type crust is generated. In modern ocean basins, inferred normal faults bound narrow downdropped blocks (grabens) along the axis of the mid-ocean ridge system, the longest continuous geologic feature on Earth. Topographically, these structural troughs form deep valleys along the ridge crest. Individual troughs range up to 30 kilometers wide and are filled or partly filled with sediment. The Mid-Atlantic Ridge, a mountain range along the midline of the Atlantic basin, extends northward from Antarctica to Iceland. In Iceland, measurements across the boundary between the North American and Eurasian plates indicate that the crustal blocks are currently moving apart at the rate of a few centimeters per year.

In the Middle East, along the Red Sea, steeply dipping normal faults are associated with a large dome or uplift over a plumelike hot spot in the Earth's mantle. Near the Afar region of Ethiopia, the uplift has been subdivided by three radial fault systems that intersect at angles of about 120 degrees. These systems are characterized by large, high-angle normal faults that initially formed a series of downdropped blocks, or grabens. The three-pronged structural feature represents a "triple junction" that separates the African, Arabian, and Indian-Australian lithospheric plates. The East African rift system, which consists of both east and west zones, forms the second prong; it trends northward for nearly 5,000 kilometers and is marked by a series of elongate lakes. The maximum displacement on the bordering faults is nearly 2,500 meters at some places. The east-trending third arm of this large feature is a rift that is partly occupied by the Gulf of Aden.

Earth scientists have also been able to identify historical divergent lithospheric plate boundaries from regional geologic and geophysical (application of physics to geological problems) studies. In the modern Appalachian mountain chain in the eastern part of the United States, a series of elongate structural troughs (grabens and half-grabens) occur along the axis of the range. These structural features, which extend from Nova Scotia in Canada southwestward to North Carolina, contain thick deposits of Triassic (period of geologic time ranging from about 200 to 245 million years ago) sedimentary rocks with associated igneous rocks. Internally, steeply dipping normal faults divide the troughs into narrow tilted blocks that range up to 10 kilometers wide. Some of the border faults were active during Triassic deposition and have a cumulative displacement of nearly 4,000 meters. The formation of these troughs probably marked the separation of North America and Europe about 200 million years ago.

ECONOMIC IMPORTANCE

Normal faults are also economically important. These faults serve as traps for hydrocarbons at many places. Migrating oil and gas moving updip from a place of origin, usually a sedimentary basin, are trapped against the fault, which acts as an impermeable barrier or seal. If the fault is not completely sealed, however, it may serve as an avenue for fluids to move to a higher level. Most commercial hydrocarbon deposits occur on the upthrown side of the fault where "rollover" of the rock layers has provided a suitable site for the accumulation of hydrocarbons. The faults are also the locus of metallic mineral deposits. Mineralization may occur in the openings along the fault or in the adjacent fractured rock. Drag ore, related to fault drag, occurs at some places. Also, rich ore bodies are moved downward along younger normal faults. A classic example is at the United Verde extension mine near Jerome, Arizona. There, a rich copper deposit was displaced more than 500 meters vertically.

Donald F. Reaser

CROSS-REFERENCES

BIBLIOGRAPHY

Allison, I. S., R. F. Black, J. M. Dennison, R. K. Fahnestock, and S. M. White. *Geology: The Science of a Changing Earth.* 5th ed. New York: McGraw-Hill, 1974. An interesting and well-written book on physical geology that is designed for high school or introductory college students. An excellent section on geotectonics that details features associated with normal faults.

Billings, M. P. *Structural Geology.* 3d ed. Englewood Cliffs, N.J.: Prentice-Hall, 1972. A popular college textbook that presents basic concepts and structural features in a clear, concise, and understandable manner. Emphasizes a field approach to recognizing and solving geological problems. Suitable for upper-level undergraduate geology students.

Buck, Roger, et als., ed. *Faulting and Magmatism at Ocean Ridges.* Washington, D.C.: American Geophysical Union, 1998. This collection of essays covers topics including seismology, magmatism, active faults, and seafloor spreading. The articles lean toward the technical but are illustrated with charts, maps, and graphs. Bibliography and index.

Cloos, E. "Experimental Analysis of Gulf Coast Fracture Patterns." *American Association of Petroleum Geologists Bulletin* 52 (1968). This journal article describes the results of experimental work with clay and dry sand models. Model grabens bounded by normal faults as well as single normal faults accompanied by fault drag are produced by applying tensional forces in a pressure box. These model fractures are compared to the Texas Gulf coast fracture pattern. Suitable for high-school-level readers and college students who are interested in geological models.

Coble, Charles R., E. C. Murray, and D. R. Rice. *Earth Science.* Englewood Cliffs, N.J.: Prentice-Hall, 1986. A general textbook designed for junior high school students and interested laypersons. In the structural section of the text, challenging scientific questions are presented for the reader. Also, the activities of specialists in the structural career field are summarized.

Davis, G. H. *Structural Geology of Rocks and Regions.* New York: John Wiley & Sons, 1984. This book is very readable and takes a practical approach to regional tectonics. Basic concepts and principles of structural geology are emphasized. Suitable for upper-level undergraduate geology students.

De Sitter, L. U. *Structural Geology.* New York: McGraw-Hill, 1959. This book effectively relates geological theory with practice; it compares similar geological phenomena, both on a small and large scale, from different parts of the world. An advanced book designed primarily for students with a good background in geology.

Doyle, Hugh A. *Seismology.* New York: John Wiley, 1995. A good introduction to the study of earthquakes and the Earth's lithosphere. Written for the layperson, the book contains many useful illustrations.

Jacobs, J. A. *Deep Interior of the Earth.* London: Chapman and Hall, 1992. Deals in detail with all aspects of the Earth's inner and outer core. The origin of the core, its constitution, and its thermal and magnetic properties are discussed in detail. Well suited to the serious science student.

Judson, Sheldon, Marvin E. Kauffman, and L. Don Lee. *Physical Geology.* 7th ed. Englewood Cliffs, N.J.: Prentice-Hall, 1987. An interesting and well-written book on physical geology designed for introductory college students. Includes a section on geotectonics that details features associated with normal faults. An excellent glossary of technical terms.

McClay, Kenneth R. *Thrust Tectonics.* London: Chapman and Hall, 1992. This collection of papers was presented as part of the Thrust Tectonics Conference held at Royal Holloway and Bedford New College, University of London, in 1990. The advanced nature of the collection makes this most useful for the college student.

Mitra, Shankar, George W. Fisher, et al., eds. *Structural Geology of Fold and Thrust Belts.* Baltimore: The Johns Hopkins University Press, 1992. A good discussion of physical geology focusing on the structure and processes of thrust faults and folds. Suitable for the college reader. Illustrations, bibliography, and index.

Park, R. G. *Foundations of Structural Geology.* New York: Methuen, 1983. An excellent reference for high school or college students specifically interested in structural geology. Good coverage of the relationship between geologic structures and plate tectonics.

Shelton, J. W. "Listric Normal Faults: An Illustrated Summary." *American Association of Petroleum Geologists Bulletin* 68 (1984). This article provides the reader with details about the characteristic features of a specific type of normal fault. The paper also discusses the general geometry, causes, and occurrences of most types of normal faults. A number of illustrations, mostly cross sections, are included in the text as an aid to understanding the concepts presented.

FAULTS: STRIKE-SLIP

Strike-slip faults separate portions of the Earth's crust that have moved horizontally past each other. They can be thousands of kilometers in length, with offsets of hundreds of kilometers across them. Many of the most devastating earthquakes occur along strike-slip faults.

PRINCIPAL TERMS

DIP: a measure of slope, the dip is the angle between a plane and the horizontal, measured in the vertical plane perpendicular to the strike of the plane

FAULT: a fracture in the Earth's crust across which there has been measurable movement

PLATE TECTONICS: a theory that holds that the surface of the Earth is divided into about twelve rigid plates that move relative to one another, producing earthquakes, volcanoes, mountain belts, and trenches

SLIP: the relative motion across the surface of a fault

STRIKE: the orientation of a horizontal line on a plane; the strike is measured using compass directions and represents the angle between the horizontal line on the plane and a horizontal line in the north direction

CHARACTERISTICS OF STRIKE-SLIP FAULTS

A fault is a surface within the Earth's crust across which displacement has occurred. A surface can be curved, flat, tilted, vertical, or horizontal. A curved surface may be subdivided into smaller pieces, each of which can be considered to be a plane. At any location, therefore, a fault can be thought of as a planar element. The orientation of this element is referred to by two angles: its "strike" and its "dip." The strike is the orientation of a horizontal line on the plane, given as a compass direction. The dip is the angle at which the plane tilts down into the ground.

Stipulating that a fault occurs within the Earth's crust implies something about the scale of the process. Although some may consider fractures on a centimeter or meter scale to be faults, in general faults are expected to be tens of meters or kilometers in size, and some can be hundreds of kilometers long.

Displacement across the fault, or "slip," refers to the amount and direction of relative motion of the blocks of rock on opposite sides of the fault. If this motion is entirely horizontal and neither block has moved up or down, the slip will be in the direction of the strike on the fault surface; such faults are called "strike-slip" faults. If, on the other hand, the motion is in the direction of the slope on the fault plane, it is called a "dip-slip"

fault. There are two kinds of dip-slip faults. When the motion is such that the block above the fault plane slides down the fault, it is a "normal" fault. When the motion is such that the block above the fault plane slides up the fault, it is a "reverse" fault.

For many reasons, some mechanical and some geological, most faults on Earth are one of these three types. Less common are faults with significant components of both dip-slip and strike-slip motions, which are called "oblique-slip" faults. In addition, a surface may be a strike-slip fault over most of its length, but portions may be dip-slip or oblique-slip. If the relative motion between the blocks of rock on either side is predominately horizontal, it is considered to be a strike-slip fault.

TYPES OF STRIKE-SLIP FAULTS

Because their motion is restricted to being horizontal, only two kinds of strike-slip faults are possible. These are defined on the basis of how the block of rock that is across the fault from the observer appears to move. It is not necessary to know which block actually moved; only their relative motion is important.

Consider two buses parked next to each other with their drivers' sides adjacent and facing in opposite directions. If either bus were to move forward, passengers in one bus would see the other bus appear to move forward. They might not even be certain whether it was their bus moving or the other one. To passengers looking out the side windows, the view would be of a bus moving to the left, past the windows. This relative motion could be called "left-lateral" or "sinistral." Similarly, if either bus were to move in reverse, passengers in either bus, looking out the side windows, would see the other one move to their right. The relative motion between the buses could be called "right-lateral" or "dextral." Such relative motion has nothing to do with the absolute motions (movement across the ground) of either bus. If one bus were to move in reverse and the other bus were to move forward somewhat more rapidly, relative motion would be left-lateral or sinistral, even though both buses were moving across the ground in the same direction.

Geologists can study features that are offset across a strike-slip fault to determine the sense of relative motion and then name the fault either right-lateral or left-lateral. As an example, if a fence is offset across an east-west-trending fault such that the fence north of the fault has moved horizontally to the east (or the fence south of the fault has moved horizontally to the west), then it is a right-lateral strike-slip fault.

This concept was successfully applied for decades but needed modification when seafloor spreading was discovered. Offsets along mid-ocean ridges, where new ocean crust is being manufactured, appear to have a sense of relative motion that is opposite the sense of motion observed along the strike-slip faults connecting them. This can be demonstrated by considering twelve fast-growing trees planted in pots. Imagine that they are all in two parallel north-south lines initially. Next, imagine that the six at the north end of the lines (three in each line) are moved a few meters to the east. This produces an offset in the lines with a right-lateral sense. Finally, imagine that all of the trees are knocked down: Those on the western lines fall to the west, while those on the eastern lines fall to the east. Temporarily ignoring some tenets of biology, imagine that all the trees continue to grow taller, but because they are lying down, this growth is horizontal. Along the offset, trees to the north will be growing to the west, and trees to the south will be growing to the east. This motion is left-lateral, opposite the offset in the pots. Mid-ocean ridge segments are thus not good

indicators of the sense of motion on the strike-slip faults connecting them because new crust is created along them. The term "transform" fault refers to this situation and to similar manifestations of strike-slip faults.

RECOGNIZING STRIKE-SLIP FAULTS

Two approaches are used to find and delineate strike-slip faults: identifying features that have been juxtaposed or offset by the fault, and detecting features in the landscape that are known to have been produced by the faulting process. If an area has experienced fault displacements after human development, obvious offsets of anthropogenic structures are generally easy to find, measure, and date. Highways, railroads, fence lines, pipelines, and buildings have been studied over the years to decipher recent displacement histories on a great number of faults. Other, less obvious, features include surveys of real estate bound-

A high-angle reverse fault in Woburn, Quebec. (Geological Survey of Canada)

aries, such as town and village borders. Although less tangible, these surveys are usually done with high precision and provide good estimates of regional deformation associated with faulting.

Faulting is episodic by nature, and movement on a particular fault may recur on a time scale of centuries. Such lengthy recurrence intervals are too long to be reflected in offsets of most cultural features, and so topographic characteristics are also examined. Offsets in the courses of rivers and streams, interruptions in hills and valleys, and other topographic changes can reveal relative displacements along a fault. Unlike anthropogenic structures, such alterations of a natural topography can be difficult to date with great precision. By extending the time scale back thousands of years, however, they often provide important data.

Interruptions in the deposition in lakes and swamps near strike-slip faults can reveal considerable detail about the timing and intensity of former earthquakes. In an area called Pallet Creek, along the San Andreas fault in California, evidence for eight earthquakes has been dated using radiocarbon techniques. The earliest occurred at about 750 C.E., and the most recent in 1857. Unfortunately, the slip associated with each of these earthquakes is not easily ascertained from the sedimentary strata.

During an earthquake, a fault may slip several meters in a matter of seconds, but, averaged out over millions of years, most faults have displacement rates on the order of a few centimeters a year. Over the course of 1 million years or so, the displacement across most faults will be on the order of a few tens of kilometers. The San Andreas fault is thought to have been in existence for about 29 million years. Hundreds of kilometers of displacement are possible over this vast period of time. Topographic features persist for thousands, perhaps even hundreds of thousands of years, but few could be expected to exist after millions of years. Here, the juxtaposition of rock types and the records left in the magnetic minerals of the ocean floor can be used to delimit the time of inception and the subsequent movement history of a fault.

The faulting process itself produces clues, distinct from offsets or juxtapositions, that can reveal the presence of a fault. In general, the broken rock in the vicinity of the fault will weather and erode more easily than sound rock some distance away. This results in the formation of long, linear valleys along fault lines.

By grinding up the rock in the immediate vicinity of the fault, faulting often modifies the groundwater system. Sometimes rock that was impermeable has its permeability increased by the new fractures produced. Other times, a permeable rock is rendered less permeable because the conduits that permitted water to flow through it are disrupted by the fracturing process. These changes may result in the formation of springs or lakes called "sag ponds" directly above the fault.

Finally, some faults are revealed by the earthquakes that occur on them. By studying the seismographic data obtained from an earthquake, geophysicists can determine the location of the earthquake and the direction in which the blocks of rock moved, even if the fault involved lies kilometers beneath the surface.

OCCURRENCE

The surface of the Earth is made up of twelve or so tectonic plates that are roughly 100 kilometers thick and persist with little deformation within them for hundreds of millions of years. Plates diverge from each other along ridges (generally beneath the oceans but occasionally running through a continent, such as the African rift valley). Connecting ridge segments, which may be separated by hundreds of kilometers, are strike-slip faults. Plates converge, with one plate moving beneath the other, along subduction zones. When, as is often the case, the direction of convergence is not perpendicular to the subduction zone (oblique convergence), a strike-slip fault may develop to accommodate the horizontal component of relative motion. Sometimes the horizontal stresses produced by plate convergence are sufficient to extrude a wedge-shaped piece of a plate to the side, which results in strike-slip faults. If plates move past each other without diverging or converging, the motion is horizontal and is accomplished along strike-slip faults.

The majority of strike-slip faults connect ends of ridge segments. In the ocean floor, such ridge segments are often called offset spreading centers (OSCs), and the reason for their existence is not well understood. Active faulting occurs along the transform faults between ridge segments, resulting in an age difference across a line that extends the fault beyond the offset region. Because the ocean floor cools, contracts, and sinks as it gets older, this age difference is often expressed by significant topographic relief, with cliffs stretching out for hundreds of kilometers from the transform fault as an oceanic fracture zone.

Oblique convergence is frequently accommodated by a partitioning of relative motion between a dip-slip subduction zone and a strike-slip fault parallel to the subduction zone. Some examples of strike-slip faults that are parallel to subduction zones are the Great Sumatran fault on the island of Sumatra and the Denali fault in Alaska. In 1995, the Hyogoken-Nanbu earthquake near Kobe, Japan, occurred on one of these faults, killing more than five thousand people and causing more than $200 billion of damage.

Horizontal stresses perpendicular to the trend of a subduction zone can also produce strike-slip faults. Sometimes called "watermelon seed" or "horizontal extrusion" tectonics, this process occurs when a segment of a plate is wedged off to the side by plate convergence. The motion of this plate segment is similar to that of a watermelon seed when it is squeezed between the thumb and forefinger. Mechanical engineers say that this type of deformation is caused by a "rigid indenter," and the faults produced may be called "indent-linked" strike-slip faults.

The subcontinent of India has acted as a rigid indenter as it has pushed northward into the Eurasian plate. subduction produced the Himalaya Mountains and the Tibetan plateau; at the same time, huge, wedge-shaped pieces consisting of most of Southeast Asia have been moved off to the side. As much as one-half of the convergence has been accommodated by this eastward extrusion along the Altyn Tagh, Haiyuan, and related strike-slip fault systems, resulting in many devastating earthquakes. The North Anatolian fault in Turkey is another indent-linked strike-slip fault. In this case, much of Turkey is being extruded to the west as the Arabian and Eurasian plates converge. Movement on this fault brought Turkey 1.2 meters closer to Europe during a devastating 1999 earthquake.

If the relative motion between plates has little convergence or divergence, it may be taken up almost entirely along strike-slip faults. In California,

motion between the North American plate and the Pacific plate occurs largely across the San Andreas fault, for example. Although it is presently a variety of transform fault that connects ridge segments in the Gulf of California to the Mendocino triple junction (where it meets a trench and another strike-slip fault), the San Andreas fault has such a complex history that it is often best to consider it as a plate boundary where most of the relative motion has a right-lateral strike-slip sense.

FORMATION AND SECONDARY FEATURES

There are often long periods of time between episodes of motion on strike-slip faults. During these periods, soil and other unconsolidated sediments can accumulate over the fault region. When motion again occurs on the fault, the offset in these new layers at the surface can be seen to develop in a complex but systematic way. Similar processes have been observed in laboratory models involving clay cakes being offset above moving plates.

Initially, many small offsets develop above the fault in a parallel, offset geometry resembling the slats on a venetian blind. These are called Riedel shears, conjugate Riedel shears, or P shears, depending on their angular relationship to the underlying fault. Complex technical issues are involved in their formation, but of particular interest is the fact that many minor faults form initially, and only later do the principal displacement shears develop. These are the strike-slip faults across which most of the movement occurs. Study of strike-slip faults in bedrock often reveal the complexities introduced by the early shears.

The complicated geometry of strike-slip faults means that motion across some parts of them will not occur in a simple, strike-slip sense. In some places, the fault surface will bend, or be offset, resulting in either extension or compression. Imagine an east-west-trending right-lateral strike-slip fault with an offset across which the eastern side has been offset to the south relative to the western side. In the vicinity of the offset, the crust will be extended. Sometimes called "transtension," this stretching may result in normal faults bounding a down-dropped block of crust, producing a "pull-apart" basin. The Salton Sea in Southern California and the Dead Sea in Israel are examples of these.

Using the same geometry, but this time considering the offset to be of a left-lateral strike-slip fault, the vicinity of the offset would be in compression. This "transpression" might be expected to produce buckling and mountain ranges. The Transverse Ranges of California, occurring near the "big bend" of the San Andreas fault, probably owe their existence, in part, to these compressive stresses.

SIGNIFICANCE

Many of the horizontal movements on the surface of the Earth occur along strike-slip faults. If a complete set of slip data for all of the strike-slip faults on the planet could be constructed, it would reveal most of what is known about tectonics. Most strike-slip faults extend to the surface, providing exposures where they can be studied in detail. Knowledge of the geometry, offset history, and earthquake recurrence intervals on strike-slip faults is therefore more developed, and based on better data, than similar knowledge for normal or reverse faults.

Because they often cut through the continental crust, strike-slip faults are likely to traverse populated regions. Their effects on topography may even encourage development of the most earthquake-prone areas. Disastrous earthquakes are common on strike-slip faults today, as they have been in the past, and will certainly be in the future. Learning more about these faults and the earthquakes that occur along them is likely to help in predicting those earthquakes and in mitigating their negative consequences.

Otto H. Muller

CROSS-REFERENCES

Creep, 185; Cross-Borehole Seismology, 190; Discontinuities, 196; Earthquake Distribution, 277; Earthquake Hazards, 290; Earthquakes, 316; Elastic Waves, 202; Experimental Rock Deformation, 208; Faults: Normal, 213; Faults: Thrust, 226; Faults: Transform, 232; Himalaya, 836; Lithospheric Plates, 55; Notable Earthquakes, 321; Ocean Ridge System, 670; Paleoseismology, 1139; Plate Margins, 73; Plate Tectonics, 86; San Andreas Fault, 238; Seismic Observatories, 246; Seismic Tomography, 252; Seismometers, 258; Stress and Strain, 264; Transverse Ranges, 858.

BIBLIOGRAPHY

Davidson, Jon P., Walter E. Reed, and Paul M. Davis. *Exploring Earth: An Introduction to Physical Geology.* Upper Saddle River, N.J.: Prentice Hall, 1997. Chapter 10, "The Conservative Boundary: Transform Plate Margins," provides an easily understood treatment of strike-slip faulting that covers the common transform types. However, it does not deal with indent-linked strike-slip faults. Profusely illustrated with colored maps and diagrams, it is suitable for high school readers.

Davis, George H., and Stephen J. Reynolds. *Structural Geology of Rocks and Regions.* 2d ed. New York: John Wiley & Sons, 1996. Provides a thorough, comprehensive treatment of faults and faulting, as well as a great deal of information about the strength of rock, the accumulation of strain, and other related aspects of geology. More technical than the other references cited, it is not suitable for high school readers.

Fowler, Christine Mary Rutherford. *The Solid Earth: An Introduction to Global Geophysics.* Cambridge: Cambridge University Press, 1990. This book provides an outstanding treatment of how offsets between plates can be determined, excellent descriptions of the detailed structure of oceanic transform faults, and a useful discussion of the extrusion tectonics associated with the formation of the Himalaya Mountains. Although not suited for high school readers, its treatment of these topics requires no mathematics beyond algebra and trigonometry.

Keller, Edward A., and Nicholas Pinter. *Active Tectonics: Earthquakes, Uplift, and Landscape.* Upper Saddle River, N.J.: Prentice Hall, 1996. Looking at earthquakes from the perspective of how they alter the landscape, this 338-page book gives careful, easily understood descriptions of how topographic features can reveal a great deal about recent and historic movements on faults. Illustrated in black and white, this book is suitable for high school readers.

Lundgren, Lawrence W. "Earthquake Hazards." In *Environmental Geology.* 2d ed. Upper Saddle River, N.J.: Prentice Hall, 1999. Excellent treatment of earthquakes and faulting from a perspective of hazard mitigation and reduction. Includes case studies of the Loma Prieta earthquake, which occurred on the San Andreas fault in 1989, and the Hyogoken-Nanbu earthquake, which occurred near Kobe, Japan, in 1995.

Yeats, Robert S., Kerry Sieh, and Clarence R. Allen. *The Geology of Earthquakes.* New York: Oxford University Press, 1997. A thorough and detailed exploration of all aspects of earthquakes, with emphasis on the geological evidence used to study them, this book is an excellent resource. Although some of the treatment may be too detailed for a beginner or casual reader, these areas can be skimmed over easily. Concepts are explained well, and great care is taken to keep terminology concise and understandable. Profusely illustrated with black-and-white maps and diagrams, it also has a very useful index.

FAULTS: THRUST

Thrust faults are the result of compressional forces that exceed the natural strength of rocks and cause them to break and move. They can trigger earthquakes, create mountain ranges, and serve as natural traps for gas and oil deposits.

PRINCIPAL TERMS

DIP: the angle between a fault plane and a horizontal surface

FAULT: a fracture or zone of breakage in a mass of rock that shows evidence of displacement or offset

FOOTWALL: the block of rock that lies directly below the plane of a fault

HEAD WALL: the block of rock that lies directly above the plane of a fault; it is also known as a hanging wall

REVERSE FAULT: the same thing as a thrust fault, except that its fault plane dips at more than 45 degrees below the horizontal

SCARP: a steep cliff or slope created by rapid movement along a fault

SLIP: a measure of the amount of offset or displacement across the plane of the fault, relative to either the dip or the strike

STRIKE: the orientation of a fault plane on the surface of the ground measured relative to north

THRUST FAULT PRODUCTION

A mass of rock below the surface of the Earth usually cracks and fractures when it loses its resistance to an applied force. Rocks break when their ability to store energy is exceeded; when a rock shows some evidence of movement or displacement along the zone of breakage, a fault is created. Thrust faults are commonly the result of strong compressional (squeezing) forces acting on relatively brittle, older subsurface rock that has moved upward and over or on top of a mass of younger, adjacent rock. It is a particular kind of fault and one of many types that exist.

The zone of breakage between the once-united masses of rock is known as the fault plane. The motion of the rocks on either side of this plane and the plane itself are usually parallel to each other. The blocks of rock on both sides of a fault plane are known as walls, a term that comes from the days of the early prospectors, who were really the first field geologists. Because the presence of a fault marks a zone of weakness in the ground, either mineral-rich groundwater or hot fluid magmas will eventually find and follow this path of least resistance toward the surface and deposit ores, minerals, or gemstones. Prospectors would seek out faults, as they knew that a fault was likely to be the home of some valuable material. Once a fault was located, a mine shaft would be dug to follow the trace of the fault below ground.

The head wall, or hanging wall, was the wall above the miner's head; the footwall was the wall below his feet. Head walls and footwalls exist only in faults that are not vertical. In terms of the overall structure of the fault, the hanging wall is the block that occurs above the fault plane and the footwall is the rock below the fault plane. In thrust faults, the head wall always moves relatively upward and the footwall moves relatively downward. The term "relatively" is used because it is usually very difficult for a geologist to determine exactly which block has moved. For example, both blocks could have moved upward but the hanging wall moved farther; both blocks could have dropped but the footwall dropped farther; the hanging wall could have remained stationary while the footwall dropped; or the footwall could have remained stationary while the hanging wall moved upward.

FAULT ORIENTATION

The orientation of the fault relative to the Earth's surface is of great importance. It allows the fault to be located and mapped as a place to avoid during construction, especially if it is an active fault or one with the potential for continued movement. A fault's orientation, or strike, is measured by the trace of the fault plane as it would appear on a horizontal plane and is measured in degrees from the magnetic North Pole. The plane's angle of tilt, or dip, is measured from a horizontal position down to the fault plane. The dip direction is always perpendicular to the strike direction.

A fault can be straight in form and consist of one sharp, clean break, or it can have a highly irregular form and be composed of multiple breaks. Thrust faults of the latter type may be so closely spaced as to form a highly complex zone that may be hundreds of meters wide. Fault planes can also be curved, adding to the complexity. Geologists have located and mapped many small thrust faults at very shallow depths below the surface and have found some large thrust faults that extend down to a depth of 700 kilometers.

In general, the total displacement, or offset, in a rock along a thrust fault may be large or small, and horizontal, vertical, or oblique, depending on the strength of the compressional force and the rock type involved. An important factor in determining this displacement is the angle of the fault plane. Steeply dipping planes will show a small vertical uplift on the order of a few meters or less; shallowly dipping planes may exhibit a long horizontal displacement extending for many kilometers. Displacements are also described in terms of their relative motion. A "dip slip" occurs when the movement of the rock is parallel to the dip direction of the fault plane, a "strike slip" indicates motion

parallel to the strike, and an "oblique slip" occurs somewhere between the strike and dip.

FAULT VARIETIES AND AGES

Three types of faults involve movement of the hanging wall upward in respect to the footwall: Thrust faults are characterized by fault planes that dip at angles less than 30 degrees; overthrust faults

In Montana, a low-lying slab of rock known as the Lewis Overthrust shows a horizontal displacement of about 24 kilometers with a fault plane that dips at an angle of less than 3 degrees. Shown here also is Chief Mountain, a peak of Precambrian rock faulted onto young rock. (© William E. Ferguson)

are thrust faults with very large displacement; and reverse faults dip at angles greater than 45 degrees from horizontal. In large displacement thrust faults, the fault surface may be quite irregular; therefore, the fault plane may approach horizontal or even reverse direction. Overthrusts are very common in areas of intense folding where there is shortening, and usually thickening, of the crust. The offset of rock along the path of the fault is usually greatest near its middle and decreases at either end, or the thrust may terminate in strike-slip faults.

The age of rocks may be used to determine the relative age of faults. Since a fault cuts through a block of rock, the fault must be younger than the rock. Therefore, if a rock is found to be 100 million years old and is cut by a fault, then the fault must be younger than 100 million years old.

LOCATING THRUST FAULTS

Structural geologists study faults to try to understand what was happening to the Earth's crust at the time of faulting and to determine the origin of the force that caused the rocks to break and move. To study a thrust fault, the geologist must first accurately locate and measure the fault in the field. It is sometimes difficult to determine where a fault exists, especially if it is very old. The geologist, acting as a detective, must rely on direct physical evidence that he or she can gather from above and below the surface.

When a fault intersects the surface, it usually forms a fault line. The existence of this line is commonly indicated by some noticeable feature. Photographs of the land taken from high-altitude aircraft reveal these features as offsets or disruptions in rows of planted crops or trees; sharp breaks in the channels of streams; unusual linear alignments of springs flowing at the base of a mountain or along a valley; raised sections of land, such as beach and river terraces; and fault scarps. A fault scarp is a recent, sharp break in the surface of the ground that has a straight and very steep slope. The height of a scarp is directly related to the amount of upward motion along a fault. Unless the fault is still active, however, it usually is not exposed at the surface but is buried beneath the cover of more recently deposited sediments. In this case, a geologist must rely on subsurface evidence.

EVIDENCE OF THRUST FAULT MOTION

The subsurface location of a thrust fault is not always easy to find. The field geologist must rely on direct or indirect evidence that is not usually visible on the surface. Direct subsurface evidence of the existence of a thrust fault can be obtained from the examination of rocks within mine shafts, highway tunnels, or excavation sites dug for building foundations. Similarly, "roaducts" (highway excavations that run through mountains and valleys) and natural outcroppings of rock at the surface may indicate the presence of a thrust fault. In these situations, a geologist would look for any evidence that suggests that massive blocks of rock have moved relative to one another.

When huge blocks of rock are broken and continue to rub against one another, certain physical features are produced as a result of the friction between these moving blocks. Sure indicators of faulting are slickensides, which are recog-

A fault scarp near Red Canyon Creek, Gallatin County, Montana, 1959. (U.S. Geological Survey)

nized as highly polished and finely scratched, or striated, surfaces of rock along the fault plane. The direction of the striations is parallel to the direction of the last movement along the fault. Depending on the amount of friction generated between these blocks, their inward-facing surfaces may be crushed into a fine, soft, claylike powder known as "gouge" or may become fault breccias, rocks consisting of small, angular fragments. Microbreccias are formed when the crushed fragments along the fault plane are microscopic. Mylonites are a special type of solid microbreccia; they have a streaked appearance in the direction of motion. Pseudotachylites are a kind of microbreccia that does not appear streaked but exists as a thin glassy film because of the melting of rock from frictional heat. Other evidence of thrust fault motion would be the overlapping or repetition of the same rock units (like overlapping shingles on a roof) and an abrupt termination of a rock unit along its trend. One or more of these faulting criteria may be present, or evidence may be completely missing.

This view of the southern end of Scapegoat Mountain in Montana shows folds in strata. (U.S. Geological Survey)

DRILLING FOR EVIDENCE

If there are no accessible outcrops or underground viewpoints, the field geologist must turn to the expensive direct method of evidence-gathering: drilling. Drilling into the ground with specialized drilling rigs allows access to the subsurface. The examination of small broken rock samples brought up by a diamond drill bit or solid "rock cores" brought up by hollow drill bits is a direct means of studying rocks that do not exist at the surface. These samples can be compared with those taken from well-known nearby regions, and any disruption or missing units of rock may indicate the presence of a fault. If a subsurface geologic map exists, the geologist can predict what rocks exist below the drill site and use this information in conjunction with the collected rock samples.

If the evidence from drilling does not match the regional geology, then one or more faults may have been at work shuffling the sequence of rocks below. Sometimes large blocks of rock get caught between faults when they move and become bent, folded, or twisted, creating a highly complex pattern that is not easy to understand using drill core data alone; the uncertainty of what lies below ground increases.

EVIDENCE FROM GEOLOGIC MAPS

Indirect evidence of thrust faulting can be had from a geologic map of an area. Geologic maps show the distribution, thickness, age, and orientation of the various types of rock that would be seen at the surface of the Earth were all the soil covering removed. Any older rock formation that sits on top of a younger rock formation was most likely moved to its location by thrust faulting. Rock formations that appear out of place with the overall sequence of rocks probably suffered a similar fate.

Several notable large thrust faults exist in the United States. They can be seen in the Rocky Mountain region as a series of sharp, parallel ridges similar to the teeth on a saw—hence the term "sawtooth," as in the peaks of the Sawtooth Mountain Range in Montana. In the same state, a low-lying slab of rock known as the Lewis Overthrust shows a horizontal displacement of about 24 kilometers with a fault plane that dips at an angle of less than 3 degrees. Thrust faults occur in most mountain ranges on the Earth such as the Appalachian Mountains, the European Alps, and the Himalaya.

ENGINEERING AND ECONOMIC APPLICATIONS

A fault is a zone of weakness in a rock; therefore, it may continue to move over a long period. Information about the rate of rock movement is very valuable, especially when the motion along a thrust fault (or any fault) is rapid enough to trigger the release of stored energy within a rock, causing an earthquake. Accurate mapping of thrust faults is also important, since the potential for future earthquakes must be carefully evaluated, especially in highly urbanized areas. A complete study of thrust faults is not easy, however, because there are many variables to consider: the fault's exact location and total horizontal extent, the orientation and strength of the rocks relative to the fault, the direction and amount of movement, and the fault's previous earthquake history. These factors are critical to decisions of where to construct nuclear power plants, dams, housing projects, and cities.

Thrust faults have great economic potential, since they have the ability to act as "traps," or reservoirs, for deposits of migrating oil and natural gas. In such a case, one impervious rock type is brought into fault contact with a petroleum-bearing rock. The impermeable rock now acts as a barrier to any further upward fluid migration and allows oil to accumulate beneath it. Similarly, in mineral exploration, large thrust faults have been known to harbor exploitable quantities of radioactive and other rare minerals, needed for use in industry and medicine, that were either deposited by igneous activity or precipitated by circulating, mineral-rich groundwater. Thrust faults also serve as natural underground pipelines, allowing circulating groundwater easier access to the surface of areas that otherwise might have been deserts.

Steven C. Okulewicz

CROSS-REFERENCES

Continental Structures, 590; Creep, 185; Cross-Borehole Seismology, 190; Discontinuities, 196; Earthquake Distribution, 277; Earthquake Engineering, 284; Earthquake Hazards, 290; Earthquake Locating, 296; Elastic Waves, 202; Experimental Rock Deformation, 208; Faults: Normal, 213; Faults: Strike-Slip, 220; Faults: Transform, 232; Fold Belts, 620; Folds, 624; Land-Use Planning, 1490; Lithospheric Plates, 55; Mountain Belts, 841; Petroleum Reservoirs, 1728; Plate Margins, 73; Plate Tectonics, 86; San Andreas Fault, 238; Seismic Observatories, 246; Seismic Reflection Profiling, 371; Seismic Tomography, 252; Seismometers, 258; Stress and Strain, 264; Subduction and Orogeny, 92; Thrust Belts, 644.

BIBLIOGRAPHY

Billings, Marland P. *Structural Geology.* 3d ed. Englewood Cliffs, N.J.: Prentice-Hall, 1972. This is an introductory college-level textbook for all aspects of structural geology. Chapters 8 through 12 discuss the variation, classification, and recognition criteria for faults. Chapter 10 is devoted to thrust faults. The book is well written and clearly illustrated with many line drawings. Also included are black-and-white photographs of faults as they appear in the field and a useful end-of-chapter bibliography. The bible of structural geologists.

Lahee, Frederic H. *Field Geology.* 6th ed. New York: McGraw-Hill, 1961. Within this thick book are described the various techniques for the measurement, mapping, relative age determination, and interpretation of rock formations in the field. A large part of chapter 8 is devoted to faulting, with a description of applied field mapping techniques that are used in the construction of surface and subsurface maps. The chapter is easy to read, and the fault types are illustrated with either line drawings or block diagrams. Although somewhat dated, this work is still a classic and the field book carried by most practicing geologists.

Lutgens, Frederick K., and Edward J. Tarbuck. *Earth: An Introduction to Physical Geology.* 6th ed. Upper Saddle River, N.J.: Prentice Hall, 1999. This college text provides a clear picture of the Earth's systems and processes that is suitable for the high school or college reader. In addition to its illustrations and graphics, it has an accompanying computer disc that is compatible with either Macintosh or Windows. Bibliography and index.

McClay, Kenneth R. *Thrust Tectonics.* London: Chapman and Hall, 1992. This collection of papers was presented as part of the Thrust Tectonics Conference held at Royal Holloway and Bedford New College, University of London, in 1990. The advanced nature of the collection makes this most useful for the college student.

Mitra, Shankar, George W. Fisher, et al., eds. *Structural Geology of Fold and Thrust Belts.* Baltimore: The Johns Hopkins University Press, 1992. A good discussion of physical geology focusing on the structure and processes of thrust faults and folds. Suitable for the college reader. Illustrations, bibliography, and index.

Parker, Sybil P., ed. *McGraw-Hill Encyclopedia of Earth Sciences.* 2d ed. New York: McGraw-Hill, 1988. A source that covers every aspect of geology. Topics are arranged alphabetically. The entry on faults and fault structures discusses fault movement, procedures for locating faults, stress conditions, and examples of various types of fault. Includes references to other entries and a bibliography. Contains some equations, but they are not essential to the reader's understanding.

Spencer, Edgar W. *Introduction to the Structure of the Earth.* 2d ed. New York: McGraw-Hill, 1977. This college-level structural geology textbook was written from the "plate tectonics" point of view and covers the entire Earth. Chapters 7 through 10 deal with various faults that were formed as a result of colliding continents and ocean basins and are indicators of intense rock deformation. Chapters 8 and 9 describe the many types of thrust fault. Many regional examples are provided and clear line drawings are included. There is a thorough end-of-chapter bibliography.

Tarbuck, Edward J., and Frederick K. Lutgens. *The Earth: An Introduction to Physical Geology.* 2d ed. Columbus, Ohio: Merrill, 1987. This basic geology textbook contains a section on rock deformation. It provides diagrams to illustrate the various types of fault and the concepts of strike and dip, and it includes a photograph of an overthrust formation. Review questions and a list of key terms conclude the chapter. For high school and college-level readers.

Thornbury, William D. *Principles of Geomorphology.* 2d ed. New York: John Wiley & Sons, 1968. Geomorphology is the study of landforms on the Earth's surface. This basic college-level textbook describes the many types of landform that are produced by various geologic processes. Chapter 10 is an informative, highly readable, and well-illustrated chapter that discusses the landforms created by fault activity and how they are recognized.

FAULTS: TRANSFORM

Transform faults occur along fracture zones found at the mid-oceanic ridges. The ridges are areas of erupting ultramafic lavas, which cause seafloor spreading, which, in turn, drives the moving lithospheric plates. Geologists have found that the offset ridges do not move in relation to each other and have been essentially fixed relative to the spreading seafloor.

PRINCIPAL TERMS

CURIE TEMPERATURE: the temperature below which minerals can retain ferromagnetism

DIVERGENT BOUNDARY: the boundary that results where two plates are moving apart from each other, as is the case along mid-oceanic ridges

FERROMAGNETIC: relating to substances with high magnetic permeability, definite saturation point, and measurable residual magnetism

FRACTURE ZONE: the entire length of the shear zone that cuts a generally perpendicular trend across a mid-oceanic ridge

HYPOCENTER: the initial point of rupture along a fault that causes an earthquake; also known as the focus

MAGNETIC ANOMALIES: patterns of reversed polarity in the ferromagnetic minerals present in the Earth's crust

MID-OCEAN RIDGE: a long, broad, continuous ridge that lies approximately in the middle of most ocean basins

RIFT VALLEY: a region of extensional deformation in which the central block has dropped down with relation to the two adjacent blocks

SEAFLOOR SPREADING: the hypothesis that oceanic crust is generated by convective upwelling of magma along mid-ocean ridges, causing the seafloor to spread laterally away from the ridge system

TRANSCURRENT FAULT: a fault in which relative motion is parallel to the strike of the fault (that is, horizontal); also known as strike-slip fault

TRANSFORM FAULT OCCURRENCE

Faults are regions of weakness or fractures in the Earth's crust along which relative movement occurs. The simplest type of fault is the so-called normal fault, in which one block of crust is displaced vertically downward with respect to the other. A reverse or thrust fault is one in which the block is driven upward with respect to the other block. Yet another type of crustal displacement takes place horizontally, as one block slides laterally with respect to the other. These strike-slip or transcurrent faults are related to the most complex class of faults, known as transforms.

Actually regions of crustal transformation, transform faults are found almost exclusively along the mid-oceanic ridges that nearly encircle the globe. The ridges are the sites of newly forming crustal materials composed of very dense magmas of relatively high iron and magnesium content. As the new crust forms, lava flows act to push the oceanic crust laterally away from the spreading ridge, as the seafloor spreads out at a rate of a few centimeters a year.

MID-OCEANIC RIDGE SYSTEM

If one were able to view the mid-oceanic ridge system from orbit, it would quickly become apparent that the spreading centers do not occur along a smoothly continuous line but rather are broken into scores of offset ridges. The offset is marked by a fracture zone, which serves as the border between two spreading centers. Because the ridges are displaced, it was first believed that these fracture zones were simply transcurrent faults, along which right or left lateral displacement would be observed from opposing sides. The ridges are fixed with relation to each other and appear to have been so for long periods of geologic time. Clearly, a new type of faulting was being observed.

If one were to voyage to the bottom of the mid-

Atlantic Ocean to view one of these faults, its highly unusual nature would become clear only after one had traveled hundreds of miles along its entire length. Starting at the west end of the transform fault, one would find that the crust is slowly moving toward the west on either side of the fracture zone. Because the crust is essentially moving in the same direction, earthquakes are rare events in this region.

Most transform faults are oriented at right angles to the mid-oceanic ridges. Typically, they extend from ridge to ridge (the active part of the fault); if spreading has been taking place for long periods of geologic time, fracture zones extend out from the ridge systems. These transforms and fracture zones are thought to be ancient areas of weakness in the crust that formed when the ocean basin (such as the Atlantic) began forming. The discontinuous nature of the ridge system is believed to be structurally ancient. When viewed globally, the spreading axis is offset from the Earth's rotational axis, with the ridges corresponding to lines of longitude and the transforms roughly approximating latitudinal lines. Spreading rates are greatest at the equatorial regions of the globe.

VOLCANIC ERUPTIONS AND EARTHQUAKES

The underwater ridge mountains are marked by a distinctive rift valley, centered along the range. Volcanic eruptions emanating from the rift create pillow-shaped lavas. The ridge is a boundary between the Earth's lithospheric plates; it is a divergent boundary, for the seafloor is spreading laterally in opposite directions. A view of the ridge along the transform fault would reveal that the line of mountains is broken and offset. An observer passing over the high ridge on the north side of the fault would suddenly notice that the fault had taken on the appearance of a transcurrent fault. Movement on the south side of the fault is right lateral, and although the crust is traveling west, on the north side of the fault, the movement is toward the east. An observer on either side of the fault would see right lateral displacement.

Because crustal movement is in opposing directions between the offset ridges, earthquakes occur frequently in this area. The crust's relative motion is horizontal so that the focus or hypocenter (the actual point of rupture in the rock) of transform earthquakes is shallow—typically less than 70 kilo-

meters deep, whereas trench earthquakes can be up to 700 kilometers deep. Magmatic eruptions and earthquakes offer convincing evidence that the Earth's crustal plates are far from stationary. Past the offset ridge, crustal motion is once again in the same direction, and no lateral motion would be observed on opposite sides of the fracture. This transformation of crustal displacement along the shear zone is the derivation of the term "transform fault."

While the Atlantic Ocean is a basin in which new crust is forming, causing the rifting of the continents and the growth of the ocean basin at a rate of a few centimeters per year, the opposite is the case in the Pacific Ocean basin. There, the transform faults are a bit more complex. An example would be the New Zealand fault, which terminates on both ends at subduction trenches rather than at ridges. New Zealand is seismically active, because the transform fault passes directly through both of the large islands.

SAN ANDREAS FAULT

The most famous of all transform faults forms a distinctive type of lithospheric plate boundary. Extending from the East Pacific Rise off the coast of western Mexico to the Mendocino fracture zone and the Juan de Fuca ridge system off the coast of Washington State is the 960-kilometer-long, 32-kilometer-deep system of strike-slip faults known as the San Andreas. The San Andreas fault forms a boundary between the northward-trending Pacific plate and the North American plate. Horizontal displacement along this huge fault system is estimated at 400 kilometers since its inception nearly 30 million years ago. The rate of displacement has been measured at an average of 3.8-6.4 centimeters annually.

How did this impressive plate boundary form, and how does it fit into the transform fault model? About 60 million years ago, the east Pacific was also home to a third plate, the Farallon, which eventually was subducted under the North American plate. The Farallon plate was pushed eastward into the North American plate by a spreading ridge system and trench that were eventually overridden by the continental plate. The remnants of this ridge system are found at the ends of the San Andreas fault. Because the Pacific plate's motion was northward, the ridge system was converted into a transform fault, characterized by its right lateral strike-slip displacement.

TRIPLE JUNCTIONS

The complexity of three plate interactions, or triple junctions, explains the complex nature of transform fault systems such as the San Andreas. The remnants of the doomed Farallon plate are presently found as the Juan de Fuca plate, which is bounded by spreading ridges and transform faults to the west and a subduction zone where the plate is being inexorably pushed into the Earth's upper mantle, under the North American plate. Inland, the Cascade volcanoes, most notably including Mount St. Helens, have their volcanic fires fueled by a rising magma plume that is generated by the melting of the subducted oceanic plate.

North of the Juan de Fuca plate is another large transform system similar to the San Andreas fault, called the Queen Charlotte transform. An extension of the San Andreas system, the Queen Charlotte fault begins to the south at the Juan de Fuca ridge but becomes inactive at its northern end, whereas the San Andreas is a ridge-ridge transform that is bordered by the East Pacific Rise to the south and Gorda Ridge to the north.

The great ridge system that was overridden by the continental plate is responsible for the faulted structure of the Basin and Range Province, and the triple junction of the plates may have given rise to the magma plume or "hot spot" that is responsible for the Snake River plain volcanics and the thermal activity at Yellowstone National Park. Clearly, understanding the evolution of transform faults such as the San Andreas and the Queen Charlotte is pivotal to the understanding of the complex mountain scenery of western North American.

EVIDENCE FROM THE SEAFLOOR

A revolution in the Earth sciences had its germination in the ideas of the German meteorologist Alfred Wegener, who argued that the continents had once been joined in one supercontinent and had since drifted apart. Wegener's theories of continental drift were not taken seriously until evidence from the seafloor forced a rethinking of modern geology in the 1960's. Oceanographic research in the 1950's led to a new picture of the ocean basins. Far from the featureless abyssal plains they were once thought to be, the basins proved to

be marked by dramatic mountain ranges characterized by rift valleys. In 1960, a Princeton University scientist, H. H. Hess, suggested that convection currents in the Earth's upper mantle were driving volcanic eruptions along the rifts, causing the seafloor to spread and the continents to move apart.

The real breakthrough in understanding the ocean floor came as a result of numerous cross-Atlantic voyages by research vessels towing submerged magnetometers in order to measure the strength of the crustal rock's residual magnetism. As the basaltic magma erupts at the spreading rift, magnetite freezes out of the melt at 578 degrees Celsius, and the Earth's magnetic field orientation is frozen into it. This temperature is known as the Curie temperature.

MAGNETIC ANOMALIES

When the first magnetic maps of the seafloor were produced, scientists groped to explain the alternating nature of the field's polarity, which changed in a random pattern over short distances. Two research students at the University of Cambridge, Fred Vine and Drummond Matthews, solved the riddle of the magnetic anomaly stripes by proposing that the stripes represented reversals in the Earth's magnetic field over time. Because the anomaly patterns were symmetrical with respect to the axis of the spreading centers, this was proposed as evidence in favor of the Hess model of seafloor spreading.

The huge fracture zones that appeared to offset the mid-oceanic ridges were mapped with magnetic anomaly measurements, and seismic data indicated that seismic activity was concentrated between the offset ridges along the fracture zones. Networks or seismic instruments also enabled geophysicists to study the direction of the spreading crustal movement, through geometrical solutions known as first motion studies or fault-plane solutions.

In 1965, J. Tuzo Wilson dubbed the faults "transforms" because of the changing of relative motion along the length of the fracture. He reasoned that the faults were not causing the offset of the spreading ridges, but that the offset is what caused the appearance of a transcurrent fault between the ridges and an inactive segment, in which the seafloor was spreading in the same direction beyond the zone of offset.

Geophysicists mapping the magnetic anomalies were aided by the work of terrestrial geologists, who were able to identify magnetic reversal patterns in terrestrial lavas that were identical to those found in oceanic rocks. Radiometric dating established a magnetic anomaly time scale that would allow scientists to determine the ages of rocks laterally out from the ridges and hence to deduce the rates of seafloor spreading. Armed with the paleomagnetic data and with Wilson's notion of transform faults as directional guides, scientists were able to show that the Atlantic Ocean had indeed been opening at a rate of a few centimeters per year and that the landmasses of western Europe and eastern North America were once essentially in contact.

IDENTIFYING TRANSFORM FAULTS

Wilson and others went on to study the more complex faults of the eastern Pacific. The San Andreas fault had been identified long before as a transcurrent fault. The new paradigm of plate tectonics placed it in a global tectonic scale; the fault is now accepted by most as a kind of transform fault that connects to spreading ridge centers to the north and south.

While faults such as the New Zealand Alpine and San Andreas are fairly accessible to scientists, most of the planet's transform faults lie below thousands of meters of ocean. Echo sounding, radar, sonar seismographic, and magnetic anomaly maps have helped scientists to locate the Earth's transform faults. The 1978 Seasat mission radar mapped the ocean floor from space, producing detailed information on the globe's mid-oceanic ridge and fault system. In addition, deep-sea submersibles have been piloted to the rifts, where eruptions of lavas and fault displacements have been observed directly.

PLATE TECTONICS

The theory of plate tectonics in the 1960's caused a revolution in the study of geology, geophysics, and even paleontology. Post-World War II oceanographic research led to the discovery that the mid-oceanic ridges are sites of newly forming crust and resulted in J. Tuzo Wilson's explanation of transform faults as ancient fracture zones offsetting the spreading ridges.

With a vast supply of paleomagnetic, seismic,

geologic, and other evidence at their disposal, geologists have been able to reconstruct Earth history and the relative motions of the continents, using transform faults as directional guides and magnetic reversals to determine the rates of plate movement. Like any successful theory, plate tectonics is elegantly simple; it explains nearly all the Earth's diverse landforms and rock formations. Changes in continental distribution may also be linked to the extinction events that punctuate the geologic time scale.

Geologists' understanding of the interaction between ridge systems, subduction trenches, and transforms has led to a better understanding of the paleogeography of the American West and how that complex mountainous region of active faults and recent volcanic activity came into being. Aside from contributing to the fundamental understanding of Earth processes, plate tectonics theory explains forces that can influence human lives. Active plate boundaries are the sites of earthquake and volcanic activity. In California, millions of people live and work astride one of the world's largest transform faults, the San Andreas. Residents of Vancouver and Seattle are similarly threatened by the offshore Queen Charlotte Islands fault. On the other side of the Pacific Ocean, New Zealand is nearly bisected by the New Zealand Alpine fault, making it a seismically active region. Whenever human beings decide to build their homes and cities near these moving regions of the Earth's crust, the possibility of geologic catastrophe is very real.

David M. Schlom

CROSS-REFERENCES

BIBLIOGRAPHY

Condie, Kent C. *Plate Tectonics and Crustal Evolution.* 4th ed. Oxford: Butterworth Heinemann, 1997. An excellent overview of modern plate tectonics theory that synthesizes data from geology, geochemistry, geophysics, and oceanography. Of special interest is chapter 6, on seafloor spreading, and chapter 9's treatment of the Cordilleran system, including a discussion of the evolution of the San Andreas fault. A very helpful tectonic map of the world is enclosed. The book is nontechnical and suitable for a college-level reader. Useful "suggestions for further reading" follow each chapter.

Cox, Allan, and R. B. Hart. *Plate Tectonics: How It Works.* Palo Alto, Calif.: Blackwell Scientific, 1986. A valuable treatment of the geometrical relationships and movements of the Earth's lithospheric plates. Designed for the reader who has a basic qualitative knowledge of plate tectonics but who wishes to learn more, particularly about quantitative analysis of plate movements. Filled with easy-to-follow exercises that demonstrate plate motions, particularly those associated with transform faults.

Doyle, Hugh A. *Seismology.* New York: John Wiley, 1995. A good introduction to the study of earthquakes and the Earth's lithosphere. Written for the layperson, the book contains many useful illustrations.

Lambert, David, et al. *The Field Guide to Geology.* New York: Facts on File, 1988. For the beginning student of geology, this reference work is filled with marvelous diagrams that make the concepts easy to understand. Chapter 2 deals with plate tectonics and has a clear, concise treatment of seafloor spreading and transform faults. Suitable for any level of reader, from high school to adult.

Lutgens, Frederick K., and Edward J. Tarbuck. *Earth: An Introduction to Physical Geology.* 6th ed. Upper Saddle River, N.J.: Prentice Hall, 1999. This basic geology textbook contains a section on rock deformation. It provides diagrams to illustrate the various types of fault and the concepts of strike and dip, and it includes a photograph of an overthrust formation. Review questions and a list of key terms conclude the chapter. For high school and college-level reader. In addition to its illustrations and graphics, it has an accompanying computer disc compatible with either Macintosh or Windows.

McClay, Kenneth R. *Thrust Tectonics.* London: Chapman and Hall, 1992. This collection of papers was presented as part of the Thrust Tectonics Conference held at Royal Holloway and Bedford New College, University of London, in 1990. The advanced nature of the collection makes this most useful for the college student.

Mitra, Shankar, George W. Fisher, et al., eds. *Structural Geology of Fold and Thrust Belts.* Baltimore: The Johns Hopkins University Press, 1992. A good discussion of physical geology focusing on the structure and processes of thrust faults and folds. Suitable for the college reader. Illustrations, bibliography, and index.

Redfren, Ron. *The Making of a Continent.* New York: Times Books, 1983. Richly illustrated with dramatic photographs, this book is a lucid discussion of plate tectonics with respect to the continent of North America. Contains excellent explanations of seafloor spreading and transforms, along with a section on the San Andreas fault. Suitable for a general audience.

Shea, James H., ed. *Plate Tectonics.* New York: Van Nostrand Reinhold, 1985. A collection of classic and key scientific papers, mostly from the 1960's, that together constitute a sweeping overview of plate tectonic geology. Of special interest are papers by Fred Vine and J. Tuzo Wilson on the magnetic anomalies off Vancouver Island and a paper by L. R. Sykes on transform faults at the mid-oceanic ridges. With chapter introductions by the editor, the work is suitable for a college-level reader with an interest in the history of plate tectonics theory.

Shepard, Francis P. *Geological Oceanography.* New York: Crane, Russak, 1977. Chapter 2 addresses seafloor spreading and faulting of the oceanic crust. Photographs, diagrams, and supplementary reading lists augment the text, which is suitable for a beginning geology or oceanography student.

Sullivan, Walter. *Continents in Motion.* New York: McGraw-Hill, 1974. Dedicated to Harry Hess and Maurice Ewing, two late pioneers of plate tectonics theory, this is the classic popular work on moving crustal plates. Well-written explanations of transform faults and their roles in seafloor spreading and a discussion of the San Andreas fault are included in the highly readable text.

Wilson, J. Tuzo, ed. *Continents Adrift and Continents Aground.* San Francisco: W. H. Freeman, 1976. Selected readings from *Scientific American* are introduced with commentary by Wilson, a leading figure in the history of plate tectonics theory. Chapter 2 deals with seafloor spreading and transform faults with a classic article by Don L. Anderson on the San Andreas fault. Suitable for a general audience. Contains a bibliography.

Wyllie, Peter J. *The Way the Earth Works: An Introduction to the New Global Geology and Its Revolutionary Development.* New York: John Wiley & Sons, 1976. Wyllie's book has a very informative section on transform faults and earthquake studies. An extensive list of suggested readings augments the text, which is suitable for a college-level reader.

Young, Patrick. *Drifting Continents, Shifting Seas.* New York: Franklin Watts, 1976. A good entry-level discussion of plate tectonics theory, written by a journalist with a knack for simplifying complex concepts. Contains a brief glossary, indexed with a bibliography. Suitable for high school and lay readers.

SAN ANDREAS FAULT

The San Andreas fault has been recognized as a major geologic feature of California and of North America for nearly a century. It is hoped that research into this seismically active fault will help geologists to develop a simple, characteristic model to explain the behavior of the fault and possibly to forecast potentially destructive earthquakes.

PRINCIPAL TERMS

EPICENTER: the point on the Earth's surface directly above the focus of an earthquake

GEODETIC SURVEYING: surveying in which account is taken of the figure and size of the Earth and corrections are made for the Earth's curvature

GEODIMETER: an electronic optical device that measures ground distances precisely by electronic timing and phase comparison of modulated light waves that travel from a master unit to a reflector and return to a light-sensitive tube; its precision is about three times as great as that of a tellurometer

RIGHT-SLIP, or RIGHT-LATERAL STRIKE-SLIP: sideways motion along a steep fault in which the block of the Earth's crust across the fault from the observer appears to be displaced to the right; left-slip faults are exactly the opposite

SEISMIC: pertaining to an earthquake

TECTONICS: a branch of geology that deals with the regional study of large-scale structural or deformational features, their origins, mutual relations, and evolution

TELLUROMETER: a portable electronic device that measures ground distances precisely by determining the velocity of a phase-modulated, continuous microwave radio signal transmitted between two instruments operating alternately as a master station and remote station; it has a range up to 65 kilometers

TRIPLE JUNCTION: a point on the Earth's surface where three different global plate boundaries join

CALIFORNIA'S EARTHQUAKE HISTORY

With a total length of more than 1,600 kilometers, the San Andreas is the longest fault in California and perhaps the longest in North America. Because about three-quarters of California's population live within 80 kilometers of the fault, its existence—and its potential for unannounced major earthquakes—is responsible for formulating many public policy decisions and for establishing the design criteria of many engineering projects in the state. In spite of much being known about the fault, the geoscientists who have studied it readily admit that their knowledge remains far from complete.

California's earthquake history is short. The earliest recorded earthquake in the state was on July 28, 1769. It was experienced by the Spanish explorer Gaspar de Portolá while camped on the Santa Ana River southeast of Los Angeles. One of the state's earliest recorded large earthquakes occurred on the Hayward fault, a branch of the San Andreas system, in the East Bay on June 10, 1836, during which surface breakage occurred along the western base of the Berkeley Hills. Another large earthquake on October 21, 1868, also centered on the Hayward fault, caused thirty fatalities and ground breakage for about 32 kilometers, accompanied by as much as 1 meter of right-slip between San Leandro and Warm Springs. Other notable nineteenth century earthquakes along the San Andreas system in the Bay Area occurred in June, 1838, on October 8, 1865, and on April 24, 1890. Unfortunately, there are no seismograph records of these early earthquakes because it was not until 1887 that the first seismographs were installed in the United States.

Since 1850, California has experienced three great earthquakes of Richter magnitude 8 or greater. (The Richter scale, which was introduced in 1935, defines earthquakes of magnitude 2 as "felt," those of magnitude 4-4.5 as "causing local damage," those of magnitude 6-6.9 as "moderate," those of magnitude 7-7.5 as "major," and those exceeding magnitude 7.5 as "great.") Two of these three great earthquakes, the San Francisco earthquake of 1906 and the Fort Tejon earthquake of 1857, resulted from movement of the San Andreas fault.

RECOGNITION OF SAN ANDREAS FAULT

It was not until after the great San Francisco earthquake of April 18, 1906, that the San Andreas was first recognized as a continuous regional geologic structure of California. This earthquake caused sudden right-slip of up to 5 meters, ground rupturing for 420-470 kilometers, an estimated seven hundred fatalities, and damage estimated at between $350 million and $1 billion. The earthquake is generally considered as having attained magnitude 8.3. Much of the property loss in the San Francisco disaster was caused by the extensive fires following the earthquake that resulted from ruptured gas lines and lack of water from broken water mains. The strongest earthquake in the Bay Area associated with the San Andreas fault since the 1906 event was the magnitude 7.1 earthquake of October 17, 1989, which caused more than $1 billion in damage.

The other important California earthquake of magnitude comparable to the 1906 San Francisco disaster caused by the San Andreas fault was the Fort Tejon earthquake of January 9, 1857. This earthquake had an estimated magnitude of 8.3 and caused ground breakage for 360-400 kilometers on the southern part of the San Andreas fault zone, from the Cholame Valley in the Coast Ranges to the Transverse Ranges as far south as the present site of Wrightwood. Ground motion was felt from north of Sacramento to Fort Yuma on the lower Colorado River. The ground accelerations during this earthquake were so great that mature oak trees in the Fort Tejon area were toppled, buildings collapsed, fish in a lake near the fort were tossed out of the water, and the Los Angeles River was thrown out of its banks. (A wonderfully informative, and sometimes entertaining, account of the effects of the 1857 earthquake, taken from correspondence and newspaper accounts of the time, was published by D. Agnew and Kerry Sieh in 1978.) Right-slip movement ranging from 4.5 to 4.8 meters occurred in the 1857 event. Although the 1857 earthquake had

the same estimated magnitude as the 1906 earthquake, the 1857 earthquake was potentially more destructive than the 1906 disaster because the ground shaking is reported to have lasted from 1 to 3 minutes. Compare this time interval with the short 10-30 second shaking accompanying the moderate 1971 San Fernando earthquake, magnitude 6.3, in which sixty lives were lost, and the 1989 San Francisco earthquake of 7.1, which lasted for approximately 15 seconds and whose damage, although severe in certain local areas, was minor in comparison to that of the larger-magnitude, longer-lasting earthquakes. (Evidence, although inconclusive, points to the possibility that the 1989 event was a strike-slip earthquake along the San Andreas.) Such long duration and low seismic frequencies associated with events such as the 1857 earthquake occurring in California's highly populated regions are more likely to cause serious damage to large buildings and claim more lives than are moderate earthquakes.

DEFINITION OF SAN ANDREAS FAULT

The name "San Andreas" was first used in 1895 by Andrew C. Lawson, a geologist at the University of California, Berkeley. Later, it was Lawson who headed the California Earthquake Investigation Commission, whose monumental report on the San Andreas system was published in 1908. This report describes surface ruptures that developed in the 1906 earthquake that extended for more than 420 kilometers from Humboldt County to San Juan Bautista in San Benito County and continued to follow the fault southward through the Coast Ranges into San Bernardino County. The report used the term "San Andreas rift" for the rift-valley surface expression in the San Andreas Lake area on the San Francisco Peninsula, and from this term have evolved the terms "San Andreas fault," "San Andreas fault zone," and "San Andreas system." Curiously, in spite of the ample evidence of right-slip motion described for the segment that ruptured in the 1906 earthquake, the 1908 report maintained that the dominant characteristic movement on the San Andreas fault throughout its geologic history had been vertical (up and down) by as much as thousands of meters. This idea of vertical motion became the prevailing idea of motion for the San Andreas for almost fifty years. It has since been agreed, however, that activity along the fault has shown dominantly horizontal motion, with the North American plate moving southward relative to the Pacific plate. This kind of horizontal motion is termed right-lateral strike slip, or, more simply, right-slip, because the sideways displacement of the block across the fault from the observer appears to be to the right. Thus, if the historically observed motion is typical of the movement of the San Andreas fault in the geological past, rocks that once faced each other across the fault should now be separated. Furthermore, because it is possible to establish the ages of many rocks that were once a single unit and once the amount of displacement of the faulted rock unit is known, scientists are able to determine the movement rate (called slip rate) during the geologic history of the fault.

Current usage defines the San Andreas fault proper as the strand of the San Andreas zone that reveals surface rupturing produced by recent movements within the zone. The definition of the term "San Andreas zone" incorporates numerous parallel to subparallel related fractures that may be separated by 10-15 kilometers. Many of these subordinate fractures record that the locus of movement in the zone has shifted from one branch to another through time. Some examples of active or formerly active subordinate strands are the Pilarcitos, Calaveras, and Hayward faults in northern California and the Punchbowl, San Gabriel, San Jacinto, and Banning faults in southern California.

EXTENT OF FAULT

On land, the San Andreas fault extends from Shelter Cove in Humboldt County southward, crossing the Golden Gate west of the Golden Gate Bridge, and under southwestern San Francisco, where the fault trace is obscured by housing tracts in Daly City; it appears that city planners, developers, and residents have ignored the hazards of living on this infamous fault. The fault continues southward along the length of the Santa Cruz Mountains of the San Francisco Peninsula, where it forms the rift valley occupied by San Andreas and Crystal Springs reservoirs, through the Coast Ranges past its intersection with the Garlock fault at Frazier Mountain near Tejon Pass. Beginning here, the fault turns more eastward—geologists call this segment of the fault the big bend—and

marks the approximate northern boundary of the San Gabriel Mountains. It passes through Cajon Pass separating the San Gabriel and San Bernardino mountains and continues southeastward along the southern margin of the San Bernardino Mountains to the vicinity of San Gorgonio Pass. Here, the San Andreas divides into the North and South branches, the North Branch joining the Mission Creek fault and the South Branch joining the Banning fault. North of Indio, the branches rejoin, forming what some geologists call the southern big bend, and the San Andreas fault resumes its more southerly trend to border the eastern edge of the Salton Sea trough and continue south into the Gulf of California.

Near the big bend at the Frazier Mountain intersection of the San Andreas and the Garlock faults are also the ends of the San Gabriel and Big

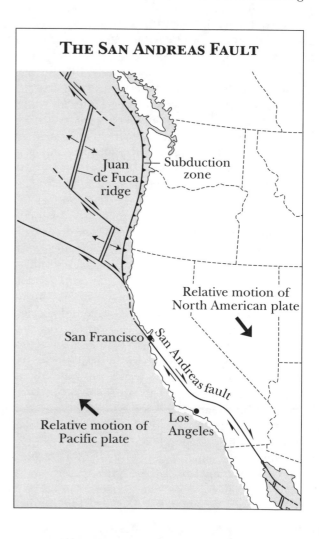

THE SAN ANDREAS FAULT

Pine faults and the Frazier Mountain thrust. This area is also known to geologists as the "knot," where both right-slip faults, the San Gabriel and San Andreas, are intertwined with left-slip faults, the Garlock and Big Pine. An important parallel branch of the San Andreas system is the San Jacinto fault. It splays off the San Andreas at the northeastern end of the San Gabriel Mountains and runs southeastward under San Bernardino, San Jacinto, and Hemet to El Centro.

The northern and southern ends of the San Andreas fault terminate in ways that are still not entirely clear. At its northern end, the San Andreas disappears into the Pacific Ocean at Shelter Cove near Point Mendocino, apparently to join the Mendocino fracture zone and the Juan de Fuca oceanic trench at a point known to geologists as a transform-transform-trench triple junction. The southern end of the San Andreas disappears beneath the waters of the Gulf of California, where it appears to develop into a number of northwest-trending transform faults that offset the East Pacific Rise. The San Andreas system, therefore, links the East Pacific Rise to the Juan de Fuca Ridge as a transform fault and marks the approximate position of a very major structural feature of the Earth's crust, the sliding boundary between the Pacific and North American crustal plates; most of California rests on the North American plate, but much of southern California is on the Pacific plate.

SEISMOLOGY AND FIELD STUDY

Virtually every technique available to geoscientists has been used in studying the San Andreas fault. Most prominent, perhaps, is the work of seismologists, geophysicists who study the waves of energy generated in earthquakes. Seismologists in California study the activity of the state's fault systems by monitoring all seismic activity in the state with seismometers, instruments that record the passage of earthquake waves. The data from these instruments allow seismologists to determine the epicentral location, depths of foci (hypocenters), surface area, and movement directions of faults responsible for given earthquakes.

The geologic history of the San Andreas fault zone has been learned by fieldwork, which began in earnest following the San Francisco earthquake of 1906 and has continued since. Fieldwork is the process of working in the field by foot, car, or aircraft in order to locate and plot on aerial photographs and topographic base maps the locations of the rock units, faults, and folds in the area of interest. A primary goal of field mapping is correlation, the determination of age relationships between rock units or geologic events in separate areas. For decades following the 1906 disaster, field mapping accumulated data on the structures and rock units exposed along the fault. For many years, the field data were interpreted to support a model that the movement history of the San Andreas was primarily vertical. A tentative proposal of 38 kilometers of right-slip motion for the fault was advanced in 1926, but it was not taken seriously.

In 1953, however, in what may be the most important benchmark paper ever presented on California geology, two California geologists, Mason L. Hill and Thomas W. Dibblee, Jr., suggested that as much as 560 kilometers of movement had taken place along the San Andreas fault in a period of 150 million years. Their proposal resulted from careful analysis of field observations that allowed correlation of rocks and fossils in separate exposures on opposite sides of the fault. They illustrated that not only had the oldest rocks studied been offset by 560 kilometers, but younger rocks, rocks that had not existed long enough to be offset as much, had been displaced by progressively lesser distances. For example, they showed that the correlation of uniquely similar rocks in the Gabilan and San Emigdio ranges illustrated 282 kilometers of offset in about 25 million years, that offset beds of gravel in the Temblor Hills and San Emigdio Mountains revealed 22 kilometers of offset in less than 1 million years, and that the Big Pine fault had been displaced from its eastern extension, the Garlock fault, by 10 kilometers in the last 200,000 years.

Few geologists at the time could accept the idea of such great right-lateral displacement along the San Andreas fault. Nevertheless, so much interest was aroused by the tentative conclusions of Hill and Dibblee that many geologists began to conduct their own investigations. Consequently, subsequent field studies clearly confirmed the thesis. The work of Hill and Dibblee has proved to be the fundamental framework to which explanations of the geology of the Coast, Transverse, and San Bernardino ranges; the Salton Trough; and the Great Central Valley must conform. In retrospect, a large cumulative right-lateral offset by the San Andreas fault is a requirement of the global tectonic theory.

FORECASTING THE NEXT "BIG ONE"

Estimates are that a repeat of the 1906 San Francisco earthquake would cause 11,000 deaths, and $35-40 billion in damage could be expected. A repeat of the 1857 event would jolt southern California with a similarly destructive earthquake, which could cost 12,000 or more lives, cause $15 billion in damage, and leave 100,000 homes unfit for habitation. The questions asked of scientists involve when and where the next "Big One" will occur. To arrive at answers to such questions, a different type of fieldwork was undertaken in the 1960's and 1970's. Studies by Tanya Atwater, then at Scripps Institute of Oceanography, showed that plate tectonics, which was at the time considered to be strictly an ocean-bound concept, was sometimes critical in understanding the deformation of crustal rocks within a continent. The San Andreas fault plays a very important role in understanding the geology of the western United States in the light of the new global tectonics. Within the framework of plate tectonics, the San Andreas fault is a transform fault along which strain is occurring between the two moving plates. In order to measure strain across the San Andreas fault, precision geodetic surveying began in the 1970's, using tellurometers, geodimeters, and long baseline interferometry (LBLI), a "space-geodetic" technique developed in the late 1970's. LBLI relies on extraterrestrial reference points, such as quasars, to measure distances to an incredible precision of 1 centimeter or less in 1,000 kilometers. Precision geodetic surveying and LBLI reveal that the relative motion between the Pacific and North American plates is about 5.5 centimeters per year, right-slip. About 4 centimeters of this motion is taken up each year by the San Andreas fault, the remaining motion being accommodated by the slippage of other faults and in other ways within the San Andreas zone.

Thus, seismology, geologic field mapping during the first half of the twentieth century, large-scale tectonic and geodetic evidence of the 1970's and 1980's, and the recorded history of earth-quakes along the San Andreas fault all suggest that Californians may expect the fault to move again—but questions remain about when and where and with what frequency great earthquakes occur along the San Andreas.

DETERMINING AVERAGE RECURRENCE INTERVAL

Researchers who estimate the frequency of earthquakes on an active fault use a still different kind of fieldwork: logging and mapping the walls of 20-foot-deep trenches dug across a fault, in association with carbon 14 isotopic age dating and statistics, in an effort to establish an average recurrence interval (RI) of earthquakes for that particular fault. The reason for attempting to establish the history of a fault's activity is that much of geological interpretation is based on the premise that geologic events occurring today result from the same processes that caused them in the past (uniformitarianism). Thus, the activity of the San Andreas fault in the geological past is the best clue to forecasting its behavior in the future.

In order to develop a regional overview of the San Andreas fault, the fault has been divided into four segments—northern, central, south-central, and southern—based on existing knowledge of seismic activity. At least two recorded large earthquakes have occurred on the northern segment, the 1838 and the 1906 events. Although little detailed work has been done on the northern segment, it is believed that the frequency of earthquakes here is similar to that on the south-central segment. The central segment has a historic record of as much as 3 centimeters of offset per year. The strain along this segment of the fault, however, appears to be released slowly by aseismic creep (very slow, incremental movement along the fault that is unaccompanied by earthquake activity). The southern segment, from Indio south, is considered dormant, with no record of seismic activity since 1688.

Using this information as a starting point, Kerry Sieh and his associates at the California Institute of Technology have trenched three sites on the south-central and southern segments of the San Andreas fault in order to obtain the information necessary to determine the average RI. The first of these sites is at Pallett Creek near Palmdale, about 55 kilometers northeast of Los An-

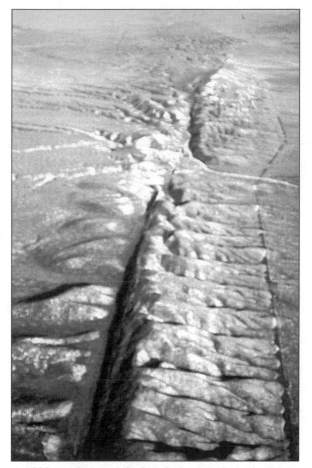

A 1983 aerial view of the San Andreas fault on the Carrizo Plain, San Luis Obispo County, California, looking southeast. A fence with tumbleweeds piled against it (right) parallels the fault. (U.S. Geological Survey)

geles; another is just north of the big bend at Wallace Creek on the Carrizo Plain west of Bakersfield; and the third locality is on the southern segment near Indio. Ray Weldon, of the University of Oregon, has studied a fourth area astride the south-central segment at Lost Lake in Cajon Pass.

RESULTS OF RECURRENCE INTERVAL STUDY

The results of the Pallett Creek work reveal the RI for ten episodes of faulting in a period of about 2,000 years to be about 132 years. Because the Pallett Creek area was affected by the great southern California earthquake of 1857, Sieh's research seems to suggest that this segment of the San Andreas should break again soon. Sieh is quick to point out, however, that the site's average RI could

have limited meaning because of newly perfected methods of carbon 14 dating, developed by Minze Stuiver of the University of Washington and statistician David Brillinger of the University of California at Berkeley, that allows much tighter precision in age dating than has been possible in the past. Consequently, age determinations of the ten events reveal that the earthquakes are clustered in bunches of two or three, with individual earthquakes within the clusters separated by tens of years. Five of the intervals between clusters are less than a century, and three of the remaining four intervals span from 200 to 330 years. From this field, laboratory, and statistical information, Sieh's interpretation of the Pallett Creek record is that it is probably near the middle of a two-hundred-year-long dormant period. Thus, if the pattern of earthquakes at Pallett Creek is truly representative of the fault's behavior of the south-central segment, it might be eighty years or more before Los Angeles could expect another great earthquake along this portion of the San Andreas fault.

Weldon's investigation at Lost Lake, about 40 kilometers southeast of the Pallett Creek site, reveals six earthquakes in the last thousand years with an average RI of 150-200 years. Thus, both the Pallett Creek and Lost Lake sites reveal relatively short RI values that appear to be at odds with what is known about the frequencies of earthquakes on the segments of the San Andreas to the north and south. Sieh's work at Wallace Creek, about 100 kilometers northwest of the Pallett Creek site and north of the big bend, results in a longer average RI of 250-450 years, and investigations on the southern segment show that this segment has not had a major, or great, event since about 1688, even though the RI for this segment is about 220 years. Comparing the results from the Indio, Lost Lake, Wallace, and Pallett creeks, Weldon and Sieh have developed several alternate models of behavior of the south-central and southern segments of the fault. Two of their models suggest that the southern segment is the most likely site for the next large earthquake. Fortunately this is a sparsely populated region at present, about 120 kilometers east of Los Angeles, far enough that a great earthquake on the southern segment is currently not a serious threat to the residents of Los Angeles.

D. D. Trent

BIBLIOGRAPHY

Agnew, D., and K. Sieh. "A Documentary Study of the Felt Effects of the Great California Earthquake of 1857." *Bulletin of the Seismological Society of America* 68 (1978): 1717-1729. An informative collection of eyewitness accounts and newspaper articles about the great 1857 southern California earthquake.

Clarke, Thurston. *California Fault: Searching for the Spirit of State Along the San Andreas.* New York: Ballantine Books, 1996. Clarke traveled the length of the San Andreas fault collecting first-hand accounts from earthquake survivors and predictors. Along with the entertaining stories, Clarke provides historical and scientific information about the fault.

Collier, Michael. *A Land in Motion: California's San Andreas Fault.* San Francisco: Golden Gates National Parks Association, 1999. Filled with beautiful color photographs that accompany text intended for the nonscientist, *A Land in Motion* gives the reader excellent insight into earthquakes and their aftermaths. There are also many diagrams and graphs that explain subduction, faults, and orogeny.

Condie, Kent C. *Plate Tectonics and Crustal Evolution.* 4th ed. Oxford: Butterworth Heinemann, 1997. An excellent overview of modern plate tectonics theory that synthesizes data from geology, geochemistry, geophysics, and oceanography. Of special interest is chapter 6, on seafloor spreading, and chapter 9's treatment of the Cordilleran system, including a discussion of the evolution of the San Andreas fault. A very helpful tectonic map of the world is enclosed. The book is nontechnical and suitable for a college-level reader. Useful "suggestions for further reading" follow each chapter.

Fradkin, Philip L. *Magnitude 8: Earthquakes and Life Along the San Andreas Fault.* Berkeley: University of California Press, 1999. Written for the layperson, *Magnitude 8* traces the seismic history, mythology, and literature associated with the San Andreas fault.

Hill, M. L. "San Andreas Fault: History of Concepts." *Bulletin of the Geological Society of America* 92 (1981): 112-131. A historical account of the recognition, mapping, and changing interpretations of the San Andreas fault, written by one of the key authorities on the geology of the fault.

Iacopi, Robert, ed. *Earthquake Country: California.* Menlo Park, Calif.: Sunset Books/Lane, 1971. Probably the best single source of information for the layperson on the San Andreas fault. Because of its photographs, maps, and diagrams, it can be well used by the traveler as a guide in finding the many fault-formed features along the fault's length.

Jordan, T. H., and J. Minster. "Measuring Crustal Deformation in the American West." *Scientific American* 259 (August, 1988): 48-58. Describes the method of measuring the actively deforming continental crust in the western United States by the use of "space-geodetic" techniques that rely on extraterrestrial reference points, such as quasars, to a precision of 1 centimeter or less in 1,000 kilometers.

Kerr, R. A. "Take Your Choice: Ice Ages, Quakes, or Impacts." *Science* 243 (January 27, 1989): 479-480. A brief overview of the December, 1988, meeting of the American Geophysical Union that includes an overview of Kerry Sieh's research and the state of knowledge in forecasting the next great earthquake on the San Andreas fault. Easily understandable by the layperson with minimal technical background.

Norris, R. M., and R. W. Webb. *Geology of California.* New York: John Wiley & Sons, 1976. The definitive textbook on the geology of California. Written for use by college students who have a year of freshman introductory geology courses as a background. A full chapter is devoted to the San Andreas fault and should be readily understandable to the interested layperson. An excellent glossary.

Plummer, Charles C., and David McGeary. *Physical Geology*. Boston: McGraw-Hill, 1999. A college-level introductory geology textbook that is clearly written and wonderfully illustrated. An excellent sourcebook of basic information on geologic terminology and fundamentals of geologic processes. The chapters on structural geology and global plate tectonics are especially relevant to understanding the San Andreas fault in the context of large-scale geologic processes. Contains a CD-ROM. Three full pages are devoted to the San Andreas fault. An excellent glossary.

Powell, Robert, R. J. Weldon, Jonathan Matti, et al., eds. *The San Andreas Fault System: Displacement, Palinspastic Reconstruction, and Geological Evolution*. Boulder, Colo.: Geological Society of America, 1993. This book provides a clear description of the power and size of the San Andreas fault. It details the history of the fault, as well as its constant evolution. Illustrations and folded leaves; includes bibliography and index.

Sieh, K. E., M. Stuiver, and D. Brillinger. "A More Precise Chronology of Earthquakes Produced by the San Andreas Fault in Southern California." *Journal of Geophysical Research* 94 (January 10, 1989): 603-623. Although this article is published in a technical journal, the section comparing the results of the investigations at the Pallett Creek site with other sites on the south-central and southern segments of the San Andreas fault and the conclusions are relatively nontechnical and worth reading.

Walker, Bryce. *Earthquake*. Alexandria, Va.: Time-Life Books, 1982. Features a nontechnical essay on the San Andreas fault that includes magnificent illustrations.

SEISMIC OBSERVATORIES

A global network of seismic observatories detects and records seismic waves produced by earthquakes and other energy releases. Their data can be used to locate the sources and determine the magnitudes of earthquakes, to interpret the Earth's internal structure, to delineate seismically active zones and thus assess earthquake hazards, to study dynamic processes in the Earth's crust, and to monitor nuclear tests.

PRINCIPAL TERMS

EPICENTER: the spot on the Earth's surface directly over the focus of an earthquake

FOCUS: also called the hypocenter, the region in the Earth's crust or upper mantle where an earthquake begins; larger earthquakes have a focus several tens of kilometers in size

SEISMIC WAVES: oscillatory vibrations generated by an earthquake that travel outward in all directions through the Earth (as body waves) or along and near the Earth's surface (as surface waves)

SEISMOGRAM: a recording, by ink pen, film, or digital data stored on magnetic tape or disk, that measures the train of seismic waves arriving by the variation in ground motion as time advances

SEISMOGRAPH: a sensitive instrument that mechanically or electromagnetically detects and records arriving seismic waves, usually by measuring the motion of the ground with respect to a relatively fixed mass

TIME OF ORIGIN: the time of an earthquake's occurrence in local time or—more conveniently for analysis of worldwide events on a standard time scale—in Coordinated Universal Time (CUT, or Greenwich Mean Time)

MONITORING SEISMIC WAVES

Earthquakes are one of nature's most sudden and terrifyingly unexpected phenomena; they are dangerous to human life and cause the destruction of buildings and property. However, they do not occur randomly in time and space. Seismology—the study of earthquakes and the outgoing seismic waves they produce—has yielded much understanding of the Earth and its dynamic physical environment, including knowledge, since the 1960's, of plate tectonics. Large tectonic plates in the Earth's lithosphere (crust and upper mantle) move at rates of several centimeters per year. It is at the margins of these plates, where they spread apart, slip sideways, or collide in great processes of mountain-building or deep-ocean trench subduction (diving down into the Earth's mantle), that earthquakes have been most common in the past and are most likely to occur in the future.

The seismic wave vibrations generated by earthquakes travel through the entire Earth, yielding information about the Earth's interior structure, physical properties, and likely composition. Study of seismic waves can also provide understanding of other sudden energy releases, such as volcanic eruptions, landslides, meteorite impacts, and large subsurface explosions such as nuclear weapons tests.

All this is possible because of the detection, recording, and analysis of seismic waves at seismic observatories. By sharing data with other observatories that have appropriate equipment, scientists can determine a seismic event's time of origin, size (energy release), geographic location, and depth. By considering many such events, scientists can investigate wave travel through the Earth and interpret its structure and properties; map the distribution of earthquakes over the Earth's surface and over time; delineate seismically active zones, major plate tectonic boundaries, and local faults (where crustal rock masses are slipping and crunching past one another); and assess seismic hazards.

Advances in seismology were made possible by the invention of seismographs in the 1870's; the establishment of progressively larger, more coordinated, and more sensitive global networks of seismic observatories beginning in the early twentieth century; and the introduction of computers in the

1960's. Computers have allowed researchers to deal with massive amounts of data and complicated mathematical analysis procedures, as well as subtle and difficult problems of interpretation.

EARTHQUAKES AND SEISMIC WAVES

An earthquake occurs when rock masses in the crust and upper mantle suddenly break and shift along a plane or zone called a fault. This can occur when stress has built up, as from plate tectonic movement in the lithosphere, and exceeds the breaking strength of the rock. The accumulated strain energy is released as frictional heating on the fault zone, plus seismic waves (vibrations) that are transmitted rapidly out from the focus in all directions through the Earth. These waves can range from barely perceptible to catastrophically destructive.

There are different types of waves produced by an earthquake. "Body" waves are generated by the faulting, termed either P (for "primary," or compressional, having forward-and-back motion like coils moving in a slinky spring) or S (for "secondary," or shear, having sideways motion like a wave in a rope snapped sideways at one end). P waves always travel faster than S waves and thus arrive at a distant point sooner. When body waves travel up to the Earth's surface, they can produce surface waves that travel out across the Earth's surface like large ripples from a stone dropped into a pond. Surface waves are termed L waves (for "large"), since they are typically larger in amplitude (extent of oscillation) than body waves. Because of their amplitude and the nature of ground shaking, they can be quite destructive.

When earthquakes are detected and analyzed by seismic observatories, information on their time of origin, epicenter, magnitude (energy release), and depth of focus can be catalogued and mapped at a central data repository. Two prime sites for global data collection are the U.S. Geological Survey's National Earthquake Information Center (USGS/NEIC) in Golden, Colorado, and the International Seismological Center in Cambridge, England. The data are publicly available and are easy to access with a computer over the Internet. One example would be a map that shows seismicity around the Pacific Rim. This margin around the Pacific Ocean basin is defined by the major plate tectonic collision and subduction as

the Pacific seafloor spreads apart on the East Pacific Rise and is forced to dive down upon collision with the other plates rimming the Pacific. The pattern of earthquake epicenters effectively outlines these active plate margins.

The amplitude (height, or deflection, of the measured vibration) of seismic waves can be used to calculate a "magnitude" for the earthquake. This is a measure of the energy released by the event, and thus its size. Several slightly differing versions of magnitude exist, depending on the wave type analyzed and the distance to the earthquake epicenter. The Richter magnitudes vary numerically from about 1 (very minor earthquake) to about 9 (extremely large and rare) in a logarithmical increasing scale. Each year there are several hundred thousand earthquakes that are instrumentally detectable, at least locally. Many of these are too small to be felt by people. A "major" earthquake has a Richter magnitude of at least 7; there are typically several of these worldwide each year.

When incoming seismic waves arrive at a seismological observatory, they are detected and ultimately recorded for display on a seismogram—a line trace related to the variation of ground motion from the passing wave train as time advances. The seismogram, in conjunction with other seismograms from other observatories, can be analyzed by computers and by experienced human interpreters to give the precise arrival times of various waves. The waves include those that have traveled directly from the focus, with travel paths shaped by the wave velocities and properties of the layers in the Earth's interior, and those that have been multiply reflected inside the Earth. Major earthquakes produce so many waves that they often continue to arrive with enough energy to be detected for several hours.

The travel velocity of seismic waves in the Earth is several kilometers per second, depending on the type of wave, the composition of the material at depth, and the temperature and pressure at depth. A seismic station at the antipode (the point on the Earth's far side opposite the epicenter, 12,740 kilometers straight through the earth) would detect the first-arriving P wave about twenty minutes after the earthquake occurred. It would be followed by a wave train of many waves having different travel paths.

SEISMOGRAPH EQUIPMENT

The collection of useful data at seismological observatories is only as good as the quality and technological ingenuity of the equipment. A seismograph is a sensitive device designed to perform this function. It consists of a seismometer, which mechanically or electromagnetically detects arriving seismic waves (or other ground vibrations) by measuring the movement of the ground; an amplifier, which uses electrical or optical systems to magnify the small vibrations being sensed by a factor of several thousand; and a recording system, which makes a record, for storage or display (on a seismogram), of the arriving seismic waves as a function of time advancing. Recording methods include an ink pen on a rotating paper drum, a light spot on photosensitive paper, or analog or digital data stored on magnetic tape or disk.

A common seismometer technique is to measure the movement of the ground—at the surface, in a buried vault, or down a borehole or mine—with respect to a suspended mass such as a pendulum. The mass, not being fixed rigidly to the ground, briefly "hangs back" as the ground suddenly moves, because of its inertia. This indicates the relative motion.

The world's first seismometer is attributed to Zhang Heng in China in the year 132 C.E. during the Han dynasty. It consisted of a bronze kettle with eight dragon heads holding balls over eight toads. Inside was probably an inertial pendulum with linkages to the dragons' heads. When earthquake ground motion moved the kettle sideways, the pendulum hung behind and opened one or more dragons' mouths so that those balls dropped into the open mouths of the toads below. This ingenious seismoscope (measures the occurrence of a vibration but not a record of its duration and behavior) indicated the relative intensity of the effects felt (by the number of balls dropping) and possibly the direction from which the waves came. The principle was neglected until Europeans again addressed the problem of earthquake detection and measurement during the nineteenth century.

John Milne, an English professor of mining and geology, arrived in Japan to teach in Tokyo in 1875. Interested in the frequent earthquakes there, he began a scientific study by developing seismographs. He, along with John Gray and James Ewing, who were also visiting Tokyo, designed swinging pendulums that detected and recorded wave motion by scratching a trace on a moving smoked glass plate. The instruments measured three components of motion, which, when combined, could give the net ground motion. Milne left Japan for England in 1894; by 1900 his efforts had encouraged the installation of seismographs at a couple of dozen seismic observatories around the world. Thus began the systematic collection of global seismic data.

Other work to develop or refine seismographs was done in Germany by Ernst von Rebeur Paschwitz and Emil Wiechert (who introduced a damped pendulum in 1898 so motions could be more quickly recovered) and in Russia by Boris Galitzin (who introduced moving-coil electromagnetic recording of motion in the early twentieth century).

In practice, seismographs must be designed to accurately sense a wide range of wave amplitudes and periods. The period of a wave is the time (in seconds) between successive oscillation, or vibration, peaks (from crest to crest). The frequency is the inverse: that is, the number of peaks passing per unit time (in cycles per second, or Hertz). A seismograph might be designed, or tuned, to emphasize quarry blasts with wave periods of less than one second (frequency greater than 1 Hertz), nearby earthquakes with body wave periods of one to ten seconds and surface wave periods of ten to sixty seconds, distant earthquakes (called teleseisms) with arriving waves having longer periods (the shorter-period energy is progressively absorbed during travel through the earth), or Earth oscillations and even Earth tides having periods of hours.

A typical seismological station would have six seismographs. Three would be short-period (about one-second response, for nearby earthquakes and vibrations) and would measure the three components (north-south, west-east, and vertical). Three would be long-period (about twenty to thirty seconds response, for teleseisms and surface waves). All would record continuously. Apart from the need to detect and record a wide range of periods, it may also be necessary to have both low- and high-sensitivity instruments for global data collection and interpretation. This is because a high-sensitivity instrument, designed to respond to small local events or moderate distant events, would be deflected or driven "off scale" (exceeding its reading limits) by a large event.

A variety of filtering, damping, and data-analysis techniques can assist in yielding a useful data set. One innovation is to restrain the relative motion of the inertial mass with a "forced feedback" mechanism. The restraining force is measured as a signal. This allows a greater range of sensitivity, and, without large excursions of the inertial element, the device can be more compact. An example is the Wielandt-Streckeisen STS-1 leaf-spring seismometer, developed in 1986, in which the mass is on a flexible strip rather than on a pendulum.

GLOBAL NETWORK OF OBSERVATORIES

In the early twentieth century, a worldwide network of seismological stations that used the fairly rudimentary seismographs available at the time was established for the purpose of studying earthquakes and the interior structure of the planet. The 1906 earthquake in San Francisco was recorded, for example, at dozens of seismological stations around the world, including in Japan, Italy, and Germany. By 1960, about seven hundred seismic observatories were operating around the world using various types of seismographs and standards, which led to incomplete data exchange and analysis. Most were operated by government agencies and universities.

What was needed was a standardized global network of calibrated instruments, coordinated accurate timing, and central data collection—an expensive and vexing organizational problem. However, scientists recognized that seismology and the mapping of earthquakes would contribute to the emerging study of global plate tectonics and the dynamic processes at the plate boundaries during the 1960's.

By convenient coincidence, the impetus and funding for these Earth-centered interests came from the new military and international political need to monitor underground nuclear testing. The 1963 Test Ban Treaty prohibited atmospheric, oceanic, and space testing—all of which could be monitored fairly directly—and nations now needed to test underground. Relations between the United States and the Soviet Union necessitated a program of remote surveillance, and the

techniques of seismology—to distinguish small, shallow earthquakes from buried nuclear explosions—were needed and quickly applied. In the early 1960's, the United States began a program that deployed a series of stations (eventually about 120) in the World Wide Standardized Seismograph Network (WWSSN). The stations used Benioff short-period seismographs (period of one second) and Sprengnether long-period seismographs (period of fifteen to sixty seconds) with moving-coil electromagnetic seismometers and galvanometers to record on seismogram drums.

During the 1970's, seismic recording began a conversion from analog recording to digital recording onto magnetic tape. The latter samples the seismic signal at short time intervals and stores the data. It can retain greater dynamic range and is convenient for direct processing by computer. The combination of new digital seismic observatories and some of the WWSSN stations that were upgraded to digital recording had a major impact on global seismology from the mid-1970's to the mid-1980's.

In 1986, seismographs based on the forced-feedback mechanism to restrain motion and increase range of response, along with digital recording of signal onto magnetic tape, began being used to form the Global Seismic Network (GSN). This is a joint effort of a consortium of universities (the Incorporated Research Institutions for Seismology, or IRIS), the U.S. Geological Survey (USGS), and upgraded stations run by European countries, Canada, Australia, and Japan. Funding comes from the U.S. National Science Foundation and the USGS. The GSN's plan was to have 128 seismic stations when the network was complete. The one hundredth station was installed in 1997. This high-quality global network, designed to replace the WWSSN, is an example of international scientific collaboration. Within hours of an earthquake, data are automatically collected and made available to government and university scientists, as well as on the Internet.

REGIONAL NETWORKS

In addition to the coordinated global collection and analysis of seismic data, a variety of regional, or more localized, networks also exist. These are set up for more detailed and rapid analysis of local earthquakes or other events, such as for detection of patterns of small earthquakes or detection of foreshocks prior to a larger event in hopes of possibly predicting it.

Such networks tend to be set up by more technologically advanced nations in regions at high risk of suffering seismic hazards, such as the United States (particularly California, the Pacific Northwest, and the New Madrid area of southeast Missouri), Japan, and Canada. The proposed U.S. National Seismograph Network will use sensitive forced-feedback seismometers with satellite telemetry of data back to the USGS/NEIC in Golden, Colorado. Plans included the establishment of stations in the continental United States, Alaska, Hawaii, Central America, and the Caribbean region. The network will be designed to detect, locate, and analyze earthquakes of magnitudes as small as 2.5 but will also have seismographs with high dynamic range to handle large earthquakes.

The occurrence of tsunamis (seismic sea waves) generated by some large earthquakes or subsea volcanic eruptions, particularly in the Pacific Ocean, prompted the United States set up the Pacific Tsunami Warning System in 1948. This was in response to a destructive tsunami that hit Hawaii in April, 1946. It traveled south from an earthquake off the coast of the Alaskan peninsula, killing 179 people in Alaska and Hawaii with waves that rose to crests up to 30 meters high on Unimak Island, Alaska, and up to 10 meters high in Hawaii. The system is administered by the National Oceanic and Atmospheric Administration (NOAA), with coordination, data processing, and alerts issued from a warning center located in Honolulu, Hawaii. The system quickly activates when any of its thirty or so participating seismographic stations, located around the Pacific Rim and on Pacific islands, detects an earthquake or other disturbance that could potentially generate a spreading tsunami wave. Another seventy-eight stations have tide gauges for monitoring unusual changes in sea level, to detect a tsunami as it passes by. If this is the case, an alert can be issued, with prediction of tsunami arrival times at Pacific-bordering nations or islands. Coastal people can be evacuated inland, and ships or fishing boats can be taken out to sea to ride out the much more subdued waves offshore.

Portable seismograph systems can be used for rapid and temporary setup of local seismic networks. These can be deployed to monitor precursory earthquakes before a suspected major earthquake or aftershocks after a large one has occurred. They can also be used around volcanoes before or after eruptions, or in anticipation of an underground nuclear test or some other energy-releasing disturbance. Local temporary arrays have also been used during the seismic probing of deep geological structures and to study local seismicity patterns.

Robert S. Carmichael

CROSS-REFERENCES

Creep, 185; Cross-Borehole Seismology, 190; Discontinuities, 196; Earthquake Distribution, 277; Earthquake Locating, 296; Earthquake Magnitudes and Intensities, 301; Elastic Waves, 202; Experimental Rock Deformation, 208; Faults: Normal, 213; Faults: Strike-Slip, 220; Faults: Thrust, 226; Faults: Transform, 232; San Andreas Fault, 238; Seismic Tomography, 252; Seismometers, 258; Stress and Strain, 264.

BIBLIOGRAPHY

Lane, N., and G. Eaton. "Seismographic Network Provides Blueprint for Scientific Cooperation." *EOS/Transactions American Geophysical Union* 78 (September, 1997): 381. The authors discuss attempts to install a global network of modern seismological stations.

Lay, T., and T. C. Wallace. *Modern Global Seismology.* San Diego: Academic Press, 1995. Chapter 5 discusses the development of seismographs, as well as their deployment in regional and global networks for interpreting the Earth's internal structure, earthquake characteristics, and nuclear weapons testing.

Natural Resources Canada/Canadian National Seismology Data Center Website: www.seismo.nrcan.gc.ca. Database of earthquakes as recorded in Canada.

Simon, R. B. *Earthquake Interpretations: A Manual for Reading Seismograms.* Los Altos, Calif.: W. Kaufmann, 1981. Instructions for reading earthquake seismograms, with examples.

U.S. Geological Survey Earthquake Website: http://quake.wr.usgs.gov. Contains useful information about earthquakes and seismicity.

U.S. Geological Survey/National Earthquake Information Center Website: www.neic.cr.usgs.gov. Database of earthquakes in the United States and around the world, as located and assembled by NEIC in Golden, Colorado. Includes data and listings on worldwide seismological stations, earthquake listings by geographical region, and listings of significant historical earthquakes.

Walker, Bryce. *Earthquake.* Planet Earth Series. Alexandria, Va.: Time-Life, 1982. Chapter 3, "Secrets of the Seismic Waves," discusses seismology, recording instruments, and historical developments.

SEISMIC TOMOGRAPHY

Seismic tomography is a technique for constructing a cross-sectional image of a slice of the Earth from seismic data. Measurements are made of seismic energy that propagates through or reflects from subsurface geological materials, and the measured time of travel and amplitude of this energy are used to infer geometry and physical properties of the geological materials, from which an image of the inside of the Earth is generated.

PRINCIPAL TERMS

AMPLITUDE: the maximum departure (height) of a wave from its average value

ATTENUATION: a reduction in amplitude or energy caused by the physical characteristics of the transmitting medium

IMAGING: a computer method for constructing a picture of subsurface geology from seismic data

INVERSION: the process of deriving from measured data a geological model that describes the subsurface and that is consistent with the measured data

LITHOLOGY: the description of rocks, such as rock type, mineral makeup, and fluid in rock pores

RESOLUTION: the ability to separate two features that are very close together

SEISMIC REFLECTION METHOD: measurements made of the travel times and amplitudes of events attributed to seismic waves that have been reflected from interfaces where changes in seismic properties occur

SEISMOMETER: an instrument used to record seismic energy; also known as a geophone or a seismic detector

TRAVEL TIME: the time needed for seismic energy to travel from the source into the subsurface geology and arrive back at a seismometer

METHOD PIONEERED

Seismic tomography is a means of making an image of a slice of the Earth using seismic data. "Tomography" is derived from a Greek word meaning "section" or "slice." Since the 1970's, seismic techniques have been used to create subsurface pictures. Although some methods that have been used in exploration geophysics for a number of years can be classified as tomographic, it is only since the mid-1980's that seismic tomography has been specifically developed for geophysical exploration and exploitation. The current interest in and promise of seismic tomography in geophysical exploration and global seismology are the product of many factors, including interaction between different scientific disciplines, along with advances in seismic field-data acquisition, imaging and inverse-problem theory, and computing speed.

The basic idea of tomography is to use data measured outside an object to infer values of physical properties inside the object. This method was pioneered by J. Radon in 1917. Radon showed that if data are collected all the way around an object, then the properties of the object can be calculated. In fact, Radon derived an analytical formula that relates the object's internal properties to the collected data.

METHOD APPLIED

Since the mid-1970's, the ideas of tomography have been applied to a number of fields of study. Applications of tomographic techniques are found in fields as diverse as electron microscopy and astronomical imaging. In medicine, the process of computed tomographic scanning ("CT" or "CAT" scanning) has developed particularly rapidly since its inception in the early 1960's, and its use has been very important in diagnostic medicine.

Geophysicists have been attempting to apply techniques similar to those of medicine to geophysical problems. Since the early 1970's, interest and applications have been rapidly expanding, and a number of papers have been presented and written on the applications of seismic tomography since 1984. These range from attempts to estimate the internal velocity structure of the subsurface to formulations that provide a complete image of the subsurface geology. Since the late 1980's, tomographic reconstruction has become a standard technique in analyzing data between drill holes (crosshole analysis). Thus, while tomography is relatively new to exploration geophysics, it is a broad, powerful concept that is making a significant impact. Seismic tomography has led to many useful new applications as well as insightful reinterpretations of some existing methods.

Seismic tomography is a type of "inverse problem"—that is, measurements are first made of some energy that has propagated through and reflected from within a medium (in this case, the earth). The received travel times and amplitudes of this energy are then used to infer the values of the medium through which it has propagated. The parameters that are extracted are velocities and depths; therefore, a gross model of the Earth's structure can be derived. Initially, this was considered the ultimate goal of seismic tomography, but it has become obvious that an accurate set of measurements can be used effectively for other purposes, such as constructing an accurate depth image of the subsurface.

In CT scanning, an X-ray source and a number of X-ray detectors are used to acquire data around the human body. The X-ray source sends out X rays, and the receivers record the transmitted X-ray intensity. This intensity is related to the attenuation of the X rays along their ray paths inside the object. In turn, the amount of attenuation is related to the density of the object encountered by the X rays. Thus, a CT scan is an actual estimate of the density distribution within a body. CT scans can be done over various parts of the body, and these scans can be put together to form a three-dimensional image. This kind of image can show with great clarity the structure or damage inside the body. Interpreting three-dimensional images of the body's interior is similar in many ways to interpreting the interior of the Earth from three-dimensional seismic data.

SEISMIC SURVEYS

The similar problem in geophysical exploration is determining from seismic data the velocities with which sound propagates through a section of the Earth, as well as other properties of the Earth, such as density and compressibility. Classical tomography is typically associated with transmitted energy and requires a distribution of sources and receivers around the object to be imaged. In imaging techniques, such as medical X-ray tomography, the source and receiver rotate all the way around the object to be imaged. In contrast, by far the most pervasive seismic measurement is the surface reflection survey. Its measurements are made on just the upper boundary of the medium of interest.

Since the first seismic detectors (seismometers or geophones) were placed on the surface of the Earth near the end of the nineteenth century, seismic waves have been used to locate remote objects. The first applications involved the location of earthquake epicenters in faraway regions. Efforts during World War I to locate heavy artillery by seismic means evolved later to the first exploration methods for oil and gas. The imaging technique in exploration seismics has been improved ever since. At first, it merely involved the interpretation of travel times of observed seismic pulses in terms of the depth and slope of reflecting surfaces. Beginning in the 1970's, complete seismic records were used, and imaging methods were developed that are firmly based on sophisticated mathematical techniques.

In a seismic survey, geophysicists typically arrange seismic detectors along a straight line and then generate sound waves by vibrating the Earth. Earthquakes release the large amounts of energy needed to probe the deep layers (mantle and core) of the Earth. Other methods can produce seismic waves that can be focused on the geologic features closer to the Earth's surface. These waves can be generated by explosions, such as a charge of dynamite, or by dropping a weight or pounding the ground with a sledgehammer. To eliminate environmental risks associated with the use of explosives, a system called "vibroseis" is used. In this system, a huge vibrator mounted on a special truck repeatedly strikes the Earth to produce sound waves. A seismograph records how long it takes the sound waves to travel to a rock layer, reflect, and return to the surface. The recorded data display the amplitudes of the reflected sound waves as a function of travel time. Such a graphic record is called a "seismogram." The equipment is then moved a short distance along the line, and the experiment is repeated. This procedure is known as the seismic reflection profiling method.

METHOD ADAPTED TO SEISMIC DATA

Since seismic waves traveling in the Earth readily spread, refract, reflect, and diffract, classical tomographic methods must be adapted to produce realistic seismic pictures, and effective software and interactive graphics are required to process the seismic data into a relevant image. It has taken some time for tomographic concepts to spread to the seismic experiment, for appropriate data to be acquired, and for effective processing and interpretation techniques to be developed. Using a variety of computer programs, seismograms are processed to yield seismic sections that represent the Earth's reflectivity in time. Geologists, though, would really like to have a lithologic picture illustrating such features as rock velocity, seismic wave attenuation, and elastic constants of the rocks as a function of depth. Reflectivity is a property associated with interfaces between rocks; a rock sample held in one's hand does not have an intrinsic reflectivity. Therefore, reflectivity is not an actual rock property, and it must be converted to some parameter that really describes the rock. In addition, seismic time data must be converted to depth measurements in the imaging process. Consequently, conventional reflection sections are being greatly improved by the use of tomographic techniques to produce subsurface images as a function of depth and to estimate rock properties from some of the images.

The basic procedure of seismic tomography is an extension of the notion of transmission tomography; this process can also be classified as a generalized linear inversion of travel times. The procedure is to locate reflected events on the raw seismograms, associate these events with the structure of a proposed or guessed geological model, use the laws of physics to trace ray paths of seismic waves through the proposed model from given seismic sources down to a particular reflector and back to the seismic detectors, compare the ray-traced travel times through the model with the travel times recorded on the seismogram, and update the geological model to make the ray tracing consistent with the observed data. Seismic tomography is distinct from classical tomography in

that only reflected waves are used, and the source-detector coverage of the object or area of interest is far from complete. These aspects of the problem create difficulties, but the tomographic velocity determination is still very useful, especially in areas of significant lateral velocity variations. By including all available well data, as well as any other available geophysical data, in the tomographic process, the resolution and certainty of subsurface images can be greatly improved.

RESOURCE EXPLORATION

Various survey geometries and tomographic constructions are used to assist in solving geophysical problems. Geophysicists can use the velocities of seismic waves recorded by a seismograph to determine the depth and structure of many rock formations, since the velocity varies according to the physical properties of the rock through which the wave travels. In addition, seismic waves change in amplitude when they are reflected from rocks that contain gas and other fluids. Sometimes the fine details of seismic records can be used to infer the type of rocks (lithology) in the subsurface. Some tomographic studies have used subsurface velocities determined from the inversion of seismic travel times to construct geological cross sections of the geology inside the Earth, while other studies have used reflection amplitudes for the same purpose. Based on the characteristic geometries and amplitudes for oil and gas traps and for mineral ore deposits, these tomographic images are used to predict where oil, natural gas, coal, and other resources such as groundwater and mineral deposits are most likely to be found in the subsurface. The tomographic cross sections constructed from seismic data make the odds of finding such resources much greater than would be the case if exploration were based on mere random drilling.

In seismic tomography, various source and detector geometries are used, such as drill-hole-to-drill-hole, surface-to-drill-hole, and surface-to-surface. The greater the degree of angular coverage around the rock mass of interest, the greater will be the reliability of the constructed tomographic image. By making numerous measurements from various source-detector positions and analyzing the travel times and amplitudes from a number of source-detector locations, the velocity and attenuation of the intervening rock can be

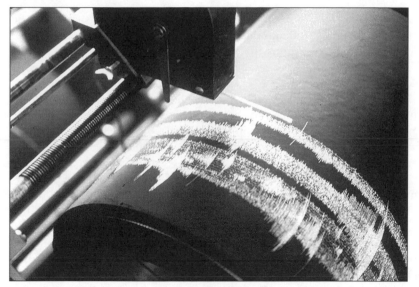

A smoke-drum seismograph record of seismic acitivity on Mount Kilauea's east rift, island of Hawaii. (U.S. Geological Survey)

calculated from the recorded energy (reflected or transmitted). This technique has found applications not only in locating subsurface natural resources but also in areas such as siting of nuclear-waste dumps and monitoring of stream floods, which are used to help produce hydrocarbons from a reservoir.

EXPLOITATION OF HYDROCARBON RESERVOIRS

In many hydrocarbon development areas, adjacent drill holes may be available. In these situations, it is desirable to have a very high-resolution description of the rock mass between the drill holes. It is often effective to adopt crosshole tomography. A seismic source, such as dynamite caps or downhole air guns, is placed in one drill hole, and appropriate detectors are placed in an adjacent drill hole. The source is fired, and the resulting seismic energy propagates through the rock and is detected in the other drill hole. The travel times and amplitudes of seismic waves that have been reflected or transmitted through the rock mass between the drill holes are recorded. The source and detectors are then moved to another position, and the process is repeated. This procedure is continued until the region of interest is adequately covered by the propagating energy. Seismic crosshole tomography has been used for a number of applications of different kinds. Among these are mineral exploration in mines, fault detection in coal seams, stress monitoring in coal mines, delineation of the sides of a salt dome, investigation of dams, and mapping of dinosaur-bone deposits. The resolution of crosshole tomography is typically better than for surface reflection tomography.

The broad objective of geophysics is to produce images that represent the subsurface geology as accurately as possible, and tomography-based imaging algorithms provide seismic depth sections that are consistent with drill-hole data in regions of resource exploration and exploitation. Integrating drill-hole and surface seismic reflection data in a tomographic approach can provide a better, less ambiguous subsurface picture. This correlation holds considerable promise to increase knowledge of the subsurface. The resulting seismic depth sections assist in interpreting the structure (geometry), stratigraphy (depositional environment), and lithology (rock and fluid types) of potential and established hydrocarbon reservoirs and mineral deposits.

The geologic detail needed to develop most hydrocarbon reservoirs substantially exceeds the detail required to find them. For effective planning and drilling, a complete understanding of the lateral extent, thickness, and depth of the reservoir is absolutely essential. This can be found only with detailed seismic interpretation of three-dimensional seismic reflection surveys integrated with drill-hole data. A common practice in three-dimensional seismic reflection surveying is to place the seismic detectors at equal intervals and collect data from a grid of lines covering the area of interest. Based upon integrated seismic tomographic imaging of the drill-hole and seismic reflection data, more wells are drilled in the area, and the three-dimensional data volume evolves into a continuously utilized and updated management tool that influences reservoir planning and evaluation for years after the seismic data were originally acquired and imaged.

GLOBAL SEISMOLOGY

Imaging in global seismology (whole-Earth geophysics) has lagged behind the developments in exploration geophysics for several reasons. In contrast to artificial sources, earthquakes are uncontrolled, badly placed sources of seismic-wave energy, and the Earth is only sparsely covered with seismometers. In addition, instrument responses were for a long time widely different, and recording was not in digital form. Thus, seismologists were faced with the paradox that the available data, despite the enormous volume, often contained crucial gaps.

In global seismology, the whole three-dimensional Earth is considered as an object to be imaged. Seismic energy generated by earthquakes travels through the Earth and is recorded by a distribution of seismic detectors, such as the World Wide Standardized Seismographic Network. By examining the travel times of the propagating energy for a number of earthquakes and stations, researchers can construct a model representing the velocity structure inside the Earth. Likewise, by measuring the shapes and sizes of the amplitudes of the recorded energy, they can estimate a seismic attenuation model of the Earth. Based upon these models, a three-dimensional tomographic image of the Earth can be constructed.

Global seismic tomography has been used to image convective flow within the mantle. Changes in seismic wave velocity have been used to identify sinking (cold) and rising (warm) mantle materials. Estimates can be made of the variation of seismic velocities inside the Earth using seismic tomography, and these variations in turn depend upon the variations in composition, structure, and temperature of the materials inside. Mantle regions that are relatively hot have lower velocities compared to cooler regions at the same depth. This is because the higher temperatures reduce the values of the elastic constants of the mantle material. Seismic tomography supports a hybrid convection theory that postulates the existence of shallow, small-scale convection currents as well as of deep, large-scale convection currents in the mantle. Convection of the mantle is the primary driving force of plate tectonics. In general, tomographic results show a strong correlation at shallow depths with present plate boundaries, such as fast movement under cold, old shields and in subduction zones, and slow movement under hot, spreading ridges and other volcanically active areas. Three-dimensional images of the Earth's interior reconstructed with seismic tomographic procedures have therefore had a major effect on the understanding of the structure and dynamics of the Earth.

Alvin K. Benson

CROSS-REFERENCES

BIBLIOGRAPHY

Bording, R. P., et al. *Principles of Seismic Traveltime Tomography.* Tulsa, Okla.: Society of Exploration Geophysicists, 1986. Discusses the basic principles of tomography and how they can be applied to seismic data to create a velocity model of the Earth from recorded traveltimes.

Iyer, H. M., K. Hirahara, et al., eds. *Seismic Tomography: Theory and Practice.* London: Chapman and Hall, 1993. Various essays explore all aspects of seismic tomography and the inversion of seismic data.

Lines, L. R. "Cross-Borehole Seismology." *Geotimes* 40 (January, 1995): 11. Discusses applications of seismic tomography to the shallow subsurface.

Lo, Tien-When. *Fundementals of Seismic Tomography.* Tulsa, Okla.: Society of Exploration Geophysicists, 1994. Lo provides a fine introduction to seismic tomography for students with little knowledge on the subject.

Lutgens, Frederick K., and Edward J. Tarbuck.

Earth: An Introduction to Physical Geology. 6th ed. Upper Saddle River, N.J.: Prentice Hall, 1999. This college text provides a clear picture of the Earth's systems and processes that is suitable for the high school or college reader. In addition to its illustrations and graphics, it has an accompanying computer disc that is compatible with either Macintosh or Windows. Bibliography and index.

Nolet, Guust, ed. *Seismic Tomography.* Boston: D. Reidel, 1987. Describes the methods and reliability of seismic tomography. Contains many qualitative discussions that will be useful to the general reader as well as more technical discussions for those with the appropriate background. Primarily discusses applications of tomography to whole-Earth geophysics, with some discussion of applications to exploration geophysics.

Poupinet, Georges. "Seismic Tomography." *Endeavour* 14, no. 2 (1990): 52. Good description of seismic tomography as it is applied to the study of the deep structure of the Earth by integrated analysis of seismic wave patterns generated from earthquakes.

Russell, B. H. *Introduction to Seismic Inversion Methods.* Tulsa, Okla.: Society of Exploration Geophysicists, 1988. Discusses techniques used for the inversion of seismic data, including principles of seismic tomography. Good illustrations.

Stewart, R. R. *Exploration Seismic Tomography.* Tulsa, Okla.: Society of Exploration Geophysicists, 1991. Recounts the historical development of tomography. Reviews the fundamentals of seismic tomographic techniques and discusses applications of seismic tomography, mainly to exploration geophysics.

SEISMOMETERS

A seismograph is a device that detects, measures, and records the ground motion at a point. The sensor that detects and, in part, measures the motion is called a seismometer. The recorded ground motion, a seismogram, is the output of a seismograph.

PRINCIPAL TERMS

EARTHQUAKE: a sudden release of strain energy in a fault zone as a result of violent motion of a part of the Earth along the fault

NATURAL PERIOD OF VIBRATION: the period at which structures undergo oscillation if they are set in motion by an impulse

PARTICLE MOTION: the motion of a particle in a material volume when it experiences the passage of seismic waves

SEISMIC WAVES: the propagation of a disturbance in the form of energy release in a solid me-

dium; the released energy propagates in the solid from one region to another by setting individual particles in motion in a particular direction

SEISMOGRAM: the recorded output of a seismograph

SEISMOGRAPH: a device that detects, measures, and records the ground motion

WAVE VELOCITY: the velocity at which a particular seismic wave travels through a medium

HISTORY AND DEVELOPMENT

According to the historical records, the first seismometer was invented by Chinese astronomer and mathematician Zhang Heng (78-139 C.E.), during the Han Dynasty. The records indicate that it was a bronze vessel containing a suspended pendulum. The pendulum was connected to an eight-spoked wheel, and each spoke terminated in the mouth of one of eight externally mounted dragon heads with movable jaws. Each mouth contained a bronze ball, and eight open-mouthed frogs were located around the base. During an earthquake, the ground motion would displace the pendulum laterally, ejecting one of the balls into a frog's mouth. The ejected balls would give some idea of the direction of the traveling waves and the source of the earthquake.

The first truly precise seismometer was developed by an English mining engineer, John Milne, in the late 1800's. This instrument was improved by J. J. Shaw in the early 1900's, when the Milne-Shaw seismograph was introduced. In the United States, the first seismographs were installed in 1887 at the University of California, Berkeley. The 1906 San Francisco earthquake was the first large event in the United States to be recorded.

In the 1960's, largely as a result of the develop-

ment of the atomic bomb and concern over the Cold War, the improvement and development of seismographs took a large leap forward. There was a national security need to be able to discriminate between an underground nuclear explosion and natural events. Such effort led to the construction and the deployment of 120 seismic stations in sixty countries, called the World Wide Standardized Seismograph Network (WWSSN). This period also marks a turning point in bringing the science of observational seismology to the forefront of the physical sciences. At the same time, countries such as the Soviet Union, France, and Canada modernized their earthquake observation systems.

MECHANICS

Currently, many types of seismographs are available. Most of them incorporate similar physical principles, utilizing some sort of spring or pendulum to detect and measure the ground motion. The exception is a strain seismograph, invented by Hugo Benioff, which uses a long horizontal bar (20-30 meters long) to measure the ground deformation between two points. A simple seismometer can be considered as a pendulum attached to a rigid frame, which is anchored to a horizontal ground surface. When an earthquake occurs, seismic waves are radiated from the earthquake source in all directions through the Earth. For a sufficiently strong earthquake, the seismometer site experiences ground vibration. Assuming that there is no slippage between the ground and the rigid frame of the seismometer, the frame experiences the same motion as does the ground. If the pendulum could stay motionless, which is the ideal case, the relative motion of the frame with respect to the pendulum would be the true ground motion. Yet, that is not the case; the pendulum undergoes motion, and because of its resistance to motion (inertia), it tends to lag the motion of the frame (ground). This lag results in a complex differential motion between the pendulum and the ground.

Pendulums and springs have the characteristic that, if they are set in motion, they oscillate with period independent of the amplitude of motion as long as the amplitude does not become extreme. This characteristic period or frequency (frequency is the inverse of period) is called the free-period or the natural frequency of a pendulum or spring, respectively. A seismometer pendulum or spring can be designed to have a high- or a low-period sensitivity. The relative frequency content of the seismic waves with respect to the natural frequency of the pendulum or spring-mass determines in part the nature of the recorded seismograms. The deflection of a pendulum that experiences ground motion with frequencies much higher than its natural frequency is proportional to the ground displacement. The deflection, however, is proportional to the ground acceleration if the frequency content of the ground motion is much lower than the natural frequency of the pendulum, and the seismograph is called an accelograph. Finally, if the pendulum has a natural frequency close to the frequency content of the ground motion, the deflection is proportional to the ground velocity. Nevertheless, displacement, velocity, and acceleration are mathematically related. For example, ground velocity and displacement could both be determined from an accelogram.

The relative motion of the pendulum with respect to the ground must be magnified, allowing very small wave amplitudes to be distinguished. The most sensitive seismometers, at a quiet observatory, can detect ground displacement of a thousandth of a millionth of a meter, which is in the order of atomic size. The record of such ground motion might show an amplitude of 1 centimeter, a magnification of 10 million. The ratio of the largest to the smallest amplitudes which are undistorted is called the dynamic range of the seismograph. Conventional visible seismographs (those that produce records directly on paper or film) have a dynamic range of approximately one hundred to three hundred.

COMPLICATIONS AND LIMITATIONS

Unfortunately, there is an unwanted complication in detecting useful seismic signals. The ground is always in motion. This motion, microseismicity, is caused by natural sources such as wind, storms, and human activities. Seismographs should be designed so that the recorded seismograms are least affected by the unwanted microseismic noise—to get an optimum signal-to-noise ratio. Most microseismic noise has a period in the range of 5-8 seconds. These noises are effectively avoided in the commonly used short-period (1 hertz) seismographs. The Moon is the ideal place for readings from advanced seismographs because of the absence of wind, ocean waves, and human-made noise. The lunar seismographs left behind by the Apollo mission astronauts can detect the seismic waves generated by a 1-kilogram meteor striking anywhere on the Moon's surface.

Borehole seismographs have been developed which can operate within deep boreholes. The advantage of a downhole recording is that the noise level at depth is much lower than at the surface, and, with certain geometrical arrangements, the recording of seismic waves that travel desired paths could be made. Seismologists have employed such data-acquisition techniques in conjunction with mathematical models similar to those used in medical tomography to construct a three-dimensional picture of the surveyed volume of the Earth.

Using mechanical vibration as the only means of measuring and recording motion, although relatively simple, has a few drawbacks. Such instruments have limited magnification and sensitivity, and the recording is mechanical, with inherent friction. In older seismographs, very large pendulum masses were used to reduce the friction, for example, 1,000 kilograms in many Wiechert seismographs. Some Wiechert seismographs built with a pendulum mass of roughly 20 tons were installed in the beginning of the century in several places in central Europe. The older seismographs all use mechanical methods for transferring the motion of the pendulum to the recording pen. The recording pen can have several variations; the most common are a stylus on a rotating smoked paper drum and a reflected light beam onto a photographic paper. The highest measurable frequency of mechanical recording depends upon the speed of the recording medium relative to the stylus pen or light spot. The typical recording speed ranges from 0.1 millimeter per second for frequencies at roughly 0.05 hertz, up to 10 millimeters per second for frequencies as high as 20-50 hertz. The dynamic range of these systems is approximately one hundred to three hundred.

INTRODUCTION OF ELECTROMAGNETICS

In 1906, Boris Golitsyn elegantly utilized the principles of electromagnetics to translate mechanical motion into electrical voltage. It is known that if a conducting loop (a coil) crosses lines of magnetic flux, there will be an induced electromotive force that will generate an electric current proportional to the velocity at which the coil crosses the magnetic field. This idea has had a significant effect on the later development of seismographs. A system of this kind consists of a coil of wires and a magnet: One of the components is fixed to the rigid frame and undergoes the same motion as the ground, while the other element is suspended by a spring from the frame. The relative motion of the coil with respect to the magnet will produce electromotive force between them, with voltage proportional to the velocity of motion. The current generated in the coil is sent through a sensitive galvanometer (an apparatus to measure electrical current) that makes a continuous record on photographic paper, a mirror, or a hot stylus. In most cases, the coil is fixed to the frame and the magnet is suspended, but there is a wide variety of possible arrangements. The sensitivity of an electromagnetic seismometer depends upon the magnet's strength, the number of turns of wire in the coil, and the geometry of the crossing of the magnetic flux by the coil. Modern seismometers have become quite small as magnetic

materials with greater strengths have become available.

The introduction of the electromagnetic principles in seismic instrumentation seismographs has been a great step forward in the development of seismographs. The virtual elimination of friction also eliminates the need for the large pendulum mass used in the mechanical systems to overcome friction. Also, a much higher recording magnification with respect to the mechanical models is possible.

DAMPING

When the motion of a pendulum has started, the oscillation will continue at its natural frequency for a time, depending upon the rate of energy dissipation. In order to measure and detect the arrival of various waveforms separately, the motion of the pendulum must be damped to prevent its free-period swinging. If the damping is small, any seismic impulse will set the suspended spring into motion with an oscillatory or "ringing" output at a period of roughly the pendulum free-period. This problem can be pronounced for long-period seismometers. The damping of a pendulum or a spring-mass can be visualized by pulling a horizontal mass-spring, resting on a smooth oily surface, to one side by 1 centimeter and then releasing it. The distance that the spring overshoots the original rest position determines the damping ratio; for example, an overshoot of 0.1 centimeter corresponds to a damping ratio of 0.1, or 10:1. In older seismographs the damping was achieved in various ways, such as air damping in Wiechert seismographs or oil damping in the early version of the Benioff seismograph. Damping in modern-day seismographs takes advantage of the same electromagnetic principles as discussed above. As the coil moves with respect to the suspended magnet, an electric current is induced in the coil, which in turn generates a magnetic force opposing the motion; this resisting force acts as the damping mechanism.

SEISMOGRAPH DESIGN

In designing a seismograph (a seismometer and the recording device), one must address issues such as the frequency content of the seismic waves of interest, the direction of the ground motion (vertical or horizontal), the smallest and the largest amplitudes of motion to be measured accurately, the plausible magnification of the pendulum motion relative to the ground motion, the method to be used for relative or absolute timekeeping, the recording of the measured ground motion (a visual recording on paper or recording on a magnetic tape), and the means of recording (an on-site or a telemetry recording). The nature of a particular application will determine the type of seismometer required, the frequency range of interest, the accuracy, and the resolution of the produced record.

Seismic waves generated by earthquake or explosive sources can have a broad frequency band of about 100-0.00033 hertz. Most of the primary earthquake-generated seismic waves have periods of between 0.05 and 20 seconds (frequencies of 20-0.05 hertz). Large earthquakes, such as the 1960 Chilean earthquake with a magnitude of 8.3, can put the whole Earth into an oscillatory motion with period of roughly one hour. Such an unusually broad frequency band creates a need for seismographs to record waveforms of different frequency bandwidth. A good seismological station such as WWSSN often houses six seismometers for obtaining a complete description of ground motion: one vertical and two horizontal (north-south and east-west) short-period seismographs, which are sensitive to the arrival of waveforms in the 0.05- to 2.0-second period range, and three long-period instruments with wave periods in the 15- to 100-second range. Seismographs are usually designed so that the output has a relatively constant magnification for displacement, velocity, or acceleration over some design frequency range. For example, a Wood-Anderson seismograph, the instrument whose seismograms are used to define the Richter magnitude scale, is a horizontal displacement seismograph with a flat response to ground displacement over the frequency range of greater than approximately 1.25 hertz with a magnification factor of 2,800. An earthquake with such a ground displacement at the distance of 100 kilometers is defined to have magnitude of 3.0 on the Richter scale.

TIMING SYSTEMS

Without exact time, a seismic record is not very useful. Seismologists use the Greenwich Mean Time (Universal Time) for timing seismograms.

The timing system can be based either on internal or external clocks or on a combination of the two. The internal timing system is a precise crystal oscillator, with temperature compensation, for keeping the time base. It is a low-cost and low-power-drain clock that has a drift rate as low as 0.1 second per month. The timing system of this type of oscillator can be as simple as a series of minute and hour pulses for visible records, or as complex as a digital time code with day, hour, minute, and second information repeated once per second. The external timing system is used to keep the internal clock correct to 1 millisecond by continuous reception of a radio time signal. The most common radio signal is the time signals transmitted by a standard world time service (such as WWV in the United States). These short wave signals can usually be received, with adequate equipment, anywhere in the world. The reception of this signal, however, is not of good quality. Low-frequency time broadcasts in the 15-100-kilohertz range, such as WWVB, can be received reliably enough for continuous recording over a large area.

RECORDING AND CALIBRATION SYSTEMS

In the 1970's, the development of analogue and digital circuits, coupled with increasing access to computers, resulted in the development and use of magnetic tapes for seismic recording. The recording on magnetic tapes can be either analogue or digital. In analogue, the motion is directly converted into proportional magnetization on the tape, although the analogue data cannot be directly used by computers and have to be digitized. In digital recording, the analogue ground motion is magnified and digitized at some selected time intervals, and the measured signals are recorded onto a magnetic tape.

A more recent development in seismic recording systems has been integrated circuit modules, allowing a huge amount of information to be held in memory on a single printed circuit board. That enables the recording system to screen the data regarding their significance and, thus, to make the decision if the data should be recorded. Such "self-editing" systems with the use of low-cost cassette tapes have made digital recording very attractive.

Seismologists are interested in the true amplitude of the ground motion. Therefore, seismographs must be calibrated in order to construct the true ground motion from the recorded seismograms. It is impossible to do a theoretical calibration because of the difficulty in modeling the physical behavior of each component. A direct method of calibration is to use shaking tables, where a seismometer is placed on a table, which is set into harmonic motion with a known amplitude and period. The record of such motion gives direct information on the magnification at that period, assuming that the period is not contaminated by the free-period of the shaking table. The experiment is repeated for different periods.

EARTHQUAKE HYPOCENTER AND MOTION DETECTION

Earthquake prediction and the design of earthquake-resistant structures have important social and economic value. A successful program would save lives and billions of dollars; the key to success is understanding the physical mechanisms of earthquakes, if there is to be hope of predicting the time, the location, and the magnitude of future earthquakes. Considering that the majority of earthquakes are caused by the jerky motion of a volume of Earth along a fault plane, their spatial distribution could be used to delineate the fault zone. Seismologists use the relative travel times of the primary seismic waves to determine the location (hypocenter) and the time of occurrence of earthquakes. To accomplish that, seismographs are distributed around the expected seismogenic region to record the earthquake-generated seismic waves. The accuracy of pinpointing the hypocentral locations depends upon the accuracy in the relative timekeeping and the location of seismographs. For example, an error of 0.1 second in the travel time of a P wave (pressure wave) could translate to some 600 meters in error in the hypocentral location (P-wave velocity in upper crust is roughly 5,000-6,000 meters per second).

Seismologists determine the direction of motion along the fault surface from studying the relative direction, upward or downward, of the initial motion of the P wave at sites surrounding an earthquake. The mechanism is similar to moving two blocks in opposite directions in a sandbox. The upward or downward motion of the sand particles on the surface can be directly related to the direction of the motion of the blocks.

Perhaps the most important single factor in the development of seismographs has been the need to detect and differentiate underground nuclear tests from earthquakes. At the present time, seismographs are the main source of information for monitoring the time, the location, and the size of such tests.

EARTHQUAKE-RESISTANT STRUCTURAL DESIGN

For the earthquake-resistant design of structures, it is necessary to determine the expected level of ground shaking from future earthquakes. Seismologists use the recording of past earthquakes to obtain relationships between the maximum ground motion, the magnitude of earthquakes, and the source-site distance for earthquakes in various regions (attenuation equations). Such relationships are used to predict the maximum ground acceleration, velocity, or displacement of future earthquakes of given magnitudes and distances for a desired site.

Another aspect of the seismic design of structures is the determination of the structural response to strong ground shaking. For this purpose, accelographs are mounted in various levels of structures to record the vibration. From the study of such records, earthquake engineers determine the frequency-dependent magnification factors of typical structures. Such information, coupled with the knowledge of the natural period of the building and the site (the site period is the period at which ground motion is magnified), is critical for the design of structures safe against future earthquakes.

Seismic waves from large earthquakes travel thousands of kilometers through the center of the Earth. These waves carry much information about the physical properties of the Earth and have been used to construct a picture of the interior of the Earth. Any discontinuity in the Earth's material properties that crosses the traveling path of seismic waves reflects part of the seismic energy and transmits the rest into the adjacent region. Examples of major discontinuities are the Earth's crust-mantle and mantle-core discontinuities. The reflected waves reach the recording stations as distinct waveforms; seismologists identify these waveforms on seismograms and determine the location of the discontinuity in the Earth by modeling the travel times of such waveforms.

Mehrdad Mahdyiar

CROSS-REFERENCES

BIBLIOGRAPHY

Bath, Markus. *Introduction to Seismology*. New York: John Wiley & Sons, 1981. Chapter 2, "Seismometry," gives a very good and comprehensive description of seismometers and seismic recording systems. Well written and includes much interesting information.

Berlin, G. Lennis. *Earthquakes and Urban Environment*. 3 vols. Boca Raton, Fla.: CRC Press, 1980. Chapter 2, "Earthquake Descriptors," gives very easily read information on the history and the fundamentals of seismic recording systems. A well-written book worth reading.

Bullen, K. E., and B. A. Bolt. *An Introduction to the Theory of Seismology*. Cambridge, England: Cambridge University Press, 1985. Chapter 9, "Seismometry," gives relatively technical details of the principles of seismometers and the recording systems. One may ignore the equations and still get some good information from this chapter.

Carlson, Shawn. "The New Backyard Seismology." *Scientific American* 274 (April, 1996). This article provides easy-to-follow instructions on building a rudimentary seismograph.

Dobrin, M. B. *Introduction to Geophysical Prospecting*. New York: McGraw-Hill, 1976. Chapter 3, "Seismic Recording Instruments," gives a general description of the principles of the electromagnetic seismometers and the analogue and digital recording systems.

Garland, G. D. *Introduction to Geophysics: Mantle, Core, and Crust*. 2d ed. Philadelphia, Pa.: W. B. Saunders, 1979. Chapter 5, "Seismometry," gives a general background on seismometry. The text is semitechnical. Also discusses seismic arrays. The whole work is a good introductory book on the subject of geophysics.

Plummer, Charles C., and David McGeary. *Physical Geology*. Boston: McGraw-Hill, 1999. A college-level introductory geology textbook that is clearly written and wonderfully illustrated. An excellent sourcebook of basic information on geologic terminology and fundamentals of geologic processes. Contains a CD-ROM. An excellent glossary.

Reynolds, John M. *An Introduction to Applied and Environmental Geophysics*. New York: John Wiley, 1997. An excellent introduction to seismology, geophysics, tectonics, and the lithosphere. Appropriate for those with minimal scientific background. Includes maps, illustrations, and bibliography.

Wiegel, R. L., ed. *Earthquake Engineering*. Englewood Cliffs, N.J.: Prentice-Hall, 1970. Chapter 6, "Ground Motion Measurements," is written by D. E. Hudson and concentrates more on the development of accelographs. The whole book is highly recommended, as it contains valuable information on the subject of earthquake engineering and seismic hazards.

STRESS AND STRAIN

Stress and strain have to do with why and how a solid body deforms. Each point within a body under a load will have a set of stresses associated with it, varying in direction, magnitude, and the planes on which they act, according to the intensity of the forces acting within the body at that point. Each point within a deformed body will have a set of strains associated with it that indicate the translation, rotation, dilatation, and distortion experienced by the material at that point during the deformation.

PRINCIPAL TERMS

DILATATION: the change in the area or volume of a body; also known as dilation

DISTORTION: the change in shape of a body

NORMAL STRESS: that component of the stress on a plane that is acting in a direction perpendicular to the plane

ROTATION: the change in orientation of a body

SHEAR STRESS: that component of the stress on a plane that is acting in a direction parallel to the plane

TRANSLATION: the movement of a body from one point to another

PHYSICS OF RIGID BODIES

Stress and strain are concepts that help to explain how and why rocks deform. Strain describes the deformation, and stress pertains to the system of forces that produce it. In considering strain, first it will be helpful to review some aspects of the physics of rigid bodies in order to appreciate the significance of these concepts.

When dealing with many problems in mechanics, it is common to assume that the bodies involved are perfectly rigid; that is, they do not deform. Such problems usually involve the balancing of forces, or, if forces do not balance, determining the resulting accelerations. Movement of a rigid body can involve translation, rotation, or both, but the individual points within the body do not move relative to one another. In a translation, all points within the body move the same linear distance and in the same direction. In a rotation, all points within a body rotate through the same angle around the same center of the rotation. Any rigid body motion can be described in terms of a translation plus a rotation. Deformation introduces further complications. A volume-conserving, shape-changing deformation is called distortion. A change in volume, without a change in shape, is called dilatation (or dilation). Strain combines all four of these possibilities: translation, rotation, distortion, and dilatation.

NET STRAINS

If the beginning and ending locations and orientations of a rigid object are known, it is easy and straightforward to determine the net translation and the net rotation of the object and of every point within the object. For example, if an airplane begins in New York facing north as it is loaded and ends up in Madrid facing northwest as it is unloaded, then it can be said to have moved 5,781 kilometers to the east and rotated 45 degrees counterclockwise. Similarly, every piece of luggage on that airplane had a net translation of 5,781 kilometers to the east and a net rotation of 45 degrees counterclockwise. Very little information is needed to determine such net displacements and rotations, but the path that the airplane took is not well represented by them. The plane probably flew along a great circular route, changing its bearings constantly, and it very likely circled a bit after taking off and again before landing. To describe the path of the plane, one would need much more data. These data might consist of a series of translations and rotations taken at one-minute intervals. Each item of luggage, rigidly fixed within the hold of the aircraft, would move through an exactly identical series of translations and rotations. Furthermore, by applying the basic laws of mechanics, one could attribute each acceleration (linear or angular) to the forces resulting from the interplay of the thrust of the engines, the force of gravity, air resistance, prevailing winds, and other relevant factors.

In much the same way that net translations and rotations can be determined by knowing original and final locations and orientations, net distortions and dilatations can often be determined relatively easily when initial and resultant shapes and volumes are known. Analysis is simplified if the area of study can be divided into subareas such that straight, parallel lines within each area remain straight and parallel after deformation. Such deformation, called homogeneous strain, is often assumed in the study of strain. Under these conditions, initially circular objects deform into ellipses.

STRAIN PATHS

Determining the strain path requires a series of known translations, rotations, distortions, and dilatations, while in order to tie the strain to the series of forces and stresses that produced it, one needs to know the strain path. Just as there are an infinite number of ways to fly from New York to Madrid, there are an infinite number of strain paths that could result in identical net strains.

As an indication of the problem, consider a circle 1 centimeter in radius that deforms into an ellipse with a semimajor axis of 2 centimeters and a semiminor axis of 0.5 centimeter. Although there is no net dilatation, the deformation may have consisted of stretching in one direction and shrinking in the direction perpendicular to it. Alternatively, this deformation could have been produced entirely by distortion, as can be seen by drawing a circle on the edge of a deck of cards and then moving each card slightly to the right of the one below it. Each card will have two spots on it, one from each side of the circle. Since the distance between the spots on an individual card does not change, and the number of cards does not change, the area inside the resulting ellipse will remain constant. Continuing to deform the deck in this manner (a process called shearing) will result in the ellipse getting longer and thinner.

STRESS MEASUREMENT

Strains are produced by stresses somewhat as movements of rigid bodies are produced by forces. More specifically, unbalanced forces acting on a rigid object cause it to accelerate. The amount of acceleration can be calculated if the net force and the mass of the object are known. The intensity of the forces acting within a body causes it to deform, and this force intensity is called stress. The units used to measure stress are the same as those used to measure pressure and are given in terms of force per unit area. Data may be presented in terms of atmospheres, pounds per square inch, bars, or similar units. The appropriate unit (based on the International System of Units) is the pascal, defined as 1 newton per square meter, or 1 kilogram per meter-second-squared. It is important to note that stress measurements contain an area term, and therefore they cannot be added, subtracted, or resolved as if they were forces. By multiplying a stress by the area over which it is applied it can be converted to a force, which can then be treated like any other force. It is customary to resolve it into forces parallel and perpendicular to a plane of interest. Finally, by dividing by the area of this plane, stresses can be obtained once again, yielding the shear stress and normal stress, respectively.

These factors can be demonstrated with a simple case in which a cube, 1 square meter on a side, has two forces acting on it in the vertical direction: One force of 10 newtons is pushing down on the top, another of 10 newtons is pushing up on the bottom. The forces balance, so there will be no acceleration. Any horizontal plane within this cube will have an area of 1 square meter. It is subject to stresses of 10 pascals, perpendicular to the plane, acting on each side of it. A diagonal plane through this cube, cutting the cube in half from one edge to the other, has an area of 1.414 square meters. The component of the vertical downward force acting perpendicular to this plane (the normal force) will be 7.071 newtons, and there will also be a component of the vertical downward force acting parallel to this plane (the shear force) of 7.071 newtons. When these forces are divided by the area over which they act, it is apparent that there will be a normal stress of 5 pascals and a shear stress of 5 pascals acting on the upper surface of this plane. Similar stresses can be shown to exist on the lower surface of the plane. Planes with different orientations will have other combinations of normal stresses and shear stresses, even though the forces responsible for those stresses remain the same. There are equations to manipulate the general situation, which give the normal and shear stresses acting on any plane as functions of the size and directions of the boundary forces. These result in what is called the stress ellipse, in two dimensions, or the stress ellipsoid in three. A graphical way of representing these equations (and the equations for strain also) was developed by Otto Mohr in 1882 and is now called Mohr's Circle.

MODES OF DEFORMATION

The results obtained above may be compared with those for a hydrostatic condition, where stresses are the same in every direction. If stresses of 10 pascals are acting on all six sides of the cube, no matter what plane one considers inside the cube, there will be normal stresses of 10 pascals acting on each side of it. The stress ellipses and ellipsoids one might construct would be circles and spheres, and there would be no shear stresses anywhere.

Different modes of deformation are favored by different combinations of stresses. Movement on a fault plane, for example, is favored by low normal stresses and high shear stresses on that plane. Through the simple analysis described above, it becomes clear that faulting is much more likely to occur along diagonal planes than along horizontal ones.

STRESS AND STRAIN FIELDS

The study of stresses often involves determining the stress field in a particular area, either at present or at some time in the past. After some simplifying assumptions are made concerning the geometry, mechanical properties, and boundary stresses of the area, a model is constructed that will indicate certain aspects of the stress field. Sometimes the model can be a physical one, produced from photoelastic plastic, for example. Such models can display the magnitudes of shear stresses when viewed appropriately with polarized light. More often, though, the model is constructed on a computer, and the stress ellipses are calculated for points of interest throughout the area. If geological stress indicators exist, such as the igneous sheet intrusions called dikes, the results of the model can be compared with the observed indicators, and the model can be adjusted until it fits the observations as closely as possible.

Determining at least parts of the strain field is in some ways more direct. Objects are sought in those rocks of the area that have net strains that can be determined. Frequently the distortion experienced by such an object can be easily observed and measured. A fossil that is elliptical but is known from its appearance in other areas to have been circular when it was alive provides a simple example. Strain ellipses showing the distortion of such objects can be constructed by measuring the shapes of these objects in the field.

These ellipses can then be plotted on a map. (A map with ellipses on it is a way of representing a tensor field, for example, a stress field or a strain field.) Most of the time, however, the initial size, location, and orientation of the objects are not known. One cannot tell whether a particular fossil is small because it never grew very large or whether it was once large and became smaller by deformation. If all the deformation occurred within a limited period of time, this map will represent the distortion part of the net strain field for that deformation. When similar data obtained from rocks deformed at different times are combined, a partial strain path can be obtained. The effects of more recent deformations are removed from the effects of earlier ones to isolate the earlier distortions.

The next step is to determine the stresses responsible for each increment of strain observed. To do so, it is necessary to know how each of the rocks responded to stress. Such data on mechanical behavior come from studies of experimental rock deformation. With these data, estimates can be made of the stress field present at different times in the history of the area. Finally, all this work can be applied to the known geological history of the area, permitting quantitative assessments of the various forces thought to have been active in the past.

APPLICATIONS FOR STRUCTURAL GEOLOGY

The stresses in a body are a function of the geometry of the body and the distribution of loads acting on it and within it. Determining the distribution of stresses is usually considered to be an exercise in statics, a branch of mechanical engineering, but it also plays a significant role in the Earth sciences. Earthquakes occur when rock fails suddenly. Mine collapses, landslides, and dam failures are other catastrophes that occur when stress exceeds the strength of the material involved.

When a load is placed on a solid, the distribution of stresses within that solid is usually uneven. If the solid deforms, the deformation will usually also be uneven. To measure such deformation, one examines strain, which includes movements and changes in size and shape. As with stress, strain usually varies throughout the region being deformed.

A structural geologist is often concerned with determining how a region of the crust of the Earth became deformed, and then why it deformed that way. Rocks often contain objects that are presently deformed but whose original shapes are known; such strain indicators include fossils, raindrop impressions, and bubbles. Using these indicators, a geologist seeks to reconstruct the strain field that existed at some time in the past. With enough indicators, along with dates for each,

it may be possible to construct a strain history for the area in question. The next step is to guess, using the known mechanical behavior of the rocks involved, what the stresses were that produced the reconstructed strains. A final goal might be to seek causes for those stresses in terms of a larger picture of Earth history, perhaps involving plate tectonics.

Otto H. Muller

CROSS-REFERENCES

BIBLIOGRAPHY

Davis, George H. *Structural Geology of Rocks and Regions.* New York: John Wiley & Sons, 1984. Chapter 4, "Kinematic Analysis" (44 pages), deals with strain, and chapter 5, "Dynamic Analysis" (30 pages), deals with stress. The treatment is the most descriptive and least technical of the references listed here, although some knowledge of stereonets is assumed. The double subscripts often used in the field are not employed here, and calculus is carefully avoided. Suitable for the general reader.

Hatcher, Robert D., Jr. *Structural Geology: Principles, Concepts, and Problems.* 2d ed. Englewood Cliffs, N.J.: Prentice Hall, 1995. This undergraduate textbook covers folds and stress in three chapters. Intended for the more advanced reader, it is accompanied by charts and illustrations.

Hobbs, Bruce E., Winthrop D. Means, and Paul F. Williams. *An Outline of Structural Geology.* New York: John Wiley & Sons, 1976. The first section, "Mechanical Aspects" (71 pages), covers stress, strain, and the response of rocks to stress. The approach is largely descriptive,

with no calculus, but double subscripts and some linear algebra are used. Although presumably written by Means, this section is sufficiently different from his book to complement it. Suitable for college students.

Johnson, Arvid M. *Physical Processes in Geology.* San Francisco: Freeman, Cooper, 1984. Chapter 5, "Theoretical Interlude: Stress, Strain, and Elastic Constants" (43 pages), develops a fairly rigorous treatment of stress and strain. Includes partial differential equations and other elements of calculus. Suitable for technically oriented college students.

McEvily, A. J., Jr., ed. *Atlas of Stress-Corrosion and Corrosion Fatigue Curves.* Materials Park, Ohio: ASM International, 1990. An excellent reference guide filled with illustrations and graphs. Includes sections on stress-strain curves, corrosion fatigue, and stress corrosion. Bibliographical references.

Marshak, Stephen, and Gautam Mitra, eds. *Basic Methods of Structural Geology.* Englewood Cliffs, N.J.: Prentice-Hall, 1988. Chapter 15, entitled "Analysis of Two-Dimensional Finite Strain," by Carol Simpson (26 pages), presents an overview of strain, with easily understood analogies, followed by an excellent survey of the techniques that have been used to measure strain in the field. The approach is definitely "how-to," with each step of each method clearly spelled out. Suitable for college students.

Means, W. D. *Stress and Strain*. New York: Springer-Verlag, 1976. This 339-page text is intended to be used for self-study by college undergraduates. Each of the twenty-seven chapters is followed by a number of problems, and solutions to the problems are provided. Although several of the chapters go into partial differential equations and other technical subjects, most of the book is easily read by nonspecialists. Suitable for college students.

Plummer, Charles C., and David McGeary. *Physical Geology*. Boston: McGraw-Hill, 1999. A college-level introductory geology textbook that is clearly written and wonderfully illustrated. An excellent sourcebook of basic information on geologic terminology and fundamentals of geologic processes. Contains a CD-ROM. An excellent glossary.

Ragan, Donal M. *Structural Geology: An Introduction to Geometrical Techniques*. 3d ed. New York: John Wiley & Sons, 1985. Chapter 7, "Stress" (18 pages), chapter 9, "Concepts of Deformation" (24 pages), and chapter 10, "Strain in Rocks," introduce the concepts in a mathematically straightforward way. The use of a card deck as a means of demonstrating strain was pioneered by Ragan and is included in this text. Suitable for college students.

Ramsay, John G. *Folding and Fracturing of Rocks*. New York: McGraw-Hill, 1967. The classic book on strain and strain analysis. Chapter 3, "Strain in Two Dimensions" (71 pages), chapter 4, "Strain in Three Dimensions" (64 pages), and chapter 5, "Determination of Finite Strain in Rocks" (70 pages), have provided the background in these subjects to generations of geologists. Although much of what is presented is fairly technical, the abundant photographs and line drawings and some of the descriptions in the text are useful to the general reader. A quick browse through this book furnishes valuable insight into what is involved in strain analysis. Suitable for college students.

Reynolds, John M. *An Introduction to Applied and Environmental Geophysics*. New York: John Wiley, 1997. An excellent introduction to seismology, geophysics, tectonics, and the lithosphere. Appropriate for those with minimal scientific background. Includes maps, illustrations, and bibliography.

Suppe, John. *Principles of Structural Geology*. Englewood Cliffs, N.J.: Prentice-Hall, 1985. Chapter 3, "Strain and Stress" (33 pages), presents an elegant development of the subject. It is the only reference listed here that uses the Einstein summation convention and relies heavily on matrix manipulations. Thus, mathematically, this work may be above the level of most nontechnical college students. The eight photographs and the discussion of strain versus displacement are excellent, however, and useful to the general reader.

6
EARTHQUAKES

DEEP-FOCUS EARTHQUAKES

Deep-focus earthquakes occur at depths ranging from 70 to 700 kilometers below the Earth's surface. This range of depths represents the zone from the base of the Earth's crust to approximately one-quarter of the distance into the Earth's mantle. Deep-focus earthquakes provide scientists information about the planet's interior structure, its composition, and seismicity. Observation of deep-focus earthquakes has played a fundamental role in the discovery and understanding of plate tectonics.

PRINCIPAL TERMS

AFTERSHOCK: an earthquake that follows a larger earthquake and originates at or near the focus of the latter; many aftershocks may follow a major earthquake, decreasing in frequency and magnitude with time

BRITTLE FRACTURE: rock that fractures at less than 3 to 5 percent compressional or tensional stress

DUCTILE FRACTURE: rock that is able to sustain, under a given set of conditions, 5 to 10 percent deformation before fracturing or faulting

EPICENTER: the point on the Earth's surface that is directly above the focus of a earthquake

FOCUS: the place within the Earth where an earthquake commences and from which the first P waves arrive; also called the hypocenter

LITHOSPHERE: the solid portion of the Earth used in plate tectonics as a layer of strength relative to the underlying plasticlike asthenosphere; encompassing the Earth's crust and part of the upper mantle, it is about 100 kilometers in thickness

PRIMARY WAVE (P WAVE): the primary or fastest wave traveling away from an earthquake through the solid rock; P waves also are capable of moving through liquids

SECONDARY WAVES (S WAVES): the secondary wave that travels more slowly through solid rock than the P wave; S waves cannot penetrate a liquid

SHEAR STRESS: stress that causes different parts of an object to slide past each other across a plane

STRIKE-SLIP FAULT: a fault along which movement is horizontal only; the movement is parallel to the trend of the fault

SUBDUCTION: the process in which a dense lithospheric plate descends into the mantle beneath another, less dense, plate in a subduction zone

WADATI-BENIOFF ZONE: a narrow zone of earthquake foci that seismically delineates an inclined subduction zone; they are generally tens of kilometers thick

EARTHQUAKES AND EARTH'S INTERIOR

Because direct physical access to the Earth's interior is restricted, most knowledge about it is derived from the study of earthquake waves that travel through the planet and vibrate the Earth's surface at some distant point. Earthquakes recorded on seismograms allow scientists to accurately measure the time required for earthquake waves (seismic waves) to travel from the focus of an earthquake to a seismographic station. Primary waves (P waves) travel the fastest, and this information is used to determine distances to earthquake epicenters. The time required for P waves

and secondary waves (S waves) to travel through the Earth depends on the physical properties encountered as the waves pass through the subsurface. Seismologists, therefore, search for variations in travel times that cannot be explained by differences in the distance traveled. These differences correspond to changes in the properties of the subsurface Earth material encountered. Changes in rock properties indicate that the Earth has four major layers: the crust, a thin outer layer; the mantle, a rocky layer beneath the crust with a depth of 2,885 kilometers; a 2,770-kilometer outer core exhibiting characteristics of a mobile liquid composition; and a solid inner core, a metallic sphere with a radius of 1,216 kilometers.

EARTHQUAKE ZONES

Most earthquakes occur in narrow zones that globally connect to form a continuous seismic network. Characteristic surface features of seismic zones are rift valleys, oceanic ridges, mountain belts, volcanic chains, and deep-ocean trenches. These global seismic zones represent the boundaries of the lithospheric plates. The interior regions of the lithospheric plates are largely free of earthquakes. The lithosphere is made up of twelve rigid plates that cover the entire globe. The depths of the lithospheric plates range between 60 and 100 kilometers. They are composed of either the entire continental crust and a portion of the upper mantle or the entire oceanic crust and a portion of the upper mantle. The lithospheric plates are in constant motion relative to each other and can diverge away from each other, forming ridge axes and new oceanic crust material; converge toward each other, where one plate subducts under the leading edge of its neighbor plate; or transform, where plates slide horizontally past each other.

Within the global seismic network, there are four types of seismic areas that are recognized by their form, structure, and geology. The first type is represented by narrow zones of high surface-heat flow and basaltic volcanic activity along the axes of mid-ocean ridges where the earthquake focus is shallow (70 kilometers deep or less). Here, molten rock material is welling up from the mantle area and is emplaced on either side of the ridge, adding to the ocean crust. Mid-ocean ridges are active areas of seafloor spreading and are found in all ocean basins. A primary example is the Mid-Atlantic Ridge trending north to south in the Atlantic Ocean basin. This ridge rises above sea level in Iceland and delineates the boundary between the South American and African lithospheric plates.

The second type of seismic zone is identified by large surface displacements occurring parallel to the fault, shallow-focus earthquakes, and absence of volcanoes. Excellent examples of this seismic zone are the San Andreas fault in California and the Anatolian fault in Turkey, both of which demonstrate strike-slip movement between plate boundaries. The third type of seismic zone is a widespread continental area ranging from the Mediterranean Sea to Myanmar (Burma); it is associated with high mountain ranges created by converging plate margins. Although this zone is usually characterized by shallow-focus earthquakes, earthquakes of intermediate focus (70 to 300 kilometers deep) have occurred in the Hindu Kush and Romania; deep-focus earthquakes (300 to 700 kilometers deep) have been recorded in a few places north of Sicily under the Tyrrhenian volcanoes.

The fourth type of seismic zone is physically connected to volcanic island-arc-trench systems, such as the Japan Trench and the Kermadec-Tonga Trench in the South Pacific Ocean. Earthquakes associated with trenches can be shallow, intermediate, or deep focus based on where they are located in the steeply converging lithospheric plate adjacent to the trench. The earthquake foci define the plate being carried into the Earth's interior and away from the trench. These inclined earthquake zones, called Wadati-Benioff zones, underlie active volcanic island arcs and have an assortment of complex shapes. The Wadati-Benioff zone also marks the downflowing portion of the convection cell as identified by geophysicists through calculations of moving oceanic plates.

SUBDUCTION OF THE LITHOSPHERE

New lithosphere is created at mid-ocean ridges by upwelling and cooling of magma from the Earth's mantle. With new lithospheric material constantly created and no measurable expansion of the Earth occurring, some of the lithosphere must be removed globally. Older oceanic lithosphere is removed by subduction of the oceanic plate into the Earth's mantle at island-arc-trench systems. The originally rigid plate slowly descends and heats; over millions of years, it may be fully absorbed into the Earth's mantle. The subduction of lithospheric plates accounts for many of the processes that shape the Earth's surface, such as volcanoes and earthquakes, including nearly all the earthquakes with deep and intermediate foci. Deep earthquakes cannot occur except in subducting plates, so the presence of a deep-focus earthquake implies the presence of subducting oceanic plate material.

The proposed driving force for movement of the lithospheric plates is a convection cell arrangement where molten mantle material rises into a crustal rift zone from the subsurface mantle zone. Portions of the molten mass that do not move into the central rift zone move to either side of the rift and travel away from it, below and parallel to the lithospheric plate. The convection current travels beneath the plate for some distance. It is in this zone that the convection current is thought to move and carry the lithospheric plate with it. As the convection cell cools because of its interaction with the cool oceanic plate, the outer area of the cell sinks into the mantle to be rewarmed by mantle convection currents. Again, the downflowing portion of the convection cell is represented by the subducting oceanic plate.

ISLAND-ARC-TRENCH SYSTEMS

Major island chains are known to geologists as island-arc-trench systems. The island chains are a surface expression of the oceanic subduction process, and the associated deep trenches are a surface expression of the seaward boundary of subduction zones. Examples of deep trenches appearing in connection with island arcs are the Java and Tonga Trenches.

When foci of earthquakes near island arcs and ocean trenches are compared, a notable pattern emerges that is well illustrated in the Tonga island arc in the South Pacific. To the east of the volcanic islands of Tonga lies the Tonga Trench, which is approximately 10 kilometers deep. Beginning at the Tonga Trench and moving from east to west, the earthquake foci lie in a narrow but well-defined zone, which slopes from east to west from the Tonga Trench to beneath the Tonga Island Arc at an angle of about 45 degrees. The earthquake record in the arc-trench zone reveals that the earthquake foci are extremely shallow at the trench; however, moving to the west away from the trench, the earthquake foci register at a depth of 400 kilometers. An additional move to the west reveals deep-focus earthquakes at below 600 kilometers. In other regions of deep earthquake activity, some variation in the angle of dip and distribution of foci is recorded, but the common feature, that of a sloping seismic zone, is characteristic of island arcs and deep-ocean trenches. Other regions associated with island arc and deep-ocean trenches are the Kurile Ridge and Trench and the Mariana Islands and Trench.

What the deepening earthquake foci are defining at island-arc-trench systems, and especially the Tonga Trench and Island Arc system, is the movement of the descending oceanic plate (the downgoing slab) into the inclined seismic zone, known to scientists as the Wadati-Benioff zone. In the process of the plate's gravity-driven downward movement, additional force is created, causing further deformation and fracturing and deep-focus earthquakes. At mantle depths of 650 to 680 kilometers, either the plate may have been absorbed into the mantle interior, or its properties may have been altered to the extent that earthquake energy cannot be released.

OCEANIC AND CONTINENTAL CRUST SYSTEMS

Major mountain belts, such as the Andes in South America, have been raised by the convergence and subduction of oceanic lithospheric plates beneath continental lithospheric plates. As the two plates converge, the denser oceanic plate is subducted under the less-dense continental plate (South America). The zone of subduction is identified by an oceanic trench, the Peru-Chili Trench. Initially, in this setting, the angle of descent of the plate into the subduction zone is low, but the angle gradually steepens into a downward curve as revealed by intermediate- and deep-focus earthquakes. On June 8, 1994, a magnitude 8.2 deep-focus earthquake occurred 600 kilometers below Bolivia in the Andes Mountains. This earthquake released a tremendous amount of energy, but because of its deep focus, the seismic waves were slow to reach the Earth's surface, preventing damage to populations and structures.

Deep-focus earthquakes have proven especially useful because they do not produce many surface seismic waves, and they provide information about the density of the Earth's mantle—a factor that ultimately controls how convection currents of rock inside the planet move the continents around the surface. Because the core and mantle remain hidden from view, researchers interested in these deep regions must wait for large deep earthquakes to provide indirect information about the planet's interior. The 1994 Bolivian earthquake was well-recorded because of global seismic detectors placed by scientists some twenty years in advance. The seismic equipment was placed in an effort to fully capture deep-focus earthquake activity as a result of a large deep-focus earthquake under Colombia in 1970.

THERMAL PROPERTIES OF THE DESCENDING LITHOSPHERE

Temperatures near the surface of the Earth increase rapidly with depth, reaching about 1,200 degrees Celsius at the depth of 100 kilometers. Here the minerals in peridotite, an olivine-rich major mineral constituent of the upper mantle, begin to melt. The temperatures then increase more gradually to 2,000 degrees Celsius at approximately 500 kilometers depth.

As the lithospheric plate descends into the mantle, it is heated primarily by heat flowing into the cooler lithospheric plate from the enclosing hotter mantle. Since the conductivity of the rock increases with temperature, conductive heating becomes more efficient with increasing depth, further warming the subducting plate. Heat within the Earth is generated by the energy released when minerals in the mantle change to denser phases or more compact crystalline structures with the higher pressures present at depth. Additional heat sources in the mantle are radioactivity and the heat of compression, which is activated by increasing pressure at depth.

Despite the elevated temperatures at depth, the interior of the descending plate remains cooler than the surrounding mantle until the plate reaches a depth of about 600 kilometers. As the plate continues to subside into the deep mantle interior, it may heat rapidly because of efficient heat transfer by radiation. At 700 kilometers, however, the lithosphere plate is difficult to decipher as a separate structural unit. Complicating the detection of the plate properties is the low number of earthquakes at or below the 700-kilometer depth, which could possibly reveal information about any remaining plate material. The subduction zone under the Japanese island of Honshu, under the Kuriles, and under the Tonga-Kermadec area (north of New Zealand) represent almost ideal subductions zones without major complications caused by the age or thermal properties of the plate.

However, not all subduction zones behave ideally. The descending oceanic lithospheric material can be assimilated before reaching deep mantle zones: A slow-moving plate may achieve thermal assimilation before reaching 700 kilometers, such as the Mediterranean plate under the Aegean Sea. In younger subduction zones, as found in the Aleutian Islands and Trench, the descending plate may have penetrated far less than 700 kilometers owing to the warmer, more buoyant oceanic plate.

STUDY OF DEEP-FOCUS EARTHQUAKES

The deep-focus earthquake problem has been one of the leading scientific problems of solid-Earth geophysics since the 1920's, when Kiyoo Wadati, a Japanese seismologist, demonstrated that some earthquakes occur hundreds of kilometers beneath the Earth's crust. Laboratory experiments attempting to replicate the temperatures and pressures of the Earth's interior have confirmed that rock under stress at the higher temperatures and pressures of 70 kilometers fails suddenly by brittle fracture—that is, shallow-focus earthquakes fail by brittle fracture. At even higher temperatures and pressures, similar to what is present in the deep interior of the Earth (300 to 700 kilometers), shear stress should deform rocks by ductile flow, even in the colder regions of the mantle beneath subduction zones. Yet seismic data indicate that rocks at these depths are apparently also failing by brittle fracture.

Research has been directed toward understanding the triggering mechanism for brittle fracture in deep-focus earthquakes. The first area of research concerns the mechanism for brittle failure on intermediate-depth earthquakes. Research seems to indicate that the brittle failure is driven by water subducted along with the oceanic lithospheric plate or by water released from the molecular structure of oceanic crustal minerals. Thus, down to a depth of 350 kilometers or so (many scientists consider any earthquake below 70 kilometers a deep-focus earthquake), the triggering mechanism is much like that of shallow earthquakes, with the available water reducing the normal stresses on faults at depth and allowing failure (an earthquake) to occur.

The second research position holds that olivine, a primary constituent mineral of the Earth's upper mantle, transforms under stress to spinel, a denser mineral in which the atoms are more closely packed and display a rearranged crystal structure. The newly formed spinel is deposited in many elongated, beadlike structures that eventually become thin, shear faulting zones. Faulting occurs along planes of greatest stress in these thin, shear zones.

However, olivine could remain cool because of its presence within the cooler subducting plate and not transform until some greater depth. At this greater depth, the newly formed, fine-grained spinel could be the lubricating material that makes deep-focus earthquake faulting possible. As olivine undergoes the transformation to spinel, heat is released that may augment the catastrophic faulting.

Neither of these two positions has been proven conclusively, nor are they accepted by all scientists. Yet most research focuses on exploring the detailed geophysics and geochemistry of each position. For example, thermal calculations suggest that olivine may remain present even at great depth in subduction zones where the plate is old and cold (cooling over time) and therefore subducts rapidly. These findings might explain why there are deep-focus earthquakes in Tonga and the Kuriles where the plate material may be older, whereas warmer, younger subducting slabs may manifest only intermediate-depth events as in the Aleutians.

Another factor that may need consideration is that analysis of seismograms from intermediate or deep-focus earthquakes reveals little or no gross differences between the seismic properties for intermediate and deep earthquakes. In fact, the only gross mechanical difference between shallow earthquakes and deeper ones is that aftershocks are much rarer for deep and intermediate earthquakes. Therefore, if the seismograms are similar, then should not the source for intermediate- and deep-focus earthquakes be similar? Thus, one of the two premises may be the triggering mechanism for all earthquakes below 70 kilometers—or perhaps the triggering mechanism is a combination of the two positions.

SIGNIFICANCE

Deep earthquakes are significant for at least four reasons. First, they are exceedingly common, constituting almost 25 percent of all earthquakes occurring during the period of 1964 to 1986. Second, deep earthquakes most often occur in association with deep-ocean trenches and volcanic island arcs in subduction zones. One of the great achievements of twentieth century geophysics was the recognition that the occurrence of deep earthquakes in Wadati-Benioff zones apparently delineates not only the subduction of the lithospheric plate but also the cold downflowing cores of convection cells in the uppermost mantle. Third, scientists use seismograms of deep earthquakes rather than those of shallow earthquakes to investigate core, mantle, and crustal structures. Deep-focus earthquakes are mechanically different from shallow-focus earthquakes because their body wave phases traverse the uppermost mantle only once from the focus to the seismic station, thus producing simpler seismographs. Finally, information on how deep-earth materials process and handle stress can be derived from the seismograms of deep earthquakes.

Mariana Rhoades

CROSS-REFERENCES

BIBLIOGRAPHY

Bolt, Bruce, A. *Earthquakes.* Rev. ed. New York: W. H. Freeman, 1995. A concise account of earthquake knowledge for general readership. Readers will be able to answer questions about the cause, location, and occurrence of earthquakes. A glossary and bibliography are provided, along with information on the Northridge (1994) and Kobe (1995) earthquakes; an earthquake quiz with answers is included.

_____. *Earthquakes and Geological Discovery.* New York: Scientific American Library, 1993. This well-illustrated text may give the impression of a simpler format, but the text is technical and deserves thoughtful reading and study. The sections on predicting and forecasting earthquakes and reducing seismic risk are excellent.

_____. *Inside the Earth: Evidence from Earthquakes.* Rev. ed. San Francisco: W. H. Freeman, 1994. Includes a full explanation of seismology; the temperatures, densities, and elastic properties of Earth; and the interiors of Earth, the Moon, and Mars. For college-level readers.

Gubbins, David. *Seismology and Plate Tectonics.* Cambridge: Cambridge University Press, 1990. Although highly technical, the text portions are easily read and understood, especially the section on ocean trenches and plate subduction. Describes seismograph interpretation techniques and includes practice seismograms for the reader to locate earthquakes and determine their source mechanisms. Suitable for college-level science courses.

Wilson, J. Tuzo, ed. *Continents Adrift and Continents Aground.* San Francisco: W. H. Freeman, 1976. Wilson brings together seventeen articles from *Scientific American* that describe the plate tectonic revolution of the late 1960's. Articles are grouped under four topics: mobility in the Earth; seafloor spreading, transform faults, and subduction; plate motion; and applications of plate tectonics. Illustrations and maps are clear and concise.

Yeats, Robert, Kerry Sieh, and Clarence Allen. *The Geology of Earthquakes.* New York: Oxford University Press, 1997. The authors provide an interdisciplinary approach to the study of well-known earthquakes. Although it welcomes a wide range of readers, the text was written for engineers, geophysicists, and government planners. Includes a comprehensive glossary, a bibliography, and a global appendix that lists more than three hundred historical earthquakes with surface rupture. Well illustrated with photographs, maps, and cross sections.

EARTHQUAKE DISTRIBUTION

For approximately a century, seismologists have been monitoring global earthquake activity. These studies have led to an understanding of earthquake frequency and distribution that contributed dramatically to the confirmation of plate tectonics theory.

PRINCIPAL TERMS

EPICENTER: the point on the Earth's surface directly above an earthquake's focus

FOCUS: also known as the hypocenter, the focus is the actual place of rupture inside the Earth's crust

P WAVE: the primary or fastest wave traveling away from a seismic event through the rock and consisting of a series of compressions and expansions of the Earth material

PLATE BOUNDARY: a region where the Earth's crustal plates meet, as a converging (subduction zone), diverging (mid-ocean ridge), transform fault, or collisional interaction

S WAVE: the secondary seismic wave, traveling more slowly than the P waves and consisting of elastic vibrations transverse to the direction of travel; S waves cannot propagate in a liquid medium

SEISMIC BELT: a region of relatively high seismicity, globally distributed; seismic belts mark regions of plate interactions

SEISMIC WAVE: an elastic wave in the Earth usually generated by an earthquake source or explosion

SEISMICITY: the occurrence of earthquakes as a function of location and time

SEISMOGRAPH: an instrument used for recording the motions of the Earth's surface caused by seismic waves, as a function of time

SUBDUCTION ZONE: a dipping ocean plate descending into the Earth away from an ocean trench

PLATE TECTONICS

Although seismic instruments can record them from virtually anywhere on the globe, earthquakes occur primarily along active tectonic regions of the Earth's crust where mountain building, folding, and faulting are occurring. More temporal and often less severe earthquakes also accompany volcanic activity.

The key to understanding earthquakes lies in the powerful theory of plate tectonics. The Earth is far from a geologically "dead" world like its moon. The Earth's crust is broken into several large slabs of crust, or lithospheric plates, and convection currents caused by the planet's internal heat drive the plates into motion, like a bunch of small rafts crowded onto the surface of a boiling pot of viscous jelly. At the mid-ocean ridges, new seafloor is created by magmatic eruptions—pushing two plates away from each other. These divergent boundaries are characterized by the Mid-Atlantic Ridge, where the North American and European continents (riding on the lithospheric plates) are moving away from each other. Along the Alpine belt, two continents are literally crashing into each other as Africa is pushing into and subducting under the Eurasian plate. The subduction zone is marked by a complex series of transform and thrust faults, which give the region its high seismicity.

Earthquakes, then, are predominantly distributed along plate boundaries. Another converging boundary is found along the Hindu Kush and Himalayan mountain ranges, where the subcontinent of India is crashing into and thrusting under the huge Eurasian plate at the rate of some 5 centimeters per year. The collision has caused the throwing up of the world's highest mountains and an earthquake-prone region.

SEISMIC BELTS

Mapping earthquake epicenters, scientists are able to map the seismicity (earthquake activity as a function of time) of the planet. Most earthquakes occur along three main belts: the Mid-Atlantic Ridge system; the Alpine Tethys belt, which extends from the Mediterranean Sea through Turkey and Armenia all the way into Asia, where it merges with the third main belt; and the infamous circum-Pacific "Ring of Fire." The least threatening of these is the mid-ocean ridge system such as that found in the Atlantic Ocean, along which new ocean crust is being created. As the seafloor spreads from the volcanic activity occurring at the spreading ridges, earthquakes occur along transform faults that bound the offset ridges. Although population centers are sparse along the mid-ocean ridges, Iceland, the Azores, and other small mid-Atlantic islands are regions of potential quake hazard. Owing to the steady rate of spreading (a few centimeters annually), earthquakes occurring along the ridge offsets tend to be frequent and of relatively low magnitude.

A far more dangerous region of earthquake activity is the Alpine Tethys belt, extending across southern Europe and Asia. A listing of only a few of the major earthquakes along this belt reads like a litany of human suffering and huge losses of property: Persia in 1505; Calabria in southern Italy in 1509, 1783, and 1832; Lisbon in 1755; the Neapolitan in Italy in 1857; numerous quakes in recent decades in Yugoslavia, Romania, Greece, and Turkey; and the tragic 1988 disaster in Soviet Armenia. Volcanic activity also occurs in the region, with notable examples including Mounts Etna and Vesuvius and the island of Thera.

Perhaps the most seismically active region of the world lies on the eastern end of the Mediterranean-Himalayan belt. Stretching across Tibet and into China, this colossal zone of high seismicity threatens all who live along its 4,000-kilometer length. More than a dozen earthquakes of Richter magnitude 8.0 or greater have caused well in excess of a million human lives to be lost in this notorious region. An estimated 830,000 casualties in the Shenshi region earthquake of 1556 makes that event what is believed the worst earthquake in historic times. The Kansu earthquake of 1920 led to 200,000 deaths and adds credibility to the claim.

"RING OF FIRE"

The trans-Asian earthquake belt passes through Burma and Indonesia, ending in the southern Philippines. This transitional region marks the border of the greatest of the Earth's seismic belts—the circum-Pacific, or Ring of Fire. A region of complex plate interactions, the Pacific Rim is no stranger to earthquakes and volcanoes. Perhaps no region characterizes the circum-Pacific belt better than the islands of Japan. In geologic terms, the Japanese islands are an island arc, formed by a subduction zone off the coast of the present landmass. As the seafloor spreads from the ridge systems, it collides with the Asian continental plate. The dense, water-soaked seafloor is subducted at a deep ocean trench. As the oceanic plate descends, the slab grinds and shudders in resistance before finally being swallowed by the mantle. Accounting for 90 percent of the world's earthquakes, trench subduction zones have a seismic fingerprint of ever-

View southwest along Hanning Bay fault scarp, reactivated during the 1964 Alaska earthquake. The 10- to 15-foot-high bedrock scarp trends from the right foreground to the left background. The uplifted wave-cut surface to the right is coated with desiccated marine calcareous organisms. (U.S. Geological Survey)

deepening quakes that can be very severe shocks.

Seafloor earthquakes can cause tsunamis (sometimes incorrectly called "tidal waves") along the coastline. A 7.7-magnitude shock hit the Oga Peninsula in 1983 and brought on a tsunami that caused extensive damage. The Fukui earthquake of 1948 and the Great Tokyo earthquake and fire (magnitude 8.2) of 1923, in which 143,000 perished, are stark testimonies to the danger of living near active plate subduction zones. In addition to the trench quakes, the volcanic islands are crisscrossed by numerous faults. Thousands of quakes have been recorded by Japanese historians, dating to well before the birth of Christ.

To the north and east along the circum-Pacific belt, the Aleutian Island arc reaches into the North American continent in Alaska. A complex system of faults and a subduction zone off the coast make Alaska a region of dangerous seismicity. In 1964, one of the most severe earthquakes ever recorded, at 8.6, struck near the port of Valdez and generated a terrifying series of tsunamis that wracked the coast. On Alaska's southern coast is the Fairweather fault, a transform fracture on which a 1958 temblor shook loose 90 million tons of rock, which cascaded into Lituya Bay, raising a wave exceeding 500 meters in height. The Fairweather fault is a northern extension of a system of transform faults (so named because the fault is transformed into a ridge or trench at its ends) that includes the most famous of all—the San Andreas fault.

SAN ANDREAS AND ALPINE FAULTS

Extending from the Mendocino fracture zone 700 miles south to Mexico, the San Andreas and its attendant system of faults make California earthquake country. A unique type of plate boundary, the San Andreas represents a fracture line along which the oceanic Pacific plate is slowly but inexorably sliding north with respect to the North American continent at a rate of roughly 5 centimeters per year. In places where the fault is displacing smoothly, small tremors regularly shake the California landscape. Yet, in regions where the fault is believed locked, major quakes are impending, placing the large population centers of San Francisco and Los Angeles at risk. On October 17, 1989, one such strong earthquake shook the San Francisco Bay area, causing extensive damage and a number of deaths and injuries. In addition to the San

Andreas, the Hayward fault in the north and a multitude of southern California faults further increase the state's seismicity. The Whittier, San Fernando, and Inglewood-Newport faults are among those that threaten Los Angeles.

As seismically active as California is, its seismicity pales by comparison with Latin America. Mexico City's tragic quake of 1985 was a grim reminder of the subduction zone off the coast of western Mexico. Central America is in a particularly precarious position, lying sandwiched between the Cocos, North American, South American, and Caribbean plates, and is thus a zone of active volcanoes and numerous severe deep-trench earthquakes. El Salvador, Guatemala, and Nicaragua are among the nations in greatest earthquake danger. As the Nazca plate bumps into South America's plate, the oceanic plate is subducted and the melting slab has caused the volcanism and massive uplifting of the towering Andes. While the eastern part of South America is seismically quiet, the west coastal regions are known for severe quakes, especially in Peru and Chile.

The circum-Pacific belt continues along the South Pacific through New Zealand. Analogous to the San Andreas, the majestic scenery of New Zealand is regularly shaken by tremors occurring along the Alpine fault. Continuing up through Indonesia, the ring of earthquake and volcanic activity completes its loop.

VOLCANIC ACTIVITY

Other earthquake regions of note include the American Northwest's Cascade volcanic chain, where a series of tremors can indicate pending eruptions. The 1980 explosive eruption of Mount St. Helens in Washington was caused by an earthquake that triggered a landslide, initiating the lateral blast, or nuée ardente.

In California, the San Andreas is not the only region of earthquake activity. An area with an explosive volcanic past in recent geologic times, the Mammoth-Mono Lake region was hit by four magnitude 6.0 temblors in 1980, occurring along the northern perimeter of the Long Valley caldera. South of the possible site of future volcanic and certain seismic events, the Sierra Nevada mountain range is still undergoing periodic spasms of uplift, like the one that caused the Owens Valley quake of 1852.

In the Caribbean, earthquakes and volcanic activity are an ever-present threat along the borders of the Caribbean plate. This seismic belt is actually an extension of the Pacific belt, although it lies on the Atlantic side of the Americas. Examples of major shocks and activity include Port Royale, which plunged 50 feet underwater following a major earthquake in 1692, and Mount Pelée, which destroyed the town of Martinique with a nuée ardente in 1902. Regions of hot spot volcanism are also zones of especially high seismicity. The Hawaiian Islands lie on top of a mantle plume of magma, and the same forces that built the island chain are working on the main island of Hawaii today. A similar region lies beneath the North American continent, the site of Yellowstone National Park in Wyoming, Montana, and Idaho. Both Hawaii and Yellowstone are earthquake-prone regions.

CONTINENTAL INTERIORS

For the most part, continental interiors, especially the Precambrian metamorphic basement and its thin veneer of sedimentary rocks that make up the craton, are regions of low seismicity. Such regions include parts of the United States and Canada in the Great Lakes region, virtually all of South America except for the Pacific coast and Andes belt, and most of Africa.

While continental interiors are seismically quiet compared with the active plate margins, there are exceptional regions. Although most earthquakes in the United States take place west of the Rockies, the Mississippi Valley has been the site of some of the most severe earthquakes ever, the New Madrid quakes of 1811-1812. New England and South Carolina have also experienced powerful shocks in the past. The western two-thirds of Africa is seismically inactive, but the East African rift valley is a zone of earthquake and volcanic activity where a geologically new plate boundary is rifting the eastern edge of the continent apart. Curiously, the only continent on Earth that is seismically quiet is Antarctica.

EARLY RECORDS

In a sense, the study of where earthquakes occur traces back to the very roots of western and eastern culture roughly four thousand years ago in Mesopotamia and in Asia. The Old Testament and more ancient Middle Eastern documents are filled with accounts of earthquakes toppling cities. The most complete records of seismic activity are, appropriately enough, those of Japan and China. Chinese earthquake records date back thirty centuries, with exhaustive accounts of tragic earthquakes striking the Asian mainland. Japan, which experiences up to one thousand noticeable shocks per year, has been keeping detailed earthquake records on Tokyo's tremors since 818 C.E.

Modern scientific views on earthquake distribution perhaps began in response to the tragic All Saints Day earthquake and tsunamis that wrecked Lisbon in 1755. While a horrified western civilization reeled at the scope of the disaster, which killed many at Lisbon's numerous downtown churches, one of the more insightful minds of the scientific revolution looked at the tragedy more objectively. Immanuel Kant advised that learning about where and why earthquakes occur was a more reasoned approach than blaming the disaster on divine causes.

GLOBAL SEISMIC STUDY

Some one hundred years after Lisbon's fateful quake, Irishman Robert Mallet published a study of the Neapolitan earthquake of 1857 in which he produced a seismographic map of the world that (except for the mid-ocean ridge systems) is accurate today. Teaching in seismically active Japan, British geology professor John Milne invented the modern seismograph. By the time of the 1906 quake in San Francisco, global seismic observatories were in place and recorded the jolts. Since the distance to an epicenter (the surface site above the actual fracture or focus of the quake) could be determined, a set of three properly placed seismographs, Milne reasoned, could pinpoint an epicenter anywhere on the globe.

Seismic waves generated by an earthquake produce different types of waves. The primary wave, or P wave, is compressional, while the secondary, or shear, waves (S waves) cause the shaking motions that occur during the most destructive part of the quake. By measuring the ratio of the arrival time of the primary and secondary waves and the size or amplitude of the waves on the seismograph recording, Charles Richter, in 1935, was able to establish a scale for measuring the energy released in a quake, its magnitude. Richter and his colleague at the California Institute of Technology Beno Gutenberg published some of the best maps of worldwide earthquake distribution in 1954.

In the 1950's and 1960's, the United States helped to organize enough of the world's seismic observatories to establish a global monitoring network called the World Wide Standardized Seismograph Network. Data from decades of shocks recorded by the network and oceanographic research vessels led to the revolutionary theory of plate tectonics and its acceptance by the vast majority of Earth scientists.

IDENTIFYING REGIONS OF SEISMIC HAZARD

Although plate tectonics theory and, specifically, plate boundaries are invoked to explain the vast majority of earthquakes, scientists are still puzzled by earthquakes far from active margins. Examples are the Mississippi Valley region and Charleston, South Carolina, which shook violently in 1886. In these regions, seismologists (well aware of the seismicity of the regions) are alarmed by public perception that the land east of the Rockies is "solid bedrock." The most plausible cause of the Mississippi Valley quake activity is the enormous weight of sediments the great river system has deposited on a weak part of the continental crust. South Carolina is a region riddled with faults, yet it is far from an active plate margin. Seismologists use historical accounts and recent seismic records to predict regions of earthquake hazard. One theory of seismic hazard involves identifying regions where earthquakes have not occurred along active fault regions. Such seismically quiet regions are called "seismic gaps" and represent regions of accumulated strain along which a major rupture may be anticipated.

Using seismicity data, seismologists produce maps that indicate seismic hazard. Not only useful for scientific purposes, such maps help public agencies to create building codes and other earthquake prevention methods appropriate to the earthquake hazard of the region. Studies conducted by the United States Geologic Survey and the National Oceanic and Atmospheric Administration have concluded that the areas of San Francisco and Los Angeles, California; Salt Lake City and Ogden, Utah; Puget Sound, Washington; Hawaii; St. Louis-Memphis, Tennessee; Anchorage and Fairbanks, Alaska; Boston, Massachusetts; Buf-

falo, New York; and Charleston, South Carolina, are at greatest seismic risk in the United States. Recently, the worldwide network of seismic stations was upgraded with new instruments that record directly onto digital tapes, which enable computers to analyze the seismic waves more swiftly, thus improving observation of the planet's moving plates.

DANGERS TO DENSELY POPULATED REGIONS

Approximately once every 30 seconds, a million times a year, the Earth's crust shivers. Most of these tremors are perceptible only by sensitive instruments, but more than three thousand are strong enough to be felt by those nearby. Roughly twenty quakes a year are strong enough to do catastrophic damage to populated areas. By tragic coincidence, some of the Earth's most active seismic regions are also among its most densely populated. The lands bordering the Mediterranean Sea and Pacific Rim, the mountainous Middle East, India, China, and Japan are all familiar with the havoc of a major shock. In China alone, the death toll exceeds 13 million lives down through the ages. Earthquakes and related phenomena claim up to fifteen thousand lives annually in these regions of dangerous seismicity.

All told, millions have been killed by seismic activity, with staggering loss of property through the half-tick of geologic time comprising human history. Seismologists still lack the capability of precise prediction of activity, but areas of high seismicity and the ominous seismic gaps warn of quake hazard. In cities such as Tokyo, Los Angeles, San Francisco, and Anchorage, citizens must be prepared for the next big quake, which is as sure to come as the slow but steady movement of the Earth's crustal plates.

David M. Schlom

CROSS-REFERENCES

BIBLIOGRAPHY

Bolt, Bruce A. *Earthquakes: Revised and Updated.* 2d ed. New York: W. H. Freeman, 1987. A revision of the University of California, Berkeley, seismologist's classic, *Earthquakes: A Primer.* Chapter 1 deals with earthquake distribution, and also of special interest are the appendices on world earthquakes and seismicity rates and lists of important earthquakes in the Western Hemisphere. Suitable for the lay reader.

Collier, Michael. *A Land in Motion: California's San Andreas Fault.* San Francisco: Golden Gates National Parks Association, 1999. Filled with beautiful color photographs that accompany text intended for the nonscientist, *A Land in Motion* gives the reader excellent insight into earthquakes and their aftermaths. There are also many diagrams and graphs that explain subduction, faults, and orogeny.

Condie, Kent C. *Plate Tectonics and Crustal Evolution.* 4th ed. Oxford: Butterworth Heinemann, 1997. An excellent overview of modern plate tectonics theory that synthesizes data from geology, geochemistry, geophysics, and oceanography. Extensive coverage of plate boundary interactions and earthquake distribution. An excellent "Tectonic Map of the World" is enclosed. Nontechnical and suitable for a college-level reader. A useful "Suggestions for Further Reading" is provided at the end of the chapters.

Doyle, Hugh A. *Seismology.* New York: John Wiley, 1995. A good introduction to the study of earthquakes and the Earth's lithosphere. Written for the layperson, the book contains many useful illustrations.

Eiby, G. A. *Earthquakes.* New York: Van Nostrand Reinhold, 1980. Written by an experienced seismologist, this text is filled with charts, maps, and photographs that help demystify the science of seismology. Two chapters address seismic geography of the world. Especially useful for readers interested in the seismicity of New Zealand. A lucid account, suitable for a college-level reader.

Halacy, D. S. *Earthquakes: A Natural History.* Indianapolis: Bobbs-Merrill, 1974. Excellent treatments of historical earthquakes and an extensive discussion of world seismicity patterns. Written for a lay audience, the book is a lively discussion of all aspects of seismology and earthquakes, along with volcanoes and tsunamis.

Jacobs, J. A. *Deep Interior of the Earth.* 1st ed. London: Chapman and Hall, 1992. Deals in detail with all aspects of the Earth's inner and outer core. The origin of the core, its constitution, and its thermal and magnetic properties are discussed in detail. Well suited to the serious science student.

Lambert, David. *The Field Guide to Geology.* New York: Facts on File, 1988. An excellent reference for the beginning student of geology, it is filled with marvelous diagrams that make the concepts easy to understand. Several sections address earthquake distribution and related topics. Suitable for any level of reader from high school to adult.

Plummer, Charles C., and David McGeary. *Physical Geology.* Boston: McGraw-Hill, 1999. A college-level introductory geology textbook that is clearly written and wonderfully illustrated. An excellent sourcebook of basic information on geologic terminology and fundamentals of geologic processes. The chapters on structural geology and global plate tectonics are especially relevant to understanding the San Andreas fault in the context of large-scale geologic processes. Contains a CD-ROM. Three full pages are devoted to the San Andreas fault. An excellent glossary.

Shepard, Francis P. *Geological Oceanography.* New York: Crane, Russak, 1977. Chapter 2 addresses seafloor spreading and faulting of the oceanic crust, along with trenches and associated earthquake activity. Supplementary reading lists at the ends of chapters, photographs, and diagrams augment the text, which is suitable for a general audience.

Sullivan, Walter. *Continents in Motion.* New York: McGraw-Hill, 1974. Dedicated to Harry Hess and Maurice Ewing, two late pioneers of plate tectonics theory, this book is the classic popular work on moving crustal plates. Lucid explanations of seismic evidence for plate motions and historical vignettes on seismology and earthquakes. A highly readable book.

Verney, Peter. *The Earthquake Handbook.* New York: Paddington Press, 1979. Superb historical accounts of humanity's struggle to understand earthquakes with easy-to-follow discussions of seismology and important sections on earthquake safety and preparedness. Extensive discussion on the causes and distribution of seismic events.

Walker, Bryce. *Earthquake.* Alexandria, Va.: Time-Life Books, 1982. A volume in the Planet Earth series, this book is filled with color photographs and diagrams, with an essay entitled "Grand Design of a Planet in Flux" addressing plate tectonics's role in earthquake distribution. Contains an index and bibliography and is suitable for all readership levels.

Wilson, J. Tuzo, ed. *Continents Adrift and Continents Aground.* San Francisco: W. H. Freeman, 1976. Selected readings from *Scientific American* magazine that are introduced with commentary by Wilson, a leading figure in the history of plate tectonics. Includes discussion of trench earthquakes, transform faults, and collisional boundaries. A classic post-San Fernando earthquake (1971) article by Don L. Anderson on the San Andreas fault will interest students of California's seismicity. Suitable for a general audience.

EARTHQUAKE ENGINEERING

Earthquake damage and injury are aggravated by the fact that neither the time nor the location of major tremors can be precisely predicted. Damage to human-made structures may be lessened, however, through the use of proper construction techniques. Earthquake engineering studies the effects of ground movement on buildings, bridges, underground pipes, and dams and determines ways to build future structures or reinforce existing ones so that they can withstand tremors.

PRINCIPAL TERMS

EPICENTER: the central aboveground location of an Earth tremor; that is, the point of the surface directly above the hypocenter

FAILURE: in engineering terms, the fracturing or giving way of an object under stress

FAULT: a fracture or fracture zone in rock, along which the two sides have been displaced vertically or horizontally relative to each other

HYPOCENTER: the central underground location of an Earth tremor; also called the focus

NATURAL FREQUENCY: the frequency at which an object or substance will vibrate when struck or shaken

NATURAL PERIOD: the length of time of a single vibration of an object or substance when vibrating at its natural frequency

SHEAR: a stress that forces two contiguous parts of an object in a direction parallel to their plane of contact, as opposed to a stretching, compressing, or twisting force; also called shear stress

UNREINFORCED MASONRY (URM): materials not constructed with reinforced steel (for example, bricks, hollow clay tile, adobe, concrete blocks, and stone)

SOIL CONDITIONS

Earthquake engineering attempts to minimize the effects of earthquakes on large structures. Engineers study earthquake motion and its effects on structures, concentrating on the materials and construction techniques used, and recommend design concepts and methods that best permit the structures to withstand the forces.

One might logically expect that the structures nearest an earthquake fault would suffer the most damage from the earthquake. Actually, structural damage seems to bear little direct relation to the faults or to their distance from the structure. It is true that buildings near the fault are subject to rapid horizontal or vertical motion and that if the fault runs immediately beneath a structure (which is more likely in the case of a road or pipe than a building) and displaces more than a few inches, the structure could easily fail. The degree of damage, however, has more to do with the nature of the local soil between the bedrock and the surface. If the soil is noncohesive, as is sand, vibrations may cause it to compact and settle over a wide area. Compaction of the soil raises the pressure of underground water, which then flows upward and saturates the ground. This "liquefaction" of the soil causes it to flow like a fluid so that sand may become quicksand. Surface structures, and even upper layers of soil, may settle unevenly or drop suddenly. Sinkholes and landslides are possible effects.

NATURAL FREQUENCY

Ground vibration and most ground motion are caused by seismic waves. These waves are created at the earthquake's focus, where tectonic plates suddenly move along an underground fault. The waves radiate upward to the surface, causing the ground to vibrate. Wave vibrations are measured in terms of frequency—the number of waves that pass a given point per second.

Much earthquake damage depends on what is known as natural frequency. When any object is struck or vibrated by waves, it vibrates at its own frequency, regardless of the frequency of the incoming waves. All solid objects, including build-

284

ings, dams, and even the soil and bedrock of an area, have different natural frequencies. If the waves affecting the object happen to be vibrating at the object's frequency, the object's vibrations intensify dramatically—sometimes enough to shake the object apart. For this reason, an earthquake does the most damage when the predominant frequency of the ground corresponds to the natural frequency of the structures.

At one time it was thought that earthquake motion would be greater in soft ground and less in hard ground, but the truth is not that simple. Nineteenth century seismographers discovered that the natural frequency of local ground depends on the ground's particular characteristics and may vary widely from one location to another. The predominant frequency of softer ground is comparatively short, and the maximum velocity and displacement of the ground are greater. In harder ground, the predominant frequency is longer, but the acceleration of the ground is greater. When the ground is of multiple layers of different compositions, the predominant frequency is quite complex.

FREE- AND FORCED-VIBRATION TESTS

In order to determine the effect of vibrations on a building, an engineer must do the obvious: shake it. Whereas the effects on a very simple structure such as a pipe or a four-walled shack may be computed theoretically, real-life structures are composed of widely diverse materials. By inducing vibrations in a structure and measuring them with a seismograph, one can easily determine properties such as its natural frequency and its damping (the rate at which vibrations cease when the external force is removed).

The simplest type of test is the free vibration, and the oldest of these is the pull-back test. A cable is attached to the top of the test structure at one end and to the ground or the bottom of an adjacent structure at the other. The cable is pulled taut and suddenly released, causing the structure to vibrate freely. Other tests cause vibrations by striking the structure with falling weights or large pendulums or even by launching small rockets from the structure's top.

Forced-vibration tests subject test structures to an ongoing vibration, thereby giving more complete and accurate measurements of natural frequencies. In the steady-state sinusoidal excitation

test, a motor-driven rotating weight is attached to the structure, subjecting it to a constant, unidirectional force of a fixed frequency. The building's movements are recorded, and the motor's speed is then changed to a new frequency. Measurements are taken for a wide range of different frequencies and forces. Surprisingly, the natural frequencies for large multistory buildings are so low that a 150-pound person rocking back and forth will generate measurable inertia in the structure, thereby providing an adequate substitute for relatively complex equipment.

Another useful device is the vibration table: a spring-mounted platform several meters long on each side. Although designed to hold and test model structures, some tables are large enough to hold full-scale structural components—or even small structures themselves. Useful forced vibrations are also provided by underground explosions, high winds, the microtremors that are always present in the ground, and even large earthquakes themselves.

EARTHQUAKE-RESISTANT DESIGN

Structures can be designed to withstand some of the stresses put upon them by large ground vibrations. They must be able to resist bending, twisting, compression, tension, and shock. Two approaches are used in earthquake-resistant design. The first is to run dynamic tests to analyze the effects of given ground motions on test structures, determine the stresses on structural elements, and proportion the members and their connections to restrain these loads. This approach may be difficult if no record exists of a strong earthquake on the desired type of ground or if the research is done on simplified, idealized structures.

The other approach is to base the designs on the performance of past structures. Unfortunately, new buildings are often built with modern materials and techniques for which no corollary exists in older ones. It follows that earthquake-resistant design is easier to do for simple structures such as roads, shell structures, and one-story buildings than for complex skyscrapers and suspension bridges.

BASIC CONFIGURATION OF STRUCTURE

The first concern in examining a structure is its basic configuration. Buildings with an irregular floor plan, such as an "L" or "I" shape, are more likely to twist and warp than are simple rectangles and squares. Warping also tends to occur when doors and windows are nonuniform in size and arrangement. Walls can fail as a result of shear stress, out-of-plane bending, or both. They may also collapse because of the failure of the connections between the walls and the ceiling or floor. In the case of bearing walls, which support the structure, failure may in turn allow the collapse of the roof and upper floors. Nonstructural walls and partitions can be damaged by drift, which occurs when a building's roof or the floor of a given story slides farther in one direction than the floor below it does. This relative displacement between consecutive stories can also damage plastering, veneer, and windowpanes.

Lateral (sideways) crossbracing reduces drift, as do the walls that run parallel to the drift. Another way to avoid drift damage is to let the nonstructural walls "float." In this method, walls are attached only to the floor so that when the building moves laterally, the wall moves with the floor and slides freely against the ceiling. (Alternatively, floating walls may be affixed only to the ceiling.) Windows may be held in frames by nonrigid materials that allow the frames to move and twist without breaking the panes. The stiffness and durability of a wall can be improved by reinforcement. For reinforcement, steel or wooden beams are usually embedded in the wall, but other materials are used as well. If the exterior walls form a rectangular enclosure, they may be prevented from separating at the top corners by a continuous collar, or ring beam.

STRUCTURAL ELEMENTS

Frame buildings are those in which the structure is supported by internal beams and columns. These elements provide resistance against lateral forces. Frames can still fail if the columns are forced to bend too far or if the rigid joints fail. Unlike bearing walls, frame-building walls are generally nonstructural; the strength of the frame, however, can be greatly enhanced if the walls are attached to, or built integrally into, the frame. This method is called "infilled-frame" construction. Roofs and upper floors can fail when their supports fail, as mentioned above, or when they are subjected to lateral stress. An effective way to

avoid such failure is to reduce the weight of the roof, building it with light materials.

Another danger to walls is an earthquake-induced motion known as pounding, or hammering, which can occur when two adjacent walls vibrate against each other, damaging their common corners. The collision of two walls because of lateral movement or the toppling of either is also called pounding. Columns and other structural elements may also pound each other if they are close enough; in fact, the elements pounding each other may even be on adjacent buildings. If the natural vibrations of the two structures are similar enough, the structures may be tied together and thus forced to vibrate identically so that pounding is prevented. Because such closeness in vibration is rare, the best way to avoid pounding is simply to build the structures too far apart for it to occur.

An earthquake-resistant architecture that employs crisscross bracing in a tall office building in San Francisco, California. (© William E. Ferguson)

Shell structures are those with only one or two exterior surfaces, such as hemispheres, flat-roofed cylinders, and dome-topped cylinders. Such shapes are very efficient, for curved walls and roofs possess inherent strength. For this reason, they are sometimes used in low-cost buildings, without reinforced walls. When failure does occur, it is at doors and other openings or near the wall's attachment to the ground or roof, where stress is the greatest.

Much earthquake damage could be prevented if the stresses on a structure as a whole could be reduced. One of the more practical methods of stress reduction uses very rigid, hollow columns in the basement to support the ground floor. Inside these columns are flexible columns that hold the rest of the building. This engineering technique succeeds in reducing stress, but the flexible columns increase the motion of the upper stories. More exotic methods to reduce stress involve separating the foundation columns from the ground by placing them on rollers or rubber pads. Structures with several of these lines of defense are much less likely to collapse; should a vital section of crossbracing, bearing wall, or partition fail, the building can still withstand an aftershock. Overall, the earthquake resistance of a structure depends on the type of construction, geometry, mass distribution, and stiffness properties. Furthermore, any building can be weakened by improper maintenance or modification.

ALTERNATIVES TO UNREINFORCED MASONRY

Buildings using unreinforced masonry (URM) or having URM veneers have a poor history in past earthquakes. Because URM walls are neither reinforced nor structurally tied to the roof and floors, they move excessively during an earthquake and often collapse. Similarly, ground floors with open fronts and little crosswise bracing move and twist excessively, damaging the building. URM chimneys may fall to the ground or through the roof.

Buildings with URM bearing walls are now forbidden by California building codes, but URM is still common in many less developed areas of the world. There are several low-cost earthquake-resistant alternatives to such construction. Adobe walls may be reinforced with locally available bamboo, asphalt, wire mesh, or split cane. Low-cost buildings should be only one or two stories tall and should have a uniform arrangement of walls, partitions, and openings to obtain a uniform stress distribution. The floor plan should be square or rectangular or, alternatively, have a shell shape such as a dome or cylinder. Roofs and upper floors should be made of lightweight materials—wood, cane, or even plastic, rather than mud or tile—whenever possible, and heavy structural elements should never be attached to nonstructural walls.

The Center for Planning and Development Research at the University of California at Berkeley noted certain features of modern wood-frame houses that make them especially susceptible to damage from strong ground motion. In addition to URM walls or foundations, such houses may have insufficient bracing of crawl spaces, unanchored water or gas heaters, and a lack of positive connections between the wooden frame and the underlying foundations. Porches, decks, and other protruding features may be poorly braced. Most of these deficiencies can be corrected.

PROTECTION AGAINST INJURY AND PROPERTY DAMAGE

The earthquake is arguably the most destructive natural disaster on the planet. No other force has the potential to devastate so large an area in a very short time. Not only can it not be predicted, but there is also even less advance warning for the earthquake than for other types of disaster. A hurricane can be seen coming by radar; a volcano may belch smoke before it erupts. An earthquake simply happens.

Yet the magnitude of the earthquake is not solely responsible for the destruction. Property damage and injury to humans also depend on the type and quality of construction, soil conditions, the nature of the ground motion, and distance from the epicenter. The tremor which struck Anchorage, Alaska, in 1964 measured 8.3 on the Richter scale and killed eleven people; on the other hand, one that hit San Fernando, California, in 1971 measured only 6.6—less than a tenth of the force of the Anchorage quake—and fifty-nine people died. Most of the San Fernando deaths occurred in one building: a hospital that collapsed. It seems likely that the hospital had not been adequately constructed to withstand the stresses to which it was suddenly subjected. The higher damage toll resulted from the soil characteristics in

San Fernando and an underground fault that had previously been unmapped.

The only protection earthquake-zone residents have against property damage is that given by the engineers who design and build their buildings, railway structures, dams, harbor facilities, and (especially) nuclear power plants and by the public officials who regulate them. Now that engineers can learn how ground movement affects engineering structures and can design new ones accordingly, many of the earthquake-prone regions have building codes for resistant construction. Laws and programs exist to determine which buildings are unsafe and how they may be made resistant. Unfortunately, not all quake regions have such rules and programs in place, because of apathy, high cost, or other reasons. The high costs of recovery after major quakes, however, certainly provide a compelling rationale for better preparation.

Shawn V. Wilson

CROSS-REFERENCES

BIBLIOGRAPHY

Center for Planning and Development Research. *An Earthquake Advisor's Handbook for Wood Frame Houses.* Berkeley: University of California, 1982. This slim book was a result of the Earthquake Advisory Service Project at the Center for Planning and Development Research at the University of California, Berkeley. The book is addressed both to the homeowner and to public policy personnel who want to plan an earthquake advisory project. Includes a long, diagrammed how-to section on repairs. Suitable for all readers.

Coburn, Andrew. *Earthquake Protection.* New York: Wiley, 1992. Coburn examines earthquake engineering and prediction, along with safety procedures. He also offers a historical look at seismic activity and earthquake hazard analysis.

Collier, Michael. *A Land in Motion: California's San Andreas Fault.* San Francisco: Golden Gates National Parks Association, 1999. Filled with beautiful color photographs that accompany text intended for the nonscientist, *A Land in Motion* gives the reader excellent insight into earthquakes and their aftermaths. There are also many diagrams and graphs that explain subduction, faults, and orogeny.

Kanai, Kiyoshi. *Engineering Seismology.* Tokyo: University of Tokyo Press, 1983. This book begins with a discussion of the seismograph and proceeds logically through seismic waves, ground vibrations, and their effects on structures. Concise, but the calculus is difficult in spots. Suitable for college-level students.

Kramer, Steven Lawrence. *Geotechnical Earthquake Engineering.* Upper Saddle River, N.J.: Prentice Hall, 1996. Intended for a person with some civil engineering background, this book can be quite technical at times. Kramer provides clear but advanced descriptions of the many aspects of seismic engineering.

Newmark, Nathan M., and Emilio Rosenblueth. *Fundamentals of Earthquake Engineering.* Englewood Cliffs, N.J.: Prentice-Hall, 1971. A highly technical, mostly theoretical graduate textbook and reference manual. Addresses basic dynamics, earthquake behavior, and recommended design concepts. Text is occasionally unclear, even where the language is not especially technical. Some diagrams are given inadequate explanation.

Okamoto, Shunzo. *Introduction to Earthquake En-*

gineering. New York: Halsted Press, 1973. This textbook has chapters on earthquakes, earthquake-resistant design procedures, and earthquake resistance of roads, tunnels, railways, bridges, and various types of dams. The chapters on seismicity and on historical earthquakes are based on data from Japan, one of the world's most active seismic zones. Written for engineers and college-level students, but the many nontechnical passages are not too difficult to follow.

Powell, Robert, R. J. Weldon, Jonathan Matti, et al., eds. *The San Andreas Fault System: Displacement, Palinspastic Reconstruction, and Geological Evolution.* Boulder, Colo.: Geological Society of America, 1993. This book provides a clear description of the power and size of the San Andreas fault. It details the history of the fault, as well as its constant evolution. Illustrations and folded leaves; includes bibliography and index.

Reps, William F., and Emil Simiu. *Design, Siting, and Construction of Low-Cost Housing and Community Buildings to Better Withstand Earthquakes and Windstorms.* Washington, D.C.: U.S. Department of Commerce/National Bureau of Standards, 1974. A report prepared for the U.S. Agency for International Development on the construction of small buildings in earthquake- and windstorm-prone areas of the world. Explains concisely and accessibly the forces put on small buildings, the effects they have, and construction techniques to prevent them. It would be a good primer on earthquake engineering in general were it not for its necessary focus on small, low-cost buildings and ways to build them using inexpensive and locally available materials. Written for a general audience.

Tierney, Kathleen J. *Report of the Coalinga Earthquake of May 2, 1983.* Sacramento, Calif.: Seismic Safety Commission, 1985. This manual focuses on the Coalinga earthquake to show how an earthquake can affect a community physically, financially, and socially and how local and state governments can deal with the aftereffects. For all readers.

Wiegel, Robert L., ed. *Earthquake Engineering.* Englewood Cliffs, N.J.: Prentice-Hall, 1969. A large and comprehensive volume on the subject. Based originally on lectures given in a University of California course on engineering. Includes data on earthquake causes, ground motion and effects, mathematical modeling and theory, tests and effects on structures, and design of earthquake-resistant structures. Aimed at students and professionals.

EARTHQUAKE HAZARDS

Over the past four thousand years, about 13 million persons have died as a result of earthquake activity, and an unknown amount of property damage has occurred as well.

PRINCIPAL TERMS

CREEP: the very slow downhill movement of soil and rock

EPICENTER: the point on the surface of the Earth directly above the focus

FAULT: a fracture or zone of breakage in a rock mass which shows movement or displacement

FOCUS: the point below the surface of the ground where the earthquake originates and its energy is released

INTENSITY: an arbitrary measure of an earthquake's effect on people and buildings, based on the modified Mercalli scale

LANDSLIDE: the rapid downhill movement of soil and rock

LIQUEFACTION: the loss in cohesiveness of water-saturated soil as a result of ground shaking caused by an earthquake

MAGNITUDE: a measure of the amount of energy released by an earthquake, based on the Richter scale

SUBSIDENCE: the sinking of the surface of the land

TSUNAMI: a seismic sea wave created by an undersea earthquake, a violent volcanic eruption, or a landslide at sea

EARTHQUAKE OCCURRENCE

Earthquakes are the result of the rapid motion and vibrations caused by movement of the ground along a fracture in a rock or along a fault. Movement occurs when rocks are unable to store any more stress, at which time they reach their breaking point, release energy, and create an earthquake. The point of origin of an earthquake below the surface where its energy is released is known as the focus. The focus can be located at either a shallow or a deep depth. The point on the surface of the Earth directly above the focus is called the epicenter; it is the spot frequently cited by the news media as the location of the earthquake.

Throughout the Earth's surface, numerous faults and fault systems exist, but the larger faults are usually confined to specific areas. Earthquakes and faults are not hazardous in themselves, but they can become hazardous when they directly endanger humans and their immediate environment. Each year, the Earth is subjected to at least 1 million earthquakes. Only a few, however, are strong enough to cause major structural damage to cities and to kill or injure thousands of persons. The major hazards directly created by an earthquake are ground shaking, ground rupture, and tsunamis. The major indirect hazards are fires, floods, building collapse, disruption of public services, and psychological effects.

GROUND SHAKING

Ground shaking occurs as energy released by the earthquake reaches the surface and causes the materials through which it passes to vibrate. The intensity of these vibrations and of the shaking at the surface depends on several factors: the amount of energy released, the depth of the focus, and the type of material through which the energy is moving. The closer the focus is to the surface, the more powerful the earthquake. Also, the denser the material, the more the vibrations will be felt and the stronger will be the resulting ground motion. There are a few documented cases in which very strong ground motion actually caused parked cars to bounce along the road, the surface of the land to move in rippling waves, and trees to become uprooted and snap. Yet, damage to open, uninhabited land is usually minimal. The amount of damage to buildings subjected to strong ground motion is controlled by many complex and interacting factors: the buildings' method of construction, the types of building material used, the depth of the bedrock, the distance from the epicenter, and the duration and intensity of the shaking. Buildings constructed on thin, firm soil and solid bedrock will fare much better than those on thick, soft soil and deep bedrock; however, if the shaking's duration is great, even the most well-constructed building will be destroyed. Such a situation was observed in the earthquake that struck Mexico City in 1985, in which more than one thousand buildings were destroyed and ten thousand persons were killed. This city was built on ancient lake-bed deposits of sand and silt that rapidly lost rigidity as a result of intense shaking, causing tall buildings to collapse vertically, one floor on top of the other.

Four other very important elements determine the amount of destruction: the degree of compaction of the soil or bedrock on which the buildings' foundations are resting, the amount of water saturation of this material, its overall chemical composition, and the buildings' physical structure. If construction took place on or within solid bedrock, then the structures would move as a unit and would suffer much less damage. Some buildings may be able to withstand severe shaking for a few seconds, although prolonged shaking will completely destroy them. Ground shaking in the 1964 Alaskan earthquake lasted for about four minutes, causing major damage to the sturdiest of buildings. On the other hand, a particular building may easily withstand the effects of shaking but be destroyed by other factors. In Soviet Armenia during December of 1988, about twenty-five thousand persons were killed because of the effects of multiple aftershocks that shook apart poorly designed structures and buildings that were designed to withstand a lesser degree of ground motion. Moreover, very strong ground motion can knock a building completely off its bedrock foundation, rendering it unusable, or a building may fall prey to other types of ground failure triggered by an earthquake.

GROUND FAILURE

Ground failure includes landslides, avalanches, fault scarps, fissuring, subsidence, uplift, creep, sand boils, and liquefaction. Areas such as mountain valleys and regions surrounding an ocean bay can be subjected to these kinds of failure, since they usually consist of recently deposited, fine-grained materials that have not yet been completely compacted or that show variable degrees of groundwater saturation.

Landslides occur when unstable soil and rock move rapidly downslope under the influence of gravity; they usually are started by an earthquake. They commonly occur on steep slopes but can also move down gentle inclines. An avalanche is similar to a landslide but consists of snow and ice mixed with rock and soil. In either case, these masses move with great rapidity and force, sometimes filling, burying, or excavating the land along their path. Some earthquake-induced avalanches have been clocked at velocities of more than 320 kilometers per hour. The Tadzhik Soviet Socialist Republic, in late January of 1988, was hit with an earthquake of a 5.4 magnitude on the Richter scale that shook the ground for almost forty seconds, unleashing a massive landslide that was 8 kilometers wide. It buried the nearby village of Okuli-Bolo with mud to a depth of 15 meters, killing between six hundred and one thousand persons.

The collapsed Cypress viaduct on Interstate 880 after the 1989 Loma Prieta earthquake, Oakland, California. (U.S. Geological Survey)

Fault scarps are created when a fault intersects the surface of the Earth and large chunks of the ground are uplifted or dropped. Within these chunks, deep ground cracks known as fissures appear. Film portrayals of earthquakes notwithstanding, there is no chance that the Earth will open, close, and "swallow" anything during an earthquake. Deep ground cracks commonly remain open, since the forces that created them operate only in one direction. Sometimes, however, animals fall into these cracks and appear to have been swallowed. In many places, buildings, roads, and other structures are constructed across an old fault scarp, and they undergo extensive structural damage from vertical or horizontal ground displacement. The largest measured vertical displacement along a fault scarp is 15 meters; the largest horizontal displacement that occurred at one time is 6 meters. The rate of displacement is variable, however, and some faults can show a slow but accumulated horizontal displacement of several kilometers.

The subsidence or depression of the surface of the land may occur when underground fluids such as oil or water are removed or drained by a nearby fault; the land sinks, creating water-filled sag ponds. This process occurs over a long period but does not result in any loss of life. Another slow form of ground failure is earthquake creep, which occurs more or less continuously along a fault. Creep is really an earthquake in slow motion; stored energy is gradually released in the form of very small earthquakes, or microearthquakes, causing the land to move in opposite horizontal directions. It results in the slow bending and breaking of underground pipes and railroad tracks and causes concrete building foundations to crack. The Hayward fault near San Francisco Bay, California, is a prime example.

Under the worst conditions—a low degree of compaction, thick and fine-grained sandy or silty soil, a high degree of water saturation, and intense ground shaking—solid land loses its cohesiveness and strength. It then begins to liquefy and flow by a process known as liquefaction. Tall buildings slowly move as if they were built on shifting sand. An unusual, localized ground failure event related to liquefaction is a sand boil. Rapid ground motion can cause a pressurized mixture of sand and concentrated groundwater to make its way toward the surface and create a small volcano-like mound of sand that spouts sandy water.

TSUNAMIS

A spectacular and extremely hazardous coastal event is the seismic sea wave, or tsunami (sometimes incorrectly called a "tidal wave"). Tsunamis are usually produced by undersea earthquakes; the seafloor undergoes rapid vertical motion along one or more active faults, and energy is transmitted directly from solid rock into the seawater. They can also originate with massive undersea landslides or the violent eruption of an oceanic volcano. The catastrophic eruption of the volcanic island of Krakatau in 1883 created a series of waves more than 40 meters high that drowned an estimated thirty-six thousand persons who lived along the low-lying coastal areas of Java and Sumatra. The exact cause of all these large waves is not completely understood by geologists and oceanographers. Regardless of their cause, the

Close-up of the damage to roads by the 1971 San Fernando earhquake in California. (U.S. Geological Survey)

waves begin to radiate away from their point of origin in a manner similar to when a stone is tossed into a quiet body of water. Generally, tsunamis have a wavelength, or distance between successive wave crests, of greater than 161 kilometers and move at a velocity of more than 966 kilometers per hour; however, these large and fast-moving walls of water are not observable as such in the deep ocean, becoming visible only as they enter shallow water. Here the trough of the wave encounters the sea floor and begins to slow down, allowing the crest to build in size. At this point, water is rapidly sucked out of inland bays and harbors to feed the increasing mass, as it comes roaring into the mainland, drowning people and smashing buildings—in some cases as far inland as 3 kilometers. For example, after tossing trains and fishing boats almost 1 kilometer inland, the seismic sea waves produced from the 1964 Alaskan earthquake traveled south many thousands of kilometers as far as Crescent City, California, where local surfers decided to challenge the now 6-meter-high waves. The east coast of Japan, the Hawaiian Islands, the west coast of the United States, Alaska, Chile, Peru, and most other Pacific coastal regions have suffered damage from these powerful waves. They are rare in the Atlantic Ocean; the last major Atlantic tsunami occurred in 1755, when an offshore earthquake (the All Saints Day earthquake) generated a series of waves that hit the coast of Portugal, killing an estimated fourteen thousand persons.

Not as spectacular, but similar to tsunamis, are seiches. Although they also have several origins, they can be produced by an earthquake. Seiches are small, oscillating waves that may travel for several hours, back and forth within the enclosed basin of a lake or reservoir, sometimes causing flooding and minor structural damage to nearby buildings.

INDIRECT HAZARDS

One of the most deadly indirect hazards from an earthquake is fire. Fire has claimed the most number of victims and caused more property damage than all the direct hazards combined. In 1906, the San Francisco earthquake, better known as the San Francisco fire, killed five hundred people, destroyed twenty-five thousand buildings, and burned 12.2 square kilometers of the city. Fires are started by the sparking of downed electrical wires that re-

sults in the ignition of ruptured gas lines. They are difficult, if not impossible, to control, since water pressure in hydrants may be low or nonexistent because of the breakage of underground waterpipes. Flooding is another indirect hazard. Although the risk of flooding is usually minimal in a seismically active area, the potential failure of large concrete or earthen dams and reservoirs poses a great threat to nearby life and property. During the 1971 San Fernando earthquake in Southern California, for example, the lower part of the Van Norman earthen dam partially gave way, threatening the eighty thousand people who lived in the surrounding area. If the shaking had continued for another minute, it would have been disastrous for the local community.

The danger of being trapped in a collapsing building during an earthquake is real. No building is "earthquake proof," but steps are being taken to make existing buildings more secure. Most concrete and steel-reinforced buildings built on solid bedrock and most one- or two-story wooden frame houses suffer little or no damage in an earthquake, provided that the ground motion is not too protracted or severe. Generally, older buildings suffer much more damage than newer ones. One of the safest places to be during an earthquake is in a doorway.

STRUCTURAL DESIGN

Although earthquakes cannot be prevented, the dangers that they pose may be either eliminated or reduced. Engineers and city planners want to design buildings that will withstand a certain amount of ground shaking. As structural design improves, the loss of life decreases.

Yet, it is very difficult to predict the effect of ground motion on a building design. To design a more flexible building, engineers must perform tests on scale models using simulated or actual samples of the local bedrock and the construction materials. In the state of California, for example, legislation requires the removal of overhanging ledges and the reinforcement of key structural supports in buildings, bridges, and highway overpasses. The installation of numerous cutoff valves on gas, water, and sewage lines can localize and minimize service disruptions, while above-ground pipes, roads, and power lines built across active faults have been designed to anticipate fault motion.

LAND-USE POLICY

The geologist, however, has the primary responsibility for the gathering of geologic information needed for the accurate assessment of the seismic risk for an area. This information can be obtained by the drafting of a geologic map of the region that includes an accurate tracing of all known faults and fault-derived topographic features, such as scarps and sag ponds, and the identity of the rock types involved. Such a map can help geologists to predict where an earthquake will most likely occur, but it cannot predict the earthquake's intensity, frequency, or time of occurrence without further data. This information may be gained through studies of the long-term movement history of existing faults, the determining of their relative ages, the monitoring of current fault motion, and the detection of previously unknown faults. Once the information is assimilated, an estimate of a possible earthquake's magnitude on the Richter scale, its epicenter, its intensity on the modified Mercalli scale, and the amount of horizontal and vertical ground movement can be made.

If the geologic data indicate that an area may suffer an unacceptable amount of destruction, alternative land-use policies for this area should be adopted. Such policies involve the establishment of a fault hazard easement, whereby construction is restricted to a certain minimum distance from the nearest fault trace or fault zone. Geologic hazard zoning would also identify areas affected by past landslides, floods, and seismic sea waves. In many regions, however, much urban or industrial development already exists in hazardous areas, simply because the danger was not recognized at the time of construction.

Since the recurrence interval between large earthquakes is very long or poorly known, the largest potential hazards exist in areas that have suffered little or no seismic activity. Local governments in areas such as New York City, South Carolina, and Missouri have given little thought to earthquake disaster planning. The New Madrid, Missouri, earthquake of 1811-1812 was felt over a sparsely populated area of 2.6 million square kilometers; large sections of the ground were uplifted, others sank, deep cracks appeared in the ground, and bells were caused to ring in church towers as far away as Boston. It was the most powerful earthquake ever to strike the eastern half of the United States, and it occurred in an area thought to be earthquake-free. If a major earthquake were to strike the East Coast cities today, the property damage and loss of life would be tremendous.

HUMANKIND'S ROLE IN EARTHQUAKES

As the population of the world continues to grow and compete for living space, more and more areas, once considered hazard-free because of their lack of development, are becoming inhabited and may suffer damage from potential movements of the Earth. Therefore, research in the area of earthquake control and prediction is growing more important.

Humans' ability to trigger an earthquake was discovered during the early 1970's at the Rocky Mountain Arsenal in Denver, Colorado. There, liquid wastes were disposed by high-pressure injection into wells that were drilled to 3,600 meters below the surface. These liquids reduced the pressure along deep faults, allowing them to slip and causing an increase in the number of minor earthquakes in the area. When the pumping stopped, the number of earthquakes decreased, and when pumping was continued, the tremors began again. A careful study proved that the number of recorded earthquakes was statistically higher than that which would normally be expected. Human-made earthquakes were also created through underground nuclear explosions and the filling of water reservoirs behind major dams. Some geologists believe that these processes may relieve the pressure along faults and prevent a major earthquake from occurring.

EARTHQUAKE PREDICTION

The ability to predict an earthquake's time and place is based entirely on the quality of geologic data available for a given area. Geologists look at historic evidence of a fault's movement and at its current activity. The problem lies in the brevity of the record-keeping period. The first seismometer was built around 1889, so the science of seismology is relatively new, and data have been collected on some faults for less than one hundred years. Many more faults have been discovered since that time, and not enough information is available to give a reliable estimate of the seismic activity of a region.

Most earthquakes are preceded by warning

signs. Some of these indicators are local ground swelling, an increase in the number and frequency of minor tremors, an increase in the amount of radioactive radon gas in water wells, and unusual animal activity. The problem is that not all earthquakes have these precursors, and they are not always reliable. Moreover, new hazards are created by the prediction of an earthquake. A short-term prediction may cause major traffic jams, panic, riots, and looting. Long-term predictions may cause property values to drop, create a decrease in tourism, and cause the gradual abandonment or economic depression of nearby cities. If the prediction is wrong, the public is unlikely to trust any future predictions, and major lawsuits may arise from injuries or damages resulting from evacuation.

Steven C. Okulewicz

CROSS-REFERENCES

Dams and Flood Control, 2016; Deep-Focus Earthquakes, 271; Earthquake Distribution, 277; Earthquake Engineering, 284; Earthquake Locating, 296; Earthquake Magnitudes and Intensities, 301; Earthquake Prediction, 309; Earthquakes, 316; Elastic Waves, 202; Faults: Normal, 213; Faults: Strike-Slip, 220; Faults: Thrust, 226; Faults: Transform, 232; Land Management, 1484; Land-Use Planning, 1490; Land-Use Planning in Coastal Zones, 1495; Landslides and Slope Stability, 1501; Notable Earthquakes, 321; Plate Margins, 73; Plate Motions, 80; Plate Tectonics, 86; San Andreas Fault, 238; Seismometers, 258; Slow Earthquakes, 329; Soil Liquefaction, 334; Thrust Belts, 644; Tsunamis and Earthquakes, 340.

BIBLIOGRAPHY

Cargo, David N., and Bob F. Mallory. *Man and His Geologic Environment.* 2d ed. Reading, Mass.: Addison-Wesley, 1977. Chapter 10 of this college textbook covers earthquakes and volcanoes.

Clarke, Thurston. *California Fault: Searching for the Spirit of State Along the San Andreas.* New York: Ballantine Books, 1996. Clarke traveled the length of the San Andreas fault collecting first-hand accounts from earthquake survivors and predictors. Along with the entertaining stories, Clarke provides historical and scientific information about the fault.

Collier, Michael. *A Land in Motion: California's San Andreas Fault.* San Francisco: Golden Gates National Parks Association, 1999. Filled with beautiful color photographs that accompany text intended for the nonscientist, *A Land in Motion* gives the reader excellent insight into earthquakes and their aftermaths. There are also many diagrams and graphs that explain subduction, faults, and orogeny.

Griggs, Gary B., and John A. Gilchrist. *Geologic Hazards, Resources, and Environmental Planning.* 2d ed. Belmont, Calif.: Wadsworth, 1983. Chapter 2 thoroughly covers geologic hazards related to earthquakes and faulting. Supplemented by an extensive bibliography and

many photographs, charts, and maps. Highly recommended.

Howard, Arthur D., and Irwin Remson. *Geology in Environmental Planning.* New York: McGraw-Hill, 1978. Chapter 8, "Earthquakes and the Environment," contains a wide-ranging discussion of earthquakes and their distributions, hazards, prediction, and warning signs. Supplemented with many photographs and line drawings relating to famous earthquakes.

Keller, Edward A. *Environmental Geology.* 5th ed. Columbus, Ohio: Merrill, 1988. Chapter 8 covers earthquakes and related phenomena in a well-illustrated and well-written discussion.

Sharpton, Virgil L., and Peter D. Ward, eds. *Global Catastrophes in Earth History.* Boulder, Colo.: Geological Society of America, 1990. A compilation of fifty-eight papers on various natural disasters. Accessible to readers with some scientific background. Bibliography and index.

Utgard, R. O., G. D. McKenzie, and D. Foley. *Geology in the Urban Environment.* Minneapolis: Burgess, 1978. Chapter 12, "Seismic Hazards and Land-Use Planning," is a brief outline of earthquake hazards. Chapter 13, "The Status of Earthquake Prediction," presents a quick overview of research in the field. Both chapters are written for the general reader.

EARTHQUAKE LOCATING

Earthquake locating requires a network of seismographs. The tremors felt at the Earth's surface originate at depth by sudden jerky motions of faults under great pressures. Finding earthquakes requires measuring depths and determining geographic locations of the source zones within the Earth where the rocks ruptured.

PRINCIPAL TERMS

CORE: the innermost portion of the Earth's interior; it measures 2,900 kilometers in diameter

CRUST: the upper 30-60 kilometers of the Earth's rocks in which shallow earthquakes occur; it is thickest under continents, thinnest beneath oceans

EPICENTER: the point on the Earth's surface directly above the focus of an earthquake

FAULT: a crack in the Earth's crust where one side has moved relative to the other

FOCUS: the point beneath the Earth's surface where rocks have suddenly fractured along a fault zone, generating a train of seismic waves that travel through the Earth and that are experienced at the surface as an earthquake

MANTLE: the portion of the Earth's interior between the Moho and the core, from 30-60 kilometers to 2,900 kilometers deep

MOHOROVIČIĆ DISCONTINUITY (MOHO): the lower boundary of the crust, 30-60 kilometers deep on the average; the velocities of P

waves and S waves both increase sharply just below the Moho

P WAVE: a type of seismic wave generated at the focus of an earthquake traveling 6-8 kilometers per second, with a push-pull vibratory motion parallel to the direction of propagation; "P" stands for "primary," as P waves are the fastest and first to arrive at a seismic station

S WAVE: a type of seismic wave generated at the focus of an earthquake traveling 3.5-4.8 kilometers per second, with a shear or transverse vibratory motion perpendicular to the direction of propagation; "S" stands for "secondary" because S waves are usually the second to arrive at a seismic station

SEISMOGRAPH: a sensitive instrument that detects vibrations at the Earth's surface and records their arrival times, amplitudes, and directions of motion

DETERMINING EARTHQUAKE SOURCE REGION

Even though tremors from a single earthquake may be felt for hundreds of kilometers and recorded throughout the world, they always begin at a point or very small region within the Earth. Earthquakes are caused when pressures build up to the point that the rocks cannot withstand any more stress, and they snap and move to adjust to the stress. As they rupture, they either form a crack or move along an existing fault, with one side moving with crushing force against the other. This violent internal rubbing creates vibrations that propagate away from the disturbance and ripple across the surface of the Earth.

When a violin bow is rubbed against a violin string at a point, the whole string vibrates, and as it vibrates, energy is transmitted to the surround-

ing air, sending out sonic, or sound, waves. Similarly, when rocks of the Earth's interior rapidly rub against one another at a point, all the neighboring rocks vibrate. As they vibrate, they transmit energy upwards and through the Earth. This energy sends out seismic waves felt and measured as tremors of an earthquake throughout a region. Sometimes, if an earthquake is strong enough, seismographs in every area of the world will measure it.

Locating an earthquake means to find the region within the Earth where the rocks ruptured and the vibrations started. For small quakes, the source region is no more than a few meters in size. For very large earthquakes, the source region may be hundreds of meters and even a kilometer or more in dimension. In the case of large quakes, however, the entire area of disturbed fault motion does not move at once. The earthquake still starts at some point in the fault region, then the disturbance moves away from point to point in a chain reaction until the entire stressed region has adjusted. The actual fault motion may occur in one or two seconds for small quakes or extend over a minute or more for the largest quakes.

FOCUS AND EPICENTER

The point within the Earth where the fault began its motion is called the focus, or hypocenter, of the earthquake. The focus is where the rocks break. The geographic location of the point vertically above the focus on the Earth's surface is called the epicenter. The epicenter is thus the point on the Earth's surface nearest to the focus. It is also the place where the most intense ground motions are usually experienced—but not always. Sometimes, because of the peculiarities of underlying geology, the most violent vibrations experienced above ground are displaced a few kilometers from the epicenter. For this reason, an array of sensitive instruments called seismographs is needed to locate the true epicenter and focus. Surface expressions of an earthquake can be very misleading. The earthquake of September 19, 1985, that devastated Mexico City actually had its epicenter some 400 kilometers away beneath the Pacific Ocean. Because of the geologic peculiarities in that part of the world, Mexico City, at a considerable distance from the source, was more severely shaken than was Acapulco, which was much closer.

Earthquakes are generated from a point or small restricted region within the Earth, but they send out waves that can be felt for hundreds of kilometers and recorded by seismographs throughout the entire world. For example, the great New Madrid, Missouri, earthquakes of 1811-1812 were felt from the Rocky Mountains to the Atlantic seaboard, but their epicenters were in the Mississippi River valley of the midwestern United States.

SEISMIC STATIONS

Among seismologists, "earthquake locating" means finding the focus and epicenter, which requires seismic instrumentation. Finding the regions of strongest shaking and heaviest damage is another kind of investigation in which seismologists also engage, but it does not require seismographs.

Earthquakes can be located by well-calibrated seismographs by noting the times that various seismic waves arrive at different stations. By knowing the speeds at which P waves, S waves, and other seismic waves travel through the Earth and the precise times they arrive at several stations, distances and directions can be calculated and the earthquake epicenter and depth determined. A minimum of three seismic stations scattered around an epicenter are needed to determine a location. At least one station near the epicenter is needed to estimate the depth accurately. For truly reliable and precise measurements within less than 0.1 kilometer, a dozen or more stations are needed at varying distances surrounding the focal region.

All seismic stations are timed to universal time referenced to the zeroth meridian passing through Greenwich, England. This time is broadcast by shortwave radio stations, such as WWV radio in Boulder, Colorado, which broadcasts every second of time, twenty-four hours a day. The time is accurate to within billionths of a second. Seismic stations are equipped with special radios to receive this information and to set their seismographs on a daily basis.

Networks of seismographs can be found throughout the world, including a worldwide net received by radio, satellite, and telephone lines in Golden, Colorado, by the U.S. Geological Survey's National Earthquake Information Service. Smaller, more local nets exist around the major faults and seismic zones in the country, including the San Andreas fault system of California; the Puget Sound area of Washington State; the Wasatch fault of Utah; the New Madrid fault of Arkansas, Missouri, and Tennessee; the Charleston, South Carolina, seismic zone; and several seismic areas of New England. Locating earthquakes is a cooperative effort between many seismic stations and scientists throughout the country and the world, including those associated with universities, government facilities, and private corporations. In earthquake seismology, the whole world is the laboratory. Sharing of information—irrespective of state, provincial, or national boundaries—is required to understand, study, and locate earthquakes.

P–S DELAY TIME TECHNIQUE

The easiest way to locate an earthquake is when three stations are located in a triangle around its source regions. For example, imagine that an earthquake occurs at exactly 06:00 hours universal time. At the instant the fault rips, two kinds of waves are generated: P waves and S waves. Velocities of P waves in the upper crust of the Earth are about 7 kilometers per second, while those of S waves are typically 40 percent less, or 4.2 kilometers per second. When the P wave has traveled 70 kilometers in 10 seconds, the S wave has traveled 42 kilometers. To reach the point the P wave reached in 10 seconds, the S wave will take 16.7 seconds, or 6.7 seconds longer. If a seismic station were at this site, it would record both waves and note that the S wave lost the "race" by 6.7 seconds. Another station 100 kilometers away in another direction would also note the arrival of the P wave and S wave. To travel 100 kilometers, the P wave would take 14.3 seconds, while the S wave would take 23.8 seconds, or 9.5 seconds longer than the P wave. If, in a third direction from the earthquake, another seismic station were situated 160 kilometers away, it would receive both the P and S waves at later times: 06 hours 22.9 seconds and 06 hours 38.1 seconds, respectively. The third seismic station would note a "P minus S" (P – S) arrival time difference of 15.2 seconds. Although the three seismic stations recording the arrivals of P and S waves would not identify where the earthquake had occurred nor the time of its origin, seismologists could calculate the exact time of arrival in universal units to the nearest tenth of a second or better. Even more useful, they would know precisely the difference in arrival times of the P and S waves, which could be read from each station's seismograms.

Each seismographic observatory has sets of empirically determined travel times for P and S waves for various distances from its station. The first station in the example, at 70 kilometers' distance, would refer to the travel time tables and see that for a difference of 6.7 seconds between P and S waves, the event had to be about 70 kilometers away. Seismologists would not know, however, in which direction the waves had come. The staff at the station could draw a circle on a map 70 kilometers in radius, with the location of the station at the center, knowing that somewhere on that circle the earthquake had occurred. By contacting the station located 100 kilometers from the epicenter, the staff at the first station would learn that the

seismograph at the second station noted a 9.5-second P – S difference. The staff could then refer to the travel time charts to discover that a 9.5-delay time corresponds to 100 kilometers, thus defining a second circle, of that radius, centered on the second station. Similarly, the staff at the first station could contact the third station and obtain another P – S delay time and find that the 15.2-second difference observed there implied a 160-kilometer distance of that station from the earthquake focus.

DETERMINING DEPTH

Plotting the three circles centered on the three station locations with the appropriate radii, the seismologists would find that all three intersect at a single point. That point would be the epicenter. To obtain the depth requires some analysis of the P – S delay times and how closely the three circles intersect precisely at a point. By assuming a depth, calculating the arrival times to the stations, and comparing them to the real arrival times, the seismologist can judge how realistic the assumed depth was. By assuming various depths and repeating the calculations until a close correspondence with real arrival times results, the depth can be said to have been determined.

The best way to determine a depth is by having a station very near the epicenter. In this way, the distance from the focus at depth to the station recording at the surface is an approximation of the depth below the epicenter. The most common way to determine depth, however, is by inputting measurements from many stations into a computer and repeating trial calculations until they best resemble the data for a depth determination. Epicenters are also located by the mathematical convergence of data from many stations, even though theoretically only three stations are required.

LIMITATIONS OF P–S DELAY TIME METHOD

The P–S delay time technique is the original and most simple way to locate earthquakes. This method does not work in all cases, however; even when it does work, beyond 500 kilometers, the curvature of the Earth becomes a contending factor. The simple triangulation method works best for shallow quakes less than one-quarter of the circumference of the world away, or less than 10,000 kilometers away.

A shallow earthquake is one that occurs in the

crust less than 60 kilometers deep. Most active faults do not extend visibly up through the surface (as does the San Andreas fault in California) but lie buried beneath layers of rock and soil. The deepest earthquakes are 300-700 kilometers deep; those between 300 and 60 kilometers deep are termed "intermediate" by seismologists. Deep earthquakes only occur in certain places, mostly around the perimeter of the Pacific Ocean. For a deep earthquake, even a station at the epicenter would be hundreds of kilometers away, and the seismic energy might take an entire minute to propagate from the focus to the surface.

There are several reasons the P – S delay time triangulation method cannot always work. One reason is that the core of the Earth acts like a liquid, not a solid. Liquids do not allow S waves to propagate. Hence, beyond a certain distance around the Earth, direct S waves cannot be received. The core also bends and diffracts P waves, even though P waves do propagate readily through the core. Because of the inner structure of the Earth, receiving P and S waves directly from the focal source zone at a seismic recorder is not possible. Instead of P waves and S waves being the first and second waves to arrive at a station, other kinds of seismic waves—including P and S waves that have been reflected, refracted, and diffracted along complex pathways—will arrive first. When the direct P and S waves are not received by a seismic station, other waves can be used to determine the epicentral distance in an analogous fashion. Hence, seismologists have tables and charts not only for direct P and S wave travel times to their station but also for more than a dozen other ray paths and wave types. By reading all such data from their seismograms and applying multiple travel time calculations, seismologists can determine more precisely depths and epicentral distances.

In the early days of seismology, near the beginning of the twentieth century, earthquake epicenters were found by using large spherical globes, thumb tacks, and pieces of string to strike off intersecting radii. Presently, epicenters and depths are found by sophisticated computer programs that consider the data of numerous stations and numerous phases of the various kinds of seismic waves and their possible wave paths through the Earth.

APPLICATION OF SEISMOGRAPHIC DATA

The science of locating earthquakes, as it has developed over the past 100 years, has been responsible for providing most of what is known about the Earth's interior. The Moho, the thickness of the crust, the Earth's mantle, the liquid core, and even the inner solid core floating within the fluid outer core—all of these have been deduced from seismograms taken from all over the world. The motions of the Earth's crustal plates are also observed by the analysis of seismograms. Locating an earthquake by a seismogram determines much more than just when and where the rocks ruptured. The amount of energy released and the relative directions of motion that occurred on both sides of the fault can also be determined.

The installation by the United States of a worldwide network of seismographs has made it possible over the decades to monitor the underground nuclear experiments of the Soviet Union and other countries. An underground nuclear blast has many of the earmarks of an earthquake and can be located by the same methodologies, but it also has distinctively different seismic characteristics. Studying seismograms of nuclear blasts has helped refine understanding of the Earth's inner makeup. Another practical application of earthquake locating is measurement of the tremors associated with volcanic magma moving toward the surface prior to an eruption. Volcanic eruption predictions are based on such data.

David Stewart

CROSS-REFERENCES

BIBLIOGRAPHY

Bolt, Bruce A. *Earthquakes.* New York: W. H. Freeman, 1988. An elementary treatment of earthquakes in general. Could be used by lower-level college students who are nonscience majors.

_____. *Inside the Earth.* New York: W. H. Freeman, 1982. An excellent introduction to seismic waves and how they behave within and define the interior of the Earth. College level.

Clarke, Thurston. *California Fault: Searching for the Spirit of State Along the San Andreas.* New York: Ballantine Books, 1996. Clarke traveled the length of the San Andreas fault collecting first-hand accounts from earthquake survivors and predictors. Along with the entertaining stories, Clarke provides historical and scientific information about the fault.

Collier, Michael. *A Land in Motion: California's San Andreas Fault.* San Francisco: Golden Gates National Parks Association, 1999. Filled with beautiful color photographs that accompany text intended for the nonscientist, *Land in Motion* gives the reader excellent insight into earthquakes and their aftermaths. There are also many diagrams and graphs that explain subduction, faults, and orogeny.

Fradkin, Philip L. *Magnitude 8: Earthquakes and Life Along the San Andreas Fault.* Berkeley: University of California Press, 1999. Written for the layperson, *Magnitude 8* traces the seismic history, mythology, and literature associated with the San Andreas fault.

Garland, G. D. *Introduction to Geophysics: Mantle, Core, and Crust.* Philadelphia: W. B. Saunders, 1971. Excellent and thorough treatment of the Earth's interior and how its character is deduced from seismic wave behavior as measured by seismographs. Understandable to college science majors.

Lutgens, Frederick K., and Edward J. Tarbuck. *Earth: An Introduction to Physical Geology.* 6th ed. Upper Saddle River, N.J.: Prentice Hall, 1999. This college text provides a clear picture of the Earth's systems and processes that is suitable for the high school or college reader. In addition to its illustrations and graphics, it has an accompanying computer disc that is compatible with either Macintosh or Windows. Bibliography and index.

Richter, Charles F. *Elementary Seismology.* San Francisco: W. H. Freeman, 1958. The author of this classic 768-page text, who was a seismologist for many years at the California Institute of Technology, developed the Richter scale for measuring the magnitude of earthquakes. Judging from his book, Dr. Richter must have been an excellent teacher. Even though this source is outdated, its lucid explanations of basic principles make it a worthwhile reference. Contains excellent and detailed chapters on the complexities of earthquake locating, along with examples, charts, diagrams, and travel-time curves. Some sections using differential equations would be for upper-level college students, but most of the book, including the parts on earthquake locating, would be quite readable to any advanced high school student.

Simon, Ruth B. *Earthquake Interpretation.* Golden: Colorado School of Mines, 1968. Basic primer on seismogram interpretation. Aimed at lower-level college science students.

Tarbuck, E. J., and F. K. Lutgens. *Earth Science.* 5th ed. Westerville, Ohio: Charles E. Merrill, 1988. Freshman college text. Covers spectrum of Earth sciences with many full-color, excellent illustrations.

Verhoogen, John, et al. *The Earth.* New York: Holt, Rinehart and Winston, 1970. Provides introductory Earth science material. Freshman college level.

EARTHQUAKE MAGNITUDES AND INTENSITIES

The measurement of earthquake intensity (based on observed effects of earthquakes) and magnitude (based on instrument readings) is useful not only for scientists wishing to study and predict earthquakes but also for land-use planning and other aspects of public policy.

PRINCIPAL TERMS

AMPLITUDE: the displacement of the tracings of the recording pen (or light beam) on a seismogram from its normal position

DEEP-FOCUS EARTHQUAKES: earthquakes whose focus is greater than 300 kilometers below the surface

EPICENTER: the point on the Earth's surface directly above the focus of an earthquake

FOCUS: the point within the Earth that is the source of the seismic waves generated by an earthquake

HIGH-FREQUENCY SEISMIC WAVES: those earthquake waves that shake the rock through which they travel most rapidly

LOW-FREQUENCY SEISMIC WAVES: those earthquake waves that shake the rock through which they travel most slowly (also called long-period waves)

SEISMOGRAM: an image of earthquake wave vibrations recorded on paper, photographic film, or a video screen

SEISMOGRAPH: the mechanical or mechanical-electrical instrument that detects and records passing earthquake waves

SHALLOW-FOCUS EARTHQUAKES: earthquakes having a focus less than 60 kilometers below the surface

INTENSITY SCALES

Magnitude is a numerical rating of the size or strength of an earthquake, based on instrument readings. Intensity is a different kind of numerical rating having to do with the actual effects of an earthquake on people, buildings, and the landscape. Magnitude rating values allow comparison between earthquakes on a worldwide basis, whereas intensity ratings are more useful for comparing relative effects in areas surrounding the epicenter.

Because only human judgment is required for an intensity rating, intensity scales were developed first. Many different scales of earthquake effects have been devised in different countries since Renaissance times. The earliest known scale was developed in Italy and had only four rating values. The earliest widely accepted intensity scale, in use after 1883, was the Rossi-Forel scale, which had ten value ratings. In the United States, a revision of a later European scale, the 1931 modified Mercalli scale, has become the standard. It has twelve possible values of intensity, ranging from I, which barely would be felt, to XII, which would be the most violent. Each earthquake has a zone of maximum intensity, surrounded by zones of successively lesser intensity.

INTENSITY RATINGS

Intensity I on the modified Mercalli scale would not be felt except by very few. It might trigger nausea or dizziness if it occurs in the marginal zone of a large earthquake. In an area experiencing intensity II effects, ground vibration may be felt by some persons at rest, especially on upper floors, where building motion may exaggerate ground motion. Regions experiencing level III intensity are characterized by a brief period of vibration like that of a passing loaded truck. Many do not recognize it as an earthquake. Zones of intensity IV are felt indoors by many, outdoors by few. Buildings may shudder slightly; windows and doors of older homes may rattle, and glassware in cupboards may start clinking. In an intensity V earthquake, which is widely felt, some people may be awakened. A few may be frightened; some run outside. Windows and glassware may break, and cracks may appear in plaster.

Intensity VI earthquakes are felt by all; many

people are frightened, and many run outside. Some plaster may fall, some brick chimneys may be damaged, and some furniture moves. Objects are often thrown from shelves, and trees shake noticeably. An intensity VII earthquake frightens all. Its strong shaking may last for many seconds, causing considerable damage to older brick buildings and slight to moderate damage to well-built wood- or steel-frame structures. These quakes are noticed by persons driving vehicles. In an intensity VIII earthquake, damage to buildings is considerable. Specially designed earthquake-resistant structures can hold up, but many older brick buildings collapse totally. Branches and trunks of trees may break. Damage to most buildings—ranging from collapse to being thrown out of plumb or off the foundation—is caused by intensity IX earthquakes. Conspicuous ground cracks appear, and buried water and gas pipes break. Intensity X earthquakes cause most buildings to collapse partially, some totally. Railroad tracks are bent, and buried pipes buckle or break. Landsliding along riverbanks and steep slopes is triggered; obvious ground cracks are widespread. Strong shaking may last for many tens of seconds. After intensity XI quakes, few structures remain standing. Broad fissures appear in the ground. The earthquake may cause a large tsunami if it occurs near a coastal area. The strong

shaking may last a minute or more. Finally, in an intensity XII earthquake, objects are thrown in the air. Waves are "frozen" in the ground surface. Fewer than a half-dozen earthquakes have been rated at this level of intensity.

ASSIGNING INTENSITY RATINGS

Assignment of the lower values on the modified Mercalli scale is possible only if people are present to experience the effects. In the middle and upper values, effects on structures are a primary basis for the assignment of ratings, although earthquakes of greatest intensity may produce long-term geologic effects on the landscape, including ground fissures, fault scarps (clifflike features visible at the Earth's surface), landslides, and sandblows (small volcano-like mounds of sediment that erupt from water-saturated ground as a result of severe shaking). Thus, earthquakes that occur in uninhabited areas of the Earth cannot be rated unless the shaking was sufficiently strong to produce geologic effects; similarly, rating is impossible for quakes occurring below large areas of the ocean with few or no populated islands, although in some cases, intensity can be estimated if a tsunami is generated.

Because effects on buildings are an important means of differentiating between ratings in the middle and upper part of the scale, the nature of construction becomes an important differentiating criterion. In earthquake-prone California, for example, earthquake-resistant design practices mandated by law have made the average new building less susceptible to damage or destruction than are older buildings. Unless this factor is taken into account, equal-sized earthquakes would be rated lower over time because of less damage as older buildings are replaced.

ISOSEISMAL LINES

Despite such problems, the modified Mercalli scale is still quite useful. There are many more people (each of whom is

Looking west along the Motagua fault after the 1976 Guatemala earthquake. At intensity IV on the Mercalli scale, conspicuous ground cracks appear, and buried water and gas pipes break. (U.S. Geological Survey)

COMPARISON OF MAGNITUDE AND INTENSITY

Magnitude	Intensity (Mercalli)	Description
1.0-3.0	I	Not felt except by a very few under especially favorable conditions.
3.0-3.9	II-III	II felt only by a few persons at rest, especially on upper floors of buildings. III felt quite noticeably by persons indoors, especially on upper floors of buildings. Many people do not recognize it as an earthquake. Standing motor cars may rock slightly. Vibrations similar to the passing of a truck. Duration estimated.
4.0-4.9	IV-V	IV felt indoors by many, outdoors by few during the day. At night, some awakened. Dishes, windows, doors disturbed; walls make cracking sound. Sensation like heavy truck striking building. Standing motor cars rocked noticeably. V felt by nearly everyone; many awakened. Some dishes, windows broken. Unstable objects overturned. Pendulum clocks may stop.
5.0-5.9	VI-VII	VI felt by all, many frightened. Some heavy furniture moved; a few instances of fallen plaster. Damage slight. VII. Damage negligible in buildings of good design and construction; slight to moderate in well-built ordinary structures; considerable damage in poorly built or badly designed structures; some chimneys broken.
6.0-6.9	VIII-IX	VIII. Damage slight in specially designed structures; considerable damage in ordinary substantial buildings with partial collapse. Damage great in poorly built structures. Fall of chimneys, factory stacks, columns, monuments, walls. Heavy furniture overturned. IX. Damage considerable in specially designed structures; well-designed frame structures thrown out of plumb. Damage great in substantial buildings, with partial collapse. Buildings shifted off foundations.
7.0+	X-XI	X. Some well-built wooden structures destroyed; most masonry and frame structures destroyed with foundations. Rails bent. XI. Few, if any (masonry) structures remain standing. Bridges destroyed. Rails bent greatly. XII. Damage total. Lines of sight and level are distorted. Objects thrown into the air.

SOURCE: U.S. Geological Survey, National Earthquake Information Center; URL: http://neic.usgs.gov/neis.

a potential observer of earthquake effects) distributed around the world than there are earthquake-recording instruments. Earthquake intensity ratings begin to be gathered soon after earthquakes large enough to be felt by more than a few people. A government agency (for example, the U.S. Geological Survey) sends questionnaires to people in the area where the earthquake was felt. The forms contain questions related to the specific location and activity of the observer at the time of the earthquake as well as details of what happened before, during, and after the quake. The responses are then rated, using the modified Mercalli scale, and the ratings are plotted on a map. Lines separating the various values, or "isoseismal lines," can then be drawn to show areas having equal inten-

sity. Frequently an irregularly shaped bull's-eye pattern emerges, with the highest rating zone in the middle. This zone of maximum intensity usually contains the earthquake's epicenter.

The size and shape of the pattern of isoseismal lines can be invaluable in land-use planning or zoning of areas that experience frequent earthquake activity. The pattern may give clues about the distribution of land that would make a poor foundation for buildings because of greater susceptibility to seismic shaking. Certain types of sediment overlying bedrock can actually amplify seismic vibrations.

The highest intensity value determined for a specific earthquake is only a rough indicator of the quake's real strength. Quite large shocks can

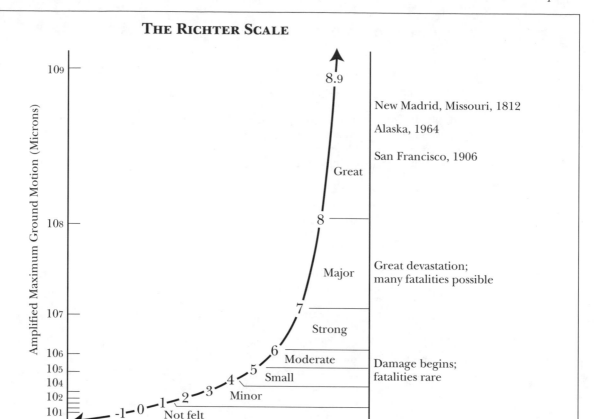

THE RICHTER SCALE

occur at many hundreds of kilometers of depth (deep-focus earthquakes), but because their most damaging vibrations are largely absorbed by rock before arriving at the surface, their maximum intensity ratings are low.

MAGNITUDE: THE RICHTER SCALE

Unlike earthquake intensity determinations, which vary with distance from the epicenter and depend on factors such as the depth of focus and soil depth, the magnitude rating of an earthquake is reported as a single number that is *usually* calculated from an instrumental recording of ground vibrations. ("Usually" is emphasized because an alternative method for magnitude determination has been developed, known as moment magnitude.) Every earthquake large enough to be detected and recorded by seismographs can be assigned a magnitude value. The Richter scale of

earthquake magnitude was the first to be used widely. It is named for its originator, Charles F. Richter, professor of seismology at the California Institute of Technology.

When an earthquake occurs, it generates ground vibrations that radiate out to surrounding areas, much as ripples do when a small pebble is dropped into a quiet pond. A seismograph is an instrument designed to detect and record, over time, even very small ground vibrations. It does this by amplifying the motion of the ground on which the seismograph sits when seismic waves pass through it. For some seismographs, the amplified recording (seismogram) is simply a piece of paper wrapped around a cylindrical, clock-driven drum, on which the ground vibrations are recorded as a zigzag line drawn by a recording pen. Larger ground vibrations result in larger zigzags or displacement on the paper. The amount of displacement from the normal or average position is called the amplitude, and it can be measured accurately with a simple fine-scale ruler.

In 1935, Richter first published his method for determining the relative size of the many earthquakes for which he had recordings in southern California. Most of these were small and moderate-sized shallow-focus earthquakes. Eventually, after various modifications, the Richter scale came to be used around the world to measure and compare earthquakes.

MAGNITUDE RATINGS

Richter's original definition of the magnitude of an earthquake is as follows: "The magnitude of any shock is taken as the logarithm of the maximum trace amplitude, expressed in microns, with which the standard short-period [Wood-Anderson] seismometer . . . would register that shock at an epicentral distance of 100 kilometers." Simply stated, the Richter scale magnitude of an earthquake is determined by measuring the greatest displacement from the average of the recording-pen tracing on the seismogram of a standard instrument at a standard distance from the earthquake epicenter (earthquake ground motions weaken as they travel out from the source). The height of the largest amplitude is measured in microns (thousandths of a millimeter). The logarithm of this number (to the base 10) is calculated and becomes the rating number "on the Richter scale." If an earthquake produced a maximum recording-pen displacement of 1 centimeter, that would be equivalent to 10,000 microns (which is equal to 10^4 microns). The logarithm of 10^4 is 4 (the exponent); thus, the earthquake would be rated at 4 on the Richter scale. If the maximum displacement had been 10^5 microns, it would be rated 5.

Because of the logarithmic nature of the Richter scale, each higher value in whole numbers actually represents a tenfold increase in the amount of ground motion recorded on the seismogram. Compared with an earthquake rated at 4, one rated at 5 involves 10 times more ground-shaking displacement; one rated at 6 would involve 100 times more displacement than 4, and 7 would be 1,000 times more. In other words, the largest earthquakes produce more than 1,000 times more ground motion than those that just begin to damage average American homes.

Another way to compare earthquakes is to determine the amount of energy they release. The energy released by earthquakes is about 30 times greater for each higher Richter scale value. For example, an earthquake rated at 6 would be about 900 times (30×30) more energetic than one rated at 4.

OTHER MAGNITUDE SCALES

Methods for magnitude determination have advanced substantially since the original magnitude scale was developed by Richter. His original scale was valid only for shallow-focus earthquakes occurring in the local region of southern California, because it depended on the Wood-Anderson type seismograph. That instrument is "tuned" to pick up the higher-frequency vibrations typical of the small and medium-sized earthquakes of southern California. Later work by Richter and others extended his original scale for use with other instruments tuned to pick up earthquakes that occurred at greater distance (more than 1,000 kilometers from a seismograph station) and had a deeper focus. These magnitude extensions were designed to coincide, as much as possible, with the values of the original scale. The extensional scales, however, use bases that are different from those of the original. One extensional scale, for example, requires a seismograph tuned to pick up only low-frequency vibrations. As a result, magnitude values differ somewhat and are not strictly "on the Richter scale." Technical literature actually limits the use of the term "Richter scale" to magnitude values determined essentially according to the original specifications.

For the very largest earthquakes, even the extensional scales become inadequate for ranking accurately the relative strength of earthquakes, because the sensitive instruments are said to become "saturated"—essentially thrown off scale. To solve this problem, a "moment magnitude" scale was developed; it is based not on instrument readings but on data obtained in the field, along the earthquake-generating fault. The average amount of fault offset, the length and width of slippage along the fault, and rock rigidity data are used to calculate the moment magnitude. Because of its greater accuracy, the moment magnitude scale is becoming more commonly used, particularly for medium-sized and larger earthquakes. Although the values of the moment magnitude scale essentially merge with those of the Richter scale for medium-sized earthquakes, at higher values there can be substantial differences. Thus, the largest earthquake rated by moment magnitude is 9.5.

PREDICTION OF EARTHQUAKE RECURRENCE

A key element to prediction of earthquake recurrence, particularly for the less frequent but larger and more destructive earthquakes, is a detailed record of the relative sizes of earthquakes occurring along a particular fault (or fault segment) through time. The most accurate and consistent indicator of size is the magnitude value. Because instruments needed for magnitude determination have been in existence less than one hundred years, magnitudes of earlier events must be estimated. This can be done using several approaches.

Maximum intensity values correlate with different but known magnitude values in areas where the depth of focus is thought to be consistent through time and where the nature of bedrock absorption of seismic waves is known. For determining maximum intensity values for unrecorded earthquakes of the past, archives and historical documents sometimes yield useful data. For prehistoric earthquakes, newly developed techniques are proving successful in defining the occurrence of large earthquakes and, under the right circumstances, even of their relative sizes. Such old events may be judged by the nature and extent of geologic traces preserved in radiocarbon-datable buried sediments.

Another avenue of study of earthquake recurrence is based on the amount of stored-up energy that is released by earthquakes worldwide in a year. A curve of the energy release can be compared to the occurrences of every possible recurrent earthquake-triggering mechanism—for example, tidal forces. To determine energy released, the moment magnitude must be determined for as many earthquakes as possible, but especially for the largest ones, because they release much more energy overall than do the more numerous smaller ones. One study, for example, revealed a strong correlation between times of higher-than-average earthquake activity (around 1910 and again around 1960) and the extent of "wobbling" of the Earth's axis of rotation.

DETERMINATION OF SEISMIC RISK

In earthquake-prone areas, seismic risk must be considered in urban and regional planning. The effects of ground shaking at any location are dependent primarily on magnitude, distance from the source of the earthquake waves, and the nature of the bedrock and the type and thickness of materials above it. After historical patterns of earthquake recurrence are determined, particularly their characteristic (or most typical) size, it is possible to make estimates of probable intensity patterns in the vicinity of faults. Detailed maps have been prepared of the areas along the San Andreas fault and for many miles on either side of it, outlining the zones of greatest potential seismic risk. Such maps can be of great value when decisions are being made regarding sites for potential secondary earthquake hazards such as nuclear power plants, fuel storage depots, and dams.

DETECTION OF UNDERGROUND NUCLEAR EXPLOSIONS

Additionally, seismic detection and characterization of distant underground nuclear explosions is of considerable political importance. One method of attempting to discriminate between a nuclear explosion and a natural seismic event is by analysis of its magnitude as recorded by several types of seismographs, each "tuned" to pick up different frequencies of ground vibrations. Ratios between such magnitude values appear to be quite useful for this purpose.

SIGNIFICANCE OF NUMBER ON RICHTER SCALE

Nearly every time a news broadcast makes reference to a damaging earthquake somewhere in the world, a number on the Richter scale is mentioned to give the listener some idea of the relative size of the event. If the earthquake has happened in a remote part of the globe and has not caused significant damage, it becomes merely another of the many facts that are soon forgotten. If the earthquake has happened where one's relative or a friend lives, however, that number on the Richter scale becomes extremely important because it is one of the first available indicators of possible severity. It can be determined within minutes after earthquake waves have been detected at seismological observatories, whereas direct communication from and damage assessment at the site of the earthquake may be very slow in coming. Exactly what does a value of 7.3, for example, mean to those who were near the epicenter? How does that figure compare to the value associated with damage to homes? (That number is about 5 on the Richter scale.)

Many of the world's large cities are located close to active earthquake-generating faults. What effect will a major earthquake in such a city have on the economy of that area, and how in turn will its misfortune affect the rest of the world? What would happen if the flow of goods between the United States and Japan were suddenly disrupted for a long period because of an 8.5-magnitude earthquake near Tokyo or Los Angeles? The initial answers to these and many other questions may someday hinge on that critical number on the Richter scale.

VALUE OF SEISMIC MAPS

Aspects of earthquake intensity are less likely to be mentioned in the media except, perhaps, when covering local aspects of seismic risk along a certain fault or the risk of earthquakes in various areas of the United States. Maps identifying zones of potential seismic hazard are likely to become more common as their need becomes more apparent. The lack of such knowledge and of the will to act on it could be costly in terms of lives and property.

One of the most instructive illustrations of seismic intensity patterns is a map comparison of the 1906 San Francisco earthquake and a similar-sized earthquake in southern Missouri. The seismic wave absorptive properties of the bedrock of the western margin of North America is much greater than that of the central and eastern states. As a result, except in California, little damage is expected to occur from earthquakes in the United States—even from major ones. A great earthquake in the New Madrid fault zone of southern Missouri, however, is likely to have a wide zone of maximum intensity, resulting in severe damage in cities hundreds of miles from the epicenter.

Valentine J. Ansfield

CROSS-REFERENCES

Continental Rift Zones, 579; Deep-Focus Earthquakes, 271; Earthquake Distribution, 277; Earthquake Engineering, 284; Earthquake Hazards, 290; Earthquake Locating, 296; Earthquake Prediction, 309; Earthquakes, 316; Elastic Waves, 202; Faults: Normal, 213; Faults: Strike-Slip, 220; Faults: Thrust, 226; Faults: Transform, 232; Notable Earthquakes, 321; Plate Motions, 80; Plate Tectonics, 86; San Andreas Fault, 238; Seismometers, 258; Slow Earthquakes, 329; Soil Liquefaction, 334; Subduction and Orogeny, 92; Thrust Belts, 644; Tsunamis, 2176; Tsunamis and Earthquakes, 340.

BIBLIOGRAPHY

Bolt, Bruce A. *Earthquakes.* New York: W. H. Freeman, 1988. An authoritative introduction to most aspects of earthquakes, including magnitude and intensity. Includes lists of important earthquakes, a glossary, a bibliography of titles suitable for the general reader, and an interesting "earthquake quiz" with answers. Suitable for high school or college-level readers.

Doyle, Hugh A. *Seismology.* New York: John Wiley, 1995. A good introduction to the study of earthquakes and the Earth's lithosphere. Written for the layperson, the book contains many useful illustrations.

Eiby, G. A. *Earthquakes.* Auckland, New Zealand: Heineman, 1980. A relatively technical discussion of earthquakes and seismology. Suitable for college-level readers.

Fradkin, Philip L. *Magnitude 8: Earthquakes and Life Along the San Andreas Fault.* Berkeley: University of California Press, 1999. Written for the layperson, *Magnitude 8* traces the seismic history, mythology, and literature associated with the San Andreas fault.

Gere, James M., and Haresh C. Shah. *Terra Non Firma: Understanding and Preparing for Earthquakes.* New York: W. H. Freeman, 1984. Similar in many respects to Bolt's book, described above. Includes a table relating maximum intensity values of the modified Mercalli scale to Richter scale magnitude values (page 87) and a table relating the duration of strong motion to Richter scale values (page 173). Suitable for high school or college-level readers.

Monastersky, Richard. "Abandoning Richter." *Science News* 146 (October, 1994): 250-252. The author describes why the Richter scale of measuring earthquake size is outdated. Current reports of earthquake magnitude are based on a number of factors quire different from that of the simpler Richter scale as originally proposed.

Nance, John, and Howard Cady. *On Shaky Ground*. New York: William Morrow, 1988. One of the best-written and easiest to understand accounts of the actual effects of some of the most significant earthquakes of various magnitudes. Based on interviews with survivors as well as with top researchers in seismology. Strongly recommended.

Richter, Charles F. *Elementary Seismology*. San Francisco: W. H. Freeman, 1958. The author of this classic 768-page text, who was a seismologist for many years at the California Institute of Technology, developed the Richter scale for measuring the intensity of earthquakes. Judging from his book, Dr. Richter must have been an excellent teacher. Even though this source is outdated, its lucid explanations of basic principles make it a worthwhile reference. Contains excellent and detailed chapters on the complexities of earthquake locating, along with examples, charts, diagrams, and travel-time curves. Some sections using differential equations would be for upper-level college students, but most of the book, including the parts on earthquake locating, would be quite readable to any advanced high school student.

Walker, Bryce. *Earthquake*. Alexandria, Va.: Time-Life Books, 1982. A well-written popular account of earthquakes; includes many good illustrations. Suitable for high school-level readers.

EARTHQUAKE PREDICTION

Predicting the location and timing of earthquakes is an active area of research in many countries throughout the world. Although significant progress has been made in understanding the causes and consequences of earthquakes, scientists are still unable to predict with sufficient accuracy the occurrence of major temblors.

PRINCIPAL TERMS

CRUST: the outermost layer of the Earth, which consists of materials that are relatively light

ELASTIC REBOUND THEORY: the theory that states that rocks across a fault remain attached while accumulating energy and deforming; the energy is released in a sudden slip, which produces an earthquake

FAULTING: the process of fracturing the Earth such that rocks on opposite sides of the fracture move relative to each other; faults are the structures produced during the process

LITHOSPHERE: the Earth's rigid outer layer, which is composed of the crust and uppermost mantle

SEISMICITY: the temporal and spatial distribution of earthquakes

SEISMOLOGY: the study of earthquakes and their causes

STRESS: a force acting in a specified direction over a given area

EARTHQUAKE OCCURRENCE

Chinese scientists pioneered the study of earthquakes hundreds of years ago. Since that time, predicting the location and time of major earthquakes has been an important part of seismology. Earthquakes occur, with varying frequency, in diffuse belts in nearly every region of the globe. The distribution of earthquakes is explained easily by the modern theory of plate tectonics, which holds that the surface of the Earth is composed of a mosaic of interlocking rigid plates that move relative to one another at speeds up to 12 centimeters per year. Motions along the boundaries of the plates produce earthquakes; if the plates are not able to accommodate the motions easily, then large earthquakes may accompany the relative motion between the plates. The widespread occurrence of earthquakes makes them important to everyone, and for people living near plate boundaries, earthquakes play an even more potentially destructive role in shaping the environment.

Earthquakes are generated when some portion of the Earth's rigid outer layer, called the lithosphere, ruptures catastrophically along a sharp discontinuity or fault. This creates significant ground motion near the source of the rupture. Earthquakes occur most commonly at the three types of boundaries of lithospheric plates, which are known as convergent or destructive, divergent or constructive, and transcurrent. The largest number of earthquakes are at divergent plate boundaries located along mountain ridges in the ocean basins. Because these earthquakes are small and far from population centers on the continents, little effort is expended to predict ruptures along divergent plate boundaries. In contrast, earthquakes at convergent or transcurrent plate boundaries, although fewer in number, are larger. Most convergent and transcurrent boundaries coincide with continental margins along which the majority of the global population lives. For this reason, earthquake prediction research focuses on convergent and transcurrent plate boundaries such as those in Japan and California.

Slip on fault planes of large earthquakes is on the order of 10-20 meters, and the forces responsible for faulting are simply the result of the relative motions of the plates at the plate boundaries. Rocks near a region of an impending earthquake may accumulate motion and change volume and shape for hundreds of years prior to causing a rupture. When the lithosphere does finally break, energy stored by the rocks is released suddenly as seismic waves that travel through and around the surface of the Earth. These waves generate the intense vibrations associated with an earthquake.

For great earthquakes, the rupture may extend for as much as 1,000 kilometers, and it may propagate at speeds in excess of 10,000 kilometers per hour.

EARTHQUAKE CATEGORIZATION

Seismologists categorize earthquakes by several different features, but the two most important for earthquake prediction are an earthquake's location and its magnitude or size. The location is described by an epicenter, which is the projection of the earthquake's focus within the lithosphere onto the Earth's surface.

The magnitude is a number from 1 to 10 on a scale devised by Charles Richter that describes the relative changes in ground motion recorded on a seismometer. The so-called Richter scale is logarithmic; an increase from a value of 1 to 2 corresponds to a tenfold increase in ground motion and to an approximately thirtyfold increase in the amount of energy released during the rupture. The Richter scale is based on a "standard seismometer" placed at a "standard distance" from the epicenter of the earthquake. The traditional Richter scale magnitudes are denoted by "M" to distinguish them from other more recent magnitude scales. Richter originally devised his magnitude scale to be most appropriate for describing moderate local earthquakes in California. Unfortunately, despite its widespread use, the scale is not a good measure of the energy released from very small or very large earthquakes.

POTENTIAL TRIGGERS

Seismologists have learned much about the rupture process that causes earthquakes by studying the ground motion close to and far away from the source. The development of modern seismological instruments and procedures in the early twentieth century led seismologists to the discovery that different rupture mechanisms are at work in different plate tectonic settings. Early attempts at earthquake prediction used analysis of the frequency of major earthquakes in specific regions of the globe to determine whether any significant pattern was apparent. This approach proved to be fruitless. With the acceptance of elastic rebound theory to describe the rupture mechanism for earthquakes, seismologists shifted their attention away from statistical analysis of earthquake occurrences toward developing methods to understand the "trigger" of major earthquakes. Most seismologists agree that the energy necessary to produce a major earthquake is accumulated slowly relative to the time it takes for a rupture to occur. If no trigger were involved in the rupture process, prediction of earthquakes would be extremely difficult, if not impossible. Modern seismologists interested in earthquake prediction primarily rely on developing methods to understand any precursory phenomena that would enable them to predict at least several days or weeks in advance the location of large (M ≥ 6.0) to great (M > 8.0) earthquakes. Of course, if an impending earthquake is far removed—for example, more than 1,000 kilometers—from any population center, the need to alert the public is minimal.

Over the years, investigators have suggested a variety of potential triggers to earthquakes. These include rapidly changing or severe weather conditions; variations in the gravitational forces among the Moon, Sun, Earth, and other planets in the solar system; and volcanic activity. Scientists have searched historical seismicity records, including extensive catalogs for California, for relationships between the suggested triggers and the occurrence of earthquakes, without much success. For example, every 179 years, the planets of the solar system align. Some researchers suggested that this alignment would increase the gravitational forces acting upon Earth and thereby trigger an increase in seismicity. The last such planetary alignment was in 1982, and no significant increase in Earth seismicity occurred.

PRIOR CHANGES IN PHYSICAL PROPERTIES

Because the research on earthquake triggers has been largely unsuccessful, seismologists have turned their attention away from potential triggers of major earthquakes and toward the role of changing physical properties prior to an earthquake. Some promising properties that are currently being studied include shifts in ground elevation near the site of an impending earthquake; variations in the velocity of certain types of seismic waves as they traverse regions that may produce a major quake; increased escape of radon, helium, and other gases from vents and cracks in the Earth's surface prior to the earthquake; changes in the electrical conductivity of rocks near the region of impending rupture; and fluctuations in pore fluid pressure in the rocks near major fault zones. In addition, seismologists also have focused on recognizing certain premonitory swarms of smaller quakes, called foreshocks, that may foreshadow a major earthquake. Another technique is assessing the time between major earthquakes in a specific region. If an area that is expected to produce earthquakes is seismically quiet—that is, a gap exists in its seismic activity—then the area may be more likely to experience an earthquake in the near future. This is referred to as "seismic gap" theory. Finally, strange animal behavior has been linked to periods of several days to several hours prior to an earthquake. Some researchers have claimed that cats and dogs tend to run away from home or exhibit unusual behavior, such as seeking out special hiding places, before the onset of an earthquake. Individual reports of odd animal behavior are substantiated by the increase in advertisements for lost pets in local newspapers during the days before a major earthquake. Scientists have suggested that some animals are sensitive to minute changes in their environment, which allows them to "sense" an earthquake prior to onset of severe ground shaking. Research in this area is actively pursued in China and Italy. Most workers, however, are interested in developing instruments that would be able to measure the same effects that disturb animals. Although many of these effects are known to occur prior to a major earthquake, scientists still must develop highly sensitive devices that will alert them in enough time to evacuate or prepare the region near an impending earthquake.

Seismogram Analysis

Several methods are used by seismologists to study earthquake prediction. The techniques include analysis of seismograms to identify either foreshock or aftershock patterns that signal an impending large earthquake, examination of active fault zones in the field to determine the frequency of great earthquakes over the last tens of thousands of years, and investigation of deep boreholes to characterize the orientations and magnitudes of stresses associated with active faults. In addition, elaborate arrays of sophisticated instruments are frequently deployed near active faults to collect geophysical data that may shed light on earthquakes.

The energy carried by seismic waves is recorded on seismometers or seismographs, which are instruments that monitor ground motion. Seismometers are composed of a mass attached to a pendulum. During an earthquake, the mass remains still, and the amount that the Earth moves around it is measured. Ground motion is recorded on a chart as a series of sharp peaks and valleys that deviate from the background value, measured during times of no earthquake activity. The arrival of the waves at different times at different places allows seismologists to calculate the epicenter of an earthquake. The height or amplitude relative to the background noise of the first peak in a long series of peaks associated with a particular earthquake is an estimate of the magnitude of that earthquake.

Since the early twentieth century, seismometers have recorded hundreds of thousands of earthquakes per year worldwide. Seismologists have carefully cataloged many seismograms, the actual paper records of ground motion from a particular location, so that they may be easily compared. Examining these records in detail has allowed seismologists to deduce certain characteristics of major earthquakes. They have noted, for example, that most large earthquakes are followed by a series of smaller earthquakes in the same region. These smaller earthquakes are called aftershocks, and they allow seismologists to constrain the orientation and dimensions of the rupture or fault plane that produced the main earthquake. With the development of modern seismometers and digital recording networks in the 1970's and 1980's, seismologists began to recognize certain precursory seismicity patterns in addition to aftershock sequences. These precursory phenomena are referred to as foreshock sequences and, as yet, are poorly understood. Seismologists hope that with enough data on the overall seismicity of an area, they will be able to note deviations from normal patterns that would signal the onset of a major earthquake.

Field Analysis

Information about prehistoric seismic activity must be obtained by examining ancient fault zones exposed at the surface of the Earth. Large motions between two rock masses produce characteristic features that may be identified in the field.

Geologists are now examining the recent rock record near the San Andreas fault in California. Careful mapping of areas that have been excavated across the fault zone has yielded evidence for large earthquakes prior to historical and seismological records. The now well established technique of carbon 14 dating was applied to organic material trapped in the fault zone to determine the approximate age, location, and intensity of several ancient earthquakes. The data, although sparse, seem to suggest that great earthquakes occur every fifty to three hundred years. In addition, there is some indication that great earthquakes may occur closely spaced in time with significant periods of quiescence between them. This type of analysis is similar to the seismic gap theory, where catalogs of seismograms are examined to determine what known faults or regions that have been active previously are currently inactive and perhaps are ready for renewed activity.

Many countries are carrying out elaborate experiments in areas of repeated seismic activity. One example is in central California on the San Andreas fault near the town of Parkfield. There, geophysicists arrayed a variety of instruments, including seismometers, tiltmeters, gravimeters, and laser surveying equipment, to measure ground motion, elevation changes, gravity variations, and minute amounts of slip on the fault. Based on seismic records, scientists discovered that earthquakes with magnitudes of approximately 5.0 M occur with predictable frequency. The experiments are designed to learn as much as possible about the changes that occur in the region prior to, during, and after an earthquake of moderate size. Scientists hope that these data will allow them to know what features to monitor to predict much larger earthquakes in other areas.

INVESTIGATION OF DEEP BOREHOLES

Another method used by scientists to understand precursory phenomena associated with earthquakes is drilling deep boreholes into the Earth's crust near major fault zones. One such hole is in Fort Cajon, California. At this site, researchers examined changes in pore fluid pressure and electrical conductivity in the borehole. In addition, instruments measured the orientation of fractures in the borehole's walls. These fractures are related to the forces acting on the

rocks deep in the crust, and some researchers attempted to relate these manifestations of stress to earthquake fault orientation. Geophysicists hope that the newest techniques will measure these stresses in real time, so they will be able to compare these data to those obtained from studies of seismicity. Understanding the behavior of a major fault zone at depth may prove useful in predicting earthquakes in the future.

PROGRESS IN PREDICTION

Many countries are actively involved in earthquake prediction research. Since the early 1960's, effort has been particularly high in Japan, the People's Republic of China, and the United States. The overall goal of these research efforts is to attain the same level of reliability in earthquake prediction as in weather prediction. Although the majority of effort has been focused on predicting the exact time and place of a major earthquake, an equally important, though often overlooked, aspect of earthquake prediction is an assessment of the severity of ground shaking for a specific site. This information is crucial for public policy discussions on the location of dams, hospitals, schools, and nuclear reactors, all of which may be at significant risk during a major earthquake.

Despite the numerous uncertainties that enter into forecasting an earthquake, some have been successfully predicted. The most spectacular was the Haicheng earthquake of northeast China in February, 1975. Five hours before the earthquake, warnings were issued and several million people from towns in the vicinity of the predicted epicenter were evacuated. Devastation was widespread, but loss of human life was minimal. Scientists who later visited the area estimated that hundreds of thousands of lives were saved. Unfortunately, the Chinese were only able to predict that a great earthquake was to strike the Tangshan region within five years. In August, 1976, a very strong earthquake struck this area without warning, leaving 700,000 people dead.

MINIMIZING EARTHQUAKE HAZARDS

Earthquake prediction remains critical to modern society because most of the world's population lives along convergent plate boundaries and, thus, within the destructive reach of a great earthquake. The purpose of earthquake prediction, then, is to prepare a society for any earthquakes with magnitudes capable of disturbing normal life. This may include warning and evacuation of the local population or assessing the risks of severe ground motion on current or future structures. In addition to strong ground motion, earthquakes may be responsible for hazards such as tsunamis, avalanches, and fires. In both the great San Francisco earthquake of 1906 and the Tokyo earthquake of 1923, many of the fatalities attributed to the earthquakes were actually the result of the subsequent fires that consumed the cities. Another danger of earthquakes is soil liquefaction, which occurs when the seismic waves cause the soil to lose rigidity and slide away. When this happens, the soil can no longer support structures. Although the structures may be strong enough to withstand the shaking associated with an earthquake, their foundations may be undermined, causing the buildings to topple.

Perhaps the most promising aspect of earthquake prediction is the development of stringent building codes. After each major earthquake in Southern California, for example, municipal, county, and state statutes are changed to reflect new data concerning the behavior of building materials during strong ground motion. Engineers now know that unreinforced masonry buildings are likely to be destroyed in even a moderate earthquake. Because the seismic risk is high in Japan and California, these areas now have the most stringent building codes in the world. That these codes can prevent much unnecessary loss of property and human life unfortunately can be demonstrated by a comparison of the 1971 San Fernando and 1988 Armenian earthquakes. The earthquakes were of similar magnitude—slightly greater than 6.0 M—yet the San Fernando earthquake resulted in about fifty deaths, most of which were in one wing of an old hospital building, while tens of thousands perished in the Armenian earthquake because of the collapse of poorly constructed masonry buildings.

Humans will never be able to prevent earthquakes and their potentially devastating effects. Understanding how, why, when, and where earthquakes occur, therefore, is extremely important to society. Earthquake prediction, like weather prediction, is one way that society seeks to minimize harmful effects of these complex natural phenomena.

Glen S. Mattioli and Pamela Jansma

CROSS-REFERENCES

Deep-Focus Earthquakes, 271; Earthquake Distribution, 277; Earthquake Engineering, 284; Earthquake Hazards, 290; Earthquake Locating, 296; Earthquake Magnitudes and Intensities, 301; Earthquakes, 316; Elastic Waves, 202; Faults: Normal, 213; Faults: Strike-Slip, 220; Faults: Thrust, 226; Faults: Transform, 232; Land Management, 1484; Land-Use Planning, 1490; Land-Use Planning in Coastal Zones, 1495; Landslides and Slope Stability, 1501; Notable Earthquakes, 321; Plate Motions, 80; Plate Tectonics, 86; San Andreas Fault, 238; Seismic Reflection Profiling, 371; Seismometers, 258; Slow Earthquakes, 329; Soil Liquefaction, 334; Subduction and Orogeny, 92; Thrust Belts, 644; Tsunamis and Earthquakes, 340.

BIBLIOGRAPHY

Berlin, G. Lennis. *Earthquakes and the Urban Environment.* Vols. 2 and 3. Boca Raton, Fla.: CRC Press, 1980. These books are two volumes of a three-part series written by a geographer who is primarily concerned with effective land-use planning in seismically active areas. Volume 3 concentrates on strategies to minimize the effects of great earthquakes, such as disaster planning and improved building codes. Many of the social aspects of earthquakes in an urban environment are presented, including human response and insurance. Volume 2 addresses earthquake prediction and building codes. Both volumes contain an extensive reference list of more than 1,400 articles and books. These volumes can be quite technical and are recommended for college-level readers.

Bolt, Bruce A. *Earthquakes and Geological Discovery.* New York: Scientific American Library, 1993. As the title suggests, an excellent introductory text on earthquakes. Earthquake prediction is discussed extensively in one chapter. The illustrations and photographs of the effects of earthquakes add considerably to the text. Anyone interested in earthquakes will find this an invaluable source.

Doyle, Hugh A. *Seismology.* New York: John Wiley, 1995. A good introduction to the study of earthquakes and the Earth's lithosphere. Written for the layperson, the book contains many useful illustrations.

Eiby, G. A. *Earthquakes.* New York: Van Nostrand Reinhold, 1980. A reference aimed at beginning college-level students. Well illustrated and addresses all topics relevant to earthquakes and seismology. Earthquake prediction and the effect of large earthquakes on human-made structures are discussed in two separate chapters.

Farley, John E. *Earthquake Fears, Predictions, and Preparations in Mid-America.* Carbondale: Southern Illinois University Press, 1998. This book examines seismic activity, the practice of predicting earthquakes, and the hazards associated with them, focusing on the American Midwest. Bibliography, charts, and index.

Iacopi, Robert. *Earthquake Country.* 3d rev. ed. Menlo Park, Calif.: Lane Books, 1971. Part of the Sunset Book series, this source is directed toward the lay reader. Discusses California geology in relation to seismic risk in a straightforward and nontechnical way. Contains many photographs of the effects of earthquakes on both human-made structures and the natural landscape. Also has a foreword by Charles F. Richter, the inventor of the famous Richter scale.

Lomnitz, Cinna. *Fundamentals of Earthquake Prediction.* New York: John Wiley & Sons, 1994. Lomnitz examines the principles and mythologies of earthquake prediction. Illustrations, maps, bibliography, and index.

Lutgens, Frederick K., and Edward J. Tarbuck. *Earth: An Introduction to Physical Geology.* 6th ed. Upper Saddle River, N.J.: Prentice Hall, 1999. This college text provides a clear picture of the Earth's systems and processes that is suitable for the high school or college reader. In addition to its illustrations and graphics, it has an accompanying computer disc that is compatible with either Macintosh or Windows. Bibliography and index.

Mogi, Kiyoo. *Earthquake Prediction.* San Diego, Calif.: Academic Press, 1985. A comprehensive and highly technical text that discusses most aspects of earthquake prediction. The majority of prediction experiments that are described are from Japan. The reader is expected to have a considerable background in Earth science and mathematics. The text is suitable for senior college-level students.

National Academy of Sciences. *Earthquake Prediction and Public Policy.* Washington, D.C.: Government Printing Office, 1975. This book was prepared by a National Research Council panel on the public policy implications of earthquake prediction. The panel was composed of outstanding scientists and engineers involved in all fields of earthquake research, in addition to sociologists and other public figures. They specifically evaluate seismic risk in certain regions of the United States and propose action to prepare these regions for significant earthquakes. Guidelines for earthquake prediction research are discussed. This text is suitable for anyone.

Press, Frank, and Raymond Siever. *Understanding Earth.* 2d ed. New York: W. H. Freeman, 1998. One of the finest illustrated introductory texts on geology. The book has chapters focusing on plate tectonics, seismology, and earthquakes. A map of the major plates is on the inside back cover. The glossary is huge and indispensable. Senior high school and college-level students should find this text suitable for general background information.

Rikitake, Tsuneji. *Earthquake Forecasting and Warning.* Norwell, Mass.: Kluwer Academic Publishers, 1982. Focuses on advances in earthquake prediction in Japan, California, the Soviet Union, and the People's Republic of China. The major programs for earthquake study in each of these regions are discussed. Case studies in Japan are presented in detail. The reader is required to have an excellent understanding of geophysics. Suitable for senior college-level students.

Uyeda, Seiya. *The New View of the Earth.* Translated by Masako Ohnuki. San Francisco: W. H. Freeman, 1978. The historical development of the theory of plate tectonics is presented, in addition to an excellent explanation of the theory itself. The text is well illustrated and is nontechnical. Designed primarily for the nonscientist.

EARTHQUAKES

An earthquake is the sudden movement of the ground caused by the rapid release of energy that has accumulated along fault zones in the Earth's crust. The Earth's fundamental structure and composition are revealed by earthquakes through the study of waves that are both reflected and refracted from the interior of the Earth.

PRINCIPAL TERMS

CRUST: the uppermost 5-40 kilometers of the Earth

DEFORMATION: a change in the shape of a rock

ELASTIC REBOUND: the process whereby rocks snap back to their original shape after they have been broken along a fault as a result of an applied stress

LITHOSPHERE: the solid part of the upper man-

tle and the crust where earthquakes occur

MANTLE: the thick layer under the crust that contains convection currents that move the crustal plates

STRAIN: the percentage of deformation resulting from a given stress

STRESS: a force per unit area

STRESS

Earthquakes are sudden vibrational movements of the Earth's crust and are caused by a rapid release of energy within the Earth. They are of critical importance to humans, first, because they reveal much about the interior of the Earth, and second, because they are potentially one of the most destructive naturally occurring forces found on Earth.

The outermost skin of the Earth, called the crust, is in constant motion as a result of large convection cells within the upper mantle that circulate heat from the interior of the Earth toward the surface. The crust of the Earth is about 5 kilometers thick in the oceanic basins and about 40 kilometers thick in the continental masses, while the upper mantle is about 700 kilometers thick. Because the crust is relatively thin compared to the upper mantle, the crust is broken up into several plates that float along the top of each convection cell in the upper mantle. Most earthquakes occur along the boundaries separating the individual plates and are represented by faults that may be thousands of kilometers long and tens of kilometers deep. Although the vast majority of earthquakes occur along these plate boundaries, some also occur within the plate interior. The rocks on either side of the fault fit tightly together and produce great resistance to movement. As the blocks of rock attempt to move against one another, the

resistance of movement causes stress, which is a force per unit area, to build up along the fault. As the stress continues to build, the rocks in the immediate vicinity slowly deform, or bend, until the strength of the rock is exceeded at some point along the fault. Suddenly, the rocks break violently and return to their underformed state, much as a rubber band snaps to its original shape when it breaks. This rapid release of stress is called elastic rebound. The point at which the stress is released is called the focus of an earthquake, and that point at the Earth's surface directly above the focus is called the epicenter.

SEISMIC WAVE MOTION

The release of energy associated with elastic rebound manifests itself as waves propagating away from the focus. When these waves of energy reach the surface of the Earth, the land will oscillate, causing an earthquake. These waves move through the Earth in two ways. P (primary) waves move in a back-and-forth motion in which the motion of the rock is in the same direction as the direction of energy propagation. This type of wave motion is analogous to placing a spring in a tube and pushing on one end of the spring. The motion of the spring in the tube is in the same direction as is the motion of the energy. These waves are called primary because they move through the Earth faster than do other waves—up to about 25 kilometers per second.

Thus, P waves are the first waves to be received at a seismic recording station. Because the individual atoms in a rock move back and forth along the direction of energy movement, P waves can move through solids and liquids and, for this reason, do not tell geologists much about the state (solid or liquid) of a given rock at depth. In contrast to P waves, for S waves, the rock motion is perpendicular to the direction of energy propagation. Guitar strings vibrate in a similar manner: Each part of the guitar string moves back and forth while the energy moves along the string to the ends. S waves are the second waves to be received at a seismic recording station and derive their name from this fact. Unlike P waves, S waves cannot move through liquids but can move through solids. Thus, when a P wave is received by a seismic station but is not followed by an S wave, seismologists know that a liquid layer is between the focus of the earthquake and the receiving seismic station. Both S and P waves are bent, or refracted, as they move in the Earth's interior. This refraction occurs as the result of the increase in density of rocks at greater depths. Furthermore, both types of waves are reflected off sharp boundaries, representing a change in rock type located within the Earth. Thus, by using these properties of S and P waves, geologists have mapped the interior of the Earth and know whether a given region is solid or liquid.

Although S and P waves represent the way seismic energy moves through the Earth, once this energy reaches the Earth's surface, much of it is converted to another type of wave. L (Love) waves move in the same manner as do S waves, but they are restricted to surface propagation of energy. L waves have a longer wavelength and are usually restricted to within a few kilometers of the epicenter of an earthquake. These waves cause more damage to structures than do P and S waves because the longer wavelength causes larger vibrations of the Earth's surface.

EARTHQUAKE INTENSITY

The amount of energy released by an earthquake is of vital importance to humans. Many active fault zones, such as the famous San Andreas fault in California, produce earthquakes on an almost daily basis, although most of these earthquakes are not felt and cause no damage to human-made structures. These minor earthquakes indicate that the stress that is accumulating along some portion of a fault is continuously being released. It is only when the stresses accumulate without continual release that large, devastating earthquakes occur. The intensity of an earthquake is dependent not only on the energy released by the earthquake but also on the nature of rocks or sediments at the Earth's surface. Softer sediments such as the thick muds that underlie Mexico City will vibrate with a greater magnitude than will the very rigid rocks, such as granites, found in other parts of the world. Thus, the great earthquake that devastated Mexico City in 1985 was in part the result of the nature of the sediments upon which the city is built.

For a given locality, earthquakes occur in cycles. Stress accumulates over a period of time until the forces exceed the strength of the rocks, causing an increase in minor earthquake activity. Shortly thereafter, several foreshocks, or small earthquakes, occur immediately before a large earthquake. When a large earthquake occurs, it is usually followed by many aftershocks, which may also be rather intense. These aftershocks occur as the surrounding rocks along the fault plane readjust to the release of stress by the major earthquake. The cycle then repeats itself with a renewed increase in stress along the fault. Although seismologists can usually tell what part of the seismic cycle a region is experiencing, it is difficult to predict the duration of each of these cycles; thus, it is impossible to predict precisely when an earthquake will occur.

SEISMOGRAPHS

Seismographs are the primary instruments used to study earthquakes. All seismographs consist of five fundamental elements: a support structure, a pivot, an inertial mass, a recording device, and a clock. The support structure for a seismograph is always solidly attached to the ground in such a fashion that it will oscillate with the Earth during an earthquake. A pivot, consisting of a bar attached to the support structure via a low-friction hinge, separates a large mass from the rest of the seismograph. This pivot allows the inertial mass to remain stationary during an earthquake while the rest of the instrument moves with the ground. The recording device consists of a pen attached to the inertial mass and a roll of paper that is attached to the support structure. Finally, the clock records the exact time on the paper so that the time of arrival of each wave type is noted. When an earthquake wave arrives at a seismic station, the support structure moves with the ground. The inertial mass and the pen, however, remain stationary. As the paper is unrolled, usually by a very accurate motor, the wave is recorded on the paper by the stationary pen. Modern seismographs, however complex in design, always contain these basic elements. The clock, which each minute places a small tick mark on the recording, is calibrated on a daily basis by a technician using international time signals from atomic clocks. The recording pen often consists of an electromagnet that converts movement of the inertial mass relative to the support structure to an electrical current that drives a light pen. The light pen emits a narrow beam of light onto long strips of photographic film that are developed at a later date.

RICHTER AND MERCALLI SCALES

Seismologists have adopted two widely used scales, which are called the Richter and Mercalli scales, to measure the energy released by an earthquake. The Richter magnitude scale is based on the amplitude of seismic waves that are recorded at seismic stations. Because seismic stations are rarely located at the epicenter of earthquakes, the amplitude of the seismic wave must be corrected for the amount of energy lost over the distance that the wave traveled. Thus, the Richter magnitude reported by any seismic station for a given earthquake will be approximately the same. Richter magnitudes are open-ended, meaning that any amount of seismic energy can be calculated. The weakest earthquakes have Richter magnitudes less than 3.0 and release energy less than 10^{14} ergs. These earthquakes are not usually felt but are recorded by seismic stations. Earthquakes between magnitudes 4.0 and 5.5 are felt but usually cause no damage to structures; they release energy between 10^{15} and 10^{16} ergs. Earthquakes that have magnitudes between 5.5 and 7.0 cause slight to considerable damage to buildings and release energy between 10^{18} and 10^{24} ergs. Earthquakes that are greater than 7.5 on the Richter scale generate energy up to 10^{25} ergs—as much as a small nuclear bomb. The Mercalli intensity scale is based not on the energy released by an earthquake but rather on the amount of shaking that is felt on the ground; it rates earthquakes from Roman numerals I to XII. Unlike the Richter scale, the Mercalli scale provides descriptions of sensations felt by observers and of the amount of damage that results from an earthquake. Thus, an earthquake of Mercalli intensity I is felt only by a very few persons, while an earthquake of intensity XII causes total destruction of virtually all buildings.

Both the Mercalli and Richter scales have advantages and disadvantages. The Mercalli scale provides the public with a more descriptive understanding of the intensity of an earthquake than does the Richter scale. The damage caused by an earthquake is a function not only of the energy released by such an event but also of the nature of the sediments or rocks upon which the buildings in the vicinity are constructed. The Richter scale is best used to study specifically the amount of energy release by an earthquake. Finally, the Richter scale, which is purely quantitative, does not rely on subjective observations such as those required by the Mercalli scale.

TRIANGULATION TECHNIQUES

The exact location of an earthquake epicenter can be deduced from three seismographic stations using triangulation techniques. Because the P and S waves travel at different velocities in the Earth, seismologists can determine the distance from the station to the epicenter. They calculate the difference in time between the first arrival of the P and S waves, respectively, at the station. They then divide this time difference by the difference in wave velocities to obtain the distance to the epicenter. The earthquake must have occurred along a circle whose radius is the distance so calculated and whose center is the seismographic station; any three stations that record the event can be used to draw three such circles, which will intersect at a single point. This point is the epicenter.

EARTHQUAKE PREDICTION

Earthquakes are one of the most important processes that occur within the Earth because they have such a profound effect on how and where people should develop cities. Geologists understand how and where earthquakes occur, yet despite their best efforts, they still cannot accurately determine when an earthquake will happen. They are merely able to predict that a large earthquake will occur in a particular region "in the near future." Very great earthquakes of magnitude 8 or greater, such as the San Francisco earthquake of 1906, occur about every five to ten years throughout the world. Industrialized societies, such as Japan, the United States, and many European countries, have developed buildings that are capable of withstanding devastating seismic catastrophes, but other countries are not as fortunate. Furthermore, some great earthquakes occur in regions that are not considered seismically active. The great Charleston, South Carolina, earthquake of 1886 and the Tangshan, China, earthquake of 1976 are examples of seismic events that could not have been easily predicted using modern technology. In such regions, buildings are not designed to withstand devastating earthquakes. Finally, many regions of the world do not experience earthquakes on a daily basis and, thus, their governments lack the motivation to plan adequately for such potentially catastrophic events.

A. Kem Fronabarger

CROSS-REFERENCES

Continental Crust, 561; Continental Rift Zones, 579; Creep, 185; Deep-Focus Earthquakes, 271; Earth's Structure, 37; Earthquake Distribution, 277; Earthquake Engineering, 284; Earthquake Hazards, 290; Earthquake Locating, 296; Earthquake Magnitudes and Intensities, 301; Earthquake Prediction, 309; Elastic Waves, 202; Experimental Rock Deformation, 208; Faults: Normal, 213; Faults: Strike-Slip, 220; Faults: Thrust, 226; Faults: Transform, 232; Heat Sources and Heat Flow, 49; Lithospheric Plates, 55; Notable Earthquakes, 321; Plate Margins, 73; Plate Motions, 80; Plate Tectonics, 86; San Andreas Fault, 238; Seismometers, 258; Slow Earthquakes, 329; Soil Liquefaction, 334; Stress and Strain, 264; Subduction and Orogeny, 92; Thrust Belts, 644; Tsunamis, 2176; Tsunamis and Earthquakes, 340.

BIBLIOGRAPHY

Bolt, Bruce A. *Earthquakes and Geological Discovery.* New York: Scientific American Library, 1993. As the title suggests, an excellent introductory text on earthquakes. Earthquake prediction is discussed extensively in one chapter. The illustrations and photographs of the effects of earthquakes add considerably to the text. Anyone interested in earthquakes will find this an invaluable source.

Doyle, Hugh A. *Seismology.* New York: John Wiley, 1995. A good introduction to the study of earthquakes and the Earth's lithosphere. Written for the layperson, the book contains many useful illustrations.

Farley, John E. *Earthquake Fears, Predictions, and Preparations in Mid-America.* Carbondale: Southern Illinois University Press, 1998. This book examines seismic activity, the practice of predicting earthquakes, and the hazards associated with them, focusing on the American Midwest. Bibliography, charts, and index.

Hodgson, John H. *Earthquakes and Earth Structure.* Englewood Cliffs, N.J.: Prentice-Hall, 1964. This source provides the reader with an understanding of how earthquakes are used to determine the structure and composition of the interior of the Earth.

Lutgens, Frederick K., and Edward J. Tarbuck. *Earth: An Introduction to Physical Geology.* 6th ed. Upper Saddle River, N.J.: Prentice Hall, 1999. This college text provides a clear picture of the Earth's systems and processes that is suitable for the high school or college reader. In addition to its illustrations and graphics, it has an accompanying computer disc that is compatible with either Macintosh or Windows. Bibliography and index.

McKenzie, D. P. "The Earth's Mantle." *Scientific American* 249 (September, 1983): 66-78. This article, written at the college undergraduate level, is a very complete description of current scientific understanding of the interior of the Earth.

Nichols, D. R., and J. M. Buchanan-Banks. *Seismic Hazards and Land-Use Planning.* U.S. Geological Survey Circular 690. Washington, D.C.: Government Printing Office, 1974. The effect of earthquakes on human-made structures is discussed in this short bulletin.

Written explicitly for the layperson by the United States government, it provides additional sources of information for land-use planning.

Press, Frank. "Earthquake Prediction." *Scientific American* 232 (May, 1975): 14-23. Press's article details geologists' current understanding of earthquake prediction. Also provides a discussion of the methods by which earthquakes can be predicted. Written at the college undergraduate level.

Press, Frank, and Raymond Siever. *Understanding Earth.* 2d ed. New York: W. H. Freeman, 1998. This text includes one of the most complete descriptions of the causes of earthquakes, their measurement, where they occur, how they can be predicted, and how they affect humans. A map of the major plates is on the inside back cover. The glossary is huge and indispensable. Senior high school and college-level students should find this text suitable for general background information.

United States Department of the Interior. *Earthquake Information Bulletin.* Washington, D.C.: Government Printing Office. This bimonthly bulletin provides the reader with a concise understanding of where earthquakes occur in the United States and which regions are likely to be affected in the future. Also lists other sources of information on earthquakes. For general and specialized readers.

NOTABLE EARTHQUAKES

Most of the casualties from great earthquakes occur from building collapse, fire, landslides, and tsunamis. Modern concepts of plate tectonics can account for the location of most great earthquakes, and sound planning can do much to minimize the dangers from them.

PRINCIPAL TERMS

AFTERSHOCKS: earthquakes that follow a major earthquake and have nearly the same focus; they are caused by residual stresses not released by the main shock

EPICENTER: the point on the surface of the Earth directly above the focus of an earthquake

FAULT: a fracture within the Earth along which opposing masses of rock slip to produce earthquakes

FOCUS: the area or point within the Earth where an earthquake originates

INTENSITY: the strength of shaking that an earthquake causes at a given point; intensity is generally strongest near the epicenter of an earthquake

MAGNITUDE: a measure of ground motion and energy release in an earthquake; an increase of one magnitude means roughly a thirtyfold increase in energy release

PLATE TECTONICS: the crust of the Earth consists of a number of moving plates; most earthquakes occur at plate boundaries where moving plates are in contact

SEISMOGRAPH: an instrument for recording motion of the ground in an earthquake; most seismographs are pendulums that remain static as the ground moves

TSUNAMI: a large sea wave caused by coastal earthquakes, probably generated by submarine landslides; not all coastal earthquakes result in tsunamis

MEASURING STRENGTH OF EARTHQUAKES

Two measures are used for describing the strength of earthquakes: intensity and magnitude. Intensity, generally rated on the twelve-point modified Mercalli scale, is the degree of shaking noted at a given point. Intensity depends on the distance to the focus, the local geology, and the observer. Customarily expressed in Roman numerals, intensity ranges from I (felt by only a few observers) to XII (total destruction; ground motion powerful enough to throw objects into the air).

Magnitude, usually expressed in terms of the Richter scale, is a measure of ground motion as measured on seismographs and is related to the total energy of an earthquake. The scale is defined so that an increase of one magnitude corresponds to a tenfold increase in ground oscillation, or approximately a thirtyfold increase in energy release. Earthquakes of magnitude 3 are often unnoticed, those of magnitude 5 produce widespread minor damage, those of magnitude 7 are considered major, and those above 8 are considered great. The greatest magnitude ever recorded is 8.9. Contrary to popular misconception, there is no upper limit to the Richter scale; it appears, however, that the crust cannot store enough elastic energy to generate earthquakes of magnitudes greater than 9.

OCCURRENCE OF STRONG EARTHQUAKES

Clear relationships exist between plate tectonics and the occurrence of great earthquakes. The magnitude of an earthquake generally corresponds to the area of fault surface where slippage occurs. The larger the slippage area, the greater the energy required to overcome friction. Because the ocean basins have a thin crust (about 5 kilometers thick), great earthquakes are rare in the ocean basins. Most great earthquakes are associated with continental crust, which has an average thickness of 40 kilometers.

The greatest earthquakes (magnitude 8.5 and higher) occur where plates converge, such as in Japan or on the west coast of South America. In these regions, one plate dips beneath the other at a shal-

A Time Line of Notable Earthquakes

Date	Magnitude	Location
1450 (c.)	?	Crete
Nov. 1, 1755	?	Lisbon, Portugal
Nov. 18, 1755	?	Off eastern Massachusetts
Jan. 24, 1556	?	Shaanxi, north-central China
Dec. 16, 1811	7.7	New Madrid, Missouri
Jan. 23, 1812	7.6	New Madrid, Missouri
Feb. 7, 1812	7.9	New Madrid, Missouri
Jan. 9, 1857	7.9	Fort Tejon, California
Apr. 3, 1868	7.9	Ka'u District, Island of Hawaii
Mar. 26, 1872	7.8	Owens Valley, California
Aug. 11, 1886	7?	Charleston, South Carolina
Feb. 24, 1892	7.8	Imperial Valley, California
Sept. 4, 1899	8.2	Near Cape Yakataga, Alaska
Sept. 10, 1899	8.2	Yakutat Bay, Alaska
Oct. 9, 1900	7.9	Kodiak Island, Alaska
Jan. 31, 1906	8.8	Ecuador
Apr. 18, 1906	7.7	San Francisco, California
Oct. 3, 1915	7.7	Pleasant Valley, Nevada
Jan. 31, 1922	7.3	West of Eureka, California
Nov. 11, 1922	8.5	Argentina
Sept. 1, 1923	8.3	Japan
Apr. 11, 1927	7.3	West of Lompoc, California
Feb. 1, 1938	8.5	Indonesia
Nov. 10, 1938	8.3	East of Shumagin Islands, Alaska
Aug. 15, 1950	8.6	India
July 21, 1952	7.5	Kern County, California
Nov. 4, 1952	9.0	Russia
Dec. 16, 1954	7.3	Dixie Valley, Nevada

Date	Magnitude	Location
Mar. 9, 1957	8.8	Alaska
July 10, 1958	8.3	Lituya Bay, Alaska
Nov. 6, 1958	8.7	Kuril Islands
Aug. 18, 1959	7.3	Hebgen Lake, Montana
May 22, 1960	9.5	Chile
Mar. 9, 1964	9.2	Prince William Sound, Alaska
Mar. 28, 1964	9.2	Alaska
Feb. 4, 1965	8.7	Rat Islands, Alaska
May 31, 1970	7.1	Peru
July 28, 1976	8.2	Tangshan, China
Oct. 28, 1983	7.3	Borah Peak, Idaho
Sept. 19, 1985	8.0	Mexico City, Mexico
May 7, 1986	8.0	Andreanof Islands, Alaska
Nov. 30, 1987	7.9	Gulf of Alaska, Alaska
Dec. 7, 1988	8.0	Armenia
May 23, 1989	8.2	Macquarie Islands region
Oct. 17, 1989	7.1	San Francisco, California
June 28, 1992	7.6 MS	Landers, California
Aug. 8, 1993	8.0	South of Mariana Islands
Jan. 17, 1994	6.7	Northridge, California
June 9, 1994	8.2	Northern Bolivia
Oct. 4, 1994	8.3	Kuril Islands
Jan., 1995	7.2	Kobe, Japan
Apr. 7, 1995	8.0	Tonga Islands
July 30, 1995	8.0	Near Coast of Northern Chile
Oct. 9, 1995	8.0	Near Coast of Jalisco, Mexico
Feb. 17, 1996	8.2	Irian Jaya Region, Indonesia
Mar. 25, 1998	8.8	Balleny Islands Region
Nov. 29, 1998	8.3	Ceram Sea
Aug. 17, 1999	7.8	Izmit, Turkey
Jan. 28, 2000	6.7	Kuril Islands

NOTE: Widely differing magnitudes have been computed for some of these earthquakes; the values differ according to the methods and data used. For example, some sources list the magnitude of the 8.7 Rat Islands earthquake as low as 7.7. On the other hand, some sources list the magnitude of the February 7, 1812, New Madrid quake as high as 8.8. Similar variations exist for most events on this list, although generally not so large as for the examples given. In general, unless otherwise noted (as in the case of MS, or surface-wave magnitude), the magnitudes given in the list above have been determined from the seismic moment, when available. For very large earthquakes, this moment magnitude, MW, is considered to be a more accurate determination than the traditional amplitude magnitude computation procedures. Note that all of these values can be called "magnitudes on the Richter scale," regardless of the method used to compute them.

SOURCE: United States Geological Survey, National Earthquake Information Center, http://neic.usgs.gov/neis.

low angle, resulting in a very large area of fault surface. Rifts, where continental crust is pulled apart, and transcurrent faults, where one block of continental crust slides horizontally past another, also have produced earthquakes above magnitude 8. The most famous transcurrent fault is the San Andreas fault of California. A few great earthquakes have also occurred well within plates. Some are reasonably well understood: Most of the earthquakes of China and central Asia are a response to the collision of India with Asia. Others, such as the Charleston and New Madrid earthquakes in the United States, are poorly understood.

EARLY EARTHQUAKES

Little is known of great earthquakes of the distant past. The casualty figures reported for ancient earthquakes are often unreliable. Nevertheless, it can usually be assumed that earthquakes that devastated large areas also inflicted great casualties. Even for modern earthquakes, destruction is often so great that casualty figures can only be estimates; different sources frequently list casualty figures differing by many thousands.

Perhaps the earliest great earthquake to have major historical impact struck the Minoan civilization on Crete about 1450 B.C.E. During this earthquake, it appears that all the major place complexes on Crete were destroyed. This earthquake possibly was related to the catastrophic eruption of Thera (Santorini), a volcano in the Aegean Sea approximately 120 kilometers north of Crete. The exact order of events is still uncertain.

The first earthquake to be well described destroyed the Greek city of Sparta in 464 B.C.E., killing a reported 20,000 people. In A.D. 62, the city of Pompeii in Italy was severely damaged by an earthquake. Pompeii is famous for being buried by an eruption of Vesuvius seventeen years later. An earthquake on July 21, 365, devastated Alexandria, Egypt, killing 50,000 and destroying the Pharos, or lighthouse—one of the Seven Wonders of the World.

The greatest killer earthquake in history struck Shaanxi in north-central China on January 24, 1556. In this region of China, many traditional dwellings were dug into hillsides of loess (wind-deposited silt). Collapse of these cave homes and landslides triggered by the earthquake reportedly killed 830,000. The area of devastation was so large that the death toll was certainly in the hundreds of thousands.

MODERN EARTHQUAKES

The catalog of famous modern earthquakes begins with the Lisbon earthquake, also known as the All Saints Day earthquake, of November 1, 1755. The city of Lisbon, Portugal, was demolished by three shocks between 9:30 and 10:00 A.M., with additional major aftershocks at 11:00 A.M. and 1:00 P.M. Approximately 70,000 people were killed by building collapse, fire, and a tsunami. Considerable damage also occurred in nearby Morocco. The Lisbon earthquake is sometimes listed as one of the greatest earthquakes of all time, producing widespread destruction as far away as Algeria and even being felt in the West Indies. In reality, the earthquakes in Algeria and the West Indies were separate events unrelated to the Lisbon earthquake. The earthquake produced effects far beyond the region where the shock was actually felt. Lake oscillations (seiches) were noted all over Western Europe, clocks stopped, and church bells rang. Many of these phenomena were noted and recorded carefully; observations showed that a wavelike disturbance had traveled outward from Lisbon. The Lisbon earthquake was thus the first earthquake to be studied systematically by modern scientific methods.

By coincidence, one of the strongest earthquakes ever to strike New England occurred off eastern Massachusetts on November 18, 1755. This event, which occurred just before news of the Lisbon shock reached America, heightened American consciousness of the Lisbon disaster. The first earthquakes to be recorded in the United States were those that struck the New Madrid, Missouri, area on December 16, 1811, January 23, 1812, and February 7, 1812. These events are among the few recorded earthquakes of intensity XII, and they took place in a region not generally considered earthquake-prone. Surface effects in the epicentral area were profound. The Mississippi River was churned into turmoil, and large tracts of unstable ground were affected by surface cracks and subsidence. The shocks were felt as far away as New Orleans and caused church bells to ring in Boston. Because of the sparse population in the New Madrid area at that time, only one death was reported. The New Madrid earthquakes, despite the vast area over which they were felt, were not of extremely large magnitude: They probably had a magnitude between 7.5 and 8. In the central United States, flat-lying and uniform rock layers transmit seismic waves with high efficiency, so that an earthquake at New Madrid is felt over a much larger area than an equally powerful earthquake in a geologically complex region such as California.

On January 9, 1857, a major earthquake (probably magnitude 8) struck Southern California. At least 60 kilometers of the San Andreas fault ruptured near Fort Tejon, north of Los Angeles. A strong earthquake (possibly magnitude 7) struck Charleston, South Carolina, on August 31, 1886. The earthquake was felt over most of the east coast of the United States and killed approximately 110 people. This earthquake was the first in the United States to receive wide scientific attention.

EARLY TWENTIETH CENTURY EARTHQUAKES

For many Americans, the word "earthquake" is synonymous with the San Francisco earthquake of April 18, 1906. The earthquake, with a magnitude of 8.3, was officially reported to have killed about 700, but later estimates have placed the death toll as high as 2,500. In October, 1989, an earthquake of magnitude 7.1, known as the Loma Prieta earthquake, would again leave the city with fatalities. The 1906 earthquake triggered fires that could not be fought because of ruptured water mains. As a result, a large area of the city was burned. From a scientific standpoint, the earthquake is important because it revealed the extent of the San Andreas fault. North of San Francisco, fence lines and roads were offset as much as 6 meters by the fault. The fault ruptured for at least 280 kilometers, possibly as much as 400.

On September 1, 1923, an earthquake known as the Kwanto earthquake, of magnitude 8.3, destroyed much of Tokyo and Yokohama, Japan. This earthquake is notable for the devastating fire that followed it. The earthquake struck when thousands of open cooking fires were in use all over Tokyo. Traditional Japanese construction, which relies extensively on wood and bamboo, is very resistant to collapse in earthquakes but is also very combustible. The earthquake set thousands of fires that coalesced into a firestorm—a self-sustaining whirlwind in which updrafts above the fire draw air in from the outside and keep the fire supplied with oxygen. About 140,000 people died. Forty thousand of those who died had taken refuge in an open square and suffocated from lack of air.

LATER TWENTIETH CENTURY EARTHQUAKES

A little-known earthquake (magnitude 7.9) in southeastern Alaska on July 9, 1958, is remarkable for creating the highest wave ever recorded. The earthquake triggered an avalanche into one arm of Lituya Bay, sending the water 530 meters over a ridge on the other side of the bay. Anchorage, Alaska, was damaged by a magnitude-8.3 earthquake on Good Friday, March 27, 1964. Much of the damage to Anchorage was the result of liquefaction of an unstable layer of clay a few meters below the surface. When the seismic shaking liquefied the clay, the ground above broke up, tilted, or collapsed. A tsunami, reaching up to 30 meters in height, devastated the nearby coast. Of the 131 people killed in Alaska, 122 were killed by the tsunami. The tsunami swept down the coast of North America, causing little damage in most places. At Crescent City, California, however, the bottom topography of the harbor focused the wave, which swept into the center of town, killing twelve people. Surveys of the epicentral region showed that almost 300,000 square kilometers of crust had been measurably deformed. Some points on the coast moved seaward by 20 meters; shorelines were uplifted by 15 meters in places. These motions are among the greatest ever documented for any earthquake.

A magnitude 7.7 earthquake in Peru on May 31, 1970, killed about 70,000 people, including the victims of one of the worst landslide disasters in history. The earthquake triggered a rock and ice avalanche from the summit of 6,768-meter Huascaran, the highest peak in Peru. A portion of the landslide rode over a 250-meter ridge and buried the town of Yungay, killing approximately 20,000 people. This earthquake was the worst earthquake disaster in the Southern Hemisphere.

The greatest earthquake disaster of the twentieth century in terms of loss of life—and the second greatest in history—took place on July 28, 1976, when a magnitude 8.2 earthquake struck Tangshan in northeastern China, an urban area with about 10 million people. According to the most widely accepted estimate, 600,000 people were killed.

The worst earthquake to strike North America killed 20,000 people in Mexico City on September 19, 1985. The epicenter of the magnitude 8 earthquake was actually on the Pacific coast, some 400 kilometers from Mexico City, yet damage on the coast was light. Buildings on the coast were generally modern, well built, and with foundations on bedrock. Mexico City, in contrast, is built on an ancient lake bed. Unconsolidated sediment shakes badly in earthquakes, accounting for the great damage in Mexico City. Many modern steel-frame buildings were undamaged, while poor-grade masonry suffered badly.

On December 7, 1988, an earthquake measuring magnitude 8 killed an estimated 80,000 people in Soviet Armenia. This event was notable for its political impact, because it happened at a time when the Soviet Union appeared to be moving toward greater political openness. For the first time in many years, the Soviet Union accepted foreign relief efforts after a natural disaster and permitted foreign news coverage at a disaster scene.

On October 17, 1989, a magnitude 7.1 earthquake centered 20 miles from downtown San Francisco at Loma Prieta caused greatest damage in the San Francisco Marina District, which is built upon unstable, water-saturated landfill. The earthquake caused widespread damage to the road system, including collapse of the I-280 Skyway, many landslides along the coastal highway, and at least sixty-three deaths. On January 17, 1994, a magnitude 6.7 earthquake on a previously unknown fault rocked Northridge, California, in the San Fernando Valley for 40 seconds. Damage was estimated at $15-30 billion with 63 dead, thousands injured, nine freeways destroyed, and 250 ruptured gas lines. Power was cut to 3.1 million people, and 40,000 were left without water. A magnitude 7.2 earthquake rocked Kobe, Japan, in January, 1995. Although it lasted only 20 seconds, it caused more than 5,000 fatalities, 25,000 injuries, and at least $30 billion in damage.

On August 17, 1999, a magnitude 7.8 earthquake near Izmit, Turkey, 55 miles east of Istanbul lasted for 45 seconds, flattened 60,000 buildings, caused up to $6.5 billion in direct property loss, and killed more than 30,000 people. Nearly 300 aftershocks rocked the region in the next 48 hours.

STRONG-MOTION STUDIES

Great earthquakes present special problems and opportunities for geologists. Because of their great energy release, earthquakes are detected clearly by instruments all over the planet; these records frequently reveal details of Earth's structure that cannot be detected on the records of smaller earthquakes. The infrequency and unpredictability of great earthquakes, however, mean that instruments and observers are rarely close by when the event occurs, and instruments that are close by are often destroyed.

Ground motion during great earthquakes can be measured by special seismographs called strong-motion seismographs. Strong-motion studies require that instruments be set up in locations that might experience major earthquakes. These instruments are left in place, possibly for years. After remaining dormant for a long time, the instruments must work properly when the earthquake occurs. The need to place and periodically tend instruments that may never record an event makes strong-motion studies expensive.

It is possible to simulate the effects of earthquakes on buildings. During the planning stage, models of the proposed building can be tested on a vibrating table or through computer modeling. Existing buildings can be shaken artificially. The apparatus for testing buildings consists of a set of large, rotating, off-center weights. Sensors at critical points in the building can detect motion without subjecting the building to destructive vibrations. Corrective measures might include reinforcing weak portions of the structure or redesigning connecting wings so that they can vibrate independently.

LONG-TERM SEISMIC STUDIES

Short-term earthquake prediction on the lines of severe weather warning is probably not achievable in the near future. Geologists are pursuing a variety of studies aimed at assessing the long-term likelihood of great earthquakes. One obvious and low-cost approach is simply to compile all historical records of earthquakes. China and the Middle East, areas with the longest and best-written records, show variations in intensity and location of earthquakes on a time scale of centuries. The short historical record of the United States is insufficient for long-term seismic studies.

One way to extend the record of great earthquakes is to look for geological changes created by ancient events. In Japan, uplifted shorelines have been identified with specific historical earthquakes. At Pallett Creek, north of Los Angeles, trenches across the San Andreas fault have revealed evidence of earthquakes over the last 2,000 years. Each earthquake ruptured sediment layers below the then-existing ground surface. Radiocarbon dating (using radioactive carbon in the sediment as a geologic clock) establishes the age of each fault break. The average interval of great earthquakes in this area is approximately 140 years, but actual intervals have ranged from 75 to 300 years.

EARTHQUAKE HAZARDS

Most of the casualties from great earthquakes result from a few basic causes. Building collapse is a major cause of loss of life. Wood-frame buildings, which are flexible, and steel-frame buildings, which are very strong, are the safest kinds of buildings during earthquakes. Nonreinforced masonry and adobe (mud brick) are the most dangerous; unfortunately, these construction styles are very common in underdeveloped nations. Fire is another major threat in urban areas. Earthquakes overturn stoves and furnaces, rupture gas lines, and create electrical short circuits. At the same time, ruptured water mains and streets blocked with rubble impede fire-fighting efforts. Earthquake-induced landslides are a hazard in mountainous areas and have caused tremendous loss of life.

Tsunamis are a threat in coastal areas. Believed to be generated by submarine landslides, tsunamis are waves of low height and long length that travel at up to 600 kilometers per hour. Because of their breadth and low height, they are entirely unnoticed by ships at sea but can cause great damage when they reach shore, sometimes thousands of kilometers away. Whether a tsunami causes damage depends greatly on its direction of travel, on local tide and weather conditions, and particularly on the bottom topography near shore. Tsunami warnings are routinely issued after large earthquakes.

EARTHQUAKE MYTHS

There are a few misconceptions about great earthquakes. After a newsworthy earthquake, people often wonder if earthquakes are becoming unusually frequent. The reverse was true in the twentieth century: There were about two earthquakes per year of magnitude 8 on the average, in contrast to an annual average of eight during the years 1896-1907. One apparent pattern is real, however: Great killer earthquakes are becoming more common. The reason is demographic rather than geologic. Many seismically active regions are in underdeveloped nations where populations, especially in cities, are growing explosively and where construction standards are often poor. The population at risk from earthquakes is steadily increasing.

There are a few geologic misconceptions about earthquakes. Earthquakes frequently cause ground subsidence in areas underlain by poorly consolidated materials, often causing cracks to open on the surface, but stories of fissures opening and engulfing people, buildings, or even entire villages are unfounded. Most of these stories are probably inspired by landslides. Earthquakes and volcanoes tend to occur in the same geologic settings, and there are some recorded cases of major earthquakes associated with the eruption of a nearby volcano. As a general rule, though, earthquakes do not trigger volcanic activity. Also, the earthquakes that accompany volcanic eruptions are generally not very large.

Steven I. Dutch

CROSS-REFERENCES

BIBLIOGRAPHY

Anderson, D. L. "The San Andreas Fault." *Scientific American* 224 (February, 1971): 52. A description of the most famous North American fault, particularly good for its block diagrams showing the complex southern portion of the fault system. *Scientific American* is written for nonspecialists at a college reading level.

Boore, D. M. "The Motion of the Ground in Earthquakes." *Scientific American* 237 (December, 1977): 68. A summary of how earthquakes occur, the types of motions they cause, and their effects on structures.

Clarke, Thurston. *California Fault: Searching for the Spirit of State Along the San Andreas.* New York: Ballantine Books, 1996. Clarke traveled the length of the San Andreas fault collecting first-hand accounts from earthquake survivors and predictors. Along with the entertaining stories, Clarke provides historical and scientific information about the fault.

Coffman, Jerry L., Carl A. Von Hake, and C. W. Stover. *Earthquake History of the United States.* U.S. Department of Commerce Publication 41-1. Washington, D.C.: National Oceanic and Atmospheric Administration and U.S. Geological Survey, 1982. The most detailed general reference on earthquakes in the United States. Contains lists of events by geographic area, descriptions of all widely felt earthquakes, maps of earthquake epicenters, and detailed references. Written at a nontechnical level. A must for any student of earthquakes.

Collier, Michael. *A Land in Motion: California's San Andreas Fault.* San Francisco: Golden Gates National Parks Association, 1999. Filled with beautiful color photographs that accompany text intended for the nonscientist, *A Land in Motion* gives the reader excellent insight into earthquakes and their aftermaths. There are also many diagrams and graphs that explain subduction, faults, and orogeny.

Fradkin, Philip L. *Magnitude 8: Earthquakes and Life Along the San Andreas Fault.* Berkeley: University of California Press, 1999. Written for the layperson, *Magnitude 8* traces the seismic history, mythology, and literature associated with the San Andreas fault.

Molnar, P., and P. Tapponier. "The Collision Between India and Eurasia." *Scientific American* 236 (April, 1977): 30. The collision between India and Eurasia causes faulting and great earthquakes over all of China and central Asia. Simple mechanical models duplicate the behavior of the crust remarkably well.

Nash, J. R. *Darkest Hours.* Chicago: Nelson-Hall, 1976. A nontechnical encyclopedia of disasters. Descriptions of events are generally accurate, but some errors in geological terminology were noted. Has extensive reference lists for each type of disaster, mostly popular books and periodicals. Individual articles lack references, and specific events can be hard to find. For example, the article on the great 1923 Tokyo earthquake is titled "Japan."

Reasenberg, Paul A., William Bakun, William Prescott, et al., eds. *The Loma Prieta, California, Earthquake of October 17, 1989: Aftershocks and Postseismic Effects.* Washington, D.C.: Government Publications Office, 1997. A detailed account of the 1989 Loma Prieta earthquake in the San Francisco Bay Area.

Richter, Charles F. *Elementary Seismology.* San Francisco: W. H. Freeman, 1958. The author of this classic 768-page text, who was a seismologist for many years at the California Institute of Technology, developed the Richter scale for measuring the intensity of earthquakes. Judging from his book, Dr. Richter must have been an excellent teacher. Even though this source is outdated, its lucid explanations of basic principles make it a worthwhile reference. Contains excellent and detailed chapters on the complexities of earthquake locating, along with examples, charts, diagrams, and travel-time curves. Some sections using differential equations would be for upper-level college students, but most of the book, including the parts on earthquake locating, would be quite readable to any advanced high school student.

Wesson, R. L., and R. E. Wallace. "Predicting the Next Great Earthquake in California."

Scientific American 252 (February, 1985): 35. A summary of the major active faults in California, their history, and an assessment of the likelihood of activity in the near future. The most likely location for the next great earthquake is the southern San Andreas fault or one of its branches in the Los Angeles basin.

Woods, Mary C., Ray Seiple, et al., eds. *The Northridge, California, Earthquake of January 17, 1994.* Sacramento: California Department of Conservation, Division of Mines and Geology, 1995. A look at the 1994 Northridge earthquake and its effects on the San Fernando Valley and Los Angeles.

SLOW EARTHQUAKES

The Earth's outer shell, or crust, is composed of huge blocks called plates that regularly move small distances, usually several inches each year, causing them to collide with other plates. Slow earthquakes—the movements of the Earth that result from small plate movements—are barely felt but occur regularly in some areas.

PRINCIPAL TERMS

ASTHENOSPHERE: flexible rock in layers beneath the Earth's brittle crust

FAULT: a deep fissure in the Earth's surface into which rock moves

RICHTER SCALE: one of several scales used to measure an earthquake's magnitude

SEISMIC WAVE: a wave of energy released during an earthquake

SEISMOLOGIST: a scientist, often a geophysicist, who specializes in studying earthquakes

TECTONIC PLATE: any one of about ten enormous pieces that form the Earth's outer layer

KINDS OF EARTHQUAKES

Most people who think of earthquakes immediately visualize huge, destructive, earth-shattering movements of the Earth's surface that last only seconds but that bring down buildings, rupture gas and water mains, crush people under tons of rubble, and often are followed by fires. When such movements occur, newspaper headlines are filled with statistics about the amount of damage they have done and about the numbers feared dead.

Among the most severe earthquakes in modern history are the one in the Kansu Province of China in 1920, which killed 180,000 people; the Japanese earthquake of 1923, which killed some 143,000 people in Tokyo and Yokohama; the 1935 earthquake in Quetta, India, which killed more than 60,000 people; the 1970 earthquake in Peru, which killed more than 60,000 people; and the colossal 1976 earthquake in the Hopeh Province of China, which killed 750,000 people. Severe earthquakes that occur in heavily populated areas result in heavy casualties. This is particularly true of those that strike less affluent countries in which buildings are often badly constructed, resulting in their collapse when the Earth shakes violently.

Another type of earthquake, often as severe as those felt on land, are deep earthquakes beneath the ocean's surface. The ocean floor has been drastically changed by such earthquakes. If they result in casualties on land, it is usually from the tsunamis, or enormous waves, that they generate.

These waves can attain heights of more than 30 meters. When they hit developed and heavily populated areas of a shoreline, they can crush everything in their paths and leave behind incredible destruction and thousands of casualties.

Slow earthquakes, also called silent or quiet earthquakes, are undramatic and receive no headlines in the press. They occur with considerable frequency, although most people are unaware of their existence even if they are in an area where considerable seismic activity is taking place. Seismographs may record their occurrence, but few people feel threatened by slow earthquakes because they do not cause the Earth to tremble, books to fall from library shelves, or walls of canned goods to fall on the floors of supermarkets. Their destructive force is cumulative; it takes place over substantial periods of time, causing such minute changes on a day-to-day basis that these changes are not apparent to the naked eye.

STRUCTURE OF THE EARTH

Regardless of what kind of earthquake one is considering, all earthquakes have similar underlying causes. In order to understand these causes, one must consider how the Earth is constructed. The planet is composed of three basic parts. The one with which people are most familiar is the crust, the outer layer that, below the surface, consists of solid rock. The crust has an average thickness of about 32 kilometers beneath the Earth's seven continents, although it is considerably thinner beneath the sea, where its thickness averages about 5 kilometers.

Underlying the crust is the mantle. It, like the crust, is composed of solid rock, but this rock is extremely hot. The mantle is thick, extending in many places more than 3,000 kilometers below the crust. Its rigid upper portion is called the lithosphere. Beneath it is a weaker area of the mantle, the asthenosphere, which, being closer to the Earth's molten core, is much hotter than the lithosphere. The lithosphere may be nearly 100 kilometers thick beneath the continents and some oceanic areas, but it shrinks to just 8 to 10 kilometers in thickness beneath submerged ridges in mid-ocean.

Inward toward the Earth's center from the lithosphere and asthenosphere are the two major parts of the Earth's core, the liquid core and the solid core. The liquid core has a radius of about 2,300 kilometers and consists of molten iron and nickel whose temperature averages about 5,000 degrees Celsius. The solid core, which is at the Earth's very center, has a radius of just under 1,300 kilometers and is composed of solid iron and nickel.

Ancient people had various quaint explanations about what caused earthquakes. The ancient Greeks thought that the titan Atlas carried the world on his shoulders and that every time he shifted the weight of this great burden, the Earth moved, causing earthquakes. In India, people conceived of the Earth as an object balanced on the head of an elephant riding on the back of a huge tortoise. Whenever either animal moved, an earthquake resulted. Other theories viewed the world as being carried by giant catfish, whales, or oversized gods who rode in sleds pulled by dogs. Some religions, including Judaism and Christianity, once viewed earthquakes as expressions of God's anger and as the punishment he visited upon humankind for its transgressions.

Modern science has been slow to offer rational physical explanations for earthquakes. John Mitchell, in 1760, suggested that they are caused by the movement of subterranean rocks. Few accepted this explanation, and those who did thought that such movement was caused by gigantic explosions deep inside the Earth. It took another hundred years before Robert Mallet, an engineer from Ireland, contended, in 1859, that the causes of earthquakes were strains in the Earth's crust. After 1960, most seismologists accepted the theory of plate tectonics as the cause of earthquakes.

PLATE TECTONICS

According to the theory of plate tectonics, which has gained wide acceptance, the Earth was once a solid landmass surrounded by a great sea. The planet began to cool after its fiery formation some 5 billion years ago. As it cooled, its surface cracked. Over hundreds of millions of years, parts of the once-solid landmass drifted away, forming seven large and twelve small islands, all with ragged edges. The large islands are the Earth's seven continents. These islands, or tectonic plates, float and are in constant but often quite limited motion. For example, the two plates that exist on the western part of the North American continent, the Pacific plate and the North American plate, hardly move at all. The Pacific plate drifts north at the barely perceptible rate of about 5 centimeters per year. The North American plate drifts southwest at a similar rate.

Despite the slow movement of the North American plate, it has been estimated that over hundreds of millions of years, the North American continent could, through continental drift, collide with Australia. Between the Pacific and North American plates lies the San Andreas fault, a gash in the Earth's surface that runs more than one-half the length of California. The movement of these plates results in slow or silent earthquakes.

When the edges of the two plates collide, however, as they did in the San Francisco earthquake of 1906 and the Northridge earthquake of 1994, the result is a major earthquake. The Northridge quake, which registered 6.8 on the Richter scale, revealed the existence of a hidden thrust fault and a horizontal fault that had previously gone undetected. The Northridge earthquake was followed by more than one thousand aftershocks as the Earth beneath the area resettled. Many slow earthquakes preceded the Northridge disaster as foreshocks and followed it as aftershocks. Slow earthquakes often presage the coming of major earthquakes. As increased knowledge about slow earthquakes evolves, seismologists are beginning to understand more fully how to interpret the often subtle signals they send. The interpretation of these signals can help predict future deep earthquakes.

Signs of impending earthquakes exist—probably quite often in the form of slow earthquakes—that cause animals to react in anticipation of severe, deep earthquakes. The behavior of animals in zoos in the hours preceding a severe earthquake shows clear signs that they are disquieted and sense something, perhaps minute subterranean vibrations, that humans are not able to perceive. The fields of plate tectonics and seismology are becoming more and more sophisticated as technology produces ever more sensitive instruments for detecting seismic activity.

TYPES OF SLOW EARTHQUAKES

Most people think of earthquakes as cataclysms in which a crack breaks through the Earth's crust at a speed of several kilometers per second, causing a violent shaking of the ground, severe damage to structures, and injury to living things in the quake zone. In many parts of the world, however, the development of a crack along a fault line occurs at a speed of less than 1 meter per second, with some slips even measured in millimeters per year. Three faults in California—the Hayward, San Andreas, and Calaveras Faults—demonstrate the great variety of seismic activity, ranging from the ordinary earthquakes that occur from rapidly developing breaks in the Earth's crust that suddenly release waves of stored elastic-strain energy, to a variety of smaller tremors.

Among the types of earthquakes that geologists and seismologists have discovered and named are slow earthquakes, defined as having speeds of hundreds of feet per second; silent earthquakes, defined as having speeds of tens of feet per second; strain migration events, measured at speeds of centimeters per second; and creeping earthquakes, with speeds measured in millimeters per second. These varieties are not always measurable on typical seismographs. Few of them attract attention as they are taking place.

Slow earthquakes can, at times, cause rapid ruptures that produce high-frequency sound waves, but more often they take a much longer time to rupture through the Earth's crust than ordinary earthquakes of comparable magnitude. Some slow earthquakes occur in oceanic transform faults, as happened on June 6, 1960, in the Chilean transform fault, which ruptured for about one hour as a series of small, barely detectable breaches.

Silent earthquakes have been so named because they are never accompanied by the high-frequency sound waves that most seismographs need to register seismic events. Some researchers have employed delicate instruments that measure tectonic strain to detect silent earthquakes. Such instruments also have revealed creeping movement of about 10 millimeters per second in parts of California's San Andreas fault. The low-frequency waves of a silent earthquake moving about 0.3 meters per minute were recorded shortly before the 1976 earthquake in Fruili, Italy, and again in 1983 before a severe earthquake hit the Japan Sea. Seismologists think that the occurrence of some slow and some silent earthquakes may be warning signs that, if heeded, could prevent substantial loss of life when an ensuing ordinary earthquake is on the brink of shattering a region. The stick-slip earthquake, with its jerky, sliding motion at a fault, usually comes after a slip with propagation speeds of 20 to 200 meters per second. Not all such slips can be detected on the typical seismographs that most geophysicists and seismologists use. Close to the earthquake, silent earthquakes can be recorded geodetically and by using strainmeters. Only digital, broadband seismographs are able to record seismic waves of such low frequencies.

Pacific Rim nations are shaken yearly by thousands of tiny earthquakes resulting from the collision of oceanic and continental plates. Some of these are slow or silent earthquakes that relieve the pressure that is built up in subterranean rocks when one tectonic plate rams into another. Severe earthquakes are the result of several years of pressure buildup, but slow and silent quakes relieve the pressure more gradually and possibly act as the safety valves that prevent the Earth from experiencing more numerous major earthquakes than it does.

SLOW CREEP AT WORK

Hollister, California, about 160 kilometers southeast of San Francisco, lies close to the San Andreas fault. Studies of major fault lines near Hollister have revealed a gradual slippage beneath the Earth's surface, although it has not been possible to measure this slippage with total accuracy. A winery constructed in Hollister in 1939 is located almost on top of the San Andreas fault. In 1956, the winery began to experience damage that could not be easily explained but that could no longer be ignored. Strong reinforced concrete walls and floors in one of the warehouses were gradually crumbling. None of the local people remembered any overt seismic action that could explain the phenomenon.

Finally, because of the winery's location close to the known fault line, the owners engaged seismologists to assess the situation. They found that an active branch of the fault zone ran directly below the building. They discovered that the two sides of this fault line were moving past each other at an estimated rate of 1.3 centimeters every year. Although such movement does not attract immediate attention, over fifty years the distance involved is more than 0.6 meters, which causes damage readily observable by anyone who looks at the building.

As such movement continues, structures are weakened and are finally felled by it. Because the San Andreas is a right-lateral fault, the winery's west side was steadily moving north of its east side. When cracks appeared in the floor and walls, they were patched up. Sagging walls were reinforced. Since 1956, the situation has been monitored carefully. It has been determined that the creep continues, as it surely will do in the foreseeable future.

This sort of creep is related to slow earthquakes in that it does not involve a dramatic underground upheaval that happens in a matter of minutes, although creep is thought not to be entirely gradual. It often occurs in a matter of seven to ten days at a time, after which there is a period of quiescence for weeks or months. It can, however, continue for decades and be barely detected in areas that are neither built up nor heavily populated. In 1960, recorders clocked an earthquake in Hollister in which instant creep of about 0.3 centimeters occurred. The frequency with which earthquakes occur may have an influence on periods in which seismic activity takes place. An earthquake in the area near Hollister in 1939 separated adobe walls from side walls and pulled girders from their brick moorings. Rents appeared in the ground around the winery. A major jolt in 1960 severely shook the winery, causing damage to it.

The slow creep beneath the winery and in other areas in the Hollister area is under the constant scrutiny of seismologists, who are trying to determine why slow creep occurs along some areas of the fault but is not observed in other nearby areas. A tunnel for the Los Angeles Aqueduct that was constructed in 1911 and crosses the San Andreas fault has remained intact for its entire existence, with no signs of seismic activity.

R. Baird Shuman

CROSS-REFERENCES

BIBLIOGRAPHY

Beroza, G. C., and T. H. Jordan. "Searching for Slow and Silent Earthquakes Using Free Oscillations." *Journal of Geophysical Research* 95 (1990): 2485-2510. The authors relate how free oscillations, which ring like a bell, were recorded over a decade, most of them caused by large, ordinary earthquakes. In some instances, the earthquake involved was not big enough to cause free oscillation, suggesting that they were slow earthquakes. Of the 1,500 free-oscillation earthquakes, 164 were not accompanied by a recorded earthquake.

Bolt, Bruce A. *Earthquakes.* Rev. ed. New York: W. H. Freeman, 1993. This comprehensive overview of earthquakes is easy to understand and highly informative. Its material on seismic waves and seismography is of great significance to those interested in various types of earthquakes and in where and how they occur. Bolt writes clearly and with authority in this field.

Ebert, Charles H. V. *Disasters: Violence in Nature and Threats by Man*. Dubuque, Iowa: Kendall/Hunt, 1988. Chapter 1, "Earthquakes," and Chapter 4, "Tsunami Waves and Storm Surges," should prove of particular interest to readers interested in earthquakes, although parts of other chapters also contain relevant information. Chapter 3, for example, relates how earthquakes can trigger landslides and avalanches.

Koyhama, Junji. *The Complex Faulting Process of Earthquakes*. Dordrecht, Netherlands: Kluwer, 1997. This well-documented, carefully researched study, though quite technical, is excellent in the scope of its coverage. Koyhama explains how various faults have developed and what courses they have taken.

Levy, Matthys, and Mario Salvadori. *Why the Earth Quakes: The Story of Earthquakes and Volcanoes*. New York: W. W. Norton, 1995. Written with general readers in mind, this volume is exceptionally clear. Its illustrations, both verbal and graphic, add to the accessibility of what the authors are saying. Chapter 9, which focuses on California's San Andreas fault, should be of particular interest to American readers, illustrating as it does how various forms of seismic activity can occur simultaneously along the fault.

Rundle, John B., Donald L. Turcotte, and William Klein, eds. *Reduction and Predictability of Natural Disasters*. Reading, Mass.: Addison-Wesley, 1996. Among the nine contributions to this book on earthquakes, "Thoughts on Modeling and Prediction of Earthquakes," by S. G. Eubanks, and "A Hierarchical Model for Precursory Seismic Activation," by W. I. Newman, D. L. Turcotte, and A. Gabrielov, are the most relevant to the topic of slow earthquakes. The contributions to this volume, while remarkably significant, are highly specialized and may be difficult for beginners.

Schenk, Vladimir, ed. *Earthquake Hazard and Risk*. Dordrecht, Netherlands: Kluwer Academic Press, 1996. The twenty contributions to this volume, each written by acknowledged specialists in the field, focus on the prediction and management of earthquakes. The volume covers the field thoroughly but would be more useful if it contained an index. The essays seem more specialized than most beginners can easily handle.

Walker, Sally M. *Earthquakes*. Minneapolis: Carolrhoda, 1996. Written with the young adult audience in mind, Walker's account is lively, interesting, and accurate. The illustrations are colorful and cogent. Various easily comprehended charts and tables add considerably to the substance of the book's engaging text. The glossary is of especial value to readers, as is the index.

Yeats, Robert, Kerry Sieh, and Clarence R. Allen. *The Geology of Earthquakes*. New York: Oxford University Press, 1997. This comprehensive study of the geology of earthquakes, written essentially as a textbook, is well presented and thorough. Its material on slow, silent, creeping, and strain migration occurrences, though brief, is as solid as any in the field. The writing style, even in the presentation of the more technical material, is extremely appealing.

Zebrowski, Ernest, Jr. *Perils of a Restless Planet: Scientific Perspectives on Natural Disasters*. New York: Cambridge University Press, 1997. Chapter 1, "Life on Earth's Crust," and chapter 6, "Earth in Upheaval," are the most useful to those seeking more information about types of earthquakes. Zebrowski obviously has a strong background in ancient Greek and Roman mythology, as well as a scientist's grasp of the mechanics of earthquakes. The illustrations are well chosen, and the appendices include a great deal of technical information in charts that make it easily understandable.

SOIL LIQUEFACTION

Soil liquefaction is the group of processes by which otherwise solid soil particles are shaken apart by earthquakes or collapse away from edge-to-edge contact with one another and become temporarily supported by the pore water contained within them. The resulting fluid mush can allow buildings to sink; this phenomenon is responsible for considerable loss of property and life.

PRINCIPAL TERMS

BEARING CAPACITY: the ability of granular soil materials to support the weight of building structures

CLAY MINERALS: the diverse group of very finely crystalline mineral structures that are dominantly composed of silicon, aluminum, and oxygen and that have very different water retention capabilities

COHESION: molecular attraction by which the particles of a body are united throughout the mass, whether alike or unlike

ELECTROSTATIC CHARGE: the fundamental atomic force in which objects that have a similar electric charge repel each other whereas those with unlike charges attract each other

FLOCCULATION: the sedimentation process by which a number of individual minute suspended particles are held together in clotlike masses by electrostatic forces

REMOLDING: the property of some sensitive clays upon disturbance to reorient their particles, which softens them, and to flow in a liquid form

SENSITIVE (also QUICK): describes fine-grained deposits that are characterized by considerable strength in the undisturbed condition, but upon disturbance their ability to support themselves declines dramatically

VAN DER WAALS FORCE: a weak electrostatic attraction that arises because certain atoms and molecules are distorted from a spherical shape so that one side of the structure carries more of the charge than does the other

TYPES OF LIQUEFACTION FAILURE

Soil liquefaction is the abrupt and temporary change of seemingly solid soil materials into liquid mush. The soil can lose all of its cohesion and bearing capacity in a matter of moments. Once the process begins, most commonly when an earthquake strikes, ground that had supported a hospital or a high-rise apartment can suddenly become a fluid into which the buildings sink like rocks into quicksand. Anything built upon such materials can slip or sink into the new liquid. Buried gasoline or septic tanks can suddenly become buoyant and float to the surface.

A number of different sorts of liquefaction failure are recognized; several occur on land and a few are subaqueous. Quick condition failure (not to be confused with sensitive or quick-clay landslips) is the complete loss of bearing capacity caused by liquefaction of sands and silts so that structures sink or rise in material that appears otherwise solid. Flow landslips, or flowslides, can occur on moderate slopes on land or beneath water and involve either sands or clays. These types tend to retrogress, or work their way backward, slice by slice as the material beneath them liquefies. Lateral spread landslips can occur on gentle, or nearly flat, slopes. Many lateral spreads are in quick clays. In extreme cases, the liquefaction slides result from spontaneous subaqueous liquefaction that propagates, or spreads, in all directions.

WATER-SATURATED SAND LIQUEFACTION

Two classes of soil materials can liquefy: water-saturated sands and silts and the unusual sensitive, or quick, clays. In the first example, layers of loosely packed, well-sorted, fine- to medium-grained sands and coarse silts are subject to liquefaction where groundwater tables are within 10-15 meters of the ground surface. When this water-saturated sediment starts to shake apart in an earthquake or other vibration, the grains temporarily lose rigid contact with one another and collapse inward. Much of the pore water is then superfluous but does not escape at once, so that the rearranged grains cannot fit close to one another. The particle weight is thereby transferred to the pore water, its pressure increases (which reduces friction between the grains), and the soil becomes liquid for a short time.

When water-saturated sands liquefy during the course of an earthquake, sand boils erupt muddy water and sand from ground fractures and turn the surface into a quagmire. Sand boils tend to be roughly circular in plan and can have a depressed center like a volcano. After a very large earthquake, sand boils may form as much as several hundred meters across and several meters high. Their presence in an area is a reliable guide to past earthquake activity.

Water-saturated earth-fill dams are also subject to soil liquefaction, with obvious disastrous consequences. For example, during the 1971 San Fernando Valley earthquake in California, the Lower Van Norman dam collapsed. The 40-meter-high dam had been built with a core of clay surrounded by a fill of water-saturated sand. After about twelve seconds of strong shaking, a large, wedge-shaped segment of this water-saturated sand fill liquefied. Eight large blocks on the upstream side of the dam slid into the reservoir as parts of the sediment fill liquefied. Before the quake, the crest of the dam was a safe 10 meters above the water level of the reservoir. Afterward, however, only a thin barrier of 1.5 meters was left above the water level. Had this minor amount of remaining dam surface moved down but a fraction, the dam would have quickly failed, because water would have rushed down and eroded the downstream side. The 80,000 people who lived directly below the reservoir thus escaped disaster by a very slim margin.

QUICK-CLAY LIQUEFACTION

In the quick-clay type of soil liquefaction, because of a rather special geologic history, certain clays develop the ability to liquefy, or to become quick, as with quicksand. Two main theories have been advanced to account for the unusual distribution of these quick clays. The saltwater theory holds that sensitive clays are first formed where glaciers erode very fine-grained clay mineral platelets from the soils and bedrock over which they ride. Where the weight of such glaciers depresses the surface of the land close to the sea, the small platelets can be deposited directly into the salty waters.

The small clay mineral platelets of a sensitive clay carry a negative charge, and as they settle in ocean water they tend to pick up positively charged sodium ions from the salt in solution. These oppositely charged, mutually attractive forces act as a glue to cause the particles to clump together into a sort of coagulated honeycomb structure. The platelets thus develop an edge-to-edge, or "house of cards," structure held together by the electrostatic forces of the salt ions attached to the platelet edges.

A sand boil (sand volcano) during the Peruvian earthquake of 1970. Water-saturated mud and sand squirted from the two pits. (U.S. Geological Survey)

Glacial marine clays may remain quite stable as long as the salt water remains in the pore spaces of the open card-house structure. These deposits can later be raised above sea level by rebound of the land following removal of the glacier ice load. Then the salty pore waters may be flushed out of the clays by the fresh waters that normally occur above sea level. Removal of the mutually attractive electrochemical charges then sets the stage for eventual liquefaction.

A different theory explains the freshwater type of quick clays. In this case, glacial erosion also provides extremely fine particles of other than clay minerals. Instead, the fine particles (mostly quartz minerals) are so small that the normally weak van der Waals attraction is sufficient to hold them together. When the particles are small enough, the ratio of these weak attractions to the weight of the particles is greater, and the material will be cohesive, strong but brittle, and sensitive. When the cohesive bonds are broken through shaking, the short-range forces are ineffective, and a total loss of strength results. If there is sufficient pore water, the material will liquefy and flow on gentle slopes.

An example of a well-documented failure in quick clay is the St. Jean-Vianney failure of May 4, 1971, in Quebec, Canada. This landslip was located in the middle of a much larger prior failure that was at least five hundred years old. At about 7:00 P.M., instability first began with no warning on the steep west bank of the small Petit Bras tributary to the Saguenay River. The first actual failure, however, did not occur until 10:15 P.M., when, in fifteen minutes, the failure surface retrogressed 150 meters west into the slope. The resulting debris moved into the river and blocked both the river valley and the opening to the failure. The dam blocked the outflow of the liquefied material from the crater until the pressure became too high. Finally, the dam burst and allowed the remolded fluid to flow down the river valley with a wavefront about 18 meters high and traveling about 26 kilometers per hour. The mass carried with it thirty-four houses, one bus, and an undetermined number of cars. Thirty-one lives were lost as well.

SOIL STRENGTH

The strength, or load-carrying capacity, of all soils varies considerably. In addition, the strength of any specific soil can vary under different conditions of moisture and density. Sensitive soils have natural water contents already above the liquid limit, which is the moisture content at which a soil passes from a plastic to a liquid state. Plasticity is a characteristic of clayey soils that allows them to be squeezed and easily deformed without disintegration, whereas being above the liquid limit, as the name implies, allows the soil to flow easily and remold itself.

Sensitive clays are generally considered to be thixotropic in the sense that strength diminishes upon disturbance and is regained when disturbance ceases. Once collapse of the card-house matrix is initiated and flow begins, the clays seem to be remolded by lining up of the platelets and particles in a parallel or linear fashion. Remolded clays slip easily over one another in the watery mixture. Once these instabilities develop, they can spread rapidly throughout an area of retrogressive failure. The regaining of strength is not well understood but is thought to result from the gradual rearrangement of particles into positions of increasing mechanical stability under the action of new bonding forces. Perhaps the compaction and gradual expelling of excess water from the moving mass also are responsible for some restoration of strength. The artificial addition of new salts to such materials can be used in certain cases to regain strength and thereby stabilize hazardous areas.

REMOLDING

The typical lateral spread and flow landslips in sensitive clays have certain characteristics that make them distinctive. The processes are essentially a gravitational remolding that transforms the clay into a viscous slurry, or mixture of liquid and solid. Overlying sediments can break into strips or blocks that then become separated. The cracks between the blocks fill with either soft material squeezed up from below or detritus (loose material resulting from disintegration) from above.

Such failures commonly start in the lower part of a slope as a result of local oversteepening through stream or other erosion. After the initial slip, the failure spreads retrogressively, slice by slice, farther and farther into the bank. Movement generally begins quickly, without appreciable warning, and proceeds with rapid to very rapid velocity. A large, bowl-shaped crater commonly results.

SOIL LIQUEFACTION STUDY

Analysis of soil liquefaction is not extensive, because the phenomenon has been known as an important hazard for a relatively short period of time and is an unusual event in any case. The rarity of the phenomenon means that scientists have had trouble installing instruments and making measurements of the ground in places that would eventually liquefy and provide a well-documented record. In most cases, studies had to be done on ground that had liquefied in the past but had since become stable or in the rather artificial situations of the laboratory.

After soil liquefaction was first discovered following catastrophic ground failure, interviews with survivors provided most of the early information. Borehole cores and measurements of displaced ground showed types of sediments and characteristic depths and amounts of water involved. In many engineering applications, the undisturbed cores are subjected to various stresses to determine their behavior when shaken or loaded. Stirring of undisturbed quick clays to a liquid state in the lab, followed by resolidification upon addition of salt, is a means of analyzing the condition. Such studies allow the hazardous condition of certain areas to be represented on maps of liquefaction susceptibility that indicate places in which the phenomenon is likely to occur.

Success in observing soil liquefaction in action has occurred where instruments were previously installed in the ground to measure pore-water pressures at various depths during carefully measured earthquakes. Surprisingly, and in contrast to laboratory experiments in which liquefaction occurs at the same time as does strong shaking, pore-water pressures rose only slowly as the shaking intensified, and sandy layers completely liquefied only well after strong earthquake motions had ceased. The delay appears to occur as uneven pore-water pressures are redistributed in the ground.

EVER-PRESENT EARTHQUAKE HAZARD

The significance of soil-liquefaction potential is enormous: Many of the world's major cities are partly built upon weakly consolidated, water-saturated sediments. Understanding the mechanisms of liquefaction is also important in analyzing earthquakes long past; old sand boils have been used to date and to estimate magnitudes of prehistoric earthquakes. Major earthquakes in Niigata, Japan, and Anchorage, Alaska, brought about a heightened appreciation of the importance of soil liquefaction as a general geologic processes. Following these events, geologists, engineers, and planners have come to recognize soil liquefaction as an ever-present earthquake hazard and have associated it with almost all major earthquakes since.

In June, 1964, for example, a magnitude 7.3 earthquake occurred 55 kilometers from the city of Niigata on Japan's west coast. In fifty seconds of shaking, the city of 300,000 was subjected to dramatic soil liquefaction that affected thousands of dwellings and industrial structures. Much of the city was built originally upon sand deposits about 100 meters thick along the Shinano River and upon younger lowland sediments and reclaimed riverfront land. During the earthquake, subsurface sand and water flowed up and out of cracks in the ground. The liquefaction caused major destruction of highways, bridges, railroads, utilities, oil refineries, and harbor facilities. There were 3,018 houses destroyed outright and 9,750 damaged moderately or severely because of cracking and unequal settlement of the ground; much of this damage occurred on the newly reclaimed areas. In a most spectacular occurrence, a number of large apartment buildings, which had been designed to be earthquake-resistant, tipped over on their sides to settle at angles of as much as 80 degrees, though the structures themselves remained intact. People were able to escape by walking down the sides of the buildings. Several of these apartment houses were later jacked up, reinforced, and opened for reoccupation.

The Good Friday (March 27) earthquake of 1964 in Alaska was, at 8.4-8.6 magnitude, the largest ever recorded in North America. The Turnagain Heights landslide in Anchorage, Alaska, took place in flat terrain along the steep coastal bluffs that border the Knick Arm of the Cook Inlet there. Before the earthquake, the bluffs rose steeply some 30-35 meters above sea level. Marine clays and silts with layers and lenses of sand of the Bootlegger Cove Clay were exposed in the bluffs. A 6-meter thickness of sand and gravel on the flat terrace above had provided an apparently fine place to build homes. During the 1964 earth-

Anchorage, Alaska, was hit by one of the most damaging earthquakes in history, on Good Friday, March 27, 1964, with a magnitude of 8.3 (higher, by some estimates). Much of the damage to Anchorage was the result of liquefaction of an unstable layer of clay a few meters below the surface. (U.S. Geological Survey)

selves commonly have surface slope angles of as small as 3-4 degrees. Such a failure occurred at Borssele in 1874 and involved nearly 2 million cubic meters of sand.

A liquefaction process also seems to be significant in the fine silts and clays of the Mississippi Delta region, particularly in the formation of collapse depressions and elongate flows. In these sediments, the pore-water pressures are extremely large, and the pore spaces also contain large amounts of methane gas from decaying organic matter. Following initial failure of these materials, softening of the highly pressured clay/water/gas system causes remolding and strength loss similar to a type of liquefaction or quick behavior. Collapse of offshore structures and sinking of pipelines and seafloor monitoring equipment vertically into the sediment may result. Increased pore pressures during storms also cause bottom movement, collapse, and indications that the sediment can become active as a fluid. On very low-angle slopes (0.1-0.2 degree), distinct collapse depressions are formed, whereas on slightly steeper slopes (0.3-0.4 degree), more elongate flows can result.

John F. Shroder, Jr.

quake, giant blocks of Bootlegger Cove Clay and the overlying sands and gravels were set in motion toward the sea as the sand layers in the clay formation were liquefied. The first movements of the blocks began about two minutes after the onset of intense earthquake shaking. In the next five minutes, the previously flat terrain was transformed into a jumble of blocks capped by tilted trees and broken buildings. Seventy-five homes and three lives were lost in the breakup of the ground. More than seven huge blocks became widely separated as they moved toward the sea.

INCIDENCES OF SUBAQUEOUS FAILURE

In a subaqueous example of failure, the sand beds along the coast of Zeeland in the Netherlands periodically liquefy. The coast is located on a thick layer of fine quartz sand that consists of rounded quartz grains. The slope of the beach is only about 15 degrees. Once every few decades, especially after exceptionally high spring tides, the structure of the sand breaks down beneath a short section of the coastal belt. The sand flows out and spreads with great speed in a fan-shaped sheet over the bottom of the adjacent body of water. The tongue of such a flowslide is always much broader than is the source; the flowslides them-

CROSS-REFERENCES

BIBLIOGRAPHY

Dennen, W. H., and B. R. Moore. *Geology and Engineering*. Dubuque, Iowa: Wm. C. Brown, 1986. This volume is one of the most nontechnical but accurate books available on general geotechnical subjects. The sections on soil liquefaction and related phenomena are not long but are easy to understand.

Doyle, Hugh A. *Seismology*. New York: John Wiley, 1995. A good introduction to the study of earthquakes and the Earth's lithosphere. Written for the layperson, the book contains many useful illustrations.

Holzer, T. L., T. L. Youd, and T. C. Hanks. "Dynamics of Liquefaction During the Superstition Hills, California, Earthquake." *Science* 244 (April 7, 1989): 56-59. This article reports the details of the first record of a natural liquefaction event after an earthquake. An array of instruments had been installed up to 12 meters deep that recorded excess pore pressures generated once horizontal ground acceleration from the earthquake exceeded a certain threshold value.

Lundgren, Lawrence. *Environmental Geology*. Englewood Cliffs, N.J.: Prentice-Hall, 1986. A general book, with several excellent sections and photographs of soil liquefaction phenomena. Not too difficult or technical for the average interested reader.

Penick, J. L., Jr. *The New Madrid Earthquakes*. Rev. ed. Columbia: University of Missouri Press, 1981. This well-written account covers the most intense earthquakes ever to strike the North American continent and details their effects upon people, animals, waterways, and land. Some of the largest sand boils from soil liquefaction ever recorded are described here. The vivid description of the devastation wrought upon the face of the land gives a picture of the dramatic changes caused by the upheaval of natural forces.

Pipkin, Bernard W., and Richard J. Proctor. *Engineering Geology Practice in Southern California*. Belmont, Calif.: Star Publications, 1992. This book provides a detailed description of geological engineering in areas prone to earthquakes, including attempts to prepare for the destructive effects of soil liquefaction. Illustrations and bibliographic references.

Plummer, Charles C., and David McGeary. *Physical Geology*. Boston: McGraw-Hill, 1999. A college-level introductory geology textbook that is clearly written and wonderfully illustrated. An excellent sourcebook of basic information on geologic terminology and fundamentals of geologic processes. Contains a CD-ROM. An excellent glossary.

Spangler, M. G., and R. L. Handy. *Soil Engineering*. New York: Harper & Row, 1982. This book is a detailed engineering text with many equations. Nevertheless, readers will find the discussions on soil liquefaction useful even if they disregard the numerical equations.

Terzaghi, K., and R. B. Peck. *Soil Mechanics in Engineering Practice*. New York: John Wiley & Sons, 1948. This text is the classic work on soil liquefaction. Although it includes many equations, the authors have also provided plentiful written descriptions that are easy to understand.

TSUNAMIS AND EARTHQUAKES

A tsunami is a series of traveling ocean waves of extremely long length and depth generated by violent submarine disturbances associated primarily with earthquakes. The mass movement of water represented by a tsunami poses a significant danger to low-lying coastal regions, particularly along the Pacific Rim.

PRINCIPAL TERMS

CONTINENTAL SHELF: the gently sloping surface that extends between the shoreline and the top of the continental slope, which slopes steeply to the deep ocean bed

EARTHIC CRUST: the thin, outermost layer of the Earth that varies in thickness from about 30 to 50 kilometers

OCEAN WAVE: a disturbance on the ocean's surface, viewed as an alternate rise and fall of the surface

RICHTER SCALE: the measurement of the magnitude of seismic disturbance (an earthquake)

using the amplitude of seismic waves

SEISMIC SEA WAVE: an enormous wave in the ocean generated by an earthquake under the floor of the ocean or along the seacoast

SUBDUCTION ZONE: the zone, at an angle to the Earth's surface, down which an upper layer of oceanic or continental plate descends

WAVE REFRACTION: the process by which the direction of waves moving through shallow water is altered by local submarine conditions

WAVELENGTH: the distance between the crest of one water wave to the crest of the next

UNDERSEA ENVIRONMENT

While other catastrophic events can trigger a tsunami, undersea earthquakes are responsible for the majority of them. Many tsunamis occur around the Pacific Rim in regions designated by geologists as subduction zones, where the dense earthic crust of the ocean floor slips below the lighter continental shelf crust and into the Earth's mantle. subduction zones are most common along the west coasts of North and South America and the coasts of Japan, eastern Asia, and the Pacific island chains. Another subduction zone is located in the Caribbean Sea but is not considered as active as those in the Pacific.

Scientists believe the sudden movement in the seafloor during an earthquake disrupts the equilibrium of the overlying expanse of water, raising or lowering enormous amounts of it all the way from the seafloor to the surface. Once the seafloor settles into its new position, it has nowhere to go until the next disturbance. However, the water mass above it is still subject to the downward pull of gravity. Consequently, as the swell of water returns to its original position, the water around it is pushed up, creating a rippling effect or series of

waves called a tsunami. The primary factor in determining the size of the tsunami is the amount of seafloor uplifted during the undersea disturbance. A side-to-side movement of the seafloor is unlikely to cause a severe tsunami. For example, the San Andreas fault does not generate tsunamis, since its primary movements are horizontal.

Tsunami waves are not to be confused with tidal waves, which are simply the movements of water associated with the rise and fall of tides generated by the gravitational pull of the Sun and Moon. Since they often are the result of a sudden movement in the Earth's crust, tsunamis often are referred to as seismic sea waves. This can be somewhat misleading because the term "seismic" indicates an earthquake-related event, when other natural phenomena, such as landslides, volcanic eruptions, or even meteor strikes, can also generate tsunamis, though on a much less frequent basis.

WAVE MOVEMENT

The waves that form a tsunami are different from the surface waves observable from a beach. The latter are produced by winds blowing over the surface of the sea, and their size is directly dependent on the strength of the wind that creates them. The distance between these waves can range from a few centimeters to nearly 300 meters, though the normal separation is about 9 to 18 meters. The speed of the common surface wave can range from a few kilometers per hour to more than 90 kilometers per hour, with the lapse of time between two successive waves running from about five to twenty seconds.

Unlike the common surface wave, the tsunami wave is categorized as a shallow-water wave because of its extensive wavelength, which can stretch up to 500 kilometers and run for a period of ten minutes to two hours at speeds up to 800 kilometers per hour, depending on the depth of water in which it is traveling. A wave is classified as a shallow-water wave when the ratio between the water depth and wavelength becomes very small. The formula for making the determination is as follows: The speed of a shallow-water wave is equal to the square root of the product of the acceleration of gravity (98 centimeters per second per second) and the depth of the water. This formula enables seismologists to alert coastal communities about the potential of a tsunami following an earthquake. The formula also points out another significant distinction between a tsunami wave and common surface wave. The rate at which a wave runs out of energy is inversely related to its wavelength. Because of its long wavelength, a tsunami can travel thousands of kilometers across ocean waters without dissipating; in contrast, the average surface wave begins to lose its energy after a distance of a few kilometers. In 1960, an undersea earthquake off the coast of Chile generated a tsunami that had enough lasting power to kill 150 people in Japan, following an earlier strike on the Hawaiian Islands, where it killed 51 people.

Tsunamis not only can travel great distances but also can reverberate through an ocean for extended periods, as they bounce back and forth between continents. At speeds of up to 800 kilometers per hour, a tsunami can travel across an ocean in about the same time it takes a jet aircraft to cross. The swell of the waves is of such magnitude that it takes only a few surges and collapses for it to span the sea. The tsunami is barely discernible to the naked eye, since the crest of one of its waves represents only the tip of the vast amount of water that extends deep into the ocean. The depth of the tsunami is the reason that its course can be altered by undersea mountain ranges, valleys, or other landforms that stand in its way.

WAVE TRANSFORMATION

As a tsunami approaches land, it undergoes a major transformation. While its total energy level remains constant, it begins to slow from the friction it encounters in the shallow waters above the continental shelf. The process is called "wave refraction," and it occurs when the tsunami's wave train travels in shallow water and begins to move at a slower pace than the portion still advancing in deeper water. As its speed decreases, the wave's height grows because of a "shoaling" effect in which the trailing waves pile onto the waves in front of them. At this stage, the tsunami takes on a more visible appearance, as its waves often reach up to 9 meters or more in height. The incoming waves approach much like an incoming tide, except at a much faster pace. The maximum vertical height a wave reaches in relation to the sea is called a "run-up." The maximum horizontal distance attained by a wave is termed an "inundation." The contours of local reefs, bays, mouths of rivers, and undersea features, as well as the angles of beaches, can have a significant effect on the shape and impact of the tsunami as it nears land.

Tsunami waves normally do not curve and break like common surface waves. Survivors of the waves often describe them as walls or plateaus of water being driven by what appears to be the whole weight of the ocean behind them. Although greatly drained of their energy, the waves retain sufficient momentum to wash away nearly everything in their path, including buildings, houses, and trees. Tsunamis can cause enormous erosion, stripping beaches of sand and vegetation that has taken years to accumulate. The fact that a tsunami consists of a series of waves poses a hidden danger to coastal residents. In some cases, curious onlookers and homeowners have returned to an exposed area following the initial wave, only to be overwhelmed by a succeeding one.

Damage from tsunamis generally falls into three categories: inundation of coastal structures resulting from rapid flooding, destruction of buildings and beaches caused by water velocities, and a combination of the two in which velocity and flooding result in a complete tidal-like inundation. The most destructive form of tsunami is one that transforms itself into a bore, a concentrated wave of great force created when the tsunami moves from deep water into a well-defined shallow bay or river. This was the case during the 1960 Chilean tsunami when one of its waves struck Hilo Harbor in Hawaii as a high-velocity bore.

Destructive tsunamis strike somewhere in the world an average of once or twice each year, with most occurring in the Pacific basin. During the four-year period from 1992 to 1996, seventeen recorded tsunamis in the Pacific claimed seventeen hundred lives. Thirteen major tsunamis hit the Hawaiian Islands during the twentieth century. All were generated by earthquakes along the Pacific basin. The largest recorded wave heights were nearly 16 meters on the islands of Hawaii and Molokai in 1946, as a result of an earthquake off the Aleutian Islands that registered 7.1 on the Richter scale. The waves from this tsunami crested at about fifteen-minute intervals.

HISTORICAL RECORD

Since scientists are unable to predict exactly when earthquakes will occur, they cannot determine the precise moment when a tsunami will be generated. However, with the aid of historical records and numerical models, they can predict where they are most likely to occur. In addition, scientists have deployed sensors on the floor of the Pacific Ocean along the Aleutian Islands and the Pacific Northwest to enhance advance warning systems. The Pacific Tsunami Warning System (PTWS) in Hawaii, established after the 1946 tsunami devastated many coastal areas of the islands, monitors seismological and tidal stations throughout the Pacific basin to evaluate potential tsunami-causing earthquakes for the purpose of determining their direction and issuing adequate advance warnings.

For confirmation of past tsunamis, scientists have turned to geological evidence. Based on a technique called dendrochronology (the dating of trees by counting the ring patterns in their trunks), researchers discovered in the 1980's that a cataclysmic earthquake that struck the Pacific Northwest was the triggering mechanism for a giant tsunami that inundated the coastal Japanese island of Honshu in 1700. For some time scientists had suspected that an earthquake, centered some-

where along the Cascadia subduction zone—a fault that stretches from British Columbia, Canada, to Northern California—was to blame. They found their evidence in the marshlands along the Washington and Oregon coasts. Traces of a thin, unbroken sheet of sand were detected nearly 1 kilometer inland, indicating that a wave at least 9 meters high was likely to have hit the local coast. Calculations indicate that it would have taken an earthquake up to a magnitude 8, striking at high tide, to have produced such a wave. However, the energy of an earthquake of at least magnitude 9 would have been necessary to send waves of sufficient size across the Pacific to inflict substantial damage in Honshu. The link between the two events was established more firmly when researchers discovered that annual growth rings of drowned cedar trees and damaged spruce revealed they had died or were damaged around the same time of the Pacific Northwest waves and Honshu floods. The trees were located in regions where scientists have discovered geological evidence of previous earthquakes.

In 1998, an earthquake off the north coast of Papua New Guinea, registering a magnitude of 7.1, produced a tsunami that killed close to three thousand people at Sissano Lagoon with waves reaching 12 meters in height. Earthquakes of this size normally do not generate significant tsunamis, but researchers concluded that the earthquake probably occurred in relatively shallow water near the ocean floor and thus was able to induce a tsunami larger than normal. A later hypothesis was that the earthquake likely produced an undersea landslide that helped generate the giant waves. The lagoon is separated from the ocean by a narrow strip of land and represents an especially vulnerable section of the coast. Since the lagoon blocked the route inland, the families of the fishermen living on the sand bar had no way of escaping the waves. The entire coastal strip was swept clean by the tsunami, except for some stilts that the residents used to raise their houses off the sand.

EARTHQUAKE MAGNITUDE

Many questions remain to be answered regarding the relationship between earthquake activity and the formation of tsunamis. For a long time it was assumed that the magnitude of the earthquake, as registered on the Richter scale, was the major factor in determining the size of the tsunami. It was also believed that the shock would have to register somewhere near the 7.4 range to have a major impact on wave generation. However, scientists have learned that the largest earthquakes do not always create the greatest seismic waves. Some earthquakes can release energy in subterranean settings where the Earth's crusts react slowly and with less violent convulsions than might be the case elsewhere.

Another assumption long held by scientists was that the highest waves of a tsunami were generated immediately following the disturbance; in 1992, however, the highest tsunamis measured along the Northern California coast arrived nearly six hours after a nearby earthquake. During the same year, an earthquake with a magnitude of 7 produced a series of giant waves that devastated a 320-kilometer stretch of the Nicaraguan coast, killing nearly 170 people and injuring 500 others. The earthquake generated only moderate shaking but shifted an estimated 193-kilometer stretch of seafloor nearly 1 meter in about two minutes. A slow-paced event of this type is very efficient in producing great amounts of water to supply the tsunami. The unusually large waves in the Nicaraguan incident resulted from the relatively shallow depth of the disturbance and an accompanying subterranean landslide. The waves arrived at some coastal regions only twenty minutes after the earthquake and struck in the evening, when most of the fishing boats were in dock, destroying or damaging many of them. Numerous homes and two schools also were destroyed. As in the Papua New Guinea disaster, the initial waves were relatively weak, misleading local residents about the potential danger.

An example of a landslide-induced tsunami occurred in Skagway, Alaska, in 1994, when a large accumulation of sediment along the eastern edge of the harbor was loosened by a drop in the tide. An estimated one-third of the landmass involved in the landslide was situated above the water. The collapse produced a wave nearly 12 meters in height at the shore, killing a construction worker and causing extensive damage to a railroad dock.

Perhaps the most destructive of the volcanic-induced tsunamis was the one that occurred in 1883, when the volcano Krakatau, located between the islands of Java and Sumatra in Indonesia, erupted in a series of violent undersea explosions.

The upheaval destroyed Krakatau and created gigantic waves as high as 35 meters that inundated towns and villages on nearby islands, killing more than thirty-six thousand people. In a desperate maneuver, the crew of the ship *Loudon* aimed the boat's bow into the oncoming tsunami and managed to ride out the waves. Scientists have debated whether it was the submarine explosions, the slumping of the cone into the crater, or the occasional surges of matter falling into the water that caused the event. Waves from the Krakatau tsunami were recorded as far away as South America and Hawaii. Nearly nine hours after the event, the waves smashed boats resting in the harbor at Calcutta, India. When the tsunami reached Port Alfred in South Africa, it was still nearly 0.5 meters high. The English Channel even recorded a minor surge.

SIGNIFICANCE

By exploring the relationship between earthquakes and tsunamis, scientists continue to learn more about the conditions that lead to a tsunami. The bottoms of the oceans, where most tsunamis originate, are largely unexplored. Only in the last few decades of the twentieth century did researchers begin to accumulate data in sufficient amounts to assess the destructive potential of the phenomenon, particularly in regard to the near-shore tsunamis similar to the one that struck Papua New Guinea.

Scientists already have determined that the hazard to American shorelines, especially along the heavily populated West Coast, is greater than previously thought. The likelihood that the earthquake that triggered the Papua New Guinea tsunami also generated an underwater landslide carries great significance for coastal settlements located close to the offshore seismic faults stretching from Northern California to the Aleutian Islands and near Hawaii. These areas feature canyons that slope deep into the ocean, making them vulnerable to the kinds of underwater landslides that occur in conjunction with moderate earthquakes. How such relatively small disturbances generate such large waves remains a key area for research. The fact that they take place only minutes from shorelines underscores the importance of the public's being made aware of the tsunami's potentially destructive force, if damages are to be limited.

Researchers have concluded that the technology

utilized in early warning systems, from monitoring seismographs to measuring water-level changes at tide-gauging stations, cannot significantly alter the advance warning time for near-coastal tsunamis as it can for the disturbances that originate in deeper seas. Instead, with the aid of computer-generated models and historical records, researchers seek to provide a basis for educational programs designed to alert the general public about the regions where tsunamis are most likely to occur and about the impact that can be expected.

William Hoffman

CROSS-REFERENCES

Continental Crust, 561; Continental Structures, 590; Deep-Focus Earthquakes, 271; Earthquake Distribution, 277; Earthquake Engineering, 284; Earthquake Hazards, 290; Earthquake Locating, 296; Earthquake Magnitudes and Intensities, 301; Earthquake Prediction, 309; Earthquakes, 316; Notable Earthquakes, 321; Ocean Basins, 661; Ocean Tides, 2133; Ocean Waves, 2130; Oceanic Crust, 675; Oceans' Structure, 2151; Slow Earthquakes, 329; Soil Liquefaction, 334; Tsunamis, 2176.

BIBLIOGRAPHY

Associated Press. *The Associated Press Library of Disasters.* Danbury, Conn.: Grolier, 1998. Volume 1 of this work covers earthquakes and tsunamis and provides a compilation of news reports on the major events. Entries are arranged chronologically and are accompanied by supporting texts that place each event in context and explain any inaccuracies in the original reports. A note at the end of each entry indicates the outcome of the event.

Dudley, Walter, and Min Lee. *Tsunami!* 2d ed. Honolulu: University of Hawaii Press, 1998. This work offers firsthand accounts and photographs of major tsunamis, including the 1946 event in the Aleutian Islands and Hawaii and the 1998 occurrence in Papua New Guinea. Other sections cover the development of the early warning system and its effectiveness in tracking actual tsunamis.

Folger, Tim. "Waves of Destruction." *Discover* 15 (May, 1994): 66-73. Folger offers an account of the surprisingly large tsunami that hit the coast of Nicaragua in 1992 following an offshore earthquake of moderate size. The article reveals how the relatively low intensity of an earthquake's vibrations can mislead coastal residents into thinking the potential danger of a tsunami is minimal.

Montastersky, Richard. "Waves of Death." *Science News* 154 (October 3, 1998): 221-223. Provides a close look at the tsunami that struck Papua New Guinea in 1998 as a result of a moderate undersea earthquake. The data retrieved from studies of this event highlight the differences between near-shore and deep-sea earthquakes and their potential effects on similar coastal formations elsewhere along the Pacific Rim.

Satake, Kenji, and Fumihiko Imamura, eds. *Tsunamis 1992-1994: Their Generation, Dynamics, and Hazards.* Boston: Birkhauser, 1995. Reviews the unusually active period between 1992 and 1994, when six destructive tsunamis occurred with waves up to 12 meters high and casualties totaling close to fifteen hundred people. The book reports on the findings of geologists, seismologists, oceanographers, and specialists from other disciplines who were sent to investigate the circumstances surrounding the events.

Tsuchlya, Yoshito, and Nobuo Shuto, eds. *Tsunami: Progress in Prediction, Disaster Prevention, and Warning.* Advances in Natural and Technological Hazards Research Series 4. Norwell, Mass.: Kluwer Academic Publishers, 1995. This work contains a series of studies conducted by scientists, engineers, and specialists on all aspects of the tsunami phenomenon. The articles are divided into three major categories: tsunami generation, propagation and inundation (including prediction and simulation), and observation and warning systems.

7
EXPLORING EARTH'S INTERIOR

DEEP-EARTH DRILLING PROJECTS

Deep-earth drilling projects represent the newest and most ambitious attempts by Earth scientists to investigate the origins, structure, and nature of planet Earth. The projects already under way and those proposed for the future promise to reveal information concerning the planet unobtainable by other methods.

PRINCIPAL TERMS

CORE DRILLING: a method of extracting samples of the materials being drilled through in a deep-drilling project

CRUST: the outer layer of the Earth, averaging 35 kilometers in thickness on land and 5 kilometers on the ocean bottoms

MANTLE: the area of basaltic rocks separating the Earth's crust from its core; it is estimated to be about 2,900 kilometers thick

MOHOROVIČIĆ (MOHO) DISCONTINUITY: an area of undetermined composition and depth between the Earth's crust and mantle

ROTARY DRILLING: a method of drilling holes to great depths using a rotating drill bit

ORIGINS OF SCIENTIFIC DRILLING PROJECTS

The idea of drilling deep holes in the Earth's crust in order to determine its nature and history originated with nineteenth century geologists and naturalists. Charles Darwin was among the first scientists to call for a deep drilling project for purely scientific purposes. In 1881, Darwin proposed that a shaft be sunk to a depth of 150-180 meters on a Pacific atoll to test his theory concerning the origins and growth of coral islands.

Eighteen years later, the Royal Society of London financed the drilling of a 348-meter-deep hole on one of the Ellice Islands (Tuvalu) in the South Pacific to test Darwin's theories. That was perhaps the first deep-earth drilling project ever undertaken for purely scientific purposes.

THE MOHOLE PROJECT

Since Darwin's time, many Earth scientists have proposed deep-earth drilling projects to advance scientific knowledge of the Earth's interior; those proposals have usually met with indifferent success because of the great expense and the technological problems involved. In the 1950's, a number of well-known geologists, geophysicists, and oceanographers from several countries began corresponding with one another. A major topic of their correspondence was a deep-drilling project that might capture the imagination of the public (which they perceived to be over-focused on space research) and consequently allow Earth scientists to claim a larger share of research funds from government agencies. The correspondents eventually formed an unofficial organization they called the American Miscellaneous Society (AMSOC), which held informal meetings during official scientific conferences at which they discussed the desirability of a deep-drilling project that would penetrate the Mohorovičić (Moho) discontinuity into Earth's mantle. Ideally, they agreed, there would be two deep-drilling projects, one on land, the other on the ocean bottom. In 1958, the members of AMSOC became the Deep Drilling Committee (DDC) of the National Academy of Sciences (NAS), a private organization chartered by then-president of the United States Abraham Lincoln in 1865 to act as an adviser to the federal government on scientific matters.

NAS, on the recommendation of the DDC, proposed in 1958 that the federal government of the United States fund a deep-drilling project to penetrate the Moho. The catalyst for NAS's recommendation seems to have been rumors circulating at the time that the Soviet Union had already begun such a project and was about to "beat" the United States in learning the secrets of the Earth's interior as it had recently beaten the Americans into outer space. The original proposal envisioned a hole to be drilled on land to a depth of perhaps 10,500 meters as a training project to develop the technology necessary to penetrate the Moho at the ocean's bottom, where the crust is thinner.

The DDC subsequently scrapped the ground-hole idea when it received a grant of $15,000 for a feasibility study of the deep-sea drilling project. A DDC member summarized the result of the feasibility study in an article in *Scientific American* (April, 1959) entitled "The Mohole," which stirred immediate industrial and public interest. The interest thus generated resulted in a much-publicized government-sponsored deep-sea drilling project that failed to meet most of its objectives, cost considerably more than originally estimated, and discredited the idea of deep drilling in the minds of many members of Congress and the public. The Mohole project became a source of considerable international embarrassment for American science and may have delayed a sound deep-drilling project by two decades. It was not until 1968 that oceanographers were able to convince Congress to finance another deep-sea drilling project. Another decade passed before federal funds were forthcoming for continental deep-drilling projects.

THE KOLA PROJECT

The rumors concerning a Soviet deep-drilling project were at least partially responsible for the urgency with which the U.S. scientific community and government embraced the ill-fated Mohole project. The rumors were unfounded. Scientists in the Soviet Union did not begin such an effort for twelve years after Mohole. In 1970, Soviet geophysicists launched a deep-drilling project on Kola peninsula near Murmansk, 240 kilometers north of the Arctic Circle; the project reached a depth of more than 12,000 meters, almost twice the depth of any preceding hole. Under the direction of a Soviet government agency called The Interdepartmental Council for the Study of the Earth's Interior and Superdeep Drilling (formed in 1962), the Kola project became the first of several proposed deep holes meant to explore the structure of the Earth's crust and mantle.

The Kola drillers penetrated through almost 3 billion years of Earth's geologic history, into rock from the Archean eon. Along the way, they discovered hot, highly mineralized water in larger quantities and at greater depths than geophysicists and geologists had previously thought possible. Scientists at the Kola project concluded from this and other unexpected findings that enormous mineral deposits may be located at great depths, waiting for humankind to discover a way to reach them. Soviet scientists expected that as they developed a better understanding of the deeper layers of the Earth's crust from the Kola project, they would concurrently find ways to discover and exploit petroleum, gas, and minerals at great depths.

THE APPALACHIAN PROJECT

Spurred in large part by Soviet successes in deep continental drilling and the propaganda reaped therefrom by the Soviet government, scientists and government agencies in several Western nations began developing similar programs in the 1970's. The U.S. Geodynamics Committee of NAS held a workshop on deep continental drilling for scientific purposes near Los Alamos, New Mexico, in 1978. The members of the workshop convinced NAS to form its own permanent Continental Drilling Committee (CDC) that same year. The members of the CDC identified a number of geophysical problems that could be resolved through deep-earth drilling projects. The CDC also solicited proposals from U.S. geologists and geophysicists for specific projects that would address those problems. After examining the many proposals received, the CDC assigned highest priority to two of them: a project to drill a hole 3.7 kilometers deep through the highly mineralized area near Creede, Colorado, and a core hole in the southern Appalachian region 8-10 kilometers deep.

At its 1983 meeting, the CDC unanimously endorsed the Appalachian project as the most promising for America's first deep continental drilling project. At the same time, it endorsed drilling projects at a future time in two other areas, including Creede and Cajon Pass in California. The following year, the CDC convened a workshop in New York to consider exactly how the project should be approached. After the workshop, the committee organized the Continental Scientific Drilling Program, established with the aid of a grant from the National Science Foundation (NSF). NSF gave the grant to a management group called Direct Observation and Sampling of the Earth's Continental Crust (DOSECC). DOSECC, coordinating its activities with the U.S. Geological Survey and the Department of Energy, prioritized deep-drilling projects and issued contracts for drilling deep continental holes for scientific purposes. In 1985, the White House's Office of Science and

Technology Policy (OSTP) recommended to the NSF that it appropriate $2 million for preliminary studies of the Appalachian project. OSTP also recommended that by 1990 the various deep-earth drilling projects be funded at a level of $20 million per year. In 1988, the Cajon Pass project, overseen by DOSECC, got underway; within two years, it reached its targeted depth of 4,875 meters. DOSECC began the Appalachian project in the northwest corner of South Carolina in 1989. Its goal was a continuously cored hole 15,250 meters deep.

QUEST FOR PETROLEUM

The technology employed in deep-earth drilling projects derives directly from the petroleum industry. The modern origins of that industry date back to the middle of the nineteenth century, when petroleum fields were discovered and exploited in widely separated regions of the world, including Romania, Burma (now Myanmar), Sumatra, Iran, the Caucasus, and the United States. The oil wells in those areas were initially shallow,drilled with tools designed to drill water wells.

As the petroleum industry grew in importance, the search for petroleum went deeper and deeper into the Earth and required ever more sophisticated drilling machinery. In addition, oil companies began to employ geologists and geophysicists in their quest to meet the skyrocketing demand for petroleum of an increasingly industrialized society. These Earth scientists, in part to become more efficient in their effort to locate significant quantities of petroleum and in part as a by-product of that endeavor, began to learn more and more about the nature and history of the Earth's crust from the drilling process itself. This new knowledge led to the modern deep-earth drilling projects that seek to drill deep and superdeep holes, not necessarily to locate mineral resources (although that may well be an important spinoff of the projects) but to learn still more about the Earth.

The quest for petroleum in the twentieth century penetrated ever deeper into the Earth's crust. By the latter part of the century, wells in the United States were producing oil from depths of more than 6,000 meters (the record being a well in Louisiana, producing oil at a depth of 6,527 meters).

DRILLING METHODS

There are two basic methods of drilling deep wells: cable-tool and rotary. Cable-tool drilling utilizes a heavy drill bit and drill that are suspended by a cable and raised about a meter above the bottom of the hole, then dropped. Workers add mud and water to the hole to hold the rock chips produced by the concussion in suspension. Periodically, drilling crews extract the tools from the hole and pump out the mud, water, and rock chips. The method is slow and has been largely superseded by the rotary drill, which is much more effective in deep-earth drilling.

Rotary drilling, though much faster than the cable-tool method, is also more expensive. The rotary drilling method requires hundreds to thousands of meters of drill pipe and well casing, a derrick, drill bits of several kinds (depending on the type of rock being drilled through), drilling muds and chemicals, a power source (usually one or more diesel engines), and a sizable crew of workers. In rotary drilling, the workers attach the drill bit to a string of drill pipe that has at its top a square cross section called the kelly. The kelly passes through a square hole in a powered turntable, which rotates the drill pipe and bit. The workers add new sections of drill pipe just below the kelly as the bit progresses downward. The rock cut by the rotating drill bit is removed by pumping chemically treated mud down the drill pipe through the drill bit, then back up through the space outside the pipe (the pipe being somewhat smaller in circumference than the hole in which it rests) to a settling pit on the surface.

One drill bit developed for the rotary process, the hollow or core bit, is of particular importance for current deep-earth drilling projects. Because petroleum is often found in readily identified geological formations, companies exploring for oil found it expedient to bring to the surface intact samples of the rock being drilled through for examination by geologists. To extract the samples, technicians developed a hollow bit that would allow a cylindrical section of the rock, called a core, to be extracted from the hole without otherwise damaging it. The cores thus derived allow geologists to determine not only the petroleum potential of the rock but also the geological history of the Earth in the area of the hole. Core drilling has become an integral part of all deep-earth drilling projects.

MODIFICATION OF METHODS

The deep-earth drilling projects have required considerable modification of the methods developed for petroleum drilling. Soviet scientists at the Kola project learned that at depths exceeding 9,000 meters, conventional rotary drilling encounters virtually insurmountable problems. The conventional rotary method uses a power source to turn the drill bit by rotating the entire string of pipe connected to it and to the kelly. At 9,000 meters, the pipe weighs in excess of 800 metric tons, the rotation of which creates enormous stress at the kelly-power source interface and multiplies the friction resistance of the rock through which the scientists are drilling. The Soviets overcame this problem by developing a bottom-hole turbine to rotate the drilling bit, driven by the flow of the drilling mud being injected into the hole. The necessity of rotating the pipe is thus completely eliminated.

To reduce the weight of the huge lengths of drill pipe necessary for deep-earth drilling, Soviet scientists at Kola began to utilize a high-strength aluminum alloy pipe weighing only about half as much as conventional drill pipe. This innovation considerably reduced the burden on the derrick and the power required to lift the pipe periodically to replace the drill bits and remove the core samples.

Core sampling also underwent modification at Kola. In conventional wells, core samples several centimeters in diameter enter the hollow drill tube as the bit cuts a ring of rock at the bottom of the hole. The core remains in the tube until it is brought to the surface and removed. At depths of more than 2,100 meters, however, the rock is under such tremendous pressure that it literally bursts when the drill bit relieves the compression of overlying rock strata. Soviet drillers developed a new core-sampling device that diverted some of the mud into the core tube and caught the pieces of burst rock, then carried them to the surface in a special chamber. This technique also offered the advantage of clearing the tube for new core samples without the necessity of bringing the tube to the surface.

Undoubtedly, other modifications of present drilling technology will be necessary as continental deep-drilling projects proceed. Given the progress of such technology, it may confidently be expected that innovations will solve the problems that will certainly be encountered, provided that funding for the projects is adequate.

EXPLOITING EARTH'S ENERGY RESOURCES

The mineral and fossil fuel resources of the Earth are finite and nonreplaceable commodities. As industrialization continues to spread to the so-called Third World countries and to intensify in the regions where it already exists, the demand for those irreplaceable and limited commodities can only increase. Sometime in the twenty-first century, the presently known reserves of fossil fuels may be exhausted. If current lifestyles are to be maintained in industrialized regions of the world and living standards raised for a burgeoning population in the Third World, new reserves of fossil fuels and sources of minerals must be discovered and exploited, and/or new sources of energy must be developed. Deep-earth drilling projects offer one approach to both these necessities.

Strong evidence exists that significant quantities of fossil fuels, both petroleum and natural gas, may be found at depths that are currently unreachable or economically unprofitable. Technology developed in the deep-earth drilling program will almost certainly put those resources at the disposal of humankind.

Deep-earth drilling projects under way in the United States and elsewhere are exploring the possibility of exploiting on a large scale a relatively new way of producing energy by utilizing the Earth's internal heat. The temperature of the Earth increases by about 1 degree Celsius per hundred meters to a depth of 3 kilometers and, as shown at Kola, by 2.5 degrees Celsius thereafter. At 15 kilometers beneath the surface, the Earth's temperature is more than 300 degrees Celsius. The U.S. Department of Energy is currently funding several deep-drilling projects to investigate the possibility of commercially exploiting this source of virtually inexhaustible energy.

DISPOSING OF TOXIC WASTES

Some scientists suggest that deep-earth drilling might offer a solution to the problem of toxic waste disposal. Safe disposal of the wastes produced by modern industry and by the production of atomic energy ranks as a major concern of contemporary scientists. Before industries and atomic energy plants inject these materials into the deep crust of the planet, however, extensive studies must be conducted into the potential ecological consequences of such an action. Nevertheless, deep-earth drilling may help to solve the problem of toxic waste disposal.

Paul Madden

CROSS-REFERENCES

Earth Resources, 1741; Earth's Structure, 37; Engineering Geophysics, 353; Hazardous Wastes, 1769; Nuclear Waste Disposal, 1791; Ocean Drilling Program, 359; Ocean-Floor Drilling Programs, 365; Offshore Wells, 1689; Oil and Gas Exploration, 1699; Oil and Gas Origins, 1704; Onshore Wells, 1723; Petroleum Reservoirs, 1728; Seismic Reflection Profiling, 371; Well Logging, 1733.

BIBLIOGRAPHY

Anderson, Ian. "Drilling Deep for Geothermal Power and Science." *New Scientist* 111 (July 24, 1986): 22-23. This brief article presents a nontechnical overview of current deep-drilling projects exploring the possibility of exploiting the Earth's own heat to produce energy. The author is optimistic that many of the energy needs of the future may be met if governments are willing to make the necessary expenditures for research.

Bascom, Willard. "Deep Hole Story." *Modern Machine Shop* 54 (March, 1982): 92-111. A thorough review of the technology of deep drilling, including Soviet innovations. Assumes considerable technical knowledge on the part of the reader but very informative for those wanting to know more about the technology involved in deep-drilling projects.

_____. "Drilling the World's Deepest Hole." *Engineering Digest* 34 (June, 1988): 16-24. This article is primarily concerned with technical drilling problems. Readers should have strong technical background. Will be of interest only to those readers wanting to know more about the actual mechanics of deep-earth drilling.

_____. "Geothermal Boreholes Make Drilling History." *Machine Design* 53 (October 22, 1981): 8. This article contains a rather technical discussion of the problems encountered in deep drilling for geothermal purposes and their often ingenious solutions. The reader should have a dictionary handy.

_____. *A Hole in the Bottom of the Sea: The Story of the Mohole Project.* Garden City, N.Y.: Doubleday, 1961. Although the main topic of Bascom's book is the ill-fated Mohole pro-

ject, it contains a considerable amount of information about early proposals for scientific drilling projects and deep continental drilling projects. Also contains fascinating insights into the workings of the "scientific establishment" in the United States and the ways in which it interacts with the federal bureaucracy. Written for a nontechnical audience.

Gurnis, Michael, et al., eds. *The Core-Mantle Boundary Region.* Washington, D.C.: American Geophysical Union, 1998. This collection of articles is one volume of the American Geophysical Union's Geodynamics series. Although intended for the specialist, the essays contain plenty of information suitable for the careful college-level reader. Bibliography.

Heath, Michael J. "Deep Digging for Nuclear Waste Disposal." *New Scientist* 108 (October 31, 1985): 30. This article briefly explores the possibilities of using deep-drilling technology to solve the perplexing problem of disposing of nuclear wastes. The ideas presented are interesting but not entirely convincing, as they do not address the environmental problems that might be created by such a program.

Kearey, Phillip. *Global Tectonics.* 2d ed. Cambridge: Blackwell Science, 1996. This college text gives the reader a solid understanding of the history of global tectonics, along with current processes and activities. The book is filled with colorful illustrations and maps.

Kerr, Richard A. "Continental Drilling Heads Deeper." *Science* 224 (June 29, 1984): 1418-1420. An excellent account of the then-current status of United States deep-earth drilling projects and proposals. Offers considerable information about the organizations and individuals who are forming the policies for the United States continental drilling program.

Kious, Jacquelyne W. *This Dynamic Earth: The Story of Plate Tectonics.* Washington, D.C.: U.S. Department of the Interior, United States Geological Survey, 1996. Kious is able to explain plate tectonics in a way suitable for the layperson. The book deals with both historic and current theory. Illustrations and maps are plentiful.

Kozlovsky, Yephrim A. "The World's Deepest Well." *Scientific American* 251 (December, 1984): 98-104. The best and most complete account in English of the Kola deep-drilling project in the Soviet Union. The reader should keep a geological dictionary nearby but will find a wealth of information about the origins, purpose, scientific findings, and future of the Soviet continental drilling program.

Lutgens, Frederick K., and Edward J. Tarbuck. *Earth: An Introduction to Physical Geology.* 6th ed. Upper Saddle River, N.J.: Prentice Hall, 1999. This college text provides a clear picture of the Earth's systems and processes that is suitable for the high school or college reader. In addition to its illustrations and graphics, it has an accompanying computer disc that is compatible with either Macintosh or Windows. Bibliography and index.

Press, Frank, and Raymond Siever. *Understanding Earth.* 2d ed. New York: W. H. Freeman, 1998. This text includes one of the most complete descriptions of the causes of earthquakes, their measurement, where they occur, how they can be predicted, and how they affect humans. A map of the major plates is on the inside back cover. The glossary is huge and indispensable. Senior high school and college-level students should find this text suitable for general background information.

ENGINEERING GEOPHYSICS

Engineering geophysics involves the application of the techniques of exploration geophysics to problems of interest to the engineering and groundwater geologist, including site evaluation, resource exploration, and pollution monitoring. The research methods are focused on discovering shallow targets by seismic, electrical, magnetic, electromagnetic, gravity, and radar surveys.

PRINCIPAL TERMS

GEOPHYSICAL SURVEY ARRAY: a description of the orientation and spacing of sensors and energy sources relative to one another for a geophysical survey

GEOPHYSICAL TARGET: the object or surface that one wishes to detect by means of a geophysical survey; knowledge of the target is essential to selection of a survey type

PHYSICAL PROPERTY CONTRAST: the difference in a characteristic (density, velocity) between the object of interest and its surroundings

SITE EVALUATION: a process whereby a site is selected or rejected as a location for a particular use such as construction or mining

SURVEY LINE: a usually straight line along which points are located where geophysical measurements will be taken

SEISMIC REFRACTION AND REFLECTION

The methods of engineering geophysics include a number of surveying techniques that detect the physical characteristics of materials and objects in the subsurface. The technique may detect such phenomena as rock and soil layers, disturbed or disrupted soils, buried walls or foundations of archaeological interest, buried metal drums, buried sand and gravel deposits, and aquifers. Each of these targets has a distinct set of physical properties that may be sensed from the surface with the right kind of survey. The selection of a particular geophysical surveying instrument and method depends upon the size, depth, and other characteristics of the target, as well as upon the nature of the surrounding materials. The most commonly used methods are seismic refraction and reflection, ground penetrating radar, magnetic and gravity surveys, and electrical and electromagnetic techniques.

The seismic refraction method relies on the different wave velocities of rocks and soils and the principles of refraction, which govern all wave motion. A seismic survey is accomplished using a number of seismic sensors, or "geophones," that are arrayed in a straight line and are usually spaced at regular intervals. These phones will detect the arrival of seismic waves from an artificial energy source such as a hammer blow, a dropped weight, or an explosive charge. The additional equipment includes a digital recording system that measures the interval between the time that the energy is put into the ground and the time that it is picked up at the geophones. The velocities of the seismic waves are then calculated from the distance between the energy source and a particular geophone and the time elapsed for the wave to reach that geophone. The data are simplest to interpret where layers are flat and of constant thickness; more complex equations allow interpretation where layers are tilted and of variable thickness.

Seismic reflection profiling is used to detect surfaces at depth. In general, the seismic array of the energy source and geophones is kept under 10 meters in total length because the waves that are detected are traveling along nearly vertical paths. The elapsed time for the wave to reach a geophone is a function of the velocities of the subsurface layers and their depths. The applications of this method for shallow targets are more limited than seismic refraction because of surface noise and the difficulties in identifying individual reflections.

GROUND-PENETRATING RADAR

Ground-penetrating radar, or GPR, works on the same principles as seismic reflection but does not suffer from the same noise problems for shallow investigations. This technique measures the time necessary for a controlled pulse of radar energy to return from a reflecting surface. The equipment involves a transmitter and receiver that may be towed across an area to detect buried objects and soil layers.

While radar waves can travel great distances in air or space, they are rapidly absorbed by rocks, soils, and especially water. For this reason, GPR is used for depths of 10-15 meters. The depths of radar penetration are even less in areas underlain by clayey soils or shale, both of which contain large amounts of water in the crystal structure, or where the water table is high. For most engineering purposes, GPR is a fast and efficient survey method in relatively smooth terrain.

MAGNETIC AND GRAVITY SURVEYS

Magnetic and gravity methods involve the detection of variations in the natural magnetic and gravity fields of the Earth. Changes in the magnetic field are caused by the presence of materials that have a high magnetic susceptibility. These substances include very magnetic materials, such as metallic iron and steel, but also may include rocks and soils that have elevated quantities of the mineral magnetite. Either the very magnetic metals or the slightly magnetic rocks and soils may be detected by surveying the area of interest using a magnetometer. Magnetic surveys are often used to locate metallic targets that are quite small.

A gravity survey detects variations in the gravitational field as a result of changes in density of subsurface materials. Denser materials will cause a slight increase in gravity, while less dense objects cause a small decrease in gravity. A small object may be detected only if it has large contrast in density. If the density contrast of the target is small, as it is in most engineering applications, the object must be large to be detectable. Gravity surveys are usually employed to find large bodies of rock or soil as opposed to smaller objects. The survey techniques perhaps used most in engineering geophysics detect the electrical properties of subsurface materials. The contrasts in electrical properties of natural and human-made materials are very large, so surface surveys can detect anomalous regions easily.

ELECTRICAL AND ELECTROMAGNETIC TECHNIQUES

Electrical-resistivity surveys involve an array of electrodes planted in the ground in one of a number of patterns. Generally, two electrodes are used to introduce a current into the ground, while other electrodes measure the voltage drop between two points in the array. The pattern of the array and the spacing of the electrodes are chosen based on the character of the target and whether the investigator wishes to see a single depth or different depths. Measurements are interpreted by a series of complex equations that model the depths, thicknesses, and resistivities of the soil and rock layers.

Terrain conductivity is an electromagnetic method that measures the electrical properties in the subsurface without the necessity of placing electrodes. The method uses a pair of electrical coils, one to "broadcast" and the second to receive. The first coil is energized with an alternating current, creating an alternating magnetic field that penetrates into the Earth. Where this magnetic field encounters a good conductor in the subsurface, a secondary electrical current is generated or induced. The secondary electrical current is accompanied by a secondary magnetic field that is detected at the surface by the receiving coil. The depths of penetration of the conductivity method are related to the spacing of the two coils and to the power of the electrical system.

All the methods used in engineering geophysics involve time and expense. Nevertheless, they are used, because they may provide additional data on the continuity of conditions discovered from outcrops, drill holes, or excavations. In addition, anomalous regions within the site may become targets for more detailed study by drilling or coring.

SITE EVALUATION

Geophysical surveys are used to detect potential problems and provide additional information on a site undergoing evaluation. Engineering geology involves, for example, site evaluations for buildings, highways, well fields, dams, and sanitary or hazardous waste sites. Engineering geologists also deal with questions of slope stability, where data on the thickness of the unstable or moving part of the slide may be needed to assess the feasibility of construction or to develop plans for slope stabilization. In all these cases, information on the depth to bedrock, the lateral continuity of soil or rock layers, and the existence and orientation of surfaces within the rocks may be needed. Additionally, the geologist may have to detect human-made objects within an area. These include buried or abandoned gas and power conduits, storage tanks, and metal drums. One of the major problems requiring engineering geology today is the monitoring and correction of contaminant leakage from hazardous waste disposal sites across the country. In these situations, data are required to define geological conditions around the site or to map the extent and location of the plume of the contaminants that may be moving off the site.

For each of these needs, a geophysical survey may be useful. The first step in the selection of a specific technique depends on the identification of the target. Those physical properties of the target that contrast most with the properties of the surrounding rock and soil will indicate which geophysical technique may be most valuable. Next, the expected size, shape, orientation, and depth of the target will be guides in the selection of the best survey methods. Finally, the character of the terrain (whether it is dry or wet and boggy, topographically smooth or irregular) will be considered. Three examples may serve to elucidate this kind of planning.

LANDFILL APPLICATION

The first involves a landfill in which a large number of 55-gallon steel drums containing toxic chemicals were buried many years ago. The need now is to locate these drums and remove them from the site. The task must be accomplished carefully, for it is essential that none of the drums be split or broken. Knowledge of the exact location of the drums would greatly lessen the chances of spilling toxic waste during the process of removal. A number of geophysical methods might apply, but the geologist wishes to choose the surest method. The geologist begins by considering the physical properties and sizes of the bodies in an effort to eliminate some methods. While the density of the drums is likely to be slightly different from that of the surroundings, the drums are too

small to result in a measurable gravity anomaly. Similarly, a seismic refraction survey is not likely to detect individual objects as small as the drums even if there is a large contrast in velocity between the metal drums and the other material on the site.

Each of the remaining methods reviewed earlier might be useful. Seismic reflection and GPR would be able to detect reflections off the drums. In this situation, however, the drums are probably too shallow to use the seismic method, since surface-wave noise will be high. GPR, on the other hand, would do the job very well and has worked in other areas. The only weakness of GPR is that the survey line must cross directly over a drum to detect it. To ensure finding all the drums, a very close, and therefore expensive, set of GPR survey lines would be needed.

Electrical methods could detect the large contrast in the resistivity of the metal drums. Either a resistivity or a terrain-conductivity survey could sense the drums. Because the electrical or electromagnetic field extends out beyond the direct survey line, either method could locate drums slightly off the direct line of the survey. A drawback to the method, however, would be the possible existence of salty water or other electrolytes in the landfill. If these fluids are present, their high electrical conductivities could mask the presence of the drums. Because the drums are steel, they will cause a magnetic anomaly, and a magnetic survey will be able to detect even those drums off of the survey line. Although the magnetometer will detect other iron objects in the landfill, the method may be the fastest and least expensive method to survey the site for this particular target and should result in locating all the drums.

LANDSLIDE APPLICATION

In the second example, the geologist needs to determine the depths to the slip-surface, or base, of a landslide. In this case, the target is too deep for GPR, and the variations in density and magnetic susceptibility are expected to be too small to produce a significant anomaly. Previous studies, however, have shown that seismic refraction and electrical surveys can be quite successful in solving this kind of problem. The upper, mobile, and generally disrupted part of the landslide will have a different seismic velocity from that of the lower, undisturbed rocks or sediments. While the thickness of the upper layer is likely to be variable and the slip-surface complex, the seismic method has been useful. Electrical methods will depend either on a difference in conductivity of the disrupted upper layer and the undisturbed rock beneath or on the conductivity of the slip-surface itself. The exact nature of the conductivity contrast will depend on the local geology, the soil moisture, and the groundwater conditions.

WATER SUPPLY APPLICATION

The third example involves a common problem with water supplies in coastal areas where fresh groundwater floats typically above a deeper zone of salt water. In this case, withdrawal of fresh water from wells leads to a drawdown of the water table and also to intrusion from the sides and from below of salt water from the ocean. In many places, the supply of potable water is very limited by the occurrence of salty water beneath the thin lens of fresh water. The depth to and migration of saline water can be monitored by wells at critical places, but to obtain a more continuous set of data across a region, a geophysical survey is desirable. The greatest contrast in physical property will be the high electrical conductivity of the salt water versus the low values of the fresh water. Surface electrical surveys have been successfully used to estimate the thickness of the freshwater layer above salt water and to monitor the extent of the migration of salt water laterally into freshwater aquifers. Similar applications of electrical techniques have been used to detect acid mine drainage or to monitor movement of electrically conductive pollutant plumes from industrial or waste-disposal sites.

DATA INTERPRETATION

Engineering geophysics involves the application of general geophysical techniques for shallow investigations. The methods are appropriate for depths from a few meters to a few hundred meters. Applications of geophysics to shallow targets often involve greater complexity in interpretation than for deep targets because the roughness and irregularities of buried objects or surfaces are large compared to the depth. Similarly, many of the targets of interest to the engineer are not simple in shape. Surfaces such as bedrock buried beneath

soil or recent sediments or the fracture patterns in bedrock are inherently complex. Furthermore, the transitional character of physical properties in near-surface environments may be difficult to detect or model. Examples of such transitions include velocity changes caused by gradations from unweathered to weathered bedrock and density and electrical property changes caused by variation in the percentages of clay or sand in surficial gravels.

While geophysics is often used in a qualitative fashion, in many areas, mathematical modeling allows a quantitative measure of the depth, size, shape, and composition of the target. Seismic refraction and reflection and GPR generally give the most unequivocal information about the subsurface. Magnetic and gravity surveys yield information on the depths, sizes, and character of subsurface objects, but usually more than one interpretation will fit the data. Electrical and electromagnetic methods also supply data that can be interpreted in a number of ways. Thus, some geophysical methods are less exact than others, and all involve uncertainties in the interpretation. For these reasons, it is important to gather geophysical information in areas where other kinds of geological data are available from surface mapping, drilling, or excavation. The combination of geophysical and geological data provides an excellent basis for the evaluation of sites of engineering interest, for estimation of water or other resources, and for the monitoring of hazards.

GEOLOGICAL AND CIVIL ENGINEERING VALUE

The power of geophysics lies in its use to detect anomalous regions and to evaluate the lateral continuity from one point to another across an area of interest. In many cases, the existence of an anomaly can guide the engineering geologist to the critical area where drilling or excavation will be used to acquire data. In other cases, extensive drilling and sampling in an area may tell much about the region, but information is needed between the sample sites. In these cases, geophysical surveys may be used to show the extent to which the regions between sample sites are the same as the sample sites themselves. The nature of the array will allow the scientist to look to different depths in a single area or to improve the accuracy and precision of the data.

The use of geophysics to solve problems in geological and civil engineering has become more common as the need for more detailed information on site assessment and evaluation has become apparent. While nothing can match the factual nature of soil and rock sampling by drilling, excavation, or examination of outcrops, the need to test the continuity from one sample site to another requires additional data. Geophysical methods can supply that additional information through detection of the bulk physical properties of materials in the subsurface or the interfaces between the different materials.

The bulk physical properties that can be detected include density, magnetic susceptibility, seismic-wave velocity, and electrical conductivity or resistivity. The surfaces separating different bodies in the subsurface may be thick enough to have their own bulk properties, but generally they act as reflectors of seismic or radar wave energy.

Donald F. Palmer

CROSS-REFERENCES

Aquifers, 2005; Dams and Flood Control, 2016; Deep-Earth Drilling Projects, 347; Gravity Anomalies, 120; Groundwater Pollution and Remediation, 2037; Hazardous Wastes, 1769; Land Management, 1484; Land-Use Planning, 1490; Landfills, 1774; Landslides and Slope Stability, 1501; Ocean Drilling Program, 359; Ocean-Floor Drilling Programs, 365; Remote Sensing and the Electromagnetic Spectrum, 2802; Rock Magnetism, 177; Seismic Reflection Profiling, 371.

BIBLIOGRAPHY

Beck, A. E. *Physical Principles of Exploration Methods: An Introductory Text for Geology and Geophysics Students.* New York: Halsted Press, 1981. This text offers a mathematically rigorous, theoretically oriented introduction to geophysical prospecting of field data are derived and used in examples. The treatment of electrical and electromagnetic methods is especially good. An index is supplied. Most references in the bibliography are from geophysical journals.

Das, Braja, M. *Principles of Geotechnical Engineering.* Boston: PWS, 1998. This book explores and clarifies the principles, policies, and applications of geotechnical engineering. Graphs, charts, and maps reinforce concepts throughout the chapters.

Dennen, William H., and Bruce R. Moore. *Geology and Engineering.* Dubuque, Iowa: Wm. C. Brown, 1986. The authors have provided a nonmathematical introduction to the field of engineering geology with excellent diagrams and an easily understood text. The treatment of geophysical methods is qualitative but useful. The discussion of the accuracy and precision of analytic data is an excellent and meaningful addition. A short glossary defines many terms used in engineering geophysics.

Dobrin, Milton B., and Carl H. Savit. *Introduction to Geophysical Prospecting.* 4th ed. New York: McGraw-Hill, 1988. One of the standard texts in geophysical exploration, this reference is complete on all topics except ground penetrating radar. Derivation of equations from first principles and evaluation of problems that arise in data reduction and interpretation make this one of the best books for the serious student. Although the book does not focus on shallow targets, it offers a sound background with many applied examples.

Fetter, C. W. *Applied Hydrogeology.* 2d ed. Westerville, Ohio: Charles E. Merrill, 1988. The text is a superb introduction to all aspects of groundwater studies. The mathematical treatment is good, and graphs and figures are excellent. A chapter on field methods includes a brief review of geophysical methods of investigation with informative examples and mathematically treated solutions. A short section on ground penetrating radar includes examples of its use. An excellent glossary defines virtually all terms used in groundwater studies.

Griffiths, D. H., and R. F. King. *Applied Geophysics for Engineers and Geologists.* Elmsford, N.Y.: Pergamon Press, 1976. This text considers application of most geophysical techniques for shallow targets, with good treatments of problems of data reduction and drift corrections. Interpretation of simple and complicated geological settings is discussed. The bibliography provides many references to engineering or groundwater uses of geophysics.

Legget, Robert F. *Cities and Geology.* New York: McGraw-Hill, 1973. This book presents a complete view of the importance of geology in urban areas. The text covers historical and recent examples of engineering and planning problems and forms a good background for consideration of geophysical methods applied to geology. The references are compiled by chapter to aid in finding further readings on a particular subject. A well-annotated section offers suggestions for further reading.

McCann, D. M. *Modern Geophysics in Engineering Geology.* London: Geological Society, 1997. For the careful reader with a technical background, McCann offers a look at the science and practical applications of geophysics. Illustrations, bibliography, and index.

Maund, Julian G., and Malcolm Eddleston. *Geohazards in Engineering Geology.* London: Geological Society, 1998. An in-depth look at the policies and practices of geophysics engineering, as well as the dangers associated with the profession. Suitable for college-level readers.

Plummer, Charles C., and David McGeary. *Physical Geology.* Boston: McGraw-Hill, 1999. A college-level introductory geology textbook that is clearly written and wonderfully illustrated. An excellent sourcebook of basic information on geologic terminology and fundamentals of geologic processes. Contains a CD-ROM. An excellent glossary.

Rahn, Perry H. *Engineering Geology: An Environmental Approach.* 2d ed. Upper Saddle River, N.J.: Prentice Hall, 1996. This book offers a good introduction to the field of engineering geology, with an emphasis on environmentally sustainable practices. There are plenty of maps and graphs to illustrate key engineering concepts. References and index.

Sharma, Vallabh P. *Environmental and Engineering Geophysics.* Cambridge, England: Cambridge University Press, 1997. In a somewhat technical manner, Sharma describes the developments and policies within the field of engineering geophysics. A close look is also given to the role the environment plays in engineering decisions. Bibliography and index.

OCEAN DRILLING PROGRAM

The series of ocean-drilling efforts called the Ocean Drilling Program (ODP) has revolutionized understanding of the Earth's structure, climate, and available minerals. It has also allowed researchers to collect data about the Earth's cosmological history.

PRINCIPAL TERMS

ABYSSAL PLAINS: flat areas that make up large areas of the ocean floor

BASALT: rock formed from recrystallization of molten rock; most of the rock in the mid-ocean ridges and underlying the abyssal plains is basaltic

HYDROTHERMAL VENTS: areas on the ocean floor, typically along fault lines or in the vicinity of undersea volcanoes, where water that has percolated into the rock reemerges much hotter than the surrounding water; such heated water carries various dissolved minerals, including metals and sulfides

MANTLE: the thick rock layer between the Earth's crust and the core below

MARKER FOSSIL: a species that existed in a wide area but died out in a short time; finding such a fossil fixes the date of the strata in which it is found

METHANE HYDRATE: mineral formed when methane (natural gas) is trapped within the structure of water ice crystals; extensive ocean-floor deposits of methane hydrate might influence climate and could become a major resource

PLATE TECTONICS: mechanism that allows continental drift; the Earth's crust consists of individual shifting plates that form at mid-ocean ridges and other locations and are destroyed where plates collide and send material back into the mantle

MOHOLE PROJECT

Scientific drilling on the ocean floors began almost as a stunt, but it evolved into a long-term research program that revolutionized geology, contributed vital clues about climate change and about the extinction of the dinosaurs, discovered a major new type of subsea hydrocarbon deposit that may fuel the world, and aided in the study of rich metal ores that may also be tapped. Along the way, scientific drilling pioneered many techniques that have been used in offshore drilling for petroleum and natural gas.

Drilling samples have been a major part of geology since the second half of the nineteenth century. Drillers seeking water or oil could sample pieces of rock drilled from varying depths and log them into their drill records. For scientific purposes, drilling with circular drills allowed the cutting of a long cylinder or "core" that could be pulled up and carefully measured for position and composition. By the middle of the twentieth century, hard-rock miners were using cores to sample for minable ores. Data from these drillings were correlated into three-dimensional maps of distinctive strata showing dips and faults.

In the late 1950's, a number of geologists envied the space program with its romantic goal of flight to the Moon. In 1959, they proposed a similarly dramatic program called the Mohole project. In 1909, Andrija Mohorovičić had analyzed seismic waves from earthquakes and concluded that the rock of the crust changed significantly about 16 to 40 kilometers below the surface as it changed to partially melted mantle. There was speculation that rocks might be different at this so-called Mohorovičić discontinuity, or Moho, and researchers proposed drilling all the way to the Moho for samples. Drilling such a tremendously deep hole would be expensive and maybe impossible. Several marine geologists suggested that oceanic drilling to the Moho would be cheaper because the crust is thinner under the ocean floor. Also, a drill core through the seafloor might yield a complete fossil record of tiny marine shells.

Drilling into the ocean floor required several major innovations. The drilling locations were in deep water, so the drilling platform had to be an oceangoing vessel rather than a tower resting on the bottom. Anchoring in those depths would be difficult, so the vessel had to actively maintain position. This "dynamic positioning" had to keep the drill ship within two ship lengths of straight, or the drill string would break. While out of sight of land, the drilling vessel crew had to navigate within this small area with no landmarks whatsoever. This was managed originally by taut, moored buoys and later by satellite position-finding and acoustic beacons on the seafloor. Since the drill bit would hang as much as several thousand meters below the drill ship, bottom-hole assemblies were required for getting the drill started and for returning to the same hole. Finally, because the drill platform rose and fell with the waves, the rig needed a heave compensator.

Project Mohole started with engineering tests in 1961 in waters off California by *CUSS I*, a drilling barge developed for offshore oil drilling and built by Global Marine. (A number of famous drill ships from this company begin with the abbreviation "Glomar.") Initial tests off Baja California, Mexico, were promising, but the project eventually ended when technical difficulties and political management problems caused projected drilling costs to increase significantly. It was later discovered that the Moho outcrops at the Earth's surface in certain areas and that most areas of the ocean are comparatively young geologically, so continuation of the Mohole project would have been of less use than its backers had hoped.

However, the possibility of deep-ocean drilling had been demonstrated. Oil drillers, who had previously worked in depths of fewer than 100 meters, saw new possibilities. Project Mohole stirred interest in dynamic positioning for oil and gas exploration; the offshore drilling industry subsequently developed many technologies that passed back to the scientific drillers. Oceangoing rigs allowed exploratory drilling in deep areas of the continental shelf before making multibillion-dollar investments in production platforms. Thus, the Mohole project was a major factor in opening many offshore oil fields.

Furthermore, the Mohole demonstration had obtained one major data point. A mysterious lower layer visible on sonar graphs was found to be not another layer of sedimentary rock but rather basalt, an igneous rock formed by recrystallization of molten rock and often associated with volcanoes. This discovery suggested that the oceans were younger than expected and that the Mohole project, as originally planned, would have been able to gather minimally useful data about the geological history of the Earth. Conversely, it gave support to the then-radical theory of continental drift.

Testing continental drift required not one deep hole to the Mohorovičić discontinuity but many shorter holes surveying many areas. These shorter holes were still a tremendous advance. Before oceanic drilling, the only data about the seafloors came from dredge hauls and piston cores. Dredges only pull a jumbled mass of material from the first few centimeters of the seafloor. A piston core is essentially a weighted pipe that is allowed to fall as fast as possible to the seafloor; its weight and momentum drive it into soft sediment. Its limitations are that it penetrates only a few meters and that it cannot penetrate hard surfaces.

JOIDES

In 1964, the Joint Oceanographic Institutions for Deep Earth Sampling (JOIDES) was formed; it has become an international organization that includes universities and government research organizations. In April and May of 1965, JOIDES used the drill ship *Caldrill I* to test upgraded methods by drilling six holes on the Blake Plateau off the coast of Florida to sub-bottom depths of more than 1,000 meters. Based on that success, JOIDES proposed an eighteen-month program of scientific drilling in the Atlantic and Pacific Oceans to the U.S. National Science Foundation. The resultant Deep Sea Drilling Project (DSDP) was operated by the Scripps Institution of Oceanography. The DSDP drill ship *Glomar Explorer* was capable of drilling 760 meters in 6,100 meters of water. It began operations in July, 1968, and made ninety-six voyages (or drilling legs) for JOIDES that focused on sites on or near the mid-ocean ridges. The cores retrieved revolutionized geology by proving the theory of continental drift.

In 1978, the *Glomar Explorer* was replaced by the *JOIDES Resolution*, run by Texas A&M University as part of the one-third internationally funded Ocean Drilling Program (ODP). The ship was designed to drill in 8 kilometers of water for a total drill-string length of 9 kilometers, and it can drill 2 kilometers into the seafloor in shallower waters. Both the *Glomar Explorer* and the *JOIDES Resolution* have worked in concert with remote instruments and piloted instruments in a number of revolutionary developments.

PLATE TECTONICS

Drilling in marine strata has yielded tremendous advances in geological knowledge for several reasons. First, oceans cover more than 70 percent of the Earth, so a proportionate number of discoveries should be expected from oceanic data. Second, tens of thousands of drill cores have been logged on land, so many of the land discoveries have already been made. Third, land is often subject to erosion, so there are more gaps than in marine strata. Fourth, land strata have been subjected to more compression, heat, and chemical attack—all of which could confuse possible geological data—than seafloor strata. Fifth, some of the mechanisms that happen under several kilometers of water in the oceans happen under several kilometers of rock on land, so these mechanisms are easier to study with marine drilling. Finally, drilling on land requires disassembly and reassembly of equipment, often in difficult terrain; a drill ship simply sails to the next drill site.

In 1620, Francis Bacon noted similarities between the coastlines of South America and Africa and suggested that they had once been interconnected. In the early twentieth century, Alfred Wegener proposed the former supercontinent of Pangaea to explain similar fossils from areas now widely separated. However, these theories lacked a provable mechanism for moving entire continents through the rock underlying the ocean floor.

It was then noted that the Mid-Atlantic Ridge had bands of different magnetic orientation on either side. These bands corresponded to reversals of the Earth's magnetic field that occur every several thousand years. It appeared that new melted rock was crystallizing (and thus freezing a weak magnetic field) at the ridge and was then shoved away from the ridge by new rock. The crystallizing rock that cooled and hardened out of magma would quite likely be basalt. The Mohole finding of basalt in its deeper cores was encouraging, but conclusive proof required correlating the age from a large number of drill samples. The initial *Glomar Explorer* drill cores allowed geologists to concentrate on that dating using marker fossils and radioisotopic dating. Data from the cores showed that the ocean floor approaching the Mid-Atlantic Ridge was progressively younger. Thus, it appeared that Europe, North America, Africa, and South America were joined until sometime between 200 and 170 million years ago, when the Atlantic Ocean began opening.

This revolutionary discovery allowed development of plate tectonic theory, which proposes that large plates of connected rock exist on the Earth's surface. Plate tectonics has many implications. Because the Earth cannot expand indefinitely, the idea that new rock accretes along mid-ocean ridges means that plate material must be disappearing elsewhere. That explains the existence of deep ocean trenches (such as the Mariana Trench) where oceanic strata are diving steeply down toward the mantle and some of the heated rock spurts up, forming island arcs. One plate riding over another provides the mechanism for raising up mountains, such as the Andes Mountains, which are riding up over the edge of the Pacific plate. A more extreme example is the Indian subcontinent, which is burrowing under the Tibetan Plateau and lifting the Himalaya.

HYDROTHERMAL VENTS

Continued study of plate tectonics along the mid-ocean ridges and other plate boundaries led to the unexpected discovery of hydrothermal vents, which are somewhat like geysers and hot springs but which are found on the ocean floor; however, the scale and results of the activity surprised researchers. In 1965, a sample of rocks dredged from the seafloor yielded rock samples of two types. One was depleted in certain chemicals that were present in the other type. Also in the 1960's, it was noted that thermal readings in sediments near plate boundaries did not have as much heat as expected. Hydrothermal vents were suggested as the mechanism for moving the large amounts of heat. Seawater percolating down to the molten rock beneath the seafloor would be heated and return to the ocean as hot springs. The heated water would be much less dense than the near-freezing seawater near the ocean floor, so convection would drive the process.

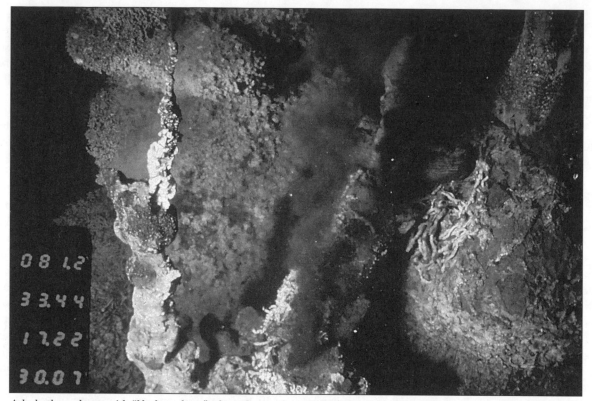

A hydrothermal vent with "black smokers," where plumes of superheated water meet colder waters and dark minerals begin to precipitate out and form "chimneys." (National Oceanic and Atmospheric Administration)

In 1977, researchers in a piloted submersible discovered an active field of hydrothermal vents with "black smokers," where plumes of superheated water meet colder waters and dark minerals begin to precipitate out and form "chimneys." This discovery revealed the importance of such vents in circulating heat from the Earth's core, circulating minerals into and out of the ocean waters, and sustaining life driven by chemosynthesis (chemical energy) rather than photosynthesis (light energy).

However, only drilling could supply data on subsurface events. For example, bacteria living by chemosynthesis have been found several hundred meters below the ocean floor, suggesting that such life-forms may be more biomass-based than photosynthesis-based life. Also, drilling in mounds built by successive chimneys revealed large amounts of carbonate salt that eventually dissolves when the area cools. Thus, old, inactive hydrothermal fields might have rich mining potential. Finally, the sulfide-enriched upper layer may be underlain by a copper-enriched layer, also with good mining

potential. More important, comparable areas of former seabed have been lifted up and become land. These areas, such as the Klamath Mountains of northwestern North America and areas of Cyprus, have already served as mining areas, and better knowledge of the mineral-formation processes will probably improve land mining long before any ocean mining.

MARKER FOSSILS

Probably the most important dividends from the ODP are carefully logged and partially analyzed cores from every ocean except the Arctic. By the end of the twentieth century, there were more than 240 kilometers of cores available for new studies and new instruments. These cores have helped provide increasingly fine chronological correlation among strata from various locations. Oceanic core data allow the identification of "marker fossils" that (ideally) lived for a short time but over a wide area. Those marker fossils can then be used to date match with other oceanic cores and land cores. This correlation is crucial for tracking individual layers of rock through folds and faults. The dated and mapped layers allow one to calculate what might have happened in the past, what might happen in the future, and where ores may exist at present.

The types of fossils in the cores are generally tiny microfossil remains of plankton that lived in the surface waters. The types of fossils indicate the climate in those waters when they were deposited. Likewise, percentages of carbon and oxygen isotopes vary with sea surface temperature. Data from isotopes and microfossils can be checked against one another and against climate data from land to calculate past climates (paleoclimatology). With better estimates of past climate, climatologists may better project possible climate change in the future.

Ocean drilling provides another climate indicator: sediments from land. Drilling confirmed massive catastrophic flows when ice dams broke at the end of the last ice age, releasing torrents of water down the Columbia and St. Lawrence River Basins. Drilling also shows the timing, direction, and quantity of windborne sand and dust from places such as North Africa and Central Asia. Cores around Antarctica have corroborated that Antarctica's continental ice sheet has existed for millions of years; however, continued studies are needed to detail fluctuations of ice level.

Drill cores also confirmed that turbidity currents (dense masses of sediment-laden water) are a major mode of deposition in the ocean. (The low relief of the abyssal plains in the Atlantic Ocean, which has a high amount of sediment compared to total area, is caused by turbidity flows filling in low areas.) Turbidities indicate past ocean currents, another paleoclimate factor.

Petroleum and natural gas traces in deep waters are another major discovery of ocean drilling. However, the *Glomar Challenger* and *JOIDES Resolution* had to stop drilling whenever they encountered such traces. These craft have room for laboratories because they were not fitted with a "riser," which collects anything coming out of the drill hole and pumps down drilling mud. The riser can be used to stop a flow of fluid from a borehole. Therefore, the riserless drill ships risked causing an uncontrolled leak into the ocean if they continued drilling in areas with any sign of petroleum or natural gas.

METHANE HYDRATES

Still another hydrocarbon deposit researched by ocean drillers is methane hydrate. Scattered reports of flammable ice and fizzing material brought up by piston cores led to a theory that methane (natural gas) might be frozen into the structure of ice in deposits beneath the deep ocean floors. These deposits could form because the waters are only slightly above the freezing point of fresh water and they have high pressure—the two requirements for methane hydrate. The methane source would probably consist mostly of the decay of biological material in sediments. Sonar readings even showed layers in seafloor sediments that might be such hydrates.

Drill cores from several sites confirmed extensive methane hydrate deposits, and their analysis led to several conclusions. Hydrates may contain several times as much energy as all other fossil fuel deposits combined. Hydrates might act as a cap rock to contain conventional gas and petroleum. It might be some time before hydrates are exploited because they are often spread thinly in sub-seafloor ocean strata. Hydrates might be a major factor in climate change because disturbances on the ocean floor or lowered sea levels might release methane, a greenhouse gas; conversely, raised sea levels would increase the area of high pressure, allowing more hydrates to form, thus decreasing the flow of greenhouse gases to the atmosphere. Finally, hydrates are stronger than unfrozen seafloor muds and oozes. Consequently, a hydrate-melting event could cause a major collapse of sediments on the continental slope and result in a major turbidite flow.

Drill cores also corroborated the asteroid-impact explanation for extinction of the dinosaurs. A thin dead zone marked the end of the Cretaceous period, the last great age of dinosaurs, and the beginning of the Tertiary. This Cretaceous-Tertiary (K-T) boundary zone was noted in a few rock outcrops in Europe, and rock from this layer had elevated levels of platinum-group metals (similar to elevated levels in asteroids) in that strata. However, the theory was doubted until multiple drill cores closed in on the probable impact site on the Yucatán Peninsula of Mexico with thicker corresponding deposits off Florida and in the Caribbean Sea. Besides the platinum-group-enriched dead layer, those deposits had large amounts of shocked quartz, suggesting cosmic impact.

Sea levels over time have also been deduced from oceanic cores. One major sea-level surprise was massive evaporite deposits in the Mediterranean Sea from when sea levels were low enough to make the Strait of Gibraltar dry land and the Mediterranean a salty inland sea. Finally, remote stations have been implanted in several oceanic boreholes. These stations take seismic data to augment land seismic stations, and they also have temperature sensors.

Roger V. Carlson

CROSS-REFERENCES

Abyssal Seafloor, 651; Atlantic Ocean, 2203; Basaltic Rocks, 1274; Continental Drift, 567; Cretaceous-Tertiary Boundary, 1100; Deep Ocean Currents, 2107; Deep-Earth Drilling Projects, 347; Deep-Sea Sedimentation, 2308; Engineering Geophysics, 353; Geochemical Cycle, 412; Hydrothermal Vents, 2117; Mediterranean Sea, 2234; Microfossils, 1048; Mid-Ocean Ridge Basalts, 657; Ocean Ridge System, 670; Ocean-Floor Drilling Programs, 365; Paleoclimatology, 1131; Plate Tectonics, 86; Seismic Reflection Profiling, 371; Turbidity Currents and Submarine Fans, 2182.

BIBLIOGRAPHY

Bascom, Willard. *The Crest of the Wave: Adventures in Oceanography.* New York: Doubleday, 1988. Chapter 8, "The First Deep Ocean Drilling," describes the technical issues faced by the Mohole project, as well as the changes in attitudes that the project prompted.

Broad, William J. *The Universe Below: Discovering the Secrets of the Deep Sea.* New York: Simon & Schuster, 1997. Provides an entertaining general overview of oceanic exploration, including descriptions of oceanic drilling, associated submersible explorations near those sites, and implications for possible future mining.

Cone, Joseph. *Fire Under the Sea: The Discovery of the Most Extraordinary Environment on Earth, Volcanic Hot Springs on the Ocean Floor.* New York: William Morrow, 1991. A detailed history of the exploration and description for the lay reader of hydrothermal vents on the ocean floor and the plate tectonics that cause them. The book is an excellent introduction to the technical articles listed in its extensive reference list.

Davies, Thomas A. "Scientific Ocean Drilling." *Marine Technology Society Journal* 32 (Fall, 1998): 5-16. Davies summarizes the history of the Ocean Drilling Program, major accomplishments, the technologies involved, and proposed work for the future.

Huey, David P., and Michael A. Storms. "Novel Drilling Equipment Allows Downhole Flexibility." *The Oil and Gas Journal* 93 (January 16, 1995): 63-68. Details the ocean-drilling innovations developed by the DSDP and ODP.

Moore, J. Robert. *Oceanography: Readings from Scientific American.* San Francisco: W. H. Freeman, 1991. Contains several classic articles written about seafloors, plate tectonics, and studies of microfossils (micropaleontology).

Zierenberg, R. A., et al. "The Deep Structure of a Sea-Floor Hydrothermal Deposit." *Nature* 392 (April 2, 1998): 485-488. The authors detail an important set of results about ocean drilling used to study metallic minerals associated with the Juan de Fuca spreading center off the coast of Seattle.

OCEAN-FLOOR DRILLING PROGRAMS

Ocean-floor drilling programs have allowed geologists and oceanographers to extend their knowledge of the Earth's history by analyzing long marine sediment cores and basement rock cores recovered from the seafloor. Data from ocean-floor drilling have provided evidence supporting the theories of seafloor spreading and plate tectonics and have permitted the investigation of the paleoclimatic and paleoceanographic history of the Earth.

PRINCIPAL TERMS

ABYSSAL PLAINS: flat-lying areas of the seafloor, located in the ocean areas far from continents; they cover more than half the total surface area of the Earth

BASALT: a dark-colored, fine-grained rock erupted by volcanoes, which tends to be the basement rock underneath sediments in the abyssal plains

CHERT: a hard, well-cemented sedimentary rock that is produced by recrystallization of siliceous marine sediments buried in the seafloor

CORRELATION: the demonstration that two rocks in different areas were deposited at the same time in the geologic past

DEPOSITION: the process by which loose sediment grains fall out of seawater to accumulate as layers of sediment on the seafloor

MID-OCEAN RIDGE: a continuous mountain range of underwater volcanoes located along the center of most ocean basins; volcanic eruptions along these ridges drive seafloor spreading

PALEOCEANOGRAPHY: the study of the history of the oceans of the Earth, ancient sediment deposition patterns, and ocean current positions compared to ancient climates

PLATE TECTONICS: a theory that the Earth's crust consists of individual, shifting plates that are formed at oceanic ridges and destroyed along ocean trenches

SEAFLOOR SPREADING: a theory that the continents of the Earth move apart from each other by splitting of continental blocks, driven by the eruption of new ocean floor in the rift

ORIGINS OF TECHNOLOGY

Most of our knowledge of the history of the Earth comes from the study of sedimentary rocks, as sediments may contain the preserved fossil remains of ancient plants and animals, while sedimentary structures record the processes of deposition. Sedimentary rocks exposed on land tend to have an incomplete record because they may be deformed by folding and faulting, which may destroy both the fossils and the sedimentary structures and may be eroded by wind, water, and ice moving across the surface. In contrast, marine sedimentary sequences contain a more complete record of accumulation because they accumulate in a lower-energy environment, which is not as affected by folding, faulting, erosion, and post-depositional alteration as are terrestrial sediments. As a result, the deep-sea sediments tend to preserve a continuous record of sediment deposition in the ocean basins.

Ocean drilling techniques were originally developed in the 1950's by petroleum exploration companies searching for shallow-water hydrocarbon and petroleum deposits located on the continental shelves. These industrial exploration methods were adapted in the 1960's to obtain long sediment cores from the seafloor in deep-water areas on the continental slopes and abyssal plains. By drilling through the entire sediment record into the harder basement rocks below marine sediments, geologists hoped to acquire the complete history of sediment deposition within an ocean basin from the time that sediments were first deposited atop volcanic basalts.

GLOMAR CHALLENGER

Preliminary attempts to drill the ocean floor included engineering tests for Project Mohole by the drilling barge *CUSS I*, which in 1961 drilled marine sediments off La Jolla, California, and at the deep-water Experimental Mohole Site east of Guadalupe Island, off Baja California, Mexico, in a water depth of 3,566 meters. Although further Project Mohole development was not undertaken because of a combination of political conflicts and increasing cost estimates for the project, in 1964 four American universities formed a consortium to initiate a program of scientific deep-sea drilling. JOIDES, the Joint Oceanographic Institutions for Deep Earth Sampling, successfully operated a drilling program in April-May, 1965, using the vessel *Caldrill I* to drill six holes on the Blake Plateau off Florida to sub-bottom depths of more than 1,000 meters, with continuous core recovery.

Following these successful trials, JOIDES proposed an eighteen-month program of scientific drilling in the Atlantic and Pacific oceans, to be called the Deep Sea Drilling Project (DSDP), operated by the Scripps Institution of Oceanography of La Jolla, California, using the drilling vessel *Glomar Challenger. Glomar Challenger* left Orange, Texas, on July 20, 1968, on Leg 1 of the Deep Sea Drilling Project. The results of DSDP drilling on the first nine cruises in the Atlantic and Pacific oceans caused the National Science Foundation to extend the drilling program beyond the initial eighteen-month period, with further drilling in the Indian Ocean and in the seas surrounding Antarctica.

When DSDP began operations, many other American oceanographic institutions joined JOIDES in support of the drilling program, and the success of DSDP also attracted scientific participation and financial support from foreign countries. The International Program of Ocean Drilling (IPOD) started in 1975 when the Soviet Union, the Federal Republic of Germany, France, the United Kingdom, and Japan joined JOIDES, with each country providing $1 million yearly to support drilling programs. DSDP/IPOD drilling activities continued through the early 1980's, leading to international scientific exchange of information between oceanographers.

Sedco/BP 471

Because the initial JOIDES proposal was only for eighteen months, it was never expected that ocean drilling would continue for fifteen years. Because of demands for ocean drilling in deeper waters and in high-latitude polar areas, JOIDES proposed in the early 1980's that a larger drilling vessel be acquired for continued drilling. The last cruise of the *Glomar Challenger,* DSDP Leg 96, ended in Mobile, Alabama, on November 8, 1983, with the retirement of the drilling vessel from service. In 1983, responsibility for scientific supervision of the international project, now called the Ocean Drilling Program (ODP), passed from the Scripps Institution to Texas A&M University, and the drilling vessel *Sedco/BP 471* replaced *Glomar Challenger.* The first cruise of a ten-year ODP drilling program began on March 20, 1985, when the *Sedco/BP 471,* informally called the *JOIDES Resolution,* left port to begin drilling on ODP Leg 101. Leg 186 was scheduled for completion in the year 2000, and planning continues for cruises up to Leg 201 in 2002.

The results of each cruise, or leg, have been published in a series of books, entitled *Initial Reports of the Deep Sea Drilling Project,* which are published by the U.S. Government Printing Office. The cores recovered from the DSDP and ODP holes represent an invaluable record of the history of ocean sediment deposition around the globe. These recovered sediment cores are studied by a variety of scientists, who are interested in the sediment type, fossil content, geochemistry, magnetic orientation and strength, shear strength, and other sedimentary properties of the samples.

DRILLING PROCEDURES

In shallow water, drilling is accomplished either by building a drilling platform directly atop the seafloor or by firmly anchoring a drilling vessel to the bottom. In deep-water ocean drilling, however, it is not possible to anchor the drilling vessel to the bottom, so the technique of dynamic positioning is used to maintain the position of the vessel above the hole being drilled. In dynamic positioning, an acoustic beacon emitting sounds at either 12.5 kilohertz or 16 kilohertz is dropped to the seafloor. Four hydrophones, located at different points on the hull of the drillship, receive the signal from the acoustic beacon at slightly different times, depending on the position of the hull relative to the beacon. The position of the ship is maintained by a shipboard computer, which interprets the information from the hull hydrophones and controls the position of the ship by driving both the main propellers and two laterally oriented propellers, or hull thrusters. If the vessel is pushed off location by waves or surface currents, the shipboard computer attempts to compensate by using the propellers and hull thrusters to maintain the ship's location relative to the seafloor beacon.

In order to drill sediment and rock samples from the seafloor, a drill bit is attached to the bottom of a 9.5-meter-long piece of hollow cylindrical stainless steel drill pipe. More individual lengths of pipe are connected on the rig floor of the drillship to make a "drill string," which extends from the vessel through the water down to the seafloor, where coring may begin. Usually, about 450-510 lengths of drill pipe are required simply to reach the bottom, and the assembly of this drill string may take twelve hours before bottom drilling may be started. Once the string reaches the seafloor, the drill pipe is rotated by hydraulic motors on the rig floor, and the rotary action combined with the weight of the drill string causes the drill bit to spiral down into the seafloor. Sharp iron carbide or diamond cutting teeth on the drill bit assist the penetration of the drill string into the sediment and the rock on the ocean bottom.

Samples of sediment and rock are retrieved from the seafloor by drilling a hole about 25 centimeters in diameter, using a drill bit with a 7.5-centimeter hole in its center. In effect, sediment is cored by "drilling the doughnut and saving the doughnut hole": Rotating the drill string grinds the outer ring of sediment to small pieces against the diamond teeth of the drill bit, while the material in the center of the drill hole is saved as a core of drilled sediment 6.6 centimeters in diameter. As the drill string is lowered deeper into the drilled hole, the core is pushed up into a plastic core liner in a steel "core barrel" within the lowest stand of drill pipe. After 9.5 meters of the seafloor has been drilled, the core barrel is pulled up to the rig floor by a cable lowered through the drill string.

ATLANTIC OCEAN DRILLING

The first three cruises of the *Glomar Challenger* provided information proving that seafloor spreading had occurred in the Atlantic Ocean: A series of DSDP holes across the Mid-Atlantic Ridge showed that the age of bottom sediments increased with distance from the ridge crest and indicated that the ages of sediments with depth cor-

relate from one hole to the next. The total thickness of sediment atop basaltic basement also increased with greater distance from the ridge crest, on both the east and west sides of the Mid-Atlantic Ridge. Further DSDP and ODP drilling has provided evidence that seafloor spreading has occurred in all the Earth's ocean basins. In addition, ocean drilling has confirmed the relative youth of the ocean basins, as predicted by plate tectonics; the oldest seafloor yet discovered is Early Jurassic in age (160 million years old), compared to continental rocks, which may be as old as 4.5 billion years.

PACIFIC OCEAN DRILLING

Glomar Challenger and *JOIDES Resolution* have operated from the Norwegian Sea to the Ross Sea off Antarctica and have drilled holes in water depths from 193 meters on the Oregon continental shelf to 7,050 meters in the Mariana Trench off Guam, in the western Pacific Ocean. The deepest hole through seafloor sediment deposits is more than 1,750 meters below the seafloor, and one site in the equatorial Pacific Ocean west of South America (DSDP Hole 504B) has been drilled through 300 meters of sediment and 1,500 meters of volcanic basement rock.

GLOBAL STRATIGRAPHIC CORRELATION

Seafloor drilling has indicated that deep-sea sediments contain long sequences of well-preserved microfossils, which may be used for global stratigraphic correlation, in contrast to the fragmentary record preserved on land, where structural deformation of sediment deposits may complicate the problem of correlating different sedimentary sequences. Analyses of these sediments has revealed the history of deposition in the different ocean basins and has provided information on ancient climates and oceanographic conditions (such as the position, strength, and temperature of past ocean currents).

Sediment cores have indicated the presence of great shifts in oceanic climate conditions during the geologic past and have demonstrated that the Antarctic continent has been covered by glacial ice caps for at least 40-50 million years, rather than the 5 million years accepted prior to DSDP drilling near Antarctica. Another startling result of ocean drilling has been the discovery that the

Mediterranean Sea dried up between 12 and 5 million years ago: Massive salt and evaporite mineral deposits below the Mediterranean basin indicate that the Strait of Gibraltar connection to the Atlantic Ocean was blocked during this time. Blockage of the Gibraltar connection allowed the water in the basin to evaporate, causing the deposition of vast salt and evaporite mineral deposits as the Mediterranean dried up.

STUDY OF SEAFLOOR BASEMENT ROCKS

Not all the information provided by ocean drilling has been concerned with the sediment column. Drilling into basement rocks has allowed geophysicists to compare the structure of seafloor basement to that of layered igneous-rock deposits that have been uplifted above sea level on the edges of certain continents. Similarly, direct drilling through these basalt and gabbro layers has allowed a comparison of the rock type to sound velocities measured by marine geophysicists. Some other results of hard-rock seafloor drilling have been the investigation of sediment and mineral deposition by hydrothermal processes at rapidly opening mid-ocean ridge segments, such as the sulfides deposited by high-temperature fluids emitted by "black smoker" and "white smoker" structures near the Galápagos Islands west of South America. Drilling of bare basement rocks along mid-ocean ridges in the Atlantic, Pacific, and Indian oceans has enabled geochemists and igneous petrologists to study the frequency at which seafloor volcanic rocks are produced at individual ridge segments and to determine whether temporal changes occur in the chemistry of basalts erupted from one location on the ridge. These studies of seafloor basement rocks may be applied to mineral exploration of marine rocks that have been uplifted above sea level and exposed on continents.

IMPROVEMENTS IN DRILLING TECHNOLOGY

In addition to scientific results, DSDP and ODP operations have resulted in improvements in drilling technology by developing the ability to reenter seafloor boreholes, by devising techniques for "bare-rock" drilling on the seafloor, and through the development of new coring bits. During DSDP Leg 1, it was discovered that existing drill bits could not penetrate hard chert beds; thus, they also would not be able to penetrate through deeper igneous rocks below seafloor sediments. The drag bits were solid, consisting of a central opening and radial curved ridges of steel or tungsten carbide, capped with industrial diamonds and designed to churn through soft sediments. As a result, DSDP began a drill-bit design program, which led to the development of roller bits capable of penetrating both chert layers and seafloor basalts. These bits consist of four conical cutting heads studded with tungsten carbide or diamond cutting teeth, situated around the central core opening in the bit.

Another important technical development of DSDP, first successfully accomplished on Leg 15, was the ability to reenter a drilled borehole on the seafloor. Even with roller bit designs, drill bits wear out from the stresses of rotary coring through seafloor sediments and rocks. When a bit fails, the entire drill string has to be "tripped," or pulled up to the vessel to replace the bit at the lower end of the string, which in most deep-ocean drilling sites requires pulling the string up not only several hundred meters from below the seafloor but also through 2,000-5,000 meters of seawater. During early DSDP legs, bit failure forced the abandonment of a hole because after the fatigued bit was replaced, it was impossible to reenter the original borehole. Successful reentry techniques were facilitated by the development of a steel reentry cone 6 meters in diameter, topped with three sonar reflectors and a rotating sonar scanner that can be lowered through the drill string. In areas where hardened sediment layers are anticipated, requiring bit replacement to complete drilling, the reentry cone is placed on the seafloor prior to drilling the initial borehole. As bits become worn, they may be replaced and the hole reentered by using the sonar scanner to locate the reentry cone (and thus the original hole).

DSDP and ODP drilling specialists have also devised methods to enable drilling in hard seafloor areas, such as mid-ocean ridges, which were not previously drillable by existing techniques. Development of a seafloor "guide base" for drilling has allowed successful drilling and reentry of boreholes in these areas and has permitted the implantation of seafloor sensing devices, such as earthquake-measuring seismometers, in these holes.

ROLE IN DEVELOPMENT OF PALEOCEANOGRAPHY

Ocean-floor drilling programs have enabled scientists to correlate apparently unconnected phenomena into the theory of plate tectonics, a global synthesis of geology and oceanography. Ocean drilling has provided verification of the seafloor spreading hypothesis as it applies to plate tectonics and has indicated that seafloor spreading has occurred in all the Earth's ocean basins.

Before long sediment cores could be acquired from the seafloor, scientists' knowledge of seafloor geology was sparse, based on limited samples available from dredging and shallow coring of the seafloor by oceanographic vessels. Prior to DSDP, global stratigraphic correlation was based on a fragmentary record preserved on land, where structural deformation of sediment deposits may complicate the problem of correlating different sedimentary sequences; DSDP drilling, however, has revealed that deep-sea sediments contain long sequences of well-preserved microfossils. Furthermore, seafloor sediment cores have revealed the history of the ocean basins and have provided information on ancient climates and oceanographic conditions (such as the position, strength, and temperature of past ocean currents). A new science, paleoceanography, has been developed based on this information from DSDP and ODP drilling. Analysis of the Earth's ancient climates may provide information to predict future shifts in the biosphere.

INDUSTRIAL APPLICATIONS

Ocean drilling has also provided evidence for the existence of deep-water hydrocarbon accumulations, which has enabled petroleum exploration companies to drill petroleum deposits on the continental slopes and may eventually lead to the discovery of significant hydrocarbon deposits in ocean basins. If future technology is developed, humans may be able to exploit these deep-water petroleum resources. Furthermore, scientific ocean drilling has enabled geologists to understand the processes controlling the deposition of "black shale" deposits and other high-productivity seafloor sediments, which may be altered by burial into source beds for the generation of petroleum hydrocarbons. Understanding of the processes affecting the formation and distribution of these sediments may assist in future exploration for fossil fuel resources. In addition, studies of seafloor basement rocks may lead to a more complete understanding of the nature of mineral deposition at mid-ocean ridges, which may be applied to mineral exploration of similar marine rocks that have been uplifted above sea level and exposed on continents.

Finally, deep-ocean drilling has led to technological innovations in the tools and techniques used to sample the seafloor. These methods have been adapted by industrial companies exploring for hydrocarbons buried beneath marine sediments and for mineral deposits on the seafloor.

Dean A. Dunn

CROSS-REFERENCES

Abyssal Seafloor, 651; Biostratigraphy, 1091; Carbonate Compensation Depths, 2101; Deep-Earth Drilling Projects, 347; Deep Ocean Currents, 2107; Deep-Sea Sedimentation, 2308; Engineering Geophysics, 353; Microfossils, 1048; Ocean Basins, 661; Ocean Drilling Program, 359; Oil and Gas Exploration, 1699; Paleobiogeography, 1058; Paleoclimatology, 1131; Plate Tectonics, 86; Sedimentary Mineral Deposits, 1637; Seismic Reflection Profiling, 371; Stratigraphic Correlation, 1153; Transgression and Regression, 1157.

BIBLIOGRAPHY

Bascom, Willard. *A Hole in the Bottom of the Sea.* Garden City, N.Y.: Doubleday, 1961. A history of the Mohole Project, which planned to drill through oceanic rocks down to the crust-mantle boundary.

Condie, Kent C. *Plate Tectonics and Crustal Evolution.* 4th ed. Oxford: Butterworth Heinemann, 1997. An excellent overview of modern plate tectonics theory that synthesizes data from geology, geochemistry, geophysics, and oceanography. A very helpful tectonic map of the world is enclosed. The book is nontechnical and suitable for a college-level reader. Useful "suggestions for further reading" follow each chapter.

Cramp, A., ed. *Geological Evolution of Ocean Basins: Results from the Ocean Drilling Program*. London: The Geological Society, 1998. Intended for the reader with some scientific background, this book offers a detailed discription of the processes and results of ocean drilling.

Davidson, Jon P., Walter E. Reed, and Paul M. Davis. *Exploring Earth: An Introduction to Physical Geology*. Upper Saddle River, N.J.: Prentice Hall, 1997. An excellent introduction to physical geology, this book explains the composition of the Earth, its history, and its state of constant change. Intended for high-school-level readers, it is filled with colorful illustrations and maps.

Hamblin, William K. *Earth's Dynamic Systems*. 8th ed. Upper Saddle River, N.J.: Prentice Hall, 1998. This geology textbook offers an integrated view of the Earth's interior not common in books of this type. The illustrations, diagrams, and charts are superb. Includes a glossary and laboratory guide. Suitable for high school readers.

Hsu, Kenneth J. *The Mediterranean Was a Desert*. Princeton, N.J.: Princeton University Press, 1983. A personal account of DSDP Leg 13 drilling in the Mediterranean Sea basin during 1970, as written by one of the two chief scientists on the drilling vessel *Glomar Challenger*.

Nierenberg, William A. "The Deep Sea Drilling Project After Ten Years." *American Scientist* 66 (January/February, 1978): 20-29. A review of the significant technical and scientific developments of the DSDP, written by the director of Scripps Institution of Oceanography.

Peterson, M. N. A., and F. C. MacTernan. "A Ship for Scientific Drilling." *Oceanus* 25 (Spring, 1982): 72-79. Summary of the technical aspects of ocean drilling on *Glomar Challenger*, written for a nonscientific audience by the director of the DSDP.

Segar, Douglas. *An Introduction to Ocean Sciences*. New York: Wadsworth, 1997. Comprehensive coverage of all aspects of the oceans and the oceanic crust. Readable and well illustrated.

Suitable for high school students and above.

Shor, Elizabeth Noble. *Scripps Institution of Oceanography: Probing the Oceans, 1936 to 1976*. San Diego, Calif.: Tofua Press, 1978. The book recounts the formation of the Scripps Institution of Oceanography and provides a history of the oceanographic research performed by its scientists. Chapter 12 covers the history of the Deep Sea Drilling Project.

Van Andel, Tjeerd H. "Deep-Sea Drilling for Scientific Purposes: A Decade of Dreams." *Science* 160 (June 28, 1968): 1419-1424. A summary of the results of scientific ocean drilling up to the start of the Deep Sea Drilling Project.

Warme, John E., Robert G. Douglas, and Edward L. Winterer, eds. *The Deep Sea Drilling Project: A Decade of Progress*. Tulsa, Okla.: Society of Economic Paleontologists and Mineralogists, 1981. A volume of scientific papers discussing the results of oceanographic research based on sediments and rocks recovered by DSDP. Best suited to those with some scientific background.

West, Susan. "Diary of a Drilling Ship." *Science News* 119 (January 24, 1981): 60-63.

_____. "DSDP: Ten Years After." *Science News* 113 (June 24, 1978). Summarizes the choices facing the DSDP in 1978: whether to continue ocean drilling with *Glomar Challenger* or to seek a larger and more sophisticated drilling vessel.

_____. "Log of Leg 76." *Science News* 119 (February 21, 1981): 124-127. These articles tell a reporter's story of seven days aboard *Glomar Challenger* during Leg 76 drilling operations in the Atlantic Ocean off Florida.

Wilhelm, Helmut, Walter Zuern, Hans-Georg Wenzel, et al., eds. *Tidal Phenomena*. Berlin: Springer, 1997. A collection of lectures from leaders in the fields of Earth sciences and oceanography, *Tidal Phenomena* examines Earth's tides and atmospheric circulation. Complete with illustrations and bibliographical references, this book can be understood by someone without a strong knowledge of the Earth sciences.

SEISMIC REFLECTION PROFILING

Seismic reflection profiling is a method of applied exploration geophysics that allows scientists to determine the location of subsurface geological structures. It is accomplished by using one of various methods of generating seismic waves. These waves are reflected to the surface from the subsurface and are received and analyzed. This analysis enables geologists to locate oil and, less frequently, mineral-bearing formations.

PRINCIPAL TERMS

ATTENUATED: becoming less intense as distance from the source increases

DISCONTINUITY: the sudden change in physical properties of rock with increased depth

FOCUS: the source of earthquake waves; the actual point of rock breakage

ELECTROMAGNETIC RADIATION: forms of energy, such as light and radio waves, that consist of electric and magnetic fields that move through space

PROPAGATED: conducted through a medium

STRATA: rock layers produced by sediment deposition in layers or beds

BODY SEISMIC WAVES

Seismic reflection profiling enables Earth scientists to determine what the Earth's subsurface looks like without actually drilling exploratory wells. This study of applied seismics is related to seismology, which is the study of earthquake waves. When an elastic body such as rock is stressed and suddenly breaks, the energy released is transferred through the material in the form of various types of waves, which is what happens during an earthquake. When stress builds up and the rock fractures, energy radiates out from the focus or zone of breakage in the form of an ever-enlarging sphere of wavefronts. As the sphere gets larger, the energy along any part of the wavefront is diminished or attenuated. This sphere will continue to enlarge and maintain its basic shape as long as the properties of the rock through which the waves are traveling remain the same.

The four types of seismic waves are classified into two major categories. Those waves that travel beneath the surface of the Earth (that is, in a three-dimensional medium) are called body waves. Waves that travel at or near the surface of the Earth (in a two-dimensional medium) are called surface waves. Body waves are caused by earthquakes that take place well beneath the surface of the Earth. Surface waves are caused by near-surface earthquakes or by human-made explosions.

The two types of body waves are P, or primary, waves and S, or secondary, waves. The P waves are compressional waves and are the faster of the two types. P waves, for example, can be generated by pinching several coils of a "Slinky" toy together and releasing them. Sound waves are an example of this type of wave; sound will travel through any medium that is capable of being compressed. In a compressional wave, the particles of the conducting medium vibrate parallel to the direction of wave propagation, or travel. A compressional wave is set up by a vibration. In the case of sound waves, the vibration may be that of the human vocal chords; in the case of seismic compressional waves, it is the breaking of rock. The particles of the conducting material (air in the case of sound, rock in the case of seismic waves) nearest the point of vibration also begin to vibrate. These particles, as they move back and forth, strike additional particles and then return to their original position—the moving forward is known as compression, the moving back to the original position as rarefaction. This next group of particles moves forward and strikes another group of particles and then returns to their original positions. This repeated succession of compressions and rarefactions is how a compressional wave travels. The S waves, which are sometimes known as shear waves, are transverse waves: The particles in the conducting medium travel perpendicular to the direction in which the wave is traveling. S waves can be generated by tying a rope to a doorknob and rapidly moving the other end up and down with a flick of the wrist. Light and other forms of electromagnetic radiation are propagated by means of transverse waves. Unlike compressional waves, the secondary seismic waves will not travel through liquids.

SURFACE WAVES

The two types of surface waves are named for the scientists who demonstrated their existence, Lord Rayleigh (John William Strutt) and A. E. Love. Both of these types of seismic wave are of the S or transverse variety but travel at a slower speed than the body type of S wave. The Rayleigh wave travels along the surface in the vertical plane not unlike the waves of the ocean. This type of wave differs from a transverse body wave in that instead of moving back and forth along a straight line perpendicular to the direction of wave motion, the particles in a Rayleigh wave move in an elliptical motion with the long axis of the ellipse usually vertical. Unlike other types of waves that may cause particles in their path to move back and forth a few times before coming to rest, the Rayleigh wave vibration lasts much longer. The particles will travel in their elliptical path many times before coming to rest.

The other type of seismic surface wave is the Love wave. It is a shear wave but, unlike the Rayleigh wave, the motion is parallel to the surface and at right angles to the direction of wave transmission. The vibration of a Love wave lasts much longer than that of a transverse body wave.

WAVE REFLECTION AND REFRACTION

By studying the paths of seismic waves generated by earthquakes, scientists have learned much about the interior of the Earth. As stated earlier, waves expand in a spherically shaped shell from the focus, or center, as long as the rock properties remain the same. Should the rock properties change—for example, the waves may encounter more or less dense material or perhaps a boundary or contact where rock properties change instantly rather than gradually—there is a significant change in the wave pattern. Some waves are reflected from that contact; others enter the new rock body and are refracted. For example, consider only the waves that a shallow earthquake will send straight down from the focus into the Earth.

Some of these waves will strike rock layers at various depths and will be reflected at various angles; some will return to the surface of the Earth. Other waves will be refracted through the rock layers at different angles and will emerge elsewhere on the Earth's surface.

The study of reflection and refraction of seismic waves has revealed that the Earth is divided into three major zones: the crust, the mantle, and the core. It has been found that secondary or shear waves will not pass through the core but are reflected from it, while the primary waves pass through easily. Based on this evidence, scientists have concluded that a portion of the Earth's core is a liquid. In addition to being able to find discontinuities or rock boundaries by studying the paths of seismic waves, from the study of the velocities of these waves, scientists can determine the density of the rock through which the waves pass.

During World War I, both the Allies and the Germans made some progress in using seismic detectors to locate the positions of large field artillery pieces. Some scientists who were involved with this study during the war became active in seismic prospecting development in the United States during the early 1920's. The method of seismic reflection was first used in an attempt to find oil-bearing rock in the area of the Texas Gulf Coast in the late 1920's. These early attempts were not very successful. The same techniques were being successfully employed in Oklahoma, however, and with this experience, seismic reflection prospecting became well established by 1931.

Later improvements in instrumentation and in field techniques enabled reflection profiling to be used under a wide range of geologic conditions. It is now used successfully in all oil-producing regions of the world.

SEISMIC SURVEYING

One of the methods that geologists and geophysicists use to study the Earth's subsurface is seismic reflection profiling. The great advantage of this technique is that Earth scientists can have a very accurate picture of subsurface geologic strata without going to the considerable expense of drilling numerous core samples or exploratory wells. The petroleum industry, more than any other concern, uses seismic reflection techniques. Seismic surveying is used to a lesser extent for construction site evaluation, groundwater exploration, and mineral exploration.

Since the early days of seismic surveying, explosives such as dynamite have played an important role. In modern times, however, dynamite has been replaced by safer types of explosives such as ammonium nitrate. The size of the charge can be varied depending on the nature of the survey. For most surveys, the explosive is placed in a hole called the shot hole. The holes may vary in depth from about a meter to a few hundred meters. These holes are drilled by a small drilling rig attached to a truck. There are always certain disadvantages to the use of explosives. First is the always-present potential for destructive side effects, and second is the inconvenience and cost of drilling shot holes. In addition, an explosion introduces an uncontrolled range of wave frequencies.

SURVEYING TECHNIQUES

In practice, seismic surveying consists of placing receivers, known as geophones, at various intervals (usually about 30-60 meters apart) and using them to detect vibrations that have been reflected back to the surface from subsurface geological features. Those vibrations may be caused by an explosive charge (impulsive technique), the explosion of a mixture of gases inside a closed steel chamber in contact with the ground (dinoseis technique), the repeated dropping of a 2-ton mass from a height of 2-3 meters (thumper technique), or a vibrator located on the surface (vibroseis technique).

When the dinoseis method is used, several canisters of gas (usually three to six) are fired simultaneously, which forces more of an energetic reaction from the surface. Like the conventional explosives technique, the dinoseis method can be used to reach and identify the location of deep structures. The thumper technique does not work well except for small engineering surveys because this method produces very little energy in the form of waves that penetrate into the surface. Much of the energy generated by dropping the weight is dissipated by surface waves. The vibroseis technique was developed in the 1950's as an alternative to the use of an impulse source. In the vibroseis system, energy is produced by a vibrating pad that is pressed firmly to the ground. The pad is attached to the underside of a truck by means of

hydraulic jacks; when these jacks are employed, most of the weight of the truck forces the pad against the ground. Unlike the case of the impulsive source, the frequency of the vibrations can be controlled. Typically, the frequency is varied or swept from 15 cycles per second to 90 cycles per second over a period of several seconds. The reverse is also sometimes used, with frequencies being changed from higher to lower.

GEOPHONES

The receiver used to detect vibrations from any one of these methods is known as a geophone. This device consists of a piece of magnetized metal attached to a container and surrounded by a suspended coil of wire. When the ground vibrates, the coil picks up these vibrations and oscillates up and down around the magnet. This action induces an electric current in the coil, which is detected: The greater the vibration of the ground, the greater the current generated. As the ground vibrates, the geophone produces a continuously varying signal.

The geophone signal is transmitted to the recording systems by means of the seismic cable. The recording device produces a record of the vibration of the ground, called a seismogram. In some seismic recorders, the signal coming in from the geophones is first amplified and then sent to galvanometers, which are devices that detect small currents. Each geophone sends a signal to a different galvanometer. This device contains a suspended coil that rotates in response to an electrical current. Attached to each coil is a tiny concave mirror that reflects light to a photosensitive paper. As the coil and mirror rotate back and forth and the paper slowly advances, an irregular line is projected on the paper. This line shows the vibrations of the ground. Marks are also placed on the chart by a timing device. It is common practice to record the signals from six, twelve, twenty-four, forty-eight, or even ninety-six different geophones simultaneously. Each geophone-amplifier-galvanometer of the system is known as a channel. With the growth of the electronics and computer industries, new types of recording devices have been developed. A modern digital seismic recording system records the incoming vibrations on magnetic tape, which can later generate the seismogram display on electrostatic paper.

SEISMIC CREW

In seismic exploration studies, the equipment that has been described is operated by a unit known as a seismic crew. Before any actual seismic work is done, the necessary permits must be obtained. Next, the land surface must be surveyed and a decision made on where the actual tests will be conducted. If explosives are to be used, shot holes must be drilled. Two or more "shooters" handle the explosives. If vibroseis is used, one to five technicians are required to operate the large vibrator truck. A ground crew consisting of a foreman and several crew members places the geophones in their proper positions and lays out and connects the seismic cables. Depending on the size of the operation, the ground crew may consist of from about six to twenty-four people. Many more may be added to the crew when operating in a rugged terrain.

Until the 1960's, seismic crews were accompanied by two or more trained seismologists who interpreted the data. Now the interpreters are found primarily at data-processing centers. Digital tapes of the field data are delivered to the centers on a regular basis.

DATA INTERPRETATION

Although the actual method of interpretation of the data is quite technical, the basic principle is rather simple. A signal (vibrations) from an explosive or another source is transmitted through the ground. When this seismic energy encounters discontinuities of various types in the Earth, part of that energy is reflected back to the surface where it is detected by geophones. The signal is then recorded. The time at which the signal was sent is known, and the time at which the reflected signal was received at the geophone can be determined. If the speed of the waves through the various types of rock is known, the depth to the reflecting boundary is a matter of velocity of the wave multiplied by the time of travel.

USE IN OIL EXPLORATION INDUSTRY

The goal of seismic reflection is to reveal as clearly as possible the subsurface structure of the Earth. Because of the great versatility of this method, some 90 percent of all seismic studies done in the world today are seismic reflection studies. The greatest use of seismic reflection continues to be in the oil exploration industry. An important advantage of this technique is that reflections are obtained from boundaries at several different depths. The depth of any reflection can be determined if the velocity of waves in that particular type of rock is known or can be determined by another method. The exact depth to oil- or mineral-bearing strata can therefore be determined without the costly drilling of test wells or drilling cores. The extent of the rock strata in question can be determined by moving the survey equipment and redoing the test as many times as necessary.

David W. Maguire

CROSS-REFERENCES

Deep-Earth Drilling Projects, 347; Earthquakes, 316; Engineering Geophysics, 353; Ocean Drilling Program, 359; Ocean-Floor Drilling Programs, 365; Oil and Gas Exploration, 1699; Petroleum Reservoirs, 1728; Seismometers, 258; Well Logging, 1733.

BIBLIOGRAPHY

Dohr, Gerhard. *Applied Geophysics*. New York: Halsted Press, 1981. A well-illustrated volume dealing with both basic principles of exploration geophysics and how these principles are actually applied to such areas as seismic, gravitational, magnetic, and geoelectrical methods. Some topics involve the use of trigonometry and differential calculus. Suitable for college-level students of physics or geophysics.

Howell, Benjamin F. *Introduction to Geophysics*. New York: McGraw-Hill, 1959. A technical volume dealing extensively with various areas in the study of geophysics. Topics such as seismology and seismic waves, gravity, isostasy, tectonics and continental drift, and geomagnetism are covered. The reader should have a working knowledge of differential and integral calculus. Suitable for college students of physics or geophysics.

Judson, Sheldon, Marvin E. Kauffman, and L. Don Leet. *Physical Geology.* 7th ed. Englewood Cliffs, N.J.: Prentice-Hall, 1987. An excellent introductory text on the principles of physical geology. Suitable for the high school or college introductory geology course.

Meissner, Rolf, et al., eds. *Continental Lithosphere: Deep Seismic Reflections.* Washington, D.C.: American Geophysical Union, 1991. Meissner provides a brief account of seismic reflections and imaging through an examination of the continental lithosphere. Illustrations and maps reinforce the ideas and concepts that Meissner describes.

Nettleton, Lewis L. *Geophysical Prospecting for Oil.* New York: McGraw-Hill, 1940. A complete volume on the methods of applied exploration geophysics by one of the pioneers in the field. There are several chapters each on gravity studies, geomagnetism, seismic methods, and electrical methods. Trigonometry and calculus are used extensively in the book. Suitable for college-level students of physics and geophysics.

Robinson, Edwin S., and Cahit Coruh. *Basic Exploration Geophysics.* New York: John Wiley & Sons, 1988. A well-illustrated volume dealing with the science of geophysics both in theory and in applications. Contains well-developed chapters on seismic, gravity, and magnetic exploration techniques. The reader should have a working knowledge of algebra and trigonometry. Suitable for college students of geology, geophysics, or physics.

Scales, John Alan. *Theory of Seismic Imaging.* Berlin: Springer-Verlag, 1995. For the reader with a strong interest in Earth sciences, Scales describes the theories and applications of the study of seismology and seismic imaging. Illustrations, charts, and maps.

Spencer, Edgar W. *Dynamics of the Earth.* New York: Thomas Y. Crowell, 1972. An introduction to the principles of physical geology, the book covers all aspects of geology, from introductory mineralogy through a study of the agents that shape the planet's surface. Concludes with chapters on global tectonics and geophysics that tend to be somewhat technical, requiring the use of algebra. Suitable for college-level geology students.

Tucker, R. H., A. H. Cook, H. M. Iyer, and F. D. Stacey. *Global Geophysics.* New York: Elsevier, 1970. A technical volume covering such topics as geodesy, seismology, and geomagnetism. The reader should have an understanding of trigonometry and differential calculus in order to comprehend the material as presented. Suitable for college students of physics or geophysics.

8
GEOCHEMICAL PHENOMENA AND PROCESSES

ELEMENTAL DISTRIBUTION

Ocean floors are composed of a dark, fine-grained rock called basalt that is more depleted in silicon and potassium and is richer in magnesium and iron than are the abundant light-colored granitic rocks on the continents. Igneous rocks that form where one oceanic plate is being thrust below another are generally intermediate in composition. Certain ore deposits occur only where certain plate tectonic processes take place, thereby enabling a geologist to focus the search for these deposits.

PRINCIPAL TERMS

ANDESITE: a volcanic rock that is lighter in color than basalt, containing plagioclase feldspar and often hornblende or biotite

BASALT: a dark-colored, volcanic rock containing the minerals plagioclase feldspar, pyroxene, and olivine

GRANITIC ROCK: a light-colored, intrusive rock containing large grains of quartz, plagioclase feldspar, and alkali feldspar

LIMESTONE: a sedimentary rock composed mostly of calcium carbonate formed from organisms or by chemical precipitation in oceans

P WAVES: the first waves from earthquakes to arrive at a seismic station; because they travel at different speeds through different types of rock, they may be used to deduce the rock types below the surface

PERIDOTITE: a dark-colored rock composing much of the Earth below the crust; it usually contains olivine, pyroxene, and garnet

PLATE TECTONICS: the theory that assumes that the Earth's crust is divided into large, moving plates that are formed and shifted by volcanic activity

SANDSTONE: a sedimentary rock composed of larger mineral grains than those forming shales, thus deposited from faster-moving waters

SEDIMENTARY ROCK: a flat-lying, layered rock formed by the accumulation of minerals from air or water

SHALE: the most abundant sedimentary rock, composed of very tiny minerals that settled out of slowly moving water to form a mud

EARTH'S OCEANIC AND CONTINENTAL CRUSTS

The surface of the Earth may be broadly divided into the oceanic crust and the continental crust. The oceanic crust is on the average "heavier," or denser, than the continental crust. Both the continental and oceanic crusts are denser than the underlying rocks in the Earth's mantle. The continental and oceanic crusts can thus be considered a lower-density "scum" floating on the denser mantle, somewhat analogous to an iceberg floating in water. Because the denser oceanic crust sinks lower into the mantle than the continental crust, much of the oceanic crust is covered by the oceans, but the less dense continental crust is mostly above the level of the oceans. Also, seismic waves from earthquakes indicate that the oceanic crust is much thinner (about 6 to 8 kilometers) than the continental crust (about 35 to 50 kilome-

ters). The density difference between the oceanic and continental crusts is related to the kinds of minerals composing them. The oceanic crust contains more of the denser iron- and magnesium-rich minerals, olivine (iron and magnesium silicate) and pyroxene (calcium, iron, and magnesium silicate), than does the continental crust. The continental crust contains much more of the less dense minerals, quartz (silica) and alkali feldspar (potassium, sodium, and aluminum silicate), than does the oceanic crust. In addition, the oceanic crust contains much of the feldspar called calcium-rich plagioclase (calcium, sodium, and aluminum silicate) than does the continental crust.

This difference in mineralogy between the oceanic and continental crusts is reflected in their average elemental composition. The oceanic crust is enriched in elements concentrated in olivine, pyroxene, and calcium-rich plagioclase, and the continental crust is enriched in those elements concentrated in quartz and alkali feldspar. Thus, the continental crust contains larger concentrations of silicon dioxide (60 weight percent in the continental crust versus 49 weight percent in the oceanic crust) and potassium oxide (2.9 versus 0.4 weight percent) and lower concentrations of titanium di-oxide (0.7 versus 1.4 weight percent), iron oxide (6.2 versus 8.5 weight percent), manganese oxide (0.1 versus 0.2 weight percent), magnesium oxide (3 versus 6.8 weight percent), and calcium oxide (5.5 versus 12.3 weight percent) than does the oceanic crust. The other major elements, aluminum and sodium, are fairly similar in concentration in both the oceanic and continental crusts.

TYPICAL COMPOSITION OF ROCKS THAT COMPOSE MUCH OF THE EARTH'S MANTLE OR CRUST

Element Oxide	Unmelted Peridotite in the Mantle	Basalt Formed at Oceanic Ridges or Rises	Andesite Formed at Subduction Zones	Granite Rock Along Continental Subduction Zones	Continental Rift Basalt	Shale	Sandstone Near the Source	Sandstone Far from the Source	Limestone
SiO_2 (silicon oxide)	45.0	49.0	59.0	65.0	50.0	58.0	67.0	95.0	5.0
TiO_2 (titanium oxide)	0.4	1.8	0.7	0.6	3.0	0.7	0.6	0.2	0.1
Al_2O_3 (aluminum oxide)	8.7	15.0	17.0	16.0	14.0	16.0	14.0	1.0	0.8
Fe_2O_3 (ferric iron oxide)	1.4	2.4	3.0	1.3	2.0	4.0	1.5	0.4	0.2
FeO (ferrous iron oxide)	7.5	8.0	3.3	3.0	11.0	2.5	3.5	0.2	0.3
MnO (manganese oxide)	0.15	0.15	0.13	0.1	0.2	0.1	0.1	—	0.05
MgO (magnesium oxide)	28.0	8.0	3.5	2.0	6.0	2.5	2.0	0.1	8.0
CaO (calcium oxide)	7.0	11.0	6.4	4.0	9.0	3.0	2.5	1.5	43.0
Na_2O (sodium oxide)	0.8	2.6	3.7	3.5	2.8	1.0	2.9	0.1	0.05
K_2O (potassium oxide)	0.04	0.2	1.9	2.3	1.0	3.5	2.0	0.2	0.3
Volatiles (water or carbon dioxide)	1.0	1.0	1.0	2.0	1.0	8.0	2.0	1.0	42.0

Note: Compositions are given as weight percentages of the element oxide in the entire rock.

The mantle is even denser than the crust, since it contains the dense minerals olivine, pyroxene, and garnet (magnesium and aluminum silicate) in the rock called peridotite. It does not contain the less dense minerals, quartz and feldspar. Thus, the mantle is even more enriched in iron oxide and magnesium oxide and more depleted in potassium oxide, sodium oxide, and silicon dioxide than are the crustal rocks.

COMPOSITION OF OCEANIC CRUST

The foregoing discussion summarizes the average characteristics of the oceanic and continental crusts, but they also vary substantially in composition. The continental crust is considerably more heterogeneous than is the oceanic crust. The oceanic crust consists of an upper sediment layer (about 0.3 kilometer thick), a middle basaltic layer (about 1.5 kilometers thick), and a lower gabbroic layer (about 4 to 6 kilometers thick). Basalts and gabbros both contain olivine, pyroxene, and calcium-rich plagioclase. They differ only in grain size; the basalts contain considerably finer minerals than do the gabbros. The basaltic and gabbroic layers are thus very similar in composition. They are also of fairly constant thickness across the oceanic floors. The gabbroic layers disappear over oceanic rises, or linear mountain chains on the oceanic floors. The basaltic rocks are believed to form at the rises by about 20 to 30 percent melting of the underlying peridotite in the upper mantle. The newly formed oceanic crust and part of the upper mantle are believed to be slowly transported across oceanic floors, at rates of about 5 to 10 centimeters per year, to where this material is eventually subducted or thrust underneath another plate.

The thickness of sediment on ocean floors varies considerably. It is nearly absent over the newly formed basalts at oceanic rises. It is thickest in basins adjacent to continents where weathering and transportation processes carry large amounts of weathered sediment into the basins. The composition of oceanic floor sediment varies considerably. It contains varied amounts of calcite or aragonite (calcium carbonate minerals), silica (silicon dioxide), clay minerals (fine, aluminum silicate minerals derived from weathering), volcanic ash, volcanic rock fragments, and ferromagnesian nodules.

Finally, a few volcanoes composed of basalt form linear chains on the ocean floor, away from the rises or subduction zones such as the Hawaiian Islands. These ocean-floor basalts are similar in composition to those at oceanic rises, except that they contain greater amounts of potassium. The amount of basaltic rocks produced by these ocean-floor volcanoes is insignificant, however, compared to the vast amounts of basalt produced at oceanic rises.

COMPOSITION OF CONTINENTAL CRUST

In contrast to the oceanic crust, the continental crust is quite heterogeneous in mineralogy and chemical composition. About 75 percent of the surface of the continents is covered by great piles of layered rocks called sedimentary rocks. The average thickness of these sedimentary rocks on the continental crust is only about 1.8 kilometers, although they may locally range up to 20 kilometers in thickness. The main kinds of sedimentary rocks on the continents are the very fine-grained shales or mudrocks (about 60 percent of the total sedimentary rocks), the coarser-grained sandstones (about 20 percent of the total), and limestones or dolostones (about 20 percent of the total). The shales or mudrocks are composed of very small grains of mostly clay minerals and quartz. The resultant composition of the shales is often high in the immobile elements, aluminum and potassium, and low in the mobile elements, sodium and calcium. Sandstones vary in composition depending on which rocks weather to form the sandstone, the distance of the sandstone from the source, and the intensity of weathering. Sandstones formed close to a source of granitic rocks may have a composition similar to that of the granitic rock: high in silicon and potassium and low in magnesium, iron, and calcium compared to basaltic rocks. Sandstones formed a long distance from the source have more time to be weathered. Thus, these sandstones may have most of the unstable minerals weathered away to clays or soluble products in water (for example, sodium), and they may be enriched in silicon because of the abundance of the stable mineral quartz. Limestones typically form in warm, shallow seas by the action of organisms to produce most of the calcium carbonate in these rocks. Thus, limestones are enriched in calcium and depleted in most other elements. The dolostones are enriched in magnesium as well as calcium. Some places, such as the Great Plains in the United States, consist mostly of alternating limestones and shales formed in ancient, shallow seas. (Thus, the average composition of the surface rocks in these areas may be considered an average of that of shale and limestone in whatever proportion they occur.) The average composition of sedimentary rocks on the continents is significantly different from that of the granitic rocks that weathered to form them. The average sedimentary rocks on continents are much more enriched in calcium (because of carbonate rocks), carbon dioxide (also because of carbonate rocks), and water (because of incorporation in clay minerals), and they are depleted in sodium (because of its solubility).

The thickness of these sedimentary rocks is still small compared to the 35- to 50-kilometer thickness of most of the continental crust. Only about 5 percent of the continental crust by volume is composed of sedimentary rocks. Most crustal rocks are igneous rocks or their metamorphic equivalents. Metamorphic rocks form in the solid state at high temperatures and pressures because of their deep burial in the Earth. A substantial percentage of these igneous rocks of the upper continental crust are either granitic rocks (quartz and alkali feldspar rock) or andesitic rocks (plagioclase-rich rock). Basaltic rocks probably compose only about 15 percent of the upper continental crust.

CONTINENTAL MARGINS AND RIFT ZONES

Most of the granitic rocks and andesites originally formed along subduction zones, where oceanic crust is being thrust or subducted below either oceanic or continental crust. There also may be some basalts formed along these subducted plates. These basalts, andesites, and granitic rocks that formed along continental margins may eventually be plastered along the edges of the continents, resulting in the gradual growth of the continents. Other basalts are formed in portions of continents, called continental rifts, that are being stretched apart much like taffy. These basalts are considerably more enriched in potassium than basalts formed on ocean floors. For example, a large fraction of the states of Washington, Oregon, and Idaho is covered with these rift basalts extruded as lavas since about 20 million years ago. The total volume of about 180,000 cubic kilome-

ters for these basalts is still comparatively insignificant; therefore, basalts make only a small contribution to the composition of the average upper continental crust.

COMPOSITION OF THE MIDDLE AND LOWER CRUST

The composition of the lower continental crust is much more difficult to determine than that of the upper continental crust because the rocks forming the lower crust are not exposed at the surface. Estimates of about 50 percent granitic and 50 percent gabbroic rocks in the lower crust have been reached. Thus, the lower continental crust is more enriched in the basaltic components, calcium, magnesium, iron, and titanium, and depleted in the granitic components, potassium and silicon, than is the upper continental crust.

The average compositions of the middle and lower oceanic and continental crusts are difficult to determine because they cannot be directly sampled. Much of the information about the nature of the crust below the surface comes from the behavior of seismic waves given off by earthquakes, from heat-flow measurements, and from the composition of rock fragments brought up by magma passing through much of the crust. In addition, there are places in the crust where rocks from the lower crust have been uplifted to the surface, so their composition can be examined in detail.

The speed of the earthquake waves through the oceanic crust is consistent with the crust being composed of a thin upper layer of sediment (indicated by P-wave velocities of 2 kilometers per second), a thicker middle layer of basalt (P-wave velocities of 5 kilometers per second), and a thick lower layer of mostly gabbro (P-wave velocities of 6.7 kilometers per second). The thicker continental crust, however, has P-wave velocities (6.1 kilometers per second) consistent with mostly granitic rocks below the overlying sedimentary rock veneer (2 to 4 kilometers per second). The lower continental crust has P-wave velocities (6.7 kilometers per second) similar to those expected for lower-silica rocks like gabbro, so there is probably more gabbro mixed with granitic rocks in the lower crust.

HEAT-FLOW MEASUREMENTS

How fast heat flows out of the Earth may also be used to limit the composition of crustal rocks.

Variation in heat flow at the surface depends on how much heat is flowing out of the Earth below the crust; the distribution of radioactive elements in the crust, such as uranium, thorium, and potassium, that give off heat; and how close magmas are to the surface. Oceanic ridges and continental rift zones, for example, have high heat flow, suggesting that magmas are close to the surface. In contrast, the heat loss from much of the ocean floor and over much of the continents with old Precambrian rocks (older than about 600 million years) is considerably lower because of the lack of magma close to the surface. It is surprising, however, that the oceanic floor and continents with old Precambrian rocks have similar low heat flow, as the abundant granitic rocks in the continents ought to be more enriched in the heat-producing radioactive elements than is the oceanic crust. That suggests that many of the granitic rocks at depth in these parts of the continental crust are depleted in radioactive elements, perhaps because of melting processes carrying away the radioactive elements in the magmas during the Precambrian. Also, that is consistent with the presence of abundant basaltic rocks depleted in radioactive elements in the lower crust.

GLIMPSES INTO EARTH'S INTERIOR

There are places on the Earth, such as the island of Cyprus in the Mediterranean Sea, that appear to be ruptured portions of the entire oceanic crust and part of the upper mantle. In Cyprus, the lower zone is composed of peridotite, olivine-rich rocks, or pyroxene-rich rocks, as are predicted to occur in the mantle. These rocks correspond to the P-wave seismic velocities of 8 kilometers per second. There is a rather abrupt change to the next overlying layer of mostly gabbros that correspond to the sharp decrease in P-wave velocities to about 6.7 kilometers per second. These rocks grade upward into basalt corresponding to the upper igneous rock layers of the oceanic crust with P-wave velocities of about 5 kilometers per second. The basalt and gabbros are also penetrated by a multitude of tabular igneous dikes that were feeders of magma to the overlying basalt at the surface. Finally, there are overlying sedimentary rocks corresponding to the upper oceanic layers with P-wave velocities of about 2 kilometers per second.

Deep drill holes provide scant information about the composition of the crust at depth. Drill cores provide mostly information about sedimentary rocks; they also give some information about the first igneous rocks just below the sedimentary cover. Unfortunately, deep drilling is costly and limited in depth and distribution. Generally, wells are never drilled deep enough to obtain samples from the intermediate and lower crust, and none reach the mantle.

Some volcanoes derive their magma from the upper part of the mantle. These volcanoes often bring up fragments of mantle material and random samples of crustal rocks thrown from the volcanic conduits walls. Although volcanic vents of this kind are not common and sample an extremely small distribution of lower continent and upper mantle material, they are extreme important.

Finally, meteorites are used as analogies of the interior of the Earth. Iron meteorites, rich in iron and nickel, are thought to approximate composition of the core of the Earth. Stony meteorites with compositions close to that of peridotite are thought to match the composition of the mantle of the Earth.

GUIDE TO ORE DEPOSITS

A knowledge of the overall distribution of rock types and the corresponding elemental compositions of these rocks over the Earth can give geologists a guide to where to look for certain kinds of ore deposits, as certain ores occur in certain kinds of rocks. The most generalized pattern is the association of certain types of ores with certain tectonic environments. Both oceanic rises and subduction zones tend to heat waters and drive the resultant metal-rich waters toward the surface. Oceanic rises often contain sulfide-rich, copper and zinc hot-water deposits. These hot-water deposits at subduction zones are often enriched in copper, gold, silver, tin, lead, mercury, or molybdenum.

One example of subduction zone deposits is the copper porphyry deposits. These important ore deposits are formed in granitic rocks that crystallized at shallow depths below the surface in areas where an oceanic plate is being subducted, or thrust below a second plate. They are especially abundant around the rim of the Pacific Ocean. The copper ores contain low copper concentrations (0.25 to 2 percent) and have some associated molybdenum and gold. These low-grade ores are often profitable to mine because of the large volume of ore (over a billion tons in some places) that can be rapidly extracted from the rock. A geologist looking for such ores designs an exploration campaign to search out only areas with active or inactive subduction zones. Also, the geologist looks for certain compositions of granitic rocks intruded at fairly shallow depths below the surface that have been exposed to erosion near the top of the intrusion, as these are the places where the copper porphyries form. Hundreds of these copper porphyry deposits have been discovered, accounting for about half of the copper ores of the world. Copper is used in wires to transmit electricity and in bronze and brass.

Robert L. Cullers

CROSS-REFERENCES

Andesitic Rocks, 1263; Basaltic Rocks, 1274; Chemical Precipitates, 1440; Continental Crust, 561; Continental Rift Zones, 579; Evolution of Earth's Composition, 386; Fluid Inclusions, 394; Fractionation Processes, 400; Freshwater Chemistry, 405; Geochemical Cycle, 412; Geothermometry and Geobarometry, 419; Granitic Rocks, 1292; Hydrothermal Mineralization, 1205; Igneous Rock Classification, 1303; Nucleosynthesis, 425; Oxygen, Hydrogen, and Carbon Ratios, 431; Phase Changes, 436; Phase Equilibria, 442; Plate Margins, 73; Plate Tectonics, 86; Sedimentary Mineral Deposits, 1637; Sedimentary Rock Classification, 1457; Siliciclastic Rocks, 1463; Subduction and Orogeny, 92; Water-Rock Interactions, 443.

BIBLIOGRAPHY

Ahrens, L. H. *Distribution of the Elements in Our Planet.* New York: McGraw-Hill, 1965. This book provides a clear summary of the composition of the solar system and the Earth. The elements are grouped in a geochemical classification. Directed to the nonspecialist.

Craig, J. R., D. J. Vaughan, and B. J. Skinner. *Resources of the Earth.* Englewood Cliffs, N.J.: Prentice-Hall, 1988. This is an excellent book describing the distribution of ore deposits on the Earth. Information is provided on the history and use of the elements, geologic occurrence, and reserves. For a nonscience major in college or interested layperson. There is a glossary of technical terms.

Emsley, John. *The Elements.* 3d ed. Oxford: Oxford University Press, 1998. Emsley discusses the properties of elements and minerals, as well as their distribution in the Earth. Although some background in chemistry would be helpful, the book is easily understood by the high school student.

Greenwood, Norman Neill. *Chemistry of Elements.* 2d ed. Oxford: Butterworth-Heinemann, 1997. An excellent resource for a complete description of the elements and their properties. The book is filled with charts and diagrams to illustrate chemical processes and concepts. Bibliography and index.

Krebs, Robert E. *The History and Use of Our Earth's Chemical Elements: A Reference Guide.* Westport, Conn.: Greenwood Press, 1998. This book defines geochemistry and examines its principles and applications. A vital resource for anyone interested in the field of geochemistry and in the Earth's elements. Illustration, charts, and bibliography.

Skinner, B. J., and S. C. Porter. *The Dynamic Earth.* New York: John Wiley & Sons, 1989. This is one of many introductory geology textbooks for college students that has a chapter on mineral and energy resources in the Earth. The interested reader with some understanding of geology can find information here about the major ore deposits and their distribution within the Earth.

Smith, D. G., ed. *The Cambridge Encyclopedia of the Earth Sciences.* New York: Crown, 1981. This reference is written for the reader with some background in science who needs to locate information on a specific Earth science topic. Chapters 4 ("Chemistry of the Earth"), 5 ("Earth Materials"), and 10 ("Crust of the Earth") might be most appropriate for further reading related to elemental distribution. There are also chapters on plate tectonics.

Utgard, R. O., and G. D. McKenzie. *Man's Finite Earth.* Minneapolis, Minn.: Burgess, 1974. This book is written as supplementary reading for college geology courses. A section on Earth resources that gives some insight on ore distribution and how it relates to public policy is suitable for a layperson.

Wedepohl, Karl Hans. *Geochemistry.* Translated by Egon Althaus. New York: Holt, Rinehart and Winston, 1971. This book gives nontech-nical descriptions of the elemental distributions within the solar system and the Earth. Some knowledge of chemistry and geology is necessary for full use of the book. Chapter 7 gives specific information on the distribution of elements in the Earth's crust.

EVOLUTION OF EARTH'S COMPOSITION

Using meteorites and some of the oldest-known crustal rocks, geochemists are trying to unravel the mysteries of the early Earth's composition. The theories about Earth's evolution are speculative, and much of the Earth's earliest history is unknown, but understanding the processes that have evolved the Earth can help to unify various Earth and biological sciences.

PRINCIPAL TERMS

ACCRETION: the process by which small bodies called planetesimals are attracted by mutual gravitation to form larger bodies called protoplanets

ARCHEAN EON: the older of a two-part division of the Precambrian, also known as the Archeozoic

CHONDRITES: stony meteorites that contain rounded silicate inclusion grains called chondrules; they are believed to have formed by crystallization of liquid silicate droplets and volatiles

DIFFERENTIATION: the process by which a planet is divided into zones as heavy elements (metals) sink to the core, while lighter elements collect near the surface

ISOTOPE: atoms of an element that have the same number of protons in the nucleus, the same atomic number, and the same chemical properties but that have different atomic masses because they have different numbers of neutrons in the nucleus

MAFIC and ULTRAMAFIC: rock-forming magmas that are high in dense, refractory elements such as iron and magnesium; oceanic basalts are examples of mafic rocks

REFRACTORY (SIDEROPHILE) ELEMENTS: elements least likely to be driven off by heating; the last elements to be melted as a rock is heated to form magma

VOLATILE ELEMENTS: elements most likely to be driven off by heating; those that are first to melt or be driven off as gas in a heated rock

ZIRCONS: mineral inclusions found in granitic rocks, zircons are often the only evidence left of early crustal rocks

EARTH'S FORMATION

About 4.5 billion years ago, scientists believe, a massive star exploded in a supernova event that shined as brightly as a whole galaxy of stars. Shock waves from the celestial fireworks overtook a cloud of gas and dust a few light-years away and triggered its contraction, simultaneously seeding the nebula with heavy elements (those heavier than iron on the periodic table). The solar nebula's collapse led to the formation of the Sun (which swept up most of the matter), and the planets formed by the accretion of small bodies called planetesimals. As the planetesimals grew into protoplanets, their gravitational fields increased, so they continued to sweep up material not garnered by the protosun. The innermost planets, Mercury, Venus, Earth, and Mars, contained the dense metals and rocks, while the outer planets were mostly made of gases and volatile ices. During the protoearth's initial accretion process, small, cold bodies collided to form a large mass of homogeneously heterogeneous composition. By the process of differentiation, the heavier metallic elements, such as iron and nickel, migrated to the core of the early Earth, while the lighter elements migrated to the outer portions of the contracting planet.

After the initial accretion of the planetesimal materials and just prior to differentiation of the lithophile and siderophile elements, the Earth's thermal history began through the process of radioactive decay. During this early thermal period, short-lived radioactive nuclides (atoms of a specific isotope, distinguished by their atomic and mass numbers) produced heating seven times greater than that of today's molten core. Most of the heating was attributable to the decay of potassium 40 as well as of the short half-lived elements such as aluminum 26. After about 100,000 years, the planet

386

separated into the iron-nickel core and magnesium-iron-silicate lower mantle. Over a longer time scale (probably more than 10 million years), the high-volatility compounds, such as lead, mercury, thallium, and bismuth, along with the noble gases, water in hydrated silicates, and carbon-based organic compounds, all condensed. This volatile-rich material migrated to the surface, where it was melted into magmas in a continuous period of crustal reprocessing that lasted for about 1 billion years.

The Earth's original inventory of gases appears to have been lost, based on the relative present abundances of the rare gases (helium, neon, argon, krypton, xenon, and radon) compared with the present silicon content of the Earth. Later periods of volcanic outgassing and perhaps impacts

with volatile-rich cosmic objects such as carbonaceous chondritic meteorites and comets may also have played a role in the evolution of the atmosphere and oceans.

EARTH'S CORE AND MANTLE COMPOSITION

Separated into three main layers—the crust, mantle, and core—the Earth is an active body, its internal heat far from exhausted. The complexity of the chemical composition increases as one examines each successive outward layer. This generalized model gives a starting point with which to examine the complex nature of Earth materials. Earth's wide range of pressure and temperature regimes helps explain why there are more than two thousand distinct minerals and numerous dif-

PROPERTIES OF SILICATE MINERALS

Mineral	Mohs Hardness	Specific Gravity	Cleaving Features	Composition, Other
Quartz	7	2.65	Fractures unevenly	Silicon dioxide
Potassium feldspar	6-6.5	2.5-2.6	Cleaves well, 2 directions	Aluminosilicates of potassium; pink or white
Plagioclase feldspars	6-6.5	2.6-2.7	Shows striations	Aluminosilicates of sodium, calcium; white or gray
Muscovite mica	2-3	2.8-3	Perfect in one direction	Aluminosilicates of potassium with water; yields flexible, colorless, transparent plates
Biotite mica	2.5-3	2.7-3.2	Perfect in one direction	Aluminosilicates of magnesium, iron, potassium, with water; yields flexible, black or brown plates; ferromagnetic
Pyroxenes	5-6	3.1-3.5	Two directions, 87° and 93°	Silicates of aluminum, calcium, magnesium, iron; black or dark green; ferromagnetic
Amphiboles	5-6	3-3.3	Two directions, 56° and 124°	Silicates of aluminum, calcium, magnesium, iron; black or dark green; ferromagnetic
Olivine	6.5-7	3.2-3.6	—	Silicates of magnesium, iron; translucent; light green; ferromagnetic
Garnets	6.5-7.5	3.5-4.3	Fractures unevenly	Aluminosilicates of iron, calcium, magnesium, manganese; red, brown, yellow

SOURCE: Data are from Harold L. Levin, *The Earth Through Time* (3d ed., Saunders College Publishing, 1988).

ferent combinations of minerals in rock types.

The core is actually composed of two basic parts: the solid inner core, with a density equal to twelve times that of water and a radius of 1,300 kilometers, and a molten outer core, 2,200 kilometers thick, with a density of about 10 grams per cubic centimeter. This model consists of an essentially iron-nickel inner core at high pressure and a metallic outer core that also contains iron sulfide and light elements such as silicon, carbon, and oxygen. As a whole, the core unit comprises about 32 percent of the Earth's mass.

Comprising the outer 68 percent of the Earth's bulk, the mantle makes the crust, atmosphere, and oceans insignificant by comparison. The mantle is rich in dense, or mafic, rocks such as olivine and pyroxene (which comes in two basic types, calcium-rich or calcium-poor), with olivine the dominant mineral.

Basic Earth materials are derived via reaction series from mafic magmas melting and settling out in the mantle's upper regions. As the temperatures drop in the melt zone, a discontinuous series (a set of discrete reactions) occurs. Magnetite, an oxide of iron and titanium, is the first to settle out, with the highest melting point at about 1,400 degrees Celsius. Olivine, a mineral whose silicate structure is a simple tetrahedron, is the next to so-

lidify out of the melt, with a density of 3.2-4.4, followed by the single chain structure pyroxene, with a density of 3.2-3.6. As temperatures in the magma drop to near 1,000 degrees Celsius, the amphibole group forms with a lesser density, 2.9-3.2. As the cooling progresses, the structures increase in complexity with the micas—biotite and muscovite, which form in planar sheets. Next in the cooling sequence would be orthoclase, or potassium feldspar, and plagioclase, or calcium feldspar, and, finally, quartz, which are all distinguished by their characteristic three-dimensional diamond shapes and varying colors. The calcic through sodic plagioclase to muscovite, biotite, and quartz occurs in a smooth, or continuous, transition rather than the stepwise, or discontinuous, reactions that characterize the formation of olivine through biotite.

EARTH'S CRUSTAL COMPOSITION

An estimate of the crustal elemental composition of the Earth indicates that only a handful of elements (oxygen, silicon, aluminum, iron, magnesium, calcium, sodium, and titanium) make up more than 99 percent of the Earth's crust. The simple oxide quartz is the most common of the silicate minerals, which account for 95 percent of the crust. With these facts in mind, one can start to hypothesize about how the continents evolved.

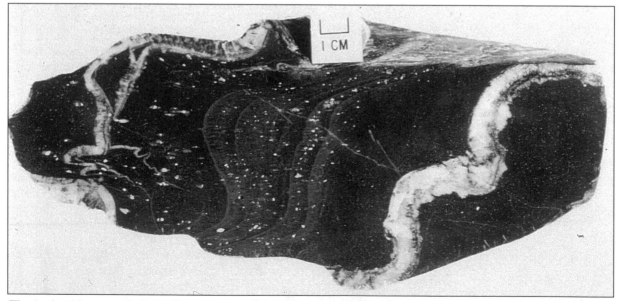

The simple oxide quartz (the lighter buckled veins in this rock sample) is the most common of the silicate minerals, which account for 95 percent of the crust. (Geological Survey of Canada)

About 700 million years after the initial formation of the Earth through accretion and differentiation, the first rocks of the Archean eon formed. They are composed of olivine, pyroxene, and anorthite (calcium-rich plagioclase feldspar), which settled out of the basaltic magma. The lighter plagioclase would rise to the surface to form a hardened crust of anorthosite, the same material that comprises the Moon's ancient highlands, which are about 3.8 billion years old.

The anorthosite formed a thick sheet that was fractured into pieces and subjected to further heating through radioactive decay, leading to an essentially granitic rock layer 10-15 kilometers thick. Extensive volcanic activity and high surface temperatures gradually diminished until the hydrosphere (water cycle) was established. The Earth's crust is divided into two main types: the dense, or mafic, oceanic crust and the lighter, or sialic (silica-aluminum), continental crust. Archean rocks (up to 3.5 billion years old) found in the stable interiors of the continents contain massive anorthosite inclusions and may be viewed as the nuclei of the continents.

SHAPING BY LIFE-FORMS

About the time of the formation of the continental nuclei, or cratons (relatively stable portions of crust), the oldest-known sedimentary rocks accumulated as the rock cycle began, eroding the parent igneous rocks into secondary types of rock. This occurrence may coincide with the beginning of plate tectonics, as the lithosphere (rock crust) of the Earth broke into plates and began its hallmark active motion. Life is thought to have arisen at about the same time, with primitive blue-green algae found in strata 2.8 billion years old. With the oceans growing in volume and salinity and the development of oxygen-releasing blue-green algae, Earth's geochemistry became more complex. Chemically precipitated rocks of calcium carbonate, commonly known as limestones, are an example of the evolving rock cycle.

Life-forms shaped the Earth's chemical composition. By the end of Precambrian time, oxygen levels had reached 1 percent of their present value. Multicelled animals in the oceans scrubbed the carbon dioxide from the atmosphere and locked it up in the carbonate rocks, forming biochemically precipitated limestones. By the late Paleozoic era, coal formations grew as a result of the first land forests being periodically inundated by ocean transgressions.

CLUES FROM METEORITES

Perhaps no other Earth science is as speculative as that of early Earth history and the geochemical evolution of the Earth. Varying models for crustal development are advanced and overturned annually. Despite the problems of extrapolating back to a time before there were solid rocks, the established models are based on some solid lines of evidence as well as on conjecture. In 1873, American geologist James D. Dana made one of the initial advances in the study of the Earth's internal chemical composition when he suggested that analogies could be drawn from the study of meteorites. Believed to be pieces of differentiated bodies that were later disintegrated into smaller pieces, meteorites come in differing types that are analogous to the Earth's interior. Stony meteorites comprise the most abundant group and are composed of silica-associated, or lithophile, elements such as those found in the Earth's crustal materials. Stony iron meteorites are composed of roughly equal parts of rock (typically olivine) suspended in a matrix of iron. Iron meteorites are composed of siderophile elements, iron being the major constituent, along with (perhaps) 10-20 percent nickel.

Iron meteorites are of particular interest to scientists attempting to model the composition of the Earth's core. The mean density of the Earth is about 5.5 grams per cubic centimeter. The mean density of crustal rocks, however, is only about 2.7 grams per cubic centimeter (water is conveniently 1 gram per cubic centimeter), which indicates a core density of ten to twelve times that of water. The only known objects approaching these densities are the iron meteorites. Because meteorite types approximate elemental distribution in the Earth, they are valuable samples for laboratory examination by scientists. Geochemists studying meteorites have derived radiometric dates of 4.6 billion years—corresponding to the initial time of accretion and differentiation of the planets.

INVESTIGATION INTO ARCHAEN CRUSTAL EVOLUTION

Geophysicists use seismic waves to study the Earth's interior. Changes in velocity and deflections of the waves passing through the Earth have revealed a differentiated Earth with a very dense core, less dense mantle, and a light crust "floating" on top. The well-established theory of plate tectonics has shown that the crust is broken into pieces, or plates, that are moving, driven by convection currents in the upper mantle. Some of the major challenges confronting Earth scientists are the questions about how the Earth's crust formed and about when plate movement began.

During the 1960's, interest in Archean crustal evolution was aroused by the discovery of Archean eon magnesium-oxygen-rich lavas similar to those found in the early Precambrian. Called komatiites, these rocks date back to 3.7 billion years ago and represent ultramafic lavas that form at 1,100 degrees Celsius. Komatiites are generally found around greenstone belts, an agglomeration of Archean basaltic, andesitic, and rhyolitic volcanics, along with their weathering and erosion derived sediments. One hundred million years older than any previously known rocks, the finds led to further exploration of Archean formations by field geologists in West Greenland-Labrador, Zimbabwe, Transvaal-Swaziland, Ontario-Quebec, southern India, Western Australia, and more recently, China and Brazil.

Important work by field geologists in these regions launched a new era in Precambrian geology. The primary targets for study are the greenstone belts and granitic-gneiss associations. An important twentieth century find included detrital zircon, discovered in Australia. An age of 4.2 billion years for the zircons was determined using precise ion microprobe analysis. The zircon find is significant because it places an approximate birth date for the continental crust, as zircon is a mineral constituent of granite (recall that oceanic crust is composed of mafic and ultramafic rocks while continental crust is granitic).

The drive to study Archean rocks was further fueled by the United States' Apollo missions to the Moon, which returned rocks of slightly older age from the lunar surface. At the same time, geochemists were able to refine their study of these ancient rocks with more sophisticated methods to determine ratios of isotopes in the samples. Instruments common in the geochemical lab today are X-ray diffraction and gamma-ray spectrometers, which probe the nuclei of atoms to determine the spectral fingerprint of elements and their various isotopes. Isotopic ratios in rocks are of particular interest to geochemists because they provide clues as to chemical cycles in nature, such as the sulfur, chemical, nitrogen, and oxygen cycles. The equilibria of these cycles, as indicated by the isotopic ratios, offer insights into volcanic, oceanic, biological, and atmospheric cycles and conditions in the past.

THEORIES OF EARLY CRUSTAL EVOLUTION

It is generally accepted by most Earth scientists that crustal formation and heat flow were substantially greater in Archean times. The question is whether this crust was broken into moving lithospheric plates as it has been for the past 900 million years. The question of plate motion during this early period has generated debate among scientists and has led to two general theories of early crustal evolution. If plate tectonics was occurring 4 billion years ago, one would expect to find formations of arc deposits and complexes similar to the Franciscan formation in California's coast range. Oölite and arc deposits are terranes that accumulate near zones of subduction, where dense mafic rocks are recycled into the mantle. Such formations have not been found to date—geologic evidence arguing against rapid plate motion.

If crustal rock production was great and yet plate tectonics minimal, what process shaped the early Earth? An answer may have emerged from one of the Earth's sister planets. Shrouded in clouds, Venus did not give up the mysteries of its geology until the radar maps generated by Soviet and American spacecraft. Like Mars, with its giant volcanoes in the Tharsis region, Venus appears to have great shield volcanoes and continentlike regions the size of Africa and Australia. Hot-spot volcanism, in which plumes of magma rising from the planetary interior erupt to form regions of volcanic activity at the surface, may indeed be the key to understanding incipient plate tectonics on the early Earth. Side scan radar images of Venus provided by the Magellan spacecraft reveal a planet surface devoid of plate tectonics but dominated by large, circular structures caused by rising and sinking mantle plumes. Several spacecraft since 1969 have imaged the giant volcanoes of Mars, including Olympus Mons, the largest volcano in the solar system. The massive volumes of lava poured out at these volcanic constructs are attributed to persistent hot spot volcanism in the absence of plate tectonics.

ADVANCES IN RESEARCH AND STUDY

Perhaps no other area of scientific study is more intriguing and controversial than that of the origin and evolution of the Earth. Geochemists have been at the forefront of the quest for understanding the Earth's present geology in terms of its past. Before the 1960's, little was known of the Earth's history during early Precambrian times. This lack is significant when one realizes that the Precambrian comprises about 87 percent of the geologic time scale.

It is likely that new techniques used to analyze rocks and minerals in the laboratory will lead to a better understanding of the formation of the Earth's crustal materials and the evolution of moving crustal plates. Precise dating of zircons from ancient rocks, isotope analysis, and high-resolution seismic data will help scientists to comprehend the relationships between the granite-greenstones and gneiss terranes (crustal blocks) that typify Archean formations.

Studying materials on other bodies of the solar system will also lead to a better understanding of the early Earth and its evolution. Space missions directed toward understanding the composition and history of the Moon and Mars are planned by several nations. The question of whether the Earth's early history was dominated by hot-spot volcanism (as on Mars with its huge volcanoes and on Venus with its complex ring structures) will receive valuable evidence from continuing studies of asteroids, meteorites, the Moon, and the planets.

David M. Schlom

CROSS-REFERENCES

Archean Eon, 1087; Atmosphere's Evolution, 1816; Earth's Age, 511; Earth's Core, 3; Earth's Core-Mantle Boundary, 9; Earth's Crust, 14; Earth's Differentiation, 20; Earth's Mantle, 32; Earth's Oldest Rocks, 516; Earth's Origin, 2389; Earth's Structure, 37; Elemental Distribution, 379; Fluid Inclusions, 394; Fractionation Processes, 400; Freshwater Chemistry, 405; Geochemical Cycle, 412; Geothermometry and Geobarometry, 419; Magmas, 1326; Nickel-Irons, 2718; Nucleosynthesis, 425; Oxygen, Hydrogen, and Carbon Ratios, 431; Phase Changes, 436; Phase Equilibria, 442; Ultramafic Rocks, 1360; Water-Rock Interactions, 443.

BIBLIOGRAPHY

Burchfiel, B. Clark, et al. *Physical Geology.* Westerville, Ohio: Charles E. Merrill, 1982. An excellent and comprehensive textbook covering all aspects of geology suitable for the lay reader or liberal arts college student. Of special interest are chapter 2 on mineralogy, chapter 7 on the Earth's interior, chapter 9 on crustal materials and mountain building, and chapter 10 on the origin and differentiation of the Earth and early geologic time.

Davidson, Jon P., Walter E. Reed, and Paul M. Davis. *Exploring Earth: An Introduction to Physical Geology.* Upper Saddle River, N.J.: Prentice Hall, 1997. An excellent introduction to physical geology, this book explains the composition of the Earth, its history, and its state of constant change. Intended for high-school-level readers, it is filled with colorful illustrations and maps.

Fyfe, W. S. *Geochemistry.* Oxford, England: Clarendon Press, 1974. Part of the Oxford Chemistry Series, this work was written for lower-division college chemistry students. Although in some respects dated, it is nevertheless a brief (about one-hundred-page) and excellent introduction to the science of geochemistry. Of special interest is chapter 9, "Evolution of the Earth." The book has a bibliography, glossary, and index.

Gregor, C. Bryan, et al. *Chemical Cycles in the Evolution of the Earth.* New York: John Wiley & Sons, 1988. A systems approach to geochemistry, this book is suitable for the serious college student. Although filled with graphs, tables, and chemical equations, sections are very readable for the layperson. Discussions of mineralogical, oceanic, atmospheric, and other important chemical cycles are extensive, and the work is well referenced.

Jackson, Ian, ed. *The Earth's Mantle: Composition,*

Structure, and Evolution. Cambridge: Cambridge University Press, 1998. Intended for the college student, *The Earth's Mantle* provides a clear and complete description of the elements that make up the Earth's mantle and the process of change that it has undergone since its formation. Includes a bibliography and index.

Jacobs, John. *Deep Interior of the Earth.* London: Chapman and Hall, 1992. This introductory geophysics textbook is formidable for the average student because there is considerable mathematics in some chapters, but it does cover many useful topics. It contains a minimum of equations but many figures and graphs.

Kroner, A., G. N. Hanson, and A. M. Goodwin, eds. *Archaean Geochemistry: The Origin and Evolution of the Archaean Continental Crust.* Berlin: Springer-Verlag, 1984. A collection of reports by the world's leading geochemists studying the geochemistry of the world's oldest rocks. Although many of the articles are technical in nature, the abstracts, introductions, and summaries are accessible to a college-level reader interested in the work of top international scientists.

Levin, Harold L. *The Earth Through Time.* 3d ed. Philadelphia: Saunders College, 1988. An excellent and very readable text dealing with historical geology. Filled with illustrations, photographs, and figures, this book is suitable for the layperson. Chapters on planetary beginnings, origin and evolution of the early Earth, and plate tectonics are of special interest. Contains an excellent glossary and index.

Lutgens, Frederick K., and Edward J. Tarbuck. *Earth: An Introduction to Physical Geology.* 6th ed. Upper Saddle River, N.J.: Prentice Hall, 1999. This college text provides a clear picture of the Earth's systems and processes that is suitable for the high school or college reader. In addition to its illustrations and graphics, it has an accompanying computer disc that is compatible with either Macintosh or Windows. Bibliography and index.

McCall, Gerald J. H., ed. *The Archean: Search for the Beginning.* Stroudsburg, Pa.: Dowden, Hutchinson and Ross, 1977. A superb collection of thirty-eight papers by outstanding geologists, arranged under topical headings. The papers are at times technical, but the editor provides an introduction and integrating commentary that helps bridge the gap for the nontechnical reader. Contains a subject index.

Plummer, Charles C., and David McGeary. *Physical Geology.* Boston: McGraw-Hill, 1999. A college-level introductory geology textbook that is clearly written and wonderfully illustrated. An excellent sourcebook of basic information on geologic terminology and fundamentals of geologic processes. Contains CD-ROM. An excellent glossary.

Ponnamperuma, Cyril, ed. *Chemical Evolution of the Early Precambrian.* New York: Academic Press, 1977. A collection of papers from the second colloquium of the Laboratory of Chemical Evolution of the University of Maryland, held in 1975. Written by experts in the field, the papers are still, for the most part, accessible to the nontechnical reader. The volume contains a subject index.

Salop, Lazarus J. *Geological Evolution of the Earth During the Precambrian.* Berlin: Springer-Verlag, 1983. A top Soviet geologist conducts an exhaustive survey of Precambrian geology. Suitable for a college-level reader with a serious interest in the subject. Contains numerous graphs and tables, with extensive references.

Tarling, D. H. *Evolution of the Earth's Crust.* New York: Academic Press, 1978. Written for the undergraduate-level college reader with some background in geology, this volume is an excellent collection of nontechnical, well-written essays covering the origin and evolution of the Earth's crust and plate tectonics. Contains references and an index.

Wedepohl, Karl H. *Geochemistry.* New York: Holt, Rinehart and Winston, 1971. An accessible and brief introduction to geochemistry fundamentals. Contains an excellent chapter on meteorites and cosmic abundances of the elements. Suitable for the nontechnical reader, with index and references. A good starting point for those unfamiliar with mineral formation.

Wetherill, George W., A. L. Albee, and F. G. Stehli, eds. *Annual Review of Earth and Planetary Sciences.* Vol. 13. Palo Alto, Calif.: Annual Reviews, 1985. Three articles of interest to the Earth history student are "Evolution of the Archean Crust," by Alfred Kroner, and "Oxidation States of the Mantle: Past, Present, and Future" and "The Magma Ocean Concept and Lunar Evolution," by Richard Arculus. Kroner's article is particularly readable for the college-level audience, with an excellent overview of the historical views on Precambrian geology. References at the end of each article.

FLUID INCLUSIONS

Fluid inclusions are small amounts of fluids trapped in minerals within rocks. They contain valuable clues regarding many geologic processes. Their study also has a number of practical applications in the exploration for mineral and petroleum resources, the study of gemstones, and the search for a storage site for nuclear wastes.

PRINCIPAL TERMS

BRINE: water with a higher content of dissolved salts than ordinary seawater

FLUID: a material capable of flowing and hence taking on the shape of its container; gases and liquids are both examples of fluids

GLASS: a solid without a periodic ordered arrangement of atoms; it frequently forms when molten material is rapidly cooled

IGNEOUS ROCKS: rocks formed by the crystallization of a magma

IMMISCIBLE FLUIDS: two fluids incapable of mixing to form a single homogeneous substance; oil and water are common examples

INTRUSIVE ROCKS: igneous rocks formed from magmas that have cooled and crystallized underground

MAGMA: molten material capable of yielding a rock upon cooling

METAMORPHIC ROCKS: rocks that have transformed from their original condition as a result of changes in physical or chemical conditions within the Earth

MINERAL: a naturally occurring, inorganic substance with a regular periodic arrangement of atoms

ORE: any concentration of economically valuable minerals

SEDIMENTARY ROCKS: rocks formed by the consolidation of material transported by and deposited from wind or water

VOLCANIC ROCKS: igneous rocks formed at the surface of the Earth

PRIMARY AND SECONDARY FLUID INCLUSIONS

During its history, almost every rock will come in contact with at least one fluid. Geologists have long recognized that fluids play an important role in shaping and altering rocks and in determining the Earth's geologic history. It has been said that the fluid phase is the critical "missing" phase in petrology (the study of rocks). Behind this statement is the widespread belief that the fluids eventually leave rock systems, so that only indirect evidence can be used to deduce their nature. There is a growing recognition, however, that small amounts of these fluids are often left behind, trapped in small cavities, as fluid inclusions. In most cases, these inclusions represent the only available direct samples of fluids active deep within the Earth or in the distant past.

Fluid inclusions can form in a variety of ways, though two are most common. Fluids can be trapped during mineral growth, to yield primary fluid inclusions. Most of these inclusions probably form when fluid fills pits on the surface of a grow-

ing crystal. New material added to the crystal grows over the tops of these cavities, trapping the fluid. These inclusions provide information about the nature of the fluids during the growth of the host minerals. Fluids trapped after the growth of their host minerals are secondary fluid inclusions. These form when fluid enters a crack within a mineral: As the ends of the crack grow together, the fluid is trapped. Secondary fluid inclusions originally form a thin envelope with a high surface-area-to-volume ratio. With its very high surface energy, this envelope is very unstable. Therefore, the host mineral will frequently recrystallize around the inclusion, causing it to break up into a swarm of smaller but thicker inclusions with a smaller surface-area-to-volume ratio. Such recrystallization occurs when the fluid inside the inclusion dissolves material from some parts of the inclusion wall and precipitates new material on other parts. Secondary fluid inclusions are sources of information about the fluids that have interacted with the rock after the growth of the host mineral.

SIZE, SHAPE, AND APPEARANCE

Fluid inclusions vary greatly in their size, shape, and appearance. Those that have diameters larger than a few tenths of a millimeter can be seen with the naked eye, but they are rare. Fluid inclusions with diameters between 1 and 100 microns are common and can be studied with a microscope. Inclusions with diameters as small as 0.01 micron have been observed with electron microscopes.

Fluid inclusions occur in almost any shape, but particularly noteworthy are negative crystals—cavities shaped like a crystal of the host mineral. This is the shape with the lowest surface energy and in many cases appears to be the final result of the recrystallization of the host mineral around the inclusion. Many inclusions contain only one phase, liquid or gas; however, gas bubbles within a liquid are also common. If these bubbles are small enough, they will move vigorously back and forth; this motion is a consequence of their relatively high surface tension. Some fluid inclusions contain immiscible liquids such as oil and water. In this situation, one of the liquids generally lines the walls of the inclusion, and the other liquid forms a droplet inside it. Solids can precipitate out of the fluids trapped within an inclusion to form tiny crystals known as daughter minerals. Trapped magma will solidify upon cooling to form either a glass or a mass of tiny mineral grains. Although such inclusions may not be fluid now, the material was trapped as a fluid, and hence they are generally classified as fluid inclusions.

TYPES OF INCLUSIONS

The composition of fluid inclusions depends on the environment in which they are trapped. Many different kinds of inclusions have been found in materials formed at the surface or in the upper levels of the Earth's crust. Glacial ice and amber (fossilized tree sap) contain gas inclusions, which serve as modified samples of the atmosphere. Minerals in evaporite deposits (rocks formed by precipitation from evaporating water) may contain samples of the concentrated brines from which they formed. Inclusions of groundwater can be found in the mineral deposits formed on walls, ceilings, and floors of caves. Fluid inclusions can be found in the cements that hold grains together in sedimentary rocks, as well as in the minerals that line the walls of vugs (roughly spherical cavities) and fractures in these rocks. Two kinds of inclusions most frequently occur in these situations. Water-bearing inclusions generally contain fairly high amounts of dissolved salts. Hydrocarbon inclusions can contain natural gas (most commonly methane) or crude oil.

Volcanic rocks can contain a number of different kinds of inclusions. Glass inclusions may represent trapped samples of the silicate liquid from which the rock crystallized. In some cases, inclusions represent immiscible liquids present as droplets in the main magma. These droplets may have consisted of another silicate-rich liquid now present as a glass or a fine-grained mixture of minerals, or they may have been a nonsilicate liquid rich in sulfur and iron, now represented by tiny sulfide mineral grains. Water-rich inclusions in volcanic rocks tend to be filled with either a concentrated brine or water vapor. Carbon dioxide is the other gas most commonly found in volcanic rocks.

COMPOSITION AT GREATER DEPTHS

Fluid inclusions are also found in rocks formed at greater depths in the Earth. The rock portions of the Earth that surround the iron core consist of the relatively dense mantle surrounded by the much thinner, less dense crust. Carbon dioxide is the most common fluid in inclusions in rocks from the upper mantle and in metamorphic rocks from the lower crust. Because these fluids are trapped at high pressures, they are often very dense, and the carbon dioxide is usually present as a liquid, which may or may not contain a gas bubble. Variable amounts of nitrogen and methane can be dissolved in these inclusions, and it appears that compositions range from pure carbon dioxide to pure nitrogen or methane. At low temperatures, dense liquid carbon dioxide and water form immiscrible liquids. Thus, when water occurs in these dense carbon-dioxide-rich inclusions, it forms a separate liquid phase that generally lines the wall of the inclusions.

Water is the most abundant fluid in inclusions in metamorphic rocks formed in the upper levels of the crust, although carbon dioxide, methane, and nitrogen also occur. The kinds and amounts of dissolved solids vary greatly in these aqueous solutions and in many cases appear to reflect the nature of the rocks in which they were trapped. In the lower crust, the most abundant inclusions are rich in carbon dioxide. This change is coincident with a decrease in the abundance of water-bearing minerals in the deeper levels of the Earth. Intrusive igneous rocks cool and solidify slowly within the Earth. Under these conditions, inclusions of the silicate liquid from which the rock forms will become not glass but rather clusters of small mineral grains. Many other kinds of fluids can be trapped during or after the crystallization of an intrusive rock. Although inclusions rich in carbon dioxide, hydrocarbons, sulfide minerals, and nitrogen have been found, the most common inclusions in these kinds of rocks are water rich and in some cases contain large amounts of dissolved solids.

CHANGES UNDERGONE BY FLUID INCLUSIONS

Fluid inclusions can undergo a variety of changes after they are trapped. Most inclusions form at temperatures significantly above those normally found on the Earth's surface. As an inclusion cools, the volume of liquid will decrease much more rapidly than will the volume of the surrounding solids. This differential shrinkage frequently results in the formation of a vapor bubble from the liquid. If the inclusion is heated, this bubble will disappear. For inclusions trapped near the Earth's surface, the temperature at which the bubble disappears will be close to the temperature at which the inclusion was trapped. For fluids trapped at elevated pressures deeper within the Earth, the temperature at which the bubble disappears (in the laboratory) will be lower than the initial temperature of trapping. As temperature decreases, the fluid may become saturated with one or more solid compounds; the result is the precipitation of daughter crystals.

The thermal expansion or contraction of solids is low enough that temperature changes have a negligible effect on the volume of most inclusions. Thus, inclusions can generally be treated as constant volume systems. Unless material leaks out of the inclusions, their density will usually not change after trapping. Dense fluids trapped at elevated pressures will exert pressure on the walls of the inclusion. If this pressure exceeds the strength of the host mineral, the inclusion will burst, and much of the fluid will leak out. Fluid inclusions may also leak slowly instead of catastrophically. One way this leakage can occur is by slow diffusion of molecules through the host mineral. Certain molecules have a greater tendency to diffuse through the host; leakage by diffusion, then, can change the composition of an inclusion. Recrystallization of the host mineral can cause changes in the shape of an inclusion, as well as cause a larger fluid inclusion to break into smaller ones. If the larger inclusion contains two or more fluids (for example, a liquid and a vapor), generally they will not split evenly between the new smaller inclusions. Thus, the compositions of the new inclusions will differ from each other and from the compositions of the original inclusion. The effects of these and many other possible secondary changes have to be carefully evaluated during any fluid inclusion study.

EXTRACTION OF FLUID FROM INCLUSIONS

Two major problems confront anyone trying to study the composition of a fluid inclusion: separating the fluid in the inclusion from the rest of the sample (including other inclusions) and obtaining an analysis of a very small sample. Most attempts to analyze fluid inclusions by conventional chemical methods have involved the extraction of fluid from many different inclusions in the same rock. One way to do this is to heat the rock. As temperature increases, so does the pressure exerted by the fluid on the walls of the inclusion. When the pressure exceeds the strength of the host mineral, the inclusion will brust open, and the fluid will escape—a process known as decrepitation. Another possibility is to crush the rock, thereby releasing the fluid. The fluid given off during crushing or decrepitation is collected, and its composition is determined by any of a number of different analytical techniques. Unfortunately, most samples contain different kinds of fluids trapped at different times. Analyses done by the methods above give the composition of a mixture of these fluids, which usually differs significantly from the composition of the fluid in any of the individual inclusions.

ANALYSIS OF INDIVIDUAL INCLUSIONS

A number of techniques have been developed which will permit a partial chemical analysis of a single inclusion. Most of these involve hitting the inclusion with a small, tightly focused beam of light—usually a laser. Under these conditions, radiation will be emitted or absorbed; its wavelengths will be characteristic of the kinds of molecules present, while the intensity of the radiation will be proportional to their concentration. Measuring the spectra, then, makes it theoretically possible to obtain the composition of the inclusion. Most attempts to do this have used the Raman spectrum, emitted when the inclusion is struck by a laser beam. Methods using the infrared spectrum have also been used.

Most information on the composition of individual inclusions has come from observations made under the microscope. For such studies, a wafer of rock is ground to a millimeter or less in thickness and then polished on both sides. Light can be transferred through many minerals at this thickness. Much information can be obtained by careful observation at room temperature.

HEATING-COOLING STAGES

Even more information can be obtained by observing changes in an inclusion as it is heated or cooled. For example, if a solid forms upon cooling and then melts at 0 degree Celsius, when the sample is heated, the inclusion contains water with no dissolved solids (0 degree is the melting point of pure water). In order to make these kinds of observations, the rock wafer is placed in the sample chamber of a heating-cooling stage, which is in turn placed onto the microscope stage. Most heating-cooling stages are capable of cooling a sample to about 180 degrees Celsius or heating it to 600 degrees while it is being observed under the microscope. Observations made over this temperature range usually allow scientists to identify the major fluid species present (for example, water, carbon dioxide, or methane) and put some limits on the amounts and kinds of dissolved solids. Moreover, heating-cooling stages are relatively cheap and easy to use and give results quickly. Finally, they provide important information on the density of inclusions. Thus, the heating-cooling microscope stage has been and is likely to continue to be the major tool of fluid inclusion studies.

At low temperatures, most fluid inclusions contain both a gas and a liquid. As the inclusion is heated, the fluid will homogenize; the volume of one of the fluids increases, and the other completely disappears. The temperature at which this homogenization occurs is a function of the density of the fluid inclusion. Thus, by observing the temperature at which homogenization occurs on the heating-cooling stage, researchers can often obtain the density of the inclusion. Such density data may supply important information about the temperatures and pressures at the time the fluid was trapped.

SEARCH FOR ECONOMIC RESOURCES

The principal use of fluid inclusions has been in the study of ancient fluids and their interaction with solid Earth materials. Although this research has made important contributions to the understanding of the Earth's geologic history, the interest has been largely academic. Nevertheless, a number of practical applications of fluid inclusion studies have also been found. Much of this application has involved the search for mineral resources and the study of petroleum migration.

Fluid inclusions have contributed to the search for mineral resources in many different ways. First, studies of these inclusions have contributed enormously to the understanding of the processes by which ore deposits form. Many ore deposits have been formed by hot, water-rich solutions (hydrothermal solutions), which circulate through the rocks, dissolving and removing elements from some areas and precipitating them as ore minerals in others. This understanding has in turn guided the selection of areas in which to search for ore deposits. Fluid inclusion studies can also contribute to knowledge of the geologic history of a specific area, including an understanding of the development of features most likely to control the emplacement of an ore body. Finally, fluid inclusions can be one of the telltale signs in the search for ore. The strategy is to increase the size of the "target"—the small area containing a valuable mineral—by looking for secondary effects that accompanied ore deposition but affected a larger area. The improvement in the chances of finding ore with this strategy can be compared to the increase in the chances of finding a nail rather than a needle in a haystack. Fluid inclusions surrounding an ore body can show special properties related to the development of the ore, and these anomalies can extend well beyond the deposits. A search for such anomalies has been used in ore exploration.

Petroleum originates in a fine-grained source rock, moves into a more permeable reservoir rock, and then migrates into a petroleum trap. Hydrocarbon inclusions in old reservoir rocks are potential clues to the process of petroleum migration. This is a relatively new area of research, but it shows good promise of increasing the understanding of the movement of oil underground, leading to improved strategies in the search for this oil.

OTHER PRACTICAL APPLICATIONS

The study of fluid inclusions has also been applied to the establishment of the authenticity of gemstones and investigation into long-term storage sites for nuclear waste. Establishing the authenticity of gemstones is one of the more important jobs of a gemologist. Many gemstones contain fluid inclusions; the nature of these inclusions often allows the expert to distinguish between synthetic and natural stones. In some cases, the fluid inclusions can be used to identify the source of the gem. Thus, for example, Colombian emeralds can often be recognized by fluid inclusions containing a concentrated brine and large daughter crystals of sodium chloride.

Because of its highly dangerous nature, nuclear waste must be isolated for periods of time ranging from thousands to hundreds of thousands of years. Most proposals of ways to bring about such isolation involve the underground burial of the waste. The principal hazard to burial is that the nuclear waste could be dissolved in and carried away by fluids circulating through the local rocks. The study of fluid inclusions has been used to evaluate this danger. These studies are based on the realization that fluid inclusions represent a partial record of the fluids that moved through the rocks in the past and hence provide some basis for extrapolating into the future. Thus, for example, studies of fluid inclusions from salt beds have indicated that some of these beds probably have been penetrated by circulating groundwater. This finding is especially significant, since salt deposits have often been mentioned as possible repositories of nuclear waste.

Edward C. Hansen

CROSS-REFERENCES

Earth Resources, 1741; Elemental Distribution, 379; Evolution of Earth's Composition, 386; Fractionation Processes, 400; Freshwater Chemistry, 405; Gem Minerals, 1199; Geochemical Cycle, 412; Geothermometry and Geobarometry, 419; Hydrothermal Mineralization, 1205; Metamorphic Mineral Deposits, 1614; Nuclear Waste Disposal, 1791; Nucleosynthesis, 425; Oil and Gas Exploration, 1699; Oxygen, Hydrogen, and Carbon Ratios, 431; Phase Changes, 436; Phase Equilibria, 442; Water-Rock Interactions, 443.

BIBLIOGRAPHY

Aharonov, Einat. *Solid-Fluid Interactions in Porous Media: Processes That Form Rocks.* Woods Hole: Massachusetts Institute of Technology, 1996. Aharonov examines the processes involved in rock formation. This is a technical book at times, but it can be understood by the careful reader.

Berner, Elizabeth K., and Robert A. Berner. *Global Environment: Water, Air, and Geochemical Cycles.* Upper Saddle River, N.J.: Prentice Hall, 1996. This book offers a clear and readable introduction to the processes that sustain life and effect change on the Earth, including a useful section on aquatic geochemistry. Color illustrations and maps.

Emsley, John. *The Elements.* 3d ed. Oxford: Oxford University Press, 1998. Emsley discusses the properties of elements and minerals, as well as their distribution in the Earth. Although some background in chemistry would be helpful, the book is easily understood by the high school student.

Hollister, L. S., and M. L. Crawford, eds. *Mineralogical Association of Canada Short Course in Fluid Inclusions: Applications to Petrology.* Mineralogical Association of Canada, 1981. This book contains some good general information about fluid inclusions but concentrates on the information that fluid inclusions can give about rocks and rock-forming processes. Contains twelve separate articles by nine different authors. Some articles may be difficult for a reader with no previous knowledge of geology.

Roedder, Edwin. "Ancient Fluids in Crystals." *Scientific American* 207 (October, 1962): 38-47. One of the few articles on fluid inclusions intended specifically for a general audience. Written by the "dean" of American fluid inclusion studies: Roedder has done more to promote interest in fluid inclusion than any among his contemporaries in the English-speaking world. Very well illustrated and clearly written, it is an excellent introduction to the subject. Although it does not cover any of the many more recent discoveries, Roedder's description of the basic phenomenon remains valid.

_____. "Fluid Inclusions as Samples of Ore Fluids." In *Geochemistry of Hydrothermal Ore Deposits*, edited by L. B. Barnes. 2d ed. New York: John Wiley & Sons, 1979. The bulk of this paper is an introduction to fluid inclusions and their interpretation, which should be comprehensible to someone without a technical background. The remaining portion of the article concentrates on fluid inclusions in ore deposits and is especially good for those with an interest in oreforming solutions.

_____. *Reviews in Mineralogy.* Vol. 12, *Fluid Inclusions.* Washington, D.C.: Mineralogical Society of America, 1984. This book is certainly one of the best and one of the most complete works on fluid inclusions that is available in the English language. Written as an introduction to fluid inclusion research for the geologist, the text may be rough going in places for those with no previous geologic knowledge.

FRACTIONATION PROCESSES

Processes as fundamental as evaporation, condensation, fluid movement through a rock, and many biological functions result in isotope fractionation. The record of these processes as they influenced the formation of various minerals is preserved in the distribution of stable isotopes within rock.

PRINCIPAL TERMS

CHEMICAL BOND: the force holding two chemical elements together as part of a molecule

DEPLETION: the process by which the light isotope is concentrated in either the reactants or the products of a chemical reaction

ENRICHMENT: the process by which the heavy isotope is concentrated in either the reactants or the products of a chemical reaction

PRODUCT: the material that results from a reactant undergoing a chemical process

REACTANT: the starting material or materials in any chemical reaction

STANDARD: a material of known isotopic composition; all enrichment and depletion is measured relative to the standard value

WEIGHT DIFFERENCE

Fractionation implies the breaking of a whole into its parts. In the case of a chemical element, the parts are the naturally occurring isotopes of that element. Some of the isotopes may be radioactive and may spontaneously decay to form another element. Most isotopes, however, are stable; they do not decay, and they differ from other isotopes only in their mass. Thus, stable isotope fractionation comprises several physical and chemical processes that can separate the stable isotopes of an element on the basis of weight difference.

The weight difference can be substantial. For example, deuterium is a stable isotope of hydrogen. The hydrogen nucleus contains one proton, whereas the deuterium nucleus contains one proton and one neutron. Because the proton and the neutron are about equal in mass (and the electron is so small it can be ignored), the atomic weight of hydrogen is 1 and the atomic weight of deuterium is 2; deuterium weighs 100 percent more than does hydrogen. In another example, the most common isotopes of carbon are carbon 12 (six protons and six neutrons in the nucleus) and carbon 13 (six protons and seven neutrons in the nucleus). Just as deuterium has one more neutron than does hydrogen, carbon 13 has one more neutron than does carbon 12. The weight difference between carbon 13 and carbon 12, however, is less on a percentage basis—only 8 percent. It is apparent, then,

that the addition of a neutron in an isotope has the greatest relative mass impact for the light elements (for example, hydrogen) and that the impact decreases as the elements become heavier. Thus, stable isotope fractionation processes are most obvious when light elements (those preceding sulfur in the periodic table) are involved.

EVAPORATION AND CONDENSATION

Fractionation includes those processes that separate light and heavy isotopes by physical means or through some chemical reaction. Evaporation and condensation are two physical processes that result in the separation of stable isotopes. For their impact on the Earth's climate, the evaporation and condensation of water are arguably the most important physical processes.

Evaporation and condensation are mirror images of each other as far as isotope fractionation is concerned. In evaporation, energy is absorbed by water. This energy absorption is reflected by an increase in the temperature of the water. The individual water molecules absorb the energy and begin to move and to vibrate faster. Eventually, the individual molecules absorb so much energy that large quantities of them change from water molecules in a liquid to water molecules in a gas. The water molecules are said to have undergone a change of phase, from a liquid to a gas. If the process is reversed and energy is removed from the gas containing water molecules, the molecules slow down and begin to clump together. In the atmosphere, this condensation creates water droplets that may eventually produce a rain shower.

Water molecules are actually not all the same. With its two hydrogen atoms and one oxygen atom, the water molecule may have any of a range of molecular weights, depending upon the range of isotopic substitution. Stable isotopes of hydrogen (hydrogen and deuterium) and stable isotopes of oxygen (oxygen 16, oxygen 17, and oxygen 18) combine to yield water molecules of varying molecular weights. Normal water has a molecular weight of 18 (two hydrogens plus one oxygen 16), and heavier water has a molecular weight of 20 (two hydrogens and one oxygen 18). The heavier water molecule is 11 percent heavier than is light water.

If a mixture of light- and heavy-oxygen water is undergoing evaporation, it will take more energy input to "lift" the heavy water out of the liquid phase and into the gas phase. This process is analogous to a weight lifter's expending more energy to raise a 200-kilogram weight than a 180-kilogram weight. In the case of evaporation, the input energy is thermal (heat) rather than mechanical (physical lifting). The lighter oxygen 16 water will evaporate more readily, leaving the heavier oxygen 18 water behind. The gas phase above the evaporating water (for example, the atmosphere above the ocean) is dominated by light water, whereas the water remaining in the liquid phase is dominated by heavier water. This is not to say that heavy water molecules never evaporate; they do. Yet if one defines a number, R, as the ratio of heavy oxygen to light oxygen and measures R before the evaporation starts and then in the water vapor and the liquid water after some period of evaporation, the value of R will change. The ratio will be larger in the liquid (concentrating heavy oxygen) and smaller in the water vapor (dominated by light oxygen).

MAGNITUDE OF ISOTOPIC FRACTIONATION

Geochemists who study isotope fractionation use an equation which uses the R values calculated above to express the magnitude of isotopic fractionation. The equation $\delta = [(R_{(sample)} - R_{(standard)}) \div R_{(standard)}] \times 1000$ defines a quantity called delta (δ), which represents the difference in the ratio of heavy oxygen to light oxygen between a sample (for example, the water vapor) and a standard (for example, seawater). This difference is multiplied by 1,000 and expressed as per mil (instead of percent).

The value of δ may be positive or negative, depending on the sign of the numerator. If the numerator is negative, R_{sample} is less than $R_{standard}$. In the evaporation example, that means that the sample has less heavy oxygen (or more light oxygen) than does the standard. The sample is depleted in the heavy isotope of oxygen and enriched in the light isotope of oxygen. In evaporation, the vapor phase is depleted in heavy oxygen and its δ value is negative. Conversely, the δ value for the remaining liquid phase will be positive (R_{sample} greater than $R_{standard}$), because the liquid phase is enriched in the heavy isotope of oxygen. When water vapor condenses to form raindrops, the first droplets to form are enriched in heavy-oxygen water (δ greater than 0), and the remaining vapor is enriched in light-oxygen water (δ less than 0).

SEAWATER AS STANDARD

Seawater makes up greater than 90 percent of the liquid water at the Earth's surface and, as such, makes a good standard for identifying fractionation effects. It should be apparent from the previous example that rainwater will be isotopically lighter (in both hydrogen isotopes and oxygen isotopes) than seawater. Similarly, evaporative waters, such as the waters of the Red Sea, will be isotopically heavier than is normal seawater.

If an oxygen-containing mineral forms in contact with water, some of the oxygen in the mineral will probably come from the water. If the water is isotopically light (from a freshwater, terrestrial environment, for example), the mineral oxygen will be relatively light also. If the mineral forms in contact with an evaporative brine, the mineral oxygen should be enriched in the heavy isotope of oxygen. Thus, the stable isotopic composition of a mineral can tell scientists about the composition of ancient waters.

BOND BREAKING

Chemical and biological reactions involve the breaking of chemical bonds. Just as it takes energy to evaporate a molecule of water, it takes energy to break the bond between hydrogen and oxygen atoms in water or between carbon and oxygen atoms in carbon dioxide. The bond between atoms can be visualized as a spring. Inputs of energy cause the bond (spring) to vibrate. The atoms at the ends of the bond move apart as the bond stretches, and they move together as the bond restores the molecule to its original shape. If sufficient energy is applied, the atoms move so far apart (stretching the "spring") that there is no force remaining to pull them back together. At this point, the bond has been broken.

Bond breaking occurs in any chemical reaction, whether it involves living organisms (a biochemical reaction) or proceeds without biological intervention (an inorganic reaction). In the case of inorganic reactions, the energy to break bonds is usually supplied by the environment in the form of heat. At low temperatures, such as normal room temperature, bonds involving heavy isotopes are less likely to be broken than are those involving only light isotopes. For the same amount of energy input, a spring with heavy weights on each end vibrates and stretches more slowly than does the same spring with lighter weights on the ends. This analogy shows why it takes more energy input to break bonds involving heavy isotopes. As the environmental temperature increases, however, the energy necessary to break bonds becomes readily available, and the degree of isotopic fractionation decreases. The slight differences in bond strength are insignificant at higher temperatures.

When living organisms are involved, the fractionation can be exaggerated. In biochemical reactions, the organism is often the source of energy for bond breaking. When the isotopes are chemically identical, it does not make sense for an organism to expend the extra energy to break heavy-isotope bonds when the same reaction with light isotopes uses less energy. Even in the case of photosynthesis, where sunlight is the primary energy source used to break the carbon-oxygen bond in carbon dioxide, isotopically light carbon dioxide is more likely to be photosynthesized than is isotopically heavy carbon dioxide. The result is that light isotopes are concentrated in the reaction products of biochemical reaction. Thus, biochemical molecules, which contain elements such as hydrogen, carbon, oxygen, nitrogen, sulfur, and phosphorus, tend to concentrate the lighter isotopes of those elements. The surrounding environment, be it lake water, seawater, or sediments, tends to accumulate the heavier isotopes.

MASS SPECTROMETRY

The mass spectrometer made it possible for scientists to identify and study the phenomenon of isotope fractionation. As its name implies, the instrument analyzes a spectrum based on mass, or weight. The rainbow is a common example of the spectrum of white light. The original light has been broken into its colorful components by passing through water droplets in the atmosphere that act like a glass prism. Each individual color represents light of a different wavelength. Water droplets and prisms, then, serve as simple wavelength spectrometers. In a mass spectrometer, a sample is analyzed for the range of masses of the elements it contains.

In an actual analysis of a geologic sample—a coal, for example—the sample is converted through chemical processes into gases suitable for analysis in the mass spectrometer. The gases produced by the sample, in this case probably carbon dioxide, water, and some nitrogen and sulfur gases, are separated and purified by passing them through a series of freezing and drying steps. When a pure gas sample is obtained, that portion of the sample is injected into the mass spectrometer. Inside the mass spectrometer, the sample first enters an ionization chamber, where one or two electrons are stripped from the molecule. The molecule is now a positively charged ion (it has lost one or more negative electrons). The ionized gas sample next approaches a negatively charged metal plate with a hole in the middle. Because the sample gas is positively charged and the metal plate is negatively charged, the sample molecules are attracted to the plate and accelerate toward it. Some of the sample molecules pass through the hole.

ISOTOPIC SEPARATION

At this point, isotopic separation becomes important. For example, in nitrogen gas (composed of two nitrogen atoms), the nitrogen molecule may contain either nitrogen 14 or nitrogen 15 isotopes. The molecule may weigh 28, 29, or 30 units. During the acceleration toward the metal plate, the lighter-isotope molecules will be moving faster than will the heavier molecules. Nitrogen gas molecules composed of two nitrogen 15 atoms are moving most slowly.

After passing through the hole, the nitrogen molecules enter a metal pipe with a slight bend. There is a vacuum in the pipe, and the pipe is surrounded by a strong magnet. As the accelerated, charged molecules enter the bent pipe, their flight path is bent by the presence of the magnetic field. The degree of bending is dependent upon how fast the molecule is moving, which, in turn, depends on the mass of the molecule and the amount of charge on the metal accelerating plate. Thus, given one set of conditions (accelerating plate voltage and strength of the magnetic field), only the nitrogen gas molecules weighing 29 units will negotiate the bend in the pipe and reach the particle detector at the other end. The lighter nitrogen would be moving too fast to make the turn and would adhere to the outer wall of the pipe. On the other hand, the heavier nitrogen molecules would be moving too slowly and would be bent sharply into the inside wall of the pipe. By varying the accelerating voltage, the scientist can focus beams of all three types of nitrogen gas onto the detector and thus analyze the nitrogen isotopic composition of the sample. The same sort of analysis is done with hydrogen isotopes, using hydrogen gas and carbon, and with oxygen isotopes, using carbon dioxide gas.

UNDERSTANDING CLIMATE AND BIOCHEMICAL PROCESSES

In general, a scientist analyzes a sample and seeks a measure of isotopic fractionation in order to determine either the temperature at which a mineral formed or whether a mineral formed at lower temperatures is inorganic or biochemically influenced. The samples analyzed may occur naturally, or they may be minerals grown in carefully controlled laboratory systems where the precise isotopic compositions of the starting materials are known.

Stable isotope fractionation is a process that occurs continuously. The processes involved are as simple as evaporation and as fundamental to human survival as are photosynthesis and respiration. Many of the natural fractionation processes studied with stable isotopes are very simple, but others are extremely complex. Among the former, precipitation in the form of snow leaves a permanent, frozen record of climate in the polar ice caps. Studies of climatic change in these polar lati-

tudes involve drilling ice cores out of the thick ice sheet and determining the stable isotope ratios in the water. Because human activities such as the burning of fossil fuel and forest clear-cutting are so extensive, some scientists believe that the Earth's surface climate is being changed. Understanding the range of climate variability over the recent geologic past through the study of isotopic variations in ice cores may tell scientists whether the Earth's climate can absorb and recover from such changes.

Biological processes would fit into the category of extremely complex processes. Scientists who study the details of biochemical processes must examine and understand each step in a series of many that make up the overall reaction. At any point in that series of reactions, if something goes wrong, the chance for development of a disease or abnormality enters. Isotopes in general, and stable isotopes especially, can be used to trace the starting chemicals through the maze of reactions to the ultimate products. Stable isotopes act to identify certain elements throughout the reaction sequence, without introducing possible radiation effects (as when radioactive isotopes are used). Knowledge of fractionation processes allows the scientist to correct for or ignore certain amounts of enrichment or depletion in the reaction products. Identifying stable isotopes as parts of nonproductive side branches of the reaction series may provide a clue to the reaction process, which, in turn, may suggest a weakness in the process. Understanding important biochemical reactions on the molecular scale may help scientists to find cures for some deadly diseases and to prevent birth defects.

Richard W. Arnseth

CROSS-REFERENCES

Bibliography

Bowen, Robert. *Isotopes and Climates*. London: Elsevier, 1991. Bowen examines the role of isotopes in geochemical phases and processes. This text does require some background in chemistry or the Earth sciences but will provide some useful information about isotopes and geochemistry for someone without prior knowledge in those fields. Charts and diagrams help clarify difficult concepts.

Criss, Robert E. *Principles of Stable Isotope Distribution*. New York: Oxford University Press, 1999. Criss describes isotopes and their properties with clarity. In addition to well-written text, the book features diagrams and illustrations that present a clear picture of the different phases of isotopes and isotope distribution. Bibliography and index.

Drever, J. I. *The Geochemistry of Natural Waters*. Englewood Cliffs, N.J.: Prentice-Hall, 1982. Nearly every general geochemistry text contains a chapter or two on isotopes, both radioactive and stable, but Drever's chapter is certainly one of the most accessible to the general college-level reader. His approach to explaining geochemistry in the text is very good, so the person without a strong chemistry background could use this text. Contains a reasonably helpful bibliography.

Faure, Gunter. *Principles of Isotope Geology*. 2d ed. New York: John Wiley & Sons, 1986. A standard text for undergraduate- to graduate-level courses in geochemistry and isotope geochemistry. The majority of the book is devoted to radioactive isotopes, but the last five chapters discuss the fractionation processes for stable isotopes of oxygen, hydrogen, nitrogen, carbon, and sulfur. In addition, the introductory chapter presents an interesting history of the development of isotope geology. Each chapter concludes with a summary section, followed by a few calculation problems and an extensive reference list.

Fritz, P., and J. C. Fontes, eds. *Handbook of Environmental Isotope Geochemistry*. Vol. 1, *Terrestrial Environment*. New York: Elsevier, 1979. Part of a five-volume set that covers the state of knowledge in isotope geochemistry. Later volumes in the series concentrate on the marine environment and high-temperature geologic environments. The discussions are highly technical and about evenly divided between stable and radioactive isotopes. Contains an extensive bibliography. The main strength of this set is that it is comprehensive and would serve as a good source for specific, detailed information. The technical nature of the presentation will limit its usefulness to college-level readers with good chemistry backgrounds.

Hoefs, J. *Stable Isotope Geochemistry*. 3d ed. New York: Springer-Verlag, 1987. A standard reference in the field of stable isotope geochemistry. The first chapter gives a thorough, though somewhat technical, introduction to the range of isotope fractionation processes and the basic principles of mass spectrometry. Chapter 2 discusses some of the fractionation processes observed for specific isotopes in geologically important processes. The remainder of the book systematically examines the range of geologic materials, from extraterrestrial dust to the composition of the mantle, and their isotopic signatures. A very comprehensive text. The reader should have some background in chemistry. An extensive bibliography makes this book especially valuable.

Rankama, K. *Progress in Isotope Geology*. New York: John Wiley & Sons, 1963. A sequel to the author's 1954 *Isotope Geology*, this book is basically a progress report on developments during the intervening decade. Though both the original and this text are old and highly technical, they do provide a wealth of information on isotopes, some of which is only rarely discussed in more recent texts. Suitable as a reference only. For college-level readers.

Schimel, David Steven. *Theory and Application of Tracers*. San Diego: Academic Press, 1993. Schimel examines the geochemistry of isotopes in plants, soils, and waters.

FRESHWATER CHEMISTRY

The unique properties of water that are so important to geological processes depend upon the distinctive polar structure of the water molecule. The chemistry of fresh waters is highly variable. It is globally influenced by the atmosphere and locally influenced by reactions with the local soils and bedrocks. Changes through time in the chemistry of natural fresh waters is one of the best indicators of changes in the Earth's surface environment.

PRINCIPAL TERMS

ACID: a substance that yields free hydrogen ions in solution

ACIDITY: the degree of a solution's being acid as determined by the quantity of base needed to neutralize the solution

ALKALINITY: the degree of a solution's being basic as determined by the quantity of acid needed to neutralize the solution

ANION GROUP: a combination of ions that behaves as a single anion

COLORIMETER: an instrument that measures the intensity of color produced when a reagent reacts with a substance in a solution; the intensity of color is used to quantify the amount of the substance in solution

COVALENT BONDING: a type of chemical bonding produced by sharing of electrons between overlapping orbitals of adjacent atoms; covalently bonded solids usually have low solubility in water

DENSITY: a property of a substance expressed in units of weight per unit of volume, such as pounds per cubic foot or grams per cubic centimeter

IONIC BONDING: a type of chemical bonding that holds the constituents of a crystal together primarily by electrostatic attraction between oppositely charged ions

TOTAL DISSOLVED SOLIDS (TDS): a quantity of solids, expressed in weight percent, determined from the weight of dry residue left after evaporation of a known weight of water

POLAR MOLECULE

Earth is a unique planet, if for no other reason than that its temperature lies within the range that permits water to exist as a liquid, a gas, and a solid. Earth is sometimes called the "blue planet" because of the brilliant blue color, seen from outer space, that results from more than three-quarters of the planet's surface being covered by water. Ultimately, this water comes from within the planet during magmatic differentiation and formation of the crust. The size of the Earth provides a gravitational field that is sufficient to keep water vapor from escaping through geologic time. Steam, the gaseous form of water, is the dominant constituent of volcanic gases, and pressure generated by steam accounts for the power of most explosive eruptions. Ice in glaciers has been a major force in influencing the topography of vast areas of the continent.

The chemical symbol for water, H_2O, refers to a molecule composed of one hydrogen ion (H^+) and one hydroxyl ion (OH^-). When combined to form a water molecule, the two hydrogen ions are arranged to one side and have an angle of about 105 degrees between them. Because the positively charged hydrogen ions lie on one side of the molecule, they impart a positive charge to that side of the molecule while leaving a negative charge of the oxygen exposed at the other end. Therefore, each water molecule is like a tiny magnet, with a positive pole and a negative pole. The molecule has polar charge distributions and thus is termed a polar molecule.

Because each water molecule is like a magnet, as a substance, water will behave somewhat like a box full of very tiny magnets. The tendency for the magnets would be to line up positive pole to negative pole, and a similar tendency occurs in water, where the ionic attraction of the hydrogen of one water molecule for the oxygen of another water molecule actually draws water molecule to water molecule and creates a type of ordering. The cohesive ordering that results from this attraction is called hydrogen bonding, and it accounts for some remarkable physical properties that belong only to water.

Hydrogen Bonding

Water is one of the few liquid substances that do not simply become denser as they cool. The hydrogen bond allows liquid water molecules to pack themselves into a tighter pattern than would be possible if the molecules were not polar. Water achieves its maximum density of 1.00000 gram per cubic centimeter at 3.94 degrees Celsius. Liquid water molecules vibrate rapidly (about 10^{12} vibrations per second), but when the water is cooled to the freezing temperature of ice (0 degree Cel-

The hydrogen bond of water permits water to become a powerful mechanical agent in weathering, as exemplified through "frost heave," a process whereby frost in soils or wedged in rock lifts the rock, as shown here in the frost-heaved blocky boulders of Shepard Formation in the Boulder Pass area of Montana's Glacier National Park. (U.S. Geological Survey)

sius), these vibrations decrease to about 10^6 vibrations per second. At that point, new covalent bonds are able to overcome the force of the hydrogen bond. The slowing of vibrations as water cools permits some rearrangement and some creation of open space to begin at just below 3.94 degrees, even though ice crystals do not form until 0 degree is reached. This accounts for the maximum density of water occurring at 3.94 degrees rather than at 0 degree Celsius.

When actual freezing occurs, water molecules arrange themselves into a covalently bonded crystal structure that has open space not present in liquid water. Water goes from a density of 0.99987 gram per cubic centimeter as liquid to 0.917 gram per cubic centimeter as solid ice. The pronounced decrease in density causes ice cover to form on the surface of lakes and rivers rather than at the bottom. The decrease in density is accompanied by an increase in volume; water expands as it freezes. When water seeps into cracks and pores of rock and soil and then freezes, the force generated by the expansion causes rock to break, soils to heave and swell, and small grains and crystals to spall away. Therefore, it is the hydrogen bond of water that ultimately permits water to become a powerful mechanical agent in weathering, as exemplified through frost heave in soils or frost wedging in rock.

Above 3.94 degrees Celsius, the density of water decreases with increasing temperature. At temperatures likely to be encountered in lakes and streams, the density of pure water can be closely approximated with a formula. The small changes in density that occur during heating, cooling, and freezing are responsible for seasonal circulation of water in freshwater lakes.

Water is unlike other substances in that its freezing point is lowered, rather than raised, by pressure, a result of a tendency at near-freezing temperatures for the covalent bonds of ice to collapse back to the hydrogen bonds of liquid water. Simple pressure is sufficient to enact the transition. When an ice skater places pressure on the blades of the skate, the pressure melts the ice at the sharp edges of the blade, and the skater glides on a thin layer of water that is created momentarily by the pressure. Such skating is not possible on other substances such as solid carbon dioxide (dry ice). This same liquefaction under pressure permits glacial ice to flow over a thin layer of water created at the base of the glacier and for this water to flow into joints and cracks, refreeze, and "pluck" out sections of the bedrock as the glacier moves.

ELECTROSTATIC BONDING

Water is a powerful solvent because the water molecule is polar. The magnetlike quality of the water molecule strongly influences substances that are bonded primarily by electrostatic (ionic) bonds. Ions bound into many solid minerals by virtue of the ionic bond may therefore be dissolved into water. In this way, water picks up dissolved materials from the soils and rock. The dissolving power provided by the polar nature of the molecules is further assisted by their vibration. As the molecules vibrate faster, they act like high-speed jackhammers against solid particles, and those solids held together with ionic bonds are particularly affected. Heating is one way to increase the vibration of molecules; therefore, fresh water tends to be a more effective solvent in warm climates.

Fresh waters contain dissolved solids and gases. The substances that are likely to occur in waters of this planet are those substances that are abundant and that occur in forms that are easily soluble. More than 99 percent of the Earth's crust can be accounted for by only twelve elements (oxygen, silicon, aluminum, iron, calcium, magnesium, sodium, potassium, titanium, hydrogen, manganese, and phosphorus—in order of abundance from greatest to least), and nearly 100 percent of the atmosphere by five elements (nitrogen, oxygen, argon, hydrogen as water vapor, and carbon as carbon dioxide).

The most common dissolved substances in fresh water are bicarbonate, calcium, sodium, magnesium, potassium, fluorine, iron, phosphate, sulfate, chloride, and dissolved silica. These substances enter natural waters both from the atmosphere and from the rocks and soils contacted by the waters. Even rainwater is not completely pure and

An example of frost heave, uplifted metasediments in the Yellowknife Supergroup, Northwest Territories of Canada. (Geological Survey of Canada)

contains small amounts of silica, sulfate, calcium, chlorine, nitrate, carbonic acid, sodium, potassium, iron, and aluminum.

Some elements are crustally abundant but are not abundant as dissolved species in fresh waters because they form insoluble compounds. For example, aluminum and iron are abundant elements, but in most waters they form nearly insoluble hydroxides—aluminum hydroxide and iron hydroxide. Silicon is extremely abundant, but it forms strong covalent bonds with oxygen that are not possible for water to break. In nature, silicon forms minerals that include silica-oxygen structures with very low solubility in water.

ACID AND ALKALINE WATERS

Although water is a good solvent because of the polar nature of the water molecule, it can be made an even more effective solvent if it can be rendered acidic. Chemists have devised the pH scale as a means for expressing the degree to which a solution is acid. The scale runs from 0 (most acid) to 14 (least acid), with a pH value of 7 termed neutral. Waters with a pH of less than 7 are termed acidic, and those with pH values greater than 7 are termed alkaline. Natural rainwater is acidic and has a pH of 5.6, because carbon dioxide in the atmosphere dissolves into the rainwater to make carbonic acid. (The atmosphere has a powerful effect on the composition of fresh water; the character of fresh water on this planet has changed through geologic time in accord with changes in the atmosphere.) Once rainwater strikes the ground, the weak acid is neutralized as it dissolves minerals in the rocks and soils. Therefore, even though natural rainwater is acidic, most fresh groundwater and surface waters are near neutral, with pH values between 6 and 8.5.

Unusually acid waters occur around volcanic vents, geysers, and fumaroles, where gases such as sulfur dioxide react with water to produce sulfuric acid. Acid waters also occur in bogs and marshes (pH 3.3-4.5), where carbon dioxide released during the decay of organic matter reacts with water to produce carbonic acid. The strongest acids occur where sulfide minerals such as pyrite and marcasite oxidize in the presence of air and water to produce sulfuric acid. Sulfuric acid is a strong acid, and its production through the weathering of sulfide minerals is so common around coal and metal mines that the waters released are termed acid mine drainage. Streams flowing from these sites often have pH values below 2. Thousands of miles of streams have become polluted by acid mine drainage. Alkaline waters occur in limestone terrains, where pH values between 8 and 9 are common. Extremely alkaline waters occur in playa lakes in contact with sodium carbonate or sodium borate, where pH values above 12 have been noted.

Extremely acid and extremely alkaline waters permit substances that remain immobile in normal waters to be dissolved. For example, aluminum is normally insoluble but does dissolve in both very acid and very alkaline waters. Acidic waters dissolve and transport metals such as iron and manganese and nonmetals such as phosphorus that would normally be present in most natural waters at low or undetectable levels.

ACIDITY VS. PH

A clear distinction needs to be made between pH and acidity, because these terms are too often misused interchangeably. In a large tank of pure distilled water at pH = 7, if a few drops of strong acid (a substance that releases free positively charged hydrogen ions into solution) are added, a low pH, perhaps pH = 3 throughout the tank, will soon be registered. If a few drops of strong base (a substance that releases free negatively charged hydroxide ions into solution) are added, however, the pH will easily rise back to 7. The measure of the amount of base required to get the tank to a pH of 7 is a measure of acidity. The tank had a low pH, but yet also a low acidity. Suppose another tank contains water with a considerable amount of dissolved solids and has a pH of 6. It may take many gallons of strong base to get the tank up to a pH of 7. In the second tank, then, there is water with a moderate pH but a very high acidity. In short, pH is the measure of the concentration of hydrogen ion present at a given time, but acidity is a measure of the amount of base needed to change the pH back to neutral. Natural waters with high acidity are usually high in dissolved solids and have a low pH. Acid mine drainage is an example.

AQUEOUS GEOCHEMISTRY

Freshwater chemistry is a part of the field of aqueous geochemistry, or low-temperature geochemistry, which is a specialty field of many geologists. Proper collection, analyses, and interpretation of data from natural waters require more than a knowledge of chemistry. Success also requires knowledge about the natural environmental system, and this is why hydrology and aqueous geochemistry fall more properly within the province of geology than that of pure chemistry.

A typical water analysis includes determination of the levels of silica, aluminum, iron, manganese, copper, calcium, magnesium, strontium, sodium, potassium, dissolved oxygen, carbonate, nitrate, bicarbonate, sulfate, chloride, fluoride, phosphate, arsenic, selenium, boron, total dissolved solids, pH, acidity or alkalinity, temperature, and conductivity. If waters are polluted with unusual substances such as pesticides, sewage, or industrial wastes, specialized tests must be undertaken for these.

As soon as a sample is taken, the water is removed from its actual environment, and changes start to occur. Dissolved gases may leave, precipitates may form, and bacteria and microscopic algae may metabolize substances or die and release substances that were formerly solids. Because some parameters change so quickly and easily, some tests must be done immediately in the field. The temperature, conductance, pH, dissolved oxygen, and sometimes acidity or alkalinity are measured in the field. Small battery-operated instruments such as specific ion meters (for nitrate, sulfate, carbonate, and a number of elements), colorimeters, and conductivity meters are made by several manufacturers to be used in field analyses.

Pure water is a poor electrical conductor, but water's ability to conduct electricity increases markedly with the amount of dissolved solids. Therefore, a conductance test in the field is a rapid method to use to make a rough estimate of the amount of total dissolved solids. Sometimes rapid colorimetric tests for sulfate, phosphate, and some metals are done in the field, but more often the analyses of those substances are done in the laboratory.

Samples to be taken to the laboratory are filtered in the field to remove any microscopic suspended solids, because it is important that the water analyses show only dissolved substances. After the water is filtered, it is acidified with a few milliliters of high-purity acid to ensure that all dissolved species remain dissolved. The water is then put into a plastic container and filled to the top so as not to admit any air. The samples are usually placed in a dark cooler and kept refrigerated until they are analyzed in the laboratory.

DANGER OF CONTAMINANTS AND POLLUTANTS

Of the vast amount of water present on the planet's surface, only about 2.5 percent is fresh water; the remainder is in oceans and inland seas. Of that 2.5 percent, 2.15 percent is locked away in glaciers and polar ice caps. About another 0.3 percent of the total is accounted for by fresh groundwater. All the surface water commonly seen in freshwater lakes and streams amounts to only .009 percent of the Earth's total water.

The small percentages do not reveal the true importance of fresh water. Fresh water is the major sculpting agent of the planet's land surface, and it is an essential substance for all terrestrial life-forms. Changes in the chemical composition of fresh water constitute one of the most sensitive indicators of changes in the environment. Changes may be local, as exemplified by the eutrophication

of a small lake that receives phosphorus from a few local septic tanks, or they may be global, as exemplified by the acidification of sensitive lakes in northern Europe and Canada as a result of the burning of fossil fuels, which releases carbon dioxide and oxides of nitrogen and sulfur into the atmosphere.

The category "fresh water" includes a group of waters from wells, lakes, and streams with diverse chemistry. Even natural substances in excessive amounts pose problems for water consumption or industrial use. In addition to the toxic content of acid mine drainage, excessive amounts of nitrate, sodium, and fluoride present health hazards. Excessive amounts of calcium and magnesium promote buildup of scale in water tanks and boilers, and waters that are high in these elements (hard waters) must have these constituents chemically removed with a water softener prior to domestic and industrial use. High phosphate contents of waters promote algal blooms and eutrophication. High boron content is important in irrigation, because boron is toxic to many crops. Of all natural resources, fresh water ranks as one of the most essential to human survival. Continual monitoring of the chemistry of fresh waters is essential to ensure that these valuable resources are not rendered unsuitable by human-made contaminants and pollutants.

Edward B. Nuhfer

CROSS-REFERENCES

BIBLIOGRAPHY

Berner, Elizabeth K., and Robert A. Berner. *Global Environment: Water, Air, and Geochemical Cycles.* Upper Saddle River, N.J.: Prentice Hall, 1996. This book offers a clear and readable introduction to the processes that sustain life and effect change on the Earth, including a useful section on aquatic geochemistry. Color illustrations and maps.

Drever, James I. *The Geochemistry of Natural Waters.* Englewood Cliffs, N.J.: Prentice-Hall, 1982. The book is intended for students in advanced courses in geochemistry or water chemistry and also for professionals working in the area of water chemistry. Thermodynamic tables, references, and an index are included. A particular strength of the book is its inclusion of case study examples, particularly in the chapters on weathering and water chemistry.

Faust, Aly. *Chemistry of Natural Waters.* Stoneham, Mass.: Butterworth Publishers, 1981. Chapter 1, "Chemical Composition of Natural Waters," of this useful reference book - requires little chemical or mathematical background. Later chapters cover thermodynamics, equilibria, reactions, and models. The final chapter deals with toxic metals in the aquatic environment.

Greenberg, Arnold, ed. *Standard Methods for the Examination of Water and Wastewater.* 16th ed. Washington, D.C.: American Public Health Association, 1985. This reference has been the major reference for water chemists. It is the essential "cookbook" for water collection and water analysis laboratory procedures.

Hem, J. D. *Study and Interpretation of the Chemical Characteristics of Natural Water.* 3d ed. U.S. Geological Survey Water Supply Paper 2254. Washington, D.C.: Government Printing Office, 1985. Written by one of the world's foremost water chemists, the usefulness of this reference has proved itself since the first edition appeared in 1959. In addition to including a condensed discussion of chemical thermodynamics, the writer provides many analyses of unusual water types and discusses, component by component, the common constituents of natural water. This section is virtually a water chemist's tour of the periodic table. The concluding sections deal with interpretation and presentation of analytical data and with the relationship of water quality to water use. Well indexed and contains an excellent list of references.

Krauskopf, Konrad B. *Introduction to Geochemistry.* 2d ed. New York: McGraw-Hill, 1979. One of the best-written texts produced for students. Chapters 1, 2, 3, 4, 8, 9, 10, and 12 provide an excellent introduction to water chemistry, particularly with respect to equilibrium and chemical thermodynamics. The book is designed as a text for undergraduate students in geochemistry but may be understood by students with good high school chemistry and Earth science courses. Includes problem sets at the end of each chapter, good references, and tables of thermodynamic properties that are especially pertinent to water chemists.

Langmuir, Donald. *Aqueous Environmental Geochemistry.* Upper Saddle River, N.J.: Prentice Hall, 1997. A thorough look into aqueous geochemistry, this book examines the geochemical cycles and processes of water systems and their interactions with other cycles on the Earth. Suitable for the careful high school reader or college student.

Nicholson, Keith. *Geothermal Fluids: Chemistry and Exploration Techniques.* Berlin: Springer-Verlag, 1993. Nicholson provides the reader with an examination of the behavior of geothermal fluids, as well as study and exploration practices. Slightly technical, this book is intended for the reader with some background in chemistry.

Sparks, Donald L., and Timothy J. Grundl, eds. *Mineral-Water Interfacial Reactions: Kinetics and Mechanisms.* Washington, D.C.: American Chemical Society, 1998. This collection of essays deals with the ongoing chemical reactions and processes that occur in aquatic systems. A technical piece intended for the person with a background in chemistry or Earth sciences.

Stumm, Werner, and James J. Morgan. *Aquatic Chemistry: An Introduction Emphasizing Chemical Equilibria in Natural Waters.* 2d ed. New York: John Wiley & Sons, 1981. This reference is indispensable for limnologists and aqueous geochemists. Thermodynamic principles are stressed from the outset, but a wealth of examples from natural systems are provided. References, a subject and author index, and a good set of thermodynamic tables make the book appropriate as a reference and as a college text.

Todd, D. K. *The Water Encyclopedia.* Port Washington, N.Y.: Water Information Center, 1970. A compilation of hundreds of pages of useful tables about water properties, water chemistry, and bodies of water of the world.

GEOCHEMICAL CYCLE

The geochemical cycle describes the movement of and changes in elements through the Earth's atmosphere, surface waters, biosphere, crustal sediments and rocks, and upper mantle. Human activities have altered the earth-surface part of the geochemical cycles of many elements, such as carbon, sulfur, nitrogen, and phosphorus.

PRINCIPAL TERMS

BIOSPHERE: all living organisms, including plants and animals

CARBONATE ROCKS: sedimentary rocks composed mainly of carbonate minerals whose structure includes a carbon atom linked to three oxygen atoms

CRUST: the outermost layer of the Earth; the continental crust is between 30 and 40 kilometers thick, of dominantly silicon-rich igneous rocks, metamorphic rocks, and sedimentary rocks, while the oceanic crust is only 5 kilometers thick, of magnesium and iron-rich rocks such as basalt

FLUX: the rate of transfer of an element from one reservoir to another

HYDROSPHERE: the waters of the Earth, including rivers, lakes, and oceans

LITHOSPHERE: the outer, rigid part of the Earth, consisting of crustal rocks and the upper mantle

PLATE TECTONICS: very slow movement of sections of the Earth's lithosphere, called plates, away from one another in some areas and toward one another in other areas

RESERVOIR: a place on or in the Earth where an element remains for a period of time

SEDIMENTARY ROCK: a rock formed by the consolidation of loose sediments deposited at the Earth's surface by water, air, or ice or precipitated from solution

SILICATE ROCKS: rocks containing silicate minerals, whose structure contains silicon linked to four oxygen atoms

UPPER MANTLE: the fairly rigid part of the Earth's interior below the crust of the Earth down to about 700 kilometers, composed of magnesium- and iron-rich rock

EARTH-SURFACE CYCLE

Geochemical cycles refer to the transfer of various chemical elements between different reservoirs on the Earth. The largest reservoirs include the atmosphere, hydrosphere, biosphere, and lithosphere (crustal rocks and sediments and the upper mantle). The basic geochemical cycle, which is often referred to as the rock cycle, has a time scale of hundreds of millions to billions of years. Subcycles within the rock cycle involve only certain of these reservoirs or small parts of them and occur on a shorter time scale. The atmospheric-hydrologic-biological-sedimentary cycle which operates on the Earth's surface will be referred to as the earth-surface cycle and has a time scale of 10 million to 100 million years. Within the earth-surface cycle are many smaller cycles, such as the oceanic cycle, which takes hundreds of years, or the biological cycle, which varies seasonally. The rock cycle is continuous, including an earth-surface cycle and a deeper, subsurface cycle. The earth-surface part of the cycle is driven by the atmospheric-hydrologic cycle, and elements are transported by water or as atmospheric gases. The biosphere, or plants and animals living on the land and in the oceans, is strongly involved in the chemical changes which occur on the Earth's surface, and it is also greatly affected by them. Geochemical cycles which involve strong biogenic interaction are often referred to as biogeochemical cycles.

Rain falling on the Earth surface erodes rock and soil and carries fragments of them to rivers. The rain, which contains atmospheric gases and dissolved chemicals, also chemically reacts with rocks and minerals in the soil in a process known as weathering. As a result, some elements become dissolved in water and new minerals are formed. Plants and bacteria living in the soil are also involved in weathering, and gases are exchanged with the atmosphere. Certain elements are taken up from the atmosphere and soil solutions and stored within the biota (the plants and animals of a region), and others are cycled by the biota and released to waters and the atmosphere. Rivers transport dissolved chemicals resulting from weathering and biological activity, as well as sus-

pended solid particles of rocks, soils, and organic matter. Some of this material may be deposited on the land surface, but most of it is ultimately carried by rivers to the ocean.

INTERNAL OCEANIC CYCLE

Upon reaching the ocean, inorganic and organic particles which have been suspended in wa-

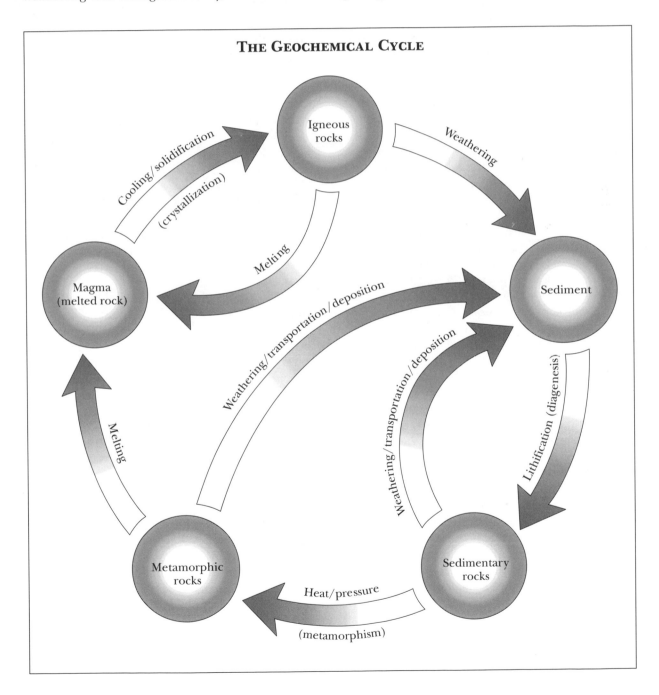

THE GEOCHEMICAL CYCLE

ter are deposited on the ocean floor. These particles make up most of the sediment on the bottom. Dissolved chemicals in the oceans along with atmospheric gases may be taken up by organisms living in the ocean surface waters, where there is light, to make cellular matter and shells. The shells accumulate on the sea floor and eventually form deposits of calcium carbonate or limestone. When the surface organisms die, most of the cellular organic matter is broken down and rapidly recycled. A small part of the organic matter falls into the deeper oceans, however, which have been isolated from the ocean surface for thousands of years. Here, organic matter may be broken down into chemicals and gases and slowly recycled to the surface. Thus, there is an internal oceanic cycle between the bottom and surface ocean waters.

Part of the organic matter which reaches the bottom survives and is buried in oceanic sediments, with a small fraction of it ultimately transformed to petroleum upon burial. Some dissolved chemicals, such as sodium chloride salt, can precipitate directly out from shallow ocean water to form sediments. Water is evaporated and other gases are released from the ocean surface into the atmosphere, where they are transported back to land and rained out again. Thus, the earth-surface cycle involves rain, erosion, weathering, river transport, oceanic sedimentation, evaporation, and atmospheric transport of water and gases.

SUBSURFACE PART OF CYCLE

The subsurface part of the rock cycle begins when oceanic sediments are buried, compressed, and lose water to form hard sedimentary rocks. These sedimentary rocks may be then uplifted to form part of the land, where they are eroded and begin the earth-surface cycle again. Other sedimentary rocks become involved in plate tectonics at the edges of continents, are buried to greater depths, and are subjected to heat and pressure to form metamorphic rocks. If the rocks are heated enough, they melt and are mixed with deeper rocks to form molten rock or magma, which then moves toward the Earth's surface. Some magma erupts on the land or sea floor to form volcanic rocks. Other magma cools beneath the surface to form crystalline igneous rocks such as granite. Gases escape from the magma and are released to the atmosphere. Deeply buried igneous and meta-

morphic rocks may be uplifted in the rock cycle into mountain ranges, where they become part of the earth-surface cycle and are eroded. Thus, over the history of the Earth, the chemical elements which make up rocks have been through many cycles of erosion and weathering, river transport, oceanic deposition, burial, metamorphism, melting, and uplift.

The rock cycle has apparently been operating for most of the age of the Earth. Over a long period of time, the rock cycle is roughly in steady state; in other words, the total amount of chemicals within the system remains constant although the locations of these elements vary. Sedimentary rocks have been found in Greenland that are 3.8 billion years old and were formed by the same sort of geochemical cycling which occurs today. The major changes between that time and the present are the evolution of life from a very primitive form, the gradual cooling of the Earth, an increase in the amount of oxygen in the atmosphere, and the growth in the size of the continents. These changes are unidirectional, and the rock cycle is essentially superimposed upon them.

CARBON CYCLE

Cycles of individual elements are included within the rock cycle, for example, the carbon cycle or the sulfur cycle. The importance and size of any reservoir, the size of the fluxes between reservoirs, and the amount of chemical change that occurs in a reservoir vary from element to element. The carbon cycle is a good example of an element cycle. Considering the earth-surface carbon cycle first, the atmosphere contains carbon in the form of carbon dioxide gas. Although the amount of carbon contained in the atmosphere is not very great, atmospheric carbon dioxide is an important reservoir because it provides a means of exchanging carbon between the other reservoirs, affects climate via the "greenhouse effect," and is part of the air that we breathe. The forests and terrestrial plants and animals also make up a carbon reservoir. Plants use carbon dioxide from the atmosphere to produce organic matter through the process of photosynthesis. When organic matter is broken down in a yearly biological cycle, carbon dioxide is released back to the atmosphere. Deforestation, the cutting and burning of tropical forests by humans, releases excess stored plant carbon to the atmosphere as carbon dioxide. Another large terrestrial carbon reservoir is carbonate rocks, such as limestone, and shales, or rocks that contain organic carbon. When carbonate rocks and other rocks containing silicate minerals, such as feldspar, are weathered, atmospheric carbon dioxide is taken up, and dissolved carbon in the form of bicarbonate is released to rivers. Rivers transport dissolved bicarbonate and organic carbon, from plant and animal life which has not been decomposed, to the oceans. Terrestrial carbon also occurs in fossil fuels, such as coal, oil, and gas. When humans burn fossil fuel for energy, carbon dioxide is released to the atmosphere and the cycling of carbon is speeded up.

The oceans are another important earth-surface carbon reservoir. Carbon occurs as dissolved bicarbonate ion and as dissolved carbon dioxide gas. The upper part of the ocean rapidly exchanges carbon dioxide with the atmosphere. Small, floating marine plants living in the surface oceans produce their organic matter by taking up atmospheric carbon dioxide through photosynthesis. When they die, most of their remains are decomposed by bacteria, and carbon dioxide is recycled to the atmosphere. A small part of this organic matter, however, is not destroyed. It falls to the bottom of the ocean to be recycled into bottom waters or buried in oceanic sediments along with organic matter carried into the ocean from rivers. Marine organisms also secrete shells and hard parts, removing bicarbonate from ocean water to form calcium carbonate and releasing part of the carbon to the atmosphere as carbon dioxide. Corals are an example of calcium carbonate secretion. Much of this carbonate material is redissolved at depth, thus recycling bicarbonate to ocean water again, but part of the calcium carbonate becomes buried in ocean sediments.

The long-term part of the carbon cycle occurs over millions of years, when the organic carbon and calcium carbonate buried in oceanic sediments are converted into rock through burial, compaction, and heating. Oceanic muds containing organic carbon are converted into shales. Sediments composed of calcium carbonate become limestone rock. When limestones and organic carbon are heated, metamorphosed, and melted at depth, carbon dioxide is released. Part of this carbon dioxide escapes to the atmosphere from vol-

canoes on the land, and some is released at midoceanic ridges. Ultimately, limestone rocks containing calcium carbonate and shales containing organic carbon are uplifted to the land surface to be weathered again, beginning the earth-surface cycle anew.

LONG AND SHORT TIME SCALE PROCESSES

In order to study geochemical cycles, geochemists combine information from various specialities, such as geology, geochemistry, geophysics, oceanography, biology, hydrology, and atmospheric sciences. They must have a broad knowledge of the long and short time scale geologic processes involved in the cycling of an element and an idea of how these processes interact on the Earth's surface with the atmosphere and biosphere. They must know the major reservoirs for various elements, the concentrations of elements within these reservoirs, processes that change the elements within the reservoirs, and the fluxes of elements between various reservoirs. The relative importance of various processes and fluxes depends on the time scale involved. Processes that are important on a short time scale are often unimportant in long-term cycling.

COMPUTER MODELING

Computer modeling is one technique that is used in the study of geochemical cycles, particularly those that involve many reservoirs and fluxes. With the carbon cycle as an example, it can be seen what type of information is needed to model geochemical cycles. The earth-surface carbon cycle has a time scale of decades to tens of thousands of years. Information about the present size of the atmospheric reservoir of carbon, in the form of carbon dioxide, comes from atmospheric science. Measurements are made of the average yearly concentration of carbon dioxide in the atmosphere and how this concentration is increasing with time. Knowledge of the amount of atmospheric carbon dioxide over the last thousands to tens of thousands of years can be obtained from the analysis of air bubbles trapped in glacial ice, which is sampled from below the surface by ice cores. Estimates of the size of the reservoir of carbon stored in the form of forests and terrestrial biota and the amount of forests that are being destroyed, with their carbon recycled to the atmosphere, come from biologists. The amount of carbon stored as fossil fuel and the amount being burned each year, releasing carbon dioxide to the atmosphere, is estimated by economic geologists, particularly those who study petroleum and gas reserves and coal reserves. The oceans are an important reservoir in the carbon cycle. The amount of carbon dioxide being taken up from the atmosphere by the surface ocean and how fast it is being transferred to the deeper oceans are important fluxes, which come from a knowledge of ocean chemistry provided by oceanographers. All this information is used by geochemists to study the short time scale carbon cycle and to predict increases in the concentration of atmospheric carbon dioxide.

MODELING LONG TIME SCALE CARBON CYCLE

In modeling the long time scale carbon cycle over millions of years, different reservoirs and fluxes become important. Geochemists use information about past conditions on the Earth, preserved in sedimentary rocks. In order to estimate the amount of carbon stored in sedimentary rocks, a geochemist must know the average carbon content of different types of sedimentary rocks, such as carbonates and shales, and also the abundance of various sedimentary rock types as a function of age, both those rocks exposed at the Earth's surface and those at depth in the subsurface. This information comes from compilations of vast amounts of data. The average chemical composition and amount of water being carried by world rivers gives a measure of how much weathering is occurring on land. Information about ancient oceans is preserved in the ratio of isotopes, or heavier and lighter forms of different elements, that are found in rocks formed from these oceans. For example, the ratio of the isotopes carbon 13 and carbon 12 in marine carbonates gives an indication of how much organic carbon was being buried in the ocean millions of years ago. Fossil records of past life preserved in the rocks give an idea of past climate conditions. For example, certain types of organisms, such as crocodiles, are favored by a warm climate; therefore, their presence in polar regions in the past suggests higher levels of atmospheric carbon dioxide, which tends to raise Earth temperatures especially at high latitudes. From modeling the carbon cycle using in-

formation of the type discussed above and other data, it has been concluded that changes in the concentration of atmospheric carbon dioxide have probably occurred on the time scale of the rock cycle, that is, millions of years. The important fluxes involved are the uptake of atmospheric carbon dioxide in the weathering of silicate rocks and in the formation of organic carbon, the burial of limestone and organic carbon in sediments, and carbon dioxide release from volcanoes and midoceanic ridges.

HUMAN ALTERATION OF GEOCHEMICAL CYCLES

Human activities such as fossil fuel burning, industry, and agriculture have altered the earth-surface part of some of the more important geochemical cycles, particularly the carbon, sulfur, and phosphorous cycles. For example, there have been large changes in the transport rate of sediments in streams, and excess carbon dioxide and sulfur dioxide have been released to the atmosphere by human activity. It is important to know the natural fluxes within these element cycles in order to determine how important human changes are and what their overall effect may be. In trying to correct human changes in our environment, it is important to understand the connections between different parts of the geochemical cycle. By studying the long-term rock cycle we can see how large changes were in the geologic past and how they were stabilized within the cycle. This gives us an idea of what the range of fluxes has been over the Earth's history and how the Earth reacts to change.

The short-term geochemical cycling of carbon dioxide gas within the earth-surface carbon cycle on a time scale of years to hundreds of years has been of particular interest recently because of the observed increase in the concentrations of atmospheric carbon dioxide and concern about the "atmospheric greenhouse effect." Both burning of fossil fuel and deforestation by humans release carbon dioxide to the atmosphere. Some of this excess carbon dioxide is taken up by the oceans and stored there. In order to predict how fast the concentration of atmospheric carbon dioxide will change over time, it is necessary to know the fluxes of carbon between the various reservoirs. Increased carbon dioxide should lead to increased temperatures on the Earth, the so-called atmospheric greenhouse effect. Long-term changes in the concentration of carbon dioxide have probably also occurred and affected past climates on the Earth. These changes can be deduced from the long-term carbon cycle.

Humans have also greatly altered the geochemical phosphorus cycle. Phosphorus is an important nutrient, a food source for plants and animals. The release of phosphorus by humans in sewage and from agriculture and industry has increased plant growth in lakes and in estuaries. When these plants die and decay, oxygen in the water is often completely used up. Both the increased plant growth and the oxygen depletion make lakes and estuaries less usable by humans.

ACID RAIN

Another topic of current interest which involves geochemical cycles is acid rain. Acid rain speeds up the weathering part of the geochemical cycle, releasing more and different ions into solution. It also changes the chemistry of and alters the geochemical cycle in lakes. The primary gases which cause acid rain are sulfur dioxide, which forms sulfuric acid, and nitrogen dioxide, which produces nitric acid. Thus, a knowledge of the at-mospheric-Earth surface geochemical sulfur cycle and the nitrogen cycle are important in understanding acid rain and its effects. Sulfur dioxide gas is produced primarily by fossil fuel burning. From studying the sulfur cycle, it is known that the amount of sulfur released into the atmosphere from fossil fuels is about equal to that formed by natural processes. Thus, humans are greatly altering the sulfur cycle. Nitrogen dioxide is produced mainly by fossil fuel burning in cars and in power plants. Another important source is forest burning, particularly to clear land in the tropics. Human sources of nitrogen dioxide add up to nearly three-quarters of the total amount released into the atmosphere. Clearly, that indicates an alteration of the geochemical nitrogen cycle.

Elizabeth K. Berner

CROSS-REFERENCES

BIBLIOGRAPHY

Berner, Elizabeth K., and Robert A. Berner. *Global Environment: Water, Air, and Geochemical Cycles.* Upper Saddle River, N.J.: Prentice Hall, 1996. This book discusses the processes that sustain life and effect change on the Earth. Topics include the hydrologic cycle; the atmosphere; atmospheric carbon dioxide and the greenhouse effect; acid rain; the carbon, sulfur, nitrogen, and phosphorus cycles; weathering; lakes; rivers; and the oceans. It is understandable to the college-level reader.

Berner, Robert A., and Antonio C. Lasaga. "Modeling the Geochemical Carbon Cycle." *Scientific American* 260 (March, 1989): 74-81. This article discusses the modeling of the carbon cycle over millions of years and should be understandable to the general reader. Changes that may have occurred in atmospheric carbon dioxide are discussed, along with a discussion of the factors important in regulating atmospheric carbon dioxide over geologic time. The illustrations, particularly of the carbon cycle, are very helpful.

Davidson, Jon P., Walter E. Reed, and Paul M. Davis. *Exploring Earth: An Introduction to Physical Geology.* Upper Saddle River, N.J.: Prentice Hall, 1997. An excellent introduction to physical geology, this book explains the composition of the Earth, its history, and its state of constant change. Intended for high-school-level readers, it is filled with colorful illustrations and maps.

Garrels, Robert M., and Fred T. Mackenzie. *Evolution of Sedimentary Rocks.* New York: W. W. Norton, 1971. This book discusses the geochemical cycle of sedimentary rocks over a long time scale. It is a college-level textbook and has detailed information about many parts of the rock cycle, particularly in chapters 4 through 11. Chapter 10 summarizes the geochemical cycling of sedimentary rocks.

Gregor, C. Bryan, Robert M. Garrels, Fred T. Mackenzie, and J. Barry Maynard, eds. *Chemical Cycles in the Evolution of the Earth*. New York: John Wiley & Sons, 1988. The prologue gives a summary for the college-level reader of the historical development of the concepts of geochemical cycling. The rest of the book covers other aspects of geochemical cycling in more detail than would interest the nonspecialist.

Lutgens, Frederick K., and Edward J. Tarbuck. *Earth: An Introduction to Physical Geology.* 6th ed. Upper Saddle River, N.J.: Prentice Hall, 1999. This college text provides a clear picture of the Earth's systems and processes that is suitable for the high school or college reader. In addition to its illustrations and graphics, it has an accompanying computer disc that is compatible with either Macintosh or Windows. Bibliography and index.

Nicholson, Keith. *Geothermal Fluids: Chemistry and Exploration Techniques*. Berlin: Springer-Verlag, 1993. Nicholson provides the reader with an examination of the behavior of geothermal fluids, as well as study and exploration practices. Slightly technical, this book is intended for the reader with some background in chemistry.

Siever, Raymond. "The Dynamic Earth." *Scientific American* 249 (September, 1983): 46-55. This article discusses for the general reader the geochemical cycle in the context of the history of the Earth. Topics covered include geologic time, the rock cycle, the earth-surface cycle, the carbon cycle, and atmospheric carbon dioxide over geologic time. There is also a discussion of evolutionary changes in the Earth over time. Excellent illustrations.

Trabalka, J. R., ed. *Atmospheric Carbon Dioxide and the Global Carbon Cycle*. Washington, D.C.: Government Printing Office, 1985. This reference provides a summary of the global carbon cycle and how it is related to changes in atmospheric carbon dioxide. It also discusses current research on these subjects. The Executive Summary and chapter 1, "The Global Cycle of Carbon," are good summaries for the college-level reader.

Woodwell, George M. "The Carbon Dioxide Question." *Scientific American* 238 (January, 1978): 34-43. This article discusses deforestation as a source of atmospheric carbon dioxide and shows how this flux is estimated. It also summarizes the present global carbon cycle and how excess carbon dioxide produced by humans can be taken up. For the general reader. The illustrations are excellent.

GEOTHERMOMETRY AND GEOBAROMETRY

Geothermometry and geobarometry use the difference in chemistry of minerals that exist together to estimate the temperature and pressure at which some geological processes occur. The estimation of temperatures and pressures is crucial to the understanding of such complex phenomena as the melting of the rocks that ultimately result in volcanic eruptions on the Earth's surface and the mechanical properties of rocks that are involved in ruptures that result in earthquakes.

PRINCIPAL TERMS

COMPONENT: a chemical entity used to describe compositional variation within a phase

IGNEOUS ROCK: any rock that forms by the solidification of molten material, usually a silicate liquid

METAMORPHIC ROCK: any rock whose mineralogy, mineral chemistry, or texture has been altered by heat, pressure, or changes in composition; metamorphic rocks may have igneous, sedimentary, or other, older metamorphic rocks as their precursors

MINERAL: a naturally occurring solid compound that has a specific chemical formula or range of composition; minerals normally have regular crystal structures such that their internal arrangement of atoms is predictable

PHASE: a chemical entity that is generally homogeneous and distinct from others in the system under investigation

PHASE EQUILIBRIA: the properties of chemical systems described in terms of classical thermodynamics; systems of specified composition are generally investigated as a function of temperature and pressure

THERMODYNAMICS: the area of science that deals with the transformation of energy and the laws that govern these changes; equilibrium thermodynamics is especially concerned with the reversible conversion of heat into other forms of energy

VOLCANIC ROCK: a type of igneous rock that is erupted at the surface of the Earth; volcanic rocks are usually composed of larger crystals inside a fine-grained matrix of very small crystals and glass

BASIC PRINCIPLES OF CHEMICAL THERMODYNAMICS

Geothermometry and geobarometry are methodologies used by geologists to determine the temperature and pressure attending igneous and metamorphic processes. The temperatures below the Earth's surface at which magma (molten rock) is generated and subsequently crystallizes (solidifies) is an example of one type of information that may be obtained through geothermometry and geobarometry. Another is the temperature at which the constituent mineral phases of large masses of sediments are metamorphosed to different compositions or structures. In order to understand how geologists have developed the methodologies of geothermometry and geobarometry, one needs to understand some of the basic principles of chemical thermodynamics and how these principles are applied to rocks.

Temperature, along with pressure and bulk composition, largely controls the macroscopic physical and chemical properties of most materials. Based on experience, scientists know that certain substances behave in predictable ways under some specified set of conditions. The substance may be as simple as a glass of water or a single grain of homogeneous iron metal, or it may be as complex as an igneous or metamorphic rock. For example, at atmospheric pressure (the pressure of the atmosphere at sea level, or 14.69 pounds per square inch), pure water will become solid ice at a temperature of 0 degrees Celsius (its freezing point) and gaseous steam at a temperature of 100 degrees Celsius (its boiling point). The behavior outlined above, however, may not be commonly observed, because pure water is rarely encountered in communities or homes. In the colder regions of the United States and Canada, many communities spread sodium chloride (common household salt) on the streets during the winter months. The introduction of salt to a pure water system changes its properties such that at atmospheric pressure, the water freezes at a temperature below 0 degrees Celsius. The new freezing point is a function of the salted water's composition and pressure. The addition of small amounts of salt to pure liquid water also raises its boiling point to a temperature above 100 degrees Celsius that, again, depends on the amount of salt and pressure. Many recipes in which food is cooked in boiling water call for small amounts of salt, allowing the liquid to attain a higher temperature, and hence heat content, before reaching its boiling point.

PHASE EQUILIBRIA

In order to discuss these concepts in greater detail, one may define a system that is the portion of the universe under investigation. In the first example described above, the system is the ice and liquid water on a road surface, where water existed in one of two distinct phases, either solid or liquid, depending on the temperature. In the ice-water system, salt spread on the road will primarily dissolve into the water phase, changing its composition from pure water to a saltwater solution. The composition is fixed by the amount of salt relative to the amount of water present, which is defined as the salt concentration in solution and is generally expressed in units of grams per liter. When the system obtains its minimum energy configuration for the specified set of conditions, it is at equilibrium.

For fixed temperature and pressure, this is characterized by a minimum in the Gibbs free energy function, which balances the heat content of the system against energy unavailable to the system to perform work on its surroundings. The type of work may be mechanical—for example, pushing a piston as a result of the rapid expansion of a gas—or electrochemical—for example, oxidation and reduction of chemical species at the positive and negative electrodes of a battery to produce an electric current. Gibbs formalism relies on the principle of chemical equilibrium, which allows scientists to relate changes in a chemical system, such as the ice-water-steam examples outlined above, to specific thermodynamic properties. The principle of chemical equilibrium states

that the chemical potential of each component in every phase in the system must be equal. Chemical potentials are used by scientists to relate the reactivity of individual chemical species at some specified temperature and pressure. At equilibrium, then, the component chemical potentials are equal and the total Gibbs free energy of the system under consideration is at a minimum. As a consequence, one would expect to observe no net change in the status of the system. A description of a well-defined chemical system in terms of classical thermodynamics is referred to as the study of the phase equilibria of that system.

When scientists analyze phase equilibria in terms of thermodynamics, they develop relationships between the composition of constituent phases and the temperature and pressure conditions under which they formed. In the salt-water system described above, if one could experimentally determine the relationship between the amount of freezing point depression and the concentration of salt in solution, one could then easily predict the temperature at which the road surface would freeze, given the amount of salt added to the system. The appropriate relationships applied to systems of geologic interest are commonly referred to as geothermometers, if they are used to determine temperature, or geobarometers, if they are used to determine pressure.

APPLICATION OF THERMODYNAMICS TO PETROLOGY

For the last 250 years, scientists from a variety of disciplines have carefully gathered experimental data in order to characterize the behavior of many materials of variable composition under a wide range of temperature and pressure conditions. These experimental data may be analyzed using a formalism called classical thermodynamics, which was largely developed by J. Willard Gibbs in the 1870's, based on the pioneering experiments and ideas of S. Carnot and E. Clapeyron. In the 1910's, N. L. Bowen, a geologist, pioneered the application of classical thermodynamics to the study of rocks. Bowen and his coworkers began a long and productive career of careful experimental characterization of geological systems. During this period, they developed experimental apparatus and techniques that are still in use presently with only moderate modification. Bowen mainly focused on examining the phase equilibria of silicate systems comprising two to three oxide components. Many of his original experimental investigations form the basis of modern igneous petrology.

Although Bowen and his coworkers did extensive experimental work to characterize the phase equilibria of many important geologic systems, the development and application of the methodologies of geothermometry and geobarometry did not gain wide acceptance until the 1960's. During the 1960's and 1970's, four common types of chemical reactions were developed and applied to rocks as both geothermometers and geobarometers: solid-solid exchange reactions, solid-liquid exchange reactions, solid-gas buffer reactions, and stable isotope exchange reactions. Each type of reaction is based on slightly different thermodynamic data and therefore has a somewhat different application to geological systems.

SOLID-SOLID GEOTHERMOMETRY AND GEOBAROMETRY

Solid-solid geothermometry or geobarometry is based on the exchange of one or more chemical components between coexisting solid phases. In most igneous and metamorphic rocks, silicate phases (solids whose structure and properties are dominantly controlled by the presence of silicon, usually coordinated by four oxygens at the apexes of a tetrahedron) represent more than 90 percent of the rock's weight. The other 10 percent is composed of oxide phases (solids whose structure and properties are dominantly controlled by oxygen, usually in a close-packed arrangement). The exchange of a magnesium component between two similar silicate phases, clinopyroxene and orthopyroxene, is the basis of a commonly applied geothermometer. The exchange of an aluminum component between orthopyroxene and garnet (another common silicate mineral) is the basis of a commonly applied geobarometer.

There are several prerequisites for the application of either geothermometer or geobarometer to rocks. First, both phases must be present in the system of interest. Second, the two phases must be in demonstrable chemical equilibrium. Often, that is quite difficult to prove; however, evidence of disequilibrium is usually discernible in the form of compositionally zoned phases. Third, the thermodynamic properties of each of the phases must

be known. Fourth, the composition of the two phases must be determined by chemical analysis. Last, to apply the geothermometer, one needs an independent determination of the pressure at which the two pyroxene phases formed, because the thermodynamic relations for the two pyroxenes are a function of pressure. Often, a geologist would use a geobarometer to obtain an estimate of the pressure. Concomitantly, to apply the geobarometer, one needs an independent determination of the temperature. Armed with these data, one may easily calculate the temperature at which the pyroxenes formed.

GEOTHERMS

Only the uppermost 10-20 kilometers of the Earth's continental crust have been directly sampled by drilling; direct observation of the oceanic crust is even less extensive because of the extreme difficulty and expense of drilling from the ocean's surface through 5 kilometers of water prior to reaching the oceanic crust. There often are pieces of rock called xenoliths, however, which are accidentally entrained in magmas that originate from depths of 5 to as great as 200 kilometers below the Earth's surface. Many of these xenoliths contain several mineral phases that are in equilibrium. The compositions of the constituent minerals are determined either by chemical analysis using traditional wet chemical methods, which require separation of individual phases, or by electron microprobe microanalysis, which allows direct determination of the composition of a spot with a diameter on the order of 1-10 microns. By examining these xenoliths in detail and applying a geothermometer and a geobarometer in concert, geologists have been able to determine temperature versus depth profiles for parts of the Earth that are inaccessible to direct observation. Temperature-depth profiles are called geotherms, and they provide the link between the region of the Earth where its composition is reasonably well known and the much deeper region where its composition and temperature are poorly constrained and subject to debate.

SOLID-LIQUID GEOTHERMOMETRY

The principles of solid-liquid geothermometry are much the same as those outlined above for solid-solid geothermometry. In contrast to the simple example offered above, in which salt only dissolved in the liquid water phase, the chemical components in silicate systems tend to be soluble in several solid phases and the liquid phase simultaneously. That complicates the thermodynamic analysis considerably, and significantly more experimental data are needed to fully characterize the system. In addition, unlike the application of the two pyroxene solid-solid geothermometer, where both phases are examined directly to obtain the temperature at which they formed, application of a solid-liquid geothermometer requires extensive assumptions about the composition and state of the liquid in equilibrium with the solid, as that liquid is no longer present in the rock. Plagioclase is a common silicate mineral found in rapidly frozen lavas that span a wide compositional spectrum. Because the plagioclase phase incorporates both sodium oxide and calcium oxide, which are both components in the liquid phase as well, the composition of plagioclase may be used to infer the composition of the coexisting liquid and the temperature at which the plagioclase formed. This method has been applied extensively to lavas that range from basaltic (silica poor) to andesitic (silica rich) in composition.

SOLID-GAS BUFFER REACTIONS

The reaction between one or more solid phases and a gas phase also yields information about temperature and the partial pressure of some gas species. One example of a solid-gas buffer reaction defines the coexistence of magnetite and hematite (oxide phases composed of iron). If both phases are pure, that is, they contain only iron and oxygen in the appropriate ratio that defines the phase, then their equilibrium with each other is directly related to their theoretical coexistence with an oxygen-bearing gas phase, which is defined by the partial pressure of oxygen (the amount of oxygen present, expressed as a percentage of the total pressure of the system).

The concept of a buffer arises because if oxygen is added to or subtracted from the system, the relative proportions of magnetite and hematite in the system will change to maintain internal equilibrium. This type of reaction does not specify the temperature explicitly, since magnetite and hematite may coexist over a range of temperatures and oxygen contents. Thus, if both pure solid phases

are present and the temperature is varied, then the oxygen partial pressure will change sympathetically. Similarly, if the oxygen partial pressure is varied, then the temperature must also change. The observed coexistence of magnetite and hematite in many silica-rich igneous systems has been used to infer either the temperature or oxygen partial pressure attending formation, depending on which parameter may be fixed by independent information.

STABLE ISOTOPE EXCHANGE GEOTHERMOMETRY

The exchange of different stable isotopes of oxygen between coexisting phases forms the basis of the last common type of reaction used as a geothermometer. The isotopes of any chemical element, for example, oxygen, have the same number of protons but a different number of neutrons in their nuclei. Oxygen has two stable isotopes; one contains sixteen neutrons, and one contains eighteen neutrons. These two isotopes are referred to as oxygen 16 and oxygen 18. Their exchange between two coexisting phases in general is a function of temperature. At present, very few solid-solid, solid-liquid, or solid-gas exchange equilibria have been experimentally calibrated as a function of temperature and isotopic concentration of oxygen, however, and in the absence of such experimental calibrations, stable isotope exchange geothermometry has seen little application to igneous or metamorphic systems.

UNDERSTANDING OF VOLCANOES AND EARTHQUAKES

Many geologic processes can affect people's daily lives. Examples include potentially devastating earthquakes and the rapid and oftentimes catastrophic eruptions of volcanoes at the surface of the Earth. By applying geothermometry and geobarometry methods to rocks, geologists have developed a much better understanding of how some aspects of these complex phenomena are initiated and how they evolve.

By determining temperature-depth profiles for the upper 200 kilometers of the Earth, geologists have gained valuable information on the composition and state of regions of the Earth that are not accessible to direct observation. This information is crucial to the understanding of how and why rocks deform during earthquakes. In fact, for rocks of fixed composition, ambient temperature is the primary variable that determines whether rocks will be able to deform in such a way as to produce an earthquake. In addition, temperature may be important in controlling the magnitude of some earthquakes.

The second and perhaps more direct application of geothermometry is in an effort to gain a greater understanding of the causes and warning signs of potentially catastrophic igneous eruptions, such as the one that occurred in 1980 at Mount St. Helens, Washington. Studies of volcanic rocks, for example, which generally integrate geothermometry, geobarometry, geochemistry, and field geology, have revealed that the processes that governed the eruption of Mount St. Helens continue to operate there as well as at other sites worldwide where oceanic crust is subducted below continental crust. In the Cascade province of the western United States, for example, Mount Shasta, Mount Bachelor, and Mount Rainier share many characteristics with Mount St. Helens, so scientists suspect that these volcanoes should erupt in a similar manner. Eruption of Mount Rainier or Mount Bachelor could prove to be extremely dangerous to the large population centers of Seattle and Portland, which are in close proximity to these volcanoes.

Glen S. Mattioli

CROSS-REFERENCES

Acid Rain and Acid Deposition, 1803; Atmosphere's Structure and Thermodynamics, 1828; Carbonates, 1182; Deep-Sea Sedimentation, 2308; Earthquake Distribution, 277; Earthquakes, 316; Elemental Distribution, 379; Eruptions, 739; Evolution of Earth's Composition, 386; Fluid Inclusions, 394; Forecasting Eruptions, 746; Fractionation Processes, 400; Freshwater Chemistry, 405; Geochemical Cycle, 412; Greenhouse Effect, 1867; Hydrologic Cycle, 2045; Magmas, 1326; Minerals: Physical Properties, 1225; Minerals: Structure, 1232; Mount St. Helens, 767; Nucleosynthesis, 425; Oxygen, Hydrogen, and Carbon Ratios, 431; Phase Changes, 436; Phase Equilibria, 442; Plate Tectonics, 86; Precipitation, 2050; Rocks: Physical Properties, 1348; Seawater Composition, 2166; Sediment Transport and Deposition, 2374; Silica Minerals, 1354; Surface Water, 2066; Water-Rock Interactions, 443; Weathering and Erosion, 2380.

BIBLIOGRAPHY

Anderson, Greg M., and David A. Crerar. *Thermodynamics in Geochemistry: The Equilibrium Mode.* New York: Oxford University Press, 1993. An exploration of geochemistry and its relationship to thermodynamics and geothermometry. A thorough resource, but it can be somewhat technical at times. Recommended for the person with some chemistry and Earth sciences background.

Blatt, Harvey, and Robert J. Tracy. *Petrology: Igneous, Sedimentary, and Metamorphic.* New York: W. H. Freeman, 1996. Undergraduate text in elementary petrology for readers with some familiarity with minerals and chemistry. Thorough, readable discussion of most aspects of geothermometry and geobarometry. Abundant illustrations and diagrams, good bibliography, and thorough indices.

Carmichael, Ian S. E., Francis J. Turner, and John Verhoogen. *Igneous Petrology.* New York: McGraw-Hill, 1974. A classic text used by most colleges for a first course in igneous petrology. Includes extensive discussions on all aspects of the formation of igneous rocks. The text is highly technical but quite readable. An extensive discussion of geothermometry and geobarometry and their application to solving geologic problems. Suitable for college-level students.

Ernst, W. G. *Petrologic Phase Equilibria.* San Francisco: W. H. Freeman, 1976. This book outlines the elements of classical thermodynamics. Also discusses experimental approaches to acquiring the thermodynamic data necessary for geothermometry and geobarometry. Recommended for readers with a college-level background in chemistry.

Gregory, Snyder A., Clive R. Neal, and W. Gary Ernst, eds. *Planetary Petrology and Geochemistry.* Columbia, Md.: Geological Society of

North America, 1999. A compilation of essays written by scientific experts, this book provides an excellent overview of the field of geochemistry and its principles and applications. The essays can get technical at times and are intended for college students.

Klotz, Irving M., and R. Rosenberg. *Chemical Thermodynamics: Basic Theory and Methods.* Menlo Park, Calif.: Benjamin/Cummings, 1974. This book is designed to accompany a first course in classical thermodynamics at the college level. Knowledge of basic chemistry is desirable. Each chapter contains worked examples as well as problems to be solved by the reader. The answers are provided.

Mortimer, Charles E. *Chemistry: A Conceptual Approach.* 3d ed. New York: D. Van Nostrand, 1975. An excellent basic chemistry text designed to accompany a first course in general chemistry, this book is suitable for advanced high school and beginning college students. Contains extensive descriptions of all basic chemical phenomena. Problems follow each chapter; a separate answer book is available. Several appendices and tables of data that the reader might find useful.

Press, Frank, and Raymond Siever. *Understanding Earth.* 2d ed. New York: W. H. Freeman, 1998. An excellent general text on all aspects of geology, including the formation of igneous and metamorphic rocks. Contains some discussion of the structure and composition of the common rock-forming minerals. The relationship of igneous and metamorphic petrology to the general principles that form the basis of modern plate tectonic theory is discussed. Suitable for advanced high school and college students.

Uyeda, Seiya. *The New View of the Earth: Moving Continents and Moving Oceans.* San Francisco: W. H. Freeman, 1971. This college-level text outlines the modern theory of plate tectonics in detail. Because the relevant observations are discussed in their historical context, the reader can learn about the personalities involved in the development of this central paradigm of the Earth sciences. Many illustrations. The text is quite easy to read.

Wood, B. J., and D. G. Fraser. *Elementary Thermodynamics for Geologists.* Oxford, England: Oxford University Press, 1976. A basic review of the thermodynamics necessary for understanding the chemistry of geologic processes. Outlines in great detail specific examples of geothermometers and geobarometers and their application to igneous and metamorphic petrology. Recommended for readers who have a college-level understanding of chemistry.

NUCLEOSYNTHESIS

Nucleosynthesis is the process by which the elements are formed in the interiors of stars during the course of their normal evolution. Hydrogen and helium are thought for the most part to have been generated at the origin of the universe itself (nucleogenesis), while all other heavier elements are synthesized via nuclear reactions in stellar cores. The heaviest elements are created during the death throes of massive stars.

PRINCIPAL TERMS

BIG BANG THEORY: the theory that the universe was created via an initial explosion that resulted in the formation of hydrogen and helium

CHARGED-PARTICLE REACTION: a nuclear reaction involving the addition of a charged particle, proton or electron, to a nucleus

DEUTERIUM: an atom built of one proton and one neutron; an essential stepping-stone in the proton-proton cycle in solar-type stars

ISOTOPE: an atom with the same number of protons as another but differing in the number of neutrons and the total weight

NEUTRON REACTION: a nuclear reaction in which a neutron is added to increase the atomic mass of the nucleus, forming an isotope

NECLEONS: positively charged protons and neutral neutrons; large particles that occupy the atomic nucleus

SUPERNOVA: a massive star that explodes after available energy in the interior is used up and the star collapses

ELEMENTAL SYNTHESIS

Two of the most fundamental questions of modern astrophysics have to do with the origin and composition of the universe's primordial matter: when it came into existence and how it relates to the Einsteinian space-time structure of the present universe. With developments in physics, the problems have resolved themselves essentially into two parts: the origin of the simplest elements, hydrogen and helium, during the initial formation of the present universe and the subsequent nucleosynthesis of the other elements in the pressure cookers known as stars.

To understand elemental synthesis, one must rely on experimental observations interpreted in the light of current theory. Such data principally have to do with abundances of nuclear species now and in the past. This data set is provided from composition studies of the Earth, meteorites, and other planets, and from stellar spectra. The distribution of hydrogen and heavier elements in stars throughout the galaxy, particularly in what are referred to as population I and II stars (younger and older, respectively), indicates how the chemical composition of the Milky Way galaxy has changed over time. From these studies, most theorists conclude that the galaxy has synthesized 99 percent of its own heavy elements and thus that nucleosynthesis occurs during the natural evolution of stars. In the light of variations observed in stars of diverse ages, scientists have formulated theories regarding the formation of elements within stellar structures. A dramatic piece of evidence along that line, for example, was the discovery of technetium, all of whose isotopes, being radioactive, are short-lived, indicating that the star in which it is found must be currently producing the element. The study of naturally occurring radioactive isotopes, long- and short-lived, not only allows for measuring the time of galactic and stellar nucleosynthesis but also provides evidence that synthesis of elements heavier than hydrogen must be occurring continuously throughout the universe.

BIG BANG THEORY

Starting with the simplest element, hydrogen, which possesses one proton and one electron and is by far the predominant element in the universe, the study of nucleogenesis has progressed to a consideration of the origin of the universe. Beginning in 1946, George Gamow and others presented the theory that the entire structure started as a gigantic explosion of an extremely dense, hot "singularity," or infinitesimally small object. The explosion would have been so intense as to provide the propellant for all subsequent motion of the outwardly expanding matter and for the creation of the elements. Such a "big bang" concept has come to be accepted almost unanimously, with certain modifications. The discovery of an isotropic microwave background radiation, corresponding to a 3-Kelvin temperature residual from the original fireball, lent support to the theory, along with the use of gigantic accelerators in the 1970's, which permits examination of the formation and interactions of the basic constituents and forces of nature.

Such physics has determined that elemental synthesis, via nuclear reactions, combining protons, electrons, and neutrons, could have occurred only when the temperature dropped to below 1 billion Kelvins about three minutes after the explosion. Before that point, the energy of motion would have been too great either to form those particles or to let them cling together in electromagnetic interactions. That period of elemental synthesis probably lasted about one hour; eventually the temperature and pressure would have dropped too low to sustain any further reactions. Because of the instability of particles with atomic masses of 5 to 8, no particle combinations beyond a mass of 4 would have been formed; thus the universe was probably composed of about 75 percent hydrogen and 25 percent helium. The formation of the helium nucleus, with a mass of 4, would have used up all the available neutrons. The reactions would have progressed in a certain order. First, neutrons and protons would combine to produce deuterium; deuterium and protons would then give helium 3; the collision of two helium 3 nuclei would produce a helium 4 nucleus and two protons, releasing energy in the process as gamma rays. This postulated process seems to be in excellent agreement with observational data and theoretical calculations.

STELLAR EVOLUTION

For roughly a million years, radiation was so intense that larger particular bodies could not form. Only after the radiation pressure became low enough would such inhomogeneities as galaxies

and stars solidify. At that state, the dominating force in the universe became gravity, the galaxies and stars forming as a result of gravitational contraction. Scientists' understanding of galactic formation remains sketchy, but stellar evolution— from birth in dust-cloud nurseries to death—is well understood through a combination of observational data, laboratory measurements of nuclear reactions and their rates, and copious amounts of theoretical work. As stated best by Subrahmanyan Chandrasekhar, the working hypothesis generally accepted by astrophysicists is that the stars are the places where the transmutation of elements occurs, all the elements beyond hydrogen being synthesized there. All the energy available to a star throughout its life span, with minor exceptions, is derived from such transformations.

As the original gas and dust in a nebula collapse and contract, they heat up enormously, until the temperature in the core reaches some 10 million Kelvins, at which point thermonuclear proton-proton reactions occur, to form deuterium and give off positrons and radiant energy. Further reactions occur, increasing the helium formed, decreasing hydrogen, and producing energy sufficient to halt the gravitational collapse of the star. For most stars, this stage probably occupies the greater part of their lifetime. The more massive the star is, however, the faster it will exhaust the hydrogen supply at the core and the shorter its time of stability will be. Stars such as the Sun are in the range for forming helium. Some interesting side reactions occur also. Some 5 percent of the helium reacts to make beryllium, boron, and lithium; in an even rarer occurrence, proton capture produces the isotope boron 8. The latter is important because it is very sensitive to temperature and therefore acts as a good test of stellar theories; the reaction produces neutrinos, which earthbound astronomers can then study.

CARBON-NITROGEN-OXYGEN CYCLE

In older stars, formed as second, third, or later generations, some heavier elements are present. In these, the so-called carbon-nitrogen-oxygen cycle proposed by Hans Albrecht Bethe works, again turning four protons to helium. Because a higher temperature is necessary to overcome the electrostatic (Coulomb) repulsion barrier, this cycle takes place only in larger stars. In either case, when a significant amount of core hydrogen is used up, with helium ash left, the star will contract. Meanwhile, the hydrogen-containing outer area expands, causing the star to become a red giant; its central temperature rises to 100 million Kelvins. At this stage, helium burns to form beryllium, forming one beryllium atom per billion helium atoms. Also produced are carbon, oxygen, and neon, the principal source of energy being the conversion of three heliums to carbon 12 plus gamma radiation. This burn, however, is short-lived, lasting only 10 to 100 million years, as compared to more than 5 billion years for the present Sun. Any further synthesis requires much higher thermal energy input than can ever occur.

Beyond this stage, in larger stars, the processes become more complex. When helium is exhausted, contraction starts again. For objects such as the Sun, this shrinkage will continue until it is halted by electron degeneracy (a mutual repulsion of tightly squeezed electrons) to form white dwarfs, small and intensely radiative bodies losing their heat into space, with no further nuclear energy available. Many become surrounded by a halo of expanding gases, the so-called planetary nebula; material from the star flows into space as a last gasp of the red giant stage. In larger stars the temperature continues to climb, to 70 million Kelvins, eventually causing new sets of elements to form, including magnesium 24, sodium 23, neon 20, silicon 28, and sulfur 32. With further contraction, until the temperature reaches 1 billion Kelvins, elements up to and including iron are created. Synthesis stops here, however, because of the energy required to form more stable nucleons bound together.

NEUTRON-INDUCED REACTIONS

Additional synthesis does not involve charged-particle reactions but rather neutron-induced reactions, which tie up neutrons and produce energy. Such reactions are called s-processes because they proceed very slowly, taking from 100 to 100,000 years per capture step. This process accounts for the heavier isotopes on even atomic number elements and the distribution of nuclides up to bismuth. This reaction, in conjunction with a p-process involving successful proton reactions, can account for all the stable isotopes up to bismuth. For higher elements, however, a more rapid

neutron-capture chain called the r-process is required; it takes place when there is an enormous neutron flux so that many captures can take place in milliseconds. Conditions perfect for such acts occur in supernovas, stars that explode with some of the greatest violence seen in the universe. Type I supernovas are from old, small stars, with masses of 1.2 to 1.5 times that of the Sun; in such an explosion the entire star is destroyed, pushing the temperature to 10 billion Kelvins. Type II supernovas occur in stars with masses greater than ten times that of the Sun. Under contraction, the temperature in the nucleus of the star rises to 5 billion Kelvins, iron and nickel nuclei photodissociate to helium and neutron particles, the helium present dissociates into nucleons, and the resulting protons capture any available electrons, forming new neutrons. The collapse, which takes one second, results in a core mass of neutrons, with explosive ejection of the outer layer and all the elements into the intersteller regions. Such explosions, which occur perhaps once in a hundred years in a galaxy, contribute all the material from which other stars, clouds, and planets such as Earth are formed.

DETERMINATION OF ISOTOPIC ABUNDANCES

Since the first theories of the processes of elemental origins were proposed, scientific understanding of nucleosynthesis has progressed greatly, thanks principally to an improved ability to determine abundances of elements particularly in stars and nebulas, and to better understanding of transformation conditions during synthesis. Nuclear physics data on reaction rates, particle formation, and interactions at diverse temperatures and energies, along with clearer notions of strong and weak force interactions, have contributed vital knowledge on both the universe's origin and the generation of elements in stellar bodies during and at the end of a star's life cycle.

Isotopic abundances can be determined from meteorites by the use of mass spectroscopy. In this experimental technique, particles are heated until they break apart into ionic forms; the bodies are then propelled, under the influence of electric and magnetic fields, through a vacuum chamber. The curved path followed depends on the mass and the charge on the elements. Collection at the end of the path allows detailed comparisons to be made, with particular attention to the anomalies that are critical to theories of nucleosynthesis.

ANALYSIS OF EXTRATERRESTRIAL OBJECTS

Spectral analysis has been the principal tool for studying extraterrestrial abundances. In such analysis, light from the observed object is passed through a prism or diffraction grating so that it is broken into all of its component colors; the resulting spectrum ranges from the blue to the red region of the visible section of the spectrum. Invariably, the background will be crossed by dark or bright lines, depending on whether it is an emission or absorption spectrum. These lines, identifiable in the physics laboratory, act as fingerprints, quickly showing such information as what elements are present and their abundances. Observation of material emitted by supernovas, for example, not only shows how heavy elements are enriched in space but also contributes greatly to theories of explosive charged-particle nucleosynthesis.

Similar analyses, using spectroscopes, telescopes, and various light-intensity enhancing instruments such as charge-coupled devices (CCDs), have been done of other objects, including medium-mass stars with s-process element formation, nova explosions, and mass flows from solar-type stars. The latter can be studied best in the Sun, by analyzing the composition of the solar wind with data returned by meteorological and scientific satellites. Detectors placed above the atmosphere can be equipped to detect charged particles such as protons or electrons. Experiments to view the universe in some region of the electromagnetic spectrum besides visible light, such as radio, gamma, infrared, or ultraviolet, also must be placed beyond the disturbing influence of the atmosphere.

ADVANCES IN THEORETICAL STUDY

Much of the information usable for the theoretical study of nucleosynthesis comes from two terrestrial sources. First, experimental studies using nuclear reactors and particle accelerator machines have provided comprehensive measurements of reaction rates and of the actions of the weak force in nature. Increasingly reliable determinations of critical cross sections, representing the space in which reaction occurs between two particles, and of the neutron-capture process, which is responsible for the bulk of nuclei more

massive than iron, have become possible with highly refined accelerators and electrical detectors. Theoretical predictions and experimental results are thus more in harmony than ever before.

The second important advance has been in computer technology, which has made possible greatly increased numerical calculations of structures and the evolution of astrophysical objects. The advent of high-speed computers has allowed much greater predictive ability for the standard model of the big band and regarding the formation of elements at various stages in the stars. Such detailed models, particularly of massive stars, in terms of hydrodynamic phases, have shown, for example, that supernovas are immensely important in the synthesis of heavy elements. Models for actions at extreme temperature and density conditions are very close to what is observed during the expansion, cooling, and mass ejection processes of the dying stars. Computer technology has helped identify further problems through capture modeling, such as the sites necessary for r-process neutron-capture nucleosynthesis.

IMPORTANCE TO ASTRONOMY

Nucleogenesis and nucleosynthesis are two of the most important topics in astronomy and hence the Earth sciences, promising to cast light on not only the evolution of stars but the ultimate origin of the universe as well. The understanding of universal origins has been advanced greatly by the advent of particle accelerators of remarkably high energies. These instruments provide physicists with clearer pictures of the elementary particle structures of the universe and of their interaction under the four forces controlling them. During the creative process of the big bang, these four forces—strong and weak nuclear, electromagnetic, and gravitational—were unified as one, separating only as initial conditions of temperature, pressure, and density changed. Under their actions, radiation and particles ultimately formed, with radiation finally dispersing enough that combinations of protons and electrons, then neutrons, could arise.

Although the modern understanding of nucleosynthesis is thought to be quite satisfactory, there are still problems unsolved. Certain elemental anomalies have not been explained by either experiments or theory; neither has the shortage of neutrinos from the Sun. Predictions of energy fluxes and solar winds from other stars, particularly red giants, represent other unsolved problems. The answer to what causes a dust cloud to begin to contract to form a star is unknown; a widely accepted notion is that the contraction is prompted by the shock wave of a supernova. Problems remain with the big bang theory itself, so that alternative theories, such as the "inflationary universe," have been proposed. The investigation of such problems of modern physics and astronomy has led to numerous insights, including the possibility that planets may be by-products of stellar formation; in such a scenario, the galaxy may be filled with planets and, possibly, life-forms. Further fine-tuning of reaction rates, mechanisms, and such experimental topics as element reactions may solve some of the deepest philosophical and scientific mysteries of modern science.

Arthur L. Alt

CROSS-REFERENCES

BIBLIOGRAPHY

Abell, George O. *Exploration of the Universe.* 4th ed. New York: Holt, Rinehart and Winston, 1982. One of the best standard textbooks on astronomy available. Covers in detail the life history of stars, particularly those in which heavy elements are formed. Separate sections on white dwarfs, neutron stars, black holes, and supernovas. Glossary and references. Excellent diagrams and pictures.

Arnett, David W. *Supernova and Nucleosynthesis: An Investigation of the History of Matter, From the Big Bang to the Present.* Princeton: Princeton University Press, 1996. This book sheds light on discoveries concerning the origin of hydrogen and other elements through the actions of stars. Deals extensively with reaction rates, primordial hydrogen, galactic chemical composition, massive stars, and supernovas. Extensive bibliography.

Bowers, Richard, and Terry Deeming. *Astrophysics.* Vol. 1, *Stars.* Boston: Jones and Bartlett, 1984. A detailed exposition on the characteristics of stars. The writer extends basic data into an understanding of how stars evolve differently depending on their original mass and brightness. Extensive sections on formation of elements during diverse stages of life cycles. Difficult reading unless one ignores the mathematics.

Brush, Stephen G. *Transmuted Past: The Age of the Earth and the Evolution of the Elements from Lyell to Patterson.* New York: Cambridge University Press, 1996. A look into modern planetary physics, this book traces the evolution of the elements and the solar system. Intended for the reader with some background knowledge in astronomy, this book is well illustrated and includes a bibliography and index.

Clayton, Donald D. *Principles of Stellar Evolution and Nucleosynthesis.* Chicago: University of Chicago Press, 1984. Arguably one of the best, most complete works on the evolution of stars, tracing the life histories of different-sized objects. Extensive dicussion of element formation and of unsolved problems in the field. Mathematics spreads throughout the work, but the advanced layperson should find it understandable. Additional references.

Hartmann, William, Pamela Lee, and Tom Miller. *Cycles of Fire: Stars, Galaxies, and the Wonder of Deep Space.* New York: Workman Publishing, 1988. A delightful work encompassing the history of stars from birth to death. Touches on a multitude of topics, including black holes, white dwarfs, binary stars, and the origin of the universe and planets. Attractive illustrations complement the well-written text. Glossary and some references.

Henbest, Nigel. *The Exploding Universe.* New York: Macmillan, 1979. An overview of the violent nature of the universe, this book deals with how atoms, elements, stars, planets, and other objects are formed and die. Detailed sections on the fundamental forces and particles of nature and their relationships to the formation of the elements. Glossary. Fairly easy reading.

Ozima, Minoru. *Geohistory: Global Evolution of the Earth.* New York: Springer-Verlag, 1987. A well-written book, deals with the origin of the Earth, its waters, atmosphere, and rocks. Addresses the issue of the internal structure of the Earth as it has changed over time. Abundant diagrams. Bibliography. For the more advanced layperson.

Pagel, B. E. J. *Nucleosynthesis and Chemical Evolution of Galaxies.* Cambridge: Cambridge University Press, 1997. An overview of the chemical composition of the universe, this book examines the origin and evolution of atoms, elements, planets, and stars, and discusses how all of these pieces work together to form galaxies. Illustrations and diagrams help explain difficult processes and concepts. Intended for the college reader.

Prantzos, N., E. Vangionu-Flam, and M. Cassae. *Origin and Evolution of the Elements.* Cambridge: Cambridge University Press, 1993. A look at how elements are formed and their processes of evolution, this book follows the life of the elements and documents each phase as it occurs. A good introduction to nucleosynthesis, it is well illustrated and includes a bibliography and index.

Rolfs, Claus E., and William S. Rodney. *Cauldrons in the Cosmos: Nuclear Astrophysics.* Chicago: University of Chicago Press, 1988. This book represents the then-current state of knowledge on the stars, pertaining to the synthesis of the elements. Starting with basic data, it tracks the lives of stars of varying masses, detailing the elements created at each step. Good pictures and diagrams; extensive bibliography. Although there is some mathematics, the text is quite clear.

Schramm, David, ed. *Supernovae.* Dordrecht, Netherlands: Reidel Press, 1977. A detailed work on supernovas, particularly the Crab Nebula. Traces how stars explode and how elements are formed in the final death throes. Presents a good overview of theories with a minimum of mathematics. Contains helpful illustrations and bibliography.

Shklovskii, Iosif S. *Stars.* San Francisco: W. H. Freeman, 1978. A basic review of stars—how they work and why they exist and die. Extensive sections on origins of elements at various stages in stellar life, focusing on the production of heavy elements in star deaths. Well written. Numerous line drawings provide clarity. Bibliography.

Taylor, R. J., ed. *Stellar Astrophysics.* Philadelphia: Institute of Physics, 1992. This multi-authored work examines advances in the study of the stars. Essays track the lives of stars of varying masses, detailing the elements created at each step and the processes that occur throughout the phases. Useful illustrations and diagrams.

OXYGEN, HYDROGEN, AND CARBON RATIOS

Oxygen 18/oxygen 16, deuterium/hydrogen, and carbon 13/carbon 12 ratios in rocks are heavy- to light-isotope ratios of oxygen, hydrogen, and carbon, respectively. These ratios can give clues to the geologic conditions under which rocks were formed.

PRINCIPAL TERMS

CALCITE: a mineral composed of calcium and the carbonate ion, which fizzes when acid is poured on it

CARBONATE: a mineral containing the carbonate ion, which is composed of one carbon atom and three oxygen atoms

GEOTHERMOMETERS: minerals whose components can be used to determine temperatures of mineral formation

ION: an atom that has either lost or gained electrons

ISOTOPES: atoms with an identical number of protons but a different number of neutrons in their nuclei

ISOTOPIC FRACTIONATION: the enrichment of one isotope relative to another in a chemical or physical process; it is also known as isotopic separation

LIMESTONE: a sedimentary rock composed predominantly of calcite

MINERAL: a natural substance with a definite chemical composition and an ordered internal arrangement of atoms

RADIOACTIVE ISOTOPE: an isotope of an element that naturally decays into another isotope

STABLE ISOTOPE: an isotope of an element that does not change into another isotope

STABLE ISOTOPES

Rocks are divisible into three major types: igneous, sedimentary, and metamorphic. Igneous rocks are formed from hot, molten rock material called magma. Sedimentary rocks are formed by the compaction and cementation of mineral grains and rock fragments, which collectively are called sediments. Metamorphic rocks are formed by the alteration of rocks caused by increased heat and pressure and interaction with water in pore spaces and fractures. The temperature of formation and the types of water that interacted with the rocks are among the geologic conditions that scientists can determine from stable isotope ratios in rocks.

Rocks are aggregates of minerals. Minerals are natural compounds made of atoms that are arranged in an ordered fashion. The atoms in minerals exist in different isotopes. Some isotopes are unstable, or radioactive; others are stable. Radioactive isotopes, or parent isotopes, change into isotopes of other elements, or daughter isotopes. This change, which is called radioactive decay, occurs at a constant rate and can be known from experimental work. The amounts of parent and daughter isotopes in rocks and the decay constant of the isotope are used by scientists to determine rocks' ages. Unlike radioactive isotopes, stable isotopes do not decay into isotopes of other elements. Thus, stable isotopes are not used for age-dating purposes. Rather, the ratio of stable isotopes—particularly of low atomic number elements, such as hydrogen, carbon, and oxygen—are used to determine the geologic conditions under which rocks are formed.

IONIC AND COVALENT BONDS

The three elements oxygen, hydrogen, and carbon are components of the minerals which make up rocks. Oxygen, the most common element in the crust, the topmost layer of the Earth, is found in all kinds of rocks. Carbon is abundant in coal. Carbon and oxygen are important components of a fairly common group of sedimentary rocks called limestones and their metamorphosed product, marble. Hydrogen and oxygen are constituents of water and can be used to characterize the types of watery solution that interact with rocks.

Oxygen, hydrogen, and carbon form bonds

that range from ionic to covalent. Ionic bonds are formed by the electrostatic attraction of adjacent atoms; covalent bonds are formed by the sharing of electrons and are stronger bonds. Scientists have found that bonding characteristics affect the behavior of isotopes in natural conditions; elements which form only one type of bond in all conditions are not useful for isotopic work.

ISOTOPIC SEPARATION

These three elements have low atomic weights. The relative mass difference between the heavy and light isotopes is large for such elements, unlike the elements of high atomic weight. For example, deuterium, the heavy stable isotope of hydrogen, is heavier than the light isotope of hydrogen by about 100 percent. In contrast, the stable strontium isotopes differ from each other by only 1.2 percent. Some conditions and processes favor the incorporation of heavy isotopes into a material; others favor light isotopes. Thus, depending on the geologic conditions, there will be a difference between the heavy- to light-isotope ratio in one mineral and the ratio of the same isotopes in another mineral or in the source material. This syndrome is called isotopic separation or fractionation. Isotopic separation is significant and detectable only for elements with low atomic numbers, such as oxygen, carbon, and hydrogen. The conditions and processes that cause separation include temperature, evaporation and diffusion, and oxidation reduction reactions.

A mineral is a compound formed by the chemical bonding of atoms. The internal energy of a mineral is controlled by factors including the vibration of atoms. The atoms of light isotopes vibrate with higher frequencies than do atoms of heavy isotopes. Since the bond strength between atoms depends on the vibration frequencies of atoms, light isotopes are weakly bonded. Such bonds can be broken comparatively easily upon dissolution or by bacterial action. Significantly, with increased temperature, the vibrational frequencies of all isotopes of the same element become nearly equal. Therefore, all other factors being equal, a mineral which is formed at low temperatures will contain a higher heavy-to-low-isotope ratio than a similar mineral which is formed at high temperatures. A change in temperature during mineral formation can cause iso-

topic separation. Consequently, scientists can use isotopic separation to determine temperatures of mineral formation; in other words, isotope ratios can be used as geothermometers. Commonly, oxygen-isotope ratios in two different oxygen-bearing minerals which were formed at the same time and from the same source are used to determine the temperature at which a rock formed.

WATER-ROCK INTERACTIONS

Applications of oxygen isotope work have provided scientists with additional insights into the formation of rocks. It is now known that groundwater interacts with magmas: Part of it is incorporated into the magma, and part of it circulates through rocks, changing their nature (metamorphosing them) in the process. Furthermore, it is now possible to determine whether a magma has been contaminated by the incorporation and subsequent melting of roof rocks.

Evaporation and diffusion also cause isotopic separation. Water is a compound of oxygen and hydrogen. During evaporation, the light isotopes of oxygen (oxygen 16) and of hydrogen break through the water surface and escape to the atmosphere in the form of water vapor, while the heavy isotopes, oxygen 18 and deuterium, concentrate in reservoirs. Similarly, molecules with lighter isotopes move across a boundary (diffuse) faster than the same molecules with heavier isotopes. Evaporation and diffusion lead to different waters' having different isotopic ratios. Scientists have determined oxygen 18/oxygen 16 and deuterium/hydrogen ratios in many kinds of water: lakes, rivers, rain, snow, oceans, and water that comes from molten rocks. When these waters enter into interconnected pore spaces and fractures of rocks and circulate through the rocks at depth, they become hot solutions. These solutions interact with the rocks, leading to the diffusion (movement) of atoms from rocks to the solutions and from the solutions to the rocks. Such water-rock interaction changes the nature of both the rocks and the solutions. From stable isotope ratios, scientists can determine the nature of the rock, the nature of solution, and the type of water-rock interaction.

OXIDATION AND REDUCTION

Oxidation reactions are another cause of isotopic separation. Atoms are said to be in an oxidized state if they have a lower number of electrons and in a reduced state if they have a higher number of electrons when compared with other atoms of the same element. For example, the element carbon can occur in the form of C^{+4}, C, and C^{-4}. C^{+4} is a highly oxidized state of carbon, a positively charged ion formed by the loss of four electrons. Such ions combine with negative ions of other elements, such as oxygen ions, to form compounds such as carbon dioxide. In a highly reduced state, carbon atoms gain four electrons, forming the negatively charged carbon ion, C^{-4}. These ions combine with positively charged ions to form compounds such as methane.

Oxidized carbon is enriched in the heavy carbon isotope carbon 13, and reduced carbon is enriched in carbon 12. Thus, if the same source material were to permit the formation of two compounds, one with oxidized carbon and the other with reduced carbon, the one with the oxidized carbon will be enriched in carbon 13. The application of this simple principle is quite involved, however, because carbon can cycle through living organisms and other environments.

Green leaves of plants photosynthesize carbon dioxide, reducing the carbon and making it part of organic compounds. This reduction is done in stages, which are different in different plants. Thus, some plant types can be distinguished by their carbon 13 values. Generally, land plants have lower values of carbon 13 than marine plants; however, marine algae have values within the range of land plants. Evaporation leads to the enrichment of seawater in carbon 13. Condensation in clouds leads to the enrichment of carbon 13 in raindrops as compared with water vapor. Consequently, repeated evaporation and rain cause seawater to be richer in carbon 13; the atmosphere is lower. Carbon dioxide in soils and groundwater has an even lower carbon 13 value, because the carbon there has cycled through decaying plants.

DIAGENESIS

It appears that freshwater limestone should have lower carbon 13 values than marine limestone. Also, marine limestone should have variable carbon 13 values, depending on the amount of carbon inherited from algae (lower values indicating higher algal content). Most limestones, however, undergo a change called diagenesis, which involves the reconstitution of carbonate minerals under conditions different from the original ones. That makes it difficult to determine precisely the original conditions under which the rock formed or the subsequent conditions that resulted in the diagenetic change. Commonly, low values of carbon 13 are obtained from the diagenesis of marine limestones.

ANALYTICAL TECHNIQUES

Scientists use many analytical techniques, including mass spectrometers, mass spectrographs, and ion microprobes, to determine isotopic ratios. All these techniques utilize the fact that different isotopes separate from each other and arrive at a detector at different speeds when they travel through a magnetic field from an ionization chamber, in which a sample containing the elements is bombarded by electrons or by (in the case of the microprobe) a beam of negatively charged oxygen atoms.

The separated isotopes are detected electronically in mass spectrometers; in mass spectrographs, they are detected by nonelectronic methods, such as photographic devices. For these methods of analysis, the elements or molecules of interest are chemically separated and introduced into an ionization chamber in a gaseous form, or they are deposited as solids on filaments that are then vaporized in the ionization chamber. In the ion microprobe method, chemical separation of the sample is not necessary, and the original sample does not have to be destroyed. A small sample of a rock is polished and then coated with gold or carbon. A beam of negatively charged oxygen is focused on the sample, over an area of less than 0.01 millimeter. That causes ionization of the sample; the ions are accelerated through a magnetic field to a detector, and the isotopes are measured by a mass spectrometer or spectrograph.

MASS SPECTROMETERS AND SPECTROGRAPHS

There are many different spectrometers and spectrographs, but the principles involved can be understood by considering one type of spectrometer. In this device, an appropriate voltage applied across a filament, possibly of tungsten, produces a stream of electrons. These electrons bombard a sample and cause the removal of electrons from the atoms of the sample. The resulting positive ions of different isotopes are accelerated by a high-voltage field and are collimated into a beam. Since the kinetic energies of isotopes of the same element are identical, the lighter isotopes travel faster than the heavier isotopes. The ion beam passes through a magnetic field which is constructed in such a way that different isotopes are separated from each other as they exit the field and enter a collector cup. The accelerating voltage and the magnetic field can be adjusted so that an ion beam of one isotope can be focused through a collector slit to enter a detector cup. The focused ions are neutralized by electrons which flow through a resistor, and the voltage difference across the resistor can be measured with a voltmeter. The ensuing electrical signals can be digitized or, more commonly, displayed on a strip-chart record. A series of peaks and valleys—each peak representing an isotope, with the peak height being proportional to the abundance of the isotope—can be recorded by adjusting the accelerating voltage or the magnetic field, which would vary the ions being focused through the collector slits. In this way, the various isotopes and their relative abundances can be determined. From those, the ratios of heavy to light isotopes can be calculated.

Modern commercial mass spectrometers, equipped with multiple collectors for the simultaneous detection of different isotopes and with digital computers, have improved both the speed of acquisition and the reliability of isotopic data.

APPLICATIONS OF STABLE ISOTOPIC GEOCHEMISTRY

Many kinds of chemical analysis can be used to determine the concentration of ions in groundwater. If different ions originate from different source areas, then these areas can be distinguished by such analysis. When the source areas produce the same ions, the isotopic ratios of the ions may be different because of different environmental conditions. In such instances, stable isotope geochemistry can be used to specify the source area, such as a groundwater-polluting factory. Stable isotope geochemistry, then, is useful in groundwater pollution studies. One of the uses of mass spectrometers is to identify elements by their isotopes, and by using a procedure called isotopic dilution, scientists can identify elements whose amounts in a sample are extremely small. The isotope dilution method provides the added advantage of identifying the groundwater pollutants.

Scientists have used stable isotopic geochemistry to study the diet of prehistoric humans. The remains of wood in ancient campfires have been analyzed for their stable isotopic ratios. Such ratios can help the researcher to distinguish between land and marine plants and between groups of land plants. The ion microprobe, with its ability to analyze particles smaller than 0.01 millimeter, has a potential application in forensic science, where trace amounts of hair and clothes are used to identify criminals. The isotopic dilution method is another procedure that can be applied to such an effort.

Habte Giorgis Churnet

CROSS-REFERENCES

BIBLIOGRAPHY

Berner, Elizabeth K., and Robert A. Berner. *Global Environment: Water, Air, and Geochemical Cycles.* Upper Saddle River, N.J.: Prentice Hall, 1996. This book discusses the processes that sustain life and effect change on the Earth. Topics include the hydrologic cycle; the atmosphere; atmospheric carbon dioxide and the greenhouse effect; acid rain; the carbon, sulfur, nitrogen, and phosphorus cycles; weathering; lakes; rivers; and the oceans. It is understandable to the college-level reader.

Ewing, G. W., ed. *Chemical Instrumentation.* Easton, Pa.: Chemical Education Publishing, 1971. An excellent publication on instrumentation written for a college-level audience.

Faure, Gunter. *Principles of Isotope Geology*. 2d ed. New York: John Wiley & Sons, 1986. An excellent book that integrates theory with practice and is easy to read.

Greenwood, Norman Neill. *Chemistry of Elements*. 2d ed. Oxford: Butterworth-Heinemann, 1997. An excellent resource for a complete description of the elements and their properties. The book is filled with charts and diagrams to illustrate chemical processes and concepts. Bibliography and index.

Hoefs, J. *Stable Isotope Geochemistry*. New York: Springer-Verlag, 1980. This useful source provides ample field examples.

Krauskopf, Konrad B. *Introduction to Geochemistry*. New York: McGraw-Hill, 1979. A section in this book provides a distillation of the nature and uses of stable isotope work.

Krebs, Robert E. *The History and Use of Our Earth's Chemical Elements: A Reference Guide*. Westport, Conn.: Greenwood Press, 1998. This book defines geochemistry and examines its principles and applications. A vital resource for anyone interested in the field of geochemistry and in the Earth's elements. Illustrations, charts, and bibliography.

Richardson, S. M., and H. Y. McSween, Jr. *Geochemistry Pathways and Processes*. Englewood Cliffs, N.J.: Prentice-Hall, 1989. A chapter in this book offers an accurate summary of stable isotope geochemistry.

Valley, J. W., H. P. Taylor, Jr., and J. R. O'Neil, eds. *Reviews in Mineralogy*. Vol. 1b, Stable Isotopes in Higher Temperature Geologic Processes. Washington, D.C.: Mineralogical Society of America, 1986. A collection of work from authorities on various aspects of the topic, with a primary emphasis on igneous and metamorphic rocks and ore deposits.

PHASE CHANGES

Phase changes among liquids, solids, or gases are important in many geologic processes. The formation of ice from water, of minerals from magma, of gases bubbling out of magma, and of halite (sodium chloride) precipitating out of a lake are examples of phase changes in nature. Many of these phase changes aid in the understanding of deposits that are of economic importance.

PRINCIPAL TERMS

GAS: a substance that can spontaneously fill its own container

IGNEOUS ROCK: a rock formed from molten rock material (magma or lava)

LIQUID: a substance that flows

MAGMA: a liquid, usually composed of silicate material and suspended mineral crystals, that occurs below the Earth's surface

METAMORPHIC ROCK: a rock in which the minerals have formed in the solid state as a result of changing temperature or pressure

PHASE: that part of nature that has a more or less definite composition and thus homoge-

neous physical properties, with a definite boundary that separates it from other phases

PRECIPITATE: the process in which minerals form from water or magma and settle out of the liquid

SEDIMENTARY ROCK: a rock that has formed from the accumulation of sediment from water or air; the sediment might be fragments of rocks, minerals, organisms, or products of chemical reactions

SOLID: a substance that does not flow and has a definite shape

POLYMORPHS

A phase is a physically distinct and mechanically separable portion of a mixture. Different phases in a mixture may be of differing chemical composition, or they may be identical in composition. Thus, the rock, granite, containing the minerals, quartz, feldspar, and mica is made up of three physically distinct components or three phases. Also, water, ice, and vapor are three distinct phases even though they share the same chemical composition.

Three different minerals in a rock—for example, quartz, plagioclase, and alkali feldspar—constitute three separate phases. The number of separate mineral grains is not the same, however, as the number of phases. There might be 231 grains of quartz, 257 grains of alkali feldspar, and 199 grains of plagioclase in a given rock, but the rock does not contain this many different phases. Instead, there are only three phases in the rock, corresponding to the three different minerals with the same composition and physical properties. Minerals of the same composition with different crystal structures are called polymorphs. Calcite

and aragonite are examples of polymorphs of calcium carbonate. Water, ice, and water vapor are polymorphs of dihydrogen oxide, and the spectacularly diverse minerals graphite and diamond are polymorphs of carbon. The calcium carbonate minerals calcite and aragonite are examples. Ice may exist as several different phases depending upon the temperature and pressure.

LIQUID PHASES

There are also many liquid phases. Water is a liquid with the same composition as ice. Ice cubes and the water in which they float may be considered as two separate phases, since the ice and water have different physical properties (ice is lighter or less dense than water, for example) and are separated by boundaries. Melted rocks form liquid rock material called magma. The magma may be considered one phase, while any minerals suspended in it are considered separate phases.

Two or more liquids may coexist as separate phases if they do not mix. Oil and water form separate layers with a boundary between them, so they are separate phases. Being less dense than

436

the water, the oil floats on top of the water. Similarly, carbonate-rich magmas may not mix with many silicate-rich magmas, and they may form separate liquid phases. Water and ethyl alcohol, on the other hand, mix in all proportions and thus form only one homogeneous phase with no boundary surfaces. In a similar fashion, two silicate magmas of somewhat different composition may mix and form a homogeneous magma of an intermediate composition.

Ice changes to water at 0 degree Celsius, and at one atmosphere pressure and 100 degrees Celsius, water changes to steam. Such phase changes may differ with changes in atmospheric pressure. At about 200 times atmospheric pressure, the boiling point of water is more than 300 degrees Celsius, and the freezing point is less than 0. At a pressure of less than 0.006 atmosphere, liquid water is not stable; rather, ice changes directly to water vapor at less than 0 degree Celsius without any intervening water phase. There is even one temperature (0.1 degree Celsius) and pressure (0.006 atmosphere), called the triple point, in which ice, water, and steam coexist.

The phase relations of water have direct application to understanding the formation of certain features on Mars. Some features appear to have been formed by a running fluid such as water. The atmospheric pressure of Mars is currently too low for the planet to have any running water. Billions of years ago, however, Mars's atmospheric pressure might have been high enough to permit stabilized water to exist there. Thus, water could have been an erosional agent on Mars early in that planet's history.

SOLID AND GAS MIXING

Though it may seem difficult to visualize, some solids of different composition may be able to mix partially or completely in all proportions. The silicate mineral olivine, for example, can accommodate any ratio of magnesium to iron into its composition; the magnesium end member is said to have a complete solid solution with the iron end member. Gases, in contrast to solids and liquids, mix in all proportions. The Earth's atmosphere, for example, is a fairly homogeneous mixture of nitrogen (the predominant gas) and oxygen. There are also small amounts of other gases, such as water vapor and carbon dioxide.

PHASE CHANGES IN METAMORPHIC ROCKS

Important phase changes occur among solids in metamorphic rocks. Metamorphic rocks were formed from other rocks by chemical reactions in the solid state because of differing temperatures and pressures. The minerals kyanite, andalusite, and sillimanite are different aluminum silicate minerals with the same composition occurring in metamorphic rocks. Phase changes among these three solids depend on temperature and pressure, as in the ice-water-steam system; no liquid or gas, however, is involved in the aluminum silicate minerals. The triple point of the aluminum silicate minerals is at about 600 degrees Celsius and nearly 6,000 times atmospheric pressure (6 kilobars), so changes among these minerals take place only deep within the Earth. Sillimanite is stable from about 600 degrees Celsius and 5-6 kilobars up to more than 800 degrees and 1-11 kilobars. In contrast, andalusite is stable at pressures up to only 6 kilobars over a wide range of temperatures (200 to 800 degrees). Kyanite is also stable over a wide temperature range but at a higher pressure for a given temperature than is the case for either andalusite or sillimanite. For a geologist, then, knowing which of these aluminum silicate minerals is present in a rock helps to show the range of temperature and pressure at which the rock formed. There are also solid-to-liquid phase changes in sedimentary systems. A variety of minerals may crystallize or precipitate from water to form sediments.

PHASE CHANGES IN IGNEOUS ROCKS

In igneous rocks, too, there are many phase changes between solids and liquids. Igneous rocks form from the crystallization of minerals from magma or melted rock, usually of silicate composition. As magma slowly cools within the Earth, it forms minerals that gradually either sink or float in the magma, depending on their density (weight in relation to volume). The minerals that are heavier or denser than the magma gradually sink, and the lighter or less-dense minerals gradually float upward. The magma composition gradually changes as the minerals are extracted, because the minerals' compositions are different from that of the magma. Magma forms by the melting of solid rock in the lower crust or upper mantle of the Earth. The magma's composition will depend on the composition of the rock melted, the pressure, and the degree of melting. Also, the magma composition will be different from that of the solid. The melting of a typical rock in the upper mantle, for example, will produce a basaltic magma. A basaltic magma will produce a dark, fine-grained rock of low silica content, called basalt, when extruded at the surface of the Earth. The melting of a silica-rich rock in the continental crust is more likely to produce magma with a high silica content. These high-silica magmas will crystallize to light-colored rocks called dacite or rhyolite when extruded at the surface.

SATURATION POINTS

The maximum amount of a mineral that may be dissolved in water is called its saturation point. Different minerals have different saturation points in water. Considerably less calcite (calcium carbonate mineral) may dissolve in water than gypsum (calcium sulfate mineral). Even more halite (sodium chloride) may dissolve in water than gypsum. If the saturation points for these minerals are exceeded, the minerals will begin to crystallize or precipitate and sink to the bottom of the water. Saturation points of minerals may be exceeded when water evaporates or when the temperature changes. If seawater is present in a bay in which evaporation exceeds the influx of new seawater, calcite, gypsum, and halite may precipitate, in that order, as the water gradually evaporates. Vast amounts of salt deposits of halite and gypsum are believed to have formed in this fashion during the geologic period called the Permian (about 250 million years ago) in Kansas and Oklahoma. Such salt deposits are not nearly as common as are limestones. Limestones are sedimentary rocks composed of mostly calcite. The calcium carbonate is believed to have precipitated in warm, shallow seas either by inorganic precipitation or by organisms forming calcite or aragonite. The precipitation of calcite or aragonite is aided by the evaporation of seawater in shallow seas and by warming of the water.

EXPERIMENTS AT ATMOSPHERIC PRESSURE AND TEMPERATURE

A variety of techniques are used to study phase changes. The technique selected to study phase changes depends on the pressure, temperature, and types of phases. The easiest phase changes to study are those involving precipitation of minerals from water solutions at atmospheric pressure and temperature. One of the intriguing problems in the study of sedimentary rocks, for example, is why among ancient rocks so much limestone that is made up of calcite and dolostone is composed of dolomite (a calcium/magnesium carbonate mineral), as modern sediments seem to be forming mostly aragonite and calcite. Little dolomite is apparently forming today. Experiments in the laboratory have helped geologists to explain such observations. The precipitation of calcite and aragonite in the laboratory is temperature-dependent. A temperature of about 35 degrees Celsius, for example, favors precipitation of needlelike crystals of aragonite. In contrast, a lower temperature of 20 degrees favors precipitation of mostly stubbier crystals of calcite.

To identify the minerals, scientists observe them under a microscope or by X-ray diffraction. X-ray wavelengths and the distances between atoms in the minerals are about the same, so the X rays will be reflected off planes of atoms in the mineral. The angle of reflection depends on the distance between the atoms and the wavelength of the X rays. Since every mineral has different spacings between atoms, the reflections of different minerals have different angles and serve as "fingerprints" for the minerals. Thus, calcite and aragonite can easily be distinguished; it is difficult, however, to produce dolomite under any conditions in the laboratory. Such laboratory observations are consistent with the observed abundance of aragonite needles in warm, shallow seas and with the greater abundance of calcite forming from cooler waters. They are also consistent with the lack of observed dolomite formation.

The only dolomite that can form in the laboratory is produced by the conversion of calcite to dolomite in contact with concentrated waters with a high magnesium-to-calcium ratio. Thus, only under special geologic conditions below the land surface will calcite convert—slowly—into dolomite. Waters high in magnesium moving through calcite-rich rocks below the surface will convert the calcite into dolomite over long periods of time. This process may be occurring in certain places presently, though it simply cannot be observed.

EXPERIMENTS AT HIGHER TEMPERATURES AND PRESSURES

Other experiments in furnaces at high pressure tell geologists that aragonite is in reality stable only at a pressure much higher than atmospheric pressure. Aragonite is unstable at atmospheric pressure, so any aragonite forming presently should slowly revert to the more stable calcite with time. This fact explains why there is no aragonite in ancient rocks.

Experiments involving phase changes at higher pressures and temperatures are more difficult to carry out because of the problem of controlling and measuring the temperature and pressure. In some experiments, a cylindrical container or hydrothermal vessel composed of a special steel alloy is hollowed out in the center so that a sample container may be placed inside of it. The sample container is composed of pure gold or platinum and is sealed at one end. The sample and some water are placed in the container, and the other end is sealed. The container is placed inside the hydrothermal vessel, and water is pumped into the container and heated to the desired temperature. As the temperature gradually rises, the water vapor must periodically be released so that the pressure does not rise too high and rupture the hydrothermal vessel. The gold or platinum container distorts easily and transmits the pressure to the sample inside the container. After the experiment has continued for the desired length of time, the container is suddenly cooled so that the sample is frozen in the state it had reached at the higher temperature and pressure. Suppose the experimenter is studying the melting of rocks. After the sudden cooling, he may find that some of the sample is glass with embedded crystals of one or more minerals. Presumably the glass represents liquid that was quickly "frozen." The minerals may be identified by observing them under a microscope or by using X-ray diffraction. The exact mineral and glass composition may be determined through the use of an electron microprobe. In this technique, a narrow electron beam is focused on a part of the material to be analyzed, causing electrons of various elements to be removed from the atoms. Other electrons take the places of the removed electrons. X rays of certain specific energies are then emitted; because these are characteristic of a given element, that element may now be identified. The number of gamma rays or their intensity depends on the amount of the element in the sample; thus, the concentration of the element may be determined.

The mineral and glass composition at a series of temperatures and pressures may be determined during the gradual solidification of a silicate liquid, for example, to understand how the crystallization of the minerals may change the composition of the liquid. These liquid changes in the experiment may then be related to the changing composition of a series of natural lavas to see whether they might have formed by a similar process.

OCCURRENCE IN FAMILIAR PROCESSES

Phase changes occur in familiar processes every day. The phase changes from ice to water to water vapor are familiar to most people. Ice is less dense than water, so the ice takes up more space than does the original water. A drink placed in a freezer may explode as a result of this phase change. This effect is avoided in automobile cooling systems when ethylene glycol is mixed with the water; the freezing point of this mixture is much lower than that of water. Ice floats in water because of its lower density. What if ice were more dense than water? Then ice would surely sink to the bottom of lakes and oceans, and it might remain there the year round. Profoundly different oceanic, lake, and atmospheric circulation and very different climates and ecosystems would be the result.

People cook with boiling water. It is generally known that the boiling point of water decreases with increased elevation as the pressure is reduced. Cooking time must thus be increased to compensate for this lowered boiling temperature at higher elevations. Alternately, salt could be added to the water to raise the boiling point.

Water vapor in the atmosphere can increase only up to a certain maximum point, called the saturation point. This saturation point varies with temperature. More water vapor may be contained in warmer air. Rainfall results when warm, saturated air rises and cools. The cooler air cannot hold as much moisture as can warmer air, so rain falls.

ROLE IN UNDERSTANDING GEOLOGICAL PROCESSES

An understanding of phase changes is essential for an understanding of geological processes. The concentration of elements in geologic systems involves one or more phase changes. The so-called fractional crystallization process, for example, involves the precipitation of minerals from a slowly cooling magma. The minerals either sink or rise in the magma, depending on whether they are heavier or lighter than the same volume of the magma. Some elements are more concentrated in the minerals than in the magma; others are more concentrated in the magma than in the minerals. Some elements may boil out of magma with water vapor and become concentrated in hydrothermal deposits. Common table salt (sodium chloride) forms vast deposits where large, saline bodies of water evaporated slowly over long periods of geologic time, much as the Great Salt Lake in Utah is doing today. Animal matter may slowly change to petroleum or natural gas when buried gradually below the surface of the Earth. As large swamps are gradually buried, they may be transformed into coal. Buried deep, some of this material may change to graphite (all carbon). The mineral diamond (also all carbon) may have existed as graphite before it was transformed to diamond at even greater pressure within the Earth.

Quartz (silica or SiO_2) is a common mineral at the Earth's surface. Silica polymorphs trydimite and cristobalite are found in volcanic rocks formed at high temperatures. Under extreme pressure quartz is converted to high-density polymorphs known as stishovite and coesite. The only known occurrence of pressure high enough to cause this change is the impact of meteors on the planetary surfaces. The presence of coesite is used as evidence that various suspect structures were caused by meteor impact. Thus, the different polymorphs of quartz provide evidence concerning the origin of the rock in which it is found.

Robert L. Cullers

CROSS-REFERENCES

BIBLIOGRAPHY

Bowen, Robert. *Isotopes and Climates.* London: Elsevier, 1991. Bowen examines the role of isotopes in geochemical phases and processes. This text does require some background in chemistry or the Earth sciences but will provide some useful information about istopes and geochemistry for someone without prior knowledge in those fields. Charts and diagrams help clarify difficult concepts.

Brownlow, Arthur H. *Geochemistry.* Englewood Cliffs, N.J.: Prentice-Hall, 1979. A variety of phase changes are discussed in this introductory text in geochemistry. Suitable for a college student who has taken introductory courses in geology and chemistry. Many illustrations.

Ehlers, Ernest G. *The Interpretation of Geological Phase Diagrams.* San Francisco: W. H. Freeman, 1972. An excellent and very detailed discussion of the principles used to interpret geological phase diagrams in geology. Suitable for college-level students with basic knowledge of mineralogy and chemistry.

Ernst, W. G. *Earth Materials.* Englewood Cliffs, N.J.: Prentice-Hall, 1969. This book is part of a series which supplements introductory textbooks in geology. Discusses mineralogy, igneous rocks, sedimentary rocks, and metamorphic rocks in more detail than do most introductory textbooks. Features good treatments of phase changes and phase diagrams in all the rock types. Accessible to the college-level student who has studied general geology.

_____. *Petrologic Phase Equilibria.* San Francisco: W. H. Freeman, 1976. A detailed treatment of phase diagrams in geologic processes. Appropriate for college students with background in chemistry and mineralogy. Illustrated.

Hamblin, William K. *Earth's Dynamic Systems.* 8th ed. Upper Saddle River, N.J.: Prentice Hall, 1998. This geology textbook offers an integrated view of the Earth's interior not common in books of this type. The illustrations, diagrams, and charts are superb. Includes a glossary and laboratory guide. Suitable for high school readers.

Krauskopf, Konrad B. *Introduction to Geochemistry.* New York: McGraw-Hill Book Company, 1979. In this well-written introductory geochemistry text a variety of phase changes are discussed in detail. College students who have taken geology and chemistry courses will find it helpful. Contains many figures.

Mason, Brian. *Principles of Geochemistry.* New York: John Wiley & Sons, 1968. An introductory college-level text in geochemistry. Includes some discussion of phase changes. Illustrated with many figures.

Thompson, Graham R. *An Introduction to Physical Geology.* Fort Worth: Saunders College Publishing, 1998. This college text provides an easy-to-follow look at physical geology. Thompson walks the reader through each phase of the Earth's geochemical processes. Illustrations, diagrams, and bibliography included.

PHASE EQUILIBRIA

The mineral assemblages in most igneous and metamorphic rocks preserve a record of the chemical equilibrium related to the initial rock-forming process. Phase equilibria studies attempt to determine quantitatively the pressure/temperature conditions of rock formation from these mineral assemblages.

PRINCIPAL TERMS

DEGREE OF FREEDOM: the variance of a system; the least number of variables that must be fixed to define the state of a system in equilibrium, generally symbolized by F in the phase rule ($P + F = C + 2$)

EQUILIBRIUM: the condition of a system at its lowest energy state compatible with the composition (X), temperature (T), and pressure (P) of the system; the smallest change in T, P, or X induces a state of disequilibrium that the system attempts to rectify

ISOCHEMICAL PROCESSES: processes that leave rock compositions unchanged; in thermodynamic terms, a system in which X remains constant even if T and P change

MOLE: the amount of pure substance that contains as many elementary units as there are atoms in 12 grams of the isotope carbon 12

PHASE: any part of a system—solid, liquid, or gaseous—that is physically distinct and me-chanically separable from other parts of the system; a boundary surface separates adjacent phases

PHASE DIAGRAMS: graphical devices that show the stability limits of rocks or minerals in terms of the variables T, P, X; the simplest and most widely used are P-T diagrams (X = constant) and T-X diagrams (P = constant)

SYSTEM: any part of the universe (for example, a crystal, a given volume of rock, or an entire lithospheric plate) that is set aside for thermodynamic analysis; open systems permit energy and mass to enter and leave, while closed systems do not

THERMODYNAMICS: the science that treats transformations of heat into mechanical work and the flow of energy and mass from one system to another, based on the assumption that energy can neither be created nor destroyed (the first law of thermodynamics)

CHEMICAL EQUILIBRIUM

The traditional methods of studying rock bodies are descriptive in nature and involve mapping large-scale outcrop features in the field and detailed microscopic observations of rock textures and mineralogy in the laboratory. These methods, successful by themselves, are supplemented by a second, more theoretical, approach wherein rocks are treated as chemical systems and the principles of phase equilibria are applied to determine the conditions of their origin. A full and complete description of a rock body is still required, but that is no longer the main goal of petrologic study. The principles of phase equilibria are simply the laws that govern attainment of equilibrium of chemical reactions such as A + B = C + D, where A and B are known as reactants, and C and D are known as products. Before exploring how these principles

cast light on rock-forming processes, the concept of chemical equilibrium must first be developed. By analogy with gravitational potential, there must exist a similar tendency in chemical systems to lower their energy state through chemical reactions. Reasoning along these lines, J. Willard Gibbs introduced the term "chemical potential" to describe the flow of chemical components from one site (of high potential) to another (of lower potential) during reactions that lead a chemical system toward its lowest energy state. The total energy available to drive a chemical reaction must therefore be the sum of the chemical potentials of each component in the system multiplied by the number of moles of each component. The usual definition of chemical potential of a phase (or pure substance) is "the molar free energy" (or free energy per mole). This simple statement leads to a workable, three-part definition of chemical equilibrium, which is central to the understanding of phase equilibria: First, if the chemical potential of a component is the same on either side of a reaction equation, the component can have no tendency to participate in the reaction; second, in a multicomponent system consisting of several phases under uniform temperature (T) and pressure (P), equilibrium must prevail when the chemical potential of each component is the same in all phases in which the component is present; third, the condition of equilibrium is one of maximum chemical stability. The second part of the definition is equivalent to saying that, for a given chemical reaction, the free energy of the reactants must equal the free energy of the products if a condition of chemical equilibrium prevails under fixed conditions of T and P. If either T or P changes, the system is no longer in equilibrium, but it will immediately adjust itself in such a way as to "moderate" the effect of the disturbing factors. The last statement is known as the moderation theorem, or Le Châtelier's principle. If the free energy on the product side of an equation is less than the free energy on the reactant side, the reaction will be spontaneous. If the opposite is true, no reaction is possible. Scientists are thus able to predict the result of any chemical reaction if the free energy of the reactants and products under the reaction conditions (T, P) is known.

EQUILIBRIUM STATE OF A SYSTEM

The most significant processes of rock formation—magmatism, metamorphism, and sedimentation—all involve large-scale flow of energy and movement of matter that produce an uneven distribution of chemical potential. Inevitably, the result must be chemical reactions tending to restore these natural systems to a state of equilibrium. The equilibrium state of a system is governed by its bulk composition (X), temperature (T), and pressure (P). For most geological processes, T and P change slowly relative to the rates of most chemical reactions, which means that most rock-forming reactions may be considered to take place under constant T and P and, if X also remains constant, most such reactions should easily attain chemical equilibrium. In geology, the major concern then is not so much with achievement of equilibrium but rather with preservation of equilibrium mineral

assemblages through hundreds of millions of years, which must follow before deep-seated rocks are finally exposed at the surface. Rocks formed at depth must clearly undergo significant reductions in T and P prior to exposure at the surface, and there are several mechanisms that may induce changes in X during this lengthy period. Geologists are acutely aware of the implications of the moderation theorem: Retrograde metamorphism, mineralogical inversions and exsolutions, hydrothermal alteration, and weathering are but a few of the processes that could trigger reequilibration in rock bodies before they are exposed for study. Fortunately, microscopic studies coupled with Gibbs's pioneering work in phase equilibria provide the means to discern whether a given mineralogical assemblage preserves a former equilibrium.

GIBBS PHASE RULE

The "phase rule," generally called the Gibbs phase rule, was initially derived by Gibbs in 1878 from the mathematical formulas of thermodynamics. The phase rule, which is fully applicable to all chemical systems, expresses the relationship between the governing variables T, P, X, and the number of phases that may coexist in a state of equilibrium. Usually the phase rule is expressed in equation form as $P + F = C + 2$, where P = number of phases, C = number of chemical components, and F = degrees of freedom possessed by the system (normally T, P, X). Phases are chemically pure, physically separable subparts of the system and may be gases, liquids, or solids. In the formal sense C is the minimum number of chemical entities needed to define completely the composition of each reactant and product phase in a given reaction.

Although the objective of phase rule applications is to determine F for major rock-forming reactions, a far simpler situation that could be experimentally verified in any high school laboratory or even in an ordinary kitchen may be considered. Pure water (H_2O) boils at T = 100 degrees Celsius at sea level (P = 1 bar; atmospheric pressure). The effect of dissolving common salt (NaCl) in water is to raise the boiling temperature approximately 0.8 degree Celsius for each mole percent of NaCl in the liquid phase. The steam given off by boiling is pure H_2O and, therefore, the salt concentration in the remaining liquid must progressively increase with temperature during boiling. To apply the phase rule to this simple system, one first must tally up the participating phases: there is steam and there is liquid salt solution, and one must conclude that P = 2. Both pure water and salt are required to form these coexisting phases and, therefore, C = 2. The phase rule for this process (boiling), under conditions of fixed pressure (P = 1 bar), tells one that

$$F = C - P + 2$$
$$= 2 - 2 + 2$$
$$= 2$$

Therefore, the system has two degrees of freedom and is said to be "divariant," which means that, because one degree of freedom is utilized by fixing P = 1 bar, only one additional variable need be known to specify completely the state of the system. That may be either T (boiling temperature) or X (composition of boiling solution); in other words, T and X are dependent variables at constant P. This T-X compositional dependence is easily determined for P = 1 bar by direct experiment. If the resulting T-X data were graphically plotted, the diagram would indicate, for example, that the boiling temperature for an 8 mole percent solution is close to 106 degrees Celsius. Conversely, if it were known only that the boiling temperature of a salt solution were 106 degrees Celsius, that would necessitate a solution concentration of 8 mole percent at P = 1 bar.

APPLICATION OF PHASE RULE

To apply the phase rule to a reaction that has some geological significance, consider the appearance of diopside in siliceous dolomite during contact metamorphism by the reaction

$$CaMg(CO_3)_2 + 2\ SiO_2 = CaMgSi_2O_6 + 2\ CO_2$$
(dolomite) (quartz) (diopside) (gas)

The reaction involves three mineral phases and a fugitive gas phase, which is necessarily lost from the rock if diopside appears; P = 4. Note that the Ca:Mg (calcium-to-magnesium) ratio is the same in the reactant phase (dolomite) and the product phase (diopside); consequently, the minimum number of components needed to define the compositions of the four phases in the reaction is

three. Therefore, C = 3. Substituting these values into the phrase rule, one obtains

$$F = C - P + 2$$
$$= 3 - 4 + 2$$
$$= 1$$

and the reaction above has one degree of freedom and is said to be univariant. For any given reaction pressure, the phases appearing in that reaction can coexist in equilibrium at one, and only one, temperature. Univariant reactions are of great interest to petrologists because it is frequently possible to estimate the depth of a rock-forming reaction and, hence, P, from field relationships. It is then a simple matter to estimate the reaction temperature from the P-T diagram for the univariant reaction involved. For example, suppose that field relationships lead to the conclusion that diopside-bearing contact metamorphic rocks formed by the reaction above at an estimated depth of 4 kilometers. The pressure equivalent of this depth is about 1,000 bars; the experimentally derived P-T diagram indicates a reaction temperature of about 450 degrees Celsius.

MINERALOGICAL PHASE RULE

The principles of phase equilibria were first applied to rocks in 1911 by V. M. Goldschmidt in his classic account of contact metamorphism in the Oslo area in Norway. Countless similar attempts have followed, with most authors concluding, as did Goldschmidt, that rocks, in general, record a state of chemical equilibrium governed by the temperature and pressure prevailing at their time of origin. The major generalization emerging from eighty years of such studies is that rocks with a large number of mineral phases tend to have a low number of degrees of freedom. Goldschmidt recognized that, at Oslo, metamorphism must be controlled mainly by T and P and that divariant equilibrium (F = 2) is the general case for isochemical rock-forming reactions. For F = 2, the phase rule reduces to P = C. Dubbed "the mineralogical phase rule," this equation cannot be mathematically derived from thermodynamic principles. It simply reflects the common case in nature where rock-forming processes are approximately isochemical and their phase equilibria are controlled by both T and P operating as independent variables. It follows that rocks recording univariant equilibria are more restricted in occurrence than those recording divariant equilibria. Similarly, the rarity of rocks recording invariant equilibria ($F = 0$) can be understood, as the phase rule requires that both a unique T and a unique P be maintained during their formation. Rocks of complex mineralogy are occasionally encountered in which the number of phases, P, exceeds $C + 2$ and, consequently, the value of F is negative. Phase rule departures, indicated by negative F values, are a sure sign of disequilibrium and serve as a reminder that not all mineral assemblages can be treated by the methods of phase equilibria.

PETROGENETIC GRIDS

The temperature and pressure range in which any mineral may exist is limited, and it is the task of experimental petrology to ascertain these limits for rock-forming minerals. The data resulting from this experimental work are utilized to construct "phase diagrams." Such diagrams show the effects of changing P, T, X values on mineral stability fields and are, therefore, simply graphical expressions of the phase rule. In metamorphic petrology, the major use of phase equilibria data has been for construction of "petrogenetic grids." A petrogenetic grid, initially conceived by N. L. Bowen in 1940, is a P-T diagram on which experimentally derived univariant reaction curves are plotted for a particular metamorphic rock type (for example, blueschists, marbles, calcsilicates, or pelites). The value of such "grids" lies in the fact that each natural equilibrium assemblage recognized in the field will fall within a definite P-T pigeonhole and thereby inform the field geologist immediately of the P-T conditions of the metamorphic terrane under study. This goal, so simple in concept, has proven elusive even after half a century of vigorous experimental, theoretical, and field effort. The problem lies in the lack of truly univariant reactions. For nearly seventy years, field geologists mapped isograds recording the "first appearance" of notable zone minerals such as biotite, garnet, staurolite, kyanite, and sillimanite under the impression that they represented the intersection of the ground surface with a plane of univariant equilibrium. Virtually all such "isograds" have proven to be the result of divariant equilibria and thus plot as a "band"—which may be rather wide—on a P-T diagram. This undesirable result has the effect of "smearing" grid boundaries and rendering them less useful as metamorphic indicators.

The general absence of univariant reactions in metamorphic rocks was eventually recognized because of theoretical and experimental advances in phase equilibria studies. The problem stems from the fact that most mineral phases participating in metamorphic reactions are solid solutions of variable composition and the commonest reactions lead to the release of a fluid, the composition of which may vary with time. Each of these effects introduces an additional degree of freedom in phase rule terms, and, as a result, virtually all important reactions are divariant. In spite of these difficulties, petrogenetic grids, based on divariant and quasi-univariant equilibria, have gradually evolved for all major metamorphic rock types. These are not the simple, quantitative grids envisioned by Bowen, but they do provide quick, reliable, and fairly narrow estimates of P-T conditions for common metamorphic mineral assemblages. Modern grids, continually subject to refinement, are phase equilibria's greatest contribution to metamorphic petrology.

APPLICATION TO IGNEOUS PETROLOGY

In the area of igneous petrology, phase equilibria methods and data have become indispensable. Natural rocks, spanning the compositional spectrum, are melted under strictly controlled laboratory conditions to determine solidus and liquidus temperatures at pressures ranging from 1 to 35,000 bars. The results of such experiments place tight constraints on depths and temperatures of magma generation. They also permit the experimentalist to explore P-T effects on partial melting (anatexis) in terms of melt composition and refractory solid phases. The resulting phase diagrams, like metamorphic grids, permit petrologists to "see" deep into the crust and upper mantle and to test hypotheses dealing with the origin of magma.

For nearly a century, igneous petrologists have studied crystal-melt equilibria of simplified, synthetic melts as models for complex, natural magmas. The objective is to reduce the number of equilibrium phases by elimination of minor components of real magmas. Studies of this type were introduced by Bowen at the Geophysical Laboratory of the Carnegie Institution in Washington, D.C. Through its many subsequent researchers, this laboratory published hundreds of phase diagrams and earned a reputation for meticulous and exhaustive experimental work.

Phase equilibrium studies have provided a rather complete understanding of two fundamentally different modes of magma crystallization. Equilibrium crystallization occurs when P-T-X conditions change so slowly that chemical reactions within the melt are able to maintain the state of chemical equilibrium. On the other hand, fractional crystallization results when changes in P-T-X conditions outpace the compensating reactions. This disequilibrium process greatly influences the behavior of natural magmas and extends the range of melt compositions that can be derived from a given parent magma. This latter type of behavior, recognized through the early phase equilibria studies of the Geophysical Laboratory, is the major factor in explaining the compositional diversity of igneous rocks.

The relatively simple phase diagrams of synthetic systems unraveled the complexities of sequential crystallization, cast light on the mechanics of crystal nucleation, and exposed the crucial role that water plays in magmatic processes. Collectively, these diagrams are the foundation of modern igneous petrology.

GOAL OF PHASE EQUILIBRIA STUDIES

The refined symbolic notation and elegant mathematical derivations of thermodynamics are likely to remain unappreciated by the majority of laypersons and geologists alike. It is precisely these formalisms, however, that place phase equilibria on a quantitative footing and permit calculation of mineral stability fields from compositional data. Future development in the area of phase equilibria will follow this theoretical line.

The qualitative form of phase equilibria is expressed in phase diagrams rather than equations. Such diagrams have been a major part of petrology since the 1950's. Historically, emphasis in phase equilibria studies has been on high-temperature igneous and metamorphic rocks, which are most likely to preserve former equilibrium mineral assemblages. This preservation is the fundamental prerequisite for any application of phase equilibria methods. For this reason, the phase equilibria approach has generally not been applied to sedimentary rocks, except saline deposits formed by intense evaporation of seawater and record chemical equilibrium.

The phase diagrams and sophisticated calculations utilized in phase equilibria studies are often imposing, but that merely reflects the compositional complexity of natural rocks and minerals. What must be appreciated is that the goal of such studies is both simple and practical: to determine how rocks form. All processes taking place on or within the Earth (as well as all other solar system bodies) involve the flow of energy and mass. If scientists wish to advance beyond simply describing these processes—that is, to understand the chemical nature of the world—the phase equilibria approach must be employed.

Gary R. Lowell

CROSS-REFERENCES

Chemical Precipitates, 1440; Contact Metamorphism, 1386; Elemental Distribution, 379; Evaporites, 2330; Evolution of Earth's Composition, 386; Fluid Inclusions, 394; Fractionation Processes, 400; Freshwater Chemistry, 405; Geochemical Cycle, 412; Geothermometry and Geobarometry, 419; Hydrothermal Mineralization, 1205; Magmas, 1326; Nucleosynthesis, 425; Oxygen, Hydrogen, and Carbon Ratios, 431; Phase Changes, 436; Plate Tectonics, 86; Regional Metamorphism, 1421; Sedimentary Mineral Deposits, 1637; Water-Rock Interactions, 443.

BIBLIOGRAPHY

Aharonov, Einat. *Solid-Fluid Interactions in Porous Media: Processes That Form Rocks.* Woods Hole: Massachusetts Institute of Technology, 1996. Aharonov examines the processes involved in rock formation. This is a technical book at times, it but can be understood by the careful reader.

Angrist, Stanley W., and Loren G. Hepler. *Order*

and Chaos: Laws of Energy and Entropy. New York: Basic Books, 1967. An elementary treatment of basic thermodynamic concepts that lie behind phase equilibria studies. One of very few books that present this topic in nonmathematical terms. Aimed at high school readers.

Best, Myron G. *Igneous and Metamorphic Petrology.* 2d ed. New York: W. H. Freeman, 1995. A widely used text for undergraduate geology majors. The treatment skillfully balances the traditional and phase equilibria approaches to petrology. Chapter 1 introduces basic thermodynamic concepts from the geological perspective. Chapters 8 and 14 summarize phase equilibria applications to igneous and metamorphic processes, respectively. For college-level readers with some background in geology and chemistry.

Blatt, Harvey, and Robert J. Tracy. *Petrology: Igneous, Sedimentary, and Metamorphic.* New York: W. H. Freeman, 1996. Undergraduate text in elementary petrology for readers with some familiarity with minerals and chemistry. Thorough, readable discussion of most aspects of phase equilibria. Abundant illustrations and diagrams, good bibliography, and thorough indices.

Bowen, Norman L. *The Evolution of the Igneous Rocks.* Mineola, N. Y.: Dover, 1956. A reprint of the 1928 classic that first brought the phase equilibria approach to American geologists. Still valuable as an introduction to igneous processes, though some parts are dated. Suitable for college-level readers who have had a course in physical geology.

Bowen, Robert. *Isotopes and Climates.* London: Elsevier, 1991. Bowen examines the role of isotopes in geochemical phases and processes. This text does require some background in chemistry or the Earth sciences but will provide some useful information about istopes and geochemistry for someone without prior knowledge in those fields. Charts and diagrams help clarify difficult concepts.

Ehlers, Ernest G. *The Interpretation of Geological Phase Diagrams.* San Francisco: W. H. Freeman, 1972. Aptly titled, this book is a "must" for those interested in the practical side of phase equilibria. A well-indexed guide to most chemical systems relevant to igneous petrology. Theoretical aspects of the subject are omitted, except for a terse presentation of the phase rule in chapter 1. For college-level readers with a knowledge of mineralogy.

Emsley, John. *The Elements.* 3d ed. Oxford: Oxford University Press, 1998. Emsley discusses the properties of elements and minerals, as well as their distribution in the Earth. Although some background in chemistry would be helpful, the book is easily understood by the high school student.

Fermi, Enrico. *Thermodynamics.* Mineola, N.Y.: Dover, 1956. A reprint of the 1937 work by one of the most prominent physicists of the twentieth century. Concepts are presented simply and without regard to applications. For college-level readers with some background in calculus; others may still find the text valuable as the emphasis on mathematical derivation is minimal compared with modern books on the subject.

Fyfe, W. S., F. J. Turner, and J. Verhoogan. *Metamorphic Reactions and Metamorphic Facies.* New York: Geological Society of America, 1958. This monumental book established the thermodynamic approach in metamorphic petrology, and its influence is seen in all subsequent texts on the subject. A major reference for any serious student of metamorphism. Will be found in all university libraries.

Greenwood, Norman Neill. *Chemistry of Elements.* 2d ed. Oxford: Butterworth-Heinemann, 1997. An excellent resource for a complete description of the elements and their properties. The book is filled with charts and diagrams to illustrate chemical processes and concepts. Bibliography and index.

Powell, Roger. *Equilibrium Thermodynamics in Petrology: An Introduction.* New York: Harper & Row, 1978. This text, as well as many others of recent vintage, treats phase equilibria as a problem-solving methodology rather than as a body of knowledge derived through field and laboratory work. Written for advanced students of geology, but chapters 1 and 2 are suitable for general readers seeking an introduction to equilibrium concepts and phase diagrams.

Press, Frank, and Raymond Siever. *Understanding Earth*. 2d ed. New York: W. H. Freeman, 1998. An excellent general text on all aspects of geology, including the formation of igneous and metamorphic rocks. Contains some discussion of the structure and composition of the common rock-forming minerals. The relationship of igneous and metamorphic petrology to the general principles that form the basis of modern plate tectonic theory is discussed. Suitable for advanced high school and college students.

WATER-ROCK INTERACTIONS

Water-rock interactions occur as fluids circulate through rocks of the Earth's crust. Isotopes of common elements are exchanged between a fluid and its host rock. As a result of these reactions, a rock may preserve a record of the fluids that have passed through its pore spaces. Studies of water-rock interactions provide information on the nature of fluid movement through rocks and on the origin of economic ore deposits.

PRINCIPAL TERMS

CONNATE FLUIDS: fluids that have been trapped in sedimentary pore spaces

DEHYDRATION: the release of water from pore spaces or from hydrous minerals as a result of increasing temperature

EXCHANGE REACTION: the exchange of isotopes of the same element between a rock and a liquid

FRACTIONATION: a physical or chemical process by which a particular isotope is concentrated in a solid or liquid

ISOTOPES: atoms of the same element with identical numbers of protons but different numbers of neutrons in their nuclei

JUVENILE WATER: water that originated in the upper mantle

MASS SPECTROMETER: a laboratory instrument that separates isotopes of a particular element according to their mass difference

METEORIC WATER: water that takes part in the surface hydrologic cycle

VOLATILES: dissolved elements and compounds that remain in solution under high-pressure conditions but would form a gas at lower pressures

STABLE ISOTOPES

Fluids that circulate within the crust of the Earth take part in chemical reactions involving an exchange of elements between the fluid and the host rock. Such reactions, referred to as fluid-rock interactions, are an important mechanism in the concentration of minerals in economically valuable ore deposits. A useful means of studying fluid-rock interactions is by measurement of the stable isotopic composition of rocks that have undergone such an exchange history.

Rocks are exposed to fluids that are diverse in their origin and composition. The most important volatile constituents of natural fluids are water, carbon dioxide, carbon monoxide, hydrogen fluoride, sulfur compounds, and light hydrocarbons such as methane. In addition, fluids contain dissolved solids derived from the crustal rocks through which they pass. As the list of volatiles makes clear, the dominant elements present in fluids are oxygen, hydrogen, carbon, and sulfur.

An element may be characterized by its atomic number and atomic weight. Atomic number refers to the number of protons present in the nucleus of an atom and is a constant value for each element. Atomic weight is determined by adding together the number of protons and neutrons contained in an atom. Because the number of neutrons present often varies within a limited range, atoms of a particular element may have several different atomic weights. These atoms are referred to as isotopes. Oxygen, with an atomic number of 8, occurs most commonly with eight neutrons but may have nine or ten. Therefore, three isotopes of oxygen occur in nature: oxygen 16, oxygen 17, and oxygen 18. Similarly, carbon occurs as carbon 12 or carbon 13. (Carbon 14 is a radioactive isotope that forms in the Earth's upper atmosphere and will not be considered here.) Hydrogen contains only one proton; however, a small fraction of hydrogen atoms also contain a neutron, which doubles the mass of the atom. These heavy hydrogen atoms are called deuterium. Sulfur has four stable isotopes: sulfur 32, 33, 34, and 36.

STABLE ISOTOPE RATIOS

For any given mineral or fluid, the relative concentrations of the isotopes of a particular element may be expressed as a stable isotopic ratio, or the ratio of the second most abundant isotopic species over the most abundant isotope. Any physical process that results in the enrichment or depletion of the concentration of a heavy isotope is referred to as fractionation. A common fractionation process is evaporation of water. Water molecules containing the lighter isotope of oxygen (oxygen 16) will preferentially evaporate so that the remaining liquid will be enriched in the heavier oxygen isotope (oxygen 18). Fractionation also occurs during the growth of minerals in either a magma or a water-rich solution. Some minerals, because of the nature of their chemical bonds and their crystal structure, tend to concentrate a greater number of heavy isotopes than do other minerals. Quartz, dolomite, and calcite are common examples of minerals that contain high concentrations of heavy oxygen, while oxides such as ilmenite and magnetite have very little of the heavy isotope. The effectiveness of fractionation during mineral growth is dependent upon temperature. Low temperatures permit minerals to be more selective in choosing atoms for growing crystal sites, resulting in large differences in isotopic ratios between different minerals. At high temperatures, the selection of atoms is a more random process, and differences between isotopic ratios become progressively smaller.

Because the heavier isotopes of elements naturally occur in such small concentrations, isotopic ratios of oxygen and carbon, for example, have numerical values that are very small and difficult to measure accurately. For this reason, isotopic ratios for a particular sample are presented as a relative enrichment or depletion of the heavy isotope of an element as compared with a defined standard. The difference between the sample and the standard is measured in parts per thousand or per million and is expressed by the Greek letter delta (δ). For carbon isotopic values, for example, the accepted standard is called PDB and is obtained from a belemnite fossil of the Cretaceous-age Pee Dee formation of North Carolina.

EXCHANGE OF STABLE ISOTOPES

As fluids migrate through rocks, reactions occur that involve the exchange of stable isotopes between the fluid and the solid. The exchange process may be pervasive, where fluid movement is diffusive and affects the entire rock mass, or localized along specific fluid channelways, such as fractures, where only the wall of the rock is altered along the route of water movement. The degree to which the isotopic composition of the rock is altered depends on the initial composition of the rock and the fluid, the temperature at which isotopic exchange is occurring, and the amount of fluid present. Typically, the oxygen and hydrogen isotopic values for crustal water are light compared to those for most rocks. Therefore, as isotopic exchange proceeds, the isotopic composition of the rock becomes progressively lighter, while the water becomes increasingly enriched in the heavier isotopes. If the fluid-rock interaction has occurred under constant temperature conditions, the final isotopic value of the rock is proportional to the volume of water that has passed through the rock.

RESERVOIRS OF CRUSTAL FLUIDS

There are four principal sources, or reservoirs, of crustal fluids. Each of these fluid reservoirs contains stable isotopic ratios that reflect the fractionation mechanisms at work and the result of chemical reactions between the fluids and their host rocks. "Meteoric water" is a term applied to fluids that take part in the surface hydrologic cycle. Water that undergoes evaporation, precipitation, and runoff to lakes and to the ocean is capable of penetrating the Earth's crust to a depth of several kilometers. This penetration is usually accomplished by fluid migration along weaknesses in the crust, such as faults or fracture systems. Natural hot springs are an example of meteoric water that has been heated deep in the crust and then reemerges at the surface. The oxygen and hydrogen isotopic ratios associated with meteoric water are controlled principally by the distillation effect of evaporation and precipitation. As a result of the general transport of air masses from the equator toward the poles, meteoric water isotopic values vary in a systematic manner, with heavier ratios found near the Earth's equator and progressively lighter isotopic values occurring toward the poles. Meteoric water that is trapped in the pore spaces of accumulating sediments is referred to as connate water. When loose sediments have lithified to form hard sedimentary rock, the enclosed pore fluids may become isolated for very long periods. Connate water reveals an isotopic trend similar to that of meteoric water, except that the oxygen values tend to be heavier because of the capacity of the lighter oxygen (oxygen 16) contained in the trapped pore fluids to be exchanged for some of the heavier oxygen (oxygen 18) of the host sedimentary rock. Isotopic exchange between connate water and the host sediments continues until equilibrium is achieved. As rocks undergo increases in temperature and pressure associated with metamorphism, water is frequently released in a process called dehydration. The escaping water may be from pore spaces in the rock or from hydrous minerals such as micas or amphiboles. Because of their origin, these dehydration fluids are also referred to as metamorphic fluids. Dehydration water has a very wide range of isotopic compositions, which reflects the diversity of the original sediments. Juvenile water, which originates in the upper mantle, escapes from ascending magma and represents the fourth important fluid source. Not all water derived from magma should be considered juvenile, as meteoric or connate water will frequently be present in sediments that undergo melting deep in the crust. True juvenile water has a very narrow range of isotopic compositions. Because of mixing of crustal fluids and exchange reactions with igneous rocks, samples of unaltered juvenile water are found very rarely.

ISOTOPIC COMPOSITIONS OF ROCKS

Isotopic compositions of rocks reflect the formation history of the particular rock type. Igneous rocks are controlled by the composition of their magmatic source area, which is usually in the lower crust or upper mantle. Other factors include the temperature of crystallization, the type of minerals, and the degree to which the magma remains isolated or mixes with other constituents. Unaltered igneous rocks typically have a narrow range of isotopic values as compared with natural fluids. Deep mantle rocks contain minerals that are low in oxygen 18 and therefore fall in a narrow range of isotopically light compositions. Crustal rocks, with a higher proportion of silicate minerals, which concentrate oxygen 18, are isotopically heavy. Sedimentary rocks have two distinct modes of formation. Clastic sedimentary rocks are made of transported particles of weathered material and therefore have isotopic ratios that reflect the individual components. Chemical sedimentary rocks

precipitate directly and often involve biological activity. The oxygen isotopic values of these rocks are usually much heavier than those of fluids. Because of fractionation factors associated with organisms, carbon isotope ratios are highly variable. Rocks that have undergone metamorphism contain the widest range of isotopic compositions, as a result of chemical reactions in the presence of fluids that may be derived from any of the reservoirs previously described.

MASS SPECTROMETRY

The most important analytical tool in the study of fluid-rock interactions is the stable isotope mass spectrometer. This instrument separates isotopes according to their mass differences, as determined by the deflection of charged ions within a magnetic field. Elements of interest are extracted from minerals through appropriate chemical reactions and then converted to a gas, which is entered into the mass spectrometer. The sample gas is bombarded by a stream of electrons, converting the gas molecules to positively charged ions. The ions are accelerated along a tube, where a powerful magnet deflects the charged molecules into curved pathways. Lighter particles are deflected more than heavier ones, so several streams of ions result, each with a particular mass. The relative proportions of each isotope are measured by comparing the induced current produced by each ion stream at a collector.

RESEARCH INVOLVING IGNEOUS ROCKS

Studies of oxygen and hydrogen isotopes have been particularly useful in research on water-rock interactions involving igneous rocks. Hydrous minerals, such as biotite and hornblende, are separated out of granitic rocks in order to extract hydrogen isotopic values. Oxygen isotopic compositions are usually measured from feldspars and quartz. Two areas of water-rock study have been of particular interest. The first concerns the origin and quantity of water responsible for the isotopic alteration of large igneous intrusions within the continental crust. The second area of investigation addresses the interaction of ocean water with ocean-crust basalt and the formation of associated sulfide ore deposits.

The initial isotopic composition of igneous rocks is predictable according to their mineral content and temperature of crystallization. Therefore, it is possible to recognize when igneous rocks have been influenced by exchange with a fluid. Many examples have been found of shallow igneous intrusions that have been depleted in oxygen 18 through interaction with large volumes of isotopically light meteoric water. Hydrogen isotope exchange is even more sensitive than with oxygen, because igneous rocks have so little hydrogen rela-

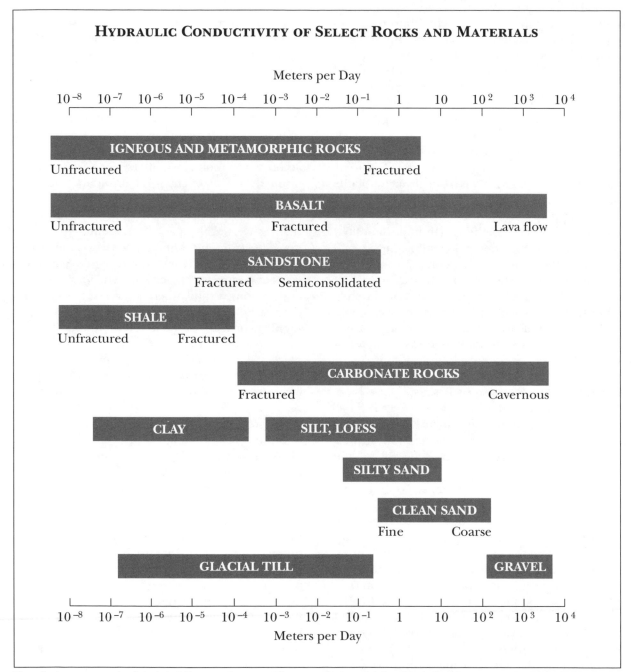

HYDRAULIC CONDUCTIVITY OF SELECT ROCKS AND MATERIALS

NOTE: Rocks vary tremendously in their ability to conduct water. The meters-per-day scale is logarithmic: Each increment to the right and left of 1 indicates a change by a power of 10. To the right, 10 meters, 1,000 meters, and 100,000 meters; to the left, 0.1 meter, 0.01 meter, 0.001 meter, and so on.

SOURCE: Ralph C. Heath, *Basic Ground-Water Hydrology*, U.S. Geological Survey Water-Supply Paper 2220, 1983.

tive to the amount in water. A very small quantity of water may produce a large isotopic shift in a rock's hydrogen value, while the oxygen is largely unaffected.

Isotopes associated with the ocean crust have been studied through cores obtained from seabed drilling and through the analysis of ophiolites, which represent portions of ocean crust exposed on land. In the vicinity of the mid-ocean ridge, where temperatures exceed 300 degrees Celsius, ocean water circulates to a depth of 3 or 4 kilometers and is responsible for depleting oxygen 18 by one or two parts per thousand.

STUDY OF METAMORPHISM AND PRECIPITATED ROCK

Fluids associated with metamorphism have also been intensively studied by stable isotopic methods. Increasing metamorphic grade is associated with progressively lighter isotopic values for oxygen and hydrogen. Areas of regional metamorphism may show two end-member types of fluid behavior. Consistent depletion of oxygen and hydrogen isotopic values throughout a large terrain points to exchange with an external source of water that flows pervasively through the region. Alternatively, only the trapped connate water is involved in exchange reactions, leading to higher isotopic composition values and to enrichment of deuterium in hydrous minerals. Because of the large difference in isotopic composition between magma and sedimentary rocks, contact metamorphism is particularly appropriate for study. Samples from intrusions indicate that the margin of the igneous rock is enriched in oxygen 18 through exchange with the country rock, while the interior of the body remains unaltered. The early stages of fluid-rock interaction are dominated by magmatic water, while meteoric water becomes increasingly important as cooling proceeds.

Sedimentary rocks that form by precipitation, such as limestone and chert, have heavy isotopic compositions as a result of the large fractionation between water and either calcite or quartz at low temperature. Clastic sandstones, which contain transported quartz grains, are characterized by the lighter isotopic values of the component particles. Sedimentary rocks selected from a large sample area frequently show a progressive change in the amount of isotopic depletion that has resulted

from water-rock exchange. As water continues the exchange process, the isotopic composition of the fluid also shifts. During the next increment of water-rock exchange, the potential amount of depletion of the rock will not be as great. By plotting isotopic values of sedimentary rocks, it is possible to determine directions of fluid motion.

SEARCH FOR ECONOMIC RESOURCES

Water-rock interactions are important primarily for their role in the formation of economically vital ore deposits. These are localized regions where metals such as gold, silver, lead, copper, zinc, and tungsten occur in unusually high concentrations and can be extracted. Water, circulating within the crust of the Earth, plays an important role in the formation of most ore deposits by leaching elements from rocks and concentrating them in zones of new mineral growth. During this process, the isotopic compositions of the host rock and fluid are progressively changed. Analysis of the resulting stable isotope ratios of the ore rocks allows geologists to understand the sources and quantity of the mineralizing fluids. Better understanding of the formation of ore deposits has led to greater success in the discovery of new mineral resources.

Stable isotope research into water-rock interaction is not limited to studying water-rich fluids associated with precious mineral deposits. Petroleum and natural gas flow from source rocks rich in organic material to porous reservoir rocks, where they may become trapped. Rocks through which hydrocarbons have migrated frequently show well-depleted carbon isotope values and thus preserve a record of fluid movement. Oil companies have used carbon data to track the migration history of hydrocarbons and to locate regions where petroleum leaks to the surface. The use of carbon isotopes has also proved successful in identifying source rocks associated with producing oil fields. This technology will become even more important as resources become increasingly scarce.

The source of fresh drinking water for almost half the population in the United States is subsurface groundwater. This crucial resource is jeopardized by contamination with common pollutants, such as pesticides, and by depletion through the withdrawing of water faster than it is replenished at recharge zones. Research involving stable isotope studies has become important in hydrology to track fluid movement within aquifers and identify sources of recharge water.

HAZARDOUS WASTE STORAGE

Another area of concern associated with water-rock interaction is the safe, long-term storage of toxic and nuclear waste. One of the most important criteria for the isolation of dangerous wastes is that the enclosing rocks be relatively dry and impermeable to water movement so that hazardous material is not transported into water supply aquifers. The record of water flow recorded by stable isotopes is sensitive to even very small fluid volumes and provides one of the means of assessing the risks associated with a toxic disposal site.

Grant R. Woodwell

CROSS-REFERENCES

Contact Metamorphism, 1386; Elemental Distribution, 379; Evolution of Earth's Composition, 386; Fluid Inclusions, 394; Fractionation Processes, 400; Freshwater Chemistry, 405; Geochemical Cycle, 412; Geothermometry and Geobarometry, 419; Groundwater Movement, 2030; Hydrologic Cycle, 2045; Hydrothermal Mineralization, 1205; Igneous Rock Bodies, 1298; Magmas, 1326; Mass Spectrometry, 483; Metamorphic Mineral Deposits, 1614; Nucleosynthesis, 425; Oxygen, Hydrogen, and Carbon Ratios, 431; Phase Changes, 436; Phase Equilibria, 442; Regional Metamorphism, 1421; Sub-Seafloor Metamorphism, 1427.

BIBLIOGRAPHY

Blatt, Harvey, and Robert J. Tracy. *Petrology: Igneous, Sedimentary, and Metamorphic.* New York: W. H. Freeman, 1996. Undergraduate text in elementary petrology for readers with some familiarity with minerals and chemistry. Thorough, readable discussion of most aspects of water-rock interactions. Abundant illustrations and diagrams, good bibliography, and thorough indices.

Bowen, Robert. *Isotopes and Climates.* London: Elsevier, 1991. Bowen examines the role of isotopes in geochemical phases and processes. This text does require some background in chemistry or the Earth sciences but will provide some useful information about isotopes and geochemistry for someone without prior knowledge in those fields. Charts and diagrams help clarify difficult concepts.

Faure, Gunter. *Principles of Isotope Geology.* New York: John Wiley & Sons, 1977. A college-level text that covers both radioactive and stable isotopes. The first five chapters are introductory in nature and include a good historical review of the development of isotope geology and mass spectrometry. The last four chapters cover stable isotopes and include figures reproduced from class research papers. Each chapter includes a detailed reference list.

Gregory, Snyder A., Clive R. Neal, and W. Gary

Ernst, eds. *Planetary Petrology and Geochemistry.* Columbia, Md.: Geological Society of North America, 1999. A compilation of essays written by scientific experts, this book provides an excellent overview of the field of geochemistry and its principles and applications. The essays can get technical at times and are intended for college students.

Hoefs, Jochen. *Stable Isotope Geochemistry.* 3d ed. New York: Springer-Verlag, 1987. Suitable for an advanced college student who seeks a detailed discussion of isotope fractionation, sample preparation, and laboratory standards. The material is introduced in three sections. The first chapter provides theoretical principles; the second chapter is a systematic description of the most common stable isotopes; and the third summarizes the occurrence of stable isotopes in nature. An extensive list of references is included at the end of the book.

Krauskopf, Konrad B. *Introduction to Geochemistry.* 2d ed. New York: McGraw-Hill, 1979. A comprehensive advanced text that covers most aspects of the chemistry of natural fluids. Radioactive and stable isotopes are briefly treated in chapter 21, while chapter 17 contains a discussion of ore-forming solutions. This resource is particularly useful for students who seek detailed information on the chemistry and interaction of crustal water. Suggestions for further reading are provided at the end of each chapter.

O'Neil, J. R. "Stable Isotope Geochemistry of Rocks and Minerals." In *Lectures in Isotope Geology*, edited by Emilie Jäger and Johannes C. Hunziker. New York: Springer-Verlag, 1979. This source provides a brief and clear introductory section on stable isotope nomenclature. The remainder of the chapter outlines major conclusions drawn from isotope analysis of igneous, metamorphic, sedimentary, and ore deposit rocks. Examples are provided from pioneering research studies. Although the text is oriented toward the college level, high school students interested in the results of isotope studies will find this chapter useful.

Smith, David G., ed. *The Cambridge Encyclopedia of Earth Sciences*. New York: Crown Publishers, 1981. Chapter 8, "Trace Element and Isotope Geochemistry," is a brief, well-illustrated summary of the occurrence of trace elements, stable isotopes, and radiogenic elements. The chapter emphasizes how trace element and isotope studies have enhanced understanding of processes such as the generation of magma and the occurrence of ore deposits. The discussion of water-rock interaction associated with the ocean crust would be accessible to advanced high school students. Few additional references are offered.

9
TECHNIQUES OF GEOCHEMISTRY

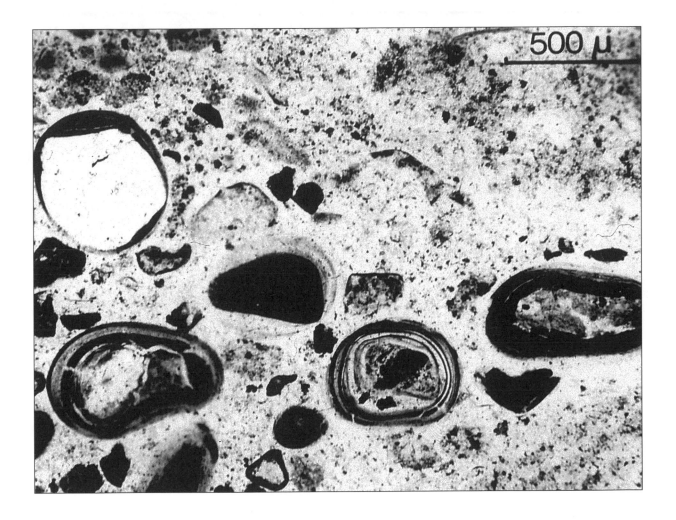

ELECTRON MICROPROBES

The electron microprobe is an analytical tool used to determine the elemental composition of Earth materials. Very small samples approximately 5 microns in diameter can be analyzed, making this instrument extremely valuable to geologists.

PRINCIPAL TERMS

DIFFRACTION: a process that allows photons of a specific wavelength to be analyzed

ELECTRON: one of the fundamental particles of which all atoms are composed; it has an electrical charge of −1

ELECTRON SHELL: a region around the nucleus of an atom that contains electrons; each electron in each shell will have a specific energy associated with it

PHOTON: a form of energy that has the properties of both particles and waves; electromagnetic (light) radiation

DEVELOPMENT OF ELECTRON MICROPROBE

Geologists often want to determine the concentration of elements in minerals and glasses. Prior to the invention of the electron microprobe, elemental analysis of geological samples was difficult and time-consuming. Furthermore, the analysis of very small samples, less than 0.1 gram, was virtually impossible. The electron microprobe provides geochemists with a rapid means of determining the elemental composition of geological samples even as small as about 5 microns in diameter. This ability to analyze extremely small samples has provided geochemists with a new understanding of the processes that form the rocks found within the Earth.

In 1947, James Hillier made the first patent application for an apparatus that would use a focused beam of electrons as a source of energy to analyze solid materials. R. Castaing and A. Guinier (1949) presented a paper in Delft, Netherlands, on the application of the electron microscope to analysis of samples, and they drew heavily on Hillier's earlier work. In 1951, Castaing developed the first usable electron microprobe as part of his dissertation research at the University of Paris. Thus, a whole new era of analytical research was born for geochemists. Although crude by today's standards, the instrument that Castaing and Guinier built contained all the fundamental elements of the modern and highly sophisticated electron microprobes.

FUNDAMENTAL PROCESS OF ANALYSIS

The electron microprobe is a complex machine, but the theory behind the analysis is relatively easy to understand. All materials are composed of atoms. These atoms contain clouds of electrons; they surround the nucleus, which is composed of neutrons and protons. The clouds of electrons are called shells and have a variety of geometric shapes. Although the exact positions of the electrons within each shell cannot be predicted, each has associated with it a discrete and known energy. These energy values are different for each element in the periodic table. Shells that are closer to the nucleus are called inner shells, and those farther away are called outer shells.

If an electron can be removed from an inner shell, then an electron from an outer shell will drop into the empty position within the inner shell. As the electron falls from an outer shell into an inner shell, it releases some energy in the form of photons. Each photon will have a characteristic wavelength associated with it, which is a function of this change in energy as the electron moves from one shell to another. Because the change in energies for all the electrons in all the elements is known, one can determine which element was responsible for the production of the photon simply by knowing the wavelength of the photon. The relative concentration of each element can be determined by counting the number of photons of a characteristic wavelength that have been generated. Thus, the type and concentration of each element in a geological sample can be determined.

Electrons are very small but can be removed by other electrons if they are moving fast enough. By accelerating a beam of electrons to a high velocity onto the sample surface, scientists can randomly "knock out" electrons of the shells surrounding the nucleus. That is the fundamental process whereby the electron microprobe performs an elemental analysis of a mineral or glass.

PARTS AND PROCESS

All electron microprobes must contain the following fundamental parts: a filament that acts as a source of electrons, an anode used to accelerate the electrons, a series of electromagnets that focus the beam of electrons, a sample holder, a crystal spectrometer used to determine the wavelength of the emitted photons, a photomultiplier tube used to determine the number of emitted photons, a vacuum chamber that contains all the previous parts, and a reporting device such as a computer.

The filament is a very thin wire that carries an electrical current that causes electrons to be emitted. These electrons are attracted to a positively charged metal plate with a hole in it through which the electrons pass. Because electrons have a negative charge, their path can be bent by electromagnets. Electron microprobes use a series of electromagnets that can adjust the focus of the electron beam from a large surface area (about 5 millimeters) to a very fine point (about 5 microns) on the surface of the sample.

The sample holder on modern instruments usually contains slots for up to ten samples and standards. The standards are used to calibrate the instrument for the material that is to be analyzed and have been subjected to rigorous elemental determinations by independent laboratories. The sample holder can be moved so that many different determinations can be made on each sample. As the sample is bombarded by electrons, photons are emitted in all directions, some of which strike a series of crystal spectrometers. These spectrometers are composed of special crystals whose structures act as diffraction gratings for the photons. Thus, they can be tuned to select specific photons of the characteristic wavelength desired for the element under analysis.

Those photons that pass through the crystal spectrometers are counted by photomultiplier tubes, also called the detectors. Each detector sends a signal to a computer that then converts the number of counts to relative concentrations for the elements being analyzed. The entire system is con-

tained in a vacuum because electrons and photons can be absorbed by the air, which will reduce the lowest concentrations that can be analyzed.

ELEMENTAL MAPPING TECHNIQUES

Within the geological sciences, the electron microprobe has become one of the most valuable tools to modern geochemists. It is used to find the composition of virtually all naturally occurring minerals and glasses to an accuracy previously unattainable. The electron microprobe can thus be used to understand the variation of the composition of naturally occurring materials over very small distances. The electron microprobe has been used to understand the origin and evolution of rocks, to determine the composition of new minerals, to develop new theories for the formation of ore deposits, and to help study how the Earth has been affected by human activities.

Prior to the invention of the electron microprobe, the best that a geochemist could hope for was the bulk composition of single large crystals found within rocks. Furthermore, older analytical techniques were, for the most part, highly destructive to the sample under investigation. Thus, once the analysis was complete, the sample was usually lost and no check on accuracy was possible. With the invention of the electron microprobe, these problems were solved. The sample is prepared by making a highly polished surface. By virtue of the very fine focus available on the electron microprobe, many points on the sample can be analyzed. This process is called elemental mapping of samples.

By using elemental mapping techniques, geochemists can determine the variation of the composition from the core to the rim of minerals. Much information on the origin and evolution of rocks can be deduced using this method of research. The composition of igneous rocks (those rocks that have solidified from a molten magma) often changes during the crystallization history. By determining the chemical variation in minerals that formed during the crystallization of the magma, geologists can understand better how and in which sequence these minerals formed. Quite often, the composition of certain minerals found in rocks is a function of the temperatures and pressures of the environment in which they formed. Thus, these two parameters can be determined using an electron microprobe. That is especially valuable for metamorphic rocks (those rocks that formed from previously existing rocks as a result of a change in temperature and pressure) as well as for igneous rocks. Sedimentary rocks (those rocks that form at or near the Earth's surface) are also good candidates for study using the electron microprobe. Many sedimentary rocks have been subjected to a variety of processes during their formation, which include erosion, transportation, and deposition. The very small changes in the composition of the minerals found in sedimentary rocks often reflect processes that formed them.

ANALYZING RARE AND VALUABLE MATERIALS

This instrument, which is used daily on samples collected from the Earth, was central to scientific study of the origin and evolution of the rocks that were collected on the Moon. Thus, for example, the modification of the Moon's surface by meteorite impacts has been understood, in part, by using the electron microprobe on samples collected by the Apollo mission astronauts. Furthermore, geologists have been able to determine the composition of the materials collected on the Moon, which, in turn, has helped them in understanding the origin of the Earth.

Geologists are continuously finding minerals that have never been described. Although the crystal structure of these new minerals can be determined using an X-ray diffractometer, the composition is best found using the electron microprobe. Most minerals that are discovered today are rare, and it is imperative that they be able to be preserved. For this reason, the electron microprobe is an ideal instrument to use.

Economic ore deposits are occurrences of rocks or minerals that may be extracted from the Earth at a profit. Exploration geologists need to understand how known ores formed so that they can find new deposits. One key part of this research is understanding what processes cause a variation in the mineral composition found in such occurrences. Thus, large mining companies often own electron microprobes as analytical tools.

APPLICATION TO ENVIRONMENTAL GEOLOGY

Finally, the Earth has been affected by the industrial activity of humankind. In an effort to repair the damage caused by this activity, geologists are called upon to study how certain activities have affected Earth materials. They are also asked to find ways to prevent further damage to the Earth.

The application of the electron microprobe is widespread in the area of environmental geology. One example involves the use of clay minerals to act as a barrier to toxic materials by absorbing them onto the clay's surface; by finding the amount of a given toxic material that has been absorbed by a particular clay, geologists can decide whether the clay is a suitable barrier in that case. Environmental scientists have needed the aid of the electron microprobe to provide quality microscopic scale analyses of geological materials to aid in resolving other issues as well. These include the amount of hazardous material released by mining operations into the environment, and the migration of toxic substances in groundwaters.

A. Kem Fronabarger

CROSS-REFERENCES

Diagenesis, 1445; Earth Resources, 1741; Electron Microscopy, 463; Experimental Petrology, 468; Geologic and Topographic Maps, 474; Geothermometry and Geobarometry, 419; Infrared Spectra, 478; Ionic Substitution, 1209; Magmas, 1326; Mass Spectrometry, 483; Minerals: Physical Properties, 1225; Minerals: Structure, 1232; Neutron Activation Analysis, 488; Petrographic Microscopes, 493; Regional Metamorphism, 1421; X-ray Fluorescence, 499; X-ray Powder Diffraction, 504.

BIBLIOGRAPHY

Birks, L. S. *Electron Probe Microanalysis.* 2d ed. New York: Wiley-Interscience, 1971. This classic book explains the history of the development of the electron microprobe, its theory, and its use. Written at a technical level, it requires a fundamental understanding of chemistry and geology. Several excellent technical articles on the subject are listed. Figures within the text are good.

Elion, Herbert A., and D. C. Stewart. *A Hand-*

book of X-Ray and Microprobe Data. Elmsford, N.Y.: Pergamon Press, 1968. This handbook is used by geochemists to interpret data obtained from the microprobe. Highly technical and intended for the advanced student. Tables and lists are extremely complete.

Heinrich, Kurt F. J., ed. *Quantitative Electron Probe Microanalysis.* U.S. National Bureau of Standards Special Publication 298. Washington, D.C.: Government Printing Office, 1968. This book is designed to introduce the student to the methods that are acceptable for obtaining quantitative results from the electron microprobe. Provides some understanding of the theory of the instrument and offers additional sources of information.

McKinley, Theodore D., Kurt F. J. Heinrich, and D. B. Wittry, eds. *The Electron Microprobe.* New York: Wiley, 1966. This series of articles discusses the state of the art of the electron microprobe during the 1960's. Although somewhat dated concerning advances in automation and standardization, it does provide a remarkably complete discussion of the theory and use of the microprobe. Some of the articles can be understood by the freshman college student with limited background. Each article has an excellent bibliography.

Murr, Lawrence Eugene. *Electron and Ion Microscopy and Microanalysis: Principles and Applications.* 2d ed. New York: M. Decker, 1991. Murr's book describes in great detail the principles, theories, and applications of modern microscopy. Sections are devoted to the field study of electrons and ions, as well as advancements in optical engineering. Written for the student with a background in the sciences. Illustrations, bibliography, and index.

Potts, Phillip J., John F. W. Bowles, Stephen J. B. Reed, Mark R. Cave, et al., eds. *Microprobe Techniques in the Earth Sciences.* London: Chapman and Hall, 1995. This book describes and illustrates many techniques and the theories concerning microprobe analysis in relation to the Earth sciences. With a strong focus on analytical geochemistry, the book can be technical at times.

Reed, S. J. B. *Electron Microprobe Analysis and Scanning Electron Microscopy in Geology.* Cambridge: Cambridge University Press, 1996. A thorough look into electron and petrofabric microprobe anaylsis, this book examines the techniques, tools, and procedures involved in scanning electron microscopy. Illustrations, index, and bibliographic references.

ELECTRON MICROSCOPY

Electron microscopy uses a "bombardment" of electron particles rather than light beams to obtain magnified images of specimen material. The process depends on carefully placed electromagnetic fields to focus the negatively charged electron particles. Much greater detail in magnification is obtained because electron particle waves are approximately four times shorter than are light waves.

PRINCIPAL TERMS

ANODE: in early cathode-ray tubes, this positively charged plate attracted negatively charged electrons that, passing through a tiny aperture in the plate, formed an electron beam

CATHODE-RAY TUBE: a tubular device, the interior chambers of which are vacuum-sealed, through which an electron beam passes, recording a dotlike image on a fluorescent screen

CONDENSER LENS: the first stage of electromagnetic focusing, which "bends" the electron beam into a tightly concentrated focal point before it passes through the specimen

ELECTRONIC LENS: electromagnetic fields inside the electron microscope that interact with the electron beam, bending the trajectory of its particles

OBJECTIVE LENS: the second, magnifying stage of electronic lens focusing, which occurs after the electron beam passes through the specimen

SCANNING ELECTRON MICROSCOPY (SEM): a later (mid-1960's) development in electron microscopy, in which surface structure images are obtained by a process that records patterns of electron emissions off the surface of the specimen as it is subjected to the impact of the microscope beam

TRANSMISSION ELECTRON MICROSCOPY (TEM): electron microscopy in which all elements of electron activity involved in contact with the specimen are transferred simultaneously to the image

DEVELOPMENT OF THE TECHNOLOGY

Electron microscopy involves the substitution of a beam of electrons and a series of electric or electromagnetic fields in the place of light beams and optical glass lenses in the common microscope. Because electron particles exhibit the same—but much shorter—wavelike movements found in light, electron beam focusing allows a much higher degree of resolution, or detailed contrast, when images of the effects of their passage through specimens are recovered, after magnified refocusing, on the fluorescent screen of an electron microscope.

A first step in what would later become standard technology came in 1895, when Wilhelm Röntgen, a German physicist, noted that solid materials released invisible rays in the first stage of what was then called a cathode-ray device (later known in the field of electron microscopy as an electron "gun"). It took some time for scientists to

discover not only that these atomic-particle rays shared the electromagnetic characteristics of light but also that they moved in wavelengths that were about four times shorter than those of light waves.

Among the earliest scientists to pioneer the field of electron microscopy was Sir Joseph John Thomson, an Englishman. In 1897, Thomson performed an experiment with electron rays that would yield the basic technology for the first television tube and, with essential adaptations which were developed by the German physicist Ernst Ruska in the 1930's, the first electron microscope. In essence, Thomson's cathode-ray device, which was not yet conceived of in terms of magnifying possibilities, created a beam of tiny negatively charged particles, or cathode rays—that is, electrons. This beam resulted from a glow discharge in a gas and an incandescent metallic filament in a vacuum. These rays (in reality, negatively charged electron particles) were caused to accelerate toward and pass through a tiny hole in a positively charged plate that served as an anode. The result was a beam of high-velocity rays, which could be recorded on a fluorescent screen. Thus was established the basic technology that would later produce the first television screen, when multiple electronic "pinpoints" would be used to create a complex image. Use of this process to obtain magnification of objects placed in the path of a ray, or electron beam, only became possible when what came to be called an electronic lens was developed.

STAGES OF MAGNIFICATION

Starting with the basic principle of a cathode-ray tube, the pioneers of electron microscopy experimented with several technological adaptations that aimed at focusing electron beams in order to obtain magnifying effects when focused rays are refocused and projected onto a fluorescent screen. In the simplest of terms, the electron lens magnifies by several stages. Each stage must take place in a vacuum.

After passing through the tiny aperture in the positively charged anode (the first phase in Thomson's 1897 cathode-ray experiment), the electron beam is acted upon by a first electronic lens, called the condenser lens. This process results from the effect of an electric or electromagnetic field, located within the tube device, that interacts with the "descending" electron beam. The effect of such an electronic lens is to "bend" electron paths toward an axis, just as converging glass lenses bend light rays toward a focal axis. Once focused by the condenser lens, the highly concentrated electron beam strikes, in a separate chamber sealed off by O-rings, a carefully prepared specimen. The beam actually passes through the object to be magnified. Because the beam is made up of particles, however, the physical effect of particle bombardment will vary according to varying levels of material density in the specimen itself. This is the key to the lighter/darker image produced in the final stage of image projection.

As the electron beam emerges (on the downward side of the specimen), two other electronic lenses (electromagnetic fields) influence its path. First, a minutely focused objective lens spreads the concentrated beam created by the condenser lens. Then, a projector lens "captures" the resultant image, which is recorded on a fluorescent screen as a picture. Details revealed in this image—as in images associated with common photography—appear lighter when highly exposed (that is, when large numbers of electrons passed directly through less dense areas of the specimen) and darker when underexposed (a substantial number of electrons were "scattered" upon impact with denser areas within the specimen). Because electron waves are approximately four times shorter than are light waves, the amount of minute detail recorded in this image (that is, the high degree of resolution obtained) promised to thrust electron microscopy to the forefront of laboratory research, especially following first stages of commercialization of the technology pioneered by Ernst Ruska in the 1930's. Ruska, along with two other scientists involved in the advanced technology of scanning tunneling microscopy, would be awarded the Nobel Prize in Physics in 1986.

SOLUTION OF TECHNICAL PROBLEMS

Eventual success in promoting electron microscopy originally hinged on the solution of several technical problems associated not with the phenomenon of electromagnetic lenses (which proved to be surprisingly easily adjustable, merely by varying the intensity of current) but with the effects of electron "bombardment" on different types of specimens. Effective solutions to several such problems were not found until several years after World War II, when not only German but also other firms producing scientific instruments would compete for commercialization of electron microscopy.

The most obvious technical problem was connected with the heat created in the process of concentrating a stream of electrons. This heat enters the specimen even if the time of exposure is very limited. Absorption of externally induced heat into the specimen caused automatic image distortion because of increased molecular agitation—a state that would not have existed prior to exposure to an electron beam. One way this problem was reduced was by use of condenser lenses to create what is called "small region radiation" techniques. Double-stage focusing reduced the area of electron bombardment to an absolute minimum and, therefore, reduced the number of electrons (that is, the intensity of the bombardment and thus the origin of heat) that were needed to obtain a detailed image. These developments increased the possibilities of using electron microscopy in research involving organic specimens.

A second technical problem involved the condensation of minor residual gases, mainly hydrocarbons, inside the chambers within which the various stages of electromagnetic focusing take place. Such condensation was particularly bothersome if it gathered on the specimen itself, causing a general darkening, and even distorting of the image produced. Research directed by the electron microscope's original inventor in Berlin, Ernst Ruska, solved this problem by "bathing" all surfaces surrounding the specimen with superchilled liquid air at the time of each experiment, which kept the specimen warmer than any of the other elements in its environment. Thus, any condensation that might occur gathers on surrounding elements rather than on the specimen itself.

SCANNING ELECTRON MICROSCOPY

What has been described up to this point applies to what is known as transmission electron microscopy (TEM), in which all elements of electron activity involved in contact with the specimen are transferred simultaneously to the image. In addition, researchers, again working mainly in Germany beginning in the mid-1930's, extended the technology of electron microscopy into a somewhat more complex domain—that of scanning electron microscopy (SEM). SEM technology relates to an entire subfield known as surface studies. Here, the image is built up

in time sequence as a far tinier electron beam than that used in TEM moves across ("scans") the specimen. The image is derived from secondary electron currents that are released from a very thin surface layer in the atomic structure of the specimen being examined. The technology needed to detect, and then to "capture," the image of such surface-layer emissions accurately took a number

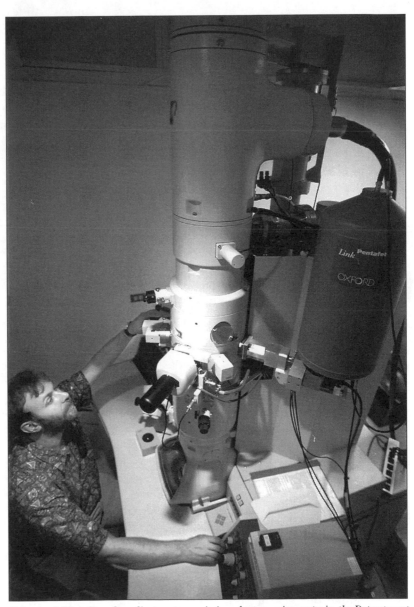

Scientist Adrian Brearley adjusts a transmission electron microscope in the Department of Earth and Planetary Sciences at the University of New Mexico. Brearley published a paper in a 1999 issue of Science *suggesting that a meteorite that fell in New Mexico in 1969 may have once had water in its composition. (AP/Wide World Photos)*

of years to develop. Even following a breakthrough in secondary electron retrieval by British researchers at the University of Cambridge in 1948, the first commercial production of an effective SEM did not come until 1965. Since that date, SEMs have been used primarily to create images that resemble three-dimensional photography of the "topography" (meaning, cross sections of molecular structures) at the surface of specimen materials.

SCIENTIFIC AND INDUSTRIAL APPLICATIONS

A number of special fields of research relating to the Earth sciences depend on diverse applications either of basic transmission electron microscopy or of scanning electron microscopy. Among them should be mentioned dark-field electron microscopy, in which the angle of the bombarding electron beam is "tilted" to produce bright spots in a darker field—a method that is particularly useful in determining whether there are crystalline structures in the specimen. The importance of scanning electron microscopy, together with computerized simulation in three dimensions, is particularly apparent in the field of crystallography and in the detailed analysis of minerals.

By far the most widely recognized application of electron microscopy to earth sciences, however, is to be found in fields relating to the petroleum industry and its by-products, either petrochemicals or related synthetic materials. In such areas, use of analytical electron microscopic techniques enables researchers to observe variations in molecular linkages that characterize "families" of synthetic products derived from petroleum. The results of such findings are critical in the search for new synthetic materials that are often much better suited to the specialized needs of modern industry and aerospace science than are natural metals and their alloys.

It is important to note that electron microscopy involves applied uses as well as invance uses. Some of the former enable alteration of the molecular or atomic structures of natural substances, especially metals and their alloys, in ways that can make them more useful for technological purposes. An example is to be found in amorphization technology, which utilizes the electron irradiation process in the electron microscope to cause a break in crystalline periodicity (regular, latticelike chains in the atomic structure of substances), creating what specialists call a "zigzag" atomic chain that alters the basic electronic and/or chemical nature of the substance in question. Without necessarily being conscious of the fact, humans are surrounded by materials, many of them originally of natural mineral or metallic origin (especially those necessary for use as conductors in high-energy-intensity situations), that have been "transformed" for a number of special functions by processes connected with electron microscope technology.

Byron D. Cannon

CROSS-REFERENCES

Electron Microprobes, 459; Experimental Petrology, 468; Geologic and Topographic Maps, 474; Infrared Spectra, 478; Mass Spectrometry, 483; Minerals: Structure, 1232; Neutron Activation Analysis, 488; Oil Chemistry, 1711; Petrographic Microscopes, 493; X-ray Fluorescence, 499; X-ray Powder Diffraction, 504.

BIBLIOGRAPHY

Beyer, George L., et al., eds. *Microscopy.* New York: Wiley, 1991. This reference tool is a wonderful introduction to microscopy. It contains descriptions of microscopes, the history of their use, and research techniques. The authors analyze microscopy procedure and protocol in a clear and understandable manner. Includes illustrations, diagrams, index, and bibliography.

Fujita, Hiroshi. "The Process of Amorphization Induced by Electron Irradiation in Alloys." *Journal of Electron Microscopy Technique* 3 (1986): 245-256. This article investigates one area in which electron microscopy proved to be more useful to metallurgists than any other method in altering the atomic structure of key alloys. What is involved here is the careful calculated removal of crystalline periodicity (regular "latticelike" chains of atoms) in key substances through irradiation.

Hajibagheri, Nasser, ed. *Electron Microscopy Methods and Protocols.* Totowa, N.J.: Humana Press, 1999. This guide book discusses advancements and protocols in molecular and electron microscopy. Many of the procedures discussed are accompanied by an illustration. Includes bibliographic references and an index.

Hunter, Elaine, Evelyn. *Pratcical Electron Microscopy: A Beginner's Illustrated Guide.* 2d ed. New York: Cambridge University Press, 1993. As the title suggests, this is an excellent guide book for the beginner. Includes step-by-step descriptions of common techniques and practices within the electron microscopy field.

Johnson, John E., Jr. "The Electron Microscope: Emerging Technologies." *Journal of Electron Microscopy Technique* 1 (1984): 1-7. The maiden article in the specialized journal that provides up-to-date terminology and changing methodologies for the entire field of electron microscopy. Of particular interest in this survey of the "state of the art" in 1984 is a discussion of analytical electron microscopy, an emerging subfield dedicated to the identification of types of atoms present in a given specimen and their probable arrangement.

Lee, W. E., and K. P. D. Lagerhof. "Structural Electron Diffraction Data for Sapphire." *Journal of Electron Microscopy Technique* 2 (1985): 247-258. This article discusses methods for using electron microscopy and computerized simulation to obtain three-dimensional images of crystalline structures. The discussion focuses specifically on sapphires, but the technology applies to a wider field of crystal structures.

Reed, S. J. B. *Electron Microprobe Analysis and Scanning Electron Microscopy in Geology.* Cambridge: Cambridge University Press, 1996. A thorough look into electron and petrofabric microprobe anaylsis, this book examines the techniques, tools, and procedures involved in scanning electron microscopy. Illustrations, index, and bibliographic references.

Ruska, Ernst. *The Early Development of Electron Lenses and Electron Microscopy.* Translated by Thomas Mulvey. Stuttgart, West Germany: S. Hirzel Verlag, 1980. An autobiographical account by (1986) Nobel laureate Ernst Ruska. Concentrates not only on Ruska's own contributions to the mid-1930's invention of the first successful electron microscope but also on other scientists' work in the field before and after.

Swift, J. A. *Electron Microscopes.* New York: Barnes & Noble Books, 1970. Probably the most accessible and easily understood guide to electron microscopy. Different techniques of specialized work with electron microscopes are surveyed, with appropriate bibliographical references for more detailed coverage.

Watt, Ian M. *The Principles and Practice of Electron Microscopy.* 2d ed. New York: Cambridge University Press, 1997. This college-level textbook is a complete survey of electron microscopy. Goes beyond mere general treatment of a number of specialized areas, especially where scanning methods—usually the object of brief discussion—are concerned.

EXPERIMENTAL PETROLOGY

Experimental petrology is the laboratory simulation of chemical and physical conditions within and at the surface of the Earth. A wide variety of apparatuses are used to obtain routinely and control precisely the range of temperature and pressure conditions known to occur up to 150 kilometers in depth. Other apparatuses allow access to the much greater temperatures and pressures of the transition zone of the Earth's mantle. Experimental data allow petrologists to interpret quantitatively the evolution of natural rocks and the Earth as a planet.

PRINCIPAL TERMS

COMPONENT: a chemical entity used to describe the compositional variation of some phase

IGNEOUS ROCK: any rock that forms by the solidification of molten material, usually a silicate liquid

METAMORPHIC ROCK: any rock whose mineralogy, mineral chemistry, or texture has been altered by heat, pressure, or changes in composition; metamorphic rocks may have igneous, sedimentary, or other, older metamorphic rocks as their precursors

MINERAL: a naturally occurring solid compound that has a specific chemical formula or range of composition; a mineral normally has regular crystal structures such that its internal arrangement of atoms is predictable

PHASE: a chemical entity that is generally homogeneous and distinct from other entities in the system under investigation; compositional variation within phases is described in terms of components

PHASE EQUILIBRIA: the investigation and description of chemical systems in terms of classical thermodynamics; systems of specified composition are generally investigated as a function of temperature and pressure

THERMODYNAMICS: the area of science that deals with the transformation of energy and the laws that govern these changes; equilibrium thermodynamics is especially concerned with the reversible conversion of heat into other forms of energy

TEMPERATURE AND PRESSURE CONDITIONS

A wide range of chemical and physical conditions is attained within and at the surface of the Earth. Experimental petrology is the simulation of these conditions in a carefully controlled laboratory environment. Temperature and pressure are the two primary physical parameters that change during geological processes. On or near the Earth's surface, temperatures ranging between 0 and 1,300 degrees Celsius are observed; such temperatures are easily attained at low pressure in the laboratory. Much higher temperatures, however, occur deep within the Earth and other planets. Pressure increases proportionally with increasing depth below the Earth's surface as a direct result of the greater mass of material overlying the material below. The pressure at the Earth's surface is that exerted solely by the overburden of the atmosphere, which defines the low pressure limit attained in geological processes.

At a depth of 3,000 kilometers below the Earth's surface, however, near the boundary between the solid silicate mantle and the liquid outer core, the pressure approaches 1.2 million times that exerted by the atmosphere on the surface. As the pressure within the Earth increases, so does the temperature, such that at the core-mantle boundary, the temperature approaches nearly 3,000 degrees Celsius. Experimental petrologists have apparatus that will routinely obtain pressures of fifty thousand times that of the atmosphere and temperatures up to 2,000 degrees Celsius, which corresponds to the physical conditions that occur at approximately 150 kilometers below the Earth's surface. Other devices available to petrologists since 1980 allow access to temperatures and pressures as great as those found at 1,000 kilometers deep.

468

NATURAL- AND SYNTHETIC-SYSTEM EXPERIMENTS

The style of experiments done by petrologists in order to quantify the conditions of formation of a particular suite of rocks or some widely occurring rock type is nearly as varied as is the number of active investigators. These experiments, however, may be broadly grouped into two major categories: The first would include all experiments done on natural rock or mineral systems, while the second would include experiments done on simpler synthetic systems that are analogous to the much more complex natural systems in some way.

Most experimental petrologists tend to work within one of these categories almost exclusively. Experiments on natural systems are necessarily more complicated and difficult to interpret because even the simplest rocks contain three to four chemical components (the simplest chemical entity that may be used to describe the system under consideration), while most rocks contain ten major and several minor components. In contrast, experiments on synthetic systems can be designed to isolate and study an individual chemical component. In such experiments it is considerably less difficult to demonstrate attainment of equilibrium and interpret the finding in terms of classical thermodynamics.

QUENCH AND GAS-MIXING FURNACES

Experimental petrologists have developed many different apparatuses that allow them to achieve conditions in the laboratory that mimic those found in the Earth. The simplest apparatuses are quench and gas-mixing furnaces. Typical working conditions for these apparatuses are pressures that range from moderate vacuums to 1 atmosphere and temperatures from 0 to 1,600 degrees Celsius. Slight variations in pressure on the order of 1 percent normally occur during the operation of these furnaces. Temperatures may be controlled and measured with a precision of several to tens of degrees depending on the particular setup.

Standard 1-atmosphere quench furnaces are set up with a vertical ceramic tube around which some resistive heating elements are either wound or placed in close proximity. The elements are then heavily insulated to prevent large heat losses to the laboratory atmosphere. The vertical geome-

try is required to enable rapid quenching of material held in the hot zone of the furnace to water at 25 degrees Celsius (or some other suitable quench medium) at the lower end of the vertical tube. Samples are generally held at the end of a ceramic rod by thin platinum loops. At the end of an experiment, the platinum loops are melted by passing a small electrical current through them, which allows the samples to fall by gravity directly into the water, where they are cooled rapidly.

Standard vertical tube furnaces may be modified to include the capability to have a gas mixture of known composition flow through the tube, replacing the static air environment. Such gas mixtures are commonly composed of species of carbon and hydrogen that, when mixed in known proportions, will fix the oxygen activity of the furnace atmosphere. The control of oxygen activity allows experimental petrologists to investigate chemical systems that contain transition metal cations, whose valence state would otherwise change in an uncontrolled and perhaps unwanted way. Samples range in size from 0.1 to 50 milliliters. One-atmosphere furnaces also may be mounted horizontally, if rapid quenching and gas mixing are not required. These apparatuses are found in virtually every experimental petrology laboratory around the world and have been used to investigate a wide variety of petrological problems.

COLD-SEAL VESSELS

Another common apparatus found in experimental petrology laboratories is the cold-seal vessel designed by O. F. Tuttle in the late 1940's. These apparatuses have been enormously important in the investigation of metamorphic and igneous processes that occur at middle to lower crustal levels. Cold-seal vessels typically are operated at pressures of several hundred to several thousand atmospheres and between 25 and 900 degrees Celsius. Modifications of the original design have allowed petrologists to obtain pressures up to 12,000 atmospheres and slightly higher temperatures. The apparatus is quite simple: The pressure vessel is fabricated from a strong alloy usually composed of nickel and chromium with smaller amounts of other metals. The rod has a small-diameter hole drilled into it to yield a container that is similar in shape to a test tube. The walls of the vessel are kept thick to support the high pressures and temperatures that occur during the course of an experiment. A pressure seal is formed by a cone-in-cone fitting at the open end of the vessel. High temperatures are obtained by placing the pressure vessel inside a simple muffle furnace, which is usually mounted vertically. The pressure seal remains outside the furnace and thus remains cold throughout the experiment. The pressure medium in most experiments is water, but to obtain pressures of more than 8,000 atmospheres, argon gas is used.

PISTON CYLINDERS

The last common apparatus found in experimental petrology laboratories is called a piston cylinder. These apparatuses were originally designed and built in the late 1950's and early 1960's. Much of the current understanding of melting relations in basaltic and ultramafic systems has been gained by using the piston cylinder. The typical operating conditions range from pressures of 5,000 to 60,000 atmospheres and temperatures between 25 and 1,800 degrees Celsius. These conditions are similar to those of the Earth's deep crust and shallow upper mantle. The apparatus consists of a small piston pressing into a cylinder, which compresses the solid materials of the furnace assembly. One end of the cylinder usually abuts a massive end load. The piston is pressed against the furnace assembly by the use of a hydraulic ram. The ratio of the areas of the piston and the ram allows one to calculate the pressure obtained inside the cylinder. Furnace assemblies consist of small graphite cylinders inside pyrex or salt outer sleeves with inner sleeves of similar materials. Sample sizes typically range between 0.01 and 0.1 milliliter, which is an order of magnitude smaller than those used in 1-atmosphere experiments. Noble metal capsules that are welded shut are commonly used to contain the sample materials and to isolate them from the rest of the furnace assembly. Temperature is generated by passing a current through the graphite furnace as a result of its finite resistance. Although this apparatus is generally quite easy to operate, careful calibration of experimental temperatures and pressures is necessary prior to its use. Large pressure corrections as a result of frictional forces may arise depending on the materials and design of the furnace assembly.

STUDYING BASALTIC ROCKS

The pioneer of modern igneous petrology, N. L. Bowen, used 1-atmosphere quench furnaces as described above to study a simple analog system for the evolution of basaltic rocks. Bowen investigated a synthetic iron-free diopside-albite-anorthite system as a function of composition and temperature. Because the system was iron-free, there was no need to control the oxygen activity during the course of the run by a gas-mixing apparatus. The system does contain sodium, however, which is notoriously volatile at high temperatures. To combat this problem, the sample charges were enclosed in platinum foil. At each bulk composition, a series of experiments was conducted to determine the onset and completion of melting. The starting materials were previously synthesized crystalline pyroxene and plagioclase.

Although this method could yield erroneous results because the melting point was only approached from the low-temperature side and not reversed from the high-temperature side, the analysis of the run products was extremely sensitive to small amounts of melting, recorded as glass. The run products were rapidly quenched and then ground to a fine powder. Gain mounts immersed in oil allowed Bowen to detect minute amounts of crystalline material and thus to determine precisely the location of the liquidus in temperature-composition space. In order to determine the liquid composition, experiments were conducted to yield only quenched liquid (glass), which was then analyzed by conventional wet chemical techniques. Today, with highly developed electron microbeam capabilities, the liquid and two solid phase compositions could all be determined simultaneously. Bowen was able to apply his experimental results to the petrogenesis of basaltic rocks. The experiments clearly demonstrated that with decreasing temperature, plagioclase composition will become more sodic and less calcic when coexisting with a diopsidic pyroxene and liquid of approximately basaltic composition. Modern experiments on basalt and peridotite systems incorporating controlled amounts of volatiles began in earnest in the 1960's and continued through the 1980's. By the 1990's a reasonably self-consistent model for the evolution of terrestrial basalt magmas was available.

STUDYING GRANITIC SYSTEMS

Granitic systems were studied extensively by Tuttle and Bowen in the late 1950's. Until the time of their experiments, many geologists did not believe that granite batholiths were the products of crystallization from silicate liquids at moderate pressure. Their experiments were conducted in cold-seal pressure vessels at temperatures below 700 degrees Celsius and pressures between 500 and 4,000 atmospheres. The experiments of Tuttle and Bowen were some of the first to use the apparatus designed by Tuttle. The solubility of water in synthetic granitic melts was determined as a function of temperature and pressure. Water was added directly to the experimental charges, which were welded shut inside platinum capsules. These experiments demonstrated conclusively that many granitic batholiths crystallized from water-bearing silicate melts.

STUDYING MAGMAS

A final example of experimental techniques for solving petrological problems is an investigation of the solubility of carbon dioxide in basaltic liquids at high pressure. In the late 1970's, D. Eggler, using a piston-cylinder apparatus similar to the one described above, demonstrated that carbon dioxide had significant solubility at pressures of 30,000 atmospheres. In order to study the effect of carbon dioxide in the melting relations of synthetic systems whose compositions approximated the Earth's upper mantle, silver oxalate was added to the experimental charges, which were then sealed by welding in platinum capsules. At high temperature, the silver oxalate decomposes to produce carbon dioxide or other carbon bearing species, depending on the composition of the liquid present and the oxygen activity during the course of the experiments. The addition of carbon dioxide decreases the solidus to lower temperatures at constant pressure and tends to favor the formation of orthopyroxene over olivine as a result of changes in the melt structure. These experiments, and others like them, are useful in helping to constrain the genesis and evolution of alkali-rich, silica-poor magmas.

STUDYING VOLCANOES AND EARTHQUAKES

Experimental petrology has provided geologists with quantitative data that allow them to under-stand many complex geologic processes. The process of magma genesis deep in the Earth's upper mantle, and its subsequent migration from depth to the surface is a process that would not be as well understood today if not for experimental petrologists and their work. The eruption of volcanoes at the surface of the Earth is merely one example of the type of process upon which experimental petrology bears. One such eruption occurred in 1980 at Mount St. Helens, Washington. Studies of volcanic rocks, which generally integrate experimental data, geochemistry, and field geology, have revealed that the processes that governed the eruption of Mount St. Helens are still operating there and at other sites worldwide where oceanic crust is subducted below continental crust. In the Cascade province of the western United States, for example, Mount Shasta, Mount Bachelor, and Mount Rainier have many characteristics in common with Mount St. Helens, and these volcanoes might be expected to erupt in a similar manner in the near future. Such an eruption of Mount Rainier or Mount Bachelor could prove to be extremely dangerous to the large population centers of Seattle and Portland.

The techniques of experimental petrology also allow geologists to develop geothermometers and geobarometers, which may be applied to pieces of the Earth that are entrained in magmatic eruptions worldwide. The compositions of the coexisting phases in these fragments of rock have permitted the petrologist to determine temperature-depth profiles for the upper 200 kilometers of the Earth, gaining valuable information on the composition and state of regions of the Earth that are not accessible to direct observation. This information is crucial to the understanding of how and why rocks deform during earthquakes. For rocks of fixed composition, ambient temperature is the primary variable that determines whether rocks will be able to deform in such a way as to produce an earthquake.

Glen S. Mattioli

CROSS-REFERENCES

BIBLIOGRAPHY

Blatt, Harvey, and Robert J. Tracy. *Petrology: Igneous, Sedimentary, and Metamorphic.* New York: W. H. Freeman, 1996. Undergraduate text in elementary petrology for readers with some familiarity with minerals and chemistry. Thorough, readable discussion of most aspects of petrology. Abundant illustrations and diagrams, good bibliography, and thorough indices.

Carmichael, Ian S. E., Francis J. Turner, and John Verhoogen. *Igneous Petrology.* New York: McGraw-Hill, 1974. Many colleges use this classic text for a first course in igneous petrology. It includes extensive discussions on all aspects of the formation of igneous rocks. The text is highly technical but quite readable. Suitable for college-level students.

Edgar, Alan D. *Experimental Petrology: Basic Principles and Techniques.* Oxford, England: Clarendon Press, 1973. The most complete book on experimental petrology presently available. All aspects, from basic thermodynamic treatment of data to starting materials and apparatus, are covered in detail with excellent illustration. The material is suitable for the undergraduate or graduate student whose interest or specialty is in experimental petrology.

Ernst, W. G. *Petrologic Phase Equilibria.* San Francisco: W. H. Freeman, 1976. This text outlines the elements of classical thermodynamics and experimental approaches to acquiring the thermodynamic data necessary for geothermometry and geobarometry. A short but well-illustrated section on experimental petrology is included. Primarily concerned with the application of data acquired by experiment to the interpretation of igneous and metamorphic rocks. Recommended for readers with a college-level background in chemistry.

Gregory, Snyder A., Clive R. Neal, and W. Gary Ernst, eds. *Planetary Petrology and Geochemistry.* Columbia, Md.: Geological Society of North America, 1999. A compilation of essays written by scientific experts, this book provides an excellent overview of the field of geochemistry and its principles and applications. The essays can get technical at times and are intended for college students.

Hall, Anthony. *Igneous Petrology.* 2d ed. Harlow: Longman, 1996. This introductory book provides a good understanding of igneous rocks and their geophysical phases. There are sections devoted to the study of petrology and magmatic processes. Well illustrated and plenty of diagram and charts to reinforce concepts, this is a good resource for the layperson.

Holloway, J. R., and B. J. Wood. *Simulating the Earth: Experimental Geochemistry.* Winchester, Mass.: Unwin Hyman, 1988. A book that describes the different philosophies, apparatus, and applications of experimental geochemistry. Although far from comprehensive, the text will give the college-level student a good idea about the nuts and bolts of experimental work.

Mortimer, Charles E. *Chemistry: A Conceptual Approach.* 3d ed. New York: D. Van Nostrand, 1975. This book aims at advanced high school and beginning college students. An excellent basic chemistry text designed to accompany a first course in general chemistry. Extensive descriptions of all basic chemical phenomena are included. Problems follow each chapter, and a separate answer book is available. Contains several appendices and tables of data that the reader might find useful.

Press, Frank, and Raymond Siever. *Understanding Earth.* 2d ed. New York: W. H. Freeman, 1998. An excellent general text on all aspects of geology. Discussions on the formation of igneous and metamorphic rocks and on the structure and composition of the common rock-forming minerals are included. Also discussed is the relationship of igneous and metamorphic petrology to the general principles that form the basis of modern plate tectonic theory. Suitable for advanced high school and college students.

Ulmer, G. C. *Research Techniques for High Pressure and High Temperature.* New York: Springer-Verlag, 1971. An edited volume of research papers that deal specifically with the finer points of how to conduct different types of experimental work. The book, although highly technical, remains quite readable and should be suitable for college-level students and above.

Uyeda, Seiya. *The New View of the Earth: Moving Continents and Moving Oceans.* San Francisco: W. H. Freeman, 1971. This college-level text provides an in-depth outline of the modern theory of plate tectonics. The author has placed all the relevant observations in their historical context, which allows the reader to become familiar with the people involved in the development of the central paradigm of the Earth sciences. The text is quite easy to read, and there are many illustrations.

GEOLOGIC AND TOPOGRAPHIC MAPS

Topographic and geologic maps are basic tools for wise management and development of the Earth's resources. Topographic maps represent on paper the Earth's surface and its various landforms. Geologic maps show the distribution of the rocks that underlie the landforms and provide a view of the Earth's surface to a depth of several thousand feet.

PRINCIPAL TERMS

CONTOUR LINES: on a topographic map, lines of equal elevation that portray the shape and elevation of the terrain

GEOLOGIC MAP: a representation of the distribution of mappable units (formations)

MAP SCALE: the scale that defines the relationship between measurements of features shown on the map compared with those on the Earth's surface

TOPOGRAPHIC MAP: a line-and-symbol representation of natural and selected human-made features of a part of the Earth's surface, plotted to a definite scale

TOPOGRAPHIC MAPS

A topographic map is a line-and-symbol representation of natural and selected human-made features of a part of the Earth's surface, plotted to a definite scale. Topographic maps portray the shape and elevation of the terrain by contour lines, or lines of equal elevation. The physical and cultural characteristics of the terrain are recorded on the map. Topographic maps thus show the locations and shapes of mountains, valleys, prairies, rivers, and the principal works of humans.

In the past, topographic maps were constructed by labor-intensive field methods, which involved detailed field measurements made with telescopic-type instruments. These data were translated in the field to actual distances and plotted by hand on field sheets for eventual office compilation and printing. More recently, however, most maps have been prepared using adjacent pairs of aerial photographs. Highly accurate, these photographs are further checked for accuracy by reference to global positioning satellites. With the advent of the global positioning satellite (GPS) network, location in respect to longitude and latitude or to Universal Metric Coordinates and elevations may be readily determined without extensive baselines and triangulation networks. In some instances, laser-beam surveys provide extremely precise control in areas that are subject to earthquakes, such as California, to monitor stress buildup. Complex stereoscopic plotting instruments are used by a trained observer to delineate contour lines and various features on a base map. Field verification of place-names and features is required before the map is printed. The maps are then compiled on a stable base (a type of plastic that does not change dimensions during temperature and humidity fluctuations). Such a base helps to ensure the map's accuracy by preventing distortions. Modern photographic and photochemical techniques are used to prepare the map for printing.

All maps must meet accuracy standards established by the government. Special standardized symbols, each with its own meaning, are used to convey a wide variety of information. Also, colors are frequently used to show the more common features. Generally, blue indicates water bodies, brown indicates contour lines, red indicates map features with special emphasis (chiefly land boundaries), pink indicates built-up urban areas, and purple indicates revisions based on new photographic information since the original map was made.

GEOLOGIC MAPS

Geologic maps are a representation of the distribution of mappable units (formations). These maps provide the data for an accurate compilation of the rock units at the surface or in the subsurface. A geologist makes hundreds of observations each day in the field. Many of these observations are recorded in a notebook or on the field sheet that eventually becomes a geologic map. Some geologic maps are prepared from aerial photographs or from remotely sensed images created through satellites. New detailed geologic mapping frequently reveals information that may require reevaluation of previously mapped areas. Large-scale (1:24,000) geologic maps require detailed examination of the area being mapped. Field investigation describes outcrops as close as a spacing of several hundred feet using a topographic map as a base. Outcrop descriptions include determinations of fossils, rock type, and mineralogy, along with descriptions of rock properties, such as color, thickness, and type of bedding units, and attitudes of the rocks. Some studies are supplemented by geophysical surveys. Drilling (cores or cuttings) is integral to many studies, including oil and gas exploration, mining, and engineering. Samples are examined later in the laboratory. Some of the laboratory work includes microscopic study of thin slices of rock to establish mineral relations, binocular study and identification of fossils, or detailed chemical analyses of the whole rock or of separated minerals, and radiometric age dating.

Geologic maps are compiled in the office. This compilation requires review of field notes and observations, laboratory data, and information from the scientific literature. Data are transferred to a topographic map base made of plastic to ensure stability during temperature or humidity variations. Finalized contacts are drawn that divide rocks of one unit from those of another. The degree of certainty of the contacts is shown by a standard set of line symbols. The orientation of the various rock units is indicated by uniform symbols. When a geologic map is complete, it is prepared for publication by conventional drafting methods or, more frequently, by digitization and computer plotting methods.

More recently, geologic and topographic maps are added to large computer databases known as Geographic Information Systems (GIS's). Through computer manipulation of layers of information such as geology and topography, informed decisions can be reached by combining these with other data layers. Individual geologic maps are issued in a numbered series by the U.S. Geological Survey and related agencies, and many of these maps can now be accessed on the World Wide Web.

MAP SCALES AND SERIES

Map scale defines the relationships between measurements of features shown on the map and measurements of features on the Earth's surface. These comparisons are numerically expressed as a ratio, for example, 1:24,000, 1:125,000, 1:250,000, 1:500,000, and 1:1,000,000 scale. Large-scale maps (1:24,000 or larger) are used when highly detailed information is required. Examples include proposed projects (roads, large construction projects, and so on) in highly developed or populated areas. Intermediate-scale maps are quite useful in land- and water-management planning projects and in resource management. The 1:100,000 scale (metric) maps have become popular for a growing number of applications, particularly environmental protection and planning. Small-scale maps (1:250,000 to 1:1,000,000) cover very large areas. They are useful mainly in regional planning.

The topographic map series of the National Mapping Program includes quadrangle and other map series published by the U.S. Geological Survey. A map series is a family of maps conforming to the same specification or having common characteristics, such as scale. Adjacent maps of the sample quadrangle series can be combined to form a single large map manually, photographically, or by computer methods. Geologic maps are prepared using existing topographic and/or planimetric base maps (maps showing boundaries but no indications of relief). Thus, the scales of geologic maps generally correspond to those of the common topographic and/or planimetric maps. In special cases, such as a major engineering project (for example, a dam or nuclear power plant), preparation of a site-specific large-scale topographic map may include detailed geologic mapping.

PRACTICAL APPLICATIONS

Topographic and geologic maps help to reveal the structure and resources of the surrounding environment. These maps are basic tools for resource management and planning and for major construction projects. They are used in the planning of roads, railroads, airports, dams, pipelines, industrial and nuclear plants, and basic construction. Both types of maps are also used in environmental protection and management, water quality and quantity studies, flood control, soil conservation, and reforestation planning. In addition, topographic maps receive wide use in recreational activities such as hunting, fishing, boating, rock climbing, camping, and orienteering.

Geologic maps provide baseline data for the identification and orderly development of the Earth materials required for modern civilization. Examples include sand and gravel, crushed stone and aggregate, clay, metal deposits, and hydrocarbon fields. Geologic maps are also used extensively in environmental monitoring and protection, in local regional planning, and in scientific studies. They help to identify areas prone to landslides or earthquakes, groundwater recharge areas, and potential sand and gravel resources. They are therefore used in land-use and planning studies to determine technically suitable and environmentally safe locations for subsurface solid, hazardous, or low-level and high-level nuclear waste repositories and excavations, waste disposal, water resources investigations, and military applications.

Jeffrey C. Reid

Cross-References

Aerial Photography, 2739; Earth Resources, 1741; Earthquake Locating, 296; Electron Microprobes, 459; Electron Microscopy, 463; Elemental Distribution, 379; Engineering Geophysics, 353; Experimental Petrology, 468; Infrared Spectra, 478; Land-Use Planning, 1490; Mass Spectrometry, 483; Neutron Activation Analysis, 488; Nuclear Waste Disposal, 1791; Petrographic Microscopes, 493; Strategic Resources, 1796; X-ray Fluorescence, 499; X-ray Powder Diffraction, 504.

Bibliography

Bohne, Rolf, and Roger Anson, eds. *Inventory of World Topographical Mapping.* 3 vols. New York: Pergamon Press, 1989-1993. This three-volume set, published under the auspices of the International Cartographic Association, provides an analysis of mapping worldwide, a guide to national mapping activities, and an inventory of topographical maps. Volume 1 covers Western Europe, North America, and Australia; volume 2 covers South and Central America and Africa; and volume 3 covers Eastern Europe, Asia, Oceania, and Antarctica.

Compton, Robert R. *Manual of Field Geology.* New York: John Wiley & Sons, 1962. This somewhat dated publication provides an excellent idea of the practical aspects of conducting fieldwork and the steps involved in making a geologic map.

Easterbrook, Don J., and Dori J. Kovanen. *Interpretation of Landforms From Topographical Maps and Air Photographs.* Upper Saddle River, N.J.: Prentice Hall, 2000. This book is a laboratory manual containing exercises that focus on developing problem-solving skills regarding the use and design of topographical maps. Assumes the reader has a basic knowledge of topographical maps and map symbols.

National Research Council. *Geologic Mapping: Future Needs.* Washington, D.C.: National Academy Press, 1988. This publication presents the results of a national survey on geologic maps. The survey was designed to identify current usage of geologic maps as well as future needs. Most important, the survey identified the relative needs for geologic maps by map scale, style of presentation, and type of user (for example, exploration, basic research, engineering, and hazard assessment).

Steger, T. D. *Topographic Maps.* Denver, Colo.: U.S. Geological Survey, n.d. This free brochure provides a concise overview of topographic maps, their production, and their use.

U.S. Geological Survey. *COGEOMAP: A New Era in Cooperative Geological Mapping.* Circular No. 1003. Denver, Colo.: Author, 1987. Single copies of this circular are free upon application to the U.S. Geological Survey. Provides an overview of how cooperative geologic mapping between state and federal geological surveys is attempting to meet the need for large- and intermediate-scale geologic maps and other types of Earth science maps.

_____. *Digital Line Graphics from 1:24,000-Scale Maps: Data Users Guide.* Denver, Colo.: Author, 1986. This free publication is a key reference to understanding how digital data are used to make topographic maps. This publication will be of value to students with an interest in computer applications.

_____. *Finding Your Way with Map and Compass.* Denver, Colo.: Author, n.d. This free brochure shows the hiker how to use a topographic map and describes the various map scales used on maps in the national topographic map series.

_____. *Large-Scale Mapping Guidelines.* Denver, Colo.: Author, 1986. This free publication provides basic information and aids in preparing specifications and acquiring large-scale maps for a variety of uses. Contains a large number of practical maps and an extensive applied glossary.

_____. *National Geographic Mapping Program: Goals, Objectives, and Long-range Plans.* Denver, Colo.: Author, 1987. This free publication provides a nontechnical overview of the use and importance of geologic maps in the United States.

_____. *Topographic Map Symbols.* Denver, Colo.: Author, n.d. This free brochure summarizes all the symbols used on large-scale maps that are in the National Mapping Program. It indicates where and how to order topographic maps.

Van Burgh, Dana. *How to Teach With Topographical Maps.* Washington, D.C.: International Science Teachers Association, 1994. This volume provides basic information about the use and development of topographical maps, as well as ideas about how they can be used in the classroom to teach map-reading and problem-solving skills.

INFRARED SPECTRA

The infrared spectrum is part of the electromagnetic spectrum that lies beyond the red color that human eyes perceive as visible light. Every body emits some energy in the infrared or near infrared. Detection of infrared radiation by special instruments can be applied to various fields, including medicine, mapping, defense, communication, and astronomy.

PRINCIPAL TERMS

ANGSTROM: a unit of wavelength of light equal to one ten-billionth of a meter

MACROSCOPIC: large enough to be observed by the naked eye

PHOTON: a quantum of radiant energy

SPECTROGRAPH: an instrument for dispersing radiation (as electromagnetic radiation) into a spectrum and photographing or mapping the spectrum

SPECTROMETER: an instrument used in determining the index of refraction; a spectroscope fitted for measurements of the observed spectra

SPECTROPHOTOMETER: a photometer for measuring the relative intensities of the light in different parts of a spectrum

SPECTROSCOPY: the subdiscipline of physics that deals with the theory and interpretation of the interactions of matter and radiation (as electromagnetic radiation)

SPECTRUM: an array of the components of an emission or wave separated and arranged in the order of some varying characteristic (such as wavelength, mass, or energy)

SPECTRUM ANALYSIS: the determination of the constitution of bodies and substances by means of the spectra they produce

ELECTROMAGNETIC SPECTRUM

The small bands of infrared radiation seeping through the atmosphere were accidentally discovered in 1800 by William Herschel; when measuring the temperatures of the visible light spectra, he found a source of greater heat and wavelength radiation beyond the color red. It was not until 1881, however, when Samuel Pierpont Langley developed the bolometer, that the first in-depth studies of the infrared were possible. Max Planck's development of the quantum theory and his discovery of photons in 1900 led to the development of quantum detectors, which further advanced the study of the infrared; these early detectors have resulted in the modern spectroscopes, spectrometers, and spectrophotometers. The electromagnetic spectrum comprises visible light and five forms of invisible radiation: radio, infrared, ultraviolet, X, and gamma rays. All spectra travel at the speed of light in waves of energy bundles called photons and can be reflected, refracted, transmitted, absorbed, and emitted.

Infrared spectrometry encompasses the study of wavelengths in the electromagnetic spectrum that range between 0.7 micron in the near-infrared photographic region and 500 microns in the far-infrared rotation region. Wavelengths of the infrared spectra are most useful for detecting certain atoms and ions visible only in the infrared, such as hydrogen, the most abundant element in the universe. Infrared rays differ from the other components of the electromagnetic spectrum: Radio waves are propagated through the atmosphere and have a wavelength of between 30,000 meters and 1 millimeter; X rays have extremely short wavelengths, approximately 1 angstrom, and are generated by a sudden change in the velocity of an electrical charge; gamma rays are similar to X rays but are of a higher frequency and penetrating power; ultraviolet rays are beyond violet in the spectrum and have wavelengths shorter than 400 angstrom.

INFRARED RADIATION

Infrared radiation is emitted in some amount by every macroscopic body in the universe that has a temperature above absolute zero (or −273 degrees Celsius). In each macroscopic body, molecules not only are moving in all directions but also are rotating, and at the same time the individual atoms within the molecule are vibrating with respect to one another. It is the interaction of molecules with radiation that is the essence of the study of infrared.

For a molecule to absorb radiation, it must have a vibrational or rotational frequency the same as that of the electromagnetic radiation. In addition, a change in the magnitude and/or direction of the dipole moment must take place. The dipole moment is a vector that is oriented from the center of gravity of the positive charges to that of the negative charges, and it is defined as the product of the size and the distance between these charges. Corresponding frequencies between radiation and molecules are possible because radiation, in addition to having a magnetic component, has an electrical component. On the other hand, a molecule has an electrical field. When the electrical field of the molecule is rotating or vibrating at the same frequency as is the incoming radiation, then it is possible for a transfer of energy to take place.

The second requirement for the study of infrared, the dipole moment change, must have something to couple the energy from the radiation to the molecule. If atoms differ in their electronegativity and they combine to form a molecule, the centers of the positive and negative electrical charges may not coincide, producing a permanent dipole moment; the energy to produce this work can come from the absorption of the incoming radiation by the molecule. A permanent dipole is also necessary for inducing rotation. Atoms rotate because the electric fields are not the same on each side, thereby allowing for a transfer of energy; when energy is transferred in this manner, a rotator will rotate faster under certain rotational frequencies, while a vibrator will not change its frequency but will increase its amplitude of vibration. Because vibrational frequencies are of the order of 10^{14} cycles per second and rotational frequencies are 10^{11} cycles, they fall within the infrared region. Absorption bands for rotational spectra are quite sharp, but the bands for vibrational spectra tend to become broader because of the rotational levels associated with each vibrational level.

INFRARED DETECTORS

Detectors, either thermal or quantum, are commonly used to study the infrared. Each type uses a different property of electromagnetic radiation to convert the infrared to an electrical signal with intensity equivalent to the amount of infrared striking the detector. A thermal detector measures heat-induced changes in a property of a material, usually electrical resistance. A quantum detector also measures change, although it uses a photon, not heat, to create successive events when it strikes a material. There are three types of quantum detectors, and each one uses a separation, or diffusion, of different types of electrons as a catalyst for an event. In brief, the photoconductive effect uses incidental radiation to increase electrical conductivity, the photovoltaic effect uses a special junction for diffusion which creates voltage from charge separation, and the photoelectromagnetic effect uses radiation falling on a semiconductor with a magnetic field. In addition to detectors, various other instruments are used in the study and application of the infrared, such as the radiometer, the comparator, the collimeter, and modulators, all of which perform unique and valuable tasks.

SPECTROMETERS

Basic for understanding how the infrared spectra works is the spectrometer. All spectrometers use certain elementary components: namely, a source of radiation, a condensing source for focusing energy onto the monochromatic (pertaining to one color or one wavelength) slit, a monochromator to isolate a narrow spectral range, a radiation detector, and some form of amplifying system and output recorder. Single-beam spectrometers record energy versus wavelength, whereas a double-beam spectrometer measures the ratio between energy transmitted by the sample and energy incident on the sample, and plot transmittance or a related quantity as a function of wavelength or wavenumber. One micron is equal to one-millionth of a meter, and wavenumber is obtained by dividing 10,000 by the wavelength in microns.

EMISSION SPECTROGRAPHY

An application of the method can be illustrated by emission spectrography, which allows the determination of major, minor, and trace elements in many materials. Approximately seventy elements can be determined in rocks and other geologic materials. When a sample of material is correctly excited by an electric arc or a spark, each element in the sample emits light of a characteristic wavelength. The light enters the spectrograph via a narrow opening and falls on a diffraction grating, which is a band of equidistant parallel lines (from ten thousand to thirty thousand or more to the inch) ruled on a surface of glass or polished metal used for obtaining optical spectra. The grating separates the reflected light of each wavelength by a different angle. The dispersed light is focused and registered on a photographic plate in the form of lines of the spectrum.

The comparator-densitometer is used to measure the intensity, or darkness, of the spectrum lines registered on the spectrograph photographic plate. Using standard films or plates for each element, scientists change the spectrograph markings to indicate the percent concentration of each element. For a visual estimate, a special screen permits a comparison of the spectrum of the sample with the spectra of standards containing known element concentrations. A direct-reading emission spectrometer is one that is tied in with a computer in which are stored electronic signals from specific parts of the spectrum during the burn of a sample. The stored signals are emitted in sequence to an electronic system that measures the intensities of the spectral lines.

Thermal infrared analytical techniques are being employed to detect the mineralogical composition of the surface of Mars. The Mars Global Surveyor sent into orbit in 1999 employed a thermal emission spectrograph (TES) to determine the composition of the Mars surface. Much of the surface varies from basaltic to andesitic-basaltic rock compositions (e.g., similar to lava flows from volcanoes of the west coast of the United States) and soils derived from these rocks. TES detected two areas of hematite-rich material and failed to detect large surface exposure of carbonates. Mini-TES instruments aboard rovers, such as the 1997 Pathfinder Mission and future rovers, will allow close-up analysis of rocks on the surface. The TES instruments offer a chance to identify mineralogy of the surface of Mars rather than elemental composition.

Use in Geochemical Research

Infrared is related to geophysics, geomorphology, structural geology, and exploration as well as to geochemistry. Specific areas in which the infrared spectra have been used in geochemical research include the study of the bonds between atoms in minerals and the gaining of unique information on features of the structure, including the family of minerals to which the specimen belongs, the mixture of isomorphic substituents, the distinction of molecular water from the constitutional hydroxyl, the degree of regularity in the structure, and the presence of both crystalline and noncrystalline impurities. For example, chalcedony, including flint, chert, and agate, has been shown by infrared spectroscopy and X-ray studies to contain hydroxyl in structural sites as well as in several types of nonstructural water that can be held by internal surfaces and pores. The content of the structural hydroxyl varies zonally in chalcedony fibers and in both natural and synthetic crystals of the same spectral type as chalcedony. The varieties of chalcedony, as well as rock crystal and amethyst formed at low temperatures and in association with chalcedony, together with crystals of synthetic quartz, show a distinctive infrared absorption spectrum in the region of 2.78-3.12 microns; natural quartz crystals formed at higher temperatures give a spectrum in this region. Structural hydroxyl is housed by different mechanisms in the two types of quartz. The fibrose nature of the low-temperature quartz may derive from the hydroxyl content and its effect on dislocations.

Use in Remote Sensing

The widest applications for the infrared spectra are in remote sensing, which is the process of detecting chemical and physical properties of an area by measuring its reflected and emitted radiation. Remote sensing has been a great aid to geologists in their study of the Earth. Thermal infrared scanning has been used to monitor and update mine waste embankment data and to locate faults and fracture zones. Landsat thematic mapper and airborne thermal infrared multispectral scanner data have been used to do surface rock mapping in Nevada. Advanced visible and infrared imaging spectrometers and other remotely sensed data have been used to locate water-producing zones beneath the surface in parts of the Great Plains region. Infrared reflectance surveys have been used to locate an extinct hot spring system in the Idaho batholith. Infrared surveys have also included quantitative measurement of thermal radiation from localized heat flow in Long Valley, California, as well as surveys of the lava dome on Mount St. Helens, Washington. Reflectance variations related to petrographic texture and impurities of carbonate rocks have been analyzed by visible and near-infrared spectra.

In addition, infrared surveys and photography of volcanic zones around Mauna Loa and Kilauea were used to obtain information impossible to gather from the ground. The effects of the 1977 earthquake in Nicaragua were surveyed by infrared photography in order to find access and evacuation routes as well as safe areas for temporary camps; in addition, the photographs provided geomorphological information for future study. Infrared surveying has been used to study California's San Andreas fault. In one section near the Indio Hills, the fault trace is not topographically distinguishable but can be located by a margin of vegetation on the northeast side of the fault. Infrared sensing verified its location by imaging a band of alluvium kept cool by the water dammed up by the displaced rock.

ADDITIONAL APPLICATIONS

On a large-scale, national defense system infrared is used in secret communications, night reconnaissance, missile guidance, and gun sitings and tracking. Weather and pollution control are accomplished by using infrared to detect levels of radiation and chemicals. The use of infrared in medicine is associated with finding hot areas on the surface of the human body as indicative of possible areas of disease. Studies have revealed positive applications in the detection of, for example, breast cancer, skin burns, frostbite, tumors, abscesses, and appendicitis. As an analyzer, infrared can pinpoint damage in semiconductors that results from overheating. It can also study electrical circuit performance while in operation, detect underrated components in regular circuits, and predict component failure or shortened lifetime: These uses are only a few that demonstrate infra-

red's analytical ability. In space technology, infrared can be used in telescopes for locating new cosmic bodies, observing star formation, studying planet temperatures, and determining the chemical and/ or physical nature of distant sources of infrared radiation. Other major uses of infrared include detecting forest infiltrations and spotting welding defects; other fields that utilize infrared include photography and organic chemistry.

Miscellaneous uses of infrared are as widely varied as the major uses already discussed. Infrared photography can reveal original charcoal sketches under oil paintings. Ecologists can study the thermodynamic world on the planet and observe the animals and plants as they adapt to the volatile thermal balance. They can also track schools of fish by mapping "warm" areas on the water's surface. Criminologists and police can use night vision to survey high crime areas in the dark without the use of a spotlight. Warm air masses associated with turbulence can be detected and avoided, resulting in smoother, safer airplane flights. Infrared studies are contributing to improved telecommunications as a result of the introduction of fiber optics, which uses glass as opposed to copper and other scarce metals. Other uses of the infrared are highly technical and are found in many fields, including geology, biology, agriculture, engineering, and defense.

Earl G. Hoover

CROSS-REFERENCES

Air Pollution, 1809; Electron Microprobes, 459; Electron Microscopy, 463; Elemental Distribution, 379; Experimental Petrology, 468; Geologic and Topographic Maps, 474; Geothermometry and Geobarometry, 419; Heat Sources and Heat Flow, 49; Land Management, 1484; Mass Spectrometry, 483; Minerals: Physical Properties, 1225; Minerals: Structure, 1232; Neutron Activation Analysis, 488; Petrographic Microscopes, 493; Remote Sensing and the Electromagnetic Spectrum, 2802; Weather Forecasting, 1982; X-ray Fluorescence, 499; X-ray Powder Diffraction, 504.

BIBLIOGRAPHY

Bernard, Burton. *ABC's of Infrared.* New York: Howard W. Sams, 1970. Bernard's book deals mostly with the theory and application of in-

frared but also includes segments on physics and optics. Written in the format of a college-level textbook, with questions at the ends of the chapters. While technical data are included, any reader can follow the carefully explained examples.

Brownlow, Arthur H. *Geochemistry.* Englewood Cliffs, N.J.: Prentice-Hall, 1979. This book is intended as an introductory text on geochemistry. Each chapter is devoted to a specific area of geochemistry. Pertinent chemical principles and concepts are reviewed. The basic data are summarized, and examples of applications to geological problems are given.

Conn, George Keith. *Infrared Methods, Principles, and Applications.* New York: Academic Press, 1960. An introduction to infrared studies, this book is divided into two parts: principles of the chief components used in studying the infrared region and practical applications. Although the text is heavily weighted in mathematics and physics, it can be understood by the layperson.

Gibson, Henry Louis. *Photography by Infrared: Its Principles and Applications.* 3d ed. New York: Wiley, 1978. This book's main topic is infrared photography. There are some aspects of indirect longwave infrared recording in which the image is formed by electronic or other means and copied onto ordinary film. Discusses how infrared photography can be used by geologists and geomorphologists.

Myers, Anne B., Thomas R. Rizzo, et al., eds. *Laser Techniques in Chemistry.* New York: Wiley, 1995. Several sections in this book cover the methods, theories, and applications of infrared and ultraviolet spectroscopy. Illustrations, diagrams, index, and bibliography included.

Roeges, Noel P. G. *A Guide to the Complete Interpretation of Infrared Spectra of Organic Structures.* Chichester, N.Y.: Wiley, 1994. An excellent handbook for exploring the field of infrared spectroscopy and its relationship to organic compunds. A good introduction for the reader without a scientific background.

Sabins, Floyd F. *Remote Sensing: Principles and Interpretation.* San Francisco: W. H. Freeman, 1978. This book is an overview of remote sensing for people with no previous training in that process. The presentation attempts to strike a balance between the physical principles that control remote sensing and the practical interpretation and use of the imagery for a variety of applications. The first chapter summarizes the important characteristics of electromagnetic radiation and the reactions with matter that are basic to all forms of remote sensing.

Salisbury, John W., et al., eds. *Infrared (2.1-25 um) Spectra of Minerals.* Baltimore: The John Hopkins University Press, 1992. A thorough look at infrared spectroscopy, this collection of essays examines the complete spectra of minerals and their properties. Includes a CD-ROM.

Siegal, Frederick R. *Applied Geochemistry.* New York: Wiley, 1975. A very good general text on applied geochemistry, with a brief discussion of emission spectrography. Written in an understandable format for the layperson, requiring a minimum background in mathematics and chemistry. Very good references.

Strahler, Alan H., and Arthur N. Strahler. *Environmental Geoscience: Interaction Between Natural Systems and Man.* Santa Barbara: Hamilton, 1973. A college text on a basic introductory level, this book contains good illustrations and general information on the Earth sciences.

MASS SPECTROMETRY

Mass spectrometry is the technique used for determining particle abundances by their mass and charge characteristics in an evacuated electromagnetic field. Its principal uses in the Earth sciences are in determining the isotope ratios of light, stable substances and in measuring the isotopic abundances of radioactive and radiogenic substances.

PRINCIPAL TERMS

ABSOLUTE DATE OR AGE: the numerical timing of a geologic event, as contrasted with relative, or stratigraphic, timing

GEOCHRONOLOGY: the study of the absolute ages of geologic samples and events

HALF-LIFE: the time required for a radioactive isotope to decay by one-half of its original weight

IONS: atoms or molecules that have too few or too many electrons for neutrality and are therefore electrically charged

ISOCHRON: a line connecting points representing samples of equal age on a radioactive isotope (parent) versus radiogenic isotope (daughter) diagram

ISOTOPE: a species of an element having the same number of protons but a different number of neutrons and therefore a different atomic weight

NUCLIDE: any observable association of protons and neutrons

RADIOACTIVE DECAY: a natural process whereby an unstable, or radioactive, isotope transforms into a stable, or radiogenic, isotope

EARLY MASS SPECTROMETRY

The golden decade of physics, from 1896 to 1906, saw the critical discoveries and experiments that resulted in the quantitative analysis of charged particles by mass. One hundred years ago, experiments with cathode rays led to their identification as streams of electrons by Eugen Goldstein, Wilhelm Wien, and Joseph John Thomson. Using the earliest application of mass analysis, Thomson identified the two isotopes of neon, neon 20 and neon 22. This work was followed in 1918 and 1919 by A. J. Dempster and F. W. Aston, who designed mass spectrographs which were used in succeeding years to determine most of the naturally occurring isotopes of the periodic table.

In 1896, Antoine-Henri Becquerel presented his discovery of the phenomenon of radioactivity to the scientific community in Paris. This finding was followed rapidly by the seminal work of Marie Curie in radioactivity, a term she coined. Her discovery of the intensely radioactive elements radium and plutonium led Ernest Rutherford to delimit three kinds of radioactivity—alpha, beta, and gamma—and, in 1900, led Frederick Soddy to formulate a theory of radioactive decay. Soddy later proposed the probability of isotopes, the existence of which was demonstrated on early mass spectrographs and mass spectrometers.

GEOCHRONOLOGIC TIME MEASUREMENT

Rutherford and Soddy's theory of the time dependence of radioactive decay, followed by breakthroughs in instrumentation for the measurement of these unstable species and their radiogenic "daughter" nuclides, caught the attention of early geochronologists and had a revolutionary effect on the study of geology. In 1904, Rutherford proposed that geologic time might be measured by the breakdown of uranium in uranium-bearing minerals, and a few years later, Bertram Boltwood announced the "absolute" ages of three samples of uranium minerals. The ages, which approximated half a billion years, indicated that at least some Earth materials were much older than had been thought, an idea developed by Arthur Holmes in his classic *The Age of the Earth* (1913). Holmes's early time scale for the Earth and his enthusiasm for the developing study of radioactive decay were not met with instant acceptance by most contemporary geologists, but eventually, absolute ages would become the prime quantitative components in the field of geology.

After the early study of the isotopes of uranium came the discovery of other unstable isotopes and the formulation of the radioactive decay schemes that have become the workhorses of geochronology. The theory of the radioactive decay of the parent, or unstable, nuclide (or the growth of the daughter, or stable, nuclide) developed in the early 1900's has not changed; it is still the basis for geochronologists' measurement of time. This field is one of the arenas for the use of mass spectrometry.

STABLE ISOTOPE FRACTIONATION

The other use of mass spectrometry in the Earth sciences results from isotopes' potential to fractionate, or change their relative abundance proportions, during geological processes, for physicochemical reasons other than radioactive decay and radiogenic buildup. Fractionation not resulting from radioactive decay (stable isotope fractionation) comes about because the thermodynamic properties of molecules depend on the mass of the atoms from which they are made. The total energy of a molecule can be described in terms of the electronic interactions of its atoms and the other energetic components of these atoms, such as their rotation, vibration, and translation. Molecules that contain in their molecular configurations different isotopes will have differing energies, because of the different energy components (usually vibrational and rotational) that are mass-dependent. The total energy of molecules also decreases with decreasing temperature; at zero Kelvin, or absolute zero, this energy has a finite value known as its zero-point energy. The vibrational component of energy, the most important factor in fractionation, is inversely proportional to the square root of its mass. A molecule with the heavier of two isotopes will have a lower vibrational energy and thus a lower zero-point energy than a similar but lighter isotope molecule. Other factors being equal, the chemical bonds of a molecule with lighter isotopic composition will be more easily broken than those of the heavier isotope analogue, and the heavier molecule thus will be less reactive chemically.

Geologic processes that result in stable isotopic fractionation are the redistribution of isotopes as a function of isotopic exchange; nonthermodynamic (kinetic) processes that depend on the amounts of the species present during a reaction; and a range of strictly physical processes, including diffusion, evaporation, condensation, adsorption, desorption, crystallization, and melting. Physical conditions such as these undoubtedly were much more intense during preaccretion events, such as star formation, than during more typical "geologic" processes, such as sedimentation or volcanism; consequently, fractionation effects are observable in primitive materials, such as some components of relatively unprocessed meteorites. These materials show extremely interesting stable isotope fractionation effects even among the heaviest elements.

ELEMENTS OF MASS SPECTROMETERS

As commonly used, the term "mass spectrometer" refers to an instrument in which beams of ionized isotopes are separated electrically. (The earlier, more qualitative mass spectrograph focused ion beams onto a photographic plate.) Mass spectrometers have three common elements: a source component, wherein elemental species are ionized so that they can be accelerated electrically; an analyzer section, where isotopic species are separated by their mass-charge ratio; and a collector assembly, where the ion beams are quantitatively measured.

MAGNETIC-SECTOR MASS SPECTOMETRY

The most common instrument in geologic use is the magnetic-sector machine, in which a uniform magnetic field is bound in a region, or sector, commonly by a stainless steel tube that can be evacuated to very low pressures to prevent sample contamination. The source region may consist of a solid source; a purified and spiked sample of a heavy element is introduced in the solid state onto a filament of purified metal such as tantalum or rhenium, and the filament is heated electrically until a sufficient percentage of the element is ionized for efficient measurement. Alternatively, the source may be a gas; in this case, the desired, commonly light, elements in a gaseous state are introduced into an evacuated region and bombarded with electrons to produce a sufficient percentage of ionized species for acceleration into the analyzer section of the instrument. The ionized species are accelerated electrically through a series of slits, onto which variable electric potentials can be applied for the purpose of acceleration and focusing, so that a well-defined, focused beam of the element or its gaseous compound is beamed into the analyzer tube.

The analyzer sector, commonly constructed so that the lowest pressure possible can be maintained and the least number of contaminant species will be struck by the focused beam, is bent at angles of 90 or 120 degrees and passed through an electromagnetic field capable of efficiently separating the ion beams by their mass-to-charge ratios. (Where the charges are uniform, as is usual in earth science research, the separation is, as desired, only by mass.)

The collector assembly commonly consists of a Faraday cup; the separated isotope beams enter, hit the metal cup, and impart unit charges to the cup as the atoms are neutralized. The resulting direct current is exceedingly low and, in many instruments, must be converted to an alternating current so that the intensity of the signal can be increased for measurement with a strip-chart recorder or, more commonly, for digital readout. Accelerating voltage in the source assembly is adjusted with the magnetic field in the analyzer sector (commonly monitored and controlled with a very precise gaussmeter) so that a beam of a unique charge-to-mass ratio (or of a unique mass, for ions of the same charge) passes through a final slit into the collector. Ions entering the collector are neutralized by electrons that flow from the ground to the metal collector cup, across a resistor whose voltage difference is amplified and measured with a digital or analogue voltmeter. These data are exhibited as a strip-chart readout or, more commonly, as digital output that is computer-collected and reduced for analysis. The collection of large numbers of highly precise isotopic ratios in computer-reduced digital form has made possible the modern use of mass spectrometry in the Earth sciences and the determination of isotopic parameters that would not otherwise have been obtainable.

ADDITIONAL TYPES OF MASS SPECTROMETRY

Many advances have been made, so it is now possible to obtain extremely precise ratio measurements of tiny pieces of material in a relatively short time. Ion probe mass spectrometers allow these measurements on *in situ* samples in thin sections that, concomitantly, can be studied petrolog-

ically. Ion probe mass spectrometry involves the combination of a microbeam probe (using ions, rather than the lighter electrons, as "bullets" for ionization) and a magnetic-sector mass spectrometer. Accelerator mass spectrometry employs the use of a particle accelerator or cyclotron as the mass analyzer; it is useful primarily to make high-abundance measurements for cosmogenic nuclides such as carbon 14. Accelerator mass spectrometry makes possible the precise measurement of cosmogenic nuclides on tiny samples.

The developing field of resonance ionization mass spectrometry holds much promise in Earth science studies because of its potentially high ionization efficiency and, therefore, sensitivity. Other possible mass spectrometric practices may include high-accuracy isotope dilution analysis utilizing a plasma ion source, and ion cyclotron resonance (Fourier transform) mass spectrometry. The supermachine of the future may combine some or even all of these potential advances.

PRINCIPAL MECHANICS OF SPECTROMETRY

Although some modern methods of determining absolute time do not involve isotopes, most do, and the standard method for their quantitative measurement is by mass spectrometry. Because the various radioactive nuclides useful in geochronology are also varied in their chemical characteristics, several instruments and techniques are involved. The principal mechanics of spectrometry, however, are mainly the same. The standard method involves placing the purified samples of the materials in question as solids on purified metal filaments and inserting the loaded filaments into a solid-source mass spectrometer. Evacuated to very low pressures, the spectrometer source regions are made so that the metal filaments can be heated to the point that the rubidium or strontium ionizes. The charged, ionized sample is accelerated through a series of collimating slits into the high-vacuum analyzing tube, where it encounters a controlled electromagnetic field. The beams of ions are separated by charge-mass ratios into beams of separated isotopes; as the charge of the elements is the same for each atom, however, the ions in this case are separated on the basis of mass only. Specific isotopic beams, controlled by the magnetic field, are channeled through more collimating slits to the collector part of the spectrometer. Commonly, a Faraday cup is used to analyze the number of atoms of each isotope by conversion of each atomic impact into a unit of charge, which is amplified, often with a vibrating reed electrometer. A digital readout is then produced. The actual output is isotope ratio measurements, which are converted by a mathematical program to the required parameters for determining time.

Scientists determine the age graphically, with the use of an isochron diagram, in which isotope ratios collected in the spectrometer are used as coordinates. A line known as an isochron connects points representing samples of equal ages. An isochron has an age value indicated by its slope on the figure; a horizontal isochron has a zero-age value, while successively greater positive slopes have increasingly greater ages. A single mineral or rock is represented by only one point in the diagram; therefore, for an isochron to be drawn, an estimate of the sample's initial isotopic composition would be necessary. Ages calculated this way are termed "model ages."

EQUILIBRIUM AND NONEQUILIBRIUM FRACTIONATION

Stable isotope fractionation, or the enrichment of one isotope relative to another in a chemical or physical process, also has Earth science applications. The two processes of this sort are equilibrium fractionation, which is useful in determining geologic paleotemperatures, and kinetic (nonequilibrium) fractionation, which is useful in establishing biologically mediated geochemical processes, such as the bacterial utilization of sulfur.

Isotopic fractionation in these processes is measured by the fractionation factor α, defined as A/B, where A is the ratio of the heavy to the light isotope in molecule A, and B is that ratio for molecule B. Although α may be calculated theoretically, in geologic use it is derived mainly from empirical data. This factor, which is largely dependent on the vibrational energies of the molecules involved, is a function of temperature; thus, it is a measure of ambient geologic processes.

ESTABLISHING PALEOTEMPERATURES OF ANCIENT SEAWATER

Many Earth science applications of stable isotope fractionation are in use, but perhaps the best-known example is the use of oxygen isotope ratios to establish paleotemperatures of ancient seawater. Surface seawater, in at least partial equilibrium with the atmosphere, contains oxygen with a characteristic isotopic composition. This composition is provided by the ratio of the most abundant species: oxygen 18 and oxygen 16. Marine plants and animals, such as foraminifera, that build their hard parts out of components dissolved in seawater, such as calcium, carbon, and oxygen (as in calcium carbonate), utilize oxygen that is isotopically characteristic of the seawater. Although this process also depends on other, incompletely understood, factors, it is primarily a function of water temperature. Therefore, the ratio of oxygen 18 to oxygen 16 in the foraminifera is a measure of the water temperature. Because the calcium carbonate does not readily reequilibrate with ambient water after it is precipitated, it retains its characteristic isotopic composition after sedimentary burial for many millions of years. Isotopic data collected from foraminifera recovered from deep-sea cores are therefore used to record water temperatures (and consequently, climate) of the geologic past. More than any other paleothermometry device, this application has been extremely useful in providing a record of global temperature changes, especially of the past glacial periods, for use in constructing and testing quantified models of the causes of climate change.

For this application, the sample is introduced into the source region of the mass spectrometer as a gas, commonly carbon dioxide. Ionization of the gas may be accomplished by bombardment of the molecules with electrons. The positively charged ions created are accelerated through collimating slits into the analyzer section of the spectrometer. In this type of gaseous analysis, use is made of the double-focusing mass spectrometer, in which the isotopic composition of the sample is determined relative to that of the standard in iterative, alternating measurements.

VALUE TO PALEONTOLOGY

The revolution in the Earth sciences, largely a result of the plate tectonics paradigm which was introduced in the early 1960's, was preceded by an even more important revolution, one that received little fanfare. In the 1940's and 1950's, Earth science began to be significantly influenced by quantitative investigations which may be considered to have provided the quantitative foundation for the plate tectonic revolution. Of these studies, none was more significant than the use of mass spectrometry for determining absolute ages of minerals and rocks and, later, for paleothermometry. Absolute-age determinations gave a firm basis for paleontology and established not only the Earth's antiquity but also a quantitative sequencing for its rocks and sediments—the geologic time scale.

E. Julius Dasch

BIBLIOGRAPHY

Barker, James. *Mass Spectrometry.* 2d ed. New York: Wiley, 1999. A college text concerning the field of mass spectrometry and its protocol and applications. There is a fair amount of analytical chemistry involved, so the reader without a scientific background may have a difficult time. Bibliographical references and index included.

Chapman, John Roberts. *Practical Organic Mass Spectrometry: A Guide for Chemical and Biochemical Analysis.* 2d ed. Chichester, N.Y.: Wiley, 1995. A good account of mass spectrometry and its use in the analysis of organic compounds. The book offers an interesting look into the relationships among organic compounds, biochemical structures, and spectrometry. Two appendices, index, and bibliography.

Duckworth, H. E. *Mass Spectrometry.* England: Cambridge University Press, 1958. An older, very technical but informative treatise on the basic principles of mass spectrometry. Suitable for college-level readers.

Faure, Gunter. *Principles of Isotope Geology.* 2d ed. New York: John Wiley & Sons, 1986. An excellent, though technical, introduction to the use of radioactive and stable isotopes in geology, including an introductory treatment of mass spectrometric principles and techniques. The work is well illustrated and in-dexed. Suitable for college-level readers.

Hoffmann, Edmond. *Mass Spectrometry: Principles and Applications.* Translated by Julie Trottier. Chichester, N.Y.: Wiley, 1996. A wonderful reference tool and introduction to the field of mass spectrometry, this book explains the history of mass spectrometry and the principles, theories, protocol, and applications associated with it. A good introduction for the college reader.

Levin, Harold L. *The Earth Through Time.* 3d ed. Philadelphia: Saunders College Publishing, 1988. Chapter 5, "Time and Geology," reviews the geologic time scale and then turns to techniques for determining absolute age. It offers a history of early attempts at geochronology, an overview of radiometric dating, and a discussion of the principal dating methods. There is a simple description and diagram of a mass spectrometer. Easy to read and suitable for high school students.

Smith, David G., ed. *The Cambridge Encyclopedia of Earth Sciences.* England: Cambridge University Press, 1981. Chapter 8 in this clearly written text contains a section on mass spectrometry and its uses. Other analytical techniques are also discussed. Includes a table of trace element abundances in common rocks and minerals. For general audiences.

NEUTRON ACTIVATION ANALYSIS

Neutron activation analysis uses a flux of neutrons to excite the nuclei of chemical elements in samples, thus causing the excited nuclei to emit characteristic gamma radiation. The technique provides a sensitive method for measuring the amount of chemical elements contained in geological samples, particularly when there is only a small amount of the element in the sample.

PRINCIPAL TERMS

CROSS SECTION: the effective area that a nucleus presents to an oncoming nuclear particle, which determines the chance that the particle will strike the nucleus, causing a nuclear reaction

GAMMA DECAY: the emission of high-energy electromagnetic radiation as a nucleus loses excess energy

GAMMA SPECTRUM: the unique pattern of discrete gamma energies emitted by each specific type of nucleus; it identifies that nucleus

HALF-LIFE: the time during which half the atoms in a sample of radioactive material will decay

ISOTOPE: atoms of the same chemical element containing equal numbers of protons whose nuclei have different masses because they contain different numbers of neutrons

NEUTRON: an uncharged particle that is one of the two major nuclear constituents having nearly equal masses and different electric charges

NUCLEAR REACTION: a change in the structure of an atomic nucleus brought about by a collision of the nucleus with another nuclear particle such as a neutron

NUCLEUS: the tiny central portion of an atom that contains all the positive charge and nearly all the mass of the atom

PROTON: a particle that carries a single unit of positive charge equal in size to that of the electron; one of the two major nuclear constituents having nearly equal masses and different electric charges

GAMMA RADIATION

In neutron activation analysis, a sample of interest to the scientist is placed in a beam of neutrons produced by a radioactive source, such as an accelerator or a nuclear reactor. The neutrons interact with nuclei contained in the sample and alter their structures, frequently leaving the nuclei of the sample with excess energy. After a predetermined time, the sample is taken out of the beam of neutrons. The altered nuclei in the sample lose their excess energy by emitting nuclear radiation that is characteristic of each individual type of nucleus. The radiation is detected and allows identification of the nucleus that emitted it. The intensity of a particular characteristic radiation is directly related to the number of nuclei of that species in the sample. Usually, the radiation studied in neutron activation analysis is gamma radiation, that is, high-energy electromagnetic radiation emitted by nuclei without altering their chemical nature. The gammas emitted by the sample are directly related to the abundance of a particular chemical element in the sample.

Neutron activation analysis grew out of the systematic study of the interaction of neutrons and nuclei conducted by nuclear physicists in order to understand the structure of the nucleus. Interpretation of the patterns of gammas emitted by a sample following irradiation by neutrons requires several types of background information. First, not all types of nuclei react in the same way with a neutron beam. The cross section for the nuclear reaction, which measures the chance that a particular nuclear species will be produced, depends not only on the structure of the nucleus involved but also on the energy of the neutrons used for bombardment of the sample. The reaction cross section must be measured in a separate experiment as the energy of the neutrons is varied. Many reaction cross sections have been measured and are tabulated in the scientific literature.

Second, the gamma radiation from a particular nuclear species has characteristic energies which can be precisely and conveniently measured using germanium-based detectors. The pattern of energies of gammas emitted by a particlar type of nucleus identifies that nucleus just as the pattern of visible light emitted by an atom—its spectrum—identifies that atom. Tables of the gamma spectra of nuclei are an important input to neutron activation analysis. Such tables, along with the cross sections for nuclear reactions, are frequently stored in the memory of a computer and automatically recalled during analysis of the gamma spectra from neutron activation analysis.

Half-Lives of Radioactive Materials

Third, emission of nuclear radiation occurs gradually in time at a rate characterized by the half-life of the given decay, that is, the time for half the nuclei in a sample to emit their gamma radiation. Half-lives for nuclear species vary from picoseconds to millions of years. Nuclear species of interest for neutron activation analysis generally have half-lives ranging from seconds to days, as the species must live long enough for the sample to be transported to the detector and must decay quickly enough so that they can be detected in a reasonable amount of time. As the neutrons interact with the sample nuclei, producing new energetic nuclei, the energetic nuclei decay with their characteristic half-lives. Thus, the number of excited nuclei in a sample depends on the half-life of the nucleus as well as on the time it has spent in the neutron beam. Half-lives of nuclear radiation are measured in separate experiments and are tabulated; the analyst keeps careful records of the exposure of the sample to neutrons.

The variations in the half-lives of excited nuclei can be used to identify nuclei present in the sample. Decays with short half-lives happen very rapidly so that the first gammas obtained from the sample are mostly those with half-lives less than about five minutes. If the sample then sits for half an hour, the short-lived nuclear species will have decayed, and the gammas from the sample will be those from nuclei with longer half-lives. Thus, in neutron activation analysis, gammas from the sample are measured at a series of carefully planned time intervals. Modern laboratories use computers to calculate the effect of half-lives on the gamma spectra that have been recorded.

Isotopes

Fourth, the number of excited nuclei produced during neutron irradiation (and thus the intensity of a particular gamma emission) depends on the concentration of a particular nuclear species in the sample. Because the chemical nature of an atom is not affected by the number of neutrons in its nucleus, most chemical elements are characterized by more than one type of nucleus or isotope, that is, nuclei with the same number of protons in them and thus belonging to the same element but with different numbers of neutrons. Different isotopes have different reaction cross sections and different gamma spectra. Thus, the analyst must know the relative amounts of each isotope of a given element in order to relate the intensity of a gamma emission to the abundance of a particular element in the sample. These data are well known and readily available.

Neutron Flux

Finally, the number of gammas at a particular energy level depends on the number of neutrons that are aimed at the sample. If the neutrons are produced by a radioactive source or an accelerator, they form a beam and are described as a particular number of particles per unit of area and time. If the sample is placed inside a nuclear reactor, the neutrons bombard it from all directions, and the irradiation is described in terms of a neutron flux, or the number of neutrons crossing a square centimeter of the sample each second. The energy distribution of the neutrons that strike the sample must also be recorded because reaction cross sections depend on energy. In an accelerator or a radioactive source, the neutrons produced usually have a well-defined single energy. In a reactor, they will have a distribution of energy levels that must be measured for an individual reactor and for a particular location in the core of that reactor.

Costs and Benefits of Method

Neutron activation analysis is expensive because it requires a neutron source and specialized detectors and counting systems, all of which are run by a computer in modern laboratories. The technique is not sensitive to the chemical state of an atom but merely determines the number of atoms of a particular element to be found in the sample; therefore, it is not suitable for determining the chemical state in which elements are present. On the other hand, neutron activation analysis is very fast compared to standard quantitative analysis, thus compensating for the expense when fast results are needed. For example, neutron activation analysis using a radioactive source to produce the neutrons has been conducted in the field to determine the copper and manganese content of ores. The technique is advantageous in that samples did not have to be transported to a laboratory for analysis, and results of the analysis could be used to guide drilling operations.

Neutron activation analysis cannot detect every chemical element in the sample, as not all chemical elements produce gammas with suitable half-lives or have large cross sections for nuclear reactions. In some cases, strong gammas from abundant elements may mask the weaker signals from less abundant species. The technique for producing neutrons may strongly influence the elements that are detected. For example, the high-energy or fast neutrons produced in accelerators using tritium targets interact strongly with oxygen and silicon, while lower-energy neutrons characteristic of nuclear reactors interact very little with these elements. Thus, fast neutrons are characteristically used for rapid determinations of the silicon and oxygen content of minerals. Frequently, samples are subjected to analysis using more than one sort of neutron source to detect different elements. Rock samples will first be subjected to fast neutron analysis to determine their content of silicon and oxygen and then to analysis using a reactor to detect about twenty-five other elements. Finally, neutron activation analysis may be supplemented by chemical separation of elements for the detection of very rare elements.

Neutron activation analysis is ideally suited for the study of trace elements, that is, relatively rare elements present in samples in small quantities. Because neutrons easily penetrate geological materials, samples for neutron activation analysis require little preparation, and the technique does not destroy the sample, which can thus be saved for display or subjected to further analysis. Therefore, the technique is often applied to samples of archaeological importance in which samples are too precious to destroy in analysis. It offers the researcher the further advantage that many elements in the same data can be identified, and thus elements may be found whose presence in the sample was not initially expected. This scanning for many elements at once is an advantage in problems such as the search for pollutants in river water. For example, extensive studies of environmental mercury in Sweden have been conducted using neutron activation analysis of minerals, coal, and plant and animal tissues.

The advent of computer-based systems has automated much of the tedious calculation needed to analyze data from neutron activation analysis. The technique is thus accessible to a much wider variety of researchers than was previously the case and promises to find increasing use as a probe of the elemental composition of samples of interest to Earth scientists.

TYPES OF EARTH SCIENCE APPLICATIONS

Neutron activation analysis provides a powerful technique for simultaneously determining the amounts of many chemical elements in a geological sample without destroying the sample. Although it requires a source of neutrons and fairly complex instrumentation, it is much faster than conventional chemical analysis and can analyze a sample for several elements at the same time with little sample preparation. This technique is also uniquely sensitive to very small amounts of particular elements and thus can often detect minute amounts of such elements present in samples that would escape all but extremely detailed and time-consuming chemical analyses using atomic spectroscopic techniques designed to search for that element.

Applications of neutron activation analysis to Earth science fall into two broad categories. The first consists of cases in which researchers take advantage of the speed of neutron activation analysis to obtain immediate results on the elemental compositions of their samples. Such work is often done in the field, using a radioactive source to produce the neutrons, and the results of the analysis guide field operations. Similarly, neutron activation analysis may be used to screen a very large number of samples rapidly on a production basis.

The second category of applications of neutron activation analysis utilizes the ability of the technique to determine rapidly very small concentrations of certain chemical elements. One example of this application has been the systematic study of trace elements in rocks of various ages. In the energy industry, neutron activation analysis has been applied to the study of trace elements in coals, thereby providing clues to the origin of particular coal beds. The quality of coal as a heat source varies widely from bed to bed. An understanding of why this variation occurs might lead to new methods of treating coals before burning them in order to reduce pollution. Finally, the ability of neutron activation analysis to scan large numbers of samples for minute quantities of chemical elements has been put to use in the study of sources of pollution, particularly by metals, in the environment. Large numbers of samples of river water or runoff near landfills can be checked for the presence of a wide variety of metals quickly and efficiently using neutron activation analysis.

A particular application of neutron activation analysis is autoradiography. In this case, the sample is irradiated and then placed in contact with a piece of film. The film is developed and records concentrations of radioactivity, showing how particular chemical elements, for example uranium and thorium, are distributed within the sample.

STUDY OF LUNAR ROCKS

Probably the most famous example of the use of neutron activation analysis in Earth science is the study of the lunar rocks brought back to Earth by the Apollo astronauts. Scientists wished to know the chemical composition of these rocks in order to obtain clues as to their history and to learn whether the Moon formed from the same original material as did the Earth. At the same time, only relatively small samples were available, as the lunar rocks had to be carried back from the Moon in circumstances where the amount of weight was critical; in addition, scientists wanted to save the lunar rocks for future analysis and for public display. Neutron activation analysis does not damage the sample it studies. Even when combined with chemical separation techniques to aid analysis for very scarce elements, the samples needed are very small—on the order of milligrams—as opposed to the gram-sized samples needed for standard chemical analysis. Thus, the lunar rock samples could be subjected to neutron activation analysis to determine their basic elemental composition and still be left intact for display or analysis by other methods. Some of the surprising results from neutron activation analysis of the lunar rock samples include the fact that lunar rocks and soils are very low in oxygen compared to their terrestrial counterparts.

One of the problems in studying the lunar samples was to determine their content of rare-earth elements such as europium, neodymium, or gadolinium. These elements are chemically very similar and thus difficult to separate by quantitative chemical analysis. On the other hand, their nuclear structures are very different; therefore, neutron activation analysis is an ideal tool for distinguishing among them. Results of the analysis showed that overall abundances of the rare Earth elements in lunar rocks were fifty to a hundred times greater than is standard for chondritic meteorites, which are meteorites believed to represent the primordial composition of the material from which the solar system formed. At the same time, lunar rocks were depleted in the element europium compared to chondritic meteorites and terrestrial rocks. Explanation of these strange patterns of elemental abundances uncovered by neutron activation analysis supports the theory that the Moon formed from a disk of material spun off from the very early Earth by a collision with a very large planetesimal.

Ruth H. Howes

CROSS-REFERENCES

BIBLIOGRAPHY

Choppin, Gregory R. *Radiochemistry and Nuclear Chemistry.* 2d ed. Boston: Butterworth-Heinemann, 1995. This widely used college text introduces the reader to the basics of nuclear chemistry and radiochemistry. It explores the theories surrounding those fields and their applications. Well illustrated with clear diagrams and figures, this is a good introduction for someone without a strong background in chemistry.

Fite, L. E., E. A. Schweikert, R. E. Wainerdi, and E. A. Uken. "Nuclear Activation Analysis." In *Modern Methods of Geochemical Analysis*, edited by Richard E. Wainerdi and Ernst A. Uken. New York: Plenum Press, 1971. A comparatively brief summary of the technique of neutron activation analysis, designed for geologists who are not familiar with nuclear physics. While not terribly difficult to read, the chapter stresses the equipment required for use of neutron activation analysis and provides examples from geology.

Keller, C. *Radiochemistry.* New York: John Wiley & Sons, 1988. A general text on the use of radionuclides, both naturally occurring and artificially formed, as in the case of neutron activation analysis. Although the section on neutron activation analysis is brief, the text provides a thorough treatment of the background material needed to understand this relatively complex analytical technique.

Kruger, Paul. *Principles of Activation Analysis.* New York: Wiley-Interscience, 1971. This volume concentrates on the experimental details and on the instrumentation needed to carry out neutron activation analysis. Written as a textbook for the scientist planning to use neutron activation analysis, who is not a specialist in the field. The details of the technique are thoroughly discussed.

Lenihan, J. M. A., S. J. Thomson, and V. P. Guinn. *Advances in Activation Analysis.* Vol. 2. New York: Academic Press, 1972. Although slightly old, this volume provides an excellent series of examples of the use of neutron activation analysis to study a variety of problems not only in the Earth sciences but also in the arts and in archaeology. It discusses some of the varied results that can be obtained by using a variety of neutron sources.

Lieser, Karl Heinrich. *Nuclear and Radiochemistry: Fundamentals and Applications.* New York: VCH, 1997. Lieser's book gives the reader a basic understanding of the practices and principles involved in the fields of radiochemistry and nuclear chemistry. Although there is quite a bit of chemistry involved in the author's explanations, someone without a chemistry background will still find the book useful. Illustrations, charts, and diagrams help clarify difficult concepts and theories. Bibliography and index.

Parry, Susan J. *Activation Spectrometry in Chemical Analysis.* New York: Wiley, 1991. Intended for someone with a background in chemistry and related fields, Parry's book contains in-depth descriptions of neutron activation and activation spectrometry. Sections deal with nuclear chemistry and radiochemistry. Bibliographical references and index.

Rakovic, Miloslav. *Activation Analysis.* London: Iliffe Books, 1970. A good introduction to the field, including details of chemical preparations and sample-handling skills. Organized by the chemical element that is being studied rather than by the technique being used. A good source for a person interested in studying a particular chemical element.

PETROGRAPHIC MICROSCOPES

The petrographic microscope is an essential tool for studying the mineral content and texture of fine-grained rocks. It also provides a rapid and accurate means for identifying minerals through their optical properties.

PRINCIPAL TERMS

ANISOTROPIC CRYSTAL: a crystal with an index of refraction that varies according to direction with respect to crystal axes

BIREFRINGENCE: the difference between the maximum and minimum indices of refraction of a crystal

CRYSTAL AXES: directions in a crystal structure with respect to which its molecular units are organized

INDEX OF REFRACTION: the ratio of the speed of light in vacuum to its speed in a particular transparent medium

INTERFERENCE: the combining of waves or vibrations from different sources so that they are in step and reinforce each other or are out of step and oppose each other

INTERFERENCE COLOR: a color in a crystal image viewed under crossed polars, caused by subtraction of other colors from white light by interference

INTERFERENCE FIGURE: a shadow shape caused by the blocking of polarized light from certain areas of a crystal image

POLARIZATION: filtration of light so that only rays vibrating in a specific plane are passed

PRINCIPAL VIBRATION DIRECTIONS: directions in a crystal structure in which light vibrates with maximum or minimum indices of refraction

RETARDATION: the progressive falling behind of part of a ray vibrating in a slower direction compared to a part vibrating in a faster direction

POLARIZED LIGHT

The crystals of many rocks and deposits are too small to be distinguished—much less identified—by the naked eye. Individual crystals can be distinguished under an ordinary microscope, but identification in this manner is still difficult. A powerful improvement was discovered in 1828 by William Nicol when he applied to the microscope his newly invented polarizing prisms cut from calcite crystals and found that different minerals have very distinctive appearances in polarized light. The modern petrographic microscope is a refinement of Nicol's discovery.

The petrographic microscope is similar to a standard biological microscope but has adaptations for use with polarized light. A polarizer (or "polar") beneath the condenser lenses polarizes the light before it passes through the specimen. A circular, rotatable stage allows the slide and specimen to be turned with respect to the polarized light. A second, removable polarizer called the analyzer, oriented crosswise to the lower polar, can be inserted into the light path in the tube, providing "crossed polars." There are other accessories used for special purposes.

A polarizing microscope adapted to view reflected rather than transmitted light is commonly used by metallurgists to study the identity, size, and texture of metal crystals produced in industrial processes. Economic geologists use the reflection microscope to identify opaque, metal-bearing minerals and to determine their abundance in samples from an ore body.

INTERFERENCE COLORS

Under the polarizing microscope, crystals show bright, distinctive colors called interference colors. As the stage is turned, these colors move and change or become dark ("extinct") in ways that can be used to identify the minerals. It is the interaction of light with the crystal structure that causes the distinctive behavior.

Light is an electromagnetic wave. As a ray of light travels along, its electric field strength oscillates back and forth transverse to the path of the ray, somewhat like the vibration that travels along a horizontal rope when it is shaken up and down at one end. The vibration direction and the travel path are perpendicular. The distance traveled by the ray between one maximum of the transverse field and the next is the wavelength of the light. Each color of visible light has its own particular wavelength, ranging from roughly 700 nanometers for red to 400 nanometers for violet (a nanometer is one-billionth of a meter). White light is a mixture of all colors.

INDEX OF REFRACTION

All light travels at the same speed in vacuum, but in a transparent medium, it is slowed by interaction with matter. The speed is characteristic of a given medium and is indicated by its index of refraction; the greater its index, the slower light moves through it. The speed in air (index 1.000) is essentially the same as in vacuum; but in quartz (index 1.54), it is only 65 percent of that, and in diamond (index 2.42), it is only 40 percent.

The atoms or molecules in crystals are arranged in a strict order, repeated over and over to make the crystal structure (or "lattice"). The structure is responsible for many characteristic features of minerals, such as the natural shapes and faces of crystals and the likelihood of breaking along flat surfaces called cleavages. The pattern of any given mineral is very distinctive, so the ways in which light interacts with the different structures can be used to identify the various minerals.

Minerals are classified as isotropic or anisotropic. In isotropic crystals, light can travel any direction at the same speed, and the index of refraction has the same value for all orientations. Isotropic minerals all have a molecular structure and spacing that are identical along each of three perpendicular crystal axes. Some common examples are halite (rocksalt) and garnet. Glass substances, which are equally disordered in all directions, are also isotropic. In anisotropic minerals, the molecular structure and spacing are different along one or more crystal axes. In such crystals, the interaction of light vibrations with matter—and, therefore, the speed of light and the index of refraction—depend on the direction of travel. The analyst, in order to identify the minerals, must know the detailed differences among the crystal structure systems and how light interacts with them. For the present purposes, however, a single fact is important: The speed of light traveling along most paths through anisotropic crystals depends on the direction of vibration. Even along the same path, light vibrating in one direction may be faster or slower than light vibrating in another (with both vibration directions being perpendicular to the travel path).

The polarizer under the stage allows the analyst to select the vibration direction. Without it, the rays of light rising vertically through the specimen vibrate parallel to the stage but with random orientation. The polarizer absorbs all these rays except those vibrating in one specific direction, which is usually fixed back-to-front or sideways in the field of view. The analyst turns the specimen on the stage to change the orientation of the crystals to the polarized light.

RETARDATION

An anisotropic crystal viewed down the microscope tube, in general, has one direction with maximum index of refraction and another perpendicular to it with minimum index. A ray vibrating parallel to the first would be the slowest, while the other would be the fastest. A ray oriented in any other direction actually separates into two parts, each part vibrating in one of the two directions but following the same vertical path. The part vibrating in the higher index direction is slower than is the other and falls progressively farther behind; it is said to be retarded. The amount of retardation depends on how far the parts travel (the thickness of the crystal) and on the difference in their speeds (and, thus, on the difference of their indices of refraction, called the birefringence). Because each color has its own particular wavelength, the amount of retardation affects whether the vibrations of the two parts of a given color are in step with each other (in phase) as they exit the crystal or determines how much they are out of step.

The analyzer (the upper polarizer) blocks all light vibrating parallel to the lower polarizer because it is oriented at right angles. Thus, the glass of the slide and any isotropic crystals appear black, as they do not alter the polarization. Similarly, the two separated parts of a ray from an anisotropic crystal, if they happen to emerge in step, recombine in the original polarization, and this light is blocked also. Although if one part is retarded out of step with the other (so that, recombined, they have a rotating "elliptical polarization"), the analyzer in effect deals with each part individually. It resolves each part once more and allows only those portions parallel to the analyzer vibration direction to pass. The passed portions of each part now vibrate in the same plane but are out of step with each other. Depending on how much they are out of step, the vibrations may reinforce each other and strengthen the color or oppose each other and weaken the color. Colors that are weakened or canceled are subtracted from the original white light, and what remains to be viewed is the complementary color. A sheet of mica placed between crossed Polaroids shows this effect well even without a microscope.

MICHEL-LEVY CHART

The interference colors that result from this process are one of the most striking features of crystals viewed under crossed polars. Because they result from the subtraction of specific colors from white light, some more and some less, they fall in a sequence that is distinctly different from an ordinary spectrum. Beginning with black when there is no retardation (the passed rays are in step), as retardation increases, the colors go through gray and white to orange and red for the first "order," then through several cycles from red through blue for higher orders, eventually merging to pinks and greens, and finally to more or less white for very high orders. The sequence is displayed on a Michel-Levy chart (which shows the sequence of colors as a function of birefringence). The colors that actually appear in a given crystal give important information about the mineral.

SAMPLE PREPARATION

Samples for microscopic examination are usually prepared either as a powder or as a thin section. The powder is made by crushing a mineral grain and screening it very fine; a small amount of the powder is then placed on a microscope slide with a drop of oil. The thin section is made by sawing a slice off of the sample, gluing it to a slide, then further sawing and grinding the slice until it is only 0.03 millimeter thick. In such thin samples, most minerals are transparent or translucent, although metals and many sulfide minerals are still opaque. Special reflection techniques can be used to examine opaque minerals.

Microscopic analysis of crushed mineral grains (powder) is the most efficient way to identify any mineral (and some nonmineral substances) whose crystals are large enough to be distinguished with a microscope. Thin sections are less efficient, but they have other advantages because they preserve the structure of the original sample; they are essential for the study of fine-grained rocks. By calculating the relative abundance of each kind of mineral, examining the shapes of grains and the ways they contact each other, and studying the distribution of grains and larger structures like bedding, the analyst can identify the rock type, estimate its properties, and interpret clues to its history. For example, a thin section of sandstone under the microscope would show the shape of the sand grains, fine details of its bedding, the amount of cement between grains, the amount of empty space, or porosity, and the presence and distribution of any mineral grains besides the quartz sand. This information could be used to estimate its mechanical strength for engineering purposes, its ability to hold water or oil, or its potential as a quarry stone, raw material for glassmaking, or ore of uranium.

MINERAL IDENTIFICATION

With either preparation, the first goal is to identify the minerals present by observing their visible properties. Features of shape, such as a characteristic crystal form, habit, cleavage, or fracture (keeping in mind that only a cross section is visible), give the first clues to identity. For example, garnets often exhibit a polygon-like cross section of their characteristic crystal form, and mica usually shows its perfect one-directional cleavage. Typical colors may be present (with polars uncrossed), although they are much fainter than in a hand sample. Some minerals, like tourmaline or biotite mica, change color as they are turned in the polarized light; these minerals are called pleochroic.

The relief of a crystal indicates the contrast between its index of refraction and that of its surroundings. A mineral with high relief appears to stand out from its background and have very distinct boundaries, while one with low relief is hard to distinguish from its background. If neighboring minerals or a medium (mounting or immersion) of known index are present, the analyst can estimate the index of refraction of an unknown mineral from its relief. The analyst can measure the index of minerals in powdered form exactly by comparison with standard index oils (called the immersion method). If the index of the mineral matches the oil closely, the grain boundary almost disappears. Anisotropic crystals require a different oil for each vibration direction. Having measured the indices, the analyst can then consult a table to identify the mineral.

If the indices cannot be measured directly, as in a thin section, the birefringence (the difference between maximum and minimum indices) gives useful information for identification. The interference colors in a crystal depend on its birefringence and its thickness (usually approximately 0.03 millimeter in a thin section). The analyst compares the highest interference colors found in a crystal to a Michel-Levy chart, determines the corresponding birefringence, and consults a table to identify the mineral.

The relationships between the vibration directions and visible features such as crystal faces and lines of cleavage give another clue to identity. At every quarter turn as an anisotropic crystal is turned on the stage, there is a point at which the crystal becomes completely dark, or extinct. Extinction occurs whenever the crystal's vibration directions are parallel to the polarizer or analyzer. The angle between an extinction direction and a crystal face or cleavage can distinguish between many otherwise similar minerals, such as the pyroxenes and amphiboles. Isotropic minerals like garnet are extinct at all positions of the stage.

INTERFERENCE FIGURES

Interference figures provide another powerful means of identifying crystals. They are shadows with distinctive shapes that appear with crossed polars and diverging light because polarized light is blocked from certain areas of the crystal image. Special lenses are used to cause the light to diverge and to change the focus of the eyepiece.

The shadow figures, which depend on the nature and orientation of the crystal, take the shapes of Maltese crosses or sweeping curves that move in distinctive ways as the stage is turned. The analyst can use them to determine many details about the crystal structure, the relationships of the vibration directions, and other features useful for identification.

COMMERCIAL AND PUBLIC-SAFETY APPLICATIONS

The petrographic microscope is an important tool for identifying many kinds of minerals and other substances that cannot easily be distinguished by ordinary physical and chemical tests. It has been used, for example, to determine the nature of corrosion products on metal surfaces; the corrosion products indicate which chemical reactions might be responsible for the damage and, therefore, how the surfaces might be protected. In another application, the microscope has been used to study the different materials traded commercially or displayed in museums as "jade." Officially, the name "jade" is applied to rocks composed of either an amphibole called nephrite or a pyroxene called jadeite, but the microscope revealed that much of what has been called jade is really composed of other minerals similar in appearance. The study showed historically significant patterns in the use of different kinds of jade in various cultures.

The petrographic microscope has many applications to areas in which geology touches on the economy or on public safety. Rock that has been sheared and fractured, as by faulting, shows distinctive texture and structure in thin section. Knowledge of these features in the rock of a given region can be important in the prediction of earthquake potential or in the evaluation of stability for engineering projects. Thin sections also show the amount of empty space, or porosity, between the grains in a rock, which is essential for estimating the potential of the rock for bearing oil, for carrying groundwater, or for allowing the passage of pollutants and radioactive waste. In the mining industry, thin sections are used to identify and evaluate the abundance of ore minerals and also to determine their grain size and how they are locked into the rock structure; all these factors determine whether the minerals can be recovered at a profit.

There are many anisotropic substances besides minerals. Whenever there is a distinct alignment of long molecules in a substance, polarized light may interact with it and reveal interference colors. Some biological tissues, structures in cells, plastics, and glasses have such anisotropic structures, and polarized light is useful for studying them. In one application, polarized light is used to study the distribution of stress in engineering structures such as machine parts and architectural members. The structure is modeled with a plastic such as Lucite and viewed through crossed polars. When the model is placed under load, the plastic develops interference colors that are concentrated at points of maximum stress.

Fiber-optic systems for transmission lines and optical switching devices developed for telephone and computer communications depend on the differences of the indices of refraction of their various parts. The polarizing microscope, which shows the differences by interference colors, is a key instrument for designing and testing such systems. New kinds of microscope systems using other kinds of radiation are becoming widely employed, but the polarizing light microscope will continue to hold a central importance both in the field of geology and outside it.

James A. Burbank, Jr.

CROSS-REFERENCES

Earthquake Engineering, 284; Electron Microprobes, 459; Electron Microscopy, 463; Experimental Petrology, 468; Fission Track Dating, 522; Fluid Inclusions, 394; Foliation and Lineation, 1390; Geologic and Topographic Maps, 474; Igneous Rock Classification, 1303; Infrared Spectra, 478; Lunar Rocks, 2561; Mass Spectrometry, 483; Metamorphic Mineral Deposits, 1614; Metamorphic Rock Classification, 1395; Metamorphic Textures, 1403; Microfossils, 1048; Minerals: Physical Properties, 1225; Minerals: Structure, 1232; Neutron Activation Analysis, 488; Rocks: Physical Properties, 1348; Sedimentary Mineral Deposits, 1637; Sedimentary Rock Classification, 1457; X-ray Fluorescence, 499; X-ray Powder Diffraction, 504.

BIBLIOGRAPHY

Barker, James. *Mass Spectrometry*. 2d ed. New York: Wiley, 1999. A college text concerning the field of mass spectrometry and its protocol and applications. There is a fair amount of analytical chemistry involved, so the reader without a scientific background may have a difficult time. Bibliographical references and index included.

Craig, James R., and D. J. Vaughan. *Ore Microscopy*. New York: John Wiley & Sons, 1981. The first three chapters (on the reflection microscope, preparation of polished specimens, and qualitative properties of minerals) give a compact overview of how the technique of reflection microscopy is used. Knowledge of basic mineralogy is assumed. Most of this college-level text is beyond the interests of the casual reader. The index is thorough, but the tables and references are technical.

Gribble, C. B., and A. J. Hall. *Optical Mineralogy: Principles and Practice*. New York: Chapman and Hall, 1993. A popular college textbook that provides a basic and comprehensible description of the theories, protocols, and applications involved in the field of optical mineralogy. Bibliography and index.

Hecht, E., and A. Zajac. *Optics*. Reading, Mass.: Addison-Wesley, 1974. A college-level text with considerable advanced mathematics, so well written and illustrated that a courageous nonmathematical reader can ignore the equations and still gain insight on many topics, especially polarization. Interesting examples and home experiments. Technical bibliography. An excellent index including many historical references.

Jambor, J., D. J. Vaughan, et al., eds. *Advanced Microscopic Studies of Ore Minerals*. Nepean, Ontario, Canada: Mineralogical Association of Canada, 1990. A collection of essays and lectures written by scientific experts in their respective fields, this volume is at times somewhat technical. Deals with topics such as mineral microscopy, the optical properties of ores, and optical microscopy. Anyone interested in petrographic microscopy is sure to find this a useful reference tool.

Kerr, Paul E. *Optical Mineralogy*. 4th ed. New York: McGraw-Hill, 1977. A college-level textbook emphasizing the identification of minerals in and interpretation of thin sections. The chapters on theory presume a knowledge of basic optics and mineralogy but provide a good summary of applications to the microscope. An ample bibliography, a selective index, excellent tables, and individual mineral descriptions make this a solid reference work for identifying common minerals.

Klein, Cornelis, and Cornelius S. Hurlbut, Jr. *Manual of Mineralogy.* 21st ed. New York: John Wiley & Sons, 1999. A college-level introduction to mineralogy. Contains a thorough discussion of crystal systems and concise descriptions of all common minerals, including essential optical data. Chapter 6 contains a summary of the construction and use of the petrographic microscope. Well illustrated and indexed, with key references after each chapter.

MacKenzie, W. S., and C. Guilford. *Atlas of Rock-forming Minerals in Thin Section.* New York: Halsted Press, 1980. The bulk of this short atlas consists of excellent color photographs of thin sections designed to show how common minerals appear with crossed and uncrossed polars. The minimal text identifies the minerals and key points of interpretation. Illustrates typical features but also shows the beauty of rocks in thin section.

Nesse, William D. *Introduction to Optical Mineralogy.* 2d ed. New York: Oxford University Press, 1991. A good introduction to mineralogy and optical mineralogy. Nesse clearly discusses the procedures and protocols of optical mineralogy and petrographic microscopy. Illustrations, bibliographic references, and an index.

Rochow, T. G., and E. G. Rochow. *An Introduction to Microscopy by Means of Light, Electrons, X Rays, or Ultrasound.* New York: Plenum Press, 1979. Thorough coverage of basic principles and the construction of various types of microscope. Very readable (high school level), with little scientific background presumed. The reader willing to cross-reference in this well-indexed volume will find technical terms carefully defined. Abundant useful illustrations, including views of thin sections. Michel-Levy chart included. Broad bibliography.

Sinkankas, John. *Mineralogy.* New York: Van Nostrand Reinhold, 1975. A wide-ranging and effective introduction to the nature and properties of crystals, written for the amateur. Very helpful illustrations. Chapter 8 contains an easy approach to many concepts needed to understand the petrographic microscope but applied instead to the simpler polariscope. Roughly one-half of the 585 pages are devoted to individual minerals. Useful tables and selected bibliography, with index.

Stoiber, Richard E., and S. A. Morse. *Microscopic Identification of Crystals.* Reprint. Malabar, Fla.: Robert E. Krieger, 1981. A compact, college-level text emphasizing the use of immersion oils for identification. The explanation of polarization and interference is clear and detailed, with many helpful illustrations, using a mostly geometrical approach.

X-RAY FLUORESCENCE

When a sample is placed in a beam of X rays, some of the X rays are absorbed and the absorbing atoms are excited, or raised to a higher energy state. X rays with energies characteristic of the particular element are emitted when these atoms decay back to their normal energy states. The intensity of these emitted, or fluorescence, X rays indicates the abundances of each element in the sample.

PRINCIPAL TERMS

ATOMIC NUMBER: the number of protons, or units of positive charge, in the nucleus of an atom

BOHR MODEL: a model of the atom in which electrons move in circular orbits around a positively charged nucleus, with orbits of only certain discrete energies being permitted

ENERGY LEVEL: the energy of an electron in one of the permitted orbits of the Bohr model of the atom

FLUORESCENCE: light emitted as the result of the decay of an atom from an excited state back to its ground state

GROUND STATE: the configuration of an atom such that all of its electrons are in the lowest energy levels that are permitted

X RAY: light in the wavelength range from 10^{-8} meter to about 10^{-10} meter, spanning the range from the ultraviolet to the gamma rays

BOHR MODEL

In the atomic model developed in 1913 by Niels Bohr, an atom consists of a positively charged nucleus surrounded by a number of negatively charged electrons that orbit around the nucleus in circles of fixed radii. Only orbits of specific radii are permitted in the Bohr model. The energy required to remove an electron completely from the atom is larger for orbits of smaller radii. Thus, the electrons in orbits of smaller radii are said to have less energy, or to be in a lower energy level. Only two electrons are permitted in each orbit, or energy level. Thus, elements heavier than helium, with an atomic number of two, must have electrons in several different energy levels. The specific energies of these levels depend on the atomic number of the atom involved.

When an atom is placed in a beam of X rays, a collision between an X ray from the beam and an electron in a lower electron energy level can result in the ejection of that electron from the atom. This action leaves a vacancy in a low energy level, which is filled by an electron from a higher energy level. The energy given up by the electron when it moves from the higher energy to the lower energy level goes into the emission of an X ray with a characteristic energy equal to the energy differ-

ence between the two electron energy levels. This process, however, leaves a vacancy at the higher energy level, which is filled by an electron from an even higher energy level. Again, an X ray with an energy characteristic of the energy difference between the two levels is emitted. This process continues until the atom that was disturbed or excited by the X ray from the incident beam has returned to its ground state, or lowest energy state. The series of X rays emitted as the atom returns to the ground state have energies characteristic of the atomic number of the atom that was excited. These emitted X rays, called fluorescence X rays, form the basis for the X-ray fluorescence method of chemical analysis.

The energy level structure of an atom depends on the charge of its atomic nucleus. Thus, each chemical element has a unique pattern of X-ray fluorescence emission. Chemical bonding into molecules generally disturbs the outer electron shell energy levels, because these electrons participate in the bonding. Because the fluorescence X rays are associated with the loss of an electron from an inner electron shell, the energies of the emitted X rays are virtually independent of the molecule in which the element is present in the sample. Thus, the X-ray fluorescence technique of chemical abundance determination is applicable to samples in their natural state and, generally, does not require significant sample preparation.

DEVELOPMENT OF X-RAY SPECTROSCOPY

The first demonstration of X-ray spectroscopy dates to around 1910, when C. G. Barkla obtained positive evidence of X-ray emission at characteristic energies by each element. By 1912, H. G. J. Moseley established the relationship between the energies of the fluorescence X rays and the atomic number of the atom responsible for the emission, laying the foundation for the identification of elements by X-ray emission analysis. This principle was used to validate the existence of the element hafnium, element 72, from its X-ray fluorescence.

Although the potential of this new technique was understood, practical difficulties limited its applicability. The early experiments used an incoming beam of electrons (not X rays) to eject an inner electron from the sample. This process required the sample to be electrically conductive, and the electron beam caused considerable sample heating. In the mid-1920's, it was recognized that the use of an incoming beam of X rays would eliminate many of the problems associated with the electron beam, but X-ray sources of high intensity and very sensitive detectors were required. By the mid-1950's, commercial instruments for X-ray fluorescence became available.

BENEFITS OF TECHNIQUE

A qualitative measure of the chemical composition of a sample can be obtained by measuring the energies of the fluorescence X rays that are emitted. These energies can be compared to tables indicating the energies expected for each element. The presence of an element in the sample at a detectable concentration is indicated if X rays of the energies corresponding to that element are seen in the emitted spectrum. Quantitative chemical analysis can be carried out on the sample by measuring the intensity, or number of X rays detected, at the wavelengths characteristic of the elements(s) of interest. This intensity is proportional to the number of atoms of that element or elements present in the sample.

The nondestructive nature of X-ray fluorescence analysis makes it the preferred technique for small or rare samples that must be preserved for other types of experimental measurements. The X-ray fluorescence technique is easily adaptable to almost complete automation; thus it is advantageous when a large number of samples needs to be analyzed, such as in mining operations to monitor ore quality.

COMPLICATIONS

In practice, corrections must be made for the efficiency of the X-ray fluorescence process, which depends on the element being detected and the energy distribution of the incoming beam of X rays. If the sample is thick enough that fluorescence X rays have a substantial probability of interaction with the sample before escaping and being detected, then corrections for this absorption process must also be made.

These complications delayed the routine application of X-ray fluorescence to the analysis of geological specimens until the early 1960's. By then, appropriate correction techniques had been developed. The most successful of these correction techniques is the preparation of a synthetic control sample of very similar composition to the rock to be analyzed and the comparison of the fluorescence intensities observed from the control to those from the rock under identical analysis conditions. Using this technique, by the mid-1960's most of the major rock-forming elements, particularly aluminum, phosphorus, potassium, calcium, titanium, manganese, and iron, as well as some minor and trace elements, could be measured as accurately using X-ray fluorescence as by the traditional wet chemical or optical spectrograph techniques. For most commonly analyzed rock specimens, the corrections are well determined, and X-ray fluorescence analyses can now be performed down to a sensitivity of about 10 parts per million with better than percent level precision.

TOOLS FOR ANALYSIS

The first requirement for X-ray fluorescence analysis is a beam of X rays to shine on the sample, usually provided by an X-ray tube or a radioactive source. A typical X-ray tube consists of a high-energy electron beam striking a heavy element target. This target then emits fluorescence X rays, which escape through a window in the X-ray tube and strike the sample. A smaller and more portable X-ray fluorescence apparatus frequently employs a radioactive source that emits X rays in its decay sequence. Some commonly used sources are iron 55, cobalt 57, cadmium 109, and curium 242. Intense X-ray beams have become available at particle accelerators, such as the National Synchrotron Light Source at Brookhaven National Laboratory, permitting X-ray fluorescence analysis of smaller samples and at lower elemental concentrations than with conventional laboratory instruments.

Two types of detectors are commonly employed to determine the number and energy of the fluorescence X rays. The first is a wavelength dispersive spectrometer, which uses a single crystal to diffract X rays of a particular energy into an electronic counter. The wavelength dispersive detector provides very high energy resolution, allowing nearby peaks from two different elements to be separated. The disadvantage of the wavelength dispersive spectrometer is that only a few X-ray energies can be measured at one time, depending on the number of counters that are placed around the diffracting crystal.

In the early 1970's, the energy dispersive X-ray detector was developed. It consists of a silicon semiconductor doped with lithium that produces an electronic pulse proportional to the energy of the X ray absorbed by the semiconductor. Thus, this device is responsive to all energies simultaneously, allowing the entire elemental composition of the sample to be determined at one time. The energy dispersive detector has two disadvantages: First, it must be operated at very low temperature, requiring liquid nitrogen to cool it; second, its energy resolution is inferior to the wavelength dispersive detector, causing the energy peaks to broaden and frequently overlap. Mathematical modeling of the peaks' shapes is then required to recover information on the number of X rays in each of the two overlapping energy ranges.

APPLICATIONS IN SCIENCE AND INDUSTRY

In mining, automated X-ray fluorescence systems are used for the continuous analysis of the zinc abundance in flowing slurries of zinc concentrates. Throughout the mining industry, the X-ray fluorescence technique is employed to analyze ores, tailings, concentrates, and drilled cores. In geology, the X-ray fluorescence method of chemical analysis has been applied to all rock types. Be-cause of its sensitivity to elements present in low abundances and its ease of application to a large number of samples, the X-ray fluorescence technique has been employed in a wide variety of geological investigations.

Recently, X-ray fluorescence has been applied to the analysis of particles collected from the air in order to determine the concentrations of toxic elements. In the analysis of airborne particles, the samples are usually collected by passing measured volumes of air through filter paper and then performing an analysis on the bulk material trapped on the filter paper. With the development of more sensitive X-ray fluorescence apparatus, it is sometimes possible to perform elemental analyses on individual dust particles. In addition, the elemental makeup of the particles is frequently useful in determining the source of the air pollution. In some cases, determination of the chemical composition of the particles has allowed identification of the source of the pollution. In agriculture and food science, the X-ray fluorescence method is used in determining the trace element content of plants and foods. This application has been used to monitor the concentrations of insecticides on leaves and fruits.

X-ray fluorescence has also been applied to problems in medicine. The sulfur content of each of the different proteins in human blood, determined by X-ray fluorescence, has proven useful in medical diagnosis. X-ray fluorescence has also been used to determine the strontium content of blood serum and bone tissue. This use was particularly important during the era of aboveground nuclear testing, when radioactive strontium absorption, particularly by children, was a major problem.

A simple X-ray fluorescence spectrometer flew on each of the two Viking spacecraft that landed on the planet Mars in 1976. These spacecraft provided the first determinations of the major element abundances in the soil of Mars. These measurements confirmed a basaltic composition for the soil, indicating that Mars had experienced planetary differentiation, that is, separation into a metallic core and a stony mantle.

ADDITIONAL APPLICATIONS

The development of the X-ray fluorescence technique for medical, chemical, geological, and industrial uses has led to a variety of additional applications. For example, X-ray fluorescence is used to observe automotive and aircraft engine wear by determining the concentrations of the metallic iron, curium, and zinc particles suspended in lubricating oils. By identifying the specific element, it is often possible to identify the actual engine component that is wearing.

The nondestructive nature of X-ray fluorescence has made it an ideal technique for the analysis and authentication of art objects and ancient coins. The elemental compositions of inks, paints, and alloys in the object are compared with the compositions in use at the alleged time of production of the object in question. The widespread availability of X-ray fluorescence apparatuses has given rise to a variety of applications for this technique that were not anticipated at the time of its initial development.

George J. Flynn

BIBLIOGRAPHY

Adler, I. *X-Ray Emission Spectrography in Geology*. New York: Elsevier, 1966. This classic 258-page text describes all aspects of the theory of X-ray fluorescence analysis as well as the practical aspects of sample preparation, sensitivity, and methods of interpreting the resulting data. Well illustrated and intended for college-level geology students, the book provides a clear introduction to the method of X-ray fluorescence analysis and a thorough bibliography.

Dzubay, T. G. *X-Ray Fluorescence Analysis of Environmental Samples*. Ann Arbor, Mich.: Ann Arbor Science, 1981. This 310-page book describes the applications of X-ray fluorescence analysis to problems of atmospheric science, particularly the chemical characterization of airborne particulate matter. Each chapter contains a reference list of scientific journal articles describing particular applications of the technique. While intended for professionals, most sections should be understandable by college-level science students.

Jenkins, Ron. *X-Ray Fluorescence Spectrometry*. 2d ed. New York: Wiley, 1999. This college-level textbook clearly describes the entire process of X-ray fluorescence analysis, beginning with a historical account of the development of the technique. The sources of X rays, the X-ray fluorescence emission process, and the various types of detectors are described in detail.

Klockenkeamper, R. *Total-Reflection X-Ray Fluorescence Analysis*. New York: Wiley, 1997. A clear description of the procedures and protocols associated with X-ray and fluorescence spectroscopy. Appropriate for the college student without much background with the field. Illustrations, index, and bibliographical references.

Liebhafsky, H. A., and H. G. Pfeiffer. "X-Ray Techniques." In *Modern Methods of Geochemical Analysis*, edited by R. E. Wainerdi and E. A. Uken. New York: Plenum Press, 1971. The process by which X rays are produced and the interaction of these X rays with matter are throughly discussed. A schematic electron-shell diagram clearly illustrates the various X-ray energies emitted in the X-ray fluorescence process. Although intended for college-level readers, this well-illustrated, 25-page chapter should be suitable for students who have completed a high school chemistry course.

Maxwell, J. A. *Rock and Mineral Analysis*. New York: John Wiley & Sons, 1968. This comprehensive textbook emphasizes the analysis of rock composition by wet chemical techniques but devotes chapter 11 to the X-ray fluorescence technique. The emphasis in this 22-

page chapter is on sample preparation, precision of the analyses, and experimental complications. Suitable for college-level readers.

Pella, P. A. "X-Ray Spectrometry." In *Instrumental Analysis*, edited by G. D. Christian and J. E. O'Reilly. 2d ed. Boston: Allyn and Bacon, 1986. This well-illustrated chapter describes the process of X-ray fluorescence and the instrumentation normally employed. Some of the complications in inferring chemical compositions from the X-ray spectra are also described. The textbook, intended for undergraduate science students, includes an extensive bibliography.

Pinta, Maurice. *Modern Methods for Trace Element Analysis*. Ann Arbor, Mich.: Ann Arbor Science, 1978. Chapter 8 of this well-illustrated book describes all aspects of the X-ray fluorescence method and its application to geological, biological, and industrial samples. An extensive reference list will direct the reader to original sources for a variety of applications of X-ray fluorescence analysis. This 53-page chapter is suitable for advanced high school students.

Robinson, J. W. *Undergraduate Instrumental Analysis*. 3d ed. New York: Marcel Dekker, 1982. This textbook, intended for undergraduate science students, includes a chapter on X-ray spectroscopy, which discusses the applications of the technique to chemical abundance determinations. The medical, industrial, and scientific applications of X-ray fluorescence are described.

Tertian, R., and F. Claisse. *Principles of Quantitative X-Ray Fluorescence Analysis*. New York: Wiley, 1982. This comprehensive 385-page text describes all aspects of X-ray fluorescence analysis. Individual chapters describe the process by which fluorescence X rays are emitted, the instrumentation employed, sample preparation, and the procedure for interpreting the observed spectra. Each chapter contains a comprehensive reference list. Suitable for college-level science students.

X-RAY POWDER DIFFRACTION

X-ray powder diffraction is a technique applied to finely powdered crystals or mixtures of crystals to identify and determine the relative amounts of the crystal phase or phases present.

PRINCIPAL TERMS

BRAGG'S LAW: the fundamental equation that relates X-ray wavelength, interatomic distances, and the angle between the X-ray beam and the lattice plane of crystals

CERAMIC: a human-made mineral, crystal, or aggregate thereof, excluding metals

CRYSTAL: a solid consisting of a regular periodic arrangement of atoms; its external form and physical properties express the repeated units of the structure

DIFFRACTOMETER: an instrument used for X-ray powder diffraction analysis

D-SPACING: the distance between successive parallel layers of atoms in a crystal

GLASS: a solid with no regular periodic arrangement of atoms; an amorphous solid

GONIOMETER: the mechanism that maintains the correct arrangement among the sample powder, the X-ray beam, and the X-ray detector in a diffractometer

MINERAL: a natural substance of fixed or narrowly limited chemical and physical properties; most minerals are also crystals

PHASE: a homogeneous, physically distinct, mechanically separable portion of matter present in a nonhomogeneous chemical system

X RAY: a photon with a much higher energy and shorter wavelength than those of visible light; its wavelength is of the same order of magnitude as the spaces between atoms in a crystal

STRUCTURE OF CRYSTALS

X-ray powder diffraction is a technique used in the analysis of fine powders. It can be used to distinguish glasses from crystals; to identify crystal and mineral phases; and to determine the relative amounts of crystal phases in mixtures, the composition of crystals that have a range of ionic substitution, and the size and shape of the unit cell of a crystal substance.

The technique is based on the structure of crystals—that is, on their orderly, periodically repeating system of atoms and molecules. For example, if the minerals quartz (silicon dioxide) and calcite (calcium carbonate) are present in a rock, then X-ray powder techniques can be used to determine their presence and relative amounts. It is not a chemical technique for determining the presence and amount of particular elements (except for those minerals with limited ionic substitution); therefore, it does not specifically determine the presence of the silicon or calcium in quartz or calcite. The procedure depends on the fact that the wavelength of X rays and the spacing between layers of atoms that make up the periodic structure of crystalline substances (d-spacing) are similar: 0.5 to 2.5 angstroms. As a consequence, when X rays are swept over a crystal lattice and geometric conditions are correct, an energy peak will be emitted that represents each lattice plane of the crystal. Every mineral has a unique set of peaks whose position and size are characteristic of its crystal structure and chemical composition.

A crystal is a homogeneous solid with an orderly, periodically repeating atomic structure. This structure is responsible for the flat faces on large crystals; the orientation of these faces relative to one another is a consequence of the internal structure of the crystal. A two-dimensional analogy to the periodic structure in a crystal lattice is the repeating pattern in wallpaper. Each design unit in the wallpaper can be envisioned as representing an atom or molecule (a cluster of atoms). The pattern in a wallpaper design obeys the same mathematical laws that pervade nature, including the structure of crystals. For simplicity, the structure of crystals is commonly envisioned as a series of points periodically repeating in a three-dimensional space. The atoms form sets of parallel planes called lattice planes, and the distance between each lattice plane is symbolized by the letter *d*. Even in a simple rectangular array of two dimensions, there are many possible lattice d-spacings (d1, d2, d3, d4, and so on). Intersecting sets of lattice planes delimit a minimal group of atoms that forms the unit cell, the fundamental building block of each crystal. There are strict mathematical laws that govern the way atoms can repeat in space, and there are only six possible crystal systems: cubic, tetragonal, orthorhombic, hexagonal, monoclinic, and triclinic. These six systems together have 230 non-identical space groups, or arrangements of points in space. When combined with the variation supplied by the ninety-two natural elements, this diversity means that no two crystals have identical structures.

GENERATION OF X RAYS

Among other ways, X rays are generated when high-energy photons bombard a metal. Copper is the metal most used in X-ray powder diffraction procedures, but many other metals can be used, including molybdenum, nickel, cobalt, iron, and chromium. An X-ray tube consists of a tungsten filament and a copper or other metal target in a vacuum. The tungsten filament supplies photons that are accelerated into the copper target by currents of 30 to 50 kilovolts and 10 to 50 milliamperes. The photons interact with the copper target in two ways. First, many photons are absorbed by a variety of processes that give rise to a broad and continuous spectrum of X-ray energy. Second, and most important in powder diffraction, X rays of a very precise wavelength will be emitted by electrons orbiting the nucleus of the copper atoms when these electrons absorb and then release a fixed amount of energy. Thus, the wavelength of X rays emitted by a copper tube consists of a broad "hump" with "spikes" at highly specific wavelengths. A typical commercial X-ray diffractometer with a copper tube is designed to allow only the X rays of wavelength 1.5418 angstroms to hit the target powder.

BRAGG'S LAW

When X rays impinge on a crystal lattice, many interactions take place. The one of importance here is diffraction. The diffraction relation between X-ray wavelength and the d-spacing between lattice planes of a crystal is expressed by Bragg's law. Bragg's law is $n\lambda = 2d \sin\theta$. In this equation, n is a whole number (usually taken as 1), λ is the wavelength of the X rays (a known quantity), d is the distance between successive parallel planes in a crystal, and θ is the angle between the direction of incoming X rays and the lattice plane of interest. (The angle is measured by a goniometer.) When the angle θ is such that the distance ABC is not a multiple of a whole wavelength, the diffracted energy will be low, because the emitted wave will be out of phase. When the angle ABC is equal to whole multiples of the wavelength, however, then the emitted waves will be in phase; they will reinforce each other, and an energy peak will be emitted. Bragg's equation can be readily solved, because the X-ray wavelength is known, the angle θ can be read on the goniometer when a peak appears, and one can assume that n is 1. Thus, the only unknown in the equation is the lattice d-spacing (d).

SINGLE-CRYSTAL DIFFRACTION

There are two X-ray diffraction techniques: single crystal and powder. Single-crystal diffraction requires one crystal of the substance of interest. The crystal must be oriented in such a way that the relation between the crystal lattice, the X-ray beam, and the detector is precisely defined. The technique is difficult, requiring precise orientation, refined analysis, and considerable mathematical skill. Single-crystal techniques are used for determination of the details of crystal structure. X-ray powder diffraction is a comparatively simple and routine analytical procedure. It depends on statistics. The fine powder is packed in its holder so that the millions of crystallites, or minute crystals, are randomly oriented relative to the X-ray beam. Usually, a powder with particles measuring 45 micrometers or less is required. The effect is as though there were one average-sized crystal present. If not already fine grained, the sample must be ground into a powder.

POWDER DIFFRACTION

There are two powder techniques: camera and diffractometer. In the camera method, a special cylindrical Debye-Scherrer camera is used. The sample powder is placed in a thin, glass tube, and the tube is placed on the axis of rotation on the centerline of the cylinder. Photographic film is placed along the inner circumference of the cylinder. X rays enter along a hole in the side of the camera, hit the rotating powder, are diffracted, and then hit the film. Peaks are recorded as lines on the film. Careful measurement of the line position relative to the hole where the X rays entered and knowledge of the geometry involved yield the location of the peaks, which in turn leads to the solution of Bragg's equation. The camera technique was the first to be developed. It is still useful when the amount of sample is very small, but it is tedious and time-consuming. A typical run may take four or five hours.

The diffractometer method is the fastest, easiest, most quantitative, and most widely used powder technique. In it, a goniometer correctly positions the X-ray beam, the surface of the sample powder, and the X-ray detector so that proper geometry for the solution of Bragg's equation is maintained. First, the proper operating conditions for the X-ray generation are set and the goniometer positioned at the desired start angle. Then, the powder is pressed into a holder, so that it has a smooth surface, and placed in the sample chamber. The goniometer scans from the starting angle to the ending angle. (In modern machines, that procedure is controlled by a computer program.) The output signal is received by a scintillation counter, electronically enhanced, and sent to an output device, typically a stripchart recorder. The output data are called the X-ray pattern, a graph of peak position and size versus angle degrees. A typical run takes thirty minutes.

Each diffraction peak corresponds to an interplaner d-spacing. The size of the peak is a function of the electron density along the "surface" of the lattice; in general, the heavier the element, the larger the peak. For identification of a single-phase powder, the three largest peaks in the low-angle range are normally adequate. Data on their position and relative size are compared to the X-ray powder diffraction card index file published by the American Society for Testing and Materials. When a match is found for the three main peaks, the identification is verified using the balance of the peaks.

If there are two or more phases in the powder,

identification becomes a matter of experience, and guesses are used until a phase is identified. Elimination of the peaks of an identified phase is followed by repetitions of the procedure until all peaks are accounted for. Computer methods are now available in which the program searches likely combinations of peaks. Where precise measurement of the d-spacings is desired, such as in the determination of the unit cell, peaks in the high-angle region are used. The d-spacings are partially controlled by composition variation. For example, when small ions substitute for larger, the d-spacing decreases. This decrease results in a shift of the peak toward higher angles.

LIMITATIONS AND BENEFITS

X-ray powder diffraction has several limitations. It is difficult to detect components that form less than 1 percent of a mixture. Furthermore, estimates of the amounts of specific minerals present in a mineral mixture are seldom more accurate than plus or minus 1 percent. On the other hand, the technique does not alter the character of the sample, and after X-ray analysis, the powder can be used for further analyses.

In fact, X-ray diffraction is widely used in the study of rocks, sediments, meteorites, and any crystalline solid in which the particles are too fine-grained for analysis by standard optical techniques. Because of its simplicity, it is also used to identify large crystals after they have been ground into a suitable powder.

One area in which X-ray diffraction is used is the classification of silicate minerals. Attempts to classify this large and diverse family of minerals on chemical grounds resulted in contradictions and confusion. X-ray diffraction is now used to divide silicates into structural groups—such as ortho-silicates, phyllosilicates (which include clay minerals), and isosilicates—giving rise to a logical and meaningful classification scheme.

ROLE IN TECHNOLOGICAL DEVELOPMENTS

X-ray diffraction procedures are vital to the technological developments that one tends to take for granted in geology, oceanography, meteoritics, ceramics, electronics, and cements. The search for a diminishing body of finite Earth resources means that geologists must look more and more closely at rocks that contain the needed materials. Many of these rocks are very fine grained, and X-ray diffraction is the only suitable analytical tool. In oceanography, the X-ray analysis of the very fine sediments and rocks on the seafloor is essential to understanding the origin and history of the ocean basins. Meteorites, rocky and metallic fragments from beyond the atmosphere, are typically fine grained. Nondestructive analysis by X-ray diffraction is essential to their classification. In the ceramics industry, the development of tougher and more durable pottery, grinding compounds, and insulators requires this analytical technique; in electronics, the development of new transistors, thermisters, and superconductors requires X-ray-based, single-crystal analysis. Cements form a complex paste of reactants that, upon curing, are best studied by X-ray powder diffraction.

In sum, whenever the samples are in the form of very fine crystallites, X-ray powder diffraction is the most powerful and easy-to-use system for mineral or crystal analysis. It is difficult to see how civilization could have reached its present technological level without this tool.

David N. Lumsden

CROSS-REFERENCES

Carbonates, 1182; Cement, 1550; Chondrites, 2660; Clays and Clay Minerals, 1187; Electron Microprobes, 459; Electron Microscopy, 463; Experimental Petrology, 468; Feldspars, 1193; Geologic and Topographic Maps, 474; Hydrothermal Mineralization, 1205; Infrared Spectra, 478; Ionic Substitution, 1209; Mass Spectrometry, 483; Metamorphic Mineral Deposits, 1614; Minerals: Physical Properties, 1225; Minerals: Structure, 1232; Neutron Activation Analysis, 488; Orthosilicates, 1244; Petrographic Microscopes, 493; Phase Changes, 436; Sedimentary Mineral Deposits, 1637; Silica Minerals, 1354; Stony Irons, 2724; X-ray Fluorescence, 499.

BIBLIOGRAPHY

Azaroff, L. V., and M. J. Buerger. *The Powder Method in X-Ray Crystallography.* New York: McGraw-Hill, 1958. This text is best suited to a second college course in X-ray diffraction analysis. Its focus is the use of powder cameras. As with all books written on this topic, the authors expect the reader to have a background in elementary physics and crystallography. They discuss the design and alignment of cameras, how to take photographs, how to interpret powder photographs in terms of unit cell size and geometry, the causes of errors, and how to overcome them.

Bowen, David Keith, and Brian K. Tanner. *High-Resolution X-Ray Diffractometry and Topography.* London: Taylor and Francis, 1998. This book examines the procedures involved with and the equipment required within the field of crystallography. Bowen and Tanner lay the foundation for a thorough look at the processes and applications of X-ray diffraction and X-ray crystallography. A somewhat technical book intended for the specialist.

Bunn, C. W. *Chemical Crystallography.* Oxford, England: Clarendon Press, 1958. A readable text for a graduate-level course in X-ray diffraction procedures. Its emphasis is on basic principles of crystallography, and it provides comparatively little information on the source and interaction of X rays. It is a valuable resource for camera techniques, both powder and single crystal. There is a minimum of math and chemistry; the author relies instead on photographs and diagrams. Includes a chapter with examples of successful solutions of crystallographic structures.

Hammond, Christopher. *The Basics of Crystallography and Diffraction.* London: Oxford University Press, 1997. Hammond offers a clear understanding of the principles and practices of crystallography and X-ray crystallography. Index and bibliography.

Jenkins, Ron. *Introduction to X-Ray Powder Diffractometry.* 2d ed. New York: John Wiley, 1996. This classic text is intended for an introductory college course in X-ray crystallography. It is a basic source for information on the principles and practice of X-ray powder diffraction as applied to inorganic materials. Jenkins discusses crystallography, X-ray production, the interaction of X rays and crystals, and the details of X-ray diffractometer design.

Jones, Christopher, Barbara Mulloy, and Mark R. Sanderson, et al., eds. *Crsytallographic Methods and Protocol.* Totowa, N.J.: Humana Press, 1996. Part of the Methods in Molecular Biology series, this volume examines the use of X-ray diffraction to determine the structure of compounds such as nucleic acids and proteins. A large portion of the book is dedicated to discussing the practices and protocols surrounding X-ray diffraction and X-ray crystallography.

Klug, H. P., and L. E. Alexander. *X-Ray Diffraction Procedures for Polycrystalline and Amorphous Materials.* New York: John Wiley & Sons, 1954. This classic text is intended for an introductory college course in X-ray crystallography. It is a basic source for information on the principles and practice of X-ray powder diffraction as applied to inorganic materials. It discusses crystallography, X-ray production, the interaction of X rays and crystals, and the details of X-ray diffractometer design. The specific diffractometers discussed are dated, but the principles remain the same.

Nuffield, E. W. *X-Ray Diffraction Methods.* New York: John Wiley & Sons, 1966. This relatively brief book combines information on powder and single-crystal techniques. It is intended as a laboratory aid for students with limited mathematical backgrounds. Discussions of elementary crystallography and X-ray generation are followed by chapters devoted to specific methods, techniques, and concepts. A good introduction to how single-crystal and powder techniques are related.

10
GEOCHRONOLOGY AND THE AGE OF EARTH

EARTH'S AGE

Determining the age of the Earth is one of the great achievements of science. Until the eighteenth century, all geological phenomena were believed to have been produced by historical catastrophes such as great floods and earthquakes. The new geology showed that the Earth was billions of years old, rather than thousands as many had previously believed, and that the Earth had the form it did because of slow uniform processes rather than catastrophes.

PRINCIPAL TERMS

CATASTROPHISM: the theory that the large-scale features of the Earth were created suddenly by catastrophes in the past; the opposite of uniformitarianism

GEOCHRONOLOGY: the study of the time scale of the Earth; it attempts to develop methods that allow the scientist to reconstruct the past by dating events such as the formation of rocks

ISOTOPE: atoms with the same number of protons in the nucleus but differing numbers of neutrons; a particular element will generally have several different isotopes occurring naturally

RADIOACTIVITY: the process by which an unstable atomic nucleus spontaneously emits a particle (or particles) and changes into another atom

SEDIMENTARY: rocks that are formed by a layering process that is generally easily visible in a cross section of the rock

UNIFORMITARIANISM: the theory that processes currently operating in nature have always been operating; it suggests that the large-scale features of the Earth were developed very slowly over vast periods of time

BIBLE-BASED CALCULATIONS

In the middle of the seventeenth century, Joseph Barber Lightfoot of prestigious Cambridge University in England penned the following words: "Heaven and Earth, center and circumference, were made in the same instant of time, and clouds full of water, and man was created by the Trinity on the 26th of October 4004 B.C. at 9 o'clock in the morning." At the time that Lightfoot wrote those words, this statement expressed the most informed opinion on the age of the Earth—namely, that it could be calculated by adding up the ages of the people recorded in the Old Testament and assuming that Adam and Eve were created at about the same time as was the Earth. This was the method that most scientists—including Nicolaus Copernicus, Johannes Kepler, and Sir Isaac Newton—used to date the Earth, and much effort was expended analyzing the first few books of the Old Testament "scientifically."

UNIFORMITARIANISM

A little over a century later, a Scottish geologist named James Hutton suggested that there was a better way to determine the past history of the Earth than by poring over biblical genealogies. Hutton believed that processes currently operating in nature could be extrapolated back in time to shed light on the historical development of the Earth. This idea—that historical processes are essentially the same as present processes—is called uniformitarianism. In 1785, he presented his new views on geology in a paper entitled "Theory of the Earth: Or, An Investigation of the Laws Observable in the Composition, Dissolution, and Restoration of Land upon the Globe." Uniformitarianism became the foundation of the newly developing science of historical geology.

Charles Lyell, who was born in the year of Hutton's death, extended these new ideas and laid the foundation for what was to become a powerful new science. The major argument was over the age of the Earth. Was it really billions of years old, as suggested by new discoveries and theories, or was it only a few thousand years old, as everyone had previously believed? The materials from which the Earth is constructed are certainly very old. Many of the atoms in the Earth date from the beginning of the universe, 15 to 20 billion years ago. The establishment of criteria by which the

511

age of anything will be determined is guided by the need for that age to be a meaningful physical quantity. The conventional definition of age for a person (number of years since birth) is meaningful; the number of years since the origin of the atoms in a person's body would not be meaningful, because it is not relevant to that particular person's duration of existence as that person. A meaningful definition for the age of the Earth can thus be formulated as follows: The age of the Earth is the time since its composite materials acquired an organization that could be identified with the present Earth.

CURRENT THEORIES OF EARTH'S FORMATION

Current theories of the formation of the Earth suggest that the atoms of the Earth and all the other members of the solar system formed a cloud of interstellar material that existed in a corner of the Milky Way galaxy several billion years ago. Under the influence of gravity, this cloud of material began to condense in those regions where the concentration of material was sufficiently higher than average. This nebular cloud, as it is called, gave birth to the Earth, the Sun, and the planets. As the material from which the Earth was forming condensed, a number of events occurred: The density increased to the point where the mutual repulsion of the particles balanced the gravity from the newly formed "planet"; the planet became hotter as friction from the now-dense material became a significant source of energy; and energy given off by materials inside the planet was unable to escape into space and was absorbed, further increasing the temperature. The early Earth was therefore very hot and existed in a molten state for many years.

There is thus no unique age for the Earth. Rather, there is a time period that can realistically be described as the "birth" of the Earth. This time period was millions of years long, and any dates given for the age of the Earth must necessarily reflect this ambiguity. Fortunately, the age of the Earth is measured in billions of years, so the uncertainties surrounding the exact time of its birth do not significantly affect measurements of its age.

RECONSTRUCTING EARTH'S HISTORY

Since the initial formation of the Earth, many processes have been taking place: Unstable (radio-active) materials have been decaying into other elements; the initial rotation rate has been declining as friction from the tides and the Moon has worked to slow the rotation of the Earth; mountains have been rising under the influence of global tectonics, and rivers have been formed from the ceaseless activities of erosion; and evolution has been transforming the planet, changing sterile compounds into organic, and barren wasteland into ecological congestion as the phenomenon of life has manifested itself over the face of the globe.

As these various physical processes traverse the Earth, they leave footprints as evidence of their passing. When these footprints are studied, the history of the Earth can be reconstructed. In some cases, this reconstruction can lead all the way back to the origin of the Earth, thus providing an answer to the question "How old is the earth?"

MEASURING NATURAL PROCESSES

Current estimates put the age of the Earth at about 4.6 billion years. This figure is firmly supported by a number of measurements—some very direct and straightforward and some rather subtle. Life itself can be used as a clock. For example, trees add distinguishable layers of growth at a rate of one a year; these are the familiar "rings" that can be counted on a stump of wood. Counting these rings provides a very accurate clock for determining the age of the tree. Giant sequoias in California are regularly dated at about three thousand years old, and the bristlecone pine has been dated at almost five thousand years. Samples of sedimentary rock, which form yearly layers called varves, can extend back as far as twenty thousand years. Unfortunately, all these annual processes that provide a direct year-by-year chronicle of Earth history provide no useful data beyond a few tens of thousands of years.

There are other, less direct, uniformitarian processes, however, that perform somewhat better in this regard. Measurements of erosion, the salinity of the ocean, the strength and direction of the Earth's magnetic field, and the internal heat of the Earth can all yield values for the "age" of the Earth, measured in millions rather than thousands of years. The validity of each of these indirect measurements requires a strict uniformitarian character for the nature of the process; this assumption, however, is not legitimate for most of these processes, which explains why the ages determined from their application are so discordant and unreliable.

RADIOACTIVE DECAY

The most consistent geological chronometer is based on radioactive decay, an atomic/nuclear phenomenon. All atoms consist of a densely packed nucleus housing a number of protons, which have a positive charge, and neutrons, which have no charge. Because the protons are all positively charged, they repel one another; an atomic nucleus would immediately explode if it were not for a different nuclear force, called the strong force, that holds them together. Every nucleus exists in a state of dynamic tension as the electrical force tries to blow it apart and the strong nuclear force tries to hold it together. Certain nuclei are frequently unstable; that is, they have a tendency to disintegrate spontaneously into other, more stable, nuclei. This disintegration is initiated by yet another nuclear force, the weak force.

Usually the protons in the nucleus of an atom are paired with a particular number of neutrons in such a way that the nucleus will be stable. For the first few elements on the periodic table, the neutron/proton ratio is equal to one, but for larger atoms, the ratio increases as the neutrons start to outnumber the protons. For almost all the elements, there are certain nuclear combinations of protons and neutrons that are stable. By definition, members of the same atomic species have the same number of protons in the nucleus and thus the same atomic number. Atoms with differing numbers of neutrons are called isotopes of that element. Carbon, for example, normally has twelve particles in the nucleus—six protons and six neutrons—and is therefore designated carbon 12. A common isotope, however, has two extra neutrons and is designated carbon 14.

The detailed structure of a particular nucleus determines its long-term stability. Most of the nuclear configurations found in nature, such as hydrogen and helium, are stable indefinitely, or at least for a time that is much longer than the age of the universe (about 20 billion years). Unstable nuclei, on the other hand, are stable for only a finite period of time, which can be either very short (a fraction of a second) or very long (billions of years), depending on the composition of the particular nucleus.

The period of stability for an unstable nucleus is known as its half-life. A half-life is defined to be the time period during which one-half of the nuclei of a given sample will spontaneously decay into another nuclear species. The half-life of carbon 14, for example, is about 5,730 years. This means that in 5,730 years, one-half of an original carbon 14 nucleus, called the "parent," will spontaneously decay into another element, nitrogen 14, called the "daughter" element. Over time, the parent element will gradually transform into the daughter. The ratio of daughter to parent can be used to determine how long the parent has been decaying and thus how old the material containing the parent is. It is important to note that the assumption of uniformitarianism for radioactive decay rates is considered very reasonable. Unlike the other processes mentioned above, there seem to be very few mechanisms in nature that can disturb the constancy of the radioactive "clock."

RADIOACTIVE DATING

A number of radioactive materials are found in nature, all with differing half-lives. Each can be used to find the ages consistent with their half-lives; that is, a material with a long half-life, such as uranium 238 (whose half-life is approximately 4.5 billion years), can be used to date objects that are billions of years old, and carbon 14 can be used to date objects that are thousands of years old.

Radioactive dating has been applied to many rocks found on the Earth. The oldest rocks be-

HALF-LIVES OF SOME UNSTABLE ISOTOPES USED IN DATING

Parent Isotope	Daughter Product	Half-Life Value
Uranium 238	Lead 206	4.5 billion years
Uranium 235	Lead 207	704 million years
Thorium 232	Lead 208	14.0 billion years
Rubidium 87	Strontium 87	48.8 billion years
Potassium 40	Argon 40	1.25 billion years
Samarium 147	Neodymium 143	106 billion years

SOURCE: U.S. Geological Survey.

lieved to have formed on the Earth are from a volcano in western Greenland and have been dated at about 3.8 billion years, using uranium 238. It is difficult to find very ancient rocks on the surface of the Earth, because most of the Earth's surface has been rebuilt many times since the Earth was born. There are probably older rocks in the deep interior of the Earth.

METEORITE DATING

The currently accepted age for the Earth, 4.6 billion years, was obtained by dating meteorites that fall to Earth from space. These meteorites are believed to have been formed at the same time as was the Earth and to have existed in the vacuum of space until they were captured by the gravity from the Earth. Similar dates have been obtained from the rocks brought back from the Moon, which is believed to have formed at about the same time as the Earth.

While many questions remain about the details of the formation of the Earth, two facts seem clear: First, the Earth owes its origin to the same processes that brought the solar system into existence; second, those processes can be dated with a high degree of confidence at between 4 and 5 billion years ago.

PRACTICAL AND SPECULATIVE ASPECTS

The problem of the age of the Earth is part of a much larger scientific question, which exists at the interface between the very practical study of the Earth and its various properties and the more esoteric question of the origin and evolution of the universe as a whole. On the practical side, knowledge of the Earth's various and occasionally delicate properties is important for the future of the human race. By knowing how long the Earth has been in existence, scientists are better able to understand the processes that have shaped the surface of the Earth into the form that it has today. Predicting earthquakes, hunting for oil, monitoring the spread of the seafloor—all these practical questions require knowledge of large-scale planetary processes, the same kind of knowledge that illuminates the question of the age of the Earth. Furthermore, knowing that the Earth is billions of years old and can easily survive for billions more should encourage human societies to take better care of the planet.

From a more esoteric or speculative point of view, the age of the Earth is important because it speaks to the most fundamental questions that are asked about the place of human beings in the universe. How old is this planet? How was it formed? In the century or so since geological science overthrew the seventeenth century notion of a much younger Earth, people have struggled with finding a new place in the universe. The argument that began centuries ago is still heard in courtrooms across the United States as "creation science" once again argues that the Earth is thousands, not billions, of years old. Legal battles rage over the issue of whether high schools across the country should teach geochronology that is based on religious dogma rather than on scientific research. Research is still being done on this very important scientific question and no doubt will continue into the foreseeable future as the human mind strives to learn more about the Earth. The growing awareness of how dependent humans are on the continued health of the Earth is a powerful incentive to learn more about their planetary home.

Karl Giberson

CROSS-REFERENCES

Earth's Oldest Rocks, 516; Earth's Origin, 2389; Elemental Distribution, 379; Evolution of Earth's Composition, 386; Fission Track Dating, 522; Nucleosynthesis, 425; Potassium-Argon Dating, 527; Radioactive Decay, 532; Radioactive Minerals, 1255; Radiocarbon Dating, 537; Rubidium-Strontium Dating, 543; Samarium-Neodymium Dating, 548; Solar System's Origin, 2607; Uniformitarianism, 1167; Uranium-Thorium-Lead Dating, 553.

BIBLIOGRAPHY

Brush, Stephen G. *Transmuted Past: The Age of the Earth and the Evolution of the Elements from Lyell to Patterson.* New York: Cambridge University Press, 1996. A look into modern planetary physics, this book traces the evolution of the elements and the solar system. Intended for the reader with some background knowledge in astronomy, this book is well illustrated and includes a bibliography and index.

Davidson, Jon P., Walter E. Reed, and Paul M. Davis. *Exploring Earth: An Introduction to Physical Geology.* Upper Saddle River, N.J.: Prentice Hall, 1997. An excellent introduction to physical geology, this book explains the composition of the Earth, its history, and its state of constant change. Intended for high-school-level readers, it is filled with colorful illustrations and maps.

Haber, Frances C. *The Age of the World: Moses to Darwin.* Baltimore: Johns Hopkins University Press, 1959. Reprint. Westport, Conn.: Greenwood Press, 1978. This interesting book does not focus on current estimates of the age of the Earth but rather on the historical controversy that emerged when nonbiblical values for the age of the Earth began to be accepted. Provides insight into the conflict between science and dogma.

Hurley, Patrick M. *How Old Is the Earth?* Garden City, N.Y.: Doubleday, 1959. One of the few full-length books on geochronology for the layperson. Even though published thirty years ago, it is still valid, as most of the material relevant to the age of the Earth has not changed appreciably since its publication.

Lutgens, Frederick K., and Edward J. Tarbuck. *Earth: An Introduction to Physical Geology.* 6th ed. Upper Saddle River, N.J.: Prentice Hall, 1999. This college text provides a clear picture of the Earth's systems and processes that is suitable for the high school or college reader. In addition to its illustrations and graphics, it has an accompanying computer disc that is compatible with either Macintosh or Windows. Bibliography and index.

Ozima, Minoru. *The Earth: Its Birth and Growth.* England: Cambridge University Press, 1981. A translation of a Japanese book that was written by a scientist whose specialty is geochronology. Written at an introductory level.

Stearn, Colin W., et al. *Geological Evolution of North America.* New York: John Wiley & Sons, 1979. Several excellent chapters discussing the age of the Earth. Contains an excellent chapter on geological time and the various ways it can be measured.

Stokes, William Lee. *Essentials of Earth History: An Introduction to Historical Geology.* 4th ed. Englewood Cliffs, N.J.: Prentice-Hall, 1982. A standard introductory text on historical geology. All the various methods for determining the age of the Earth are discussed in the first few chapters.

Stokes, William Lee, et al. *Introduction to Geology: Physical and Historical.* Englewood Cliffs, N.J.: Prentice-Hall, 1978. Textbook similar to Stokes's other book in terms of its discussion of geochronology.

Thackray, John. *The Age of the Earth.* New York: Cambridge University Press, 1989. A very short publication, about forty pages long, published by a British geological museum. Contains more pictures than text, but the pictures, most in color, are helpful and make this an interesting source.

EARTH'S OLDEST ROCKS

The oldest-known rocks on Earth have absolute (radiometric) ages approaching 3.8 billion years. Although Earth apparently has no rocks resulting from the first 7,600 million years of its history, rocks with ages ranging back to the earliest age for the terrestrial planets, about 4.56 billion years, occur for many meteorites and are closely approached in age by some rocks from the Moon.

PRINCIPAL TERMS

ABSOLUTE DATE/AGE: the numerical timing, in years or millions of years, of a geologic event, as contrasted with relative (stratigraphic) timing

GEOCHRONOLOGY: the study of the absolute ages of geologic samples and events

HALF-LIFE: the time required for a radioactive isotope to decay by one half of its original weight

ISOCHRON: the line connecting points representing samples of equal age on a radioactive isotope (parent) versus radiogenic isotope (daughter) diagram

ISOTOPES: species of an element that have the same numbers of protons but differing num-

bers of neutrons, and therefore different atomic weights

MASS SPECTROMETRY: the measurement of isotope abundances of elements, commonly separated by mass and charge in an evacuated electromagnetic field

NUCLIDE: any observable association of protons and neutrons

RADIOACTIVE DECAY: a natural process by which an unstable (radioactive) isotope transforms into a stable (radiogenic) isotope, yielding energy and subatomic particles

RADIOGENIC ISOTOPE: an isotope resulting from radioactive decay of a radioactive isotope

STRATIGRAPHIC TIME SCALE

Present knowledge of the oldest rocks on Earth developed slowly and descriptively until the 1950's, when it became possible to measure the absolute (quantitative) ages of minerals and rocks by radiometric means. These means involve the instrumental (commonly, mass spectrometric) measurement of unstable (radioactive) and stable (radiogenic) isotopes, or species of elements that differ only in their masses.

Prior to the ability of physicists, chemists, and geologists to make absolute age determinations, the oldest rocks on Earth were known only through field relations. The main field relation used is stratigraphic sequence, which involves an application of the principle of superposition: In a sequence of undisturbed layered rocks such as sedimentary layers and lava flows, the oldest rock unit—that is, the first to be deposited—is at the bottom of the sequence. Another important field principle is the manner in which one rock is cut or cuts another rock unit. The obvious chronolog-

ical conclusion is that the structure or rock that transects must be younger than the structure or rock that is transected. Through a combination of these stratigraphic and cross-cutting relationships, rock units studied and mapped can be assigned to a relative chronologic order and the geologic history of the mapped area worked out.

Accompanying the development of classical geologic principles was the understanding of the time dependence of biological evolutionary characteristics displayed by fossils found in the enclosing—primarily sedimentary—layers. Although it was early understood that fossil morphology changed through time from simpler to more complex forms, the time required for such evolutionary change could only be guessed. It is a tribute to early geologists and paleontologists that, before a quantitative measure of evolutionary scale was available, it was realized that enormous amounts of time probably were required between the deposition of rocks containing, for example, fossil collections of extinct marine animals such as trilobites and the deposition of those of containing fossils of horses.

The geologist's most important document, the stratigraphic column (the geologic time scale), was developed over the past several hundred years through the cumulative observations of field relations, paleontologic studies, and absolute dating methods. Its refinement will continue to be an important result of geologic endeavor. Correlation, the principal activity of the geologic study of stratigraphy, whereby rock units are related through their temporal and physical characteristics, enabled scientists to have some sense of the Earth's oldest rocks, long before numbers of years could be assigned to paleontologic and physical geologic phenomena. The Precambrian era, however—the vast period of geologic time that comprises more than 85 percent of the known age of the Earth—was not known to have harbored life or to have provided fossils until the past few decades; the discovery of widespread bacterial and stromatolitic fossils in Precambrian rocks has reversed this conclusion. Prior to these discoveries, the ages of the earliest rocks thus were surmised only through field relations and not through fossils.

RADIOMETRIC AGES

A misunderstanding of old rocks on Earth occurred because of the reasoning that the older the rock, the more opportunity for it to have altered, such as by weathering, tectonism (as in mountain building), or especially metamorphism. Thus, it was expected that the oldest rocks should be highly metamorphosed, as in much of the Precambrian terrain of Canada and other, commonly central continental areas of Precambrian rock (cratons or shields), and that essentially unaltered sediments and sedimentary rocks must be geologically young. Miscalculations in geologic age of billions of years occurred, owing to the incorrect correlation of rocks of similar petrology and metamorphic grade. Absolute age determinations, while not negating the essential premise of this theory, have nevertheless shown that some of the oldest rocks on Earth are not highly altered and that many young rocks may be highest-grade metamorphic and tectonized.

A major advance in geochronology has developed since radiometric ages were attached to points of the stratigraphic time scale and a quantitative framework for major fossil assemblages was

established. Once the ages of characteristic, representative fossils are quantified through absolute age determinations, the ages of sedimentary rocks containing chronologically diagnostic fossils can be assigned by comparison of these "guide" fossils with points on the stratigraphic time scale. Thus, the field geologist may establish the approximate age of sediments or sedimentary rocks in his or her area of interest (and, through field relations, the qualitative ages of associated igneous and metamorphic rocks and of geologic structures), simply through fossil identification. Although fossils, especially diagnostic fossils, are rare in many sedimentary rocks (especially those sedimentary rocks that formed prior to about 600 million years) and are absent in most igneous and metamorphic rocks, the use of paleontology as a chronologic tool is routine and in most cases quicker and less expensive than are geochemical (radiometric) age determinations.

RADIOACTIVE DECAY

Antoine-Henri Becquerel presented his discovery of the phenomenon of radioactivity to the scientific community in Paris about one hundred years ago, laying the cornerstone for scientists' present understanding of Earth's oldest rocks. The finding was followed rapidly by the seminal work of Marie Curie in radioactivity, a term she was the first to use. Her discovery of the intensely radioactive radium as well as plutonium led Ernest Rutherford to distinguish three kinds of radioactivity—alpha, beta, and gamma—and, in 1900, with Frederick Soddy, to develop a theory of radioactive decay. Soddy later proposed the probability of isotopes, the existence of which was demonstrated on early mass spectrographs and mass spectrometers.

Rutherford and Soddy's theory of the time dependence of radioactive decay, followed by breakthroughs in instrumentation for the measurement of these unstable species and their radiogenic daughter nuclides by Francis William Aston, Arthur Jeffrey Dempster, and Alfred Otto Carl Nier, among others, caught the rapt attention of early geochronologists and had a revolutionary effect on the study of geology. In 1904, Rutherford proposed that geologic time might be measured by the breakdown of uranium in U-bearing minerals and, a few years later, Bertram Boltwood published the "absolute" ages of three samples of uranium minerals. The ages, of about half a billion years, indicated the antiquity of some Earth materials, a finding enthusiastically developed by Arthur Holmes in his classic *The Age of the Earth*. Holmes's early time scale for Earth and his enthusiasm for the developing study of radioactive decay, although not met with instant acceptance by most contemporary geologists, helped to set the stage for the acceptance of absolute ages as the prime quantitative components in the study of geology and its many subdisciplines.

After the early study of the isotopes of uranium came the discovery of other unstable isotopes and the formulation of the radioactive decay schemes that have become the workhorses of geochronology, such as the rubidium-strontium, samarium-neodymium, potassium-argon, uranium-thorium-lead, and fission track methods. The formulation of the theory of radioactive decay of the parent, unstable nuclide (or the growth of the stable daughter nuclide), developed in the early 1900's, is the basis for the measurement of time, including geologic time, for any of these parent-daughter schemes used by geochronologists. Although each of the dating techniques is based on the formulation, differences occur in the kind of measurement and in the geochemical behavior of the several parent and daughter species. Thus, the geological interpretation of the data obtained is very different for the several chronometric schemes. These techniques for establishing absolute ages for minerals and rocks have been applied to the study of Earth's oldest rocks since the early 1900's and, with the development of modern mass spectrometry, more intensely since the 1950's.

DISTRIBUTION OF EARTH'S OLDEST ROCKS

Although not indigenous to Earth, the oldest rocks found on Earth (and also seen to fall to Earth) are the meteorites, many of which yield radiometric ages near 4.56 billion years, the accepted time of formation of many solar system materials. With respect to the oldest rocks indigenous to Earth, these rocks have the highest probability of being destroyed by ongoing geologic processes such as erosion, metamorphism, and subduction. It is no surprise that fewer and fewer outcrops are found as ages become older, deeper into Precambrian time. The oldest rocks are most commonly found in continental, cratonic regions, in many cases because of geologic preservative features such as their protective superjacent rocks, their location in tectonically stable continental interiors, and their low density and thus lower propensity for subduction than the more common basaltic rocks.

Histograms of rock ages thus show fewer and fewer data the further back in time. Such figures also show a feature whose significance has not been immediately apparent: the clustering of ages in rather discrete groupings. These groupings correlate with regionally defined rock/tectonic units such as those of the Grenville and Superior provinces of the Canadian Shield and indicate the intense geologic activity that resulted in these Precambrian rocks. Many scientists believe that the "magic numbers" that mark the groupings represent geologic periodicity, perhaps a result of major, discrete plate tectonic episodes. Others, however, point out that many radiometric dates fall outside these groupings and that the picture is incomplete and thus misleading. Although certainly incomplete, the available data indicate to many that there is some patterning in both the chemical and chronologic analyses of these rocks.

Early radiometric results showed some ages far back in time, near 3 billion years, and further analyses confirmed their antiquity. The oldest rock was thought to be the Morton gneiss in Minnesota, at about 3.2 billion years and questionably older, until several cratons yielded rocks with ages near 3.5 billion years. One such exposure, at North Pole, Australia, is of special significance, because of the concurrence on its age by several chronometric schemes (3.5 billion years) and especially because of its well-preserved bacterial and stromatolitic fossil assemblage, the earliest known. (Equivocal chemical evidence for organic life in even older rocks has been described; the existence of well-developed life at 3.5 billion years presupposes the existence of earlier life.) The oldest rocks, however, appear to be the well-studied Amîtsoq gneiss and contiguous, related rocks in the Godthaab area of western Greenland. Although there is incomplete agreement as to the exact range and significance of these earliest ages, several are close to or perhaps slightly greater than 3.8 billion years. Some of the disagreement with respect to these rocks, as well as for similar rocks around the world, results from incompletely known and undoubtedly variable diffusive and "freezing-in" behavior of the parent/daughter nuclides of the several chronometric systems. This varying behavior commonly results in different "ages" (dates) for the same analyzed rock specimen. A further uncertainty is whether the several isotopic systems can be completely reset, on the whole-rock scale, in metamorphic terrains that have been metamorphosed to lower physicochemical conditions.

Although not all scientists may agree, minerals of even older ages have been analyzed from Archean sandstones of Australia. Zircons (residual mineral phases from the final stages of crystallization of igneous rocks, especially granites) were separated from this stratigraphic unit and analyzed by uranium-thorium-lead dating using an innovative technique, the ion probe mass spectrometer. Although many of these zircons have been analyzed, only a few have exceptionally old ages, but these ages, ranging back to almost 4.3 billion

Three-billion-year-old gneiss in Nuuk, Greenland. (© William E. Ferguson)

years, are especially important. Because they are detrital (fragmental) in their host sandstone, they must have eroded from even more ancient rocks, perhaps granites or granitic gneisses, whose age, composition, and petrologic features are important to an understanding of the development of Earth's earliest crust. So far, their provenance (parental rocks) has not been found; apparently they have been completely eroded, altered, or buried by younger rock. If one accepts these earliest ages, crustal rocks existed on Earth less than 300 million years after Earth accreted from the solar nebula.

ANALOGY WITH EXTRATERRESTRIAL MATERIALS

It is useful to place Earth's oldest rocks within the framework of the ages of other available solar system materials, especially meteorites. Although the formation of the Earth—that is, Earth's time of accretion—is accepted by most scientists as having occurred about 4.56 billion years ago, it is obvious from the discussion above that no terrestrial rocks have ages this old. Earth's absolute age, therefore, as well as that of other solid materials of the solar system except for the Sun, is known only by analogy with meteorites. Many of the meteorites have been dated by the techniques discussed above and give formational ages near 4.56 billion years; some are thought to represent the oldest and most primitive material in the solar system with the possible exception of cometary material and cosmic dust. A few apparently unprocessed (primitive) meteorites yield radiometric age and initial isotopic composition data that suggest formational ages slightly older than 4.56 billion years. The terrestrial planets (Earth, Mercury, Venus, and Mars) are thought to have originated at the same time as did the meteorites.

A few meteorites have ages significantly younger than 4.56 billion years. These rocks are considered to have originated from parent bodies that were large enough to have maintained internal heat, and, therefore, igneous processes significantly after 4.56 billion years, as did the Earth, with its continuing volcanism and other geologic processes that have resulted in rocks of all ages from 4.56 billion years to the present. Several of these exotic rocks, collected from ice fields in Antarctica, were recognized almost immediately as pieces of the Moon, owing to scientists' familiarity with the Apollo missions' lunar rock collections. Even more spectacularly, a small collection of meteorites, long known to be different from the main collection of meteorites, was found to have crystallization ages of about 1.3 billion years, much younger than the accepted accretion age for solar system materials. These rocks must have originated from a body large enough to have maintained geologic processes between 4.56 and 1.3 billion years, unlike the Moon, which has so far yielded no rocks younger than about 3.0 billion years. This parent body is widely assumed by scientists to be Mars, a theory that is much strengthened by the compositional similarity of gases dissolved in glass from these meteorites and atmospheric compositions of present Mars, as measured from the Viking lander a decade ago.

Rocks returned from the Moon by U.S. and Soviet space programs yield ages from 0.8 to 4.54 billion years. Although the Moon is thought to have originated at the same time as the Earth, it is not massive enough to have provided a continuing internal heat source to drive volcanic or tectonic processes to the present time. Instead, significant igneous activity decreased substantially after 3.2 billion years ago until approximately 1 billion years ago. A current and popular theory is that the Moon originated by accretion in Earth orbit from material ejected from the Earth after impact with a Mars-size object. This hypothesis explains why the Moon has a composition similar to the Earth's mantle but is poor in volatile elements, and why the Moon's core is so small. Such an impact would be responsible for resetting the radiometric dates on the Moon and perhaps the Earth as well.

A widespread though not fully accepted theory for the early Moon is that it underwent a massive, perhaps global melting not long after formation (whether by nebular accretion or Earth impact). Upon cooling, plagioclase feldspar crystallized, floated, and formed the earliest lunar crust (anorthosite), which thus dates from some time after Moon accretion. If this theory is correct, no rocks older than the anorthosite will be found; this rock has yielded ages of 4.44 billion years and, arguably, somewhat older. If the Moon underwent significant or complete melting, it is possible or likely that the Earth experienced the same event, in which case there also will be no Earth rocks representing its earliest history. Finally, owing to Earth's continuing history of constructive and destructive geologic processes, it seems unlikely that significant amounts of rock will be found that date from Earth's first 200 million years.

GEOLOGIC APPLICATIONS

The absolute dating of geologic materials and events has had unprecedented influence on the evolution and understanding of geologic events on Earth, including Earth's origin and its oldest rocks, as well as other ancient minerals and rocks of the solar system. The ability of the scientist to establish events in terms of actual years, rather than in relative terms such as "older than" or "younger than," has led to a realistic knowledge of Earth's origin and its oldest rocks and has led to calibrated time scales for major geologic processes such as organic evolution. Owing to their usefulness in the precise determination of the ages of very old rocks, dating methods such as uranium-thorium-lead, rubidium-strontium, and samarium-neodymium will continue to be of major use in refining the sequence and meaning of Earth's oldest rocks and extraterrestrial materials.

E. Julius Dasch

CROSS-REFERENCES

BIBLIOGRAPHY

Ashwal, L. D., ed. *Workshop on the Growth of Continental Crust.* Technical Report 88-02. Houston, Tex.: Lunar and Planetary Institute, 1988. A technical but interesting series of articles that bear directly on Earth's oldest rocks and related material. Suitable for college-level readers.

Blatt, Harvey, and Robert J. Tracy. *Petrology: Igneous, Sedimentary, and Metamorphic.* New York: W. H. Freeman, 1996. Undergraduate text in elementary petrology for readers with some familiarity with minerals and chemistry. Thorough, readable discussion of most aspects of Earth's rocks. Abundant illustrations and diagrams, good bibliography, and thorough indices.

Faure, Gunter. *Principles of Isotope Geology.* 2d ed. New York: John Wiley & Sons, 1986. This textbook is an excellent though technical introduction to geochronology and the use of radioactive isotopes in geology and includes a thorough treatment of several dating techniques. Well illustrated and indexed. Suitable for college-level readers.

Hall, Anthony. *Igneous Petrology.* 2d ed. Harlow: Longman, 1996. This introductory book provides a good understanding of igneous rocks and their geophysical phases. There are sections devoted to the study of petrology and magmatic processes. Well illustrated with plenty of diagrams and charts to reinforce concepts. This is a good resource for the layperson.

Lutgens, Frederick K., and Edward J. Tarbuck. *Earth: An Introduction to Physical Geology.* 6th ed. Upper Saddle River, N.J.: Prentice Hall, 1999. This college text provides a clear picture of the Earth's systems and processes that is suitable for the high school or college reader. In addition to its illustrations and graphics, it has an accompanying computer disc that is compatible with either Macintosh or Windows. Bibliography and index.

Press, Frank, and Raymond Siever. *Understanding Earth.* 2d ed. New York: W. H. Freeman, 1998. An excellent general text on all aspects of geology, including the formation of igneous and metamorphic rocks. Contains some discussion of the structure and composition of the common rock-forming minerals. The relationship of igneous and metamorphic petrology to the general principles that form the basis of modern plate tectonic theory is discussed. Suitable for advanced high school and college students.

Taylor, S. R., and S. M. McLennan. *The Continental Crust: Its Composition and Evolution.* Oxford, England: Blackwell Scientific, 1985. An up-to-date though technical review of processes contributing to the formation of Earth's oldest rocks. Suitable for college-level readers.

York, Derek, and Ronald M. Farquhar. *The Earth's Age and Geochronology.* Reprint. Oxford, England: Pergamon Press, 1975. Contains good accounts of the chronologic techniques required to date rocks and Earth's age but does not include the more recent work on the oldest rocks. Technical but suitable for college-level readers.

FISSION TRACK DATING

When the isotope uranium 238 decays by fission in certain minerals, charged nuclei create a trail of damage, called a fission track. In transparent minerals, fission tracks can be enlarged by chemical etching until they are visible in an optical microscope. The age of the mineral can then be determined from the number of fission tracks and the uranium concentration in the mineral sample.

PRINCIPAL TERMS

CRYSTAL: a solid having a regular periodic arrangement of atoms

FISSION FRAGMENT: one of the lighter nuclei resulting from the fission of a heavier element

FISSION TRACK: the damage along the path of a fission fragment traveling through an insulating solid material

GLASS: a solid that has no regular periodic arrangement of atoms

ISOTOPES: two atoms of the same element having different numbers of neutrons and thus different atomic weights

SPONTANEOUS FISSION: the splitting of an unstable atomic nuclei into two smaller nuclei

NUCLEAR FISSION

The fission track dating technique is applicable to any type of rock containing minerals or glasses that record fission tracks and have sufficient uranium concentration to produce a detectable number of fission events within an appropriate-sized sample. Samples containing uranium at a concentration of one part per million can easily be dated by this technique if they are more than a hundred thousand years old. Correspondingly higher concentrations of uranium are required for younger samples, and lower concentrations are required for older ones. Because of its broad applicability, the fission track dating method has been applied to all three major terrestrial rock classes—sedimentary, metamorphic, and igneous.

Nuclear fission is a process by which an unstable atomic nucleus splits into two smaller nuclei, or fission fragments. The fission process releases a large amount of energy, causing the fission fragments to fly apart. In solid matter, each fission fragment can travel about 10-20 micrometers before coming to rest. Because of their high energy, these fission fragments do not carry along all the electrons from the original atom; thus, they have a positive electric charge during their passage through the surrounding matter. The passage of a charged particle through certain types of material gives rise to localized damage along the path of that particle. The damage caused by fission frag-

ments can be observed in some materials, making it possible to determine the number and location of fission decays that have taken place since that material was formed.

SPONTANEOUS FISSION

Spontaneous fission is a random radioactive decay process. For many elements with atomic numbers above that of lead, although alpha decay is usually the dominant mode of decay, spontaneous fission decay is also observed. The half-life (the time for half of any initial amount of an isotope to decay) for spontaneous fission has been measured for those elements having significant spontaneous fission decay.

In natural mineral samples, uranium is the only element currently present for which spontaneous fission is significant. Because uranium 235 has a much longer spontaneous fission half-life and a much lower abundance than uranium 238, the major contribution comes from the fission of the latter. In minerals that still survive from the very early era of solar system formation, plutonium 244 can have contributed many more fissions than did uranium because plutonium 244 has a very short spontaneous fission half-life. Minerals old enough to display plutonium 244 fission tracks occur in meteorites and in some ancient lunar samples, but no terrestrial rocks preserving a record of such fission have yet been found.

DETERMINING THE NUMBER OF FISSION DECAYS

The age of a natural mineral sample containing a significant amount of uranium can be calculated if the number of spontaneous fission decays since the formation of that mineral can be determined. The present abundance of uranium in the sample is first determined. The number of fission decays, along with the known spontaneous fission half-life of uranium 238 and the measured uranium abundance, is what is used to calculate the time that has elapsed since formation of the mineral.

The damage caused by the passage of each fission fragment is examined to determine the number of fission decays that have occurred in the mineral since its formation. In 1959, it was observed that the passage of fission fragments through natural silicate minerals gave rise to a population of short damage trails, which could be viewed in a transmission electron microscope. Because of the low uranium abundance in natural silicates and the high magnification required to observe these small damage trails, however, the use of the transmission electron microscope to count fission decays in such samples was not routinely practical. The major breakthrough permitting the routine fission track dating of natural minerals came in the 1960's, when a technique was developed to permit the damage trails to be observed at low magnification in an ordinary optical microscope. In these damage trails, chemical activity is greater than in the surrounding undamaged mineral. Certain chemical etches will attack the damage trail more rapidly than they will the surrounding crystal. Initially, the etch removes material along the damage trail. As the hole lengthens, however, the etch also attacks the walls, enlarging the diameter of the hole. These etched holes, or fission tracks, can then be easily counted using an optical microscope.

ION EXPLOSION SPIKE MODEL

The detailed mechanism by which a fission fragment interacts with the mineral structure to produce the damage trail has not been positively determined. The "ion explosion spike" model, however, is the generally accepted mechanism. In the ion explosion spike model, the positively charged fission fragment passes through a crystalline mineral, which consists of a periodic array of positively charged nuclei, each surrounded by or-

biting electrons. The charged fission fragment removes electrons from some of the atoms along its path, leaving a line of positively charged ions in the crystal. If the electrical conductivity of the crystal is low, a significant time elapses before the ejected electrons can migrate back to the ionized nuclei, restoring local electric neutrality. During this time, the positively charged nuclei along the path of the fission fragment repel one another electrostatically, causing displacements in the crystal structure. Once electrical neutrality is restored, the displaced atoms remain. This damage trail is visible in a transmission electron microscope as a disruption of the periodic array structure or can be enlarged by chemical etching.

The damage to the crystal structure can, however, be removed, or annealed, by heating the mineral. The temperature required to anneal fission tracks depends on both the type of mineral and the duration of exposure to heat. For time scales appropriate to most geological measurements (thousands of years or longer), the track annealing temperatures of common minerals range from less than 100 degrees Celsius to more than 600 degrees. Thus, the age actually measured by fission track dating is the time interval from the present back to the time when the mineral was last heated above its annealing temperature.

Chemical etches appropriate to reveal fission tracks have been found for more than one hundred minerals. Some etches are quite simple; for example, a boiling sodium hydroxide solution is appropriate for the common mineral feldspar. Some other minerals, such as olivine, however, require etches that are mixtures of several chemicals. Despite the fact that the ion explosion model seems applicable only to crystalline solids, fission tracks can also be revealed in most glasses when a hydrofluoric acid etch is used.

DATING OF VOLCANIC MATERIAL

The products of a volcanic eruption are frequently quite rich in uranium; thus, volcanic glasses can be dated by the fission track method. While the major terrestrial volcanoes themselves can be dated with greater precision using other techniques, volcanic debris dated by fission track techniques has proved useful in establishing the time scale for sedimentary accumulation on the ocean bottom. Wind-blown debris from major volcanic eruptions frequently accumulates as discrete layers of ash in the deposited sediments. Tiny volcanic glass fragments from the ash layers of ocean bottom cores collected by the Deep Sea Drilling Project have been dated using fission tracks, providing a chronological framework for the sediment deposition.

Fission track dating has proven to be especially valuable in the investigation of ocean-bottom, or seafloor, spreading. A model for the evolution of the floors of the ocean basins proposed that the mid-ocean ridges were sites where fresh, hot lava intrusions were deposited. After cooling, the lava would be displaced horizontally, and a new deposition would occur at the ridge. Thus, age determinations for ocean-bottom rocks at various distances from a mid-ocean ridge would constitute a direct test of the seafloor spreading hypothesis. Potassium-argon dating had been applied to such rocks, but the ages obtained were unreliable because of the way argon 40 reacts with lava. The fission track dating technique was applied to samples taken at the Mid-Atlantic Ridge and at various distances up to 140 kilometers from the ridge. The results showed material at the ridge to have an age of 10,000 years before the present and material from the most distant point to have an age of 16 million years. As the distance of the sample from the mid-ocean ridge increased, the fission track age also increased. Thus, the ocean-bottom spreading hypothesis was supported.

DATING OF METEORITES AND LUNAR MATERIAL

The fission track technique has also allowed scientists to date meteoritic impact events. Such events produce impact glass, formed of the local rock and soil that was melted by the impact event. The age of the impact glass thus dates the impact event.

Additionally, fission tracks have proven useful in determining the ages of meteoritic and lunar samples. Generally, minerals extracted from meteorites give ages of about 4.5 billion years, consistent with the age of the solar system inferred by other radioactive dating techniques. Mineral grains extracted from some meteorites, and in rarer cases grains from lunar samples, however, exhibit far more fission tracks than would be produced in 4.5 billion years by the uranium in the samples. Once all other sources of tracks were ex-

cluded, the investigators attributed these tracks to the fission of now-extinct plutonium 244, which was present in the very early solar system. Since the half-life of plutonium 244 is only 80 million years, minerals containing a substantial number of fission tracks from plutonium must have formed and cooled to below the annealing temperature within a few hundred million years of the last addition of fresh radioactive material to the solar system. Thus, grains exhibiting plutonium 244 fission tracks formed very early in the evolution of the solar system.

Plutonium 244 fission tracks have been used in the development of a technique to determine the rate at which the parent bodies of the meteorites cooled. These cooling rates then permit the sizes of the parent bodies to be inferred. Different minerals have different track annealing temperatures. The plutonium in meteorites is generally concentrated in phosphate minerals such as merrillite, which have very low track annealing temperatures (about 100 degrees Celsius for merrillite). These plutonium-rich minerals, however, occasionally occur adjacent to plutonium-poor silicate grains. Since the fission fragments have ranges of 10-20 micrometers, some fission decays near the merrillite-silicate contact surface produce fission tracks in the silicate minerals. Typical silicates have track annealing temperatures of about 300 degrees Celsius. Thus, the plutonium 244 fission fragments from the merrillite begin to produce tracks in the adjacent silicate before the tracks were recorded in the merrillite itself.

Because the plutonium 244 decay rate is known, a comparison of the number of fission tracks in the silicates with the number in the merrillite gives the time it took for the meteorite to cool from the track annealing temperature of the silicate to that of the merrillite. When the ordinary chondrite meteorite St. Severin was examined by this technique, a cooling rate of about 1 degree per million years was found. Cooling-rate data obtained on a number of ordinary chondrites suggest that these meteorites come from below the surfaces of asteroidal-sized parent bodies, no more than about 300 kilometers in diameter.

USE IN ARCHAEOLOGY

Because fission track dating measures the time interval since the last heating of the mineral above the track annealing temperature, it has proven to be particularly valuable in archaeology. Certain archaeological objects, such as pieces of pottery, are heated when they are manufactured. The age of manufacture of such an object can be determined if it or mineral grains within it record fission tracks.

The earliest application of fission track dating to an archaeological sample was to an obsidian knife blade found by I. S. B. Leakey in Kenya. The texture of the blade indicated that it had been heated after its manufacture. A small fragment of the knife blade, about one-tenth of a gram, was found to be about 3,700 years old. Fission track dating was subsequently applied to samples from the Olduvai Gorge beds, from which Leakey's team recovered the specimen of *Zinjanthropus*, a very early humanoid. Potassium-argon dating of volcanic material from this bed suggested that the actual age of *Zinjanthropus* was almost twice as great as had been inferred from the fossils associated with the bed. Fission track dating of volcanic pumice from the bed gave an age of 2 million years, consistent with the potassium-argon age. This confirmation of the age of *Zinjanthropus* resolved the controversy.

Many of the clays used in the manufacture of pottery contain crystals of zircon, a mineral rich in uranium. The high temperature the pottery reaches in the kiln erases all the tracks previously recorded in the zircons, and their high uranium concentration permits even short intervals since the heating to be established with reasonable precision. In one such study done in Japan, nine zircon-containing samples of pottery were dated, giving ages ranging from 700 to 2,300 years. This fission track technique has also been applied to many human-made glass samples, doped at high uranium concentrations (sometimes up to 1 percent) to color them. These uranium-rich glass samples can be dated by the fission track technique after only a few years of track accumulation.

COMPARISON WITH OTHER TECHNIQUES

The fission track dating technique has been applied to a wide variety of terrestrial and extraterrestrial materials. The main advantage of this technique is the large span of ages over which it can be employed, permitting the ages of objects from only tens of years old to more than 4.5 billion years old to be established. Although the ages obtained by this technique are generally not as precise as those available through radiocarbon and potassium-argon dating, fission track ages are frequently useful to confirm ages obtained by these other techniques when their applicability to the particular sample is questionable. Where such comparisons can be made, fission track dating has been shown to give correct ages ranging from less than a year to more than a billion years.

Because a large number of individual fission tracks must be counted if age is to be determined with a high degree of precision, fission track dating results are generally considered less reliable than those of techniques that use mass spectrometers to determine isotopic ratios. Fission track dating is particularly valuable for the range of ages from 40,000 years before the present, where radiocarbon dating ceases to be accurate, to about a billion years, where potassium-argon dating is relatively easy. Fission track dating is also applicable to very small samples. Individual mineral grains as small as a milligram in mass have been dated by this technique.

The fission track dating method has been adopted by many laboratories throughout the world because of its simplicity, broad applicability, and low cost. Analyses can be performed in laboratories equipped with only simple chemical etching facilities and optical microscopes. Fission tracks are recorded in a wide variety of crystals and glasses, and the technique is applicable over a wide range of sample ages.

George J. Flynn

CROSS-REFERENCES

Carbonaceous Chondrites, 2654; Chondrites, 2660; Earth's Age, 511; Earth's Oldest Rocks, 516; Geologic Time Scale, 1105; Geothermometry and Geobarometry, 419; Lunar Rocks, 2561; Ocean Ridge System, 670; Plate Tectonics, 86; Potassium-Argon Dating, 527; Radioactive Decay, 532; Radioactive Minerals, 1255; Radiocarbon Dating, 537; Rubidium-Strontium Dating, 543; Samarium-Neodymium Dating, 548; Uranium-Thorium-Lead Dating, 553.

BIBLIOGRAPHY

Fleischer, R. L. *Nuclear Tracks in Solids: Principles Applications.* Berkeley: University of California Press, 1975. This 605-page book describes all aspects of nuclear track formation and applications. Fission track dating is described in chapter 4, along with the experimental procedures to apply this technique. The authors explain the limitations imposed by track fading caused by heat and explain the methods of correcting for such track loss. An extensive bibliography is included. This well-illustrated book is suitable for college-level readers.

_____. *Tracks to Innovation: Nuclear Tracks in Science and Technology.* New York: Springer, 1998. The author explains the method of fission track dating and describes experiments that have been conducted to compare it with other dating techniques. The book also emphasizes the mechanism of track formation and the use of solid state track detectors to determine the charge and energy of each particle. Designed to acquaint geologists with the technique of fission track dating, this book is a suitable introduction for general readers.

Macdougall, J. D. "Fission-Track Dating." *Scientific American* 235 (December, 1976): 114-122. This well-illustrated account of the fission track dating technique describes applications to terrestrial and meteorite samples. A series of diagrams illustrates the mechanism by which fission fragments are believed to produce damage trails in crystal structures. Accessible to general readers.

Wagemans, Cyriel. *The Nuclear Fission Process.* Boca Raton, Fla.: CRC Press, 1991. A clear description of the process and techniques involved with nuclear fission. Wagemans discusses the applications of such procedures in relation to fission track dating. Although the subject is complicated, this book presents information in a way that a college reader without much science background can follow. Bibliography and index.

Wagner, Geunther A., and Peter Van de Haute. *Fission Track Dating.* Boston: Kluwer, 1992. This is an excellent and thorough introduction to fission track dating. The authors discuss the history, protocols, techniques, and applications of fission track dating in a manner that is understandable by the person without a strong scientific background. Illustrations, diagrams, bibliography, and index.

POTASSIUM-ARGON DATING

Radioactive decay of potassium 40 into argon 40 can be used as a natural clock to determine the age of rocks. A wide range of ages can be measured using this technique. Moon rocks brought back by the Apollo astronauts were shown to be more than 2 billion years old, and the volcanic lava that formed the island of Hawaii has been dated at less than 1 million years old.

PRINCIPAL TERMS

ATOMIC SPECTROSCOPY: a method to identify various elements by the unique spectrum of light waves that each one emits

HALF-LIFE: the time required for half of the atoms in a radioactive sample to decay, having a constant value for each radioactive material

IGNEOUS ROCKS: rocks formed by solidification of molten magma from within the Earth

ISOTOPES: atoms of the same element but having different masses as a result of extra neutrons in the nucleus

MASS SPECTROMETER: an apparatus that is used to separate the isotopes of an element and to measure their relative abundance

PHOTOMETER: a device to measure light intensity, using a light meter with a numerical output reading

RADIOGENIC: an isotope formed by radioactive decay

DISCOVERY OF ISOTOPES

The idea of using radioactivity as a clock for geology was first suggested by the physicist Ernest Rutherford in 1905. Rutherford and coworkers had shown that uranium decays into lead in a radioactive series, with helium gas formed as a byproduct. If the lead and helium are retained in the pores of a rock that contains uranium, the ratio of helium to uranium or lead to uranium can be used to calculate an age. The older the rock, the more lead and helium it should contain. There were some problems, however, that could produce incorrect results. An unknown fraction of helium may have escaped over the centuries, or extra lead may be present in a rock as the result of natural lead deposits not coming from uranium decay. Experimental uncertainties were undeniable.

An important development in 1914 was the invention of the mass spectrometer by J. J. Thomson. When neon gas was analyzed with this apparatus, Thomson was able to show that it actually consisted of a mixture of two different kinds of atoms with masses of 20 and 22 units (relative to hydrogen). His experiment constituted the discovery of isotopes. Over the next several decades, all the elements in the periodic table were analyzed. The sensitivity of the apparatus was greatly improved

so that even isotopes whose abundance is less than 1 percent could be measured with precision. In particular, Alfred Otto Carl Nier showed in 1935 that the element potassium (K) has three isotopes, of which potassium 40 is radioactive and has an abundance of less than 0.01 percent.

POTASSIUM DECAY

The atmosphere of the Earth is known to consist of about 99 percent nitrogen and oxygen, nearly 1 percent argon, and very small amounts of other gases. It was a mystery why so much argon is present in the air. Mass spectrometer data for argon (Ar) showed that argon 40 was by far the most abundant isotope. The German physicist C. F. von Weizsäcker made the suggestion in 1937 that the unexpectedly large amount of argon in the air could have come from radioactive decay of potassium in rocks. A test of this idea would be to analyze old rocks that contain potassium to see if they also contain a higher percentage of argon 40 than that found in the air. In 1948, Nier conducted this experiment on several geologically old minerals. He showed that the argon 40 isotope was indeed greatly enriched in the old rocks. The mystery of excess argon in the air had been solved: It came from potassium decay.

Many rocks contain the element potassium. All these rocks show a slight radioactivity as the potassium slowly decays into argon. The key idea of K-Ar dating is to measure accurately the relative amounts of potassium and argon. Very old rocks contain a larger amount of argon because more time has elapsed for the argon atoms to accumulate. The point at which the rock cooled to its solid form sets the starting time for the K-Ar clock. Before the rock crystallized, the argon gas could escape, but after the rock became solid, the argon gas would be retained. In some cases, corrections have to be made in the measurements because argon may have been gained or lost over long periods of time. Some minerals retain argon gas much better than others. With experience, geologists have learned to be selective in finding those applications where the K-Ar technique has its greatest validity.

Radioactivity measurements have established that potassium 40 decays into two possible end products. About 11.2 percent of the decays become argon 40, and the other 88.8 percent of the decays become an isotope of calcium, calcium 40. The calcium 40 cannot be used in dating a rock because there is too much calcium in the rock already. Only ratios are used in actual calculations. For example, after 200 million years, the ratio of potassium 40 to argon 40 is about 80 to 1. Ratios can be calculated for any age. For meteorites with an age of 4 billion years, nearly 90 percent of the original potassium 40 has decayed. For time intervals shorter than 3 million years, the amount of argon 40 becomes very small and hard to measure, even with a mass spectrometer. As the sensitivity of the apparatus has been improved, however, it has become possible to date rocks with an age as low as 100,000 years or even less.

Determining Potassium and Argon Content

To understand how the potassium-argon clock is used in geology in a quantitative way, it is necessary to look at the methods by which K and Ar are determined. For potassium, the general techniques of chemical quantitative analysis can be used. A rock sample may contain as much as 5 percent potassium or more. The sample is crushed and dissolved in acid; after unwanted elements are removed by heating and precipitation, the potassium is converted to an insoluble salt and the pre-

cipitate is collected by centrifuging. The amount of potassium can then be determined by weighing. A second method, which gives better precision (typically ± 1 percent), makes use of atomic spectroscopy. Potassium atoms emit a characteristic purple light with a particular wavelength. A standard solution with a known amount of potassium is prepared and vaporized in a burner flame. A photometer set for the proper wavelength measures the light intensity from the standard. The same process is repeated for the rock sample solution. The ratio of light intensities from sample and standard is used to calculate the potassium content of the rock. After the total amount of potassium in a rock sample has been determined, it is an easy step to calculate how much radioactive potassium 40 it contains because the relative isotope abundance is known.

For argon analysis, the mass spectrometer has to be used because the quantity is so small. A 5-gram rock sample with 5 percent potassium would contain one-fourth of a gram of potassium, but perhaps only a few billionths of a gram of argon 40. The older the rock, the more time has elapsed to allow a larger amount of argon 40 to accumulate. The rock is crushed and heated in a vacuum to collect the gases. The small amount of argon is separated; when it is run through a mass spectrometer, an electrical current is observed at mass number 40. How can one know the amount of argon that caused this current? A calibration standard using a known amount of another argon isotope, argon 38, is added. (The technique of adding another isotope is called isotope dilution.) A known mass of argon 38 is mixed with the argon sample from the rock, the gas mixture is run through the mass spectrometer, and the ratio of the electrical currents at mass numbers 38 and 40 is measured. Since the amount of argon 38 is known, the amount of argon 40 can then be calculated by simple proportion.

In the argon 40 determination, one possible source of error needs to be considered. Suppose the mass spectrometer shows the presence of another argon isotope at mass number 36, which cannot be caused by radioactive decay of potassium. Where did it come from? The most likely source of this contamination is the argon contained normally in the atmosphere. It could have gotten into the rock sample or it might be the result of residual air in the vacuum system used for analysis. In either case, the total argon 40 measured by the mass spectrometer would be the sum of radiogenic argon 40 (which comes from radioactive decay of potassium 40) plus atmospheric argon 40. Fortunately, the ratio of argon 36 and argon 40 in the atmosphere is well known, so the measured argon 36 can be used to subtract the atmospheric argon 40 from the total. The radiogenic argon 40 alone should be used for calculating the age of the rock.

AR-AR METHOD AND APPLICATION

Once the potassium 40 and the argon 40 content of a rock have been determined, all the essential data for an age calculation are available. A relatively new development in K-Ar dating is the so-called Ar-Ar method. If a rock sample is irradiated with neutrons, some of the stable potassium atoms will be converted to a new argon isotope, argon 39. With a mass spectrometer, the ratio of argon 40 to argon 39 is then determined. The argon 39 is an indirect measure of the potassium content. It is no longer necessary to do a separate potassium analysis of the rock sample. This procedure has been used to investigate possible loss of argon from the outer layers of a rock fragment during metamorphosis. The inner part of a rock would show a larger argon 40/argon 39 ratio than the outer part because the argon 40 will be retained most effectively in the interior. In favorable cases, it is possible to reconstruct a history of the rock fragment since its time of crystallization.

An application of the Ar-Ar method of dating will be described, to show its specialized applications. The astronauts brought back a remarkable orange, glass-like rock from the Moon. At first it was thought possible that it might be the product of recent volcanic activity. A sample of less than one-tenth of a gram of this glass was selected for analysis. The sample was irradiated for several days in a nuclear reactor to convert some of the stable potassium 39 into argon 39. The sample was then heated to a moderate temperature of about 650 degrees Celsius, releasing argon gas only from the outermost part of the rock. This gas was collected and analyzed in a mass spectrometer. The rock sample was heated in successive steps of one hundred degrees to about 1,350 degrees Celsius, each time collecting the argon gas that was released

from the more interior parts of the rock. The ratio of argon 40/argon 39 was plotted against the temperature of release. The resulting graph showed that the ratio varied but reached a constant plateau at higher temperatures. Using the argon ratio where it became constant, an age of 3.7 billion years was calculated. Variation in the argon ratio at lower release temperature was attributed to gain or loss of argon in the outer layers of the sample. The Ar-Ar method made it possible to discard erroneous data from the outer part of the rock when determining the age.

SUCCESSFUL K-AR APPLICATIONS

One successful application of K-Ar dating was to determine the ages of the individual Hawaiian Islands. The islands were formed by volcanic activity, and small rocks in the lava would have cooled rather rapidly after an eruption. With rapid cooling, it is possible that some atmospheric argon was trapped in the rocks, so adjustments had to be made in the calculations. The K-Ar measurements gave the following results: The island of Kauai is the oldest, at about 5 million years; Hawaii is the youngest, at less than 1 million years; and the other islands show a regular sequence of ages in between. Scientists concluded that the volcanic activity started in Kauai and gradually migrated to Hawaii, about 300 miles to the southeast, forming a chain of islands at regular time intervals over the 5 million years.

As another example, several hundred ages have been measured for granitic rock samples from the Sierra Nevada region (near Yosemite National Park) using K-Ar dating. The time when these igneous rocks formed, before they rose to the surface, is of significance for a geological understanding of the region. Samples from Half Dome, Cathedral Peak, and other sites gave initial results which were in the range of 80 million years. The data, however, had to be corrected for loss of argon as a result of reheating from later molten rock intrusions that moved upward toward the surface. With corrections, the age of the Sierra Nevada is estimated to be in the range of 140-210 million years.

Age measurements of a much longer time span have been taken for some rock samples brought back from the surface of the Moon by the Apollo astronauts in the 1970's. Radioactive dating by K-Ar, as well as uranium and rubidium decay, gave ages of about 4 billion years for some of the rocks from the lunar exploration. Four billion years is equal to about three half-lives for potassium, so almost 90 percent of the original potassium 40 would have decayed. It has been suggested that some Moon rocks may be even older than the measured 4 billion years, because a portion of the argon may have been lost as a result of a high-temperature episode in the early history of the solar system.

As a final example of the K-Ar dating method, consider an application to archaeology. In 1959, L. S. B. Leakey and his wife, Mary, discovered the fossil of a humanoid skull in Tanzania, Africa. The remains were found in an area of volcanic deposit suitable for dating by radioactivity, and rock samples lying in strata near the fossil remains were dated. At first, there was controversy about the results because some samples selected for analysis came from rock strata that were not accurately correlated with the fossils. In addition, some data had to be discarded because weathering and possible contamination by water were a problem. General agreement has now been reached that the fossil remains, dated by K-Ar of properly chosen nearby rock samples, are about 1.75 million years old. This age has been confirmed by other methods of radioactive dating. When close correlation is obtained by independent methods, one can conclude with confidence that the age determination is valid.

SCIENTIFIC AND PRACTICAL VALUE

Dating with potassium-argon has been applied to many geologically interesting questions. Among them are the early history of the solar system, the age of meteorites, the exploration of the surface of the Moon, the dates when the Earth's magnetic field reversed, the general geological history of various mountain ranges, the eruption of volca-noes, and the dates of fossil remains for archaeology. In some cases, the Ar-Ar technique using two different isotopes can be used to analyze the history of a rock sample through periods of cooling and reheating.

The problem of storing high-level radioactive wastes began to receive increased attention as more electricity was generated by nuclear power plants worldwide. Nuclear power has an advantage over coal because it does not generate acid rain or carbon dioxide. Some environmental groups now find nuclear power plants less damaging to the environment than coal or oil plants. The disadvantage of nuclear power is the radioactive waste that is produced. The information obtained by geologists from K-Ar dating has made a contribution to the development of waste storage technology because it has been shown that certain rocks can retain radioactivity for very long periods of time. Research is being done on embedding radioactivity in synthetic rocks (the SYNROC process): The radioactive waste is to be incorporated in the rock material itself, where it would be relatively impervious to water and weathering. Geological study of K-Ar dating on very old rocks can thus help scientists to find an acceptable solution to the very practical problem of nuclear waste storage.

Hans G. Graetzer

CROSS-REFERENCES

Earth's Age, 511; Earth's Oldest Rocks, 516; Elemental Distribution, 379; Fission Track Dating, 522; Geologic Time Scale, 1105; Lunar Rocks, 2561; Mass Spectrometry, 483; Nucleosynthesis, 425; Radioactive Decay, 532; Radioactive Minerals, 1255; Radiocarbon Dating, 537; Rubidium-Strontium Dating, 543; Samarium-Neodymium Dating, 548; Uranium-Thorium-Lead Dating, 553.

BIBLIOGRAPHY

Burchfield, Joe D. *Lord Kelvin and the Age of the Earth.* New York: Science History Publications, 1975. Lord Kelvin (1824-1907) was widely regarded as the greatest physicist of his era. He published some articles in the 1860's that suggested an age of 100 million years or less for the Earth. The scientific basis for and the general acceptance of this estimate by most geologists before 1900 are documented. Finally, Kelvin's conclusion was overthrown by the new technique of radioactive dating. Many original documents are cited after each chapter.

Criss, Robert E. *Principles of Stable Isotope Distribu-*

tion. New York: Oxford University Press, 1999. Criss describes isotopes and their properties with clarity. In addition to well-written text, the book also features diagrams and illustrations that present a clear picture of the different phases of isotopes and isotope distribution. Bibliography and index.

Dalrymple, G. Brent, and Marvin A. Lanphere. *Potassium-Argon Dating: Principles, Techniques, and Applications to Geochronology.* San Francisco: W. H. Freeman, 1969. The two authors are scientists with the U.S. Geological Survey, writing from their extensive personal experience with K-Ar measurements. Sample preparation, instrumentation, sources of error, and useful applications are discussed with careful attention to detail. The best technical overview of K-Ar dating, compiled into a compact volume of about 250 pages.

Durrance, E. M. *Radioactivity in Geology.* New York: Halsted Press, 1986. The author shows the wide scope of radioactivity measurements in geological investigations. Up-to-date information is presented on environmental radioactivity (including the radon hazard), heat generation, and various isotope-dating procedures. A bibliography of articles published in professional as well as popular journals follows each chapter.

Emsley, John. *The Elements.* 3d ed. Oxford: Oxford University Press, 1998. Emsley discusses the properties of elements and minerals, as well as their distribution in the Earth. Although some background in chemistry would be helpful, the book is easily understood by the high school student.

Faure, Gunter. *Principles of Isotope Geology.* 2d ed. New York: John Wiley & Sons, 1986. An excellent though technical introduction to the use of radioactive and stable isotopes in geology, including a thorough treatment of the Rb-Sr technique. The work is well illustrated and well indexed. Suitable for college-level readers.

Fleischer, Robert Louis. *Tracks to Innovation: Nuclear Tracks in Science and Technology.* New York: Springer, 1998. The author explains the method of fission track dating and describe experiments done to compare it with other dating techniques. The book also emphasizes the mechanism of track formation and the use of solid state track detectors to determine the charge and energy of each particle. Designed to acquaint geologists with the technique of fission track dating, it is a suitable introduction for general readers.

Levin, Harold L. *The Earth Through Time.* 3d ed. Philadelphia: Saunders College Publishing, 1988. This college-level text contains a brief, clear description of the Rb-Sr method. A diagram of a whole-rock isochron is included. Five other radiometric dating techniques are discussed, and background information on absolute age and radioactivity is provided. Includes review questions, a list of key terms, and references.

Parker, Sybil P., ed. *McGraw-Hill Encyclopedia of the Geological Sciences.* 2d ed. New York: McGraw-Hill, 1988. This source contains entries on radioactivity and radioactive minerals. The entry on dating methods includes a brief section on the Rb-Sr method. Includes the formula for radioactive decay and a table of principal parent and daughter isotopes used in radiometric dating. The entry on rock age determination has a longer discussion of the Rb-Sr dating and includes an isochron diagram. For college-level audiences.

Skinner, Brian J., and Stephen C. Porter. *Physical Geology.* New York: John Wiley & Sons, 1987. A widely used college-level textbook for an introductory course in geology. One chapter deals with geological time and its determination, using radioactivity and other methods. Both the K-Ar method and the more recent argon 40/argon 39 ratio are described in a readable way.

Tuniz, Claudio, et al., eds. *Accelerator Mass Spectrometry: Ultrasensitive Analysis for Global Science.* Boca Raton, Fla.: CRC Press, 1998. This book looks at the processes involved with accelerator mass spectrometry and the instruments required. There is also a substantial amount of care given to radioactive dating and its protocols, principles, and usefulness. Bibliographic references and index.

RADIOACTIVE DECAY

Radioactive decay is the release of energy by nuclei through the emission of electromagnetic energy or several types of charged particles. In geology, radioactive decay is important not only as the basis for most of the standard dating techniques and as a tracer for fluid flows and chemical reactions but also for its role in heating the interior of the Earth and changing the character of minerals.

PRINCIPAL TERMS

ALPHA PARTICLE: the nucleus of a helium atom, which consists of a tightly bound group of two protons and two neutrons

ATOM: the smallest piece of an element that has all the properties of the element

ELECTRON: a negatively charged particle that forms the outer portion of the atom and whose negative charge is equal in magnitude to the positive charge of the proton

GAMMA RADIATION: high-energy electromagnetic radiation emitted when a nucleus emits excess energy

HALF-LIFE: the time during which half the at-

oms in a sample of radioactive material undergo decay

NEUTRON: the uncharged particle that is one of the two particles of nearly equal mass forming the nucleus

NUCLEUS: the central portion of the atom, which contains all the positive charge and most of the mass of the atom

POSITRON: a positively charged electron, a form of antimatter

PROTON: the positively charged particle that is one of the two particles of nearly equal mass forming the nucleus

BEHAVIOR OF ATOMIC NUCLEI

Radioactive decay is the release of energy by the nucleus of an atom. Nuclei discharge energy either through the emission of electromagnetic radiation, a form of pure energy that does not alter the chemical nature of the atom, or through the emission of a particle that changes the atom into an atom of a different chemical element. In order to understand radioactive decay, it is necessary to understand the behavior of atomic nuclei.

Atomic nuclei occupy a very tiny central portion of the atom; if the atom were the size of a two-story house, the nucleus would be the size of the head of a pin. In spite of its small size, the atomic nucleus contains nearly all the mass of the atom. Nuclei are composed of two particles with nearly identical masses: the proton, which carries one unit of positive electric charge; and the neutron, which is uncharged. The nucleus is orbited by the electrons. Each electron carries one unit of negative electric charge equal in size to the positive charge of the proton—even though the mass of the electron is only five-hundredths of a percent of that of a proton or neutron. The atom is electri-

cally neutral; thus, the number of electrons orbiting the nucleus under normal conditions equals the number of protons contained in the nucleus. The number of protons or electrons determines the chemical element to which the atom belongs.

The particles in the nucleus are held together by the nuclear force, which is strong enough to overpower the electrical repulsion of the protons at very short distances. Just as the atomic electrons orbit the nucleus in patterns with definite energies, the nuclear particles fill states of definite energy within the nucleus. The energies of the neutron states are a little lower than those of the protons, because the neutrons are not forced apart by electrical repulsion. For elements with few protons in the nucleus, the numbers of protons and neutrons in the nucleus are nearly equal. For elements with larger numbers of protons, the electrical repulsion becomes strong enough so that neutrons in heavier atoms considerably outnumber protons. For most elements, there are several types of nuclei that contain different numbers of neutrons but have the same number of protons. Such atoms with equal numbers of protons but different numbers of neutrons are called isotopes of an element.

HALF-LIFE OF THE DECAY

If a nucleus has extra energy, it will seek to rid itself of that extra energy by emitting either electromagnetic radiation or a particle. This emission is called radioactive decay. The time at which an excited nucleus (one with extra energy) will decay is not predictable except as a probability, which depends on the time elapsed since the nucleus was formed. This probability is described in terms of the half-life of the decay, which is the time it takes for half the excited nuclei in a sample to decay. For example, if there are originally four hundred excited nuclei in a sample, two hundred of them will undergo radioactive decay during the first half-life, one hundred of them in the second, fifty in the third, and twenty-five in the fourth. After the first half-life, there will be two hundred excited nuclei left in the sample; one hundred will be left after the second, fifty left after the third, and twenty-five excited nuclei left after the fourth.

Nuclei with short half-lives decay rapidly and disappear quickly, but the large number of particles they emit may do much damage to their surroundings. Nuclei with longer half-lives exist much longer. Their radioactive decay will do less immediate damage to their surroundings but will continue to do damage over a considerable period of time.

TYPES OF DECAY

Three major types of radioactive decays are found in nature. They are called alpha, beta, and gamma decay (named for the first three letters of the Greek alphabet); the particles emitted in the decay are called alpha, beta, and gamma particles or rays. The mechanisms for the decays, their half-lives, and their effect on their surroundings differ widely.

Emission of electromagnetic radiation, called gamma decay, allows the protons and neutrons to settle into lower energy states without changing the number of protons in the nucleus that decayed. Gamma decay particles carry no charge and have no mass. Because they are electromagnetic radiation, they travel long distances through matter and do little damage to atoms through which they pass compared to the damage done by

the passage of charged particles. The half-lives of gamma decays are usually very short. Common half-lives are about a billionth of a second; it is rare to find a gamma half-life as long as a second. The energy of the gamma is characteristic of the energy levels of the protons and neutrons in the nucleus that emitted it. The pattern of emitted gammas can be used to identify a particular nuclear species.

Alpha decay occurs mostly in heavy nuclei that emit an alpha particle—the nucleus of a helium atom, consisting of two protons and two neutrons. This massive particle is believed to form a tightly bound unit inside these heavy nuclei. It takes advantage of a unique phenomenon of quantum mechanics to escape far enough from the vicinity of the nucleus that the positive electrical repulsion of the nucleus acts on the alpha's positive charge to drive it out of the atom. Because they are very massive and carry two units of positive charge, alphas heavily damage their surroundings even though they travel very short distances in matter and are easily stopped by a thin sheet of paper. Like gamma decay, each alpha decay is characterized by a unique energy determined by the energy structure of the nucleus from which is has escaped. The remaining nucleus now forms an atom of a different element, with two fewer protons and therefore two fewer electrons in the neutral atom. The half-lives of alpha decays are usually very long. Uranium 238, for example, has a half-life of 4.51 billion years, which is believed to be approximately the age of the Earth.

BETA DECAY

There are two types of beta decay: negative (the emission of an electron and changing of a neutron to a proton in the nucleus) and positive (the emission of a positron—a positively charged electron—and changing of a proton to a neutron in the nucleus). Both types of beta particles lie between alphas and gammas in the damage they do to their environment. They are typically stopped by thin blocks of aluminum and do less damage than alphas. Positrons, a form of antimatter, destroy themselves by uniting with an electron and annihilating themselves with the emission of two gamma rays that damage their surroundings. The half-lives of beta decays range from a few seconds to thousands of years.

One of the more important beta decays is the decay of the isotope of carbon with six protons and eight neutrons to the isotope of nitrogen with seven protons and seven neutrons by the emission of an electron. This decay has a half-life of 5,730 years. If a positron is emitted, the remaining nucleus has one proton fewer than it did before. Several varieties of beta decay involve phenomena such as the capture of one of the atomic electrons by a proton to turn itself into a neutron. In this case, there is no emitted positron, but an observer sees an X ray as other electrons fall close to the nucleus to replace the electron that was captured.

When they were first discovered, beta decays puzzled researchers because they did not exhibit the definite energies that characterized alpha and gamma decays. It was finally realized that nuclei undergoing beta decay emit not only an electron or a positron but also a tiny uncharged particle called a neutrino. Neutrinos carry off part of the definite energy of the nuclear charge so that the beta particles from a particular nuclear transition exhibit a statistical energy distribution. Because they are uncharged and interact very little with other atomic particles, neutrinos can pass through the mass of several earths with less than a 50 percent chance of interacting. Consequently, they are very difficult to detect. (An example of a neutrino detector is a hole in a mine the size of a ten-story building, filled with water and surrounded by detectors.) Despite their small chance of interacting, neutrinos are important to scientists' understanding of the structure of the universe, as they are believed to be produced in the nuclear reactions at the core of the Sun and other stars. Thus, they may constitute a large portion of the mass of the universe. A debate rages over whether the neutrino is massless or has a very tiny mass that has not yet been measured. The answer to this question may determine whether the universe will expand forever or will eventually stop expanding and collapse back on itself.

RADIATION DETECTORS

Radioactive decay was discovered by accident when Antoine-Henri Becquerel (1852-1908) accidentally left a piece of uranium-bearing rock on top of a photographic plate in a darkened drawer. The rock left its image on the film. The first studies of radioactive decay used film as a detector for radiation. The next generation of radiation detectors used fluorescent screens which would glow when struck by a decay particle. The screens could not be connected to electronic timers, and the flashes had to be counted through a microscope.

The Geiger-Müller counter is a gas-filled tube with a charged wire running up its center. When a gamma or a beta enters the tube, it knocks electrons off the gas atoms and makes the gas a conductor. The electrons are collected by the center wire and can be counted electronically or used to activate a speaker and make the characteristic click of a counter in the presence of radiation. Geiger-Müller counters often are not sensitive to alpha particles, as the metal used to keep the gas inside the tube also keeps alpha particles from entering the counter. The proportional counter, also a gas-filled tube, works on the same principle as does a Geiger-Müller counter except that it carefully measures the number of electrons that reach the central electrode. Because the number of electrons knocked off gas atoms is directly related to the energy of the particle that knocked them off, the number of electrons reaching the central electrode is directly related to the energy of the particle that produced them.

Scintillation counters utilize transparent materials that emit a flash of light when a particle passes through them. The amount of light is proportional to the kind of particle that passes through and its energy. The light is collected by a special tube, called a photomultiplier tube, that converts the light into a current pulse whose size is proportional to the amount of light emitted. The pulse can be used to drive an electronic counting system.

MODERN DETECTORS

Many modern studies of gamma decay use solid-state detectors which take advantage of the fact that silicon and germanium crystals can be grown with very small amounts of impurities. If the crystals are carefully prepared, they will become conductors when radiation knocks electrons off the atoms in the regions containing the impurities. Once again, the number of electrons produced depends on the energy of the gamma that originated them. The crystal is placed between electrodes with a large voltage across them, and the electrodes collect the electrons produced by the gamma.

Modern detectors produce a pulse of electrons—

an electric current—the strength of which is proportional to the energy of the radioactive decay particle that produced it. This current pulse is amplified and its size determined to study the energy of the decay particle that produced it. Such studies typically involve several stages of amplification during which the researcher must be careful not to alter the shape of the current pulses he or she is studying. The pulses are then fed into a device called a multichannel analyzer, which sorts them according to their strength and which stores each pulse as a count in a series of electronic bins. The bins can be displayed to show the number of counts received at each energy level. Called a spectrum of the decay, it is used to identify the nucleus that emitted the decay product. In modern systems, the multichannel analyzer is replaced by a dedicated computer that automatically identifies the energy of the decay and, in many cases, can tell the researcher what nucleus produced it. Such systems have memories stored with data on energies and half-lives of numerous radioactive decays that have been accumulated since World War II and carefully tabulated.

Radiation detectors and their associated counting systems have become smaller and more rugged. Studies of radioactive decays were once conducted only in laboratories; however, portable systems can now be taken into the field and are found, for example, at petroleum drilling sites. Detection systems are frequently carried into space and have been used in studies of the cosmic radiation and numerous other phenomena. The more sensitive systems used in radioactive dating are still confined to the laboratory, as they must be protected from the radiation in the environment produced by cosmic radiation and by radiation from common minerals.

RESEARCH AND EXPLORATION TOOL

Since World War II, radioactive decay has ceased to be a laboratory curiosity and has become a widely used tool. The understanding of radioactive decay has led to the development of the means to date geological and archaeological specimens. Radioactive-dating techniques take advantage of the fact that each species of excited nucleus will decay so that half of it disappears after a particular amount of time elapses. If no new excited nuclei have been added to the sample since it was formed,

one can compare the number of excited nuclei remaining in the sample to the number of nuclei in the sample formed by the radioactive decay and thus determine how long it has been since the sample was formed. The understanding of geologic time is based largely on radioactive dating.

In addition to its importance as a dating tool, radioactive decay is believed to be a major source of the heating of the Earth. This heat flow is important to the overall heat budget of the Earth and may be partially responsible for present temperatures on the Earth's surface. Radioactive decay may also contribute to driving convection currents in the Earth's interior, and it is probably at least partially responsible for the heat extracted as geothermal energy. Radioactive decay is also responsible for changing the nature of certain minerals, as in metamictization.

The presence of a large number of nuclei undergoing radioactive decay makes nuclear wastes hazardous. Although short-lived nuclei decay within a year and disappear, wastes from nuclear reactors are characterized by the presence of nuclei with long half-lives. They must therefore be stored in such a manner that they will not come in contact with the environment for tens of thousands of years. Only very stable geologic formations where wastes cannot be reached by groundwater will permit such storage. The search for suitable areas has covered many states. Radioactive decay has become a concern of many home owners with the discovery that seepage of radioactive radon gas, produced by the decay of minerals in the Earth, has raised radiation levels in some homes above levels deemed healthy.

Despite its hazards, radioactive decay has become a useful tool in many types of geological research. For example, small amounts of short-lived radioactivity have been injected into geothermal systems along with cooled water to see how long it takes the reinjected water to reach the production end of the system. Radiation detectors are inserted into boreholes during petroleum exploration to map the presence of radioactive minerals along the walls of the borehole, which assists in the identification of shaley layers within a sandstone formation. In many other cases, radioactive decay has proved to be a useful tool for research and exploration.

Ruth H. Howes

BIBLIOGRAPHY

Adolff, Jean Pierre, and Robert Guillaumont. *Fundamentals of Radiochemistry.* Boca Raton, Fla.: CRC Press, 1993. A thorough approach to radiochemistry, this book describes all aspects and applications associated with the field. An excellent source of information for someone without much knowledge or background in radiochemistry, this book is well illustrated with diagrams and charts. Bibliography and index.

Choppin, Gregory R. *Radiochemistry and Nuclear Chemistry.* 2d ed. Oxford: Butterworth-Heinemann, 1995. This widely used college text introduces the reader to the basics of nuclear chemistry and radiochemistry. It explores the theories surrounding those fields and their applications. Well illustrated with clear diagrams and figures, this is a good introduction for someone without a strong background in chemistry.

Durrance, Eric M. *Radioactivity in Geology: Principles and Applications.* New York: John Wiley & Sons, 1987. This monograph presents a very thorough review of the principles of radioactive decay and its many applications in geology. In addition to its thoroughness, this book is written for a nonspecialist in nuclear physics and is fairly readable as well as complete, discussing not only dating methods but also exploration methods, the use of radioactivity in petroleum studies, environmental radioactivity, and the role of radioactivity in heating the Earth's interior.

Keller, C. *Radiochemistry.* New York: John Wiley & Sons, 1988. In addition to covering the basic theory of radioactive decay, this volume stresses the applications of radioactive decay in the study of problems of geological interest. Unlike the preceding reference, it emphasizes radioactivity produced by humans for their own purposes and is of less general interest than is Durrance's monograph.

Lieser, Karl Heinrich. *Nuclear and Radiochemistry: Fundamentals and Applications.* New York: VCH, 1997. Lieser's book gives the reader a basic understanding of the practices and principles involved in the fields of radiochemistry and nuclear chemistry. Although there is quite a bit of chemistry involved in the author's explanations, someone without a chemistry background will still find the book useful. Illustrations, charts, and diagrams help clarify difficult concepts and theories. Bibliography and index.

Mozumder, A. *Fundamentals of Radiation Chemistry.* San Diego: Academic Press, 1999. Mozumder presents a concise introduction to radiochemistry, examining the procedures and theories associated with it. Some background in chemistry would be helpful but is not necessary. Bibliography, illustrations, and index.

Rhodes, Richard. *The Making of the Atomic Bomb.* New York: Simon & Schuster, 1986. The early chapters of this excellent history of nuclear physics are devoted to a detailed description of the discovery of nuclear decay and the development of a theory of the nucleus. The historically important experiments and ideas are presented in detail. Because the emphasis is on people and the development of a theory, Rhodes has written a very readable as well as technically accurate description of the theory of radioactive decay.

RADIOCARBON DATING

Radiocarbon dating is a means of determining the approximate time at which biological processes ceased in a once-living organism or in related organic substances. It allows scientists to estimate the ages of organic materials and the formations in which they occur.

PRINCIPAL TERMS

DENDROCHRONOLOGY: the study of tree rings; it provides a means of calibrating radiocarbon dates with absolute chronology

HALF-LIFE: the period required for half of any given quantity of a radioactive element to revert to a stable state

ISOTOPES: forms of the same element with identical numbers of protons but different numbers of neutrons in their atoms' nuclei

MAUNDER MINIMUM: the period from 1645 to 1715, when sunspot activity was almost non-existent

PHOTOSYNTHESIS: the process of fixing atmospheric carbon in organic compounds in plants with free oxygen as a by-product

RADIOACTIVE ISOTOPE: an unstable isotope that decays into a stable isotope

THERMOLUMINESCENCE: the process by which some minerals trap electrons in their crystal structures at a fixed rate and release them when heated

RADIOACTIVE CARBON 14

Carbon compounds are among the most abundant in nature. Carbon, in the form of atmospheric carbon dioxide, is continuously cycled through major environmental systems. The photosynthesis of carbon dioxide establishes carbon in various compounds in plants, which in turn may be consumed by animals. Most carbon is absorbed as carbon dioxide by the oceans or appears there as dissolved carbonate or bicarbonate compounds. Carbon occurs naturally in three isotopes: carbon 12, the most common; carbon 13, also a stable isotope; and the radioactive isotope carbon 14, which exists in minute quantities. Radiocarbon dating draws on several assumptions concerning the natural production of carbon 14, its presence in various environmental cycles, and its rate of decay, or half-life.

Radioactive carbon 14 is formed in the upper atmosphere as the result of bombardment by energetic cosmic radiation emanating from deep space. Statistically, it is most likely that free neutrons, which result from collisions between proton cosmic radiation and atmospheric gases, will shortly collide with molecules of stable nitrogen, nitrogen 14, by far the most abundant gaseous element in the atmosphere. The resulting reaction normally expels a proton from the nitrogen nucleus, producing an atom which now behaves chemically like carbon but is heavier than the stable carbon isotopes. Free carbon does not remain long in the upper atmosphere. It quickly combines with oxygen molecules to form carbon dioxide, whereupon it enters into various geological and biological processes.

BETA DECAY

The half-life of carbon 14 usually is expressed as $5,568 \pm 30$ years, though it is more likely to be on the order of $5,730 \pm 30$ years. (Once large numbers of dates had been calculated on the basis of the former figure, leading scientific journals generally preferred to stay with it.) The particular process of radioactive decay of carbon 14 is called beta decay, in which the "extra" neutron in the nucleus emits a beta particle (essentially an electron) and a neutrino (an uncharged particle), thus changing itself, in effect, from a neutron into a proton. The result is once again an atom of stable nitrogen.

As long as an organism is alive and normal biological processes are occurring, the rate of accumulation of carbon 14 is in approximate equilibrium with the rate of radioactive decay of the carbon 14 already in the organism. This level of equilibrium is extremely small, on the order of one atom of carbon 14 to every 1 trillion atoms of carbon 12. The moment that biological processes cease, however, the equilibrium is broken, and the quantity of carbon 14 in the once-living organism begins to decrease at the predictable rate of radioactive decay. By measuring the rate of beta radiation from the residual carbon 14, one may estimate the age of a material, or, more precisely, when the biological processes of the organism from which the material derives ceased to operate. In general, the age of material which gives off only half as much beta radiation as a living organism would be 5,730 years; the age of material emitting only 25 percent as much radiation would be 11,460 ($5,730 \times 2$) years; and so on.

EVOLUTION OF MEASUREMENT TECHNIQUES

Willard F. Libby, an American physicist, introduced the radiocarbon dating method in 1949 after fifteen years of research. Since Libby's pioneering work, radiocarbon dating has evolved to include several counting techniques. Normally the substance under study must be destroyed by combustion or other processes to produce gaseous carbon dioxide or a hydrocarbon gas, whose carbon 14 content is measured by a gas counter. Liquid counting techniques have also been developed, in which the carbon dioxide from the substance under study is synthesized into more complex liquid hydrocarbons, such as benzene. After 1980, direct measurement techniques, using particle accelerators and mass spectrometers, increasingly replaced gas and liquid counting.

The evolution of measurement techniques has greatly enhanced the usefulness of radiocarbon dating. The minimum acceptable mass of a sample substance is much smaller than that required in the years immediately following Libby's introduction of the method. Refined techniques and improved instrumentation also have extended the effective chronological limits of the method. At first, only materials less than about twenty thousand years old could be dated with any confidence, but by the fortieth anniversary of Libby's initial dating experiments, there was wide agreement that the effective limit was about forty thousand years and, when supporting data could be gleaned from other dating methods, possibly seventy thousand years.

RISK OF CONTAMINATION

The extremely small quantities of carbon 14, even in living organisms, demand that substances undergoing counts be prepared most carefully and that every possible source of contamination be considered. For example, fallout from atmospheric testing of thermonuclear weapons in the 1950's interfered with some of Libby's early experiments. Numerous chemical processes affect archaeological artifacts and other substances which might be candidates for radiocarbon analysis. Carbon compounds from associated soil layers find their way into dating samples. Carbonate encrustations may develop around other samples, especially in locations with abundant groundwater. In the 1980's, some radiocarbon laboratories had to consider their local environments, where, in many cases, the buildup of carbon dioxide and other carbon compounds in the atmosphere from fossil fuel combustion threatened to interfere with sample integrity and analysis. (Fossil fuels, because of their great geological age, contain virtually no carbon 14.)

Some substances are at a greater risk of contamination than others. Woody plant remains, such as charcoal or building materials carbonized by fire, produce the best results. Less dependable are textiles, fibers, and the remains of nonwoody plants, which do not live as long as woody species and, as a result, may reflect short-term or local carbon 14 anomalies. Bone is notorious for its ability to absorb carbon compounds from its surroundings. Inorganic substances which contain carbon compounds, such as eggshell and marine shell, also present serious problems of carbonate contamination. Of these substances, marine shell is preferred, since the extraneous carbonate levels can be determined on the basis of relatively constant values present in ocean water.

ADJUSTING FOR VARIABILITY

In reality, radiocarbon dating is subject to numerous variable factors and is not nearly so accurate. Radiocarbon dates are expressed as years B.P. or "before present." "Present" is actually a zero base date of 1950. (Scientists chose this year because it was close to the first experimental application of the method and because the buildup of atmospheric radioactivity from nuclear weapons tests in later years introduced complications into the measurement process.) Each date is expressed with a standard error, or standard deviation—in essence, a "confidence level"—indicated by a plus-or-minus sign. A typical radiocarbon date of, say, 2750 B.P. ± 60 indicates that there is a 66 percent probability that the true date falls within sixty years of 2750 B.P. and a 95 percent probability that the true date falls within twice the standard error—in this case, 120 years.

Another source of variability in test results was only dimly suspected before about 1970. Libby assumed that the formation rate of carbon 14 in the upper atmosphere had a constant value over time, since the process depended on cosmic radiation. Scientists now know that the formation rate varies over time and space, in part according to the strength and contours of the Earth's magnetic field. The field deflects a certain portion of cosmic radiation, so some less energetic particles never reach the atmosphere. At any time, for example, more carbon 14 is formed in the atmosphere near the poles, where the magnetic field is relatively weak, than near the equator, where it is stronger.

The strength of the magnetic field itself is affected by cyclical changes in solar activity. The Sun has roughly an eleven-year cycle of sunspot activity; the Earth's magnetic field is most energetic when sunspots are most numerous. Longer-term and even more significant changes apparently have occurred in recent solar history, as in the case of the so-called Maunder minimum (1645-1715), when sunspot activity may have been almost nonexistent. During such a period, the Earth's magnetic field would be less energetic, carbon 14 formation levels would be abnormally high, and materials from the era "younger" in radiocarbon terms than they otherwise would be.

Dendrochronology (the study of tree rings) provides a means of identifying these anomalies and thus calibrating radiocarbon dates more closely with absolute chronology. In principle, each growth ring in a tree lives only for one year. By comparing the known age of tree rings with their radiocarbon ages, scientists can locate anomalies in the carbon 14 absorption rate and, therefore, the formation rate. Wood samples from the extremely long-lived bristlecone pine provide invaluable data. The oldest living bristlecone is more than 4,500 years old, and the remains of dead specimens have yielded a calibration matrix extending almost nine thousand years into the past.

Several other dating systems have been developed which utilize the half-life of radioactive isotopes, and some provide corroboration of radiocarbon chronologies. Further support comes from closely related techniques, such as thermoluminescence. These methods, as well as the established practices of classical archaeology, strengthen the credibility of the radiocarbon technique.

APPLICATION TO ARCHAEOLOGY

Through radiocarbon dating applications, scientists have amassed a wealth of information on such matters as sea-level fluctuation, climatic change, glaciation, habits of marine life, volcanic activity, and early forms of atmospheric pollution from the combustion of fossil fuels. From time to time, results of carbon 14 analysis also attract attention by establishing the authenticity of artistic works or the ages of religious relics.

By far the most significant application of radiocarbon dating, however, has been in providing archaeologists with the means of constructing chronological relationships independently of traditional archaeological assumptions about cultural processes. The technique is especially valuable in prehistoric archaeology, for which little or no documentation is available and artifact interpretation previously could allow, at best, only relative chronological sequences. In the Western Hemisphere, Africa south of the Sahara, and other parts of the world where "prehistory"—in the sense of absence of documentation—extends nearly to the present and chronologies had been mere guesswork, carbon 14 dating was a revolutionary technological breakthrough.

The results obtained from applying carbon 14 dating to prehistoric materials, however, made the technique immediately controversial. For example, early dates for the origins of agriculture in the Near East—generally referred to as the Neolithic Revolution—were dramatically earlier than archaeologists had expected. At Jericho, one of the first sites to provide such material, scholars previously had placed the start of the Neolithic Revolution at around 4000 B.C.E., but carbon 14 dates were on the order of four thousand years earlier. Further research in the Near East has established a chronological frontier for the Neolithic Revolution at 10,000 B.C.E. or even earlier.

A second major achievement of radiocarbon

dating was the establishment of the first outlines of an absolute chronology for developments similar to the Neolithic Revolution in Mexico and Central America. Supported by pollen analysis, carbon 14 dating of the remains of early domesticated maize suggested an astonishingly long continuum for agriculture in the region extending back to 5000 B.C.E. This finding was the first real indication of the depth and complexity of Mesoamerican cultural heritage.

In sub-Saharan Africa, the immense potential of charcoal for providing carbon 14 dates led to reconstruction of the prehistory of a region which, before 1950, was unknown. Charcoal is especially plentiful at the large numbers of iron-smelting sites in Africa. Slag from the smelting operation, sometimes even iron remains, may be dated using carbon 14; carbon migrates as an impurity into the iron. Before 1950, conventional wisdom attributed the presence of iron-smelting technology in sub-Saharan Africa to cultural borrowing, either from Phoenician traders or from the Egyptians. Carbon 14 dates clearly established a thriving iron technology in West Africa as early as 1000 B.C.E., several hundred years before the earliest known incidence of smelting in Egypt itself.

CONTROVERSIAL RESULTS

Perhaps the most disturbing revelation of the first decade or so of carbon 14 dating was its support for an extended Neolithic chronology in Europe. Although archaeologists already were debating the merits of a "long" versus "short" European Neolithic, most scholars and cultural opinion favored the latter, since it accommodated the notion of cultural "diffusion" from the ancient Near East, relegating Europe to a state of barbarism until stimulated by classical civilizations. The famous megalithic structure of Stonehenge in southern England, for example, was thought to have been inspired by Mycenaean influence and was therefore dated to around 1400 B.C.E. Yet, carbon 14 results obtained from reindeer bone fragments at the site—the bones probably were digging implements used in construction—suggested dates for vital parts of the Stonehenge complex that were several centuries earlier, making it, in effect, pre-Mycenaean. These findings threatened to sever the "diffusionist" link with the Near East and

forced a reassessment of the state of prehistoric European culture.

The most severe test of the radiocarbon technique occurred when it was used to corroborate dates for Egyptian historical artifacts that already had been fairly accurately dated with conventional archaeological methods. Early results were not reassuring. The method repeatedly generated dates for materials from the second millennium B.C.E. that were several centuries too recent. Since these discrepancies were more or less consistent in scope and did not alter the *sequences* of Egyptian history, they raised serious questions about the integrity of the whole chronological framework of ancient history patiently constructed over decades by archaeological research. Conversely, many traditional scholars, confident of their work and suspicious of a technique derived from nuclear physics, preferred to reject the radiocarbon system out of hand.

These difficulties, together with the developing precision of sample handling techniques and instrumentation, by the early 1970's had led to the realization that there were variables in the carbon 14 cycle unaccounted for by Libby's earlier formulations, principally the magnetic field fluctuations, which, as noted above, result in periodic changes in the carbon 14 formation rate. Once radiocarbon dates could be calibrated on the basis of tree-ring data, they agreed very closely with previously established dates for Egyptian artifacts and therefore confirmed the work of traditional archaeologists. With that, the last serious barrier to acceptance of the technique disappeared. Some of the prehistoric dates for Europe, however, when corrected according to tree-ring data, actually pushed the Neolithic horizon even further into the past. The earliest portions of Stonehenge are now believed to be separated from what once were thought to be their Mycenaean origins by what one archaeologist has called a "yawning millennium."

In the 1980's, investigators from England, Switzerland, and the United States applied radiocarbon dating techniques to fibers from the "Shroud of Turin." Held in reverence by many as the burial shroud of Jesus, this piece of linen retains the image of a bearded man with marks consistent with crucifixion. Measurements by separate laboratories agreed that the flax from which the linen was produced grew sometime in the thirteenth or fourteenth centuries—far too recent to have been the burial shroud of Jesus.

INTERFACE BETWEEN SCIENCE AND TRADITION

Radiocarbon dating, together with similar techniques using isotopes of other elements and a variety of methods drawn from the physical and life sciences, since 1950 has elaborated a picture of human history and Earth history to a degree that could not have been conceived by earlier scholars. Among the method's practical benefits is a much more sophisticated knowledge of the scope of natural climatic change, without which it would not be possible to make useful scientific or political decisions on matters that may affect future, human-induced climatic and environmental change. Radiocarbon dating also has shattered many long-standing notions about European prehistory, Europe's historical relationship with the ancient Near East, and the antiquity and complexity of non-European civilizations, thereby undermining fundamental assumptions about the centrality of Western civilization in the human saga.

The history of the application and results of carbon 14 dating requires that one look carefully at the conditions and assumptions associated with certain dates and at the stage in the technique's development from which those dates derive. A prime example of how these matters can fuel controversy was the confusion generated by some carbon 14 dates which seemed to suggest that established dates for Egyptian artifacts of the second millennium B.C.E. were several centuries too early. Biblical scholars had long worried about an unexplained gap between the dates, derived from scientific genealogical and textual studies, for the Hebrew Exodus from Egypt and the establishment of the ancient kingdom of Israel. In terms of regarding the Old Testament as historically accurate, it would be convenient if several centuries of Egyptian history could be "erased" and earlier events moved up to fill the gap. Ironically, that is just what early carbon 14 dating of Egyptian artifacts suggested. Biblical scholars were ecstatic that science seemed to verify the biblical accounts at last; however, later calibration of these Egyptian dates using tree-ring data reinstated traditional Egyptian chronology and destroyed this temporary congruence. Clearly, the lay reader who encounters results of radiocarbon dating encounters a deviously complex interface between modern science and the most fundamental issues in the Western religious and historical tradition.

Ronald W. Davis

CROSS-REFERENCES

Carbonates, 1182; Earth's Age, 511; Earth's Magnetic Field, 137; Earth's Magnetic Field: Secular Variation, 150; Earth's Oldest Rocks, 516; Fission Track Dating, 522; Geoarchaeology, 1028; Ocean-Atmosphere Interactions, 2123; Potassium-Argon Dating, 527; Radioactive Decay, 532; Rubidium-Strontium Dating, 543; Samarium-Neodymium Dating, 548; Uranium-Thorium-Lead Dating, 553.

BIBLIOGRAPHY

Agrawal, D. P., and M. G. Yadava. *Dating the Human Past.* Pune: Indian Society for Prehistoric and Quaternary Studies. 1995. An interesting look at the use of radiocarbon dating technology in relation to anthroplogy. Suitable for the non-scientist. Illustrations.

Bard, Edouard, and Wallace S. Broecker, eds. *The Last Deglaciation: Absolute and Radiocarbon Chronologies.* Berlin: Springer-Verlag, 1992. This book deals with the radiocarbon dating processes and techniques that have been used to unravel the mysteries of glaciers and massive changes in climate. This book provides a simple explanation of radiocarbon dating and illustrates its applications.

Fleming, Stuart. *Dating in Archaeology: A Guide to Scientific Techniques.* New York: St. Martin's Press, 1976. Places the methods and results of radiocarbon dating in the context of other dating techniques, such as dendrochronology, thermoluminescence, fission track dating, pollen analysis, and chemical methods. Discusses how various of these techniques may corroborate one another in establishing chronologies. Excellent bibliography.

Lowe, J. John, ed. *Radiocarbon Dating: Recent Applications and Future Potential.* Chichester, N.Y.: John Wiley and Sons, 1996. This college-level book offers a comprehensive overview of the techniques and protocols of radiocarbon dating. Several of the essays explore possible future usage and applications. Illustrations and maps help to clarify difficult concepts. Bibliographical references.

Renfrew, Colin. *Before Civilization: The Radiocarbon Revolution and Prehistoric Europe.* New York: Alfred A. Knopf, 1973. A comprehensive discussion of the development of radiocarbon dating, the early discrepancies, and their correction using tree-ring data. Offers a thorough analysis of some alternative approaches to prehistory based on carbon 14 results from Europe.

Scott, E. M., M. S. Baxter, and T. C. Aitchison. "A Comparison of the Treatment of Errors in Radiocarbon Dating Calibration Methods." *Journal of Archaeological Science* 11 (1984): 455-466. Discusses several methods of calibration with respect to the varying degrees of accuracy required in specific applications.

Taylor, R. E. "Fifty Years of Radiocarbon Dating." *American Scientist* 88 (January/February, 2000): 60-67. This paper reviews the entire five decades of development of this remarkable technique. Taylor reviews the basic technique, deviation of C^{14} dates from true dates due to variation in the magnetic field, means of correction, and attaining increased sensitivity by the application of mass spectrometry.

_____. *Radiocarbon Dating: An Archaeological Perspective.* New York: Academic Press, 1987. An excellent treatment of the procedures and complexities involved in measuring radiocarbon content. Discusses instrumentation, sources of contamination, case studies using various substances, and the historical development of radiocarbon methodology. The bibliography covers a broad range of sources.

Vita-Finzi, Claudio. *Recent Earth History.* New York: Halsted Press, 1974. An account of the physical changes undergone by the Earth during the Holocene (modern) geologic era, presented in the form of a stratigraphical narrative based throughout on radiocarbon dates.

Wilson, David. *The New Archaeology.* New York: New American Library, 1974. Perhaps the best account for the general reader of the development and subsequent refinement of radiocarbon dating techniques. Discusses their enormous impact on the field of prehistoric archaeology and perceptions of prehistory.

RUBIDIUM-STRONTIUM DATING

Rubidium-strontium (Rb-Sr) dating is one of the most common methods of obtaining absolute (numerical) ages of geologic materials, especially older minerals and rocks.

PRINCIPAL TERMS

ABSOLUTE AGE: the numerical timing of a geologic event, as contrasted with relative, or stratigraphic, timing

GEOCHRONOLOGY: the study of the absolute ages of geologic samples and events

HALF-LIFE: the time required for a radioactive isotope to decay by one-half of its original weight

ISOCHRON: a line connecting points representing samples of equal age on a radioactive isotope (parent) versus radiogenic isotope (daughter) diagram

ISOTOPE: a species of an element having the same number of protons but a different number of neutrons and therefore a different atomic weight

MASS SPECTROMETRY: the measurement of isotope abundances by separating the isotopes by mass and charge in an evacuated electromagnetic field

RADIOACTIVE DECAY: a natural process by which an unstable, or radioactive, isotope transforms into a stable, or radiogenic, isotope

RUBIDIUM

In 1904, Ernest Rutherford proposed that geologic time might be measured by the breakdown of uranium in uranium-bearing minerals; a few years later, Bertram Boltwood published the absolute, or numerical, ages of three samples of such minerals. The ages, which approximated 500 million years, indicated the antiquity of some Earth materials, a finding developed by Arthur Holmes in his classic *The Age of the Earth* (1913). Holmes's early time scale for the Earth was not instantly accepted by most of his peers, but it helped to set the stage for the eventual use of absolute ages as the prime quantitative components in the study of geology and its many subdisciplines. After the early study of the isotopes of uranium, including uranium-series transition isotopes, came the discovery of other unstable isotopes and the formulation of the radioactive decay schemes that have become essential to geochronology, including the rubidium-strontium method.

Rubidium (Rb), an alkali-group (lithium, sodium, potassium, rubidium, cesium, and francium) element with a charge of +1, is a trace element in terrestrial and solar system materials. It is not a necessary component in any known mineral; it substitutes for the major element potassium in common, rock-forming minerals such as the alkalic feldspars and micas. Because of their similar ionic size and geochemical behavior, the ratio of potassium to rubidium is a useful petrologic parameter.

Rubidium consists of two natural isotopes, rubidium 87 and rubidium 85. Several artificial isotopes also are known. Rubidium 87 has been known to be unstable since 1940, but its use in age dating did not begin until the advent of modern mass spectrometry in the 1950's. It decays by the emission of a beta particle to the stable nuclide strontium 87. The half-life for this decay is not precisely known, because of the low energy spectrum of the emitted beta particles; values range from 4.7 to 5 billion years. Although the commonly accepted value is 4.7 billion, a comparison of Rb-Sr dates for samples of meteorites and lunar rocks with dates for the same samples yielded by other techniques indicates that a value of 4.9 billion is more correct. In any event, the long half-life and the low parent-daughter ratio mean that the radiogenic accumulation of strontium 87 in most natural minerals and rocks is very slow.

STRONTIUM

Strontium—which, along with beryllium, magnesium, calcium, barium, and radium, belongs to the alkaline-earth group of elements—has a charge of +2. Though more abundant than rubidium in materials of the solar system, it also is a trace element. It forms its own mineral phases in the form of strontianite, a strontium carbonate, and celestite, a strontium sulfate, but it occurs more significantly as a substitute for the major species calcium in common, rock-forming minerals such as plagioclase feldspar, calcium-rich pyroxenes and amphiboles, apatite, and calcium carbonate minerals. Like the potassium-rubidium ratio, the calcium-strontium ratio has use in the understanding of various geologic processes. Strontium comprises four naturally occurring isotopes, strontium 88, 87, 86, and 84. There are also several artificial isotopes, the best known being the nuclear-fission-produced strontium 89 and 90.

PURIFICATION OF SAMPLES

Concentrations of rubidium and strontium in minerals, rocks, and other natural substances can be determined by a variety of techniques, although isotopic parameters and precise elemental abundances are commonly determined by mass spectrometry. As the abundances of rubidium 85, strontium 88, strontium 86, and strontium 84 all are precise percentages of the total for each of these elements and do not vary as a function of nuclear instability, each can be calculated by taking the appropriate percentage of the total elemental concentrations, as determined by gravimetric analysis, atomic absorption spectrophotometry, X-ray fluorescence spectrophotometry, microbeam probe, or another type of analysis. Most modern work, however, consists in determining the relevant isotopic parameters by mass spectrometry, after purification by chemical techniques.

Commonly, a preliminary determination of rubidium and strontium abundance and the Rb-Sr ratio is made or estimated for the minerals, rocks, or other materials to be dated by a reconnaissance technique, such as X-ray fluorescence. Samples selected on the basis of these determinations are chosen for dating. After selection, the samples are crushed—homogenized, if necessary—and a fraction is taken which contains enough of the rubidium and especially the more critical strontium for adequate isotopic analysis. Rarer materials, such as meteorites and lunar rocks, may not afford enough sample for optimal analysis. The fraction of material—if, as is most common, it is in the geologic form of aluminosilicate compounds, perhaps

with some organic material—is dissolved in a mixture of hydrofluoric and perchloric acids and reduced by evaporation to a concentrated "mush" of material. Samples other than silicates may be dissolved in other, more appropriate solvents. The concentrated mush is spiked with an appropriate amount of purified liquids containing known amounts of rubidium and strontium of known, nonnatural isotopic composition. This material is dissolved in a small amount of hydrochloric acid and placed on calibrated ion exchange columns. The columns are washed with hydrochloric acid, and purified portions of rubidium, then strontium, of mixed natural and spiked isotopic composition, are collected and evaporated. Highest accuracy and precision requires that the smallest amounts of rubidium and strontium from the laboratory environment (contamination) be included with the completed, purified elements.

MASS SPECTROMETRY

The standard method of mass spectrometry for Rb-Sr dating involves placing the purified samples as solids on metal filaments and installing the loaded filaments in solid-source mass spectrometers. The spectrometer source regions, evacuated to very low pressures, are constructed so that the metal filaments can be heated until the rubidium or strontium ionizes. The charged, ionized sample is accelerated through a series of collimating slits into a controlled electromagnetic field, where the beams of ions are separated by charge-mass ratios into beams of separate isotopes; as the charge of the elements is the same for each atom, the ions are separated on the basis of mass only. Specific isotopic beams, controlled by the magnetic field, are channeled through more collimating slits to the collector part of the spectrometer. Commonly, a Faraday cup is used to analyze the number of atoms of each isotope by the conversion of each atomic impact into a unit of charge, which is then amplified. The actual output is isotope ratio measurements—rubidium 87/rubidium 85, strontium 88/strontium 86, and so on—which are converted using mathematical programs into the required parameters for determining time. Because strontium 87 is closest in abundance to strontium 86 in most cases, and because it differs by only one mass unit, the standard for reporting the radiogenic component is with the ratio strontium 87/strontium 86.

By mixing precise amounts of "spikes" of strontium and rubidium, whose isotopic compositions differ markedly from natural isotopic compositions, with the strontium and rubidium in the natural sample, a combined mass spectrum is obtained. From these data can be calculated the precise abundances of rubidium and strontium in the natural material (a process known as isotope dilution) and the critical isotopic composition of the natural strontium.

ISOCHRON DIAGRAMS

Although the age of the analyzed sample can be calculated using the determined Rb-Sr parameters and the decay constant for rubidium 87, it is customary and more useful to determine the age graphically, with the use of an isochron diagram. In the diagram, the actual isotope ratios collected in the spectrometer are used as coordinates. Thus, the parent, unstable component, rubidium 87, is designated by reference to a common isotope of strontium, strontium 86. The other coordinate, the measure of the radiogenic component, is strontium 87/strontium 86. A line connecting points representing samples of equal ages, an isochron, has an age value that is represented by its slope on the figure; a horizontal isochron has an age value of zero, and positive slopes of successively greater degree have increasingly greater ages. A single mineral or rock would furnish only one point on the diagram, so for an isochron to be drawn, there must be knowledge of or, more likely, an estimate of the sample's initial isotopic composition. Ages calculated in this way are termed "model ages."

Analysis of minerals of equal age but different compositions from a sample of plutonic volcanic rock would allow the construction of an isochron, whose slope would be proportional to the age of crystallization of the rock. An assumption that is justified in most circumstances is that at the time of crystallization from a magma or lava, all minerals formed have the same isotopic composition of strontium. The basis for this assumption is that, unlike isotopes of elements with a mass of less than about 40, there is no measurable fractionation of isotopes of an element at the physical-chemical conditions of the liquidus material. If the rock system has not been affected by open-system behavior—for example, by the introduction of parent or daughter species by metamorphism

or weathering—the points representing these samples will define a perfect line. In practice, uncertainties in measuring each of the parent-daughter parameters, and perhaps some open-system behavior, result in imperfect isochronism and, therefore, uncertainties in the calculated age. A benefit of the isochron method is that the isotopic composition of strontium at the time of origin of the rock, or strontium 87/strontium 86, is marked by the left-hand or lower intercept of the isochron, where rubidium 87/strontium 86 is equal to zero. As discussed above, for a model age calculated from a single mineral or rock analysis, this parameter would have to be estimated. Another benefit is that one can readily see whether one or more points are aberrant or whether a poor fit might indicate an open-system history for the rock. A wide variety of terrestrial and extraterrestrial igneous rocks have been dated by this mineral isochron method with good precision.

As with minerals from a single rock, if a series of rocks of equal age from a common parent (such as fractionally differentiated rocks from a single magma or similar rocks which resulted from different degrees of partial melting of a common source) are analyzed, their Rb-Sr isotopic parameters should yield an isochron proportional to the unique age of the rocks, or a whole-rock isochron. Consequently, one may test for ages and for co-magmatic properties in a suite of rocks. In practice, however, it is common for these properties to be unraveled only with supporting petrologic or geochemical data, if at all.

UNIQUENESS OF METHOD

The most spectacular benefit of the Rb-Sr isochron technique and one that is unique to the method is the ability, under some circumstances, to identify both the original time of crystallization of an igneous rock, such as a granite, and a later time of metamorphism resulting, for example, in a granitic gneiss. These events may be traced by constructing both mineral and whole-rock isochrons from several, chemically differing types of the gneiss. Points representing mineral components, along with points representative of their whole-rock mixtures, may form an isochron proportional to the age of metamorphism. These mineral isochrons should have the same slopes and therefore ages; however, a whole-rock isochron constructed only from the whole-rock points will be steeper, with a slope proportional to the earlier time of intrusion of the granite. This seemingly peculiar behavior results from a reequilibration of strontium isotopes at the time of metamorphism in the vicinity of the rock sampled; on a more regional scale, however, whole-rock parameters were not homogenized.

The method requires, in addition to reasonably closed-system behavior, that the later metamorphic event be capable of reequilibrating the local rock systems completely. In practice, that is accomplished either through fairly high-grade metamorphism, through the availability of sufficient water to effect the isotopic exchange, or, most likely, through both. Dry metamorphism, even if high-grade, may not result in reequilibration; the mineral isochron date obtained will therefore record only the magmatic event. Conversely, if the rocks are permeable, fine-grained, and wet, rehomogenization may be completed even under low-grade metamorphic conditions. Where conditions of economic mineralization have been sufficient to equilibrate Rb-Sr components, whole-rock isochron analysis can reveal this type of metamorphic event. If the event results in the formation of new minerals, these minerals may record the time of metamorphism. Unfortunately, in some studies, the multiple possibilities of incomplete rehomogenization, open-system behavior, and poor precision of measurement result in poor isochronism and age data of questionable or no value.

USE WITH MARINE COMPONENTS

Because of a lack of fossils with determinable ages in many sediments and sedimentary rocks, absolute age analysis by several isotopic methods, including Rb-Sr dating, has been tested. The technique works well only for certain minerals for which model ages can be calculated by assuming the ratio between rubidium 87 and strontium 86; for marine components, this assumed parameter may be reasonably determined, thanks to the effective homogenization of strontium in seawater during a given geologic time. The most commonly dated material is glauconite, although the closed-system requirement may be violated by the unavoidable inclusion of detrital components of varying strontium-isotopic composition. Other materials of limited usefulness are minerals such as phillipsite and some illite. Evaporitic minerals containing enough rubidium, such as sylvite, also have been dated by this technique, although omnipresent recrystallization may perturb the isotopic systematics.

In some cases, the isotopic composition of strontium by itself can be used to determine the age of carbonate rock. As previously stated, strontium dissolved in seawater of a particular geologic episode has the same isotopic composition everywhere in the ocean. That is because the mixing rate of marine strontium—about one thousand years—is short compared with the average "lifetime" for strontium atoms in the sea—about 2 million years. Thus, strontium of variable isotopic composition washed into the sea from rivers or other sources is well mixed. Because strontium 87 accumulates through time as a result of rubidium 87 decay, terrestrial strontium, including marine strontium, becomes more radiogenic through geologic time. Theoretically, then, marine strontium of any geologic time should have a unique value, and the time could be identified simply by measuring the ratio of strontium 87 to strontium 86. Unfortunately, from the strictly chronologic perspective, strontium is not strictly monotonic; strontium of some periods, for example, is less radiogenic than that for the preceding and succeeding periods. Careful work has delimited the marine growth curve for strontium, however, and the technique has met with considerable success.

ROLE IN UNDERSTANDING GEOLOGIC PHENOMENA

The absolute dating of geologic materials and events has had an unprecedented influence on the understanding of geologic events on Earth and of solar system minerals and rocks. The ability to establish events in terms of actual years, rather than in relative terms such as "older than" or "younger than," has led to a realistic estimate of the Earth's age and to calibrated time scales for organic evolution, geomagnetic events, and the structural development of the Earth's crust. One of the earliest and most useful chronometric schemes for the oldest rocks, the Rb-Sr technique has been of exceptional value, not only for dating but also for the use of strontium-isotopic composition as an indicator or tracer for a variety of geologic processes, such as the evolution of seawater.

Additionally, the technique has had much success in the unraveling of igneous and metamorphic processes in complex, regionally metamorphosed geologic terrains.

Because of its usefulness in the precise determination of the ages of very old rocks, the Rb-Sr method will continue to be of major use in dating the Earth's oldest rocks and extraterrestrial solar system materials, for example lunar rocks and meteorites, including meteorites from the Moon and possibly from Mars.

E. Julius Dasch

CROSS-REFERENCES

Earth's Age, 511; Earth's Oldest Rocks, 516; Fission Track Dating, 522; Geologic Time Scale, 1105; Igneous Rock Bodies, 1298; Magmas, 1326; Mass Spectrometry, 483; Potassium-Argon Dating, 527; Radioactive Decay, 532; Radioactive Minerals, 1255; Radiocarbon Dating, 537; Regional Metamorphism, 1421; Samarium-Neodymium Dating, 548; Solar System's Origin, 2607; Uranium-Thorium-Lead Dating, 553.

BIBLIOGRAPHY

Faure, Gunter. *Principles of Isotope Geology.* 2d ed. New York: John Wiley & Sons, 1986. An excellent though technical introduction to the use of radioactive and stable isotopes in geology, including a thorough treatment of the Rb-Sr technique. The work is well illustrated and well indexed. Suitable for college-level readers.

Levin, Harold L. *The Earth Through Time.* 3d ed. Philadelphia: Saunders College Publishing, 1988. This college-level text contains a brief, clear description of the Rb-Sr method. A diagram of a whole-rock isochron is included. Five other radiometric dating techniques are discussed, and background information on absolute age and radioactivity is provided. Includes review questions, a list of key terms, and references.

Parker, Sybil P., ed. *McGraw-Hill Encyclopedia of the Geological Sciences.* 2d ed. New York: McGraw-Hill, 1988. This source contains entries on radioactivity and radioactive minerals. The entry on dating methods includes a brief section on the Rb-Sr method. Includes the formula for radioactive decay and a table of principal parent and daughter isotopes used in radiometric dating. The entry on rock age determination has a longer discussion of the Rb-Sr dating and includes an isochron diagram. For college-level audiences.

Smith, David G., ed. *The Cambridge Encyclopedia of Earth Sciences.* Cambridge, England: Cambridge University Press, 1981. Chapter 8, "Trace Elements and Isotope Geochemistry," covers mass spectrometry, igneous processes, radiogenic isotopes, and radioactive decay schemes. Rb-Sr decay is used to illustrate principles of geochronology. An Rb-Sr isochron diagram is provided and explained. A more technical discussion than the one in Levin's book. Suitable for the reader with some background in science.

SAMARIUM-NEODYMIUM DATING

Sm-Nd dating is one of the more recent yet most common methods of obtaining the absolute ages of geologic materials, especially older minerals and rocks. The method depends on the natural radioactivity of one of the seven isotopes of samarium, samarium 147, which decays to neodymium 143.

PRINCIPAL TERMS

ABSOLUTE AGE: the numerical timing of a geologic event, as contrasted with relative, or stratigraphic, timing

GEOCHRONOLOGY: the study of the absolute ages of geologic samples and events

HALF-LIFE: the time required for a radioactive isotope to decay by one-half of its original weight

ISOCHRON: a line connecting points representing samples of equal age on a radioactive isotope (parent) versus radiogenic isotope (daughter) diagram

ISOTOPE: a species of an element having the same number of protons but a different number of neutrons and therefore a different atomic weight

MASS SPECTROMETRY: the measurement of isotope abundances by separating the isotopes by mass and charge in an evacuated electromagnetic field

RADIOACTIVE DECAY: a natural process by which an unstable, or radioactive, isotope transforms into a stable, or radiogenic, isotope

RARE-EARTH ELEMENTS

Samarium (Sm) and neodymium (Nd) are both rare-earth elements (REEs, or lanthanides). They occur, commonly in trace amounts, in many of the more widespread minerals and rocks and appear in high concentrations in some rare but economically important minerals, such as bastnaesite and monazite.

The sixteen REEs, their abundances, and especially their abundance patterns have achieved a remarkable usefulness in geochemistry. Their abundances, usually plotted relative to their abundances in important major reservoirs, especially the average composition of the solar system—as it is estimated from primitive meteorites, the carbonaceous chondrites—form patterns that have proved exceptionally useful as tracers for a wide variety of cosmic and geologic processes, such as the parentage of igneous, metamorphic, and sedimentary rocks. Rare-earth element concentration patterns, as tracers, are significant in a way that is similar to the importance of isotopic ratios of radiogenic nuclides such as strontium 87 and strontium 86 or neodymium 143 and neodymium 144, described herein. The reason for this similarity is that the REEs, although exhibiting almost identical geochemical properties—their charge is +3 (except for europium and cerium, which may also exist in other valence states), and they have identical outer shell electronic configurations—nevertheless vary slightly because they have slightly different ionic radii. Consequently, REEs within a given mineral or other phase act somewhat like isotopes of a given element. Because so much is known about the geochemical behavior of the REEs, the usefulness of the Sm-Nd dating and tracer techniques is great. Samarium and neodymium belong to the light REE part of the lanthanide spectrum and occur next to each other in terms of atomic number and ionic size.

SAMARIUM AND NEODYMIUM

Samarium consists of seven natural isotopes as well as several artificial isotopes. The Sm-Nd chronometer is a result of the alpha decay of samarium 147 to neodymium 143, which has an exceptionally long half-life of 1.06 billion years. Samarium 147 has been known to be unstable for many years, but because of its long half-life and the small dispersion of Sm-Nd ratios in most materials, its use in age dating did not begin until the mid-1970's, with the advent of modern mass spectrometry, high-precision instruments, and the digital collection of data. The radiogenic accumulation of neodymium 143 in most natural minerals and rocks is very slow.

The REE neodymium consists of seven natural isotopes and several unstable species. In order of isotopic abundance, the seven natural isotopes are neodymium 142, 144, 146, 143, 145, 148, and 150. The important radiogenic nuclide is neodymium 143, which forms by the alpha decay of radioactive samarium 147. Like the isotopic composition of strontium (strontium 87 and strontium 86), the abundance of the neodymium 143 is measured in a mass spectrometer relative to the neodymium isotope of closest mass and reasonable abundance, neodymium 144. Also like the isotopic composition of strontium, the isotopic composition of neodymium, specifically the neodymium 143-neodymium 144 ratio, has been of exceptional use in helping scientists to understand a variety of geologic, especially igneous, processes.

DETERMINING SAMARIUM-NEODYMIUM CONCENTRATIONS AND RATIOS

Concentrations of samarium and neodymium in minerals, rocks, and other natural substances can be measured using a variety of techniques, although isotopic parameters and precise elemental abundances are commonly measured by mass spectrometry. As the abundances of most of the samarium and neodymium isotopes are precise percentages of the total for each of these elements and do not vary as a function of nuclear instability, each can be calculated by taking the appropriate percentage of the total elemental concentrations, as determined by gravimetric analysis, atomic absorption spectrophotometry, X-ray fluorescence spectrophotometry, microbeam probe analysis, or another method. In practice, however, because of low abundances of the REEs in most minerals, the most useful techniques are neutron activation analysis or mass spectrometry. Most modern work consists of determining the relevant isotopic parameters by mass spectrometric isotope dilution after purification by chemical techniques.

A major attribute of the Sm-Nd system is that, thanks to the geochemical affinity of samarium for ultramafic and mafic minerals (minerals low in silica and rich in iron and magnesium), the method can be used to date even comparatively young low-silica rocks, such as peridotite and basalt, that cannot easily be dated by uranium-thorium-lead (U-Th-Pb) or rubidium-strontium (Rb-Sr) techniques. Additionally, in principle, the unusual geochemical behavior of the Sm-Nd ratio in magmatic crystallization or partial melting of source rocks allows contrasting isotopic trends between Sm-Nd and other ratios to be traced. For example, fractional crystallization of a magma yields residual liquid that becomes increasingly greater in terms of U-Th-Pb and Rb-Sr ratios, whereas Sm-Nd values commonly become lower. Thus, in the last formed residual rocks, strontium and lead will be more radiogenic, but samarium perhaps less radiogenic, than they were in the beginning magma. The reverse is true for the partial melting of, say, a mantle source rock such as peridotite. It has become useful, therefore, to plot time-of-crystallization (initial) isotopic ratios of lead and especially strontium against that of neodymium. For many rocks, these data form a trend of negative slope, a trend that has been called the "mantle array."

PRELIMINARY TECHNIQUES IN SM-ND DATING

Commonly, a preliminary determination of samarium and neodymium abundance and the Sm-Nd ratio is made for the materials to be dated, either by a reconnaissance technique or simply by an estimation from known concentrations of these elements in previously measured samples of the same type of mineral or rock. Samples selected on the basis of these determinations are chosen for the optimum conditions for dating. After selection, the samples are crushed, or homogenized if necessary, and a portion is taken which contains enough of the samarium and especially of the more critical neodymium for adequate isotopic analysis. Rarer materials, such as some lunar rocks or meteorites, may not afford enough sample for optimal analysis.

The sample of material, if it is the most common geologic form of aluminosilicate compounds, perhaps with some organic material, is dissolved in a mixture of hydrofluoric and perchloric acids and reduced by evaporation to a concentrated "mush." Samples other than silicates may be dissolved in other solvents. The mush is "spiked" with an appropriate amount of purified liquids containing known amounts of samarium and neodymium of known, nonnatural isotopic composition. This material is dissolved in a small amount of hydrochloric acid and placed on calibrated ion exchange columns. The columns are washed with hydrochloric acid, and purified samples of samarium, then neodymium, of mixed natural and spiked isotopic composition are collected and evaporated. Highest accuracy and precision requires that the smallest amounts possible of samarium and neodymium from the laboratory environment (contamination) be included with the completed, purified elements.

MASS SPECTROMETRY

The standard method of mass spectrometry for Sm-Nd chronology involves placing the purified samples as solids (such as oxides or metals) on metal filaments and installing the loaded filaments in solid-source mass spectrometers. The spectrometer source regions, evacuated to very low pressures, are constructed so that the metal filaments can be heated until the samarium or neodymium ionizes. The charged, ionized sample is accelerated through a series of collimating slits and into a controlled electromagnetic field, where the beams of ions are separated by charge-mass ratios into beams of separated isotopes. The charge is the same for each atom, so the ions are separated on the basis of mass only. Specific isotopic beams, controlled by the magnetic field, are channeled through more collimating slits into the collector part of the spectrometer.

Commonly, a Faraday cup is used to analyze the number of atoms of each isotope by conversion of each atomic impact into a unit of charge, which is amplified. A digital readout is produced. The actual output of the procedure is isotope ratio measurements—samarium 150/samarium 147, neodymium 143/neodymium 144, and so on—which are converted to the required parameters for determining time by mathematical programs. Because the important quantity of neodymium 143 is relatively close to that of neodymium 144 in most cases, and because it is different by only one mass unit, the standard for reporting the radiogenic component is with the ratio neodymium 143/neodymium 144.

By mixing known weights of "spikes" of samarium and neodymium, which differ markedly from the natural isotopic compositions of these elements, with the natural sample, a combined mass spectrum is obtained. From these data can be calculated the precise abundances of samarium and neodymium in the natural material (a process known as isotope dilution) and the critical isotopic composition of the natural neodymium.

ISOCHRON DIAGRAMS

Although the age of the analyzed sample can be calculated using the determined Sm-Nd parameters and the decay constant for samarium 147, it is customary and more useful to determine the age graphically, with an isochron diagram. In the diagram, the actual isotope ratios collected in the spectrometer are used as coordinates. Thus, the parent, unstable component, samarium 147, is designated by reference to the common isotope of neodymium, neodymium 144. The other coordinate, the measure of the radiogenic component, is neodymium 143/neodymium 144. A line connecting points representing samples of equal ages, or an isochron, has a value represented by its slope; a horizontal isochron has an age value of zero, and positive slopes of successively greater degree have increasingly greater ages. A single mineral or rock would furnish only one point on the diagram, so to draw an isochron, the researcher must know or, more likely, estimate the sample's initial isotopic composition. Ages determined in this way are called "model ages."

DATING ROCK CRYSTALLIZATION

Sm-Nd dating is used to determine the time of crystallization of specific types of igneous rock, the time of formation of comagmatic igneous rocks, and the time of metamorphism of a sequence of rocks of varying composition. The analysis of minerals of equal age but different compositions from a sample of plutonic or volcanic rock would form the points for an isochron whose slope would be proportional to the rock's age of crystallization. A common example is a mafic rock, such as basalt, with minerals of an increasing samarium 147-neodymium 144 ratio, such as plagioclase feldspar and pyroxene. An assumption that is justified in most circumstances is that at the time of crystallization, all the minerals formed have the same isotopic composition of neodymium. If the rock system has not been affected by the introduction of parent or daughter species by metamorphism or weathering (open-system behavior), the points representing these samples will define a perfect line. In practice, however, uncertainties about the parent-daughter parameters, and perhaps some open-system behavior, result in imperfect isochronism and, therefore, uncertainties in the calculated age. A benefit of the isochron method is that the isotopic composition of neodymium at the rock's time of origin is marked by the left-hand or lower intercept of the isochron. Another benefit is that it is easy to see whether one or more points is aberrant or whether a poor fit might indicate an open-system history for the rock. A wide array of terrestrial and extraterrestrial igneous rocks have been dated by this mineral isochron method with good precision.

DATING ROCK SERIES AND SEQUENCES

Not only can geologists use this method to date minerals from a single rock, but if a series of rocks of equal age from a common parent are analyzed, their Sm-Nd isotopic parameters should also yield a "whole-rock" isochron proportional to the rocks' age. Consequently, geologists may test for comagmatic properties and ages in a suite of rocks. In practice, however, these properties are often analyzed only with supporting petrologic or geochemical data, if at all.

The time of metamorphism of a sequence of rocks of varying composition also may be determined by the Sm-Nd technique, provided that, at the time of metamorphism, the isotopic composition of neodymium in the rocks is effectively homogenized. The geochemical behavior of the REEs, however, results in much more immobile transport for these elements as compared with, say, rubidium and strontium. Complete homogenization is accomplished through high-grade metamorphism, through the availability of sufficient water to effect the isotopic exchange, or, most likely, through both. Dry metamorphism, even if high grade, may not result in homogenization; the isochron date obtained in such a case would record only some mixture of original events or isotopic compositions. Conversely, if the rocks are permeable, fine-grained, and wet, homogenization may be completed even under low-grade metamorphic conditions. Where conditions have been sufficient to equilibrate Sm-Nd components, whole-rock isochron analysis can reveal this type of metamorphic event.

DATING MATERIALS PRECIPITATED FROM SEAWATER

The isotopic composition of neodymium by itself can in principle be used to determine the age of manganese nodules, carbonate rocks, or other materials precipitated from seawater. For this method to work, enough must be known about the isotopic composition of neodymium in that sector of the sea through time, a history that may not readily be available. Strontium dissolved in seawater of a particular geologic episode has the same isotopic composition everywhere in the ocean. Neodymium, however, does not, because the mixing rate of marine neodymium, about one thousand years, is long compared to the average "lifetime" for neodymium atoms in the sea, less than one hundred years. Thus, neodymium of variable isotopic composition washed into the sea from rivers or other sources commonly is deposited on the seafloor before it has become isotopically well mixed. Because neodymium 143 accumulates through time as a result of samarium 147 decay, neodymium on Earth becomes more radiogenic through geologic time. One might assume, then, that geologic time could be identified simply by measuring the ratios of neodymium 143 to neodymium 144; unfortunately for the chronologic usefulness of this parameter, however, neodymium is not uniform in the seas, as discussed. Therefore, marine neodymium is used more for an understanding of marine processes, such as water transport, than for chronologic studies. Future work may reveal ways to use marine neodymium as an indicator of time and as a global tectonic tracer.

ROLE IN UNDERSTANDING GEOLOGIC EVENTS

The absolute dating of geologic materials and events has had a profound effect on scientists' understanding of terrestrial and extraterrestrial geologic events. Establishing the age of events in years rather than in relative terms has led to reliable estimates of the Earth's age and to calibrated time scales for organic evolution, geomagnetic events, and the plate tectonic cycle. One of the most recent and most useful chronometric methods, the Sm-Nd technique has been extremely important, not only for dating but also for tracking of a variety of geologic processes, such as the evolution of seawater. The technique has also helped geologists to understand igneous and metamorphic processes in complex, regionally metamorphosed geologic terrains.

The Sm-Nd method will continue to be of great use in dating the Earth's oldest rocks and other solar system materials, including lunar and Martian rocks and meteorites.

E. Julius Dasch

CROSS-REFERENCES

Earth's Age, 511; Earth's Oldest Rocks, 516; Fission Track Dating, 522; Geologic Time Scale, 1105; Igneous Rock Bodies, 1298; Magmas, 1326; Mass Spectrometry, 483; Potassium-Argon Dating, 527; Radioactive Decay, 532; Radioactive Minerals, 1255; Radiocarbon Dating, 537; Regional Metamorphism, 1421; Rubidium-Strontium Dating, 543; Solar System's Origin, 2607; Uranium-Thorium-Lead Dating, 553.

BIBLIOGRAPHY

Duckworth, H. E. *Mass Spectrometry.* Cambridge, England: Cambridge University Press, 1958. An older work, but it covers well the basic principles of mass spectrometry, the major measurement technique used in conjunction with Sm-Nd dating. For readers with some background in science.

Faure, Gunter. *Principles of Isotope Geology.* 2d ed. New York: John Wiley & Sons, 1986. An excellent introduction to radioactive and stable isotopes and their use in geology. It covers the Sm-Nd technique thoroughly. The work is somewhat technical but well illustrated and indexed. Written for a college-level audience.

Parker, Sybil P., ed. *McGraw-Hill Encyclopedia of the Geological Sciences.* 2d ed. New York: McGraw-Hill, 1988. This source contains an entry on rock age determination. Not much space is devoted solely to the Sm-Nd method, but other methods are reviewed and the general principles of dating are explained. For general readers.

Smith, David G., ed. *The Cambridge Encyclopedia of Earth Sciences.* England: Cambridge University Press, 1981. A chapter in this well-written textbook covers mass spectrometry, radioactive decay schemes, and several dating methods. Both rubidium-strontium and samarium-neodymium techniques are discussed. Includes an example of an isochron diagram and a table of trace element abundances. For the reader desiring a clear yet not simplistic source.

URANIUM-THORIUM-LEAD DATING

Radioactive decay of uranium and thorium into lead can be used as a natural clock to determine the ages of rock samples. Rocks from different locations on Earth have been dated in a wide range of ages, from 100 million to several billion years. The Earth, Earth's moon, and meteorites all have a common age of about 4.6 billion years.

PRINCIPAL TERMS

COMMON LEAD: ordinary lead as it was formed at the time when all the elements in nature were created; also called primordial lead

CONCORDANT AGE: a situation in which several naturally radioactive elements, such as uranium, thorium, strontium, and potassium, all give the same age for a rock sample

DISCORDANT AGE: a situation in which several radioactive elements do not give the same age because of gain or loss of decay products from a rock sample

HALF-LIFE: the time for half the atoms in a ra-

dioactive sample to decay, having a different value for each radioactive material

ISOTOPE: atoms of the same element but with different masses as a result of extra neutrons in the nucleus, such as the two uranium isotopes uranium 235 and uranium 238

MASS SPECTROMETER: an apparatus that is used to separate the isotopes of an element and to measure their relative abundance

RADIOGENIC LEAD: lead formed from uranium or thorium by radioactive decay

RADIOACTIVE HALF-LIFE

Radioactivity was discovered by Antoine-Henri Becquerel, a professor of physics in Paris, in 1896. He found that a rock containing uranium was emitting radiation that caused photographic film to become exposed. Shortly afterward, his graduate student Marie Curie discovered a second radioactive element, radium, which is a decay product of uranium. Over the next ten years, Ernest Rutherford and other scientists were able to unravel a whole series of radioactive processes in which uranium (element 92) decayed into its stable end product, lead (element 82).

All radioactive materials have a particular half-life, which is the time for half the atoms to decay. In the early days of radioactivity, scientists wanted to see if the half-life of an element could be changed by various processes. For example, they heated or cooled the radioactive material, made different chemical compounds, or converted it to a gas at high pressure. In all cases, the half-life did not change. One can understand this result by realizing that radioactivity comes directly from the nucleus of an atom, while heat and pressure affect only the outer electron cloud. Therefore, the radioactive half-life is like a built-in clock, keeping

time at a fixed rate. No geological process, no matter how violent, can change the half-life.

Radioactivity can be applied to dating the age of rocks under certain conditions. The isotope uranium 238, for example, has a half-life of 4.47 billion years and eventually decays into lead 206. To make the situation as simple as possible, assume a rock sample contained some uranium but no lead at all when it first solidified. This phenomenon marks the starting time of the radioactive clock (time = 0). Suppose that for a very long time period, the rock remained a closed system; that is, no uranium or any of its decay products leaked out or were added from the surroundings. Because of radioactive decay, the uranium content will decrease and the lead will gradually increase. The ratio of lead 206 to uranium 238 in the rock, which will change with time, can be used to calculate the age.

For a numerical example, suppose the rock originally contained 10 grams of uranium 238 and zero lead. After 1 billion years, one can calculate from the half-life that the 10 grams of uranium 238 must have decreased to 8.56 grams because of radioactive decay. The lead 206 will have increased from zero up to 1.24 grams. (Note that

RADIOACTIVE DECAY OF URANIUM 238

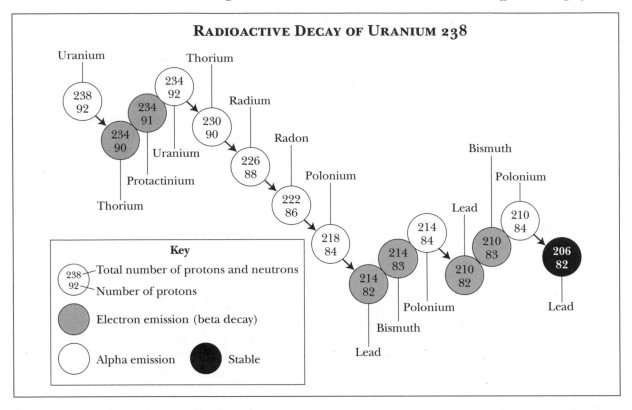

8.56 + 1.24 = 9.80 grams. The "missing" mass of 0.20 gram has gone into the creation of alpha, beta, and gamma rays.) The ratio of lead 206 to uranium 238 after 1 billion years would equal 1.24/8.56, or 0.145. In a similar way, the ratio can be calculated for any other elapsed time. The older the rock, the more lead it will contain, so the ratio of lead to uranium gradually increases.

ISOTOPES

In the early 1900's, large errors were made in age calculations because one vital item of information was missing. The idea of isotopes had not yet been discovered. The mass spectrometer, whose invention may be credited to the English physicist Sir Joseph John Thomson about 1914, was later refined by Arthur Jeffrey Dempster and others. The mass spectrometer uses a magnetic field to separate atoms of slightly different mass. Many elements were shown to be a mixture of several isotopes. Lead, for example, has four stable isotopes, with masses of 204, 206, 207 and 208. (Atom masses are expressed relative to carbon = 12.) Gradually, it became clear that uranium 238 de-

caying into lead 206 is only one of several radioactive decay processes. The situation is more complex. There are two other decay chains that produce lead as an end product: Thorium 232 decays into lead 208 with a half-life of 14 billion years, and another uranium isotope, uranium 235, decays into lead 207 with a half-life of 700 million years.

Radioactive dating requires that the individual isotopes are measured separately. Typically, three experimental ratios are measured with the mass spectrometer: lead 206/uranium 238, lead 208/ thorium 232, and lead 207/uranium 235. Each of these three ratios is combined with the half-life for that decay process to calculate an age. If all three calculations give the same result, the age is probably quite reliable and is called concordant. If the calculations disagree, the ages are discordant, and further investigation is necessary.

CORRECTING FOR DISCORDANT AGES

What can cause discordant ages when several radioactive clocks are compared? In the uranium-thorium-lead (U-Th-Pb) method (described above), one probable source of error is the possibility that the rock, when it first solidified, already contained some natural lead. In other words, the lead content of the rock is the sum of radiogenic lead (from radioactive decay of uranium and thorium) plus the primordial lead. Only the radiogenic lead should be used to calculate an age. A method is needed to subtract out the primordial lead, which did not come from radioactive decay.

Fortunately, a good method to correct for primordial lead is available. Three of the lead isotopes come from the decay of uranium and thorium; there is a fourth lead isotope, lead 204, which is not formed by a radioactive decay process. If a rock contains any lead 204, it means that it already must have contained some lead at the time it was formed. Ordinary lead contains a mixture of isotopes whose normal relative abundance has been measured. The normal ratio of lead 206 to lead 204 is about 17 to 1. Suppose a rock contains 50 milligrams of lead 206 and 1 milligram of lead 204. The investigator would subtract 17 milligrams of lead 206 from the total, which leaves a net excess of 50 − 17, or 33 milligrams of lead 206 that must have come from the decay of uranium 238. A similar correction can be made for lead 207 in normal lead.

There is an uncertainty in the correction factor. One cannot be sure that the natural lead, when it was incorporated into the rock, had the same relative abundance of lead isotopes as lead does today. It is very likely that the normal ratio of lead 206 to lead 204 was smaller than 17 to 1 at an early time in the history of the Earth because lead 206 has gradually been added as a result of decay. The most accurate age measurements are obtained if the natural lead correction is small, that is, if most of the lead in a rock sample under study came from radioactive decay.

Sometimes discordant ages can be corrected in a systematic way to calculate a consistent result. For example, some samples of the mineral zircon from the Montevideo area of southern Minnesota gave uranium-lead ages that vary from 2.6 to 3.3 billion years. Suppose a loss of lead occurred, perhaps because of temperature or weathering. It is reasonable to assume that all the lead isotopes decreased by the same percentage because they are chemically identical. One can extrapolate backward in time to show that the zircon samples must

have been formed about 3.55 billion years ago and probably had a lead loss resulting from regional heating at a later time.

LOSS OF INTERMEDIATE ELEMENTS

Another reason that discordant ages sometimes are measured may be that intermediate elements escape in the decay chain between the starting element, uranium or thorium, and the end product, lead. For example, it may happen that radium forms a chemical compound that is relatively soluble in water and is leached out from a rock sample. Also, radon gas may escape. (Scientists know this happens at least some of the time because of the radon buildup in the basements of houses in various parts of the country.)

Any loss of intermediate elements means that too little of the lead end product accumulates. Therefore, the apparent age of a rock sample would be calculated to be too short. An experienced investigator will try to select those minerals for analysis whose crystal structures are known to be relatively impervious to losses. Also, it is desirable to analyze many samples from an area. If most of the results are consistent, one may be able to reject those ages which are discordant for some reason.

DETERMINING EARTH'S AGE

Radioactive age determinations made before 1930 are considered to be unreliable because the mixture of isotopes with different half-lives was not well understood. With improvements in the mass spectrometer in the 1930's, it was shown that uranium consists of two main isotopes: about 99 percent uranium 238 and less than 1 percent uranium 235. Rutherford suggested how such data could be used to estimate at least a rough upper limit for the age of the Earth. His reasoning was as follows: The isotope uranium 235 has such a very low abundance presently because it has a relatively short half-life; most of it has decayed away. The amount of uranium 235 that existed on Earth 700 million years ago (one half-life) would have been twice as much as currently; 1.4 billion years ago, the amount of uranium 235 would have been four times as much. On the other hand, the amount of uranium 238 would have been only a little greater than at present because of its long half-life. If one calculates back far enough, the amounts of uranium 235 and uranium 238 would have been equal about 6 billion years ago; that is, uranium would have been a fifty-fifty mixture of these two isotopes. It is very unlikely that uranium 235 ever was more abundant than uranium 238 because odd isotopes in general are less abundant in nature than are even ones. Therefore, 6 billion years sets an upper limit for the age of the Earth.

A much-improved procedure to determine the age of the Earth was developed in the 1950's. As described by Harrison Brown, the lead isotopes in the Canyon Diablo meteorite (which created the famous Meteor Crater in Arizona) were analyzed. The ratio of lead 206 to lead 204 was only about 9.4, much lower than any samples on Earth. The argument is made that this ratio represented primordial lead, uncontaminated by any radiogenic lead from uranium decay. Over the history of the Earth, this ratio should gradually increase for terrestrial samples because additional lead 206 is produced from uranium, but lead 204 remains constant. Samples that are representative of modern lead on Earth contain about twice as much lead 206 as the meteorite. This amount of extra lead 206 would have required about 4.5 billion years to accumulate. The data about primordial lead in meteors, when combined with the accumulated radiogenic lead from terrestrial samples, give the most reliable result for the age of the Earth, about 4.5 billion years.

Major improvements in the sensitivity of mass spectrometers have made it possible to measure the abundance of both parent and daughter isotopes. For example, fine surface material and small rocks brought back from the Moon by the Apollo 11 astronauts in 1971 were analyzed for uranium and lead isotopes. The results were in good agreement (concordant) for the uranium 238/lead 206 and the uranium 235/lead 207 decay chains. The so-called Moon-dust was dated to be between 4.6 and 4.7 billion years old. It appears that the age of the Earth, the Moon, and meteorites all cluster around 4.6 billion years. This value would be representative for the age of the solar system.

DETERMINING EVOLUTION OF GEOGRAPHIC REGIONS

In general, rocks that solidified much later in the evolution of the Earth contain considerable uranium and thorium. Their lead content is largely radiogenic. The ratio of lead isotopes to uranium and thorium will vary greatly, depending on the time of solidification.

Uraninite is a radioactive mineral containing uranium (in the form of UO_2) and thorium. It is similar to the "pitchblende" that was used by Madame Curie in her famous experiment to isolate the new element radium. In a typical age analysis, samples of uraninite from the Black Hills of South Dakota were dated using three different isotopes, with the following results: uranium 238/lead 206 gave an age of 1.58 billion years, uranium 235/lead 207 gave 1.6 billion years, and thorium 232/lead 208 gave 1.44 billion years. The three measurements agree fairly closely and therefore are said to be concordant. The overall goal of such age measurements is to understand the stages of geological evolution for a whole geographical region on the Earth's surface.

LOCATING URANIUM DEPOSITS

Another application of U-Th-Pb dating has been to study the worldwide distribution of uranium resources. All over the surface of the Earth, the crust contains about one part per million of uranium. Because of the combined action of high temperature, chemical reactions, and water flow, concentrated uranium mineral deposits were formed when suitable geological conditions existed. Some high-grade uranium ore from Gabon, on the west coast of Africa, contains more than 20 percent uranium. This deposit took place about 2 billion years ago, according to uranium-lead dating. Much larger deposits, but with a much lower percentage yield, are located in northern Canada at Elliot Lake and at Witwatersrand, South Africa. These deposits occurred considerably earlier, about 2.5 billion years ago. In the United States, the major deposits are located in the Colorado plateau extending from Wyoming to Texas, with a relatively recent age of less than 200 million years. Such information about the age of uranium deposits is useful to understand the process of mineralization and possibly to locate new deposits.

Nuclear power plants in 1989 contributed about 16 percent of the world's electricity. Coal-burning plants generate acid rain and carbon dioxide in the environment, and the oil supply is limited, so it is likely that nuclear power will continue to be used, especially in Europe, Japan, Russia, Canada, and the United States. The location and size of the world's major uranium deposits are of great importance to supply the necessary fuel. The mining industry needs to know as much as possible about uranium ore deposits so that the present resources can be estimated accurately and exploration for new deposits can receive helpful guidance.

INVESTIGATING RADON RELEASE

Another area where the uranium-lead decay chain plays an important role is in regard to the radon hazard. Uranium in the soil decays in several steps into radium, which in turns decays into the radioactive gas radon. Because it is a gas, radon mixes with the air in small quantities and is ingested into the lungs. In the open, this natural radioactivity in the air is very dilute, so it is not a hazard. The problem comes when radon seeps into the basement of a house through cracks in the floor or through a sump hole. If the house is located in a geographic region where the soil contains considerable uranium, the radon level may be hazardous to the occupants.

The radon problem came to national attention in 1984 when an engineer at the Limerick nuclear power plant in Pennsylvania set off a radiation alarm when he entered the plant, not when he was leaving. The radioactivity was traced to the engineer's home. The radiation level in that house was found to be about one hundred times greater than exposures permitted for workers in uranium mines. Other homes in the area were also found to have relatively high levels. Surveys of radon levels have been made recently in other areas of the United States to investigate the extent of the problem. The Environmental Protection Agency (EPA) has estimated that radon in homes may be responsible for between five thousand and twenty thousand cancer fatalities per year, a cause for great concern.

Another application of radon release may be in earthquake prediction. In south-central Russia, the city of Tashkent is in a major earthquake zone. The radon content of well water was monitored in the area. A graph of the data starting in 1956 showed a low level of radon at first, increasing slowly for several years. After 1964, the rate of increase became very steep, until the earthquake came in 1966. Immediately after the quake, the radon decreased rapidly. The explanation for this phenomenon is based on the idea that stresses in the ground cause microfracturing of rocks with release of radon from the pores. This method of study is very promising, but more work needs to be done to see if the radon signal can predict the magnitude and epicenter of a quake with any quantitative accuracy.

Hans G. Graetzer

CROSS-REFERENCES

Earth's Age, 511; Earth's Oldest Rocks, 516; Fission Track Dating, 522; Geologic Time Scale, 1105; Lunar Rocks, 2561; Mass Spectrometry, 483; Meteors and Meteor Showers, 2711; Nucleosynthesis, 425; Potassium-Argon Dating, 527; Radioactive Decay, 532; Radioactive Minerals, 1255; Radiocarbon Dating, 537; Radon Gas, 1886; Rubidium-Strontium Dating, 543; Samarium-Neodymium Dating, 548.

BIBLIOGRAPHY

Allison, Ira S., and Donald F. Palmer. *Geology.* 7th ed. New York: McGraw-Hill, 1980. A college-level introductory textbook in geology that has gone through many revisions since the first edition was published in the 1930's. Chapter 5 gives a clearly written and up-to-date overview of how the ages of rocks and geologic time can be measured.

Durrance, E. M. *Radioactivity in Geology.* New York: Halsted Press, 1986. The author shows the wide scope of radioactivity measurements in geological investigations. Up-to-date information is presented on environmental radioactivity (including the radon hazard), heat generation, and various isotope-dating procedures. A bibliography of articles published in professional as well as popular journals follows each chapter.

Eicher, Don L. *Geological Time.* 2d ed. Englewood Cliffs, N.J.: Prentice-Hall, 1976. This thin volume of six chapters gives a historical overview of various methods to estimate age and time sequence in the evolution of the Earth. The evidence from heat loss, rock strata, fossils, and eventually radioactivity is described in a non-technical narrative style.

Faure, Gunter. *Principles of Isotope Geology.* New York: John Wiley & Sons, 1977. An intermediate-level book addressed to students of geology as well as to practicing geologists who may not be trained in this area of investigation. Both radioactive and stable isotope analyses are described. After each chapter, some numerical problems with actual experimental data are given. Numerous references to published scientific articles are listed.

Russell, R. D., and R. M. Farquhar. *Lead Isotopes in Geology.* New York: Interscience Publishers, 1960. A compact discussion of methodology, followed by a 120-page appendix giving specific data on many samples taken worldwide. The authors are particularly concerned about discordant age measurements and how to interpret occasional large variations in lead-isotope ratios.

Skinner, Brian J., and S. C. Porter. *Physical Geology.* New York: John Wiley & Sons, 1987. A widely used college-level textbook for an introductory course in geology. One chapter deals with geological time and its determination, using radioactivity and other physical methods. The uranium/lead and thorium/lead techniques are described in a readable way.

EARTH SCIENCE

Alphabetical List of Contents

Categorized List of Contents